Polymer Microscopy

Third Edition

Linda C. Sawyer
David T. Grubb
Gregory F. Meyers

Linda C. Sawyer
Celanese Americas, ret.
Palmyra, VA, USA

Gregory F. Meyers
Analytical Sciences
Dow Chemical Company
Midland, MI, USA

David T. Grubb
Professor
Department of Materials Science and Engineering
Cornell University
Ithaca, NY, USA

ISBN: 978-0-387-72627-4   e-ISBN: 978-0-387-72628-1
DOI: 10.1007/978-0-387-72628-1

Library of Congress Control Number: 2007928814

© 1987, 1996, 2008 Linda C. Sawyer, David T. Grubb
All rights reserved. This work may not be translated or copied in whole or in part without the written permission of the publisher (Springer Science+Business Media, LLC, 233 Spring Street, New York, NY 10013, USA), except for brief excerpts in connection with reviews or scholarly analysis. Use in connection with any form of information storage and retrieval, electronic adaptation, computer software, or by similar or dissimilar methodology now known or hereafter developed is forbidden.
The use in this publication of trade names, trademarks, service marks, and similar terms, even if they are not identified as such, is not to be taken as an expression of opinion as to whether or not they are subject to proprietary rights.

Printed on acid-free paper

9 8 7 6 5 4 3 2 1

springer.com

Preface to the Third Edition

The major objective of this text is to provide information on the microscopy techniques and specimen preparation methods applicable to polymers. The aim is to provide enough detail for the methods described to be applied by the reader, while providing appropriate references for those who need more detail than can be provided in a single text.

We recognize that scientists from a wide range of backgrounds may be interested in polymer microscopy. Some may be experienced in the field, and this text should provide a reference source and a resource whenever a new material or a new problem comes to their attention. The scientist, engineer or graduate student new to the field needs more explanation and help. The focus here is on the needs of the industrial scientist and the graduate student. Some may need to know more about the intrinsic capabilities of microscopes of all types, so there is a description of basic imaging principles and of instruments, both classical and those more recently invented. Others may know all about microscopes and little about polymers, so there is a discussion of polymer structure and properties to put the microscopy into context. A brief section on processing of polymers has also been added.

As the text has been designed to cater to this wide range of backgrounds, some of these more introductory sections will not be for every reader. However, the organization of chapter and section headings should lead the reader to the information needed, and an extensive index is provided for the same purpose.

The first edition of this book was published twenty years ago, in 1987, and a second edition was published in 1996. There were many changes between those two editions, but the advances in microscopy and polymers in the last decade have been even more significant, requiring major revision. We were pleased when Springer invited us to provide this third edition allowing us to bring *Polymer Microscopy* up to date once again. This edition follows the same basic principles as the first two, with significant editing of older work and inclusion of new material. The rapid development of Scanning Probe Microscopy (SPM) and complete conversion to digital imaging has most affected the image capabilities used for polymers. Additionally, new polymer materials such as nanocomposites have been developed that require the use of microscopy. Overall it has been an exciting decade for polymer microscopy and our goal is to provide a window to view these new technologies.

Chapter 1 provides a brief introduction to polymer materials, processes, morphology and characterization. Chapter 2 is a concise review of the fundamentals of microscopy, where many important terms are defined. Chapter 3 reviews imaging theory for the reader who wants to understand the nature of image formation in the various types of microscopes, with particular reference to imaging polymers and how instrumental parameters affect results. These chapters are summaries of large fields of science, to make this text complete, and they contain many references to more specialized texts and reviews.

Chapters 4 and 5 contain the major thrust of the book. Chapter 4 covers specimen preparation, organized by method with enough detail given to allow a reader to conduct such preparations.* Many new methods have been added, especially those developed for use with the SPM and those relating to improvements in cryo-TEM. The references are chosen to provide the best detail and support. Chapter 5 describes the application of these methods to the study of specific types of polymers. The organization is by the form of the material, as fibers, films, membranes, engineering resins and plastics, composites (including a new section on nanocomposites), emulsions, coatings and adhesives, and high performance polymers. The emphasis in this chapter is on applications, particularly where more than one specimen preparation method or microscopy technique is used.

Chapter 6 is newly named for this edition, "Emerging techniques in polymer microscopy." The change is to indicate that the chapter includes both techniques that have been recently developed and those which are not new but which have not yet been regularly applied to polymer materials. These techniques include optical, electron and scanning probe microscopy techniques. In many of these fields the techniques are still developing very rapidly and thus future improvement in practice and understanding is likely over the coming years. Chapter 7 describes how the various microscopies and other analytical techniques for investigating polymer structure should be considered together as a system for problem solving.

The selection of the authors for this text came from a desire for a comprehensive review of polymer microscopy with emphasis on methods and techniques rather than on the results obtained. The synergism provided by three authors with very different backgrounds is important. One author (LCS) has an industrial focus and a background in chemistry, while another author (DTG) is in an academic environment with a background in polymer physics. A third author (GFM), added for this edition, is a chemist with an industrial focus, and a specialist in SPM. The major contribution of David Grubb and Greg Meyers has been in Chapters 2, 3 and 6. Linda Sawyer has been responsible for Chapters 1, 4, 5 and 7, with input from her coauthors.

ACKNOWLEDGEMENTS

Special thanks go to the former Microscopy group at Celanese Americas, formerly in Summit, New Jersey, whose efforts over the years provided many of the micrographs used in all three editions of this book. Madge Jamieson (deceased), Roman Brozynski and Rong T. Chen and corporate support from Celanese Americas and from The Dow Chemical Company are also gratefully acknowledged.

Micrographs for the third edition are gratefully acknowledged from many colleagues, including: Olga Shaffer (Lehigh University, retired), Barbara Wood (DuPont), Karen Winey (University of Pennsylvania), David Martin (University of Michigan), Christian

---

*The specimen preparation methods used for microscopy of polymers involve the use of many hazardous and toxic chemicals as well as the use of instruments which can be radiation hazards. It is well beyond the scope of this text to provide the information required for the proper and safe handling of such chemicals and instruments. The researcher is strongly encouraged to obtain the required safety information from the chemicals and instrumental manufacturers before their use.

Kübel (Fraunhofer Institute), Ishi Talmon (Technion), David A. Zumbrunnen (Clemson University), G. Julius Vancso (University of Twente, the Netherlands), Javier González Benito (Universidad Carlos III de Madrid, Spain), Scot Gould (The Claremont Colleges), Gary Stevens (University of Surrey, UK), Inga Musselman (University of Texas, Dallas), Steven M. Kurtz (Drexel University), Vernon E. Robertson (JEOL, USA), Barbara Petti (Celgard LLC), Viatcheslav Freger (Ben-Gurion University of the Negev), Ron Anderson (Materials Today), Anne Hiltner (Case Western Reserve University), Francis M. Mirabella (Lyondell Chemical), Eduardo Radovanovic (Universidade Estadual de Maringa'). Greg Meyers would also like to thank Georg Bar of the Dow Chemical Company who provided technical advice on SPM content. We would also like to thank the many colleagues who supplied copies of their papers, advice and input on the content of this text. Special thanks go to Prof. James P. Oberhauser, Dept. of Chemical Engineering, University of Virginia, who enabled Linda Sawyer to have access to the UVA on-line library and obtain the electronic files needed for this new edition.

Linda Sawyer thanks a very understanding colleague and husband, David Sawyer, who, as with the earlier editions, provided technical advice and support. David Grubb would like to thank his wife, Sally, for her support over all the time spent on this project, and Greg Meyers would like to thank his wife, Debbie, for her support, patience, and understanding for the time spent on the preparation of the book.

Linda C. Sawyer
Lake Monticello, Virginia

David T. Grubb
Ithaca, New York

Gregory F. Meyers
Midland, Michigan

May 2007

Contents

Color plates appear between pages 306 and 307.

Preface to the Third Edition v

1 Introduction to Polymer Morphology 1
    1.1 POLYMER MATERIALS 1
        1.1.1 Introduction 1
        1.1.2 Definitions 2
    1.2 POLYMER MORPHOLOGY 3
        1.2.1 Amorphous Polymers 4
        1.2.2 Semicrystalline Polymers 5
        1.2.3 Liquid Crystalline Polymers 7
        1.2.4 Multiphase Polymers 8
        1.2.5 Composites 8
    1.3 POLYMER PROCESSES 8
        1.3.1 Fiber and Film Formation 9
        1.3.2 Extrudates and Moldings 11
    1.4 POLYMER CHARACTERIZATION 17
        1.4.1 General Techniques 17
        1.4.2 Microscopy Techniques 18
        1.4.3 Specimen Preparation Methods 19
        1.4.4 Applications of Microscopy to Polymers 20
        1.4.5 Emerging Microscopy Techniques 21
        References 21

2 Fundamentals of Microscopy 27
    2.1 INTRODUCTION 28
        2.1.1 Lens-Imaging Microscopes 29
        2.1.2 Scanning-Imaging Microscopes 30
    2.2 OPTICAL MICROSCOPY 31
        2.2.1 Introduction 31
        2.2.2 Objective Lenses 32
        2.2.3 Imaging Modes 32
        2.2.4 Measurement of Refractive Index 34
        2.2.5 Polarizing Microscopy 34

	2.3	SCANNING ELECTRON MICROSCOPY	35
	2.3.1	Introduction	35
	2.3.2	Imaging Signals	37
	2.3.3	Electron Sources	39
	2.3.4	SEM Types	41
	2.3.5	SEM Optimization	41
	2.4	TRANSMISSION ELECTRON MICROSCOPY	42
	2.4.1	Conventional TEM	42
	2.4.2	Scanning TEM	43
	2.4.3	Electron Diffraction	44
	2.4.4	High Resolution Electron Microscopy	45
	2.5	SCANNING PROBE MICROSCOPY	45
	2.5.1	Introduction	45
	2.5.2	Atomic Force Microscopy	47
	2.5.3	SPM Probes	50
	2.6	RADIATION SENSITIVE MATERIALS	51
	2.6.1	SEM Operation	52
	2.6.2	Low Dose TEM Operation	52
	2.7	ANALYTICAL MICROSCOPY	53
	2.7.1	X-ray Microanalysis	53
	2.7.2	X-ray Analysis: SEM versus AEM	55
	2.7.3	Elemental Mapping	55
	2.8	QUANTITATIVE MICROSCOPY	56
	2.8.1	Image Processing and Analysis	56
	2.8.2	Three Dimensional Reconstruction	57
	2.8.3	Calibration	57
	2.9	DYNAMIC MICROSCOPY	59
	2.9.1	Mechanical Deformation Stages	59
	2.9.2	Hot and Cold Stages	60
		References	60

3 Image Formation in the Microscope 67

	3.1	IMAGING WITH LENSES	68
	3.1.1	Basic Optics	68
	3.1.2	Diffraction	68
	3.1.3	Image Formation	71
	3.1.4	Resolution and Contrast	72
	3.1.5	Phase Contrast and Lattice Imaging	76
	3.1.6	Illumination Systems	78
	3.1.7	Polarized Light	80
	3.2	IMAGING BY SCANNING ELECTRON BEAM	85
	3.2.1	Probe Formation	85
	3.2.2	Probe-Specimen Interactions	88
	3.2.3	Image Formation in the SEM	92
	3.2.4	Low Voltage SEM	94
	3.2.5	Variable Pressure SEM	96
	3.3	IMAGING IN THE ATOMIC FORCE MICROSCOPE	97
	3.3.1	Microscope Components	97
	3.3.2	Probe-Specimen Interaction	100

	3.3.3 Contact Mode AFM	102
	3.3.4 Intermittent Contact AFM	105
	3.3.5 Noncontact AFM	112
	3.3.6 Practical Considerations for AFM Imaging	113
	3.3.7 Artifacts in SPM Imaging	114
3.4	SPECIMEN DAMAGE IN THE MICROSCOPE	118
	3.4.1 Effect of Radiation on Polymers	118
	3.4.2 Radiation Doses and Specimen Heating	120
	3.4.3 Effects of Radiation Damage on the Image	121
	3.4.4 Noise Limited Resolution	123
	References	124

4 Specimen Preparation Methods — 130

4.1	SIMPLE PREPARATION METHODS	132
	4.1.1 Optical Preparations	132
	4.1.2 SEM Preparations	133
	4.1.3 TEM Preparations	133
	4.1.4 SPM Preparations	140
4.2	POLISHING	142
	4.2.1 Limiting Artifacts	142
	4.2.2 Polishing Specimen Surfaces	143
4.3	MICROTOMY	146
	4.3.1 Peelback of Fibers/Films for SEM	146
	4.3.2 Microtomy for OM	147
	4.3.3 Microtomy for SEM	150
	4.3.4 Microtomy for TEM and SPM	150
	4.3.5 Cryomicrotomy for TEM and SPM	154
	4.3.6 Microtomy for SPM	158
	4.3.7 Limiting Artifacts in Microtomy	160
4.4	STAINING	160
	4.4.1 Introduction	160
	4.4.2 Osmium Tetroxide	162
	4.4.3 Ruthenium Tetroxide	166
	4.4.4 Chlorosulfonic Acid and Uranyl Acetate	173
	4.4.5 Phosphotungstic Acid	175
	4.4.6 Ebonite	177
	4.4.7 Silver Sulfide	178
	4.4.8 Mercuric Trifluoroacetate	178
	4.4.9 Iodine and Bromine	179
	4.4.10 Summary	179
4.5	ETCHING	181
	4.5.1 Solvent and Chemical Etching	181
	4.5.2 Acid Etching: Overview	183
	4.5.3 Permanganate Etching	184
	4.5.4 Plasma and Ion Etching	188
	4.5.5 Focused Ion Beam Etching	194
	4.5.6 Summary	195
4.6	REPLICATION	196
	4.6.1 Simple Replicas	197
	4.6.2 Replication for TEM	198

	4.7	CONDUCTIVE COATINGS	201
	4.7.1	Coating Devices	202
	4.7.2	Coatings for TEM	203
	4.7.3	Coatings for SEM and STM	203
	4.7.4	Artifacts	207
	4.7.5	Gold Decoration	211
	4.8	YIELDING AND FRACTURE	212
	4.8.1	Fractography	212
	4.8.2	Fracture: Standard Physical Testing	213
	4.8.3	Crazing	217
	4.8.4	*In Situ* Deformation	221
	4.9	CRYOGENIC AND DRYING METHODS	226
	4.9.1	Simple Freezing Methods	226
	4.9.2	Freeze Drying	227
	4.9.3	Critical Point Drying	230
	4.9.4	Freeze Fracture-Etching	231
	4.9.5	Cryomicroscopy	232
		References	234

5 Applications of Microscopy to Polymers — 248

5.1		FIBERS	250
	5.1.1	Introduction	250
	5.1.2	Textile Fibers	251
	5.1.3	Problem Solving Applications	260
	5.1.4	Industrial Fibers	267
	5.1.5	High Performance Fibers	270
5.2		FILMS AND MEMBRANES	276
	5.2.1	Introduction	276
	5.2.2	Model Studies	278
	5.2.3	Industrial Films	282
	5.2.4	Flat Film Membranes	294
	5.2.5	Hollow Fiber Membranes	305
5.3		ENGINEERING RESINS AND PLASTICS	308
	5.3.1	Introduction	308
	5.3.2	Process-Structure Considerations	311
	5.3.3	Single Phase Polymers	316
	5.3.4	Multiphase Polymers	321
	5.3.5	Failure or Competitive Analysis	349
5.4		COMPOSITES	354
	5.4.1	Introduction	354
	5.4.2	Literature Review	355
	5.4.3	Composite Characterization	357
	5.4.4	Carbon and Graphite Fiber Composites	365
	5.4.5	Particle Filled Composites	366
	5.4.6	Nanocomposites	370
5.5		EMULSIONS, COATINGS AND ADHESIVES	380
	5.5.1	Introduction	380
	5.5.2	Emulsions and Latexes	381
	5.5.3	Particle Size Measurements	385

Contents   xiii

5.5.4 Adhesives and Adhesion	386
5.5.5 Wettability and Coatings	388
5.6 HIGH PERFORMANCE POLYMERS	398
5.6.1 Introduction	398
5.6.2 Microstructure of LCPs	400
5.6.3 Molded Parts and Extrudates	403
5.6.4 High Modulus Fibers	409
5.6.5 Structure-Property Relations in LCPs	412
References	418

6 Emerging Techniques in Polymer Microscopy   435

6.1 INTRODUCTION	435
6.2 OPTICAL AND ELECTRON MICROSCOPY	436
6.2.1 Confocal Scanning Microscopy	436
6.2.2 Optical Profilometry	437
6.2.3 Birefringence Imaging	438
6.2.4 Aberration Corrected Electron Microscopy	438
6.2.5 Ion Microscopy	440
6.3 SCANNING PROBE MICROSCOPY	441
6.3.1 Chemical Force Microscopy	441
6.3.2 Harmonic Imaging	443
6.3.3 Fast Scanning SPM	444
6.3.4 Scanning Thermal Microscopy	445
6.3.5 Near Field Scanning Optical Microscopy	449
6.3.6 Automated SPM	449
6.4 THREE DIMENSIONAL IMAGING	451
6.4.1 Introduction	451
6.4.2 Physical Sectioning	452
6.4.3 Optical Sectioning	454
6.4.4 Tomography	455
6.5 ANALYTICAL IMAGING	459
6.5.1 FTIR Microscopy	459
6.5.2 Raman Microscopy	460
6.5.3 Electron Energy Loss Microscopy	461
6.5.4 X-ray Microscopy	462
6.5.5 Imaging Surface Analysis	464
References	468

7 Problem Solving Summary   478

7.1 WHERE TO START	479
7.1.1 Problem Solving Protocol	479
7.1.2 Polymer Structures	480
7.2 INSTRUMENTAL TECHNIQUES	480
7.2.1 Comparison of Techniques	480
7.2.2 Optical Techniques	484
7.2.3 SEM Techniques	485
7.2.4 TEM Techniques	486
7.2.5 SPM Techniques	487
7.2.6 Technique Selection	487

7.3 INTERPRETATION 488
    7.3.1 Artifacts 489
    7.3.2 Summary 492
7.4 SUPPORTING CHARACTERIZATIONS 492
    7.4.1 X-ray Diffraction 493
    7.4.2 Thermal Analysis 495
    7.4.3 Spectroscopy 496
    7.4.4 Small Angle Scattering 499
    7.4.5 Summary 500
    References 501

Appendices 505

Appendix I	Abbreviation of Polymer Names	505
Appendix II	Acronyms of Techniques	506
Appendix III	Manmade Polymer Fibers	507
Appendix IV	Common Commercial Polymers and Trade Names for Plastics, Films, and Engineering Resins	508
Appendix V	General Suppliers of Microscopy Accessories	510
Appendix VI	Suppliers of Optical and Electron Microscopes, Microanalysis Equipment, Image Analysis and Processing	512
Appendix VII	Suppliers of Scanning Probe Microscopes and Related Supplies	513

Index 515

Chapter 1
Introduction to Polymer Morphology

1.1 POLYMER MATERIALS 1
   1.1.1 Introduction 1
   1.1.2 Definitions 2
1.2 POLYMER MORPHOLOGY 3
   1.2.1 Amorphous polymers 4
   1.2.2 Semicrystalline polymers 5
      1.2.2.1 Crystallization under quiescent conditions 5
      1.2.2.2 Crystallization under flow 6
   1.2.3 Liquid crystalline polymers 7
   1.2.4 Multiphase polymers 8
   1.2.5 Composites 8
1.3 POLYMER PROCESSES 8
   1.3.1 Fiber and film formation 9
      1.3.1.1 Fiber processes 9
      1.3.1.2 Orientation methods 10
      1.3.1.3 Film processes 11
   1.3.2 Extrudates and moldings 11
      1.3.2.1 Compounding 11
      1.3.2.2 Extrusion processes 12
      1.3.2.3 Injection molding processes 12
      1.3.2.4 Other molding processes 14
      1.3.2.5 Coating processes 15
      1.3.2.6 Novel processes 15
1.4 POLYMER CHARACTERIZATION 17
   1.4.1 General techniques 17
   1.4.2 Microscopy techniques 18
   1.4.3 Specimen preparation methods 19
   1.4.4 Applications of microscopy to polymers 20
   1.4.5 Emerging microscopy techniques 21
References 21

1.1 POLYMER MATERIALS

1.1.1 Introduction

Organic polymers are materials that are widely used in many important emerging technologies of the 21st century. Feedstocks for synthetic polymers are petroleum, coal, and natural gas, which are sources of ethylene, methane, alkenes, and aromatics. Polymers are used in a wide range of everyday applications, such as in clothing, housing materials, medical applications, appliances, automotive and aerospace parts, and in communication. *Materials science*, the study of the structure and properties of materials, is applied to polymers in much the same way as it is to metals and ceramics: to understand the relationships among the manufacturing process, the structures produced, and the resulting physical and mechanical properties. This chapter is an introduction to *polymer morphology*, which must be understood in order to develop relations between the structure and properties of these materials. An introduction by Young [1], a recent book on microstructure and engineering applications of plastics by Mills [2], and a book by Elias [3] all provide a good starting point in any study of polymers. Subsequent sections and chapters of this book have many hundreds of references cited as an aid to the interested reader. The emphasis in this text is on the elucidation of polymer morphology by microscopy techniques. This first chapter provides a foundation for the chapters that follow.

Polymers have advantages over other types of materials, such as metals and ceramics, because their low processing costs, low weight, and properties such as transparency and toughness form unique combinations. Many polymers have useful characteristics, such as tensile strength, modulus, elongation, and impact strength, which make them more cost effective than metals and ceramics. Plastics and engineering resins are processed into a wide range of fabricated forms, such as fibers, films, membranes and filters, moldings, and extrudates. Recently, new technologies have emerged resulting in novel polymers with highly oriented structures. These include polymers that exhibit liquid crystallinity in the melt or in solution, some of which can be processed into materials with ultrahigh performance characteristics. Applications of polymers are wide ranging and varied and include the examples shown in Table 1.1. A listing of the names and abbreviations of some common polymers is shown in Appendix I for reference. Appendix II is a list of acronyms commonly used for analytical techniques. Appendix III provides a listing of common fibers, and Appendix IV is a listing of common plastics and a few applications. Finally, general suppliers of accessories, microscopes, and x-ray microanalysis equipment are found in Appendices V–VII.

1.1.2 Definitions

Polymers are macromolecules formed by joining a large number of small molecules, or monomers, in a chain. These *monomers*, small repeating units, react chemically to form long molecules. The repetition of monomer units can be linear, branched, or interconnected to form three dimensional networks. *Homopolymers*, composed of a single repeating monomer, and *heteropolymers*, composed of several repeating monomers, are two broad forms of polymers. Copolymers are the most common form of heteropolymers. They are often formed from a sequence of two types of monomer unit. Alternating copolymers can be simple alternating repeats of two monomers, e.g., –A–B–A–B– A–B–A–B–, or random repeats of two monomers, e.g., –A–A–A–B–B–A–B–B–A–A–B–, whereas block copolymers include long sequences of one repeat unit, e.g., –A–B–B–B–B–B–B–B–A–A–B–B–B–B–A–. There are several forms of block copolymers, including AB and ABA, where A and B each stand for a sequence of several hundred monomers. If monomer B is added while A chains are still growing, the result is a graded block copolymer. Additionally, each component of block copolymers can be amorphous or crystalline. Amorphous block copolymers generally form characteristic domain structures, such as those in styrene-butadiene-styrene block copolymers. Crystalline block copolymers, such as polystyrene-poly(ethylene

TABLE 1.1. Polymer applications

Fibers	Polyethylene, polyester, nylon, acetate, polyacrylonitrile, polybenzobisthiazole, polypropylene, acrylic, aramid
Films, packaging	Polyethylene, polyester, polypropylene, polycarbonate, polyimide, fluoropolymers, polyurethanes, poly(vinyl chloride)
Membranes	Cellulose acetate, polysulfone, polyamide, polypropylene, polycarbonate, polyimide, polyacrylonitrile, fluoropolymers
Engineering resins	Polyoxymethylene, polyester, nylon, polyethersulfone, poly(phenylene sulfide), acrylonitrile-butadiene-styrene, polystyrene
Biomedical uses	Acrylics, polyethylene, ultrahigh molecular weight polyethylene (UHMWPE), polyester, silicone, nylon
Adhesives	Poly(vinyl acetate), epoxies, polyimides
Emulsions	Styrene-butadiene-styrene, poly(vinyl acetate)
Coatings	Epoxies, polyimides, poly(vinyl alcohol)
Elastomers	Styrene-butadiene rubber, urethanes, polyisobutylene, ethylene-propylene rubber

oxide), typically form structures that are characteristic of the crystallizable component. Block copolymers that have the second component grafted onto the backbone chain are termed *graft copolymers*. Graft copolymers of industrial significance, high impact polystyrene (HIPS) and acrylonitrile-butadiene-styrene (ABS), have rubber inclusions in a glassy matrix. Many common homopolymers can also be found as repeat units in heteropolymers, such as polyethylene, as in polyethylene-polypropylene copolymer. Modified polymers, copolymers, and polymer blends can be tailored for specific end uses, which will be discussed.

There are three major polymer classes: thermoplastics, thermosets, and rubbers or elastomers. Polymers that are typical of each of these classes are listed in Table 1.2. *Thermoplastics* are among the most common polymers, and these materials are commonly termed "plastics". Linear or branched thermoplastics can be reversibly melted or can be dissolved in a suitable solvent. In some cases, thermoplastics are crosslinked in processing to provide heat stability and limit flow and melting during use. In *thermosets*, there is a three dimensional network structure, a single highly connected molecule, which imparts rigidity and intractability. Thermosets are heated to form rigid structures, but once set they do not melt upon prolonged heating nor do they dissolve in solvents. Thermosets generally have only short chains between crosslinks and exhibit glassy brittle behavior. Thermosets are used as high performance adhesives (e.g., epoxies).

Polymers with long flexible chains between crosslinks are *rubbers* and *elastomers*, which, like thermosets, cannot be melted. Elastomers are characterized by a three dimensional crosslinked network that has the well known property of being stretchable and springing back to its original form. Crosslinks are chemical bonds between molecules. An example of a crosslinking reaction is the vulcanization of rubber, where the sulfur molecules react with the double bonded carbon atoms creating the structure. Multiphase polymers, combinations of thermoplastics and elastomers, take advantage of the ease of fabrication of thermoplastics and the increased toughness of elastomers, providing engineering resins with enhanced impact strength.

The chemical composition of macromolecules is important in determination of properties. Variations in the stereochemistry, the spatial arrangement, also result in very different materials. Three forms of spatial arrangement are isotactic, syndiotactic, and atactic. The *isotactic* (i) forms have pendant group placement on the same side of the chain, whereas in *syndiotactic* (s) polymers there is a regular, alternating placement of pendant groups with respect to the chain. *Atactic* (a) polymers have disordered sequences or a random arrangement of side groups. Isotactic polypropylene (iPP) crystallizes and has major uses, whereas atactic polypropylene (aPP) cannot crystallize, is sticky (has a low thermal transition) and finds little application.

1.2 POLYMER MORPHOLOGY

In polymer science, the term *morphology* generally refers to form and organization on a size scale above the atomic arrangement but smaller than the size and shape of the whole sample. The term *structure* refers to the local atomic and molecular details. The characterization techniques used to determine structure differ

TABLE 1.2. Major classes of polymers

Crystallizable thermoplastics	Glassy thermoplastics
Polyacetal	Polystyrene
Polyamide	Poly(methyl methacrylate)
Polycarbonate	Poly(vinyl chloride)
Poly(ethylene terephthalate)	Poly(vinyl acetate)
Polyethylene	
Polypropylene	

Thermosets	Elastomers
Epoxy	Polybutadiene
Phenolic	Ethylene-propylene copolymers
Polyester (unsaturated)	
	Styrene-butadiene rubber
	Ethylene-vinyl acetate
	Styrene-butadiene copolymers

somewhat from those used to determine morphology, although there is some overlap. Examples of polymer morphology include the size and shape of fillers and additives, and the size, distribution, and association of the structural units within the macrostructure. However, as is probably clear from the overlapping definitions, the terms "structure" and "morphology" are commonly used interchangeably. The characterization techniques are complementary to each another, and both are needed to fully determine the morphology and microstructure and to develop structure-property-process relationships.

X-ray, electron, and optical scattering techniques and a range of other analytical tools are commonly applied to determine the structure of polymers. X-ray diffraction, for example, permits the determination of interatomic ordering and chain packing. The morphology of polymers is determined by a wide range of optical, electron and scanning probe microscopy techniques, which are the major subject of this text. Finally, there are many other analytical techniques that provide important information regarding polymer structure, such as neutron scattering, infrared spectroscopy, thermal analysis, mass spectroscopy, nuclear magnetic resonance, and so forth, which are beyond the scope of this text but which are summarized in Chapter 7.

Polymers are considered to be either *amorphous* or *crystalline* although they may not be completely one or the other. Crystalline polymers are more correctly termed *semicrystalline* as their measured densities differ from those obtained for perfect materials. The degree of crystallinity, measured by x-ray scattering, also shows these polymers are less than completely crystalline. There is no measurable order by x-ray scattering techniques and an absence of crystallographic reflections in noncrystalline or amorphous polymers. Characterization of semicrystalline polymer morphology can require an understanding of the entire texture. This extends from the interatomic structures and individual crystallites to the macroscopic details and the relative arrangement of the crystallites in the macrostructure. The units of organization in polymers are *lamellae* or crystals and *spherulites* [4]. Bulk polymers are composed of lamellar crystals that are typically arranged as spherulites when cooled from the melt.

The general morphology of crystalline polymers is now well known and understood and was described by Geil [5], Keller [6], Wunderlich [7], Grubb [8], Uhlmann and Kolbeck [9], Bassett [10, 11], and Seymour [12]. The work of Keller and his group has been reviewed by Bassett [13]. More recent edited books by Bassett [14], Ciferri [15], and Ward [16, 17], among others, provide excellent updates on current knowledge in the area of polymer morphology.

1.2.1 Amorphous Polymers

Amorphous polymers of commercial importance include polymers that are glassy or rubbery at room temperature. Many amorphous thermoplastics, such as atactic polystyrene and poly(methyl methacrylate), form brittle glasses when cooled from the melt. The glass transition temperature, T_g, or glass rubber transition, is the temperature above which the polymer is rubbery and can be elongated and below which the polymer behaves as a glass. Thermal analysis of amorphous polymers shows only a glass transition temperature, whereas crystalline polymers also exhibit a crystalline melting temperature.

Commercially important glassy polymers include polymers that are crystallizable but that may form as amorphous materials. These noncrystalline polymers are formed by rapid cooling of a polymer from above the melting transition temperature. They yield by forming a necked zone where the molecules are highly oriented and aligned in the draw direction. Other important polymers amorphous at room temperature include natural rubber (polyisoprene) and other elastomers. These exhibit a high degree of elasticity, stretching considerably in the elastic region and then fracturing with no plastic deformation.

Plastic deformation in glassy polymers and in rubber toughened polymers is due to crazing and shear banding. *Crazing* is the formation of thin sheets perpendicular to the tensile stress direction that contain fibrils and voids. The fibrils and the molecular chains in them are aligned parallel to the tensile stress direction.

Crazes scatter light and can be seen by eye as whitened areas if there are many of them. Crazing is often enhanced by rubber inclusions, which impart increased toughness to the polymer and reduce brittle fracture by initiating or terminating crazes at the rubber particle surface. *Shear banding* is a local deformation, at about 45° to the stress direction, which results in a high degree of chain orientation. The material in the shear band is more highly oriented than in the adjacent regions. The topic of yielding and fracture will be further explored (see Section 4.8). Overall, the mechanical behavior of amorphous polymers depends upon the chemical composition, the distribution of chain lengths, molecular orientation, branching, and crosslinking.

1.2.2 Semicrystalline Polymers

Semicrystalline polymers exhibit a melting transition temperature (T_m), a glass transition temperature (T_g), and crystalline order, as shown by x-ray and electron scattering. The fraction of the crystalline material is determined by x-ray diffraction, heat of fusion, and density measurements. Major structural units of semicrystalline polymers are the platelet-like crystallites, or *lamellae*. The dominant feature of melt crystallized specimens is the *spherulite*. The formation of polymer crystals and the spherulitic morphology in bulk polymers has been fully described by Keith and Padden [18], Ward [16, 19], Bassett [10, 11, 20, 21], and many others. Bassett [21] points out that a knowledge of morphology is an essential part of the development of polymer materials and a complete understanding of their structure-property relationships. Single crystals can be formed by precipitation from dilute solution as shown later (see Fig. 4.1). These crystals can be found as faceted platelets of regular shape for regular polymers, but they have a less perfect shape when formed from polymers with a less perfect structure. The molecular chains are approximately normal to the basal plane of the lamellae, parallel to the short direction. Some chains fold and re-enter the same crystal. In polyethylene, the lamellae are on the order of several micrometers across and about 10–50 μm thick, independent of the length of the molecule. Bulk crystallized spherulites can range from about 1 to 100 μm or larger. Small angle x-ray scattering (SAXS) and electron diffraction data have confirmed the lamellar nature of single crystals in bulk material.

1.2.2.1 Crystallization Under Quiescent Conditions

When a polymer is melted and then cooled it can recrystallize, with process variables such as temperature, rate of cooling, pressure, and additives affecting the nature of the structures formed. Two types of microstructure observed for semicrystalline bulk polymers are spherulites and row nucleated textures. Bulk crystallized material is composed of microscopic units called *spherulites*, which are formed during crystallization under quiescent conditions. The structures exhibit radially symmetric growth of the lamellae from a central nucleus with the molecular chain direction perpendicular to the growth direction. The plates branch as they grow. The molecular chains therefore run perpendicular to the spherulite radius. The crystallite or lamellar thickness in the bulk polymer depends upon the molecular weight of the polymer, crystallization conditions, and thermal treatment. The size and number of spherulites is controlled by nucleation. Spherulites are smaller and more numerous if there are more growth nuclei and larger if slow cooled or isothermally crystallized. In commercial processes, additives are commonly used to control nucleation density. When crystallizing during cooling, the radial growth rate of the spherulites is an important factor in determining their size. The morphology of isothermally crystallized polyethylene (PE) melts has revealed the nature of the lamellae [22] by a sectioning and staining method for transmission electron microscopy (TEM) (see Fig. 4.15), which will be described later.

A schematic of the spherulite structure is shown in Fig. 1.1 [16]. The structure [16, 18, 20, 23] consists of radiating fibrils with amorphous material, additives, and impurities between the fibrils and between individual spherulites. Although the shape of the growing spherulite

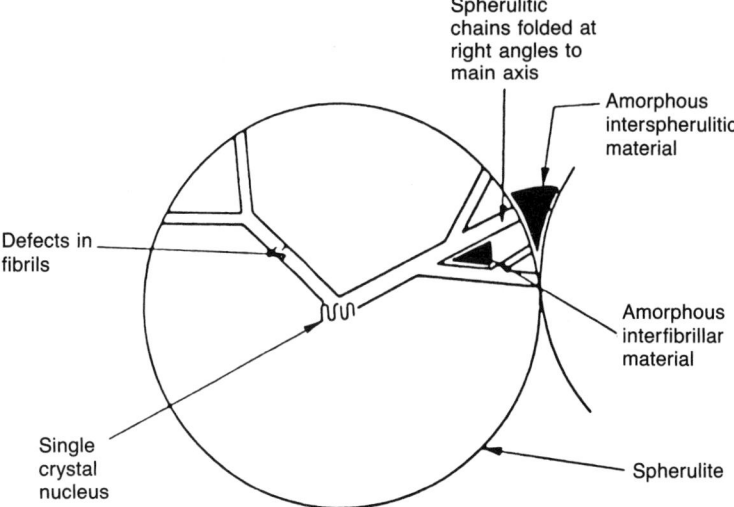

FIGURE 1.1. Schematic of spherulite structure. (From Ward [19]; used with permission.)

is round, as shown in a polarized light micrograph (Fig. 1.2), spherulites generally impinge upon one another during cooling, resulting in polyhedral shapes in the final product (Fig. 1.3). When thin melt quenched films or sections of a bulk polymer are viewed in crossed polarizers, the spherulites appear bright because they are anisotropic and crystalline in nature. Isotropic materials exhibit the same properties in all directions, whereas anisotropic materials exhibit a variation in properties with direction.

FIGURE 1.2. Polarized light micrograph of this polyoxymethylene film cooled from the melt shows recrystallization and formation of spherulites. The shape of the growing, birefringent spherulites is round.

Polarized light micrographs of a sectioned, bulk crystallized nylon (Fig. 1.3) show the size range of spherulites obtained by bulk crystallization. A more complete discussion of polarization optics will be found later (see Section 2.2.5 and Section 3.1.7), but for this discussion it is clear that the size of individual spherulites can be determined by analysis of polarized light micrographs. Average spherulite sizes are determined by small angle light scattering techniques.

1.2.2.2 Crystallization Under Flow

When a bulk polymer is crystallized under conditions of flow, a *row nucleated*, or "shish kebab," structure can be formed. Typically, the melt or solution is subjected to a highly elongational flow field at a temperature close to the melting or dissolution temperature. A nonspherulitic, crystalline microstructure forms from elongated crystals aligned in the flow direction and containing partially extended chains. At high flow rates, these *microfibers*, some 20nm across, dominate the structure. At lower flow rates, the backbones are overgrown by folded chain platelets. This epitaxial growth on the surface of the extended chain produces folded chain lamellae oriented perpendicular to the strain or flow direction [23]. At still lower flow rates, the large lamellar overgrowths do not retain this orientation. Dilute solutions of polymers stirred during crystallization are known to form this

Polymer Morphology

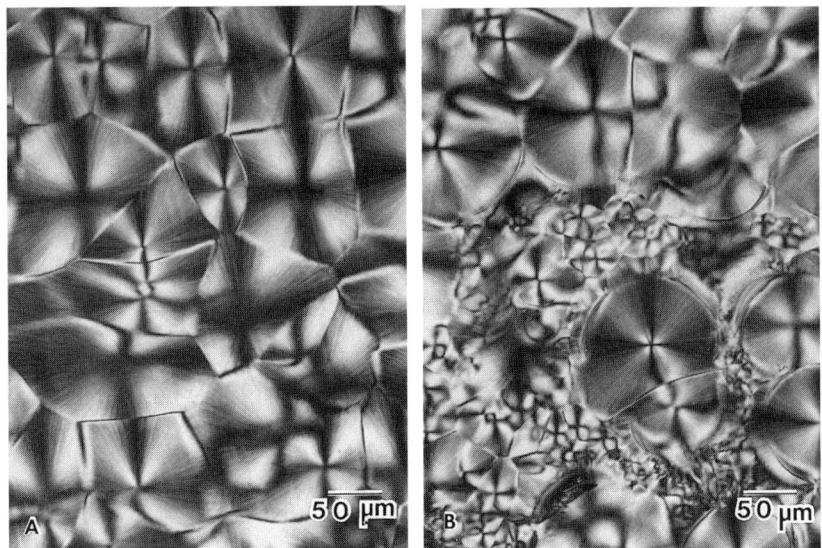

FIGURE 1.3. A thin section of bulk crystallized nylon, in polarized light, reveals a bright, birefringent and spherulitic texture. At high magnification, a classic Maltese cross pattern is seen, with black crossed arms aligned in the position of the crossed polarizers (A); the sample was isothermally crystallized, and exhibits large spherulites. The sample quenched during crystallization (B) yields large spherulites surrounded by smaller ones. (See color insert.)

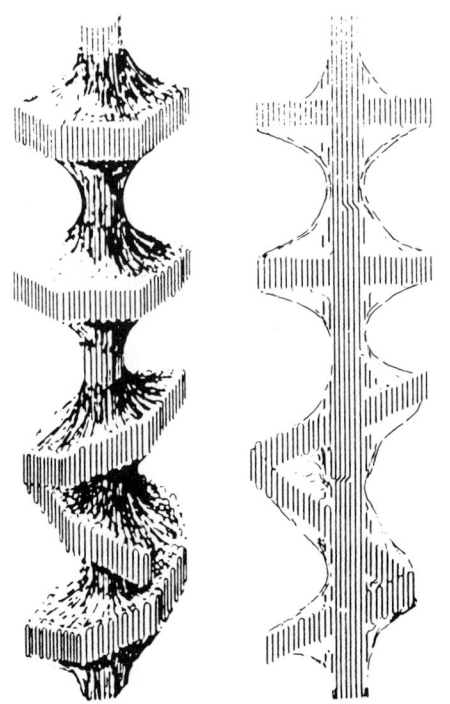

FIGURE 1.4. Schematic of a shish kebab structure. (From Pennings [24]; used with permission.)

shish kebab structure where the shish is the elongated crystals in the row structure and the kebabs are the overgrown epitaxial plates, as shown in the schematic in Fig. 1.4 [24].

High modulus fibers and films are produced from extended chain crystals in both conventional polymers, notably PE, and in liquid crystalline polymers (LCPs). The topic of high modulus organic fibers has been described and reviewed [15–17, 25] providing information on their preparation, structure, and properties. High modulus fibers are found in applications such as fiber reinforced composites for aerospace, military, and sporting applications. Industrial uses are for belts, ropes, and tire cords. Extended chain crystals can also form when polymers are crystallized very slowly near the melting temperature, but they are weak and brittle.

1.2.3 Liquid Crystalline Polymers

Rigid and semirigid polymer chains form *anisotropic* structures in the melt or in solution, which result in high orientation in the solid state without drawing. Liquid crystalline melts (*thermotropic*) or solutions (*lyotropic*) are composed of sequences of monomers with long rigid molecules. Aromatic polyamides and polyamidehydrazides are examples of two polymers that form liquid crystal solutions. Aromatic copolyesters and polyazomethines form nematic liquid crystalline melts at elevated temperature. Melt or solution spinning processing of anisotropic LCPs results in an extended chain structure in the fiber or film. Heat treatment generally improves the orientation and the high modulus and tensile strength properties of these materials.

1.2.4 Multiphase Polymers

Many amorphous thermoplastics are brittle, limiting their range of applications. Toughening with rubber is well known to enhance fracture resistance and toughness. Many major chemical industries are based on *toughened plastics*, such as ABS, HIPS, and ionomers [26–30]. Important issues in the design of fracture resistant polymers are compatibility, deformation, toughening mechanisms, and characterization. Particle size distribution and adhesion to the matrix must be determined by microscopy to develop structure-property-process relationships.

Rubber toughened polymers are usually either copolymers or polymer blends. In random copolymers, a single rubber or matrix phase can be modified by the addition of the second component. Graft and block copolymers have modified properties due to the nature of the rubber-matrix interface. In graft copolymers, grafting provides a strong bond between the rubber and matrix in the branched structure where one monomer forms the backbone and the other monomer forms the branch. Graft copolymers are usually produced by dissolving the rubber in the plastic monomer and polymerizing it to form the graft. Typical block copolymers are polystyrene-polybutadiene and polyethylene-polypropylene. They have the monomers joined end-to-end along the main chain, which results in a bonding of the two phases. New processes have resulted in novel polymer blends that will be discussed in the process section of this chapter.

1.2.5 Composites

Polymer *composites* are engineering resins or plastics that contain particle and/or fibrous fillers. Specialty composites, such as those reinforced with carbon or ceramic fibers, are used in aerospace applications, and glass fiber reinforced resins are used in automobiles and many other applications requiring enhanced mechanical properties compared with the plastic. Composite properties depend upon the size, shape, agglomeration and distribution of the filler and its adhesion to the resin matrix. It is well known that long, well bonded fibers result in increased stiffness and strength, whereas poor adhesion of the fibers can result in poor reinforcement and poorer properties. Fillers for polymers include glass beads, minerals such as talc, clay, and silica, and inorganic fillers used to strengthen elastomers. Small particles (carbon blacks or silica) are added during manufacture to change color or specific properties, such as conductivity. Reinforcements that change mechanical properties include long and short carbon and glass fibers. The toughness and abrasion resistance imparted by fillers is very important in rubber applications such as tire cords.

The theory, processes, and characterization of short fiber reinforced thermoplastics have been reviewed by De and White [31], Friedrich et al. [32], Summerscales [33], in an introductory text by Hull and Clyne [34], and in a handbook by Harper [35]. Natural fibers and composites have been reviewed by Wallenberger and Weston [36]. The introduction of new composite materials, called *nanocomposites*, has resulted in new materials that are being applied to various industrial applications. These materials have in common the use of very fine, submicrometer sized fillers, generally at a very low concentration, which form novel materials with interesting morphology and properties. Nanocomposites have been discussed in a range of texts including two focused on polymer-clay nanocomposites by Pinnavaia and Beall [37] and Utracki [38].

1.3 POLYMER PROCESSES

The growing global market for plastics has resulted in manufacturers focusing on performance improvements and novel processing to improve efficiencies and lower costs. The major point of this section is to emphasize that the structure and properties of polymers are basically a function of their chemical composition and the *process* used to make the product. The polymer manufacturing processes are not discussed here, but rather the focus is on the processes used to take neat polymers and form them into useful products. There are many processes used to manufacture polymer materials but only the basic ones will be mentioned. Important commercial processes used to manufacture polymer materials fall into three major categories: continuous (i.e., fiber spinning, extrusion, pultrusion, and calendaring); semicontinuous (i.e., injection molding and blow molding); and batch (i.e., compression molding and thermoforming) [39]. Another approach is to consider processes based on the form of the final product (e.g., fibers, films, extrudates, moldings, etc.). Development of relationships between the chemical and physical structure and properties of the polymers requires an understanding of the specific process and its effect on the resulting morphology. The objective of this section is to outline the nature of some basic processes, the important process variables, and the relation of those variables to the structure of the final product. General references on the topic of polymer processes (e.g., [2, 16, 39–45]), equipment manufacturers' Web sites, and the Society of Plastics Engineers Web site and journals should be referred to for more specific detail.

1.3.1 Fiber and Film Formation

1.3.1.1 Fiber Processes

Polymer fibers are found in textile applications (Appendix III) for clothing and household items, such as sheeting and upholstery, and also for industrial uses, such as cords, ropes, belts, and tire cords. Polymer fibers can be either short in length (staple fibers) or continuous, very long filaments. Fibers are produced by a melt or solution spinning process, as shown in the schematic in Fig. 1.5 [46]. In either case, the polymer melt or solution is extruded through a *spinneret* (or jet) to form the fiber, which cools and crystallizes and is taken up on a bobbin and may be further oriented by drawing on-line or by a post-treatment process resulting in high tensile strength and modulus. Requirements for textile fibers are that the crystalline melting temperature must be above 200°C, so that the textile fabric can be ironed, and yet below 300°C, to permit conventional, melt spinning processing. Alternatively, the polymer is dissolved in a solvent from which it can be spun by such processes as wet and dry spinning.

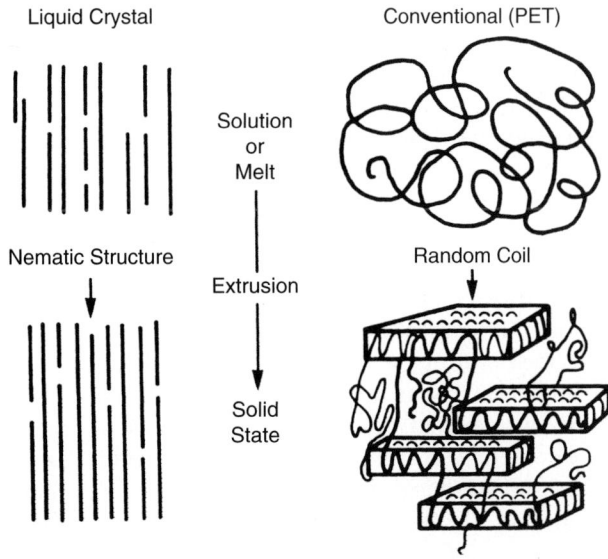

FIGURE 1.5. Schematic of fiber formation process for conventional and liquid crystalline polymers. (From Calundann and Jaffe [46]; Celanese Americas, formerly Hoechst Celanese; used with permission.)

Polymers that can be melted within this range are typically melt spun and include polyesters, polypropylene, and nylon. Polymers that are solution spun include some acetates and lyotropic LCPs. Schematic diagrams of the spinning apparatus are reviewed and shown, for example, by Griskey [39] and have been well known for decades.

In tensile drawing, stress is applied, resulting in thinning and elongation of the crystal and rotation of the molecules or bundles in the draw direction; increasing the draw ratio is known to increase the Young's modulus and the breaking strength by improving the degree of molecular alignment or extension. The diameter of the microfibrillar texture is also affected by the draw ratio with thinner microfibrils at higher draw ratios. The high speed spin-draw fiber process also yields fibers with high modulus and tensile strength when temperature, draw ratio, and speed of the process are well controlled. Heat treatments are often used to impart desired structures and properties in the fiber, and annealing of fiber forming thermoplastics yields a highly crystalline morphology. Uniaxially oriented fibers have a high degree of molecular symmetry and high cohesive energy associated with the high degree of crystallinity. Specific variables in the spinning process play an important role in the final fiber properties.

1.3.1.2 Orientation Methods

There are a variety of well known methods used to orient materials, generally by use of *tensile stress*, as reviewed by Holliday and Ward [47], Ciferri and Ward [48], and Gedde [49] and more recently by Ward [16, 50], Ciferri [15], and Chung [44]. Orientation takes place during the spinning process, as the polymer is first extruded through the spinneret, in the fiber skin and core as they solidify, and by postspinning processes. Ductile thermoplastics can be *cold drawn* near room temperature, whereas thermoplastics that are brittle at room temperature can only be drawn at elevated temperatures. Thermosets are oriented by drawing the precursor polymer prior to *crosslinking*, resulting in an irreversible orientation. Rubbers can be reversibly elongated at room temperature. The orientation in oriented rubbers is locked in place by cooling, whereas heating drawn thermoplastics causes recovery.

Cold drawing is a solid transformation process, conducted near room temperature and below the melting transition temperature of the polymer, if it is crystalline. The process yields a high degree of chain axis alignment by stretching or drawing the polymer with major deformation in the neck region. Deformation of the randomly oriented spherulitic structure in thermoplastics, such as in PE and nylon, results in a change from the stacked lamellae (ca. 20 nm thick and 1 μm long) to a highly oriented microfibrillar structure (microfibrils 10 nm wide and very long) with the molecular chains oriented along the draw direction, as shown in the schematic in Fig. 1.6 [51]. The molecules, links between the adjacent crystal plates in the spherulites, also appear to orient and yet still connect the stacked plates in the final fibrillar structure. Additionally, the drawing process

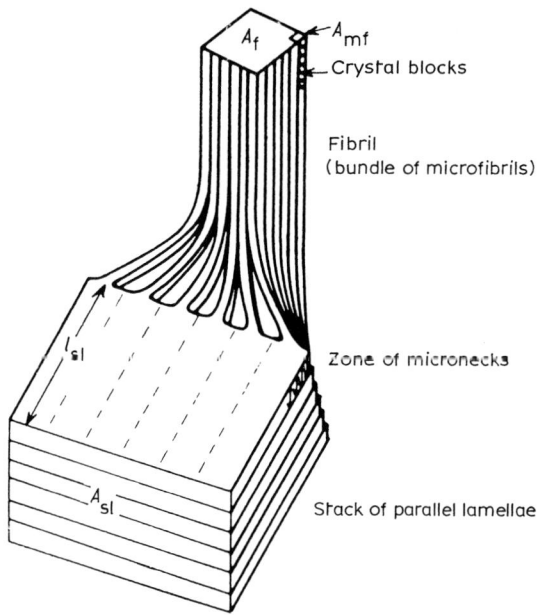

FIGURE 1.6. Schematic of cold drawing process with transformation of the lamellar texture into a microfibrillar structure. (From Peterlin [51]; used with permission.)

causes orientation in the noncrystalline amorphous component of the polymer. The degree of orientation of the crystalline component is characterized by wide angle x-ray scattering (WAXS) and birefringence. Clearly, the deformation process causes a major change in the microstructure of the polymer resulting in improved strength and tensile modulus.

1.3.1.3 Film Processes

Polymer films are used in many applications including packaging and electronic recording and in membranes for separations applications. Films are formed by similar uniaxial or biaxial processes used for fibers that impart high strength properties in either one or two directions, respectively. Processes include film extrusion, drawing, stretching, extrusion at high pressure through a die, and crystallization under flow. Extrusion is generally followed by stretching to orient the structure and also by blown film manufacture (see Section 1.3.2). Deformation processes that impart orientation to polymers can result in anisotropic mechanical properties. The increase in molecular alignment can result in increased stiffness and strength. The effects of orientation are dependent on the nature of the starting materials and on whether they are isotropic or anisotropic.

1.3.2 Extrudates and Moldings

Polymer morphology in extrudates and moldings is affected by process variables, such as melt and mold temperature, pressure, shear, and elongational flow. Process variables affect the morphology of the material thus affecting the fabricated product performance and mechanical properties. Pressure increases, for instance, can increase both the melting temperature and the glass transition temperature of a polymer, with the result that the polymer solidifies more quickly. In a crystalline polymer, the nucleation density can increase, resulting in a decrease in spherulite size with increased pressure in injection molding. In the special case of polymer blends, care must be taken to fully understand the effect of the process on the two or more polymers being used in the blend as even the size of the extruder or molding machine can cause major changes in the local orientation and stresses in the part and thus the morphology and the final properties. Many of the references cited [39, 43, 52, 53] cover this topic, but further references specific to polymer blends should be considered, for example, [26, 28, 54] and others found in Chapter 5 in the various relevant application sections.

1.3.2.1 Compounding

Many of the fabricated plastic products manufactured today include fillers and additives to modify and/or reinforce the final product properties. Fillers are blended with polymers to modify physical properties, enhance tensile strength or modulus or specific characteristics, such as wear resistance, flammability, electrical properties, color, and so forth [34, 55], or to reduce the polymer level in expensive materials. Reinforcements modify strength and modulus as the particles or fibers bear a fraction of the applied load. *Compounding* is the process of introducing fibers or particles into a resin prior to molding. Generally, the fillers or additives are added to the molten polymer and they are mixed together and extruded and cut into pellets that can be used in other processes such as molding. Process parameters, such as screw design, melt temperature and pressure, relate to the structure in the final material. High speed and pressure are known to result in a glossy surface finish, and high melt temperature is used to reduce viscosity and minimize fiber breakage. Fillers such as titanium dioxide and clay are used in paints and adhesives as pigments or toughening agents. A book edited by White, Coran, and Moet [30] examines the characteristics and methods of preparing polymer blends and compounds in batch and continuous mixing equipment.

Compounding of long fiber reinforced thermoplastics (LFRTs) is very different from compounding short fiber or particle filled composites. Long fibers enhance properties such as impact strength and are finding great utility in a number of applications in the automotive, industrial, and sports markets. Most of these products use PP or polyamide (PA), but poly(ethylene terephthalate) (PET) and other resins are also used.

Processes that are used for such compounding include an extrusion process termed *pultrusion*. The formed shapes are structural, pipes and tubing or long pellets used for molding parts. Continuous fibers, such as glass or carbon fibers, are drawn through a heated die in this process forming highly oriented extrudates. In-line or direct compounding of LFRTs can be done in which the compound with fibers and additives is formed into extrudates and directly injected into the molding process to form a part. Most of the pellets are used for injection molding but some are also used for compression molding generally for larger parts. The polymer viscosity and thermal properties are key factors in the process as are the standard process variables as noted above.

1.3.2.2 Extrusion Processes

Fundamental factors of extrusion technology have been widely reviewed, for example, in a Society of Plastics Engineers (SPE) Guide edited by Vlachopoulos and Wagner [56] and in books on polymer processing by Griskey [39], on polymer extrusion by Chung [44], and on polymer blends by Utracki [29]. As with all processing of polymers, extrusion involves the controlled melting and flow of the polymer and shaping of the material by forcing it through a die with an extruder. The goal is to form a homogeneous melt at a uniform and high rate and is part of the extrusion, blow molding, and injection molding process. Extruders are basically helical screw pumps that convert solid polymer particles to a melt delivered to a die or a mold [39]. These extruders are called single screw or twin screw devices depending on the actual number and type of screw used. Basically, the process is similar to fiber processing in that the polymer in the form of pellets is fed into a hopper and into the screw, which is driven by a motor. Heat is generated both by the heaters and by the molten polymer itself. The heated plastic is conveyed along the extruder until the pellets are melted and mixed well through a series of sections that control the melt temperature, mixing, and pressure, and thus the final product. The choices of single, twin, and multiple screws and the screw designs depend on the polymer used and the desired product. For instance, twin screws provide increased output and ability to handle materials compared with single screws. Twin screw extrusion is generally used for compounding polymers with fibers and fillers and also to fabricate nanocomposites. Process variables are used to control the mixing, uniformity, and the rate of production of extrudates.

Extruders are used in a wide range of processes to manufacture a range of products. In the simplest case, an extruded rod or pellet of material, with or without fillers, is produced and generally chopped to form pellets for various other processes such as injection molding. Extruders are also used to manufacture filaments, rod and pipe, sheet, wire coatings, blown film, and cast film. Polymer properties that influence the extrusion (and molding) process include molecular weight and molecular weight distribution, melt rheology, thermal and mechanical stability, bulk density, compressibility, and melt density. The thickness of the extrudates is important as is the extrusion speed, temperature, and viscosity, all of which affect the overall structure of the processed part. One of the key structures affected by the process variables is the size of the spherulites in the product, which in turn affects properties such as the impact strength, elongation, temperature resistance, among many others. Another structure of major importance is the *skin-core* morphology, which results from rapid cooling of the outside surface or skin of the extrudates versus the slower cooling of the central core region. This is a very simple description of a very complex process that should be well known to the engineers and scientists developing new materials and evaluating process-structure-property relations.

1.3.2.3 Injection Molding Processes

Injection molding is widely used to produce plastic parts for a broad range of industries, notably electrical and electronic, automotive, appliances, medical, and so forth, because of its flexibility to provide high rates of production on parts with tight tolerance. In injection molding, a mix of polymer and fibers, fillers and/or additives is injected into a mold at elevated temperature and pressure. These parts can be neat polymers or polymer blends, but they are more commonly chopped fiber reinforced thermoplastic composites. Molding is a semicontinuous process during which polymer pellets are fed into the heated barrel, melted or thermally softened, injected or forced into the nozzle, sprue, runner, and gate on the way to the mold cavity. Once pressure is achieved, the mold is cooled and the part is ejected. Many of the factors discussed for extrusion (in Section 1.3.2.2) are similarly important for molding, such as melt temperature, mold temperature, injection rate, and of course material properties, such as viscosity (molecular weight), chemistry, and melt temperature. The thermal and shear history, which varies with location in the mold, also affects the molecular orientation and the morphology and thus the mechanical properties of the molded article. Generally, the pellets used for molding are predried as moisture affects the viscosity and degradation of many polymers. Gates are important because they present a high resistance to flow and they allow the melt to reach the mold cavity quickly while minimizing energy and pressure losses. Mold design, especially for very fine parts, is critical to the dimensional stability of the part and to shrinkage, warpage, and thermal and chemical stability. Engineering details are found in the cited references and on the Web sites of machine manufacturers.

Macroscopic product problems that can result from poor control in injection molding include, but are not limited to: voids and sink holes on the surface generally due to poor mold filling or low pressure, incomplete mold filling, weld lines and flow marks, warping or distortion of parts, high shrinkage, and so forth.

Microstructures that are typically observed in molded parts and extrudates include anisotropic textures. The higher orientation in extrusion can result in highly oriented rods or strands, at high draw ratios and/or small diameters, or in structures with an oriented skin and a less oriented central core in thicker strands. This skin-core texture is due to a combination of temperature variations between the surface and the bulk and the flow field in both extrusion and molding processes. For instance, the flow fields in a molded part are shown schematically in Fig. 1.7 [46]. Extensional flow along the melt front causes orientation. Solidification of the polymer on the cold mold surface freezes in this orientation. Flow between the solid layers is affected by the temperature gradient in the mold, and the resulting flow effects [57, 58] result in a rapidly cooled and well oriented skin

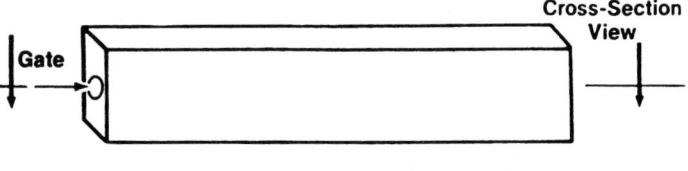
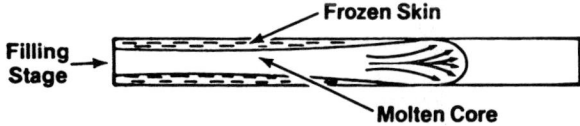
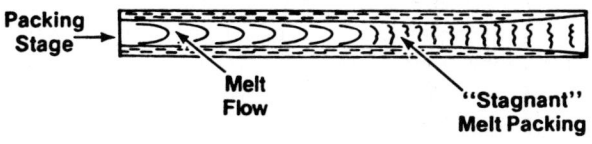

FIGURE 1.7. The pattern of the flow directions in the injection molding process is shown in the schematic diagram. (From Calundann and Jaffe [46]; Celanese Americas, formerly Hoechst Celanese; used with permission.)

structure and a slowly cooled, randomly oriented core. Extensional flow along the melt front results in molecular orientation parallel to the knit lines when two melt fronts meet. The resulting knit, or *weld line*, is a region of weakness in the molded part, and weld line fractures are commonly encountered. The local orientation is quite important as tensile strength and impact strength properties are known to be higher in the orientation direction.

Typically, in a semicrystalline polymer there are three zones within the molded part: an oriented, nonspherulitic skin; a subsurface region with high shear orientation, or a transcrystalline region; and a randomly oriented spherulitic core. The thickness of the skin and shear zone is known to be an inverse function of the melt and mold temperature with decreased temperatures resulting in increased layer thickness. In the skin, the lamellae are oriented parallel to the injection direction and perpendicular to the surface of the mold. Amorphous polymers also show a thin surface oriented skin on injection molding. When amorphous polymers are heated to the glass transition temperature and then relaxed, they exhibit shrinkage in the orientation direction and swelling in the other directions.

1.3.2.4 Other Molding Processes

There are other molding processes that will be mentioned. *Compression molding* is a very old molding process generally used for thermosetting resins. Compression molding involves the introduction of a resin and a curing or cross-linking agent into a mold followed by heating and application of pressure to cause a reaction resulting in thermosetting the material in a specific shape (e.g., [39, 52]). An overview of this topic including the underlying theory and physics is found in a text on compression molding by Davis et al. [52]. Fillers and reinforcements are used similar to those used in injection molding. The morphology of the parts is complex due to the variation in thermal properties and stress with position in the part. The process is used for large parts that do not require good tolerance compared with injection molding.

Blow molding is a very common process that is used to produce hollow objects. Products as different as food and beverage products, fuel tanks, cylinders, and blown film are formed by three different processes: extrusion blow molding, injection blow molding, and stretch blow molding. In extrusion blow molding a hollow tube of molten or thermally softened polymer, termed a *parison*, is extruded into a split cavity mold and crimped at one end. Compressed air is blown into the parison to fit the mold shape, causing the polymer to solidify. In injection blow molding, molten or softened polymer is injected into a heated mold cavity around a core pin and then the mold is opened and moved using the core pin where it is blown open and then ejected. Stretch blow molding can be by either extrusion or injection molding and results in a biaxially oriented product. Variables that are important to the morphology include thermal properties and rheology of the polymer and the transfer of heat through the part during the process. Multilayer blow molding is also increasing in demand due to potentially improved barrier properties. These processes are quite complex and this brief section only introduces the topic, which is described elsewhere (e.g., [39, 43, 53, 59]).

Thermoforming is another process used for making polymer products, such as automotive panels, underhood and fuel tank applications, chemical tanks, and packaging materials for the medical and electronics industries. A sheet formed by another process, such as extrusion, is preheated and then placed in a heated mold under pressure to form it into a part. Variations on the method include simple heating and stretching, pressure forming, contact forming, and so forth [60]. The thermoforming process is viewed as an alternative to blow molding, and it can also handle multilayered sheets such as those needed for plastic fuel tanks. Whereas an entire tank is formed by blow molding, thermoforming can involve making two halves of a tank and welding them together after all components are placed in them. As with most other processes, the final product properties are a function of the material's thermal, rheological, and chemical properties, and these are affected by the process variables, which, in turn, affect the morphology of the part. One of the advantages of thermoforming is the production of high gloss surfaces without painting, key to some automotive applications as a metal replacement. Once again, this is a brief summary of a very complex engineering process that the polymer microscopist needs to be aware of in understanding process-structure-property relationships; for more details on the process, see the many Web sites and books on processing and properties (e.g., [16, 43, 60]).

Reaction injection molding (RIM) is generally used for thermoset polyurethane, nylon, polyesters, and epoxies [39]. This process generally uses two metered reactive streams that combine and mix and then are injected into the mold. In the case of urethanes, one stream contains a polyether backbone, a catalyst, and a crosslinking agent, and the other has an isocyanate. The use of a blowing agent expands the material after mixing to fill the mold.

1.3.2.5 Coating Processes

There is a wide variety of polymer coatings and of processes to apply them to other polymers, metals, electronic devices, and for many applications in the aircraft, chemical and petroleum, food, textiles, and transport industries [43, 61, 62]. For example, polymer based coatings are used to protect steel from corrosion and extruders used in food and medical applications from wear. Adhesion is important if good protection is sought. A text by Grainger and Blunt [61] describes the many methods used for formation of surface coatings and surface modification, generally used to delay degradation and prolong the life of engineering components. The mechanisms of wear and corrosion must be understood if the coating is to provide protection. Coatings are applied by spin casting, thermal spray, electrodeposition, physical and chemical vapor deposition, laser surfacing, and various powder methods.

Licari [62] in his book on materials and processes for electronics applications describes the proper application of coating materials for the protection of electronics from environmental factors such as humidity, temperature, and high space vacuum for military and commercial applications. The chemistry and properties of a broad range of polymer coatings are discussed, including acrylics, polyesters, polystyrenes, epoxies, polyurethanes, silicones, polyimides, benzocyclobutene, fluorocarbons, polyamides, phenolics, and polysulfides. Processes discussed are spray coating, aerosol spray, electrostatic spray, dip coatings, fluidized bed coating, electrocoating, vapor deposition, spin coating, extrusion coating, and many others. Clearly, the various coating methods are affected by the polymer chemistry, thermal stability, rheology, and other factors that affect the coating thickness, adhesion, and degradation. The polymer morphology is a direct result of the chemistry and properties of the polymer and the impact of the specific process used for its application.

1.3.2.6 Novel Processes

Several processes will be discussed in this section that are not as commonly applied to polymers but have found niches for new applications. The first to be discussed is the formation of novel composites by *hot compaction* of fibers, developed by Ward and his group at Leeds (e.g., [63–65]). In this process, highly oriented fibers, such as melt and gel spun PE, PET, and LCP fibers are compacted until selective surface melting of some of the fibers permits the formation of a fiber composite with high strength and stiffness [63]. Potential applications of these materials include automotive industry, sports protection equipment, and many others [64] requiring exceptional mechanical properties. Investigation of the various process parameters showed that the time spent at the compaction temperature, termed the *dwell time* [65], was critical to formation of the composite. Molecular weight measurements showed that hydrolytic degradation occurred rapidly at the temperatures required for successful compaction, leading to embrittlement of the materials with increasing dwell time; a dwell time of 2 min was found to be optimum to have enough melted material to bind the structure together while resulting in only a small decrease in molecular weight. These studies included evaluation of mechanical properties and morphology that resulted from variations in the process variables and the polymer, such as the use of high molecular weight materials, which resulted in high impact performance of the hot compacted sheets.

Layer multiplying coextrusion is a process that has been used to fabricate one dimensional polymer blends by forced assembly of two polymers into many alternating thin layers (e.g., [66]) even to the level of nanolayer films consisting of thousands of continuous layers of two polymers with layer thickness less than 10 nm. Multilayer films have potentially improved impact strength and may also have enhanced permeability useful for many industrial applications. A schematic of the process used for microlayer and nanolayer coextrusion in Fig. 1.8 [66] shows that the viscoelastic nature of polymer melts is to repeatedly split. The process permits two immiscible polymers to come into intimate contact, and the localized mixing creates an "interphase" region, key to the properties of polymer blends [28]. A critical process variable, in the case of an amorphous polyester and polystyrene, is the extruder temperature, which was adjusted to ensure that the viscosities matched when the melts were combined in the feed block. Once the melt exited the assembly, it was spread in a film die to further reduce the layer thickness and rapidly quenched on a chill roll equipped with an air knife to freeze the melt morphology. The number of layers is dependent on the number of die elements, and the film thickness ranged from more than $10\,\mu m$ to a few nanometers [66]. The film properties were clearly a result of the morphology frozen in by the process and affected by the process variables used. An example of a film formed using this process will be described further in Section 5.2 (see Fig. 5.38).

The development of *chaotic advection*, initiated more than two decades ago by Aref and reviewed by him more recently [67], has led to its application and development for materials processing of polymer blends and clay nanocomposites by Zumbrunnen and his group at Clemson University [68–72]. This process has the exciting result of providing significant improvements to the impact properties of blends by addition of low volumes of the second phase. In one case [68], a polystyrene matrix was improved by the addition of only 9% by volume of low density PE (LDPE). The polymers were combined in the molten state within a cylindrical cavity where a quiescent, three dimensional chaotic mixing process resulted in the formation of stretched and folded minor phase domains that were interconnected and stable upon solidification. The unique microstructures were due to the process used and differed from the normally observed fine domain textures that generally result when using low volumes of a second phase polymer. The process has also been used in continuous flow to form extruded films with many layers. A batch process study of polystyrene and LDPE was used as a model binary system so that thicker layers could be formed for evaluation [69]. Figure 1.9 [71] is a schematic of the continuous chaotic advection blender (CCAB). In this study, the unique CCAB was used to investigate the influence of the morphology formed on tensile and impact toughness properties of blends of polypropylene (PP) with LDPE. Process control of the melt flow, and the stir rods, as well as the specific machinery design is important to obtaining the unique morphology that is responsible for the unexpected property

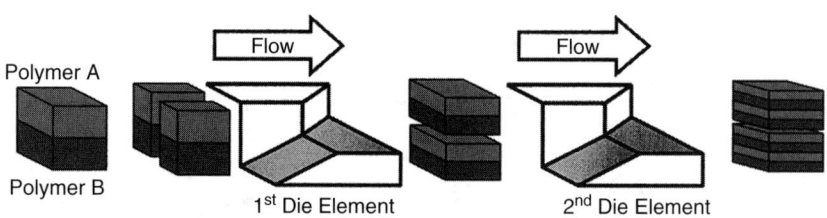

FIGURE 1.8. Schematic of layer multiplying coextrusion used for forced assembly of polymer nanolayers shows that two die elements multiply the number of layers from 2 to 8. (From Liu et al. [66], © (2004) American Chemical Society; used with permission.) (See color insert.)

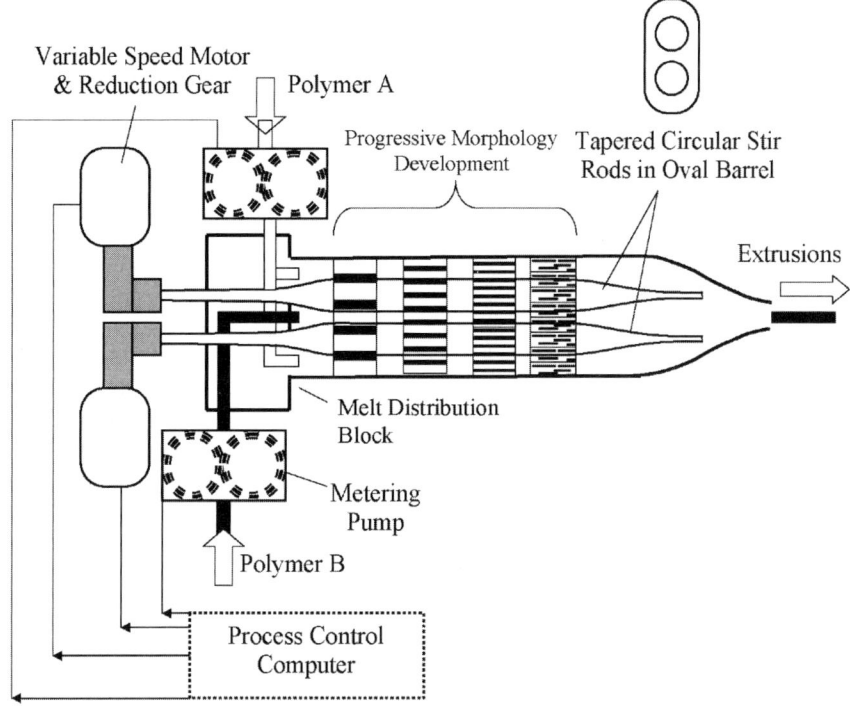

FIGURE 1.9. Schematic representation of the continuous chaotic advection blender. (From Zumbrunnen et al. [71], © (2005) Elsevier; used with permission.)

profiles. Further study currently under way involves the flow of the melt from this process into molds to maintain and control the morphology in molded parts [71]. Film formed using this process is described in Section 5.2 and an example shown in Fig. 5.39; an example of a nanocomposite formed by this process is described in Section 5.4 (see Fig. 5.109).

1.4 POLYMER CHARACTERIZATION

With the invention of scanning probe microscopes more than 25 years ago, the entire field of materials exploded into the realm of the nanoworld. The ease of imaging atoms and molecules has opened up the characterization of materials including polymers. The microscopes have changed and enabled materials to change as well, composites now include nanocomposites, and so on. Thus, the characterization of materials has changed dramatically in the past decade. Advances in many other microscopy and analytical techniques has also resulted in additional information being available about materials. The need for complementary techniques to fully understand materials continues to be required for full understanding of structure-property-process relationships.

1.4.1 General Techniques

A very wide range of analytical techniques are used to characterize polymer materials (e.g., see references on polymer physics [49], thermal analysis [73, 74], light microscopy [75, 76], Raman [77, 78], x-ray scattering [79], various spectroscopies [80, 81], and a wide range of microscopy techniques [82]). A text on polymer blends also describes many polymer characterization techniques [83]. Texts on microscopy with a focus on biological materials are often useful for the polymer microscopist (e.g., [84, 85]) as the materials have in common a tendency to be soft, to require contrast enhancement, and to suffer from radiation damage in electron beam instruments. The primary characterization of an organic material must be chemical. Elemental analysis by wet chemistry or spectroscopy may be useful in a few cases, for example to determine the degree of chlorination in chlorinated PE, but most chemical analysis is at the level of the functional group. Ultraviolet/visible spectroscopy and mass spectroscopy (MS) of fragments broken from the polymer chain are often used. Even more common spectroscopies are infrared (IR) absorption, Raman, and nuclear magnetic resonance (NMR), which is very important. All of these can distinguish specific chemical groups in a complex system. Raman may be used on small particles and inclusions much more easily than IR, but IR and in particular Fourier transform IR (FTIR) has advantages of sensitivity and precision. Nuclear magnetic resonance also gives local information, on a very fine scale, about the environment of the atoms investigated.

Once the chemistry of the molecule is known, the next important characteristic is the molecular weight distribution (unless the material is a thermoset or elastomer with infinite molecular weight). The molecular weight distribution is determined by a range of solution methods of physical chemistry, viscometry, osmometry, light scattering, and size exclusion chromatography. Chemical and physical characterization methods overlap in the polymer field, for NMR of solid samples can determine the mobility of atoms in various regions and the orientation of molecules. IR and Raman are also sensitive to orientation and crystallinity of the sample.

There are two further general types of physical characterization. They involve either scattering of light, neutrons, or x-rays or the formation of images of the polymer by microscopy, the subject of this text. Electron diffraction logically belongs in the first group but is always performed in an electron microscope, so it is associated with microscopy. This technique shares with microscopy the ability to determine the structure of a local region, whereas other scattering methods determine the average structure in a large sample volume. The impact of electron crystallography on the study of polymer materials has been reviewed by Voigt-Martin [86, 87].

The past decade has seen the emergence of analytical imaging, which is imaging using the signals from various analytical instruments, such as FTIR and Raman microscopy, x-ray microscopy, and imaging by surface analysis using secondary ion mass spectrometry (SIMS) and x-ray photon spectroscopy (XPS).

1.4.2 Microscopy Techniques

Microscopy is the study of the fine structure and morphology of objects with the use of a microscope. Resolution and contrast are key parameters in microscopy studies, which will be discussed further. The specimen and the preparation method also affect the actual information obtained as the contrast must permit structures to be distinguished. Optical bright field imaging of multiphase polymers, for instance, has the potential of resolving details less than $1 \mu m$ across; however, if the polymers are both transparent, they cannot be distinguished due to a lack of contrast. There are variations among microscopes in available resolution, magnification, contrast mechanisms, and the depth of focus and depth of field. Optical microscopes produce images with a small depth of focus, whereas scanning electron microscopes (SEMs) have both a large depth of focus and large depth of field.

There is a wide range of microscopy instruments available that can resolve details ranging from the millimeter to the subnanometer scale (Table 1.3). The size and distribution of

TABLE 1.3. Characterization techniques: Size ranges

Wide angle x-ray scattering (WAXS)	0.01–1.5 nm
Small angle x-ray scattering (SAXS)	1.5–100 nm
Transmission electron microscopy (TEM)	0.2 nm–0.2 mm
Scanning probe microscopy (SPM, AFM, STM, etc.)	0.2 nm–0.2 mm
Scanning electron microscopy (SEM)	4 nm–4 mm
Optical microscopy (OM)	200 nm–200 μm
Light scattering (LS)	200 nm–200 μm

spherulites can be observed by optical techniques, but more detailed study requires electron microscopy. Single lamellar crystals can be seen with phase contrast optical microscopy but need TEM for detailed imaging and measurements. Polarized light microscopy is an optical technique that enhances contrast in crystalline materials. Phase contrast optical techniques enhance contrast between polymers that are transparent but that have different optical properties, such as refractive index and thickness. Reflected light techniques reveal surface structures, and observation of internal textures of thin polymer slices is possible by transmitted light. Combinations of these microscopy techniques provide images of the morphology of polymer materials. Introductory texts [75, 76, 88, 89] further describe microscopy techniques and provide descriptions and definitions of microscopes and relevant principles. Books and major journal articles that describe optical microscopy (OM) and TEM advances including electron diffraction [90–93], electron energy loss spectroscopy in the TEM [92, 94], SEM and field emission SEM [95, 96], scanning probe microscopy (SPM), which includes atomic force microscopy (AFM) [97–99], and x-ray microanalysis in the SEM [96] are also important for polymer microscopists. The recent book by Hawkes and Spence [100] describes most of the known microscopy techniques, including imaging with electrons, photons, probes, and holographic techniques. Image analysis and image processing [101, 102] have taken on more importance with the advent of digital imaging with SEM, TEM and scanning probe microscopies.

1.4.3 Specimen Preparation Methods

The range of specimen preparation methods is nearly as broad as the materials that must be prepared for observation. Metals and ceramics are prepared by well known, standard methods, and biological materials have been prepared by methods specifically developed for their observation. Polymer materials are a bit newer than either of these materials, with a wide range of material forms and types and potential problems that are similar to metals and ceramics and to biological materials. Polymers, in common with biological materials, have low atomic number, display little scattering, and thus have little contrast in the TEM. In addition, they are highly beam sensitive, which must be taken into account. Like metals and ceramics, polymers can also be filled with hard inorganic materials. Overall, the types of methods developed for polymers are a composite of those methods known for metals and biological materials and are specially adapted for macromolecules. Goodhew [103, 104] described many of the metallurgical methods for preparation of specimens for optical, scanning electron, and transmission electron microscopy. Many biological methods are useful for polymers (e.g., [105, 106]). Trempler [107, 108] has an excellent two-part work on the light microscopy of plastics (with a translation available in English) that covers a broad range of methods and techniques, referenced further in succeeding chapters.

There are many characterization problems for microscopy where quite simple preparation methods are applied at least in the initial stages of morphological study. However, most polymers must be prepared with three things in mind: (1) to isolate the surface or bulk, (2) to enhance contrast, and (3) to minimize radiation damage. For surface study, simple cutting out of the specimen or more tedious replication procedures supply the specimen of interest. Bulk specimens are obtained by cutting, fracturing, polishing, or sectioning. A major problem with polymers sectioned for TEM is their inherent lack of contrast. Polymers have a low scattering power, which results in low contrast. Methods that are employed to enhance contrast include staining, etching, replication, shadowing, and metal decoration. Conductive coatings are applied to enhance contrast and to minimize radiation damage. Methods used include metal shadowing, formation of carbon support films, conductive coatings for SEM, metal coatings for optical microscopy, and gold decoration. Specimens for SEM generally need to be electrically conductive in order to produce secondary electrons and to minimize charge buildup as polymers are generally nonconductive. Low voltage or low vacuum techniques can limit the need for such preparation methods. Replication depends on the application of conductive coatings for their formation. Other methods adapted for microscopy preparation of soft, deformable materials are freeze drying, critical point drying, and freeze fracture-etching which have been primarily developed by biologists. The use of newer SPM techniques has permitted some simple methods to be used for sample preparation, but detailed morphology often requires special methods including sectioning to prepare a flat block face for study and staining and etching to enhance contrast. Methods developed specifically for polymers have been reviewed in the literature [8, 76, 109–111] and they are fully described in Chapter 4 of this book, with dozens of examples shown for polymer applications in Chapter 5.

The specimen preparation methods used for microscopy of polymers involves the use of many toxic chemicals as well as the use of instruments that can be radiation hazards. It is well beyond the scope of this text to provide the information required for the proper and safe handling of such chemicals and instruments, and the researcher is encouraged to obtain the required safety information prior to their use.

1.4.4 Applications of Microscopy to Polymers

The increased use of optical, electron, and scanning probe microscopies applied to polymer research has resulted from the widespread acceptance of these techniques combined with the need for higher performance and lower cost polymer materials. It is well known that the structures present in a polymer reflect the process variables and further that they greatly influence the physical and mechanical properties. Thus, the properties of polymer materials are influenced by their chemical composition, process history, and the resulting morphology. Morphological study involves two aspects prior to the study itself: selection of instrumental techniques and development of specimen preparation methods. Structural observations must be correlated with the properties of the material in order to develop an understanding of the material. A major issue in the application of microscopy to the study of polymers is correct image interpretation, which involves an understanding of the nature of the materials, the techniques used, the specimen preparation methods, and potential artifacts. A major issue with electron microscopes is that radiation or beam damage often changes the polymer during imaging, and the nature of the potential changes must be considered for proper image interpretation. Benefits of optical microscopy and scanning probe microscopies are that such damage does not occur, although each technique has its own benefits and potential artifacts that must be understood.

What then are the key specimen preparation methods for studying polymer materials by microscopy techniques? This topic could be organized in one of two ways; that is, by each specific microscopy technique or by each preparation method. The approach that has been chosen is to describe each specimen preparation type for all microscopies in order to minimize overlap and also to make it simpler to use for reference. Those preparation methods chosen for discussion are the typical ones found to be of major utility in the industrial laboratory. They cover the full range of study of the industrial scientist, which is everything from rapid failure analysis to process optimization studies and fundamental research. The fundamental studies must often be fitted into a limited time framework that requires good choices of methods and techniques on a wide range of materials.

Key issues in any microscopy study are that the polymer process must be understood and the structure characterized in order to develop structure-property relationships. Yet there are many questions for even the experienced materials scientist. Where do you start characterization of a new material? Are there any protocols that work most of the time in order to solve each problem by the best technique in the shortest time? How do you minimize artifacts in conducting microscopy experiments? There are no easy answers, but the approach that is described here is to understand the image formation process, to study and understand the advantages and the drawbacks of the preparation methods, to know the instruments, both theoretically and practically, and finally, to conduct the microscopy study and collect the observations and relate the characterization to the process and the physical and mechanical properties.

1.4.5 Emerging Microscopy Techniques

New microscopy techniques continue to be developed, and recent years have been a particularly active time as new microscopes have opened up the field of nanotechnology. Some new techniques are extensions and modifications of existing technology, whereas others are completely new. A 1989 issue of the *Journal of Microscopy* provides a good review of a wide range of types of microscopy available [112–116]. The reviews cover the history of the techniques and their future prospects as seen at that time. Now, many of these future prospects have become real commercialized systems, out of the hands of instrument developers and into the hands of microscopists. One major focus has been toward higher resolution in all microscopies, enabling imaging of much finer detail of many materials, especially so-called nanomaterials. These newer techniques include laser confocal scanning microscopy (LCSM) [112, 117–121], low voltage, high resolution scanning electron microscopy (HRSEM) [95, 96, 122–125], and high pressure SEM (HPSEM), also termed low vacuum SEM [126, 127]. In the latter case, a wider range of polymers can be imaged, including those that are hydrated or wet, and thus dynamic experiments may be conducted although with some loss of resolution. High resolution transmission electron microscopy (HRTEM) [128, 129] is not really new, but it is very difficult to apply the method to polymers, which has slowed its transfer to polymer microscopy. Similarly, electron energy loss spectroscopy (EELS) has been known for some time, but only recently has it been routinely applied to polymers [94, 130]. Another relatively new technique is x-ray microscopy, which has also been applied to polymers [131–133].

More recent is the invention of scanning probe microscopes, which includes the scanning tunneling microscope (STM) [134] and the atomic force microscope (AFM) [135]. These instruments can resolve individual atoms and can operate at atmospheric pressure or even underwater. The AFM requires no specimen preparation beyond the creation or exposure of a surface of interest, although many sophisticated methods, similar to those used for traditional microscopy, are used for high resolution study. With such capabilities, it is not surprising that development and commercialization has been very rapid. The new instruments have made dramatic changes in the imaging of all materials, and that includes biological and synthetic polymers [136–138]. The theory and practical aspects of these new technologies will be addressed in several chapters in this text.

References

1. R.J. Young and P.A. Lovell, *Introduction to Polymers*, 2nd ed. (Chapman & Hall, London, 1991).
2. N. Mills, *PLASTICS: Microstructure and Engineering Applications*, 3rd ed. (Butterworth Heinemann, 2005).
3. H.-G. Elias, *An Introduction to Plastics*, 2nd ed. (Wiley-VCH, Weinheim, 2003).
4. H.D. Keith, *Kolloid Z. Z. Polym.* **231** (1969) 421.
5. P.H. Geil, *Polymer Single Crystals* (Interscience, New York, 1963).
6. A. Keller, *Rep. Prog. Phys.* **31** (1968) 623.
7. B. Wunderlich, *Macromolecular Physics, Vol. 1, Crystal Structure, Morphology, Defects; Vol. 2, Crystal Nucleation, Growth, Annealing* (Academic Press, New York, 1973).
8. D.T. Grubb, in *Developments in Crystalline Polymers - 1,* edited by D.C. Bassett (Applied Science, London, 1982).
9. D.R. Uhlmann and A.G. Kolbeck, *Sci. Am.* **233** (1975) 96.
10. D.C. Bassett, *Principles of Polymer Morphology* (Cambridge University Press, Cambridge, 1981).
11. D.C. Bassett, *CRC Crit. Rev. Solid State Mater. Sci.* **12** (1984) 97.
12. R.B. Seymour, Ed. *The History of Polymer Science and Technology* (Marcel Dekker, New York, 1982).
13. D.C. Bassett, *Polymer* **41** (2000) 8755.
14. D.C. Bassett, in *Mechanical Properties of Polymers Based on Nanostructure and Morphology*, edited by G.H. Michler and F.J. Balta-Calleja (CRC Press/Taylor & Francis, Boca Raton, 2005), p. 3.

15. A. Ciferri, Ed. *Supramolecular Polymers*, 2nd ed. (CRC Press/Taylor & Francis, Boca Raton, 2005).
16. I.M. Ward, Ed. *Structure and Properties of Oriented Polymers* (Springer, London, 1997).
17. I.M. Ward and D.W. Hadley, *An Introduction to the Mechanical Properties of Solid Polymers*, 2nd ed. (John Wiley & Sons, Hoboken, NJ, 2004).
18. H.D. Keith and F.J. Padden, *J. Appl. Phys.* **34** (1963) 2409.
19. I.M. Ward, Ed. *Structure and Properties of Oriented Polymers* (Applied Science, London, 1975).
20. D.C. Bassett, *J. Macromol. Sci. Phys.* **B42** (2003) 227.
21. D.C. Bassett, *Macromol. Symp.* **214** (2004) 5.
22. D.T. Grubb and A. Keller, *J. Polym. Sci. Polym. Phys. Edn.* **18** (1980) 207.
23. A. Peterlin, in *Structure and Properties of Oriented Polymers*, edited by I.M. Ward (Halsted, Wiley, New York, 1975), p. 36.
24. A.J. Pennings, *J. Polym. Sci., Polym. Symp.* **59** (1977) 55.
25. A.E. Zachariades and R.S. Porter, Eds. *High Modulus Polymers* (Marcel Dekker, New York, 1988).
26. D.C. Bassett and A.S. Vaughan, in *Polymer Characterization Techniques and Their Application to Blends*, edited by G.P. Simon (American Chemical Society, Washington, DC, 2003), p. 436.
27. G.H. Michler, R. Adhikari, S. Henning, *Macromol. Symp.* **214** (2004) 47.
28. D.R. Paul and C.B. Bucknall, Eds. *Polymer Blends*, two-volume set (John Wiley & Sons, New York, 1999).
29. L.A. Utracki, Ed. *Polymer Blends Handbook, Volume 1 & Volume 2* (Springer, New York, 2002).
30. J.A. White, A.Y. Coran and A. Moet, Eds. *Polymer Mixing: Technology and Engineering*, (Hanser, Munich, 1999).
31. S. De and J.R. White, *Short Fibre-Polymer Composites* (Woodhead Publishing, UK, 1996).
32. K. Friedrich, S. Fakirov and A.E. Zhang, Eds. *Polymer Composites* (Springer, New York, 2005).
33. J. Summerscales, Ed. *Microstructural Characterisation of Fibre-Reinforced Composites* (CRC Press, Boca Raton, 1998).
34. D. Hull and T.W. Clyne, *An Introduction to Composite Materials*, 2nd ed. (Cambridge University Press, New York, 1996).
35. C.A. Harper, Ed. *Handbook of Plastics, Elastomers, and Composites*, 4th ed. (McGraw-Hill Professional, New York, 2002).
36. F.T. Wallenberger and N.E. Weston, Eds. *Natural Fibers, Plastics and Composites* (Kluwer Academic, Boston, 2004).
37. T.J. Pinnavaia and G.W. Beall, Eds. *Polymer-Clay Nanocomposites* (Wiley, Hoboken, NJ, 2001).
38. L.A. Utracki, *Clay-Containing Polymeric Nanocomposites* (Shrewsbury: Rapra Technology Ltd, 2004).
39. R.G. Griskey, *Polymer Process Engineering* (Chapman & Hall, New York, 1995).
40. N.P. Cheremisinoff and P.N. Cheremisinoff, Eds. *Handbook of Applied Polymer Processing Technology* (CRC Press, Boca Raton, 1996).
41. R. Beyreuther and H. Brünig, *Dynamics of Fibre Formation and Processing; Modelling and Application in Fibre and Textile Industry* (Springer, New York, 2007).
42. D.V. Rosato and D.V. Rosato, *Plastics Engineered Product Design* (Elsevier, Amsterdam, 2003).
43. D.V. Rosato and D.V. Rosato, *Plastic Product Material and Process Selection Handbook* (Elsevier, Amsterdam, 2004).
44. C.I. Chung, *Extrusion of Polymers: Theory and Practice* (Hanser Gardner Publications, Munich, 2000).
45. T.A. Osswald, *Polymer Processing Fundamentals* (Hanser, Munich, 1998).
46. G. Calundann and M. Jaffe. *Proc. Robert A. Welch Conf. on Chemical Research, XXVI, Synthetic Polymers* (Welch Foundation, Houston 1982), p. 247.
47. L. Holliday and I.M. Ward, in *Structure and Properties of Oriented Polymers*, edited by I.M. Ward (Halsted, Wiley, New York, 1975), p 1.
48. A. Ciferri and I.M. Ward, Eds. *Ultra-high Modulus Polymers* (Applied Science, London, 1979).
49. U.W. Gedde, *Polymer Physics* (Springer, New York, 1995).
50. I.M. Ward and D.W. Hadley, *An Introduction to the Mechanical Properties of Solid Polymers*, Second ed. (John Wiley & Sons, Hoboken, NJ, 2004).
51. A. Peterlin, *Adv. Polym. Sci. Eng.* (1972) 1.
52. B. Davis, A. Rios, P.J. Gramann, and T.A. Osswald, *Compression Molding* (Hanser Gardner, Munich, 2003).
53. N.C. Lee, *Understanding Blow Molding* (Hanser Gardner, Munich, 2000).

54. J.A. White, A.Y. Coran and A. Moet, Eds. *Polymer Mixing: Technology and Engineering* (Hanser, Munich, 1999).
55. S.M. Lee, in *Developments in Reinforced Plastics*, edited by G. Pritchard (Elsevier-Applied Science, London, 1984).
56. J. Vlachopoulos and J.R. Wagner, Eds. *The SPE Guide on Extrusion Technology & Troubleshooting* (SPE, Brookfield, CT, 2001).
57. Z. Tadmor, *J. Appl. Polym. Sci.* **18** (1974) 1753.
58. Z. Tadmor and C.G. Gogos, *Principles of Polymer Processing* (Wiley-Interscience, New York, 1979).
59. A. Garcia-Rejon, A. Meddad, E. Turcott and M. Carmel, *Polym. Eng. Sci.* **42** (2002) 346.
60. J.L. Throne, *Understanding Thermoforming* (Hanser Publishers, Munich, 1999).
61. S. Grainger and J. Blunt, *Engineering Coatings: Design and Application*, 2nd ed. (Woodhead Publishing, UK, 1998).
62. J. Licari, *Coating Materials for Electronic Applications: Polymers, Processes, Reliability, Testing* (Noyes Publications, Berkshire, UK, 2003).
63. I.M. Ward and P.J. Hine, *Polym. Eng. Sci.* **37** (1997) 1809.
64. I.M. Ward and P.J. Hine, *Polymer* **45** (2004) 1423.
65. P.J. Hine and I.M. Ward, *J. Appl. Polym. Sci.* **91** (2004) 2223.
66. R.Y.F. Liu, T.E. Bernal-Lara, A. Hiltner and E. Baer, *Macromolecules* **37** (2004) 6972.
67. H. Aref, *Phys. Fluids* **14** (2002) 1315.
68. Y.H. Liu and D.A. Zumbrunnen, *J. Mater. Sci.* **34** (1999) 1921.
69. O. Kwon and D.A. Zumbrunnen, *J. Appl. Polym. Sci.* **82** (2001) 1569.
70. D.A. Zumbrunnen, S. Inamdar, O. Kwon and P. Verma, *Nano Lett.* **2** (2002) 1143.
71. A. Dhoble, B. Kulshreshtha, S. Ramaswami and D.A. Zumbrunnen, *Polymer* **46** (2005) 2244.
72. C. Mahesha, D.A. Zumbrunnen and Y. Parulekar. *63rd Ann. Tech. Conf.*, Society of Plastics Engineers, Boston, 2005, p 1920.
73. B. Wunderlich, *Thermal Analysis of Polymeric Materials* (Springer, New York, 2005).
74. E.A. Turi, Ed. *Thermal Characterization of Polymeric Materials*, 2nd ed. (Academic Press, New York, 1996).
75. D.A. Hemsley, *The Light Microscopy of Synthetic Polymers* (Oxford University Press, Oxford, 1984).
76. D.A. Hemsley, Ed. *Applied Polymer Light Microscopy*, (Elsevier Applied Science, London, New York, 1989).
77. A.B. Myers, Ed. Special Issue on Raman Microscopy and Imaging. *J. Raman Spectrosc.* **27**(8) (1996).
78. J.G. Grasselli and B.J. Bulkin., Eds. *Analytical Raman Spectroscopy* (Wiley, New York, 1991).
79. B.D. Cullity and S.R. Stock, *Elements of X-Ray Diffraction*, 3rd ed. (Prentice-Hall, Englewood Cliffs, NJ, 2001).
80. J.L. Koenig, *Microspectroscopic Imaging of Polymers* (Oxford University Press, Oxford, 1998).
81. J.L. Koenig, *Infrared and Raman Spectroscopy of Polymers* (Rapra Technology, Shropshire, UK, 2001).
82. R.F. Egerton, *Physical Principles of Microscopy* (Springer, New York, 2005).
83. G.P. Simon, Ed. *Polymer Characterization Techniques and Their Application to Blends* (Oxford University Press, Oxford, 2003).
84. J.J. Bozzola and L.D. Russell, *Electron Microscopy: Principles and Techniques for Biologists*, 2nd ed. (Jones & Bartlett, Sudbury, MA, 1999).
85. M.A. Hayat, Ed. *Principles and Techniques of Electron Microscopy: Biological Applications*, Fourth ed. (Cambridge University Press, Cambridge, 2000).
86. I.G. Voigt-Martin, *Acta Polymerica* **47** (1996) 311.
87. I.G. Voigt-Martin, *Acta Polymerica* **47** (1996) 369.
88. T.G. Rochow, Tucker, Paul A., *Introduction to Microscopy by Means of Light, Electrons, X-Rays, or Acoustics*, 2nd ed. (Springer, New York, 1994).
89. M.D. Graef, *Introduction to Conventional Transmission Electron Microscopy* (Cambridge University Press, Cambridge, 2003).
90. J.C.H. Spence, *High-Resolution Electron Microscopy*, 3rd ed. (Oxford University Press, Oxford, 2003).
91. L. Reimer, Ed. *Transmission Electron Microscopy: Physics of Image Formation and Microanalysis*, Fourth ed. (Springer Verlag, New York, 1997).
92. C.C. Ahn, *Transmission Electron Energy Loss Spectrometry in Materials Science and the EELS Atlas*, 2nd ed. (John Wiley & Sons, Hoboken, NJ, 2004).
93. P.B. Hirsch, *Topics in Electron Diffraction and Microscopy of Materials* (CRC Press, Boca Raton, 1999).
94. M.R. Libera and M.M. Disko, in *Transmission Electron Energy Loss Spectrometry in Materials*

Science and the EELS Atlas, 2nd ed., edited by C.C. Ahn (Springer, New York, 2005), p. 419.
95. L. Reimer, Ed. *Scanning Electron Microscopy, Physics of Image Formation and Microanalysis,* 2nd ed. (Springer Verlag, New York, 1998).
96. J.I. Goldstein, D.E. Newbury, D.C. Joy, C.E. Lyman, P. Echlin, E. Lifshin, L.C. Sawyer and J. Michael, *Scanning Electron Microscopy and X-ray Microanalysis,* 3rd ed. (Kluwer Academic/Plenum/Springer, New York, 2003).
97. B.D. Ratner and V.V. Tsukruk, Eds. *Scanning Probe Microscopy of Polymers* (Proceedings of a Symposium at the 212th National Meeting of the American Chemical Society, held 25–29 August 1996, in Orlando, Florida.), in: *ACS Symp. Ser.* **694** (1998).
98. S. Magonov and N.A. Yerina, in *Microscopy for Nanotechnology,* edited by N. Yao and Z.L. Wang (Kluwer Academic Press, New York, 2005), p. 113.
99. B. Bhushan, H. Fuchs and S. Hosaka, Eds. *Applied Scanning Probe Methods I* (Springer, Berlin, Heidelberg, 2004).
100. P.W. Hawkes and J.C.H. Spence, Eds. *Science of Microscopy,* two-volume set (Springer, New York, 2006).
101. N. Bonnet, *Micron* **35** (2004) 635.
102. J.A. Galloway, M.D. Montminy and C.W. Macosko, *Polymer* **43** (2002) 4715.
103. P.J. Goodhew, *Specimen Preparation for Transmission Electron Microscopy* (Oxford University Press, Oxford, 1984).
104. P.J. Goodhew, J. Humphreys and R. Beanland, *Electron Microscopy and Analysis,* 3rd ed. (Taylor & Francis, UK, 2001).
105. A.M. Glauert and P.R. Lewis, *Biological Specimen Preparation for Transmission Electron Microscopy* (Princeton University Press, Princeton, NJ, 1998).
106. P.R. Lewis and D.P. Knight, Eds. *Cytochemical Staining Methods for Electron Microscopy* (Elsevier, Amsterdam, 1992).
107. J. Trempler, *Praktische Metallographie* **38** (2001) 231.
108. J. Trempler, *Praktische Metallographie* **40** (2003) 481.
109. S.Y. Hobbs, in *Plastics Polymer Science and Technology,* edited by M.D. Bayal (Wiley-Interscience, New York, 1982), p. 239.
110. E.L. Thomas, in *Structure of Crystalline Polymers,* edited by I.H. Hall (Elsevier-Applied Science, London, 1984), p. 79.
111. D.A. Hemsley, in *Applied Polymer Light Microscopy,* edited by D.A. Hemsley (Elsevier Science Publishers, Amsterdam, 1989), p. 185.
112. H.J. Tanke, *J. Microsc.* **155** (1989) 405.
113. D. McMullan, *J. Microsc.* **155** (1989) 373.
114. B. Ralph, *J. Microsc.* **155** (1989) 339.
115. P.B. Hirsch, *J. Microsc.* **155** (1989) 361.
116. A. Howie, *J. Microsc.* **155** (1989) 419.
117. T. Wilson, *Theory and Practice of Scanning Optical Microscopy* (Academic Press, London, 1984).
118. J.B. Pawley, Ed. *Handbook of Biological Confocal Microscopy* (Plenum Press, New York, 1990).
119. T. Wilson, Ed. *Confocal Microscopy* (Academic Press, London, 1990).
120. G. Cox, Ed. Special Issue on Confocal Microscopy. (Proceedings of a conference held in Heidelberg in 1999), in: *Micron* **32**(7) (2001).
121. J. Pawley, Ed. *Handbook of Biological Confocal Microscopy,* 2nd ed. (Springer, New York, 1995).
122. L. Reimer, *Image Formation in Low Voltage Scanning Electron Microscopy* (SPIE, Bellingham, 1993).
123. J.H. Butler, D.C. Joy, G.F. Bradley, S.J. Krause and G.M. Brown, in *Microscopy: The Key Research Tool* (Electron Microscopy Society of America, Woods Hole, 1992), p. 103.
124. L. Reimer, *Scanning Electron Microscopy, Physics of Image Formation and Microanalysis,* 2nd ed. (Springer, Berlin, Heidelberg, 1998).
125. L.F. Drummy, J. Yang and D.C. Martin, *Ultramicroscopy* **99** (2004) 247.
126. A.M. Donald and B.L. Thiel, *Structure and Dynamics of Materials in the Mesoscopic Domain,* Proceedings of the Royal Society-Unilever Indo-UK Forum in Materials Science and Engineering, 4th, Pune, India, Dec. 8–12, 1997 (1999) 1.
127. C.P. Royall and A.M. Donald, *Scanning* **24** (2002) 305.
128. D. Shindo and K. Hiraga, *High-Resolution Electron Microscopy for Materials Science* (Springer, Tokyo, 1998).
129. D.C. Martin, J. Chen, J. Yang, L.F. Drummy and C. Kübel, *J. Polym. Sci. B Polym. Phys.* **43** (2005) 1749.
130. L.L. Ban, M.J. Doyle, M.M. Disko and G.R. Smith, *Polym. Commun.* **29** (1988) 163.
131. H. Ade, A.P. Smith, B. Wood, I. Plotzker, B. Hsiao and S. Subramoney, *Microbeam Analysis,* Proceedings of the Annual Conference of the Microbeam Analysis Society, 29th, Breckenridge, Colo., Aug. 6–11, 1995 (1995) 141.
132. A.P. Hitchcock, C. Morin, X. Zhang, T. Araki, J. Dynes, H. Stover, J. Brash, J.R. Lawrence and

G.G. Leppard, *J. Electron Spectrosc. Relat. Phenom.* **144–147** (2005) 259.

133. S.L. Rokhlin, J.-Y. Kim and B. Zoofan, The International Society for Optical Engineering Testing Reliability, and Application of Micro- and Nano-Materials Systems, 3–5 March 2003. *Proc. SPIE* **5045** (2003) 132.

134. G. Binnig, H. Rohrer, C. Gerber and E. Weibel, *Phys. Rev. Lett.* **49** (1982) 57.

135. G. Binnig, C.F. Quate and C. Gerber, *Phys. Rev. Lett.* **56** (1986) 930.

136. G.J. Leggett, M.C. Davies, D.E. Jackson, C.J. Roberts and S.J.B. Tendler, *Trends Polym. Sci.* **1** (1993) 115.

137. G.K. Bar and G.F. Meyers, *MRS Bull.* **29** (2004) 464.

138. S. Magonov, in *Applied Scanning Probe Methods*, edited by P. Avouris, D. Klitzing, H. Sakaki and R. Wiesndanger (Springer, Berlin, Heidelberg, 2004), p. 207.

Chapter 2
Fundamentals of Microscopy

2.1 INTRODUCTION 28
   2.1.1 **Lens-Imaging Microscopes** 29
   2.1.2 **Scanning-Imaging Microscopes** 30
2.2 OPTICAL MICROSCOPY 31
   2.2.1 **Introduction** 31
   2.2.2 **Objective Lenses** 32
   2.2.3 **Imaging Modes** 32
      2.2.3.1 Bright Field and Dark Field 32
      2.2.3.2 Phase Contrast 33
      2.2.3.3 Interference Microscopy 33
   2.2.4 **Measurement of Refractive Index** 34
   2.2.5 **Polarizing Microscopy** 34
2.3 SCANNING ELECTRON MICROSCOPY 35
   2.3.1 **Introduction** 35
   2.3.2 **Imaging Signals** 37
      2.3.2.1 Backscattered Electrons 37
      2.3.2.2 Secondary Electrons 39
   2.3.3 **Electron Sources** 39
   2.3.4 **SEM Types** 41
   2.3.5 **SEM Optimization** 41
2.4 TRANSMISSION ELECTRON MICROSCOPY 42
   2.4.1 **Conventional TEM** 42
      2.4.1.1 Bright Field and Dark Field 42
      2.4.1.2 Phase Contrast 43
   2.4.2 **Scanning TEM** 43
   2.4.3 **Electron Diffraction** 44
   2.4.4 **High Resolution Electron Microscopy** 45
2.5 SCANNING PROBE MICROSCOPY 45
   2.5.1 **Introduction** 45
   2.5.2 **Atomic Force Microscopy** 47
      2.5.2.1 Contact and Associated Modes 47
      2.5.2.2 Intermittent Contact Mode. 48
      2.5.2.3 Noncontact and Associated Modes 49
   2.5.3 **SPM Probes** 50
2.6 RADIATION SENSITIVE MATERIALS 51
   2.6.1 **SEM Operation** 52
   2.6.2 **Low Dose TEM Operation** 52
2.7 ANALYTICAL MICROSCOPY 53
   2.7.1 **X-ray Microanalysis** 53
      2.7.1.1 Energy Dispersive X-ray Spectrometer 54
      2.7.1.2 Wavelength Dispersive X-ray Spectrometer 54
   2.7.2 **X-ray Analysis: SEM versus AEM** 55
   2.7.3 **Elemental Mapping** 55
2.8 QUANTITATIVE MICROSCOPY . 56
   2.8.1 **Image Processing and Analysis** 56
   2.8.2 **Three dimensional Reconstruction** 57
   2.8.3 **Calibration** 57
      2.8.3.1 OM, SEM and TEM ... 57
      2.8.3.2 AFM 58
2.9 DYNAMIC MICROSCOPY 59
   2.9.1 **Mechanical Deformation Stages** 59
   2.9.2 **Hot and Cold Stages** 60
References 60

2.1 INTRODUCTION

Microscopy is the study of the fine structure and morphology of objects with the use of a microscope. Microscopes form magnified images, magnified from a few times in an optical stereo microscope to more than a million times in microscopes that can resolve individual atoms in suitable samples. Some instruments give information about a surface and not the specimen interior, but preparation methods can create an internal surface that may be imaged. Many modern microscopes are integrated with systems that give local chemical information, adding to the structural image. Apart from this, the size and visibility of the structure to be characterized generally determines which instruments are to be used. For example, the fracture of a multiphase polymer may require a light-optical technique for the "big picture" but a study at higher resolution using electron microscopy and scanning probe microscopy to see fine details on the fracture surface. Combinations of various microscopy techniques generally provide the best insight into the morphology of polymer materials.

The name of a type of microscope generally comes from what is used to investigate the sample: *optical microscopes* use light, *electron microscopes* use electrons, *scanning probe microscopes* use a solid probe, and so on. Light and electron microscopes may operate in *transmission* where the radiation passes through the specimen and is collected on the other side, requiring a thin specimen. Alternatively, radiation can be collected from the surface that it arrives at, allowing a specimen of any thickness. In light microscopy, this mode may be called *reflection*. A more general name, *incident light*, also applies to the many cases, e.g. fluorescence microscopy, where the radiation that is detected is not the reflected incident beam. Similarly, reflection electron microscopy (REM) and its associated scanning version (SREM) are restricted to systems that detect electrons returned by a single interaction with the sample. These surface-sensitive techniques are rarely if ever applied to polymers. The scanning electron microscope (SEM) operates in a more generalized "reflection" mode, detecting whatever is emitted from the surface where electrons are incident. An electron microscope that operates in transmission will have that word in its name, thus transmission electron microscope (TEM) and scanning transmission electron microscope (STEM).

Another fundamental distinction is whether the image is formed all at once or sequentially, point by point, by scanning. In computer terms, these would be described as parallel and serial transmission of information, respectively. A television image is an everyday example of an image formed by scanning; our own eyes use a lens to form an image on the retina that is processed in parallel. At this level, the conventional optical microscope and the TEM work like an eye, using lenses to form an image all at once on a sensitive surface. This may be film or an electronic detector such as a charge-coupled device (CCD). The atomic force microscope (AFM) and other scanning probe microscopes (SPM) move a solid sharp probe over the specimen surface. These microscopes have no lenses and build up their images point by point. The SEM and STEM also form a sequential image but use a beam of electrons, focused by lenses, as the scanning probe, so some concepts of lens imaging are relevant to these microscopes. The confocal optical microscope is another intermediate type. It forms an image with lenses, but an aperture limits the area viewed at any instant to a single point and the image is built up by scanning. Table 2.1 shows some basic properties of different types of microscopes for comparison, divided into lens-imaging and scanning-imaging classes.

It is beyond the scope of this text to describe the design features and operation of specific microscopes and their attachments. Any attempt to discuss microscope operation or construction in detail would rapidly become outdated. Manufacturers' representatives or Web sites are the best source for information on their instruments. Lists of manufacturers are given in Appendices VI and VII.

Key parameters of microscope images are resolution and contrast. Two object features closer together than the *resolution* will appear as one feature, not two, in the image. The *contrast* is the fractional change in image intensity that a feature causes. Small features in the specimen that have low contrast, below about 0.05,

Introduction

TABLE 2.1. Some properties of various types of microscope

	Lens-imaging		Scanning-imaging	
	Optical microscope	Transmission electron microscope	Scanning electron microscope	Atomic force microscope
Typical lateral resolution	300 nm	0.2 nm*	3 nm*	2 nm*
Magnification	2 to 2,000	200 to 2×10^6	20 to 1×10^5	1000 to 2×10^6
Can observe	Surface, or bulk if transparent	"Bulk," but thin films, $<0.2 \mu m$	Surfaces	Surfaces
Specimen environment	Ambient, or transparent fluid	High vacuum	High vacuum (4 kPa in HPSEM)	Ambient, high vacuum or fluid
Radiation damage	None	Severe	Rarely serious	None
Specimen preparation	Easy	Very difficult	Easy	Easy
Chemical analysis	No, unless μ-Raman or IR	Yes, x-ray and electron energy loss	Yes, x-ray	No
Can detect molecular orientation	Yes	Yes	No	Sometimes

*Some instruments have higher resolutions; both TEM and AFM can resolve individual atoms if the sample is suitable. Polymers are generally not suitable, so the instrumental resolution may not be achieved.

may not be observed, even if they are larger than the resolution limit of the instrument. Image processing (see Section 2.8.1) can increase the contrast, but the microscopist has to choose to use it. Table 2.1 lists an indication of resolution but not contrast, as contrast depends on the sample.

The *field* or *field of view* is a linear measure of the area of the specimen included in the image, so the number of independent data points in the image is (field/resolution)^2. An image is made to be viewed by eye, so the number of points should match the expected final image size and human visual acuity. A person with 20/20 vision viewing a screen or print from 40 cm away can resolve details about 0.12 mm across, and a 2,000 × 2,000 array of such details would fill an area 25 cm or 10 in. square. If such a 4 megapixel image was obtained by microscopy, at for example a magnification of 1,000×, the field of view would be 250 μm and the resolution would have to be at least 0.25 μm for the finest scale of the image to be useful (see Section 3.1.4).

2.1.1 Lens-Imaging Microscopes

Electromagnetic coils are the lenses used for electron beams, and glass lenses are used for light. Electromagnetic lenses are focused by changing the current flowing through them; glass lenses are focused by changing their position. These differences and the high vacuum systems needed for electron beams tend to hide the close similarity of function and arrangement of the various components. Figure 2.1 shows a schematic of ideal image formation, using the shape of a glass lens to indicate a generic lens, which might be composed of magnetic fields.

Compare a TEM to a transmitted light optical microscope being used to take a photomicrograph, and the principles of design are very similar. Illumination of the object is very important, and in both microscopes the resolution and contrast of the image may be degraded if the illumination is not properly adjusted. The source of illumination may be a small, hot tungsten filament in both cases. The electrons or photons emitted from the filament are collected by a *condenser lens*. To increase efficiency in the TEM, an electrostatic lens (*electron gun*) is used to steer more of the flux into this lens. In the optical microscope, a mirror behind the lamp performs this function. A second condenser lens controls the transfer of this illumination to the specimen plane. In the TEM, these are simply called "condenser lens 1" (C_1) and "condenser

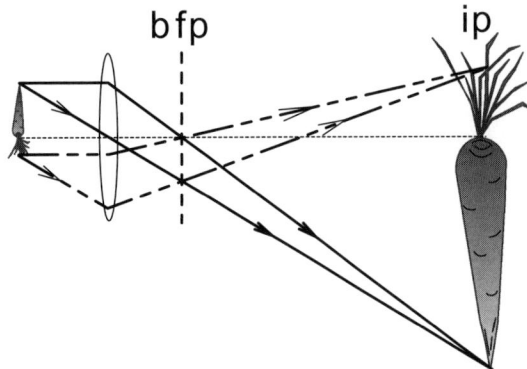

FIGURE 2.1. Schematic of image formation with a lens. Radiation from any single point on the object is brought to a focus at a point in the image plane (ip), forming an image. At the back focal plane (bfp), radiation leaving anywhere on the object in a single direction is brought to a single point.

lens 2" (C_2), but the operator's controls for these lenses may be labeled "spot size" and "brightness." In transmission optical microscopy, the second lens is called the substage condenser, or simply "the condenser." The first lens, which is built into the base unit or is part of a separate free-standing illuminator, may be called the auxiliary or lamp condenser, or the field lens.

There is an aperture associated with each condenser lens, and the apertures and lenses control the area illuminated and the angular divergence of the illumination. More details of the illumination system are described in Section 3.1.6.

After the radiation has passed through the specimen, the scattered radiation is collected by an *objective lens*. This lens is the most critical, and imperfections in it will affect the image quality directly; it is adjusted to focus the image. In the TEM, an *objective aperture* at the back focal plane of the objective lens controls the angular divergence of radiation that contributes to the image.

The *eyepiece* is the usual second imaging lens in an optical microscope; it forms a virtual image for the eye to focus on. When an optical microscope is set up for photomicroscopy, a *projector lens* that forms a real image on the camera CCD is used instead. The TEM operates in much the same way as this, but there are several lenses, *intermediate lenses*, between the objective and the projector lens that control magnification. Each lens produces a real and magnified image, as shown in Fig. 2.1. The final projector lens produces its image on the fluorescent viewing screen or on the film or CCD. More lenses are needed in the TEM to give the greater total magnification required. Modern microscopes of both types may have more lenses than described here, but these complications are usually hidden from the user.

The *depth of field* is the depth or thickness of the specimen that is simultaneously in focus. Transmission electron microscopes have a depth of field that is normally greater than the specimen thickness, so all of the transparent specimen is in focus at once. Optical microscopes have a depth of field comparable to their resolution, and this is often much less than the sample thickness. (The *depth of focus* is the depth of the image that is in focus and is not important in normal microscopy.) *Bright field* (BF) in transmission is when the direct unscattered beam is allowed to reach the image plane. An image field that contains no specimen is then bright. *Dark field* (DF) is the opposite imaging mode, where only scattered radiation is allowed to form the image. In transmission, an image field with no specimen is dark in DF.

2.1.2 Scanning-Imaging Microscopes

In scanning-imaging microscopes a small probe is passed over the specimen surface, and a signal is collected that relates to some local property of the specimen. The probe may be a sharp point or a narrow beam of electrons or photons, and the signals and detectors are very diverse. All the instruments have a common feature: a detector output signal that varies with time and is displayed like a television image. The intensity of each pixel in the image is controlled by the signal from the microscope. The magnification is then simply the linear size of the image, divided by the size of the region scanned on the specimen, the field, as shown schematically in Fig. 2.2. Magnification can thus be altered without having any effect on other imaging conditions. The spatial resolution of the microscope is determined by the size of the specimen region from which the signal is derived, which

Optical Microscopy

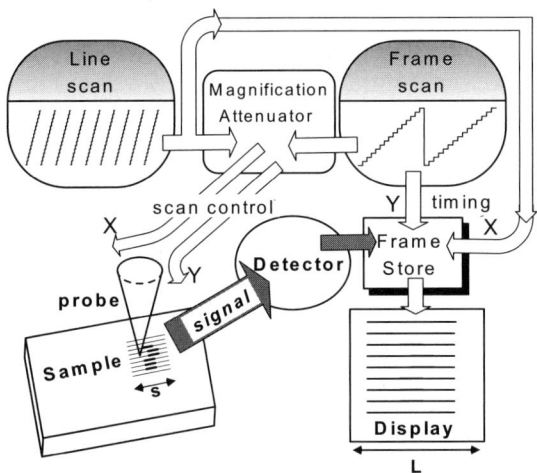

FIGURE 2.2. Schematic of image formation by scanning. The probe may be an electron beam in a SEM or a solid probe in a SPM. The detector may detect any of a wide variety of signals. The displayed magnification is simply the length scanned on the display divided by the length scanned on the object, L/s.

can be down to the atomic scale. Typical systems may scan 1,024 or 2,048 lines in each image, with the same number of distinguishable spots on each line to give 1 or 4 megapixel images.

In the original analog design the videoscan generators directly control the probe and display. Scanning in scanning microscopes is now digitally controlled, so a position of the probe corresponds to a memory location, as shown in Fig. 2.2. The display does not have to be synchronous with the probe movement. The image is thus naturally digital, while in lens-imaging microscopes an image must be captured by CCD or similar system to be digitized. This is a general advantage of scanning, but a counteracting general disadvantage is that scanning can take time. If it takes $10\,\mu s$ to collect the data at one point, a 4 megapixel image will take 40 s to form.

2.2 OPTICAL MICROSCOPY

2.2.1 Introduction

In the conventional light microscope or optical microscope (OM), an object is illuminated and light that it scatters or transmits is collected by a system of lenses to form an image. The image can reveal fine detail in the specimen at a range of magnifications from 2× to 2,000×. Resolution of about $0.5\,\mu m$ is possible, limited by the nature of the specimen, the objective lens, and the wavelength of light. At an introductory level, Spencer [1] describes the fundamental science and Rawlins [2] is more practical. There are many other texts [3–8], some concentrate on a particular topic, such as polarized light microscopy [9, 10], the identification of materials [11], the integration of microscopy and digital imaging [12], or the study of polymers [13]. More accessible are the excellent sources of information on optical microscopy on the Web (e.g., [14–16]).

The information obtained in the OM normally concerns the size, shape, and relative arrangement of visible features. Local measurement of optical constants such as the refractive index (see Section 2.2.4) and the birefringence (see Section 2.2.5) is also possible. Many techniques are used to enhance contrast and thus make more of the structure visible. Images are typically recorded with a high quality digital camera, which may be linked to a computer system for image processing and analysis.

Simple microscopes have only one imaging lens (though this may have several elements) and operate at low magnification, like a magnifying glass. The optical microscopes in the laboratory are generally *compound microscopes*, with more than one imaging lens. They operate at higher magnification and higher resolution giving more detail on smaller specimens. For visual observation, the magnification of a compound microscope is the product of the marked magnifications on the objective and the eyepiece. (40 × 12.5 = 500×, for example). Digitally stored images do not have a magnification, as they can be reproduced at any size, so it is important to calibrate them by recording an image of an object of known size under the same conditions.

Binocular stereo microscopes [17] are compound microscopes that provide two different images of the specimen through the two eyepieces. These are views from slightly different directions. The observer sees this as a three dimensional image, very useful for the examination of bulk specimens. The stereo microscope is a good starting point for investigations of the nature of the material. It also helps to identify regions of the specimen for further study. It should be noted that many people are unable to form proper three-dimensional images from stereoscopic views, using instead visual cues of size and shape. These can be misleading in microscopic images. (This is more serious in stereoscopic TEM where normal visual cues are often entirely absent from the image.) Typical compound research microscopes are also binocular, but the images in the eyepieces are identical. The two images are provided to reduce eyestrain.

2.2.2 Objective Lenses

The *objective lens* or *objective* is the most important part of the optical system. All microscope objectives have information engraved on them, and it is helpful to be able to interpret this. A modern lens may have something similar to "Pol 25/0.55 0.17/∞." The letters indicate the type of lens, here strain-free, suitable for use with polarized light. Different manufacturers may use different abbreviations, but most are readily understood. The most obscure is "Fl" or "Fluor." This means the lens is a high quality general-purpose lens, once (but no longer) made using the mineral fluorite. This is the middle quality of lenses. The best corrected and most expensive are apochromats ("Apo") and the least corrected and lowest cost are the achromats. If excellent quality color images are required, a "Plan apo" objective will be best, as "plan" means that the image appears in focus on a flat plane, so that a camera will see both center and edges of the image in focus at the same time. A plan achromat will give excellent monochromatic images.

The first two numbers on the lens are always the magnification and the *numerical aperture* (NA) of the lens. Numerical aperture is defined as ($n \sin\alpha$), where n is the refractive index of the medium in front of the objective and α is half the angular range of light that the lens can accept. The resolution of the lens, if it is perfect, is proportional to $1/(NA)$, and the depth of field is proportional to $1/(NA)^2$. If the lens is "dry," it has air in front of it, $n = 1$, and the maximum NA is 0.95. An immersion oil of $n = 1.5$ between lens and sample can significantly improve the resolution limit by increasing the possible NA to 1.3. A rule of thumb is that the highest useful final magnification is about 700–1,000 times the numerical aperture of the objective lens. Higher magnifications are "empty" as the images do not contain any extra information. An exact limit is not appropriate as it depends on the contrast of the image and the visual acuity of the observer.

The second two numbers on the lens are the correct thickness of cover glass to use between lens and specimen in millimeters and the *tube length*, also in millimeters. Cover glasses are now standardized at 0.17 mm thick, so the number is normally 0.17 or 0, indicating whether or not the objective was designed for use with a cover glass. High power (>40×) dry objectives are very sensitive to cover glass thickness variations and may have an adjustment ring that is set to suit the actual value. The tube length is not important for the user unless objectives are bought for one microscope and used with another. Very poor results will be obtained if the instruments have different tube lengths. The tube length was originally the physical length of the tube separating the objective and eyepiece lenses, but this is no longer the case. Modern microscopes commonly have a tube length of infinity, which has many advantages for modular design.

2.2.3 Imaging Modes

2.2.3.1 Bright Field and Dark Field

Bright field is the normal mode of operation of an optical microscope. In transmission, it means that the direct unscattered light is allowed to reach the image plane. In reflection, specular reflection from the surface is allowed to reach the image plane. A transparent material (in reflection a perfectly flat surface) appears bright in BF. The contrast in transmitted light is based on variations of optical density and color within the material. Carbon black agglomerates, pigment particles, and other fillers are clearly observed in polymers in bright field as the matrix polymers are typically transparent in thin sections (for examples, see Fig. 5.2 and Fig. 5.104). *Dark field*, where only scattered light is allowed to reach the image plane, is less common in transmission but has higher contrast. In reflected light, the visibility of details in polymer samples is often poor due to low surface reflectivity, scatter from within the specimen, and glare from other surfaces (for an example, see Fig. 4.4A). A metal coating on the surface (vapor deposited or sputtered) will increase the surface reflectivity. This enhances intensity and contrast in reflected light. Dark field in reflected light may be used to increase the contrast of surface roughness. If the specimen is transparent and not coated, dark field in reflection allows observation of subsurface features and details that scatter light.

2.2.3.2 Phase Contrast

Thin sections of polymer blends can give bright field images with little or no contrast between the components. Transmitted light phase contrast converts the refractive index differences in such a specimen to light and dark image regions [1, 3–9, 18]. Small differences in thickness are also made more visible. Examples of transmitted light and phase contrast images are described in Chapter 5 (see Fig. 5.70 and Fig. 5.95). Normal (Zernike) *phase contrast* requires a special condenser with an opaque center, so that illumination is limited to a ring, and a plate with a matching phase ring in the back focal plane of the objective. Unscattered light passes through the phase ring, and its phase is altered while the scattered light is unaffected. Interference between the scattered and direct rays causes changes in the image intensity. In a two-component transparent sample, light is scattered from the interfaces. More generally, local changes in thickness or refractive index are made visible.

The phase contrast image has characteristic bright halos around fine structure, due to some scattered light passing through the ring in the phase plate. A method that produces images without a halo effect is *Hoffman modulation contrast* [5, 6, 19]. In this technique, scattered light is changed in amplitude, rather than phase, by a modulator disk in the back focal plane of the objective and the illumination is limited by a slit, not a ring. Hoffman modulation contrast uses a rotatable polarizer under the condenser to control the illumination and has an asymmetric modulator disk. This makes for a very flexible system. The images are sharp and three-dimensional with shadows and textures that give an appearance of oblique illumination.

2.2.3.3 Interference Microscopy

In interference microscopy, the illumination is split into two beams with different paths, and the two beams are recombined so that they interfere. The interference pattern can be used to measure the specimen thickness in transmission, or the surface profile of the specimen in reflection. In reflected light microscopy, the beam splitter is usually a half-silvered mirror; one beam is reflected off the specimen and the other off a flat reference mirror [4–7]. This relatively simple arrangement can be built into a special objective lens such as the Mirau or the Watson designs [20]. These fit on normal optical microscopes, but they are rare; most optical profilometers are now special-purpose instruments with digital output.

Transmitted light is more complex as separate matched devices have to be used to split and then recombine the beams [3–8]. In early systems, the beam displacement was large so that the reference beam did not pass through the specimen. In the Jamin–Lebedeff system (Zeiss; last produced in the 1980s) [1, 4, 6], the reference beam is displaced by more than its diameter but both reference and measuring beam can be seen in the same field of view. The reference beam is set to pass through a featureless area of the specimen.

The modern version of interference microscopy is *differential interference contrast* (also called Nomarski contrast, or DIC). Here again, the illumination is split into two beams, one of which is displaced at the specimen plane [1, 5, 6, 21]. In DIC, the beam is displaced a very small distance, much smaller than the beam diameter. The beams remain independent because the beam-splitting device is a doubly refracting crystal, producing two beams in perpendicular polarization states (see Section 3.1.7). A region of constant properties shows no contrast in DIC because both beams see the same thing, but a sudden change of thickness or refractive index gives strong contrast. In transmission with monochromatic light, the technique is similar to Hoffman modulation contrast, containing the same information as phase contrast without the halo and with apparent relief and shadows as in oblique illumination. In white light, the interference colors can produce spectacular images in DIC. On the other hand, a birefringent specimen can ruin DIC images, whereas Hoffman contrast is unaffected. As before, in reflection a single device can both produce the polarization and separation of the beams and recombine the reflected light. In transmission a pair of matching devices are needed, one before and one after the specimen. Differential interference contrast is therefore most easily used in reflected light, where it gives a clear pseudo three dimensional image of the surface topography (for an example, see Fig. 4.4B).

2.2.4 Measurement of Refractive Index

The refractive index n of small samples can be measured in the optical microscope, and this helps identify unknown materials [8, 9, 11]. Small particles are mounted in a liquid of known refractive index and observed in transmission. The liquid can be changed until n(liquid) = n(particle), when a transparent or translucent particle will have very low contrast. To find out whether the change should increase or decrease n, the sample can be illuminated obliquely. The particles then act as rough lenses, converging lenses if n(liquid) < n(particle). Light striking one side of the particle will be diverted toward the axis, and that side will appear bright. This shading will be reversed if n(liquid) > n(particle) [22]. Alternatively, axial illumination of particles with sharp boundaries gives a narrow band of light near the edge, the *Becke line*. Light is scattered toward the side of greater n, so the line appears on this side when the focus is above the particle (overfocus) and on the other side at underfocus.

Phase contrast devices increase the contrast due to differences in refractive index and so allow a more accurate determination of n [23]. Phase and interference contrast are sensitive to the optical path length and thus to the refractive index averaged through the specimen thickness. The Becke line method is sensitive to the refractive index at the surface of the sample. Fibers often give different results by the two methods because of the variation in refractive index across a fiber cross section.

2.2.5 Polarizing Microscopy

Polarizing microscopy is the study of the microstructure of objects using their interactions with polarized light. Wood [24] gives a basic introduction to polarized light and materials, Robinson [10] introduces polarizing microscopy, and the book by Hartshorne and Stuart [9] is comprehensive. The microscopy Web sites [14–16] have good sections on polarization (e.g., [25]). The method is described in other microscopy texts [5–8, 26] and is widely applied to polymers and to liquid crystals [27]. Polarized light and polarizing microscopy is discussed in more detail in Section 3.1.7.

A *polarizing microscope* is a transmitted light microscope that has, besides all the basic features, a rotatable stage, a *polarizer* in the illumination system, and an *analyzer* between objective and eyepiece. One or both of the polarizer and analyzer must be rotatable. The polarizer and the analyzer are both *polars*, that is, devices that selectively transmit light polarized in one specific plane. Polars are made from an oriented polymer film that is *dichroic*—it selectively absorbs light of one polarization state. By far the most common arrangement in the polarizing microscope is *crossed polars*. The transmitted polarization planes of the two polars are set to be perpendicular or "crossed" so that the analyzer does not transmit light transmitted by the polarizer. With no specimen, or with an isotropic specimen, the field of view will be dark in crossed polars. The polars are usually set to transmit light polarized in the directions given by 3 o'clock to 9 o'clock and 6 o'clock to 12 o'clock, imagining a clock face on the specimen. These directions are referred to as 0° and 90°. Many examples of polarized light images of polymers are shown in Chapter 5 (for examples, see Fig. 5.3 and Fig. 5.54).

When anisotropic specimens such as fibers are rotated on the rotatable stage, they go through four *extinction positions* of minimum intensity and four positions of maximum intensity. In the extinction positions, the fiber orientation direction is aligned parallel to one polarization direction, at 0° or 90°. Maximum intensity is at the 45° positions (for an example, see Fig. 5.4). Circularly polarized light, obtained and analyzed by the addition of two crossed quarter-wave plates (marked λ/4) into the light path, one between each polar and the specimen, eliminates these extinction positions. All anisotropic specimens are bright between crossed circular polars regardless of their orientation.

The anisotropic materials are *birefringent*; they can be considered to split light that passes through them into two plane-polarized waves that vibrate in planes at right angles to one another. These transmitted waves have different velocities and refractive indices n_1 and n_2. The direction of vibration with the larger refractive index is called the *slow direction*. The sample *birefringence* is the difference between the two refractive indices, $(n_1 - n_2)$, or Δn. If the sample has a clear reference direction, it is used to define the sign of the birefringence. If the reference direction is the slow direction, the birefringence is said to be positive. Examples of reference directions are the length of a fiber and the radius of a spherulite.

Most birefringence is due to the orientation of optically anisotropic elements. These can be amorphous chains in a polymer, aligned by deformation such as drawing, or crystals aligned by deformation or by growth mechanism such as epitaxy. However, there is also *form birefringence*. This arises when the material contains at least two phases that have different refractive indices and some dimension close to the wavelength of light. Form effects can contribute to birefringence in both copolymers and semicrystalline polymers; it should be allowed for in the calculation of molecular orientation. Birefringence measurement can then be used to obtain quantitative data on the degree of molecular orientation in the sample.

In principle, birefringence can be measured directly, by measuring the two refractive indices of the sample and taking the difference. This is not accurate, so instead the specimen thickness and *retardation* are measured. Retardation is defined as ($\Delta n \times$ [specimen thickness]) and is measured using a *compensator*, which is a crystal plate of known retardation. The specimen to be measured is set to the −45° position between crossed polars, and a compensator is inserted in its slot. This is above the specimen but below the analyzer at +45°. The compensator is adjusted until the specimen is dark, when its retardation is exactly cancelled by the compensator. If this adjustment is impossible, the sample must be rotated 90° to +45°.

In white light, anisotropic structures may appear brightly colored when viewed in crossed (or parallel) polars. These *polarization colors* or *interference colors* depend on the retardation (see Section 3.1.7). An estimate of sample retardation can be made from the standard sequence of colors, published as the Michel–Levy chart in many texts [5, 8, 9, 26, 28]. Color can also be used to find the sign of a small retardation when a *first-order red plate* is inserted as a compensator in white light. Modern devices exist where the polarizing elements are electrically driven and computer controlled. These allow simultaneous measurement of retardation and orientation direction at every point on the image and thus the creation of retardation and orientation maps [29].

The polarizing microscope may have a *Bertrand lens*. When inserted, this gives a *conoscopic* view; that is, it changes the image to a view of the back focal plane of the objective (Fig. 2.1). With a high NA objective, this shows the effect of polarized light traveling through the sample in a wide range of directions. It is most often used in mineralogy to determine the optic axes (see Section 3.1.7) of crystals.

2.3 SCANNING ELECTRON MICROSCOPY

2.3.1 Introduction

The scanning electron microscope (SEM) forms an image by scanning a probe, a focused electron beam, across the specimen. The probe interacts with a thin surface layer of the specimen, a few micrometers thick at most. Scanning electron microscopy is fully described in a recent and comprehensive text [30]. Flegler et al. [31] gives a basic introduction and Reimer [32, 33] a detailed view of the physics of microscope operation. The use of the SEM for polymer studies was reviewed in 1984 [34] and in earlier editions of this text [35].

A simple analog of a scanning microscope, which may make its basic operation easy to understand, is a flashlight and a light meter in a dark room. The intensity of reflected light and thus the light meter response is large when the flashlight beam falls on a pale wall. When it falls on dark drapes, or out of a window, there is a small signal on the light meter. Scanning the spot of light systematically over the wall and recording the signal maps out the dark regions in the scanned area. In the room analog, dark regions are due to reduced reflectivity or to gaps in the reflecting surface. In the SEM, these correspond to *compositional* and *topographic* contrast, respectively, because the first depends on the composition (mean atomic number) of the sample and the second on its shape.

Figure 2.3 is a block diagram of a conventional SEM [30] showing the electron optical column with three condenser lenses used to form the probe, and the *scan generator*, the common source for display scanning and probe scanning, as in Fig. 2.2. Compared with incident light optical microscopy, the SEM has higher resolution and a much larger depth of field. The specimen chamber in the SEM is large, and samples several inches in diameter can be accommodated. Specimen preparation is generally quite simple, if the materials can withstand drying and high vacuum. Nonconductive materials, such as most polymers, require either conductive coatings, low accelerating voltages, or variable pressure to prevent them from charging up in the electron beam. Normal SEM images are easy to interpret qualitatively.

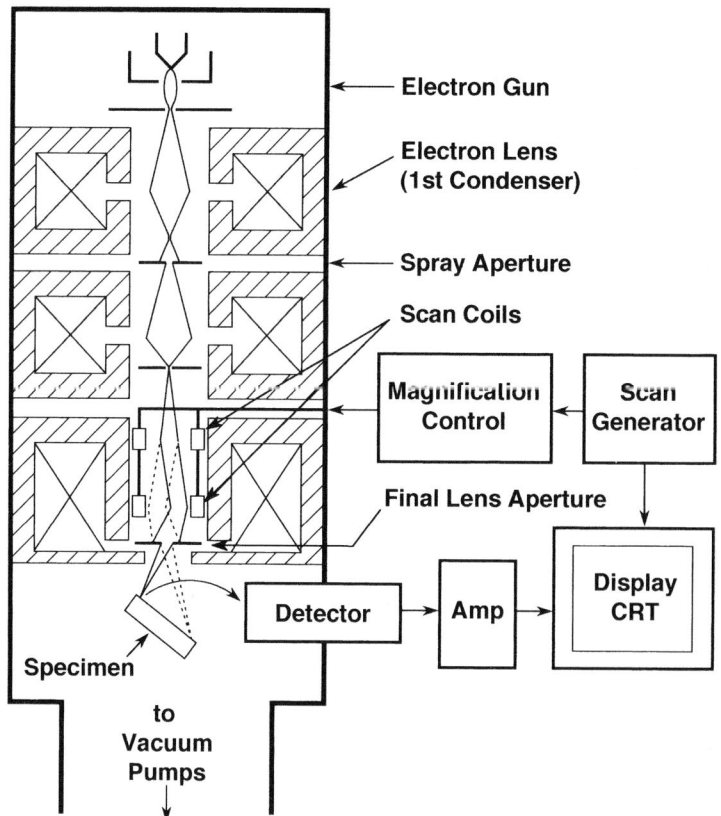

FIGURE 2.3. Schematic diagram of a scanning electron microscope. Two pairs of scan coils are shown in the SEM column. This double deflection allows the scanning beam to pass through the final aperture. Four pairs are actually used, for double deflection in both X and Y directions. (From Goldstein et al. [30], © (2003) Springer; used with permission.)

Scanning Electron Microscopy

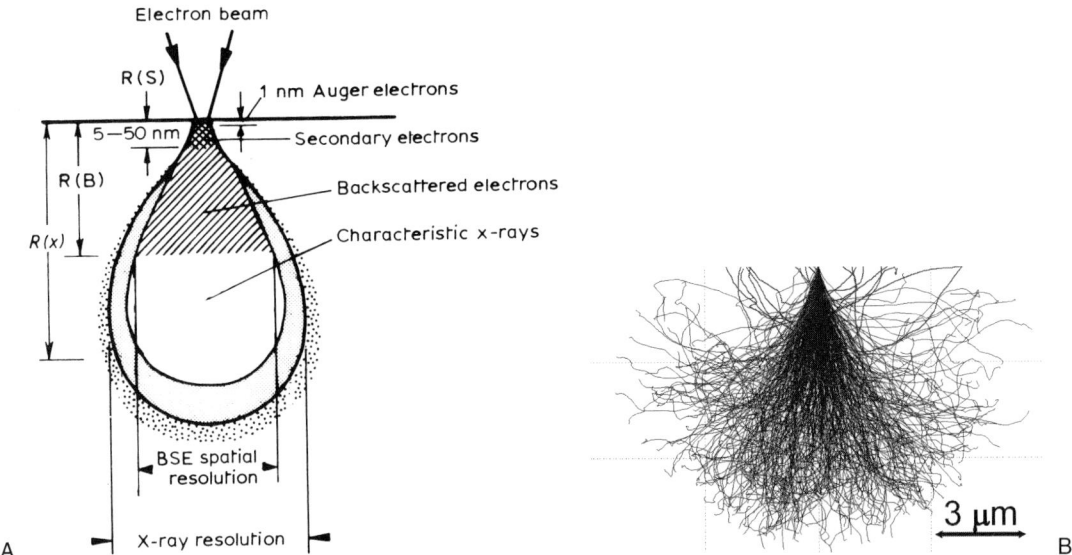

FIGURE 2.4. (A) Schematic of the interaction of an incident electron beam with a solid specimen. Backscattered electrons can escape from much greater depths R(B) than secondary electrons R(S). X-rays are produced in a larger volume and have less resolution. (B) This calculation of paths in nylon-6 for 20 keV incident electrons shows that the neat boundaries and contours in (A) are only an indication of the statistics involved. Figure drawn using CASINO [37, 38].

They appear as though the specimen is viewed from the source of the scanning beam and illuminated by a light at the detector position.

Imaging by scanning allows any radiation from the specimen, or any result of its interaction with the beam, to be used to form the image [30]. The appearance of the image will depend on the interaction involved and the detector and signal processing used. The spatial resolution, limited by the size of the specimen region from which the signal is derived [30, 36], varies considerably, as shown in Fig. 2.4. It is related to the *interaction volume*, the region where the beam interacts with the specimen (see Section 3.2.2).

The interaction volume of 20 kV electrons in poly(methyl methacrylate) (PMMA) has been shown directly by using its radiation sensitivity [39]. After exposure to a beam of electrons, the material was cross sectioned, polished, and etched. Pear-shaped holes up to 10 μm deep appear (Fig. 2.5) showing where the beam interacted with the PMMA (and reduced its molecular weight) [39]. Calculation of the interaction of the electron beam with a solid shows that the interaction volume increases at high accelerating voltage and for low atomic number and low density of the specimen [30, 32, 33].

2.3.2 Imaging Signals

2.3.2.1 Backscattered Electrons

Three signals from the specimen important for SEM are backscattered electrons, secondary electrons, and x-rays; x-rays are dealt with in Section 2.7. *Backscattered electrons* (BSE) are electrons from the beam that have been elastically scattered by nuclei in the sample and escape from the surface. The *backscattering coefficient* η—the fraction escaping—varies from 0.06 for carbon to 0.52 for gold at 20 keV [30] so backscattered electron imaging (BEI) gives strong compositional contrast. Backscattered electrons have a high energy and they can come from depths of 1 μm or more within the specimen at high beam voltages. They then leave the surface from a wide area. This means that the resolution in BEI can be low, but it

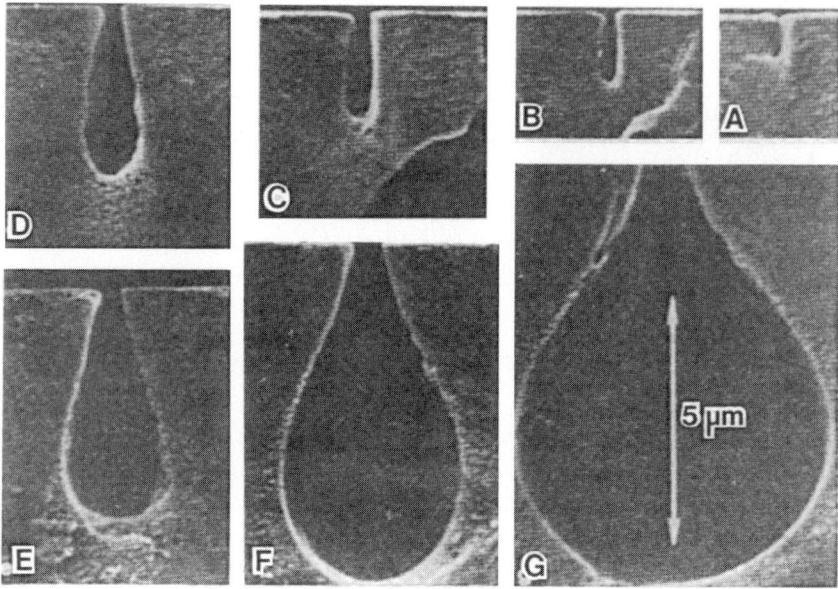

FIGURE 2.5. The electron interaction volume in poly(methyl methacrylate) for 20 kV electrons is shown directly by etching away irradiated material. The incident electron dose is the same in (A–G), but the etching time is increased, so the material irradiated less is removed [39].

depends strongly on beam voltage and sample composition. Backscattered electron imaging combined with x-ray microanalysis is a powerful method for determining the local chemical composition of a material (for an example, see Fig. 5.17).

Backscattered electrons travel in straight lines after leaving the specimen (Fig. 2.6), so a small detector placed to one side will give extreme topographic contrast, with dark regions facing away from the detector. These completely dark shadows give poor images, like pictures of the lunar landscape. A larger detector, or mixing signals from more than one detector, will produce a better BEI image. A common solution is to place large area detectors (either silicon diodes or scintillators—*Robinson detectors* [40, 41]) above the specimen and around the final lens aperture. If the sensitive area is divided into four 90° quadrants, atomic number contrast is obtained by adding the signals from all sectors, and subtraction of signals (e.g. left − right) gives topographic contrast. Because more electrons are scattered in a forward direction, a tilted surface allows more electrons to escape and increases η (see Fig. 3.20). For nylon-6 coated with 4 nm gold at 5 keV, the calculated values are 0.15 at 0° tilt, 0.20 at 30°, and 0.35 at 60° tilt [37]. Thus, even if all backscattered electrons were collected, there would still be nondirectional topographic contrast.

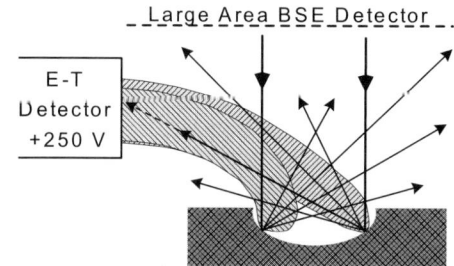

FIGURE 2.6. Shaded regions are secondary electron trajectories, drawn to the E-T detector from any part of the surface by the +250 V bias. Arrows indicate the backscattered electrons. Very few will reach the E-T detector from the surface facing away from it, so the surface facing the detector will be brighter. To collect most backscattered electrons, a large area detector is required.

2.3.2.2 Secondary Electrons

Direct *secondary electrons* (SE_1) are produced by the interaction of the primary beam with the specimen. They are emitted from the specimen with low energy, less than 50 eV, so they can only come from the top few nanometers of the material [30, 33]. If the beam falls on a tilted surface or onto an edge, more electrons interact near the surface, and so more secondaries will escape from the specimen (see Fig. 3.20). If the beam falls into a valley or pit, fewer secondaries escape; less of the interaction volume is near the surface, and more electrons are reabsorbed by the specimen.

As these direct secondaries come from an area largely defined by the beam size (see Fig. 2.4A), they can give a very high resolution topographic image. Their low energy allows a reasonable voltage to attract all of them into the detector (see Fig. 2.6), so the process is efficient, and *secondary electron imaging* (SEI) is the normal SEM image mode. However, backscattered electrons also produce secondary electrons as they leave the specimen (SE_2) and as they strike the chamber wall (SE_3) [30, 33]. These secondary electron signals will have the resolution of the BEI signal and may degrade the image.

At low beam energy, there is a single peak in the spatial distribution of secondaries. This is sharp if the incident beam is focused to a small diameter. At high energy, the SE_1 peak retains this sharpness, but there is a much broader peak of SE_2, with a resolution like that of the BSE. High-resolution information can still be obtained, but the SE_2 signal, which may be much larger than the SE_1, adds noise and reduces contrast. At low resolution, both SE_1 and SE_2 can be included. The profile at medium energy may look better, as the SE_2 peak is not so broad, but it may be worse for high resolution. The SE_2 signal is not varying slowly enough to be discarded as noise by simple image processing yet does not have the high resolution of SE_1.

The detector in the SEM that is normally used for imaging is the Everhart–Thornley scintillator/photomultiplier (E-T detector). This is a low-noise, high-speed and efficient detector that detects a small fraction of the backscattered electrons as well as the secondary electrons (see Fig. 2.6). It usually operates with a positive bias so that all the low energy electrons in the specimen chamber are attracted to it. There is therefore no contrast due to secondaries between regions of the specimen facing toward the detector and those facing away, although the edges will be well defined. The optical analog is of a surface lit by diffuse light. This detector produces good images by collecting some backscattered electrons as well as the secondaries. The result is analogous to diffuse lighting with some directional lighting to highlight the features—just what a photographer would use in a studio portrait. As we are accustomed to light shining from above, and shadows below, SEM micrographs taken with this detector system should be oriented for viewing with the detector position at the top.

In high resolution SEMs, there is little or no space between the lens and the specimen (see Fig. 3.17). In this case, the secondary electron detector is moved to a location above the final lens; the electrons are drawn there by electric and magnetic fields. When low beam voltages of only a few kilovolts are used, the great distinction between the classes of electrons is somewhat diminished; the resolution of BEI is much the same as that of SEI, and the number of secondaries emitted by the specimen increases. The beam penetration is small giving greater sensitivity to surface detail. Low accelerating voltages reduce charging effects and are increasingly used for polymer and biological samples. Scanning electron microscopy instruments have now overcome the technical problems of low accelerating voltages and can have good SEI performance at voltages down to 1 kV or less [32, 42, 43]. The low voltage SEM (LVSEM) is discussed in Section 3.2.4.

2.3.3 Electron Sources

A critical feature that controls the performance of the SEM, and is a decision made on purchase, is the type of electron gun used as a source for the electron beam. Originally, all electron microscopes used a hot tungsten filament in the form of a simple bent wire—a "hairpin." This is a thermionic emission source, so-called because the temperature is raised so high that the thermal energy of the electrons in the metal allow some to boil out into the vacuum, overcoming a potential barrier called the *work function*.

Today, more and more instruments use field emission gun (FEG) sources. Again, this may be a piece of tungsten, and in its simplest form it is an extremely sharp needle at room temperature. Now a voltage of a few kilovolts between the tip and a nearby electrode produces an electric field. The field scales as V/r, where r is the radius of the tip. The very sharp tip causes the field at the surface to reach 10 GV/m, and this lowers the potential barrier and makes it very narrow. Electrons are sucked out of the metal and tunnel through the barrier into the vacuum.

A second thermionic emission source uses lanthanum hexaboride (LaB_6). This has a much lower work function than tungsten and so will emit electrons when heated to only 1,800 K (tungsten operates at ~2,500 K). LaB_6 is reactive at its operating temperature, but the emitter is a direct replacement for a tungsten filament, requiring only very minor alteration of the instrument. Cerium hexaboride is an alternative source material of the same type.

Similarly, the field emitter can be adapted by reducing the work function. The practical device is called the Schottky field emission gun, but it is really a field-enhanced thermionic emitter. The sharp tip is coated with zirconium oxide. The field is reduced by flattening the tip and acts to reduce the potential barrier, but the electrons are thermally excited over the low barrier, not sucked through it.

Table 2.2 summarizes some important properties of these electron guns. The brightness is the current per unit area per unit solid angle (in steradians; sr)—a figure of merit for forming small probes. If two sources emit the same current, the one that has the smaller emitting area will be better, and the one that emits into a smaller solid angle will also be better as more of the flux can be captured by the lenses. The very small source size for the field emitters gives them very high brightness while their total current is less than that of the thermionic emitters. The small source size has important practical implications; it means that in a FESEM, that is, a SEM with a field emission gun, the probe diameter is always small, a few nanometers, at any condenser lens setting.

The energy spread (ΔE) depends on the temperature of the source, so it is largest for the tungsten hairpin and smallest for the cold FEG. This parameter is important at low accelerating voltages. The magnetic lenses have chromatic aberration; that is, they do not focus electrons of different energy onto the same spot. The effect scales as $\Delta E/E$ where E is the beam energy. So when E falls, the resolution is much worse for the thermionic sources. The energy spread is also important in TEM and STEM for light element analysis using electron energy loss (see Section 6.5.3).

The emitters can be poisoned by residual gases from the vacuum system adsorbing onto the emitting surfaces. High temperature devices are self-cleaning, as any condensing vapor is immediately boiled off again. Therefore, the lower operating temperature devices need better vacuum to operate in a stable and reliable manner. Ultrahigh vacuum is always expensive, and this is the main reason for the much greater cost of field emission devices. A

TABLE 2.2. Properties of electron guns

Source	Tungsten hairpin	LaB_6	Schottky field emission	Cold field emission
Brightness (MAcm^{-2} sr^{-1})*	0.1	1	100	100
Source size (μm)*	50	5	0.015	0.003
Energy spread ΔE (eV)	1–3	1–2	~0.7	0.3
Vacuum required (Pa)	10^{-2}	10^{-4}	10^{-8}	10^{-11}
Device cost	Low	Moderate	High	High

Source: Adapted from Ref. 30.
*Values are estimates for operation at 20 kV.

third type of FEG is the *thermal field emission gun*. This is like a "regular" cold FEG but the source tip is heated. This improves stability at the cost of an increase of ΔE to 1 eV.

2.3.4 SEM Types

Although SEMs are adaptable to a wide range of uses, the electron sources and some other configurations are sufficiently different to divide the instruments into different classes with different capabilities. Even so, new designs appear to try to capture the best features of all in one device. The workhorse general purpose, relatively low-cost SEM remains similar to the original instrument design. It uses a tungsten filament or a LaB_6 emitter. LaB_6 gives a greater beam current than tungsten, which might be important for x-ray analysis (see Section 2.7) or for dynamic experiments (see Section 2.9). LaB_6 costs more but needs replacement much less often. The arrangement of sample and detector is as shown schematically in Fig. 2.3 [30], allowing for large samples or different specimen stages.

The high resolution SEM (HRSEM) has the highest SEM resolution. This comes from combining a FEG with the most nearly perfect final condenser lens. Scanning electron microscopes with aberration-corrected lenses are just becoming available and should have a resolution limit well below 1 nm. Other highly perfect lenses tend to have short focal length, and so a short working distance. This may restrict the space around the specimen, limiting the specimen size and specimen stages. In the extreme case of an immersion lens, the specimen and detectors are inside the lens (see Fig. 3.17). An example of an image taken using a FESEM is found in Fig. 5.49.

If good resolution at low accelerating voltage is most important, without the need for the final nanometer of resolution, then a FEG is again required, but perhaps with a more conventional lens. Low voltage operation is often very desirable for the study of polymers as low beam voltage removes or minimizes the need to coat nonconducting samples [32,44]. High-resolution SEM images of polymer surfaces have replaced most cases of transmission electron microscopy of surface replicas [45] and compete with atomic force microscopy [46, 47]. Low beam voltage also reduces specimen damage.

Another type of SEM relaxes the condition requiring the specimen to be under high vacuum. This is the *variable pressure SEM* (VPSEM), which may be called the "high pressure" (HPSEM), "low vacuum," or "environmental" SEM (ESEM), though ESEM is a trademark of FEI. The pressure in the specimen chamber of such a SEM is 0.3–3 kPa, high only in comparison to the normal "high vacuum" value of a few $100 \mu Pa$ or less. Some distinguish "low vacuum" of 300 Pa from "ESEM" of 3 kPa, which is the vapor pressure of water at room temperature. Reaching this pressure is critical if liquid water is to be present on the sample. There is no charging even at normal accelerating voltages of 10–30 kV in the VPSEM because the positive ions from the gas neutralize surface charges on the specimen (for examples, see Fig. 5.87 and Fig. 5.98). This is particularly important for polymers and for dynamic microscopy (see Section 2.9) where new surfaces may be exposed in the microscope.

A good vacuum must be maintained at the electron source, so there are pressure-limiting apertures on the optic axis of the instrument to limit gas flow. These apertures form chambers of intermediate pressure, which can be separately pumped. The incoming beam is scattered by the gas molecules, so the focused probe is spread out. The best resolution is obtained when the specimen is close to the final aperture (a small working distance), the beam voltage is high, and the gas pressure is low. Then a resolution of about 5 nm may be obtained. At higher pressures or low beam voltages, the resolution is worse, so there is no need for field emission sources in such machines. The normal secondary electron detector is ineffective at high gas pressure, and the signal then comes from either a BSE detector or a special detector that amplifies the secondary electron signal using gas ionization [30].

2.3.5 SEM Optimization

Given a specific machine, how should it be adjusted for best results? The major issues to consider in the optimization of SEM operation for stable conducting specimens are noise, depth of field, and resolution. Parameters the operator can vary are the beam voltage, beam current, final aperture size, and working distance [48]. Standard conditions for best resolution are

1. high accelerating voltage (~20 kV);
2. small probe size, obtained with short working distance and small final aperture from thermionic emission sources (always present from field emission sources);
3. high probe current/slow scans to reduce noise, especially if contrast is low.

These conditions for high resolution imaging with secondary electrons are also conditions that result in maximum beam damage to sensitive specimens and charging of uncoated nonconducting specimens. Best conditions for these materials depends on how they interact with the beam (see Section 2.6 and Section 3.4). Reduced damage and charging in the SEM is obtained with low beam currents and low accelerating voltages. This may lower spatial resolution but improve the image, depending on the sample. The SEM may be used to study rough surfaces at magnifications below about 10,000×. In this case, the depth of field is more important than the highest resolution. Conditions for maximum depth of field are

1. long working distance;
2. small final aperture size (ca. 100 μm);
3. low magnification.

The more practical guides to SEM operation contain images showing the effect of changing these parameters (e.g., [49]).

2.4 TRANSMISSION ELECTRON MICROSCOPY

2.4.1 Conventional TEM

Conventional transmission electron microscopes (CTEM or TEM) are electron optical instruments analogous to light microscopes, where the specimen is illuminated by an electron beam. This requires operation in a vacuum because air scatters electrons. High resolution is possible because of the short wavelength of the electrons. Typical workhorse instruments made for use in biology normally use accelerating voltages from 80 to 120 kV, while high performance microscopes designed for materials science use 200 to 300 kV. Williams and Carter cover the science and the practice of TEM in four (slim) volumes [50]. Reimer [51] is an excellent reference book, but not for the beginner. There are other good recent texts [52, 53] and older reviews of the TEM of polymers [34, 54, 55]. Polymers have low atomic number and scatter electrons weakly, giving relatively poor contrast in the TEM. They can be highly beam sensitive with loss of crystalline order, mass loss, and dimensional changes occurring during observation (see Section 2.6 and Section 3.4). Increasing the accelerating voltage and cooling the specimen (for an example, see Fig. 4.50) can help to reduce the damage, and specimen preparation methods to increase the contrast (see Chapter 4) make damage less important.

2.4.1.1 Bright Field and Dark Field

Transmission electron microscope image contrast is due to elastic electron scattering. Electrons scattered to large angles by the sample do not contribute to the image in BF. In amorphous materials, the result is *mass thickness contrast*, where the image intensity depends on the local mass thickness (= thickness × density). Darker regions in the BF image are regions of higher scattering. Contrast is greater at low accelerating voltages and at small objective aperture diameters. However, a fundamental limitation of TEM is that the specimen must be very thin to allow the electrons to penetrate without losing much energy, and low accelerating voltages require even thinner samples. When a material is observed under standard conditions, the fraction of the incident beam that is transmitted can be used to measure the specimen thickness, t, using the relation:

$$I(\text{sample})/I(\text{hole}) = \exp(-Bt).$$

The constant B is determined by observing films of known thickness under the same standard conditions [51, 56, 57].

Transmission Electron Microscopy

If the sample is crystalline, the scattered intensity depends very strongly on the orientation of the crystals and on their thickness. In BF, a very thin crystal will appear dark when it is correctly oriented for diffraction. If the crystal is not perfectly flat, the contours of correct orientation will appear as dark lines, called *bend contours*. Variation of intensity such as this in crystalline specimens is called *crystallographic* or *diffraction contrast*. Many types of defects in crystals cause localized distortion of the crystal lattice. These defects change the crystal orientation locally and so cause variations in the crystallographic contrast. Detailed information on defects can be obtained by comparing the images produced by different scattered beams to theoretical predictions [50, 51, 58]. If the objects in a BF image scatter only weakly, as many polymers do, the intensity level will be high but the contrast may be too low. Dark field images normally have much higher contrast than BF images but are much weaker in intensity. Dark field images from amorphous samples are particularly low in intensity and are rarely used in CTEM. This is because the electrons are scattered in all directions, and the objective aperture can collect only a few of them. The DF arrangement in the TEM is more efficient for crystalline samples, where the scattered intensity is concentrated into a few regions of the back focal plane of the objective. To maintain symmetry of the electrons passing through the objective lens, the DF image is obtained by tilting the incident beam. Dark field of crystalline materials gives information unavailable in BF. When the DF image is formed from one spot in the diffraction pattern, bright regions in the image show only ordered areas with the correct orientation, so crystallite dimensions can be measured and their orientation determined. However, DF imaging of polymers can be difficult because the images may be unstable [54, 55, 59, 60]. An example of BF and DF images of polymers is shown in Fig. 5.144.

2.4.1.2 Phase Contrast

In *phase contrast* imaging, scattered electrons are allowed to pass through the objective aperture and recombine with the unscattered electrons to form the image. This would give no contrast if the objective lens was perfect and the specimen perfectly in focus. The lens is not perfect and often very slightly out of focus, and this causes the scattered beams to be phase shifted. When the beams recombine, this phase shift causes a change of intensity, which will depend on feature size and defocus [50, 51]. Components analogous to those in phase contrast optical microscopy (see Section 2.2.3.2) are not required but have been used [61].

If the specimen scatters strongly, there will be mass thickness contrast (or diffraction contrast if the specimen is crystalline) due to the exclusion of scattered beams by the objective aperture, and this will dominate the image. A weakly scattering object such as a thin carbon or polymer film may appear featureless at low magnification, yet have a visible structure entirely due to phase contrast at high magnification, where lens defects and defocus are more important. The contrast of this structure will be a minimum at exact focus, and this is often a good way to determine the focus condition [51, 62]. The nature of the relation between defocus and phase must be well known in order to interpret phase contrast images accurately.

Deliberate defocusing enhances phase contrast at lower magnifications but it must be used with caution. If there is only random structure in the specimen, deliberate or accidental defocus may induce clearly visible structure unrelated to the specimen—artifacts. Thomas [59] discussed this in detail for polymer microscopy, quoting several TEM studies of polymers that were dominated by phase contrast artifacts. With care, artifacts can be recognized [63, 64] and phase contrast imaging can be successfully applied to polymer systems (e.g., [65]). Phase contrast at high resolution produces lattice images (see Section 2.4.4 and Section 3.1.5).

2.4.2 Scanning TEM

In the *scanning transmission electron microscope* (STEM), as in the SEM, a fine electron beam or probe is formed and scanned across the specimen. But in the STEM the specimen is thin and the intensity of a transmitted signal is detected, amplified, and used to form an image. The image can be processed to give a wide range of structural and chemical information. Collecting the direct transmitted beam gives a BF image and an annular detector produces a DF image. The electrons collected may have been scattered to higher angles than in TEM DF, when the image mode is described as *high angle annular dark field* (HAADF), as shown in an example in Fig. 5.29.

The resolution in STEM is limited by the probe size, and fine probes are formed using high-brightness sources. A *dedicated STEM* has no imaging lenses, so the instrument is "dedicated" in that it can only operate as a STEM. Such devices generally have a cold FEG and operate at ultrahigh vacuum throughout, which is not compatible with many polymer samples. A TEM with a probe-forming condenser lens can be capable of very high resolution x-ray analysis (see Section 2.7). Add a device to scan this probe and it is also a STEM. This type of instrument may be called a TEM/STEM or AEM (analytical electron microscope). In a similar way, adding a detector below the specimen could turn a SEM into a STEM, but the relatively low beam voltage in the SEM makes this of limited use, as the specimen has to be very thin.

As with the SEM, the capabilities of a STEM are largely controlled by the electron source used. Most current instruments use some type of FEG. Older TEM/STEMs may have a LaB_6 emitter, with lower resolution and higher total current. The fine probe of the STEM can be used to obtain chemical and structural analysis from very small regions (see Section 2.7), and this is the reason for most purchases of the instrument. However, few if any polymers can withstand the radiation environment of a stationary finely focused beam of high energy electrons, which rapidly affect even stable inorganic compounds [66]. One advantage of the STEM for radiation-sensitive polymers is that only the scanned—and thus the imaged—area of the specimen is irradiated. In TEM it can be difficult to limit irradiation of adjacent areas and this is especially important for diffraction experiments. Microdiffraction can be readily conducted in a STEM with a probe that leaves adjacent regions undamaged. In polymers the microdiffraction area must be comparatively large, attainable by both instruments, but the STEM allows better control. Another possible advantage is that STEM imaging has higher resolution than the TEM in very thick films. This advantage is greatest for disordered, low atomic number materials such as polymers [67].

2.4.3 Electron Diffraction

Electron diffraction is an important technique for the study of crystalline materials [50, 68, 69]. It is regularly used to identify crystal structures and local orientation. Less frequently, it can be used to determine an unknown structure [70]. The directions in which electrons are diffracted from a specimen relate to the atomic spacings and orientation of the material (see Section 3.1.2). A crystal has a regular arrangement of atoms and so in the TEM it will produce a diffraction pattern consisting of sharp spots. Polycrystalline materials have many spots, which together form continuous rings. Small or imperfect crystals give fuzzy spots or rings.

In *selected area electron diffraction* (SAED) in the TEM, an aperture (the intermediate or selected area aperture) is used to select a region of the specimen for diffraction (for an example, see Fig. 5.146). A near-parallel beam of electrons illuminates the specimen. Generally, the region contributing to the pattern is several micrometers in diameter. This is a large area compared to that in STEM microdiffraction but very much smaller than that needed for normal x-ray diffraction.

The intermediate aperture is below the specimen, so a large area of the specimen is irradiated during SAED. This is undesirable for polymer specimens that are damaged by the beam, and other techniques must be used. The beam can be focused using the condenser lenses to limit the area irradiated, but the intensity produced by focusing the beam is too high. A strongly excited first condenser lens and a very small second condenser lens aperture will reduce the intensity. Under these conditions, a near-focused beam illuminates a small region of the specimen with a near-parallel electron beam. Here the diffraction area is "selected" by the incident beam diameter as the aperture is above the specimen.

Convergent beam microdiffraction uses a convergent rather than a near-parallel beam, and this makes it possible to limit the beam to extremely small regions. The diffracting area is limited spatially by the beam diameter, but few polymers can withstand the focused beam.

2.4.4 High Resolution Electron Microscopy

If a scattered electron beam is a sharp spot diffracted from a single crystal, the phase contrast image that forms when it is recombined with the unscattered beam is an image of the crystal lattice planes that produce the scattering by Bragg diffraction. When several beams are recombined, the result is an image of the crystal lattice. This specialized phase contrast technique has been applied to the study of atomic scale structure in many types of crystalline metals and ceramics. It is called *high resolution electron microscopy* (HREM) and allows the direct imaging of defects and interfaces on the atomic scale (for an example, see Fig. 5.22) [71–73]. The technique is difficult to apply to the study of polymeric materials because of their instability in the electron beam, as the high resolution images require high beam intensities. In 1995, Martin [74] reviewed the theory and experimental methods and Tsuji [75] reviewed HREM work done on polymers; a 2005 review by Martin [76] brings this up to date.

High resolution electron microscopy can provide information that cannot be obtained in any other way, but it requires skill and experience as well as a high resolution TEM. Lattice fringe images have been obtained from a wide range of polymers. This includes polypropylene [77], polystyrene [78, 79], poly(*p*-phenylene terephthalamide) (PPTA) [80] and poly(*m*-phenylene isophthalamide) (MPDI) [81]. Generally, it is the highly aromatic polymers and molecules with high melting points that are most suitable subjects for this method, because they are least affected by radiation damage (see Section 2.6 and Section 3.4).

Good information can also be obtained from partially ordered polymers such as poly(*p*-phenylene benzobisthiazole) (PBZT) and poly(*p*-phenylene benzobisoxazole) (PBO) [74, 82–85]. Figure 5.150A is an image from Ref. 85 that shows the details of local deformation of molecular planes in a kink band produced by plastic deformation. High resolution electron microscopy of smectic liquid crystalline polymers shows edge dislocations and other defect structures [86, 87]. The necessity of using a very thin layer of material as a sample in the TEM was used to advantage in a study of the ordering of a polyimide during imidization [88].

2.5 SCANNING PROBE MICROSCOPY

2.5.1 Introduction

A *scanning probe microscope* (SPM) is a microscope that produces an image by scanning a small solid probe on or extremely close to the surface of a specimen. Some signal is detected from the interaction of the probe with the surface, and in principle this signal can be used to form an image. In practice, this signal is used in a feedback loop, shown schematically in Fig. 2.7, to keep the signal constant by control

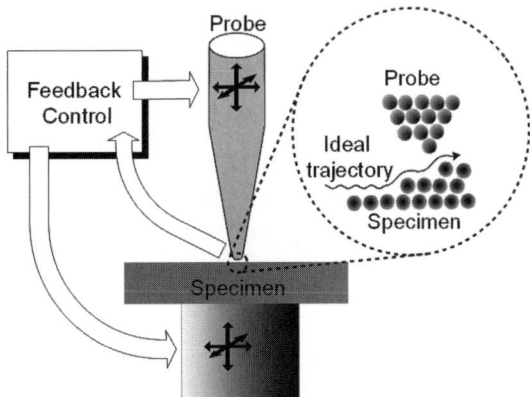

FIGURE 2.7. Schematic of an SPM. A probe is scanned in close vicinity to the sample surface while some signal that depends on some local interaction is measured. Although the signal can be measured as a function of the scanning position, in most practical applications the probe-specimen distance is continuously adjusted to keep this signal constant by a feedback control.

of the probe-specimen distance. The image is then formed from the setting of the feedback control that is required to keep the signal constant or from a second signal. For example, the amplitude of an oscillation can be controlled by feedback and the phase angle of the oscillation used to form an image (described later in Section 2.5.2.2 and Fig. 2.9).

As in the SEM, where the probe is a focused beam of electrons, the resolution of the image is controlled by the region of interaction. The part of the probe that interacts with the sample has to be very small, and in several forms of SPM it is small enough to allow atomic resolution. Extremely precise control of position is required in all SPMs both for (x, y) scanning across the surface and for the z height control, and this is accomplished by use of piezoelectric drivers. Motion control is shown schematically in Fig. 2.7 by double-headed arrows on both probe and specimen, but normally one or the other is moved, not both. Some systems split the control, moving the specimen in the x and y direction and the probe in the z direction.

Scanning probe microscopy has grown very quickly since the invention that began this field in 1982 [89]. New types of microscopes, new operation modes and new commercial designs still appear frequently, and there is a rapidly expanding range of literature. Commercial systems range from tabletop models costing little more than a top-quality optical microscope to elaborate research instruments and expensive automated high throughput microscopes for industrial quality assurance. Some instruments are optimized for the most rapid imaging. Others are optimized for the observation of (biological) samples immersed in fluid or for operation in ultrahigh vacuum systems.

The first SPM to be developed was the *scanning tunneling microscope* (STM) [89]. In the STM the probe is a conductor set at a bias voltage difference from a conducting sample and the signal is a current that passes between them. The probe-sample interaction is the quantum mechanical tunneling current that has a measurable value only when the two conductors are a very small distance apart, typically less than 1 nm. In the STM the region of interaction in the specimen can be limited to a single atom, so the microscope can show the atomic arrangement on surfaces. Because the sample must be conductive, the STM has little importance for studying polymers today, although it was the predominant SPM technique a decade ago (for an example, see Fig. 5.153).

The second type of SPM to be developed and currently the most important for the study of polymers is the *atomic force microscope* (AFM) [141]. In the AFM, the probe is a small tip mounted on a cantilever arm and the interaction is the small force (typically <10 nN) between the tip and the sample. The cantilever acts as a spring that can be deflected slightly by small forces. Its angle is generally measured with an optical lever system (Fig. 2.8) that gives an electrical signal related to the cantilever motion.

Some good general textbooks on STM and AFM include those of Magonov and Whangbo [90], Meyer et al. [91], and Weisendanger [92]. Literature reviews of SPM have been published (e.g., Bottomley [93]), but the field of research is now so highly divided among numerous subdisciplines that comprehensive literature reviews are difficult to find. Good general

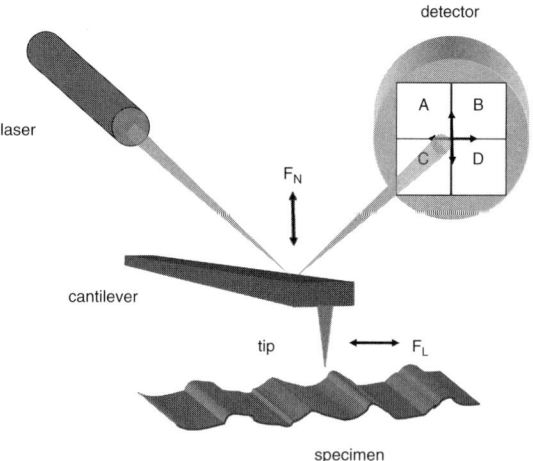

FIGURE 2.8. Schematic of the contact mode AFM. The normal and lateral forces, F_N and F_L, result in deflection and torsion of the cantilever, respectively. They can be monitored using an optical detection system and used for the purpose of feedback control and/or image contrast formation.

references for polymer SPM techniques include the introductory notes by Overney and Tsukruk [94] in the proceedings of the 1998 ACS Symposium on SPM of Polymers [95]; reviews by Reneker and Magonov in 1997 [96]; Krausch in 2001 [97]; Magonov in 2001 [98]; and Chernoff and Magonov in 2003 [99]. The published proceedings of the 2000 [100] and 2003 [101] ACS Symposia on SPM of Polymers also contain excellent collections of papers.

For information on properties of cantilevers, the second edition of Sarid's book is an excellent choice [102]. Garcia and Perez [103] have reviewed the theory behind dynamic imaging modes of intermittent contact AFM (IC-AFM) [104] and noncontact AFM (NC-AFM) [105]. The influence of the tip shape in imaging has been reviewed by Villarrubia [106]. The application of SPM to polymer ultrastructure and crystallinity has been reviewed by many including the work of Reneker et al. [107], Lotz et al. [108, 109], Hobbs et al. [110], and Magonov and Yerina [111]. Some of the ultrastructural studies have been conducted using hot/cold stages to study *in situ* crystallization [112–114].

Thermal analysis using heated tip technology [115] has been reviewed by Price et al. [116], Wunderlich [117], Pollock and Hammiche [118], and Abad et al. [119]. Reviews on mechanical deformation of polymer films by nanoindentation [120] and AFM indentation [121] as well as *in situ* tensile deformation [122] have been published. A related area of force spectroscopy to extract intermolecular and intramolecular forces between polymer chains was reviewed by Hugel and Seitz in 2001 [123]. Bushan has written extensively on the use of SPM to study tribology [124], and frictional contrast has been discussed in reviews by Mate [125], Zasadzinski [126], and Feldman et al. [127]. Spatially localized adhesion studies using chemically modified AFM tips was reviewed in 2005 by Vezenov et al. [128] and Vancso et al. [129]. Characterization of polymer film surfaces [130, 131], latexes [132, 133], multiphase polymer blends [134, 135], and nanocomposites [136, 137] has also been reviewed. Authors from the laboratories of several manufacturers of industrial engineering polymers have published reviews on the application of SPM to characterization [138–140].

2.5.2 Atomic Force Microscopy

The AFM can be operated in several modes that are sufficiently different that they need to be described separately. The basic distinction between these modes is how much the tip touches the specimen surface: *contact mode* is when the tip is in contact with the specimen all the time; *intermittent contact mode* is when the tip oscillates and touches the specimen some of the time; in *noncontact mode,* the tip does not touch the specimen at all.

2.5.2.1 Contact and Associated Modes

In contact mode AFM [141], the tip mounted on the cantilever is in contact with the sample surface during scanning. This is similar in principle to a profilometer where a stylus is dragged over a surface. In the AFM, a very small contact area and a highly sensitive cantilever allow high resolution of topographical surface features, both laterally and vertically (for an example, see Fig. 5.31C). During scanning, the local variation of vertical force for a rigid sample is due to variation in the height of the surface. Feedback control of this force causes the vertical position of the tip to trace out the surface topography. This is accomplished by varying the height of the tip or the sample so that the cantilever deflection is kept constant. Because the tip stays in contact with the surface during scanning, the cantilever experiences forces acting not only normal but also lateral to the sample surface. The origin of the lateral force is shear force in the sample at the tip position. The shear forces can be very high and cause damage to soft polymer or biological specimens. Damage to the sample surface can complicate measurement and interpretation or even make an experiment impractical and result in artifact formation.

The lateral forces result in a torsion of the cantilever, and this can be measured using the same optical detection system that measures vertical deflection (see Fig. 2.8). This signal from the tip-specimen interaction leads to another operational mode, the *lateral* or *frictional force microscope* (LFM or FFM) [142, 143]. In LFM the cantilever is designed to be relatively easy to twist, and scanning is performed perpendicular to the long cantilever axis. The variation of lateral deflection during scanning can be used to form the image. The image is correlated with the frictional force and thus with surface chemistry as well as surface topography (for an example, see Fig. 4.1B).

The forces acting between the tip and sample surface clearly play a vital role in contact mode AFM, and understanding these forces is key to image interpretation in this mode. In feedback operation, the force between tip and sample is kept constant. Controlling the force variation led to the development of the *force modulation microscope* (FMM) [144]. In this mode of operation, the vertical position of the cantilever support is modulated periodically during scanning while the tip remains in contact with the sample surface. The up and down movement results in a deflection of the cantilever that will depend on the local forces and the local mechanical properties of the sample. For example, a soft spot on the sample will allow the tip to penetrate when the force is increased by moving the cantilever support down toward the sample. This will reduce the measured deflection modulation at that point. Interpretation and experimentation in this mode can be difficult due to the contribution of nonspecific forces (e.g., the surface tension of condensed vapors) and their high values. The method is therefore not very sensitive to local properties, which may have to differ by orders of magnitude to be detectable and to result in significant contrast differences. Nevertheless, FMM has been explored on many polymeric samples [96, 137, 145].

The concept of measuring forces was further developed in an operational mode called *force spectroscopy*. Scanning is typically discontinued and the experiment is performed at a given x, y location on the surface. Again, the cantilever support is moved vertically but over a much wider range. Measuring the cantilever deflection gives a force versus distance curve, which is a plot of the force (or cantilever deflection) as a function of the tip-sample separation. In particular, this method can be used to measure the *pull-off* force (i.e. the force required to separate the tip from the sample surface). In ambient conditions the pull-off force is typically dominated by capillary forces. In a vacuum or liquid environment there are no capillary forces and pull-off can give information about the adhesive properties of the sample. Experiments that provide the local mechanical properties can be used to get new information about a sample. Force spectroscopy can also be performed in an imaging mode by collecting a data set that contains topographical information and an array of force-distance curves.

Another mode of operation derived from contact AFM that is relevant to polymer studies is the *scanning thermal microscope* (SThM) [118, 146]. In SThM the tip is a special device that has a resistive element. If a current is passed through this resistive element, its temperature depends on the heat transferred to the specimen. It thus acts as a thermal probe, as seen for example in Fig. 5.36. During scanning, thermal control of the probe can be used to generate images based on variation of either sample temperature or thermal conductivity. Filled polymers and polymer blends are candidates for this kind of study, but the resolution is relatively poor. Microfabricated thermal probes can give a resolution of 100 nm.

Two other modes derived from contact AFM that are worth noting, though they are of limited application to most polymers, are *conductive AFM* [147] and *scanning capacitance microscopy* (SCM) [148]. The conductive AFM is used to characterize conductivity variations across a sample surface. Typically, a DC bias is applied to the tip while the sample is grounded, and the current passing between the tip and the sample is used to generate the image. In SCM, an AC bias is applied to the sample, and the local tip–sample capacitance changes are measured during scanning to generate the image. Both modes have been mainly used for semiconducting materials, but contact AFM has been applied to conducting polymers [149] and composites [150, 151] containing carbon.

2.5.2.2 Intermittent Contact Mode

The application of contact mode AFM to soft materials such as polymers and biological systems is seriously limited because of the lateral forces that can damage samples, cause artifacts to form, and reduce image resolution. The invention of the intermittent contact AFM (IC-AFM), often referred to as TappingMode™ AFM (Veeco Instruments, Santa Barbara, CA) [152] or TMAFM, overcomes these limitations and has become the most important and commonly used AFM mode for the study of polymers. In IC-AFM the cantilever is forced to oscillate vertically near its resonance frequency, which is typically around 50–400 kHz. The amplitude of the free cantilever oscillation (i.e. far from the surface) is typically a few tens of nanometers. The vibrational amplitude is reduced when the cantilever is brought closer to the sample surface and begins to touch it, as shown in Fig. 2.9. In the original form of IC-AFM, the amplitude is measured and used for feedback during scanning. Topographical images are produced as the amplitude depends on the mean tip height. The lateral force and inelastic surface modification are greatly reduced because the tip makes contact with the sample surface only briefly in each cycle of oscillation ("tapping"). Examples of this mode can be seen in Fig. 5.32, Fig. 5.74 and Fig. 5.75.

As the tip is brought close to the sample surface, other vibrational characteristics of the cantilever vibration change due to the tip-sample interaction. In addition to the amplitude, these are the resonance frequency and the phase of the vibration. A further development of the IC-AFM allows shifts in phase angles of vibration to be detected when the oscillating cantilever interacts with the sample surface [153]. *Phase imaging* is the mapping of the shifts in phase angles of the oscillating cantilever, relative to the phase of the periodic signal used to drive the cantilever into oscillation. These changes often correspond in a complex way with changes in the local properties of the sample surface, such as variations in composition, adhesion and viscoelasticity. Phase imaging provides enhanced image contrast, especially for heterogeneous surfaces. It has become an important tool in polymer applications (for examples, see Fig. 5.50 and Fig. 5.73). These include mapping of different components in composite materials, differentiating variations in adhesion, locating contaminants, and high resolution imaging of surfaces. Because phase imaging is widely used to study polymer surfaces, it is important to understand how the phase signal depends on imaging conditions and sample properties (see Section 3.1.5).

2.5.2.3 Noncontact and Associated Modes

The oscillating tip, as shown in Fig. 2.9 for IC-AFM, can be influenced by long range forces originating at the sample surface without coming into contact with it—this is noncontact AFM (NC-AFM) (for an example, see Fig. 5.117). Users may sometimes think they are using the NC mode, even when there is some contact, and the mode is then really IC. In true NC-AFM the tip does not touch the surface during oscillation as the amplitude of the oscillation is kept well below the mean spacing between the tip and the sample, which is a few to tens of nanometers. Ultrahigh vacuum (UHV) conditions are usually required for

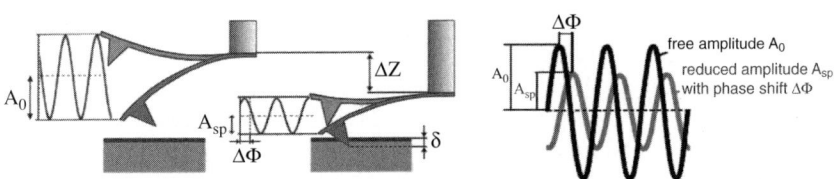

FIGURE 2.9. Schematic of the intermittent contact mode AFM: free oscillation with free amplitude A_0 far away from sample surface, and damped oscillation with set-point amplitude A_{sp} and phase shift $\Delta\Phi$ during scanning. A_{sp} is chosen by the operator, and feedback control is used to adjust tip-sample distance such that A_{sp} remains at constant value. The choice of A_0 and A_{sp} has great influence on tip-sample force interaction and image formation.

topographical imaging in NC-AFM (see Section 3.3.5). A water layer is usually present on the surface in air, and capillary forces will trap the tip in the water if it gets too close. This does not happen in IC-AFM because the amplitude of oscillation is higher. The tip can then penetrate the liquid layer and be pulled out by the spring effect of the deformed cantilever; it does not get stuck.

The *magnetic force microscope* (MFM) [154] and the *electric force microscope* (EFM) [155] work in noncontact mode. The MFM is used to map the magnetic force gradients above the sample surface using a ferromagnetic probe tip. Local variations in the magnetic field induce changes in the cantilever vibration while scanning in noncontact mode. Similarly, the EFM is used to map the electric field gradients above the sample surface. A voltage is applied between the tip and the sample (the cantilever and tip must be reasonable electrical conductors), and then variations of the electric field induce changes in the cantilever vibration. Usually the topography of the sample is first determined using IC-AFM. This topography is stored in memory and a second noncontact scan of the same area is made, using the stored height information to make the probe stay a fixed distance above the sample. This second scan is in the EFM or MEM mode. The MFM has been used for visualization of magnetic domains, and the EFM has been applied for mapping of locally charged domains, including polymeric samples [156] and filled polymers containing carbon black [157].

2.5.3 SPM Probes

Choosing the appropriate SPM probe depends on a number of factors, primarily on the intended application. For example, tips may be coated with metal for specific applications (e.g. with Au/Pt or Cr/Au for EFM, with Co or Ni for MFM) or with hard coatings such as diamond-like carbon (DLC) or SiC for extra durability. Note that these coatings generally compromise tip sharpness to some extent.

For AFM, both the tip and its supporting cantilever must be considered. The key factors to be considered are materials of construction, geometry and shape, resonance frequency, quality factor, tip sharpness, and cost. The first AFM probe used a piece of thin gold foil as cantilever, with a diamond shard glued at the end for a tip [141]. Because the instrument was operating in contact mode, the spring constant of this lever was made smaller than the spring constant between atoms to minimize damage (<1 N/m). For resonant techniques, the stiffness of the cantilevers can be much higher (~40 N/m). The cantilevers are single beam or two beams in the form of a Λ with the tip placed where they meet. The FFM requires single-beam cantilevers.

Microfabrication of probes in quantity from silicon or silicon nitride using lithography techniques [158, 159] has led to lower cost and more variety in probe shapes and sizes. If silicon is used, it may be doped to a conductivity sufficient to dissipate static charge. The tips formed by lithography have faceted cross-sections as the surfaces are crystallographic planes. If they have three sides (triangular base, tetragonal) they must terminate in a point; if there are four sides (pyramidal) they may be blunter. For conventional imaging the role of surface energetics and adhesion should be considered. Most silicon or silicon nitride probes have surfaces with oxides that are hydrophilic, whereas DLC-coated probes are hydrophobic. Nanotubes or other filaments may be attached to the tip to increase its sharpness (Fig. 2.10) [160, 161]. Cantilevers can also be fabricated out of metals (e.g. steel) when very high stiffness is required for nanoindentation.

Since a majority of commercial systems use optical deflection detection of the cantilever motion, cantilevers must be made of materials that are reflective enough to give good signal in the optical path. Gold or aluminum coatings can be deposited on the back side of the levers for improved reflectivity. This can add residual stresses if the interfaces are not carefully prepared. In the worst case, this can lead to unwanted bending or twisting of low spring constant cantilevers. Generally, higher frequency cantilevers are needed for faster response and faster scanning. Cantilever resonant frequencies can range from kilohertz to megahertz with typical values of 50–150 kHz for IC-AFM.

Radiation Sensitive Materials

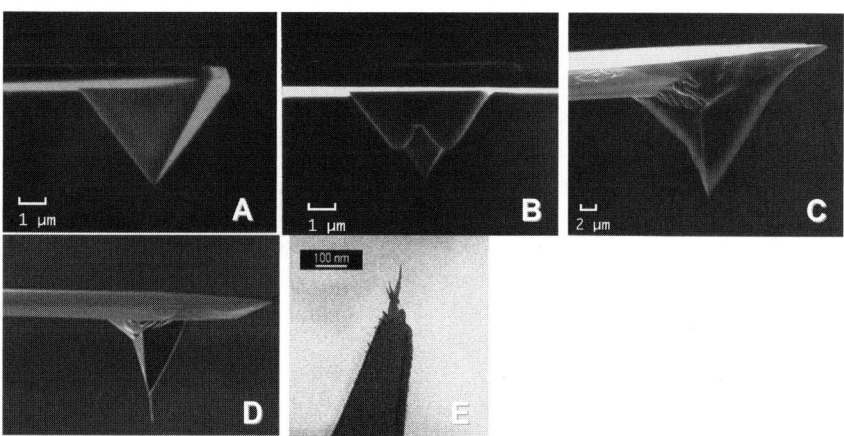

FIGURE 2.10. Scanning electron microscope images of AFM tips: (A) pyramidal Si_3N_4 contact mode tip; (B) oxide etched contact mode tip; (C) silicon tip for intermittent contact; (D) [160] same as (C) with attached carbon nanotube; (E) [161] TEM image of diamond-like filaments grown on the end of a silicon tip. (From *Nanosciences* [160, 161]; used with permission.)

The tip sharpness is usually given as a "tip radius of curvature," and this contributes directly to the ultimate resolution. The tip radius for mass-produced probes is governed by the processing conditions. The nominal radii of curvature are in the 10 nm or less range for microfabricated silicon tips used for intermittent contact or contact mode. There are many methods used to increase the tip sharpness, including e-beam deposition [162], oxidation or etching [163, 164], attachment of carbon nanotubes [165, 166], and filament growth in a plasma [167]. Some of these specialty probes tend to be expensive as they cannot be readily mass produced and the sharpness and yield can be inconsistent. Some mass-produced silicon probes may achieve the sharpness of ultrasharp tips due to random processing events. Representative tips are shown in Fig. 2.10.

Additional geometric features of tips that should be considered include the tip height; whether the tip is conical or faceted; the opening angle of the tip; the axial angle and mounting on the cantilever; and finally the location of the tip on the end of the cantilever. These geometric considerations beyond sharpness usually come into play when imaging significantly rough or textured surfaces where it is likely that the point of contact with the surface may move off of the apex and occur somewhere up the tip shaft. This may lead to artifacts in the image (see Section 3.3.7). The major manufacturers of SPMs and probes are listed in Appendix VII.

2.6 RADIATION SENSITIVE MATERIALS

This section relates to electron microscopy and particularly to TEM; a critical feature for the TEM of polymers is radiation sensitivity (see Section 3.4), and beam-sensitive materials can require very special treatment [59, 60, 168, 169]. Discussion of the optimal strategy for dealing with radiation-sensitive specimens generally deals with the choice of accelerating voltage and other tactics. First, however, the microscopist should stand back and view the overall strategy. The best way to avoid the bad effects of irradiating a radiation sensitive polymer is not to irradiate it. The following three questions should therefore be answered with a firm "NO" before considering procedures to minimize radiation damage.

1. Can information be obtained from a sample even after heavy irradiation in the microscope?
2. Can techniques other than electron microscopy be applied to the problem?

3. Can the material be made less sensitive (e.g., by staining or etching)?

Question 1. This defines the radiation sensitivity of the material in an operational way that depends very strongly on the sample and the information required from it. Thus a fluoropolymer may degrade rapidly, losing mass in the beam, and have no known staining procedure. When the microstructure of the homopolymer is studied, this will be a very sensitive material. Very limited information will be available in the TEM. The same polymer could be a component in a blend, with the TEM used to determine the size and arrangement of the component. Radiation damage may now help rather than hinder the investigation by increasing contrast between the two components. This blend sample may well not require procedures to minimize radiation damage.

Question 2. It is possible that optical microscopy or AFM or another characterization method (see Chapter 7) could give the required information. The high resolution SEM involves irradiation but the problems are less severe than in the TEM, so this may be an useful alternative.

Question 3. Etching a surface may make the arrangement of components of a blend stand out in the SEM or AFM (see Section 4.5). Stains increase the contrast of the specimen (see Section 4.4), but not all stains increase the stability of the image. Some, such as iodine, are rapidly driven off by irradiation in vacuum; others are completely stable.

A good example of avoiding radiation damage is the study of melt-crystallized polyethylene. The lamellar microstructure of a normal sample is extremely sensitive to radiation. Low dose techniques in the TEM produce low contrast images difficult to interpret. By staining (for an example, see Fig. 4.19) [170, 171] and etching (for an example, see Fig. 4.23) [172–174], completely stable, convincing high contrast TEM images of the lamellae have been produced. High resolution LVSEM [44] also shows the lamellar structure (see Fig. 5.31A) and requires only a simple specimen preparation. More recently, AFM has shown details of the lamellar structure (for an example, see Fig. 4.45), including crystal growth observed directly using the AFM with a hot stage [175].

2.6.1 SEM Operation

Electrons of even 30 keV energy do not penetrate far into the surface of the specimen. Unless the irradiated material evaporates, so that the beam drills a hole in the sample, the unaffected bulk holds the damaged thin surface layer in place and the effect on the image is limited to fine-scale details. To reduce these effects, the thickness of the surface layer that is irradiated should be reduced. Either the metal coating can be made thicker or the accelerating voltage reduced. Both reduce the total energy deposited in the polymer but both may also reduce the resolution of the image in the SEM. The optimum beam voltage for a specific specimen and instrument should be found by trial and error [48], and beam sensitivity adds another factor that may drive down the beam voltage. A beam voltage of 2–5 kV with a thin (<5 nm) layer of metal is usually satisfactory. If the resolution is sufficient, operation at 1–3 kV may be even better [32, 45, 176]. Conditions that can minimize damage in the SEM are as follows:

1. low accelerating voltage;
2. thicker surface coatings;
3. minimum beam current and exposure time that has low enough noise;
4. low magnification;
5. beam blanking between exposures.

Items (1) and (2) on this list were discussed above. Items (3) and (4) are to limit the radiation dose as much as possible by reducing the number of incident electrons and spreading them out over a large area. The last item is to make the radiation more effective by stopping it unless data is being collected.

2.6.2 Low Dose TEM Operation

Assume now that the TEM is unavoidable, that it is really necessary to form an image from some transitory feature that is destroyed by radiation. Noise is a fundamental limit, so it is most important to use all the available electrons to form an image—"make every shot count." The way to do this is to use *low dose methods* [177, 178] where focusing and other adjustments are done while the beam passes through an adjacent area of the specimen. The beam is moved between its focusing position and its exposure position with the beam deflection coils. Modern video systems allow easy viewing of the dim images. Many commercial microscopes have a "low dose" mode that automates the procedure, and software may now calculate exposures that spread the allowable dose over multiple images of the same area [179]. The main points to remember are

1. keep the beam from falling on the area to be imaged unless the image is being recorded;
2. match the maximum allowed specimen dose, the magnification, and the sensitivity of the recording medium.

When the low dose technique works well, the resolution in the image of a beam sensitive material approaches its theoretical limit (see Section 3.4). What if this is not good enough? There are a few possibilities:

1. Use image processing (see Section 2.8.1) to reduce the effects of noise.
2. Increase the maximum electron dose that the specimen can withstand;
   (a) Use *cryomicroscopy* (see Section 4.9); that is, cool the specimen to liquid nitrogen temperatures. A low temperature of irradiation increases the dose required to destroy structure in many organic materials, including polymers [180, 181].
   (b) Optimize the beam voltage; higher energy electrons are faster, so they deposit less energy in the specimen and do less damage, but contrast falls at high energy: 200–300 keV is a likely optimum.
3. Increase the contrast or the signal; for a given experiment, BF or DF, CTEM or STEM may be best, there is no clear winner that can always be recommended for the best resolution of a beam sensitive specimen.

2.7 ANALYTICAL MICROSCOPY

Chemical analysis in the light microscope has normally been performed by measuring the physical properties of the sample and comparing them to tabulated values [8, 11]. Melting point and refractive index are most commonly used. Coupling a laser Raman spectrometer to an optical microscope to make a Raman microscope has made quantitative microanalysis possible for a very wide range of materials that includes polymers [182, 183]. In TEM and STEM, the energy spectrum of electrons passing through the specimen may be used to get information about the elemental composition. The technique is called electron energy loss spectroscopy (EELS), and it is most useful for light elements [51, 184, 185]. Application of EELS to polymers has been reviewed [186] (see also Section 6.5.3). Good energy resolution for EELS or energy-filtered images to get light element mapping has been the driving force to use field emission guns in the TEM, where good image resolution can be obtained with thermionic sources, but the reduced energy spread of the FEG is critical for EELS.

More common is elemental analysis using the x-rays emitted from the specimen in the SEM [30, 33, 187, 188] and the TEM [30, 51, 189, 190]. X-ray analysis is most useful for heavier elements, so in polymers it is often applied to find out the nature of fillers and contaminants or the location of heavy elements (such as the chlorine in polyvinyl chloride) in polymer blends. Quantitative microanalysis of heavier atoms dispersed within a single phase polymer may be more difficult. The technique is limited by the sensitivity of polymers to radiation and heat damage from the intense electron beam needed for microanalysis.

2.7.1 X-ray Microanalysis

When a high energy electron beam impinges upon a specimen, x-ray photons are produced. They fall into two classes. *Characteristic* x-rays have well defined energies that are characteristic of the atoms in the specimen. These x-rays form sharp peaks in the x-ray energy spectrum and contain analytical information. They are emitted by atoms in the specimen as they return to their ground state, after an inner shell electron has been removed by an interaction with a high-energy beam electron. *Continuum* x-rays have a wide range of energies and form the background in the x-ray energy spectrum. They are produced when incident high energy electrons are slowed by scattering near the atomic nucleus and carry no useful information.

The SEM, TEM and STEM can be fitted with x-ray detectors for elemental analysis of the specimen. The electron probe microanalyzer (EPMA) or x-ray microanalyzer (XRM) is basically an SEM designed for x-ray microanalysis. In the SEM the x-ray signals are produced from almost the entire interaction volume in the specimen (see Fig. 2.4) when the beam energy is well above the energy of the x-rays. In thick specimens the interaction volume is generally much wider than the beam diameter, due to beam broadening in the specimen. The interaction volume then determines the spatial resolution of the technique, and the volume increases for low atomic number specimens and with the beam voltage. The peak to background ratio and the signal sensitivity increase with the electron beam voltage, so optimum voltage is a compromise between peak to background ratio and spatial resolution. The backscattered electron image is produced from signals coming from almost the same excitation volume as the x-ray signal (see Fig. 2.4). This is helpful, as BEI can be used to scan the specimen quickly for atomic number differences to guide the slower x-ray mapping (for an example, see Fig. 5.17).

2.7.1.1 Energy Dispersive X-ray Spectrometer

Two different types of detectors are used to measure the x-ray intensity as a function of wavelength or energy. In an *energy dispersive x-ray spectrometer* (EDS), x-rays generated from the sample enter a solid state silicon detector (a reversed bias p–n junction). They create electron hole pairs that cause a pulse of current to flow through the detector circuit. The number of pairs produced by each x-ray photon is proportional to its energy. The pulses produced are amplified, sorted by size with a multichannel analyzer, and displayed as an energy spectrum. Typically, all elements of atomic number above 11 (sodium) are detected simultaneously. There may be problems of elemental overlap in some materials. Ultrathin window (UTW) and windowless detectors have extended the range down to include carbon, atomic number 6, but contamination and lack of sensitivity are problems. Polymer samples tend to outgas and seriously contaminate the unprotected detectors.

2.7.1.2 Wavelength Dispersive X-ray Spectrometer

In a *wavelength dispersive x-ray spectrometer* (WDS), the x-rays fall on a bent crystal and are reflected only if they satisfy Bragg's law. The crystal is set to focus x-rays of one specific wavelength onto a detector and rotates to scan the wavelength detected. Only one element can be detected at a time with one crystal. The resulting WDS spectra are quite sharp and elemental overlap is minimal due to the good signal to noise ratio. In WDS, typically in an EPMA, accurate quantitative analysis is possible if the specimen is flat and standards are used for calibration. Computer analysis allows for such complicating features as the x-ray absorption and fluorescence, which depend on the elemental composition that is being determined.

Comparison of the two x-ray techniques is shown in Table 2.3. Microanalysis in the SEM is ideally conducted by a combination of these two techniques to take advantage of the strengths of both. Several spectrometers can be mounted on the microscope. Computer hardware and software are available that control both types of spectrometer and combine the data on a single system. However, EDS is much more common than WDS, so in most cases, this is not possible. Microanalysis in the TEM and STEM is conducted by EDS analysis and EELS.

TABLE 2.3. Comparison of EDS and WDS microanalysis

Energy dispersive	Wavelength dispersive
Interfaced with SEM, TEM, STEM	Interfaced with SEM, EPMA
Simultaneous detection of elements	Quantitative detection of one element at a time
Rapid analysis: about 100 s	Slow analysis: from 5 min to hours
Spatial resolution good in TEM, AEM, STEM, poor in SEM	Spatial resolution poor (1–5 μm)
Background counts from backscattered electrons reduce sensitivity	Peak/background ratio 10 to 50 times better than EDS, good sensitivity
Serious peak overlap problems; results may be ambiguous	Good energy resolution, little peak overlap
Single detector	Need several crystals to cover range of elements
Detection limit $Z > 11$ (regular window) $Z > 5$ (ultrathin window)	Detection limit $Z > 3$

2.7.2 X-ray Analysis: SEM versus AEM

Energy dispersive x-ray spectroscopy can be conducted in the SEM, STEM, and in the AEM, whereas wavelength dispersive spectroscopy is conducted only in the SEM or the EPMA. For light element analysis, from boron to sodium, the WDS technique is preferred to ultrathin window EDS for polymers, so the AEM should not be used unless its spatial resolution is required. If thin specimens are used in the AEM, high magnification images and diffraction information can be accompanied by EDS with spatial resolution about 10–100 nm. Energy dispersive x-ray spectrometry of solid specimens in the SEM normally has resolution of a micrometer or more (for examples, see Fig. 5.18 and Fig. 5.65). Just as for STEM imaging, this difference is due to the small interaction volume in thin films, where the beam does not spread out. Thin specimens also limit the need for absorption or fluorescence corrections, permitting the application of quantitative analysis techniques.

There are three problems with microanalysis of thin films in the AEM:

1. Spurious x-rays may be produced that can be detected and confused with x-rays from the specimen—this can be limited by the use of a beryllium or graphite sample holder.
2. Beam currents in the AEM are small (compared to the EPMA). This and the very small excitation volume give very low x-ray count rates. Low levels of an element present in the specimen are difficult to detect.
3. A thin film specimen must be prepared that is representative of the bulk and has a reasonably uniform thickness.

Generally, there are small amounts of heavy elements in a polymer, and these low levels are difficult to detect unless they are concentrated. Leaving the sample for long periods of time to build up the signal may not work as the material may be unstable in the electron beam. Because changes may occur in the sample under study, quantitative analysis of heavy elements in polymers in either the SEM or AEM is difficult.

2.7.3 Elemental Mapping

So far, it has been assumed that the result of microanalysis is the elemental composition of a small region of the specimen. This is obtained from the x-ray spectrum produced when the electron beam is stationary. It is often more useful to show the concentration of a specific element as a function of position on the specimen. This is *elemental mapping*. The map is formed by using the intensity of x-ray emission in a specific energy range to modulate the intensity on a display as the beam scans the specimen. The energy region, or window, is set to include the characteristic x-ray energy of the element of interest (for an example, see Fig. 5.108).

Briefly, the major issue in elemental mapping is to have a high enough count rate for good counting statistics. The higher the concentration of the element in the region scanned, the less time is required for good counting statistics. Elements present at low concentrations require long counting time during which sensitive specimens can be damaged. In principle, the spatial resolution in elemental mapping is much better in the AEM than in the SEM because of the small interaction volume. In practice, the signal is small in the AEM and only larger regions that have better statistics may be visible.

Elemental mapping using EDS in the SEM is easy to accomplish. Mapping of individual particles that are not embedded in a matrix is straight-forward, and particles below 1 μm in size can often be identified by EDS or WDS analysis. However, thick specimens with particles in a matrix or regions of differing atomic number are difficult to analyze. The two dimensional elemental maps are produced from x-ray signals from a large interaction volume that extends well below the specimen surface. The deep features contributing to the elemental map will not be visible in the image.

Current systems allow simultaneous acquisition of maps for several different elements. Digital maps with colors assigned for each element permit a more rapid and detailed analysis. Superposition of the color maps is useful in determining associations between elements. This technique is more than simple elemental mapping and approaches more definitive compositional studies. Multiple maps may be obtained in the time previously used for mapping a single element.

2.8 QUANTITATIVE MICROSCOPY

2.8.1 Image Processing and Analysis

The aim of a microscopy project is often to produce a numerical answer—the mean particle size, the volume fraction of a given phase, or some more complicated parameter involving orientation or correlation of features. *Image analysis* is the discipline that involves making these kinds of measurements [191–193]. Conceptually, image analysis is the measurement of geometric features in images, and it could be done with a ruler, given enough time and patience. Now that images are digital, such measurement methods have naturally advanced rapidly, along with available low cost computer power. The parameters that are commonly measured include particle numbers, diameters, areas, perimeters, and ferets. A flexible system can be programmed to measure anything, such as arc lengths and maximum intensity positions in diffraction patterns with texture. Analysis capability may be built into a computer that controls the microscope, but it is also convenient to have a "stand alone" system that analyzes digital micrographs from many sources.

Any problems with specimen preparation and microscopy must be solved before moving on to image analysis, as the analysis is no better than the image. The image must be carefully calibrated if it is to provide accurate data. Finally, the images analyzed must be representative. As in any statistical technique, a sufficiently large and representative population of features must be measured, so it is important to consider the number of times samples should be collected for the microscope and the number of different images taken from each specimen to obtain a random sample. For example, in a typical particle size analysis, the number of particles measured must be larger if the size distribution is broader.

Image processing is a broad field that includes computer vision and image compression, but in microscopy it is the changing of an image to make the features of interest more visible [194]. This is in principle different from image analysis, but the features must be visible before they can be measured, so the procedures are often combined into a single software suite. Elementary image processing operations such as averaging over frames to reduce noise or subtracting a background may be built into the data acquisition system. Adjusting contrast and brightness of the image is so basic it is hardly thought of as processing. The next level of processing for image improvement typically involves removing errors or distortions put in by an imperfect imaging system. This is much more complex and will likely require a stand-alone system with professional software.

An example of image processing that is not meant to lead to computer analysis is *false coloring* where a palette of colors replaces monochrome intensity as an indicator of signal level. This can be used simply to make a pretty colored picture, but it also has a more serious use. The human eye is more sensitive to changes of tint than to changes of intensity, so a false-color image can convey more information. Also, many systems can only deal with a limited number of gray levels in a monochrome image but three times that number of distinct values in a color image.

2.8.2 Three Dimensional Reconstruction

Three dimensional information on specimen microstructure is often required, but microscope images are two dimensional. They may be representations of two dimensional objects, (surfaces or thin sections of the specimen) or projections of the specimen onto a plane. Two dimensional (2D) slices can give statistical information about the three dimensional (3D) structure. *Stereology* is the field that provides the mathematical methods to go from two to three dimensions [191, 195, 196]. A simple example of the problem is a material containing spheres dispersed in a matrix. If the spheres are of uniform diameter, a random surface cut through the object will contain circular structures of varying diameter. If the spheres are of a range of sizes, a thin section will appear much the same. Analysis of the size distribution of the circles can distinguish the two cases and give the size distributions of the spheres.

A full 3D reconstruction gives exact, not statistical, 3D information. Clearly, such a reconstruction requires many images. These may be images of parallel planes, from the TEM of serial sections, or from the AFM of the serial cut faces that microtomy would leave. The confocal scanning microscope (see Section 6.2) is often used to form a 3D image by taking multiple images at different focal positions. Another possibility is to take projected images in a range of orientation, for example by tilting a small object in the TEM. The computer stacks the slices or unscrambles the projections to form a 3D image. This can be presented in projection, or any possible 2D slice can be calculated and presented as a new 2D image. Creating such slices from projections or other data is called *tomography* (see Section 6.4 and Fig. 5.22D). Statistical information about particle size distribution and so forth can be taken directly from the full 3D image. However, dealing with such objects is data intensive and the volume analyzed would have to contain enough objects for a statistically significant result [197, 198].

2.8.3 Calibration

It is obvious that the size of object features derived from image analysis and stereology depends on knowing the actual magnification. Image analysis, whether done by ruler or by computer, depends totally on knowing the relation between linear measurements on the micrograph and in the object.

Standard objects of known size are the simplest calibration tools. If an internal calibration is possible, it is often better than an external standard. For example, a common procedure is to measure latex particle diameters in the TEM. Particles with very accurately known diameters (and a narrow size distribution) are readily available. Some of these particles can be added to the sample suspension before preparation for microscopy. They are used as internal standards to correct for any changes there might be either during specimen preparation or during observation in the microscope. They must not be so close in size to the experimental particles that they may be confused, or so different that the same image magnification would not be suitable!

2.8.3.1 OM, SEM, and TEM

In optical microscopy the magnification marked on the objective lens is not very helpful in calibrating the scale of a digital image. Stage micrometers with accurate markings are required to provide calibration, and it is good practice to include an image of the micrometer in every set of micrographs. In the SEM or TEM, it is easy to read the magnification indicated on the instrument, and a micrometer marker (scale bar) may be automatically printed on the micrographs. The problem is that these may not be accurate. Magnifications should be checked after each routine maintenance of the microscope using standard diffraction grating replica specimens. Critical studies involving quantitative analysis should have calibration standards run during the study.

2.8.3.2 AFM

As the AFM measures heights, calibration of AFM images means that the displacement has to be calibrated in three dimensions (along three orthogonal axis) where x and y are in-plane with the nominal sample surface and z is along a direction normal to this plane. Microfabricated pitch and height standards are available, and their use ensures accurate dimensional measurement. Some of these standards are traceable to fundamental measurements made by standards agencies such as the National Institute for Standards and Technology (NIST). Pitch standards for lateral calibration are readily available in a range from 100 nm × 100 nm to 10 μm × 10 μm. Grating replicas can be used by both AFMs and TEMs so a common calibration is possible. Gratings or scales made in silicon by lithography can extend over the entire field of view, showing if there is any distortion of the image due to scan errors. For smaller scale pitch measurements, atomic or molecular lattices can be used and compared with lattice constants measured from diffraction or scattering experiments.

Traceable height standards are available generally as step height artifacts that range from 8 nm to several micrometers (the maximum range of common AFMs). Dimensional calibration should be done routinely as the sensitivity of actuators can change with time. For careful work, it is advisable to have dimensional standards that bracket the dimensions of real systems. Improper dimensional calibration can lead to artifacts in AFM measurement (see Section 3.3.7).

Atomic force microscopes may use sophisticated algorithms to modify the voltages applied to actuating elements (piezoelectric ceramics) that incorporate the effects of scan rate, scan size, non-linearities, and cross-talk. This approach is referred to as an open loop method. Closed loop approaches are also available. These provide an independent method of dimensional measurement (strain gauges, capacitance sensors, or optical interferometers) in order to apply real-time correction to actuator motion. Closed loop is more readily applied in x and y dimensions; recent advances include z motion. Closed loop calibration can introduce additional electrical noise, and care is required to make sure that ultimate resolution is not compromised.

The AFM is also used to measure forces, for example to quantify the load applied to the surface or to determine mechanical properties of the sample. A static applied load causes the cantilever to bend, and in the normal AFM design, this deflection moves a light spot on an optical detector (see Fig. 2.8). There are thus two steps in relating the input force to the output signal voltage. One is the cantilever spring constant and the other the sensitivity of the optical lever. The optical lever sensitivity is normally calibrated by pushing the cantilever tip against a rigid substrate. Assuming that the tip does not penetrate the surface, the known motion of the cantilever base gives the same deflection as tip motion.

Nominal values of the spring constants can be in error by as much as 50% primarily due to the difficulty in controlling the cantilever thickness during fabrication, and so vendors usually indicate only a typical value and a range. Calibration of the spring constant was reviewed in 2003 [199], and there are several methods that can be used.

1. Placing a mass on the cantilever shifts its resonant frequency [200], and the spring constant can then be calculated.
2. Pushing the tip into a substrate of known stiffness [201, 202] is also a fairly direct method.
3. The cantilever spring constant may be obtained by calculation (beam theory) from measured cantilever dimensions [102, 203].

The thickness and density are often difficult to determine, and one method replaces these inputs with the resonant frequency and damping factor [204].

4. *Thermal tuning* [205, 206] measures the fluctuations in position due to thermal excitation. A sufficiently sensitive system can determine this displacement and the mean must correspond with an energy of ½kT. If the optical lever sensitivity is known, this gives the spring constant.

Alternatively, if the spring constant is calculated, the sensitivity can be determined using thermal tuning, so that it is not necessary to push a tip against the surface and possibly damage it [207].

The AFM tip shape may need to be determined. This calibration is required to model how the topographic or height image is affected by the mixing of tip shape with surface geometry. This is especially important when trying to understand image artifacts (see Section 3.3.7). Mechanical property measurements using the projected area of contact under load in the calculation also require knowledge of the tip shape. Commercial standards have not only features designed to calibrate distances but also features designed to show the effects of probe shape. One example is an edge or spike with small radius of curvature (<10 nm). These are called imaging or tip characterizers [208]. Alternatives are direct SEM imaging of the tip or blind reconstruction [209].

2.9 DYNAMIC MICROSCOPY

Most polymer microscopy is static; that is, the specimen is not intended to change during observation. Dynamic microscopy looks at a moving or changing specimen and goes back to the observation of "animalcules" moving about in pond water. In electron microscopes and AFMs, such experiments are likely to be called *in situ* microscopy [210], and these are discussed further in Chapter 4 (see Section 4.8.4 and Section 4.9.5). The specimen is observed during deformation or under some environmental change, usually temperature change [211–213], and the data is recorded as a sequence of digital images. If the changes of interest are likely to be rapid, this requires a high frame rate. At 30 frames per second (fps), 1 MB raw images accumulate at ~2 GB/min. This requires special equipment and can still be a problem when an event occurs rapidly at some arbitrary time into the experiment, requiring storing images over a long period, then later selecting the few frames where something is happening. Here the image size is conservative; the Orius CCD camera for the TEM (Gatan Inc., Warrendale, PA) can deliver 11 megapixel frames at 14 fps. Slower data rate acquisitions are more common and at present a lot easier.

A disadvantage of conducting experiments in the microscope is that the experiments can consume a lot of time and are thus expensive. For mechanical tests, it is rare to observe a statistically significant number of specimens; also, neither stress–strain measurements nor microscopy is optimal. Nevertheless, the insight gained from direct observation of deformation and failure mechanisms outweigh these disadvantages in some cases.

Experiments involving hydrated polymers have been monitored using variable pressure SEMs (e.g., [214]), although recent evidence has shown extensive radiation damage [215, 216] occurs during these studies (see Section 5.5.2). Because the AFM can be operated in a wide range of environments, many other dynamic experiments are possible. The effect of hydration on polymer blends and membranes [135, 217] and the effect of UV irradiation in various gaseous environments on the friction of polymer films [218] are just a few examples of the hundreds of AFM dynamic studies on polymers.

2.9.1 Mechanical Deformation Stages

The common types of mechanical deformation stages are tension, compression, and bending. These types of stages designed for use in the SEM are commercially available from the SEM manufacturers or companies making accessories (Appendices V and VI). Typically, the tension and compression stages are screw-driven and symmetric, moving both of the grips that hold the ends of the specimen, so that the central region remains still and in the field of view [219]. Maximum loads of 0.1 to 1 kN are typical. Similar designs will work in the light microscope, but most polymer studies use fibers or films in transmission and there has to be good access around the specimen on both sides, for the objective and the condenser. Smaller devices suitable for use in the AFM are also available, but for continuous deformation the effects of vibration and noise coming from the driving motors have to be treated very carefully. A common use of tensile stages in the study of polymers is to study the tensile failure of fibers and yarns (see Section 5.1) [219]. Charging of polymer specimens has been a major problem in the SEM as metal coating will not work if the sample is deformed significantly or broken to expose new surfaces. Here, low voltage or variable pressure operation of the SEM is really advantageous.

Tensile stages and tensile stages combined with heating and cooling are available for use in the TEM. More detailed changes in crystallography and composition can be monitored in the TEM, but the small specimen thickness can affect the processes of deformation, and it is not possible to know the stress and strain at the region under observation. Other modes of deformation such as bending and shearing can be studied with suitably modified stages in the OM or SEM. Problems of adhesion and interfacial strength can also be studied using the same or slightly modified stages. Most SEM manufacturers and electron microscope accessory suppliers (see Appendices V–VII) provide tensile stages, and the specific design of such accessories will not be considered here.

2.9.2 Hot and Cold Stages

Hot and cold stages can easily be attached to optical, electron, and atomic force microscopes [220, 221]. Cold stages are very common in TEMs, where they may reduce specimen contamination by stopping surface diffusion. In biology, cold stages are used to keep hydrated samples frozen and stable for observation in the vacuum of the TEM or SEM (e.g., [222]), and this has some applications in polymers [223]. Cooling can also increase the lifetime of beam sensitive specimens. The use of cryomicroscopy and especially cryo-TEM for polymers has recently increased and is covered in more detail in Chapter 4 (see Section 4.9.5).

A major application of dynamic microscopy is the study of polymer structure and its development as a function of temperature in a hot stage. This type of work is rare in electron microscopes. Heating a polymer sample in an electron microscope requires caution, as not only will heating increase the rate of radiation damage, but also the polymer may outgas or degrade and evaporate, contaminating the microscope vacuum system and associated x-ray detectors.

In the past, most of these studies were done in the optical microscope [11, 212, 224, 225], whereas now they are also being done in the AFM [113, 175]. Direct observation can determine the nature of phase changes and at the same time measure the transformation temperatures. The kinetics of crystal growth can be directly observed and numerically analyzed with the aid of image processing. A liquid crystalline polymer may have several phase changes as it is heated, and changes in appearance in the polarizing microscope or birefringence can be correlated with calorimetric data [226]. The calorimetric data is acquired by differential scanning calorimetry (DSC) or differential thermal analysis (see Section 7.4). With a combined hot stage DSC, it is possible to observe the sample in the OM and simultaneously obtain the DSC trace.

References

1. M. Spencer, *Fundamentals in Light Microscopy* (Cambridge University Press, Cambridge, 1982).
2. D.J. Rawlins, *Light Microscopy* (BIOS Scientific Publishers, Oxford, 1992).
3. W.G. Hartley, *Hartley's Microscopy* (Senecio Publishing, Charlbury, 1981).
4. S. Bradbury, *An Introduction to the Optical Microscope* (Oxford University Press, Oxford, 1984).
5. T.G. Rochow and P.A. Tucker, *An Introduction to Microscopy by Means of Light, Electrons,*

References

X-rays, or Acoustics, 2nd ed. (Plenum, New York, 1994).
6. R. Telle and G. Petzow, in *Materials Science and Technology*, edited by E. Lifshin (VCH Publishers, Weinheim, 1993).
7. J.H. Richardson, *Optical Microscopy for Materials Sciences* (Marcel Dekker, New York, 1971).
8. C.W. Mason, *Handbook of Chemical Microscopy* (Wiley, New York, 1983).
9. N.H. Hartshorne and A. Stuart, *Crystals and the Polarizing Microscope* (Arnold, London, 1970).
10. P.C. Robinson and S. Bradbury, *Qualitative Polarized-Light Microscopy* (Oxford University Press, New York, 1992).
11. W.C. McCrone and J.G. Delly, *Particle Atlas* (Ann Arbor Science, Ann Arbor, MI, 1973).
12. D.B. Murphy, *Fundamentals of Light Microscopy and Electronic Imaging* (Wiley-Liss, Hoboken, NJ, 2001).
13. D.A. Hemsley, Ed. *Applied Polymer Light Microscopy* (Elsevier Applied Science, London, New York, 1989).
14. M.W. Davidson and M. Abramowitz, *Optical Microscopy Primer*. Available at http://micro.magnet.fsu.edu/primer/index.html. Accessed June 2006.
15. *MicroscopyU: The Source for Microscopy Education*. Available at http://www.microscopyu.com/. Accessed June 2006.
16. *Microscopy Resource Center*. Available at http://www.olympusmicro.com/primer/index.html. Accessed June 2006.
17. G.E. Schlueter and W.E. Gumpertz, *Am. Lab.* **Apr** (1976) 61.
18. M.W. Davidson and M. Abramowitz, *Phase Contrast Microscopy*. Available at http://micro.magnet.fsu.edu/primer/techniques/phasecontrast/phaseindex.html. Accessed June 2006.
19. R. Hoffman, *J. Microsc.* **110** (1977) 209.
20. *Principles and Applications of Two-Beam Interferometry*. Available at http://www.microscopyu.com/articles/interferometry/twobeam.html. Accessed June 2006.
21. *Differential Interference Contrast*. Available at http://www.microscopyu.com/articles/dic/dicindex.html. Accessed June 2006.
22. *Refractive Index Determination by Oblique Illumination*. Available at http://www.olympusmicro.com/primer/java/oblique/becke/index.html. Accessed June 2006.
23. *E1976–98 Standard Test Method for the Automated Determinaton of Refractive Index of Glass Samples Using the Oil Immersion Method and a Phase Contrast Microscope*, in ASTM Annual Books of Standards (ASTM International, 2003).
24. E.A. Wood, *Crystals and Light* (Dover, New York, 1977).
25. *Introduction to Polarized Light Microscopy*. Available at http://www.microscopyu.com/articles/polarized/polarizedintro.html. Accessed June 2006.
26. P. Gay, *An Introduction to Crystal Optics* (Longmans, London, 1967).
27. N.H. Hartshorne, *The Microscopy of Liquid Crystals* (Microscope Publications, Chicago, 1974).
28. P.F. Kerr, *Optical Mineralogy* (McGraw-Hill, New York, 1959).
29. R. Oldenbourg, *Nature* **381** (1996) 811.
30. J. Goldstein, D.E. Newbury, D.C. Joy, C.E. Lyman, P. Echlin, E. Lifshin, L.C. Sawyer and J.R. Michael, *Scanning Electron Microscopy and X-ray Microanalysis*, 3rd ed. (Plenum, New York, 2003).
31. S.L. Flegler, J.W. Heckman and K.L. Klomparens, *Scanning and Transmission Electron Microscopy: An Introduction* (Oxford University Press, Oxford, 1993).
32. L. Reimer, *Image Formation in Low-Voltage Scanning Electron Microscopy* (SPIE, Bellingham, WA, 1993).
33. L. Reimer and P.W. Hawkes, Eds. *Scanning Electron Microscopy, Physics of Image Formation and Microanalysis* (Springer, Berlin, 1998).
34. J.R. White and E.L. Thomas, *Rubber Chem. Technol.* **57** (1984) 457.
35. L.C. Sawyer and D.T. Grubb, *Polymer Microscopy* (Chapman and Hall, London, 1987, 1996).
36. C.J. Catto and K.C. Smith, *J. Microsc.* **98** (1973) 417.
37. Obtained using CASINO, Monte Carlo simulation software freely downloadable from [38], a site at Universitede Sherbrooke, Sherbrooke, Quebec, Canada.
38. D. Drouin, CASINO. http://www.gel.usherbrooke.ca/casino/. Accessed July 2006.
39. T.E. Everhart, R.F. Herzog, M.S. Chang and W.J. DeVore. In *Proc. 6th Intl. Conf. on X-ray Optics and Microanalysis*, edited by G. Shinoda, K. Kohra and T. Ichinokawa (University of Tokyo, Tokyo, 1972), p. 81.
40. V.N.E. Robinson, *Scanning* **3** (1980) 15.
41. R. Autrata, *Scanning Microsc.* **3** (1989) 739.
42. E.D. Boyes, *Microsc. Microanal.* **6** (2000) 307.

43. C. Gaillard, P.A. Stadelmann, C.J.G. Plummer and G. Fuchs, *Scanning* **26** (2004) 122.
44. D.L. Vezie, W.W. Adams and E.L. Thomas, *Polymer* **36** (1995) 1761.
45. T. Tagawa, J. Mori, S. Aita and K. Ogura, *Micron* **9** (1978) 215.
46. D.W. Schwark, D.L. Vezie, J.R. Reffner, E.L. Thomas and B.K. Annis, *J. Mater. Sci. Lett.* **11** (1992) 352.
47. L.C. Sawyer, R.T. Chen, M.G. Jamieson, I.H. Musselman and P.E. Russell, *J. Mater. Sci.* **28** (1993) 225.
48. S.K. Chapman, *Scanning Microsc.* **13** (1999) 141.
49. *A Guide to Scanning Microscope Observation*. Available at http://www.jeolusa.com/sem/docs/sem_guide/guide.pdf. Accessed June 2006.
50. D.B. Williams and C.B. Carter, *Transmission Electron Microscopy: A Textbook for Materials Science* (Plenum Press, New York, 1996).
51. L. Reimer, *Transmission Electron Microscopy: Physics of Image Formation and Microanalysis* (Springer, Berlin, 1997).
52. M.D. Graef, *Introduction to Conventional Transmission Electron Microscopy* (Cambridge University Press, Cambridge, 2003).
53. B. Fultz and J.M. Howe, *Transmission Electron Microscopy and Diffractometry of Materials*, 2nd Ed. (Springer, New York, 2002).
54. R.G. Vadimsky, in *Methods of Experimental Physics*, edited by R.A. Fava (Academic Press, New York, 1980).
55. D.T. Grubb, in *Developments in Crystalline Polymers*, edited by D.C. Bassett (Applied Science Publishers, London, 1982).
56. C.E. Hall, *J. Appl. Phys.* **22** (1951) 655.
57. B.D. Lauterwasser and E.J. Kramer, *Philos. Mag.* **A39** (1979) 469.
58. P.B. Hirsch, A. Howie, R.B. Nicholson, D.W. Pashley and M.J. Whelan, *Electron Microscopy of Thin Crystals* (Butterworths, London, 1965).
59. E.L. Thomas, in *The Structure of Crystalline Polymers*, edited by I.H. Hall. (Elsevier Applied Science, London, 1984), p. 79.
60. J.R. White, *J. Mater. Sci.* **9** (1974) 1860.
61. M. Tosaka, R. Danev and K. Nagayama, *Macromolecules* **38** (2005) 7884.
62. M.J. Miles and J. Petermann, *J. Macromol. Sci. Phys.* **2** (1979) 243.
63. H.P. Zingsheim and L. Bachmann, *Koll. Z. u Z. Polym.* **246** (1971) 36.
64. D.J. Johnson and D. Crawford, *J. Microsc.* **98** (1973) 313.
65. D.L. Handlin and E.L. Thomas, *Macromolecules* **16** (1983) 1514.
66. L.W. Hobbs, *Ultramicroscopy* **23** (1987) 339.
67. V.E. Cosslett, *Phys. Status Solidi* **55** (1979) 545.
68. P. Goodman, *Fifty Years of Electron Diffraction* (Riedel, Dordrecht, 1981).
69. P.E. Champness, *Electron Diffraction in the Transmission Electron Microscope* (BIOS Scientific Publishers, Oxford, 2001).
70. D.L. Dorset, *Structural Electron Crystallography* (Springer, New York, 1995).
71. P. Buseck, J. Cowley and L. Eyring, Eds. *High-Resolution Transmission Electron Microscopy and Associated Techniques* (Oxford University Press, Oxford, 1988).
72. J.C.H. Spence, *High-Resolution Electron Microscopy*, 3rd ed. (Oxford University Press, Oxford, 2003).
73. F. Ernst and M. Rühle, Eds. *High-Resolution Imaging and Spectrometry of Materials* (Springer, New York, 2003).
74. D.C. Martin and E.L. Thomas, *Polymer* **36** (1995) 1743.
75. M. Tsuji and S. Kohjiya, *Prog. Polym. Sci.* **20** (1995) 259.
76. D.C. Martin, J. Chen, J. Yang, L.F. Drummy and C. Kübel, *J. Polym. Sci. B Polym. Phys.* **43** (2005) 1749.
77. C.J.G. Plummer, R. Gensler and H.-H. Kausch, *Colloid Polym. Sci.* **275** (1997) 1068.
78. M. Tsuji, M. Fujita, T. Shimizu and S. Kohjiya, *Macromolecules* **34** (2001) 4827.
79. M. Tosaka, M. Tsuji, S. Kohjiya, L. Cartier and B. Lotz, *Macromolecules* **32** (1999) 4905.
80. M.G. Dobb, A.M. Hindeleh, D.J. Johnson and B.P. Saville, *Nature* **253** (1975) 189.
81. C. Kübel, D.P. Lawrence and D.C. Martin, *Macromolecules* **34** (2001) 9053.
82. K. Shimamura, J.R. Minter and E.L. Thomas, *J. Mater. Sci. Lett.* **18** (1983) 54.
83. W.W. Adams, S. Kumar, D.C. Martin and K. Shimamura, *Polym. Commun.* **30** (1989) 285.
84. D.C. Martin and E.L. Thomas, *Macromolecules* **24** (1991) 2460.
85. D.C. Martin and E.L. Thomas, *J. Mater. Sci.* **26** (1991) 5171.
86. I.G. Voigt-Martin and H. Durst, *Macromolecules* **22** (1989) 186.
87. I.G. Voigt-Martin, H. Krug and D.V. Dyck, *J. Phys., Paris* **51** (1990) 2347.
88. D.C. Martin, L.L. Berger and K.H. Gardner, *Macromolecules* **24** (1991) 3921.
89. G. Binnig, H. Rohrer, C. Gerber and H. Weibel, *Phys. Rev. Lett.* **49** (1982) 57.

90. S. Magonov and M.-H. Whangbo, *Surface Analysis with STM and AFM: Experimental and Theoretical Aspects of Image Analysis* (VCH, Weinheim, 1996).
91. E. Meyer, H. Hug and R. Bennewitz, *Scanning Probe Microscopy: The Lab on a Tip* (Springer-Verlag, Berlin, Heidelberg, 2004).
92. R. Wiessendanger, *Scanning Probe Microscopy and Spectroscopy: Methods and Applications* (Cambridge University Press, Cambridge, 1994).
93. L.A. Bottomley, *Anal. Chem.* **70** (1998) 425R.
94. R. Overney and V. Tsukruk, in *Scanning Probe Microscopy of Polymers*, edited by B. Ratner and V. Tsukruk (ACS, Washington, DC, 1998), p. 2.
95. B.D. Ratner and V.V. Tsukruk, Eds. *Scanning Probe Microscopy of Polymers*. ACS Symp. Ser. Vol **694** (ACS, Washington DC, 1998).
96. S.N. Magonov and D.H. Reneker, *Ann. Rev. Mater. Sci.* **27** (1997) 175.
97. G. Krausch, *Spec. Publ. Roy. Soc. Chem.* **263** (2001) 291.
98. S.N. Magonov, in *Handbook of Surfaces and Interfaces of Materials* 2, edited by H.S. Nalwa (Academic Press, San Diego, 2001), p. 393.
99. D.A. Chernoff and S.N. Magonov, in *Comprehensive Desk Reference of Polymer Characterization and Analysis*, edited by R.F. Brady (ACS, Oxford Press, 2003), p. 490.
100. V.V. Tsukruk and N.D. Spencer, Eds. *Recent Advances in Scanning Probe Microscopy of Polymers (Proc. Symp. August 2000)*, in *Macromol. Symp.* **167** (2001).
101. J.D. Batteas, C.A. Michaels and G.C. Walker, Eds. *Applications of Scanned Probe Microscopy to Polymers ACS Symp. Ser.* **897** (ACS, Washington DC, 2005).
102. D. Sarid, *Scanning Force Microscopy: With Applications to Electric, Magnetic, and Atomic Forces* (Oxford University Press, Oxford, 1994).
103. R. Garcia and R. Perez, *Surface Science Reports* **47** (2002) 197.
104. A. Schirmeisen, B. Anczykowski and H. Fuchs, in *Applied Scanning Probe Methods*, edited by B. Bushan, H. Fuchs and S. Hosaka (Springer-Verlag, Berlin, 2003), p. 3.
105. F.J. Giessibl, *Rev. Mod. Phys.* **75** (2003) 949.
106. J.S. Villarrubia, in *Applied Scanning Probe Methods*, edited by B. Bushan, H. Fuchs and S. Hosaka (Springer-Verlag, Berlin, 2003), p. 147.
107. D.H. Reneker, R. Patil, S.J. Kim and V. Tsukruk, in *Crystallization of Polymers, NATO ASI Series, Vol 405*, edited by M. Dosière (Kluwer, Dordrecht, 1993), p. 357.
108. B. Lotz, J.C. Wittmann and A.J. Lovinger, *Polymer* **37** (1996) 4979.
109. B. Lotz, *Adv. Polym. Sci.* **180** (2005) 17.
110. J.K. Hobbs, A.K. Winkel, T.J. McMaster, A.D.L. Humphris, A.A. Baker, S. Blakely, M. Aissaoui and M.J. Miles, *Macromol. Symp.* **167** (2001) 1.
111. S.N. Magonov and N.A. Yerina, in *Handbook of Microscopy for Nanotechnology*, edited by N. Yao and Z.L. Wang (Springer, 2005), p. 113.
112. S.N. Magonov and Y. Godovsky, *American Laboratory* **30** (1998) 15.
113. D.A. Ivanov and S.N. Magonov, in *Polymer Crystallization: Observations, Concepts and Interpretations*, edited by J.-U. Sommer and G. Reiter (Springer, New York, 2003).
114. C.-M. Chan and L. Li, *Adv. Polym. Sci.* **188** (2005) 1.
115. W.P. King and K.E. Goodson, *Developments in Heat Transfer* **13** (2004) 131.
116. D.M. Price, M. Reading and T.J. Lever, *J. Therm. Anal. Calorimetry* **56** (1999) 673.
117. B. Wunderlich, *Polym. Mater. Sci. Eng.* **81** (1999) 242.
118. H.M. Pollock and A. Hammiche, *J. Phys. D Appl. Phys.* **34** (2001) R23.
119. M.J. Abad, A. Ares, J. Cano, F.J. Diez, J. Lopez, C. Ramirez and L. Barral, *Recent Res. Dev. Appl. Polym. Sci.* **1** (2002) 221.
120. M.R. VanLandingham, J.S. Villarrubia, W.F. Guthrie and G.F. Meyers, *Macromol. Symp.* **167** (2001) 15.
121. V. Tsukruk and V. Gorbunov, *Probe Microsc.* **2** (2002) 241.
122. G.H. Michler, *J. Macromol. Sci., Phys.* **B40** (2001) 277.
123. T. Hugel and M. Seitz, *Macromol. Rapid Commun.* **22** (2001) 989.
124. B. Bushan, in *Applied Scanned Probe Methods*, edited by B. Bushan, H. Fuchs and S. Hosaka (Springer-Verlag, Berlin, Heidelberg, 2004), p. 171.
125. C.M. Mate, *IBM J. Res. Dev.* **39** (1995) 617.
126. J.A. Zasadzinski, *Curr. Opin. Colloid Interface Sci.* **1** (1996) 264.
127. K. Feldman, M. Fritz, G. Hahner, A. Marti and N.D. Spencer, *Tribol. Int.* **31** (1998) 99.
128. D. Vezenov, A. Noy and P. Ashby, *J. Adhes. Sci. Technol.* **19** (2005) 313.
129. G.J. Vancso, H. Hillborg and H. Schoenherr, *Adv. Polym. Sci.* **182** (2005) 55.
130. M.C. Goh, *Adv. Chem. Phys.* **91** (1995) 1.

131. V.V. Tsukruk and D.H. Reneker, *Polymer* **36** (1995) 1791.
132. J.L. Keddie, *Mater. Sci. Eng. R* **R21** (1997) 101.
133. D.C. Sundberg and Y.G. Durant, *Polym. Reaction Eng.* **11** (2003) 379.
134. J. Li, W. Liang, G.F. Meyers and W.A. Heeschen, *Polym. News* **29** (2004) 335.
135. S.E. Woodcock, W.C. Johnson and Z. Chen, *Polym. News* **29** (2004) 176.
136. J.-K. Kim and A. Hodzic, *J. Adhesion* **79** (2003) 383.
137. M. Munz, B. Cappella, H. Sturm, M. Geuss and E. Schulz, *Adv. Polym. Sci.* **164** (2003) 87.
138. G.K. Bar and G.F. Meyers, *MRS Bull.* **29** (2004) 464.
139. J. Teetsov, L. Denault, A. Alizadeh, S. Ghanti, W. Cicha and E. Hall, *Annual Technical Conference—Society of Plastics Engineers* **62nd** (2004) 2357.
140. F.M. Mirabella, Jr. and A. Weiskettel, *Polym. News* **30** (2005) 143.
141. G. Binnig, C.F. Quate and C. Gerber, *Phys. Rev. Lett.* **56** (1986) 930.
142. O. Marti, J. Colchero and J. Mlynek, *Nanotechnology* **1** (1990) 141.
143. G. Neubauer, S.R. Cohen, G.M. McClelland, D. Horne and C.M. Mate, *Rev. Sci. Instrum.* **61** (1990) 2296.
144. M. Radmacher, R.W. Tillmann and H.E. Gaub, *Biophys. J.* **64** (1993) 735.
145. H.-N. Lin, T.-T. Hung, E.-C. Chang and S.-A. Chen, *Appl. Phys. Lett.* **74** (1999) 2785.
146. A. Majumdar, J.P. Carrejo and J. Lai, *Appl. Phys. Lett.* **62** (1993) 2501.
147. F. Houze, R. Meyer, O. Schneegans and L. Boyer, *Appl. Phys. Lett.* **69** (1996) 1975.
148. R.C. Barrett and C.F. Quate, *J. Appl. Phys.* **70** (1991) 2725.
149. J. Loos, A. Alexeev and M.M. Koetse, *Ultramicroscopy* **106** (2006) 191.
150. M. Knite, V. Teteris, B. Polyakov and D. Erts, *Mater. Sci. Eng. C* **C19** (2002) 15.
151. S.-H. Cho and S.-M. Park. *208th Meeting of The Electrochemical Society, Oct. 16–21 2005*, ECS, Los Angeles, (2005), p. 2548.
152. Q. Zhong, D. Innis, K. Kjoller and V. Elings, *Surf. Sci. Lett.* **290** (1993) L688.
153. D.A. Chernoff, in *Proc. JMSA Proceedings Microscopy and MicroAnalysis 1995*, edited by G.W. Bailey et al. (Jones and Begell, New York, 1995), p. 888.
154. Y. Martin and H.K. Wickramasinghe, *Appl. Phys. Lett.* **50** (1987) 1455.
155. Y. Martin, D.W. Abraham and H.K. Wickramasinghe, *Appl. Phys. Lett.* **52** (1988) 1103.
156. A.V. Krayev and R.V. Talroze, *Polymer* **45** (2004) 8195.
157. E. Schulz, H. Sturm, W. Stark and V. Bovtoun. *Proceedings of the 1996 IEEE 9th International Symposium on Electrets, ISE 9, Sep. 25–27 1996*, Shanghai, China, (IEEE, New York, 1996), p. 334.
158. T.R. Albrecht, S. Akamine, T.E. Carver and C.F. Quate, *J. Vac. Sci. Technol. A* **6** (1990) 3386.
159. M. Nonnenmacher, J. Greschner, O. Wolter and R. Kassing, *J. Vac. Sci. Technol. B* **9** (1991) 1358.
160. Nanosciences, CNTek™ AFM Probes. Available at http://www.nanoscience.com/products/carbon_nanotube_probes.html. Accessed December 2006.
161. Nanosciences, Hi'RES AFM Probes. Available at http://www.spmtips.com/products/cantilevers/datasheets/hi-res/. Accessed December 2006.
162. D.A. Walters, D. Hampton, B. Drake, H.G. Hansma and P.K. Hansma, *Appl. Phys. Lett.* **65** (1994) 787.
163. A. Folch, M.S. Wrighton and M.A. Schmidt, *J. Microelectromech. Systems* **6** (1997) 303.
164. R.B. Marcus, T.S. Ravi, T. Gmitter, K. Chin, D. Liu, W.J. Orvis, D.R. Ciarlo, C.E. Hunt and J. Trujillo, *Appl. Phys. Lett.* **56** (1990) 236.
165. S.S. Wong, A.T. Woolley, T.W. Odom, J.-L. Huang, P. Kim, D.V. Vezenov and C.M. Lieber, *Appl. Phys. Lett.* **73** (1998) 3465.
166. J.H. Afner, C.-L. Cheung, A.T. Woolley and C.M. Lieber, *Prog. Biophys. Mol. Biol.* **77** (2001) 73.
167. D. Klinov and S. Magonov, *Appl. Phys. Lett.* **84** (2004) 2697.
168. M.S. Isaacson, in *Principles and Techniques of Electron Microscopy*, edited by M. Hayat (Van Nostrand, New York, 1977).
169. D.T. Grubb, *J. Mater. Sci.* **9** (1974) 1715.
170. G. Kanig, *Koll. Z. u Z. Polym.* **251** (1973) 176.
171. G. Kanig, *J. Crystal Growth* **48** (1980) 303.
172. D.C. Bassett and A.M. Hodge, *Polymer* **19** (1978) 469.
173. D.C. Bassett and A.M. Hodge, *Proc. Roy. Soc.* **A377** (1981) 25.
174. D.C. Bassett, *Crit. Rev. Solid State Mater. Sci.* **12** (1984) 97.
175. J.K. Hobbs, in *Polymer Crystallization: Observations, Concepts and Interpretations*, edited by J.-U. Sommer and G. Reiter (Springer, New York, 2003), p. 82.

176. L.C. Sawyer and M. Jaffe, *J. Mater. Sci.* **21** (1986) 1897.
177. R.C. Williams and H.W. Fischer, *J. Mol. Biol.* **52** (1970) 121.
178. Y. Fujiyoshi, T. Kobayashi, K. Ishizuka, N. Uyeda, Y. Ishida and Y. Harada, *Ultramicroscopy* **5** (1980) 459.
179. J. Chang, M. Marsh, F. Rixon and W. Chiu, *Microsc. Microanal.* **11** (Suppl 2) (2005) 308.
180. K. Kobayashi and K. Sakaoku, *Lab. Invest.* **14** (1965) 1097.
181. D.L. Dorset and F. Zemlin, *Ultramicroscopy* **17** (1985) 229.
182. J.G. Grasselli and B.J. Bulkin., Eds. *Analytical Raman Spectroscopy* (Wiley, New York, 1991).
183. A. Garton, D.N. Batchelder and C. Cheng, *Appl. Spectrosc.* **47** (1993) 922.
184. R.F. Egerton, *Electron Energy Loss Spectroscopy in the Electron Microscope* (Plenum Press, New York, 1986).
185. C.C. Ahn, Ed. *Transmission Electron Energy Loss Spectrometry in Materials Science and The EELS Atlas* (Wiley-VCH, 2004).
186. M.R. Libera and M.M. Disko, in *Transmission Electron Energy Loss Spectrometry in Materials Science and The EELS Atlas*, edited by C.C. Ahn (Wiley-VCH, 2004), p. 419.
187. K.F.J. Heinrich and D.E. Newbury, *Electron Probe Quantitation* (Plenum Press, New York, 1991).
188. N.J. DiNardo, in *Materials Science and Technology*, edited by E. Lifshin (VCH Publishers, Weinheim, 1994).
189. D.C. Joy, A.D. Romig and J.I. Goldstein, *Principles of Analytical Electron Microscopy* (Plenum, New York, 1986).
190. E.L. Hall, in *Materials Science and Technology Vol 2A*, edited by E. Lifshin (VCH Publishers, Weinheim, 1993).
191. J.C. Russ, *Computer-Assisted Microscopy—The Measurement and Analysis of Images* (Plenum Press, New York, 1990).
192. H.E. Exner, in *Materials Science and Technology*, edited by E. Lifshin (VCH Publishers, Weinheim, 1994).
193. D.P. Hader, *Image Analysis: Methods and Applications*, 2nd ed. (CRC Press, Boca Raton, 2002).
194. J.C. Russ, *The Image Processing Handbook*, 4th ed. (CRC Press, Boca Raton, 2002).
195. J.C. Russ and R.T. Dehoff, *Practical Stereology*, 2nd ed. (Springer, New York, 2000).
196. C.V. Howard and M.G. Reed, *Unbiased Stereology (Advanced Methods)*, 2nd ed. (BIOS Scientific Publishers, Oxford, 2005).
197. J.K. Taylor and C. Cihon, *Statistical Techniques for Data Analysis*, 2nd ed. (Chapman & Hall/CRC, Boca Raton, 2004).
198. A.C. Tamhane and D.D. Dunlop, *Statistics and Data Analysis: From Elementary to Intermediate* (Prentice Hall, Englewood Cliffs, NJ, 1999).
199. N.A. Burnham, X. Chen, C.S. Hodges, G.A. Matei, C.J.T. Roberts, M.C. Davie and S.J.B. Tendler, *Nanotechnology* **14** (2003) 1.
200. J.P. Cleveland, I. Manne, D. Bocek and P.K. Hansma, *Rev. Sci. Instrum.* **64** (1993) 403.
201. C.T. Gibson, G.S. Watson and S. Myhra, *Nanotechnology* **7** (1996) 259.
202. J.R. Pratt, J.A. Kramar, D.B. Newell and D.T. Smith, *Meas. Sci. Technol.* **16** (2005) 2129.
203. M.A. Poggl, A.W. McFarland, J.S. Colton and L.A. Bottomley, *Anal. Chem.* **77** (2005) 1192.
204. J.E. Sader, J.W.M. Chon and P. Mulvaney, *Rev. Sci. Instrum.* **70** (1999) 3967.
205. J.L. Hutter and J. Bechhoefer, *Rev. Sci. Instrum.* **64** (1993) 1868.
206. J.E. Sader, *J. Appl. Phys.* **84** (1998) 64.
207. M.J. Higgins, R. Proksch, J.E. Sader, M. Polcik, S.M. Endoo, J.P. Cleveland and S.P. Jarvis, *Rev. Sci. Instrum.* **77** (2006) 13701.
208. *E1813–96 Standard Practice for Measuring and Reporting Probe Tip Shape in Scanning Probe Microscopy*, in ASTM Annual Books of Standards, Vol. 03.06 (ASTM International, 2002).
209. J.S. Villarrubia, *J. Res. Natl. Inst. Stand. Technol.* **102** (1997) 425.
210. K. Wetzig and D. Schulze, Eds. *In Situ Scanning Electron Microscopy in Materials Research* (Akademie Verlag, Berlin, 1995).
211. A.W. Agar, R.H. Alderson and D. Chescoe, *Principles and Practice of Electron Microscope Operation*, Vol 2 of *Practical Methods in Electron Microscopy*, edited by A.M. Glauert (North-Holland, Amsterdam, 1974).
212. E.P. Butler and K.F. Hale, *Dynamic Experiments in the Electron Microscope*, Vol 9 of *Practical Methods in Electron Microscopy*, edited by A.M. Glauert. (North-Holland, Amsterdam, 1981).
213. J.A. Reffner, *Am. Lab.* **May** (1984) 29.
214. A.M. Donald and B.L. Thiel, in *Structure and Dynamics of Materials in the Mesoscopic Domain, Proceedings of the Royal Society-Unilever Indo-UK Forum in Materials Science*

and *Engineering*, edited by M. Lal (World Scientific, 1999), p. 1.
215. D.J. Stokes, J.-Y. Mugnier and C.J. Clarke, *J. Microsc.* **213** (2004) 198.
216. S. Kitching and A.M. Donald, *J. Microsc.* **190** (1998) 357.
217. P.J. James, T.J. McMaster, J.M. Newton and M.J. Miles, *Polymer* **41** (2000) 4223.
218. G.V. Lubarsky, M.R. Davidson and R.H. Bradley, *Surf. Sci.* **558** (2004) 135.
219. J.W.S. Hearle, J.T. Sparrow and P.M. Cross, *The Use of the Scanning Electron Microscope* (Pergamon, Oxford, 1972).
220. G.R. Loppnow and R.A. Mathis, *Rev. Sci. Instrum.* **66** (1989) 2628.
221. K.A. Taylor, R.A. Milligan, C. Raeburn and P.N.T. Unwin, *Ultramicroscopy* **13** (1984) 185.
222. P.K. Vinson, J.R. Bellare, H.T. Davis, W.G. Miller and L.E. Scriven, *J. Colloid Interface Sci.* **142** (1992) 74.
223. M.S. Silverstein, Y. Talmon and M. Narkis, *Polymer* **30** (1989) 416.
224. N.H. Hartshorne, *The Microscope* **23** (1975) 177.
225. S. Rastogi, in *Polymer Crystallization: Observations, Concepts and Interpretations*, edited by J.-U. Sommer and G. Reiter (Springer, New York, 2003).
226. J.A. Moore and J.-H. Kim, *Macromolecules* **25** (1992) 1427.

Chapter 3
Image Formation in the Microscope

3.1 IMAGING WITH LENSES 68
  3.1.1 **Basic Optics** 68
  3.1.2 **Diffraction** 68
      3.1.2.1 Electron Diffraction 69
  3.1.3 **Image Formation** 71
  3.1.4 **Resolution and Contrast** 72
      3.1.4.1 Limitations to Resolution 72
      3.1.4.2 Depth of Field 74
      3.1.4.3 Resolution in the TEM ... 74
  3.1.5 **Phase Contrast and Lattice Imaging** 76
  3.1.6 **Illumination Systems** 78
      3.1.6.1 Light Microscopy 78
      3.1.6.2 Transmission Electron Microscopy 80
  3.1.7 **Polarized Light** 80
      3.1.7.1 Anisotropic Materials ... 81
      3.1.7.2 Polarizing Microscopy... 83
3.2 IMAGING BY SCANNING ELECTRON BEAM 85
  3.2.1 **Probe Formation** 85
  3.2.2 **Probe-Specimen Interactions** ... 88
      3.2.2.1 Backscattered Electrons ... 89
      3.2.2.2 Secondary Electrons 89
      3.2.2.3 X-rays 91
      3.2.2.4 Interactions in the STEM 91
  3.2.3 **Image Formation in the SEM** ... 92
  3.2.4 **Low Voltage SEM** 94
      3.2.4.1 Low Voltage SEM Detectors 95
  3.2.5 **Variable Pressure SEM** 96
3.3 IMAGING IN THE ATOMIC FORCE MICROSCOPE 97
  3.3.1 **Microscope Components** 97
      3.3.1.1 Cantilever Parameters ... 98
      3.3.1.2 Cantilever Deflection ... 99
      3.3.1.3 Scanners 99
  3.3.2 **Probe-Specimen Interaction** ... 100
      3.3.2.1 Atomic Interaction 100
      3.3.2.2 Cantilever Effects 101
      3.3.2.3 Force-Separation Curves 102
  3.3.3 **Contact Mode AFM** 102
      3.3.3.1 Capillary and Electrostatic Forces 103
      3.3.3.2 Lateral Forces and Frictional Force Microscopy 103
      3.3.3.3 Force Modulation Imaging 104
      3.3.3.4 Force Volume Imaging 105
  3.3.4 **Intermittent Contact AFM** 105
      3.3.4.1 Modeling Cantilever Oscillation 106
      3.3.4.2 Real Cantilever Oscillation 108
      3.3.4.3 Imaging Conditions in IC-AFM 110
      3.3.4.4 Modeling IC-AFM 112
  3.3.5 **Noncontact AFM** 112
  3.3.6 **Practical Considerations for AFM Imaging** 113
      3.3.6.1 Feedback Loops 113
      3.3.6.2 Scan Speed 113
      3.3.6.3 Resolution 114
      3.3.6.4 Presentation 114
  3.3.7 **Artifacts in SPM Imaging** 114
      3.3.7.1 Scanner Motion Artifacts 114

3.3.7.2 Artifacts Due to Probe
Geometry 115
3.3.7.3 Tip Defects, Wear, and
Contamination 117
3.4 SPECIMEN DAMAGE IN THE
MICROSCOPE 118
  3.4.1 Effect of Radiation on
  Polymers 118
  3.4.2 Radiation Doses and Specimen
  Heating 120
  3.4.3 Effects of Radiation Damage
  on the Image 121
    3.4.3.1 Mass Loss 121
    3.4.3.2 Loss of Crystallinity ... 121
    3.4.3.3 Dimensional Changes ... 122
    3.4.3.4 Radiation Damage in
    the SEM 123
  3.4.4 Noise Limited Resolution 123
References 124

3.1 IMAGING WITH LENSES

3.1.1 Basic Optics

The basic optics of the optical microscope and the conventional transmission electron microscope (TEM) are similar. Condenser lenses illuminate the object to be imaged with a flood of radiation, and imaging lenses form the radiation leaving the object into a magnified image. The formation, contrast and resolution of images in these microscopes can be understood with classical optics, which includes geometrical (particle) and physical (wave) optics. Both electrons and light may be considered as propagating waves with an amplitude and a phase, though only the intensity which equals (amplitude)^2 can be directly observed.

A reader unfamiliar with any optics should consult a textbook for more information. Jenkins and White [1] and Hecht [2] are excellent standard college-level texts; Welford [3] is simpler, at first year college level, and Martin [4] concentrates on microscopy. Of the many that are more detailed and technical, Born and Wolf [5] should be mentioned as the professional reference book on optics. The optical microscopy texts already cited in Chapter 2 (Refs. 1–8 therein) are also useful sources. Slayter and Slayter [6] and Rochow and Tucker [7] deal with the optics of light and electron microscopy together, but most TEM texts largely assume knowledge of optics.

Matter slows both light and electrons, decreasing their wavelength λ; the *refractive index*, n is defined as:

$$n = \frac{\text{(wavelength in vacuum)}}{\text{(wavelength in material)}} \quad (3.1)$$

The *optical path length* of a wave in a material of thickness t is nt, as the material contains nt/λ wavelengths as would a path of length nt in vacuum. The *optical path difference* Δ due to the presence of this material is then $(n-1)t$, and the phase difference produced is $(2\pi/\lambda)\Delta$. Table 3.1 shows how the optical properties of polystyrene depend on the incident radiation. The refractive index of polystyrene for electrons is calculated from values for carbon, corrected for the lower density of polystyrene (see Section 3.1 of [5]).

If the phase of the wave can be calculated from its phase at nearby points and times, the radiation is *coherent*; monochromatic and parallel light, for example, is coherent. The phase and amplitude of completely *incoherent* light vary randomly in space and time. When two coherent waves of amplitudes a and b come together they *interfere* with each other. The result is a wave of intensity $(a + b)^2$ if they are exactly in phase (constructive interference) and intensity $(a - b)^2$ if they are completely out of phase (destructive interference). In general the waves must be added by vector sum rules. Incoherent waves interfere momentarily, but over any period of observation the phase effects average out. The resulting average intensity is the sum of the intensities of the two waves, $(a^2 + b^2)$.

3.1.2 Diffraction

When an object scatters coherent waves, interference between waves scattered from different

TABLE 3.1. Optical properties of polystyrene

	Green light	Electrons	
		100 keV	1 MeV
Wavelength, λ (nm)	547	0.0037	0.0009
$n - 1$	0.4	2.5×10^{-5}	3.8×10^{-6}
Film thickness (nm) for $\Delta = \lambda/4$	340	37	60

points on the object produces a variation in intensity as a function of their direction, the *diffraction pattern* of the object. An object with regular periodicity d in one dimension has a pattern with maximum intensity when the angle between incident and scattered radiation, ϕ, takes the values given by:

$$d \sin \phi = m\lambda \quad (m = \pm 1, 2, 3, \ldots) \quad (3.2)$$

A three dimensional periodicity, such as atoms in a crystal, gives maxima at different angles given by:

$$2d \sin(\phi/2) = m\lambda \quad (m = \pm 1, 2, 3, \ldots) \quad (3.3)$$

This is Bragg's law, usually written with the angle between incident and scattered radiation as 2θ instead of ϕ.

If the object has an exactly sinusoidal variation of absorption, thickness or refractive index in one dimension, diffracted beams appear only when $d \sin \phi = \pm \lambda$ (i.e., $m = \pm 1$). This is important because of Fourier's theorem, which states that any (single valued) function of a variable x can be expanded as a sum of sines and cosines of multiples of x. Thus any phase or intensity variation in the sample can be considered as a sum of sinusoidal variations of different wavelength, each giving a certain intensity at a single characteristic angle ϕ. The intensity at a point in the diffraction pattern corresponds to the strength of a variation of some sample property with a particular direction and spatial frequency. Closely spaced structures have high spatial frequencies and produce intensity at high angles in the diffraction pattern. The image is a map of the object as a function of position, and the diffraction pattern is a map of the variations in the object as a function of spatial frequency.

This concept of the diffraction pattern as a map in spatial frequency space (reciprocal lattice space and Fourier space are other names) is somewhat abstract and mathematically complex [1, 4, 5, 8–10]. It is nevertheless extremely useful, as it gives a physical insight into many facets of microscope optics.

3.1.2.1 Electron Diffraction

In electron diffraction, the spatial frequencies of interest are the distances between atoms.

Many polymers are amorphous, so in these materials the distances between atoms can have a range of values. An unoriented amorphous material will give a diffraction pattern consisting of one or two broad rings (Fig. 3.1A). Unoriented polycrystalline materials give sharp rings (Fig. 3.1C). A full analysis of the intensity of such ring patterns gives the *radial distribution function* (RDF) of interatomic spacings in the material. This analysis is infrequently performed on polymer systems; it is closely linked to modeling of amorphous phases [11–13]. When an amorphous material is oriented the ring pattern becomes arced. The *meridional* reflections are those intensifying on a line parallel to the draw direction, and they are associated with spacings along the molecular chain. Those stronger on a line perpendicular to the draw direction are called *equatorial* and are associated with intermolecular spacings (Fig. 3.1B). These associations are also normally correct for crystalline materials. It is the "off-axis" reflections (those on neither line) that prove that there is crystalline order in a fiber (Fig. 3.1D).

A large and perfect crystal scatters electrons into a diffraction pattern of sharp spots. Interpretation of this pattern requires knowledge of crystallography. There are many texts in this field: Buerger and Borchardt-Ott [14, 15] are clear and comprehensive, but beginners will find Sands or Hammond more accessible [16, 17]. Giacovazzo et al. contains useful summaries of both basic and current advanced topics [18]. Books on crystal optics [19, 20] (see Section 2.2.5) also contain a basic summary of crystallography. There are many texts on diffraction from materials [21, 22], some concentrating on electron diffraction [23, 24], and TEM books also cover electron diffraction in detail (e.g., volume II of [25]).

In simple terms we can regard the crystal as a set of lattice planes, reflecting radiation according to Bragg's law. The diffraction angle is then 2θ where $2d \sin \theta = \lambda$. For electrons $\lambda \ll d$, so θ is very small. This means that lattice planes will diffract only if they are almost parallel to the beam. The geometry of an electron diffraction pattern is thus easier to analyze than the equivalent x-ray diffraction pattern, which

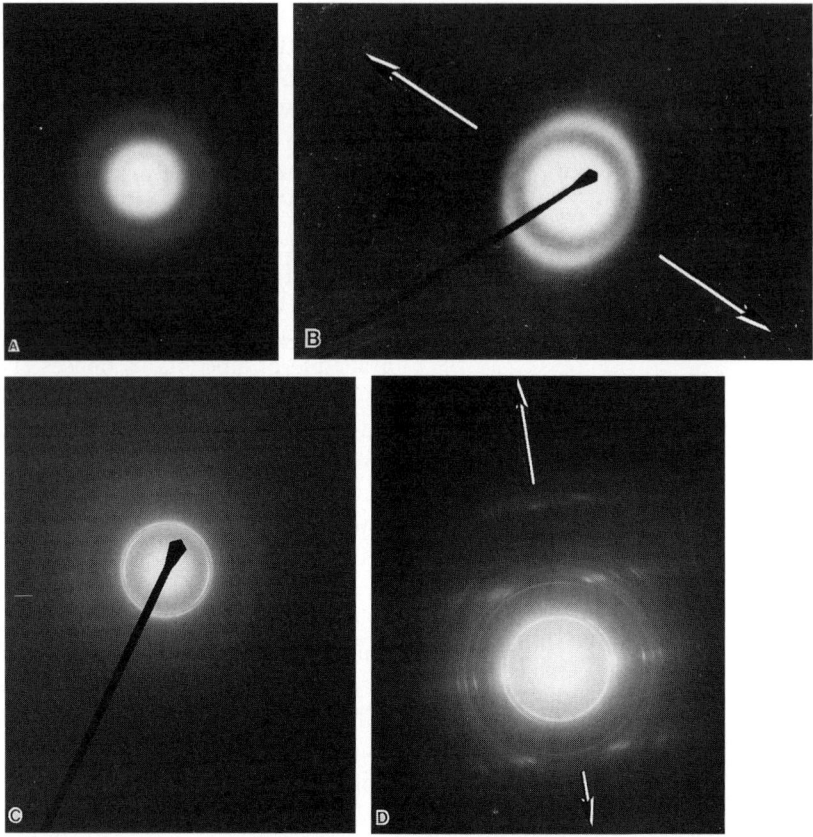

FIGURE 3.1. Electron diffraction patterns from (A) amorphous carbon, (B) oriented amorphous polystyrene (tensile direction indicated by arrows), and (C) a polycrystalline PE film. The sharpness of the rings in (C) indicates crystalline order. Highly oriented polyethylene is shown in the diffraction pattern (D) (tensile direction indicated by arrows). The off-axis spots prove the presence of three dimensional order.

involves large angles. The geometry of a single pattern is normally enough to determine orientation or to distinguish different crystallographic phases [26–28]. The unit cell and symmetry of an unknown phase may also be determined from the geometry of its diffraction patterns. Proper interpretation of the intensities of spots in a diffraction pattern gives the positions of the atoms in the crystal; that is, determines the crystal structure [29, 30]. Intensities may be difficult to measure accurately because of background from inelastic scattering of electrons. An energy filtering electron microscope (EFTEM; see Section 6.5.3) can remove the inelastic background. Figure 3.2 shows that this can make a very significant difference [31]. Here the effect is large because the micellar film spacings are large, so the diffraction spots of interest are at lower angles, where inelastic scattering is highest.

Electrons interact so strongly with matter that an electron can be scattered into one diffracted beam and then from that to another, even in a very thin crystal. This multiple scattering makes full theoretical treatment of electron diffraction complex [23, 24]. Although obtaining an accurate electron diffraction pattern from a polymer crystal may be difficult because of its instability [32], the small thickness and low atomic number of most polymer crystals can make the analysis more straightforward. The small size of available single crystals rules out the equivalent x-ray experiment. Many structural determinations have been

Imaging with Lenses

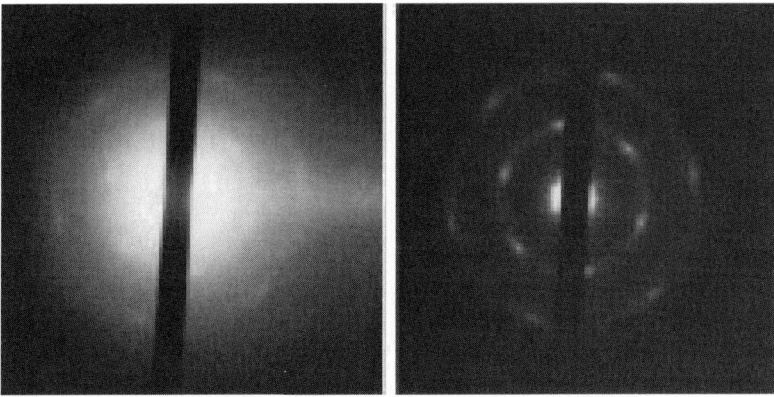

FIGURE 3.2. Electron diffraction at 120 kV from a cast micellar film with 3.5-nm spacing: at left, unfiltered; at right, filtered to include only zero-loss electrons. (From Du Chesne [31], © (1999) Wiley-VCH; used by permission.)

made of polymers and other organic crystals [29, 30, 33, 34].

Polymer crystals are frequently small and imperfect, so that the diffraction spots are fuzzy, or they are arcs from an oriented polycrystalline texture (see Fig. 3.1C). The degree of perfection of crystals of known structure can be determined from measurement of diffraction line widths and intensities. The analysis used for x-ray diffraction [21, 35, 36] can be transferred directly to electron diffraction. One can distinguish between crystal size effects and the effects of disorder within the crystals, but often a simple estimate of the mean crystal size, $0.9\lambda/$(angular breadth), is used. All electron diffraction measurements on polymers may be impacted by radiation damage (see Section 3.4).

3.1.3 Image Formation

A typical TEM objective has a focal length of about 2 mm and forms a real image with magnification $M = 20\times$ to $50\times$ at image distance 50–100 mm. An optical objective of $M = 50\times$ has a similar focal length, and older optical objectives had a similar image distance; newer ones form an image at infinity. Figure 3.3 shows the geometry of image formation by the objective lens. It is schematically shown as a thin lens, though in both optical and TEM instruments the lens is actually longer than its focal length.

The semi-angular divergence of rays from a point on the object is shown as α, limited by an aperture in the back focal plane. The divergence on a point on the image, $\alpha' = \alpha/M$. This demonstrates a general principle in geometric optics,

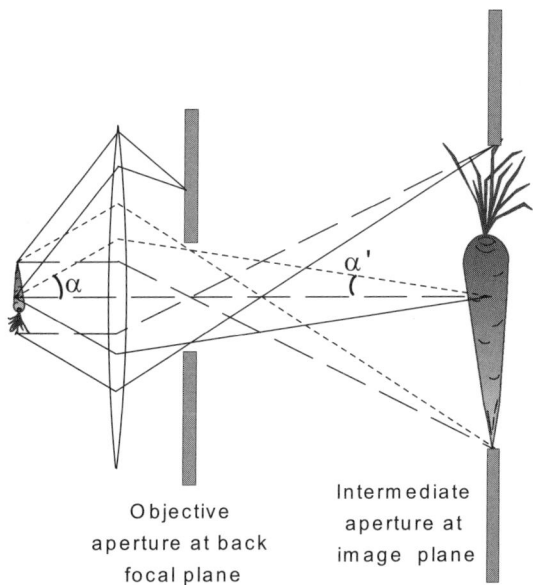

FIGURE 3.3. Image formation by the objective lens. Rays leaving the specimen in a given direction meet at a point in the *back focal plane* of the lens. The objective aperture there limits the angular acceptance of the imaging system to an angle α. The intermediate aperture is used to select the specimen area that contributes to the diffraction pattern.

that divergence angles go as $1/M$. The effects of nonideal behavior of the lens increase with ray angle (see Section 3.1.4.1), so imperfections in a second lens have less effect. The objective lens properties are most important in controlling image formation.

In the TEM the aperture in the back focal plane is called the objective aperture, and its size is always important in image formation. In the light microscope some objectives have an adjustable aperture, but in most it is a fixed part of the lens construction. The second aperture shown in the figure, the intermediate aperture, is used in the TEM to limit the area that contributes to a diffraction pattern. Figure 3.3 shows that the diffraction pattern will appear at the back focal plane.

3.1.4 Resolution and Contrast

In all microscopy a primary concern is the spatial resolution that can be obtained. If the details are points, the resolution may be specified as *point-to-point* resolution. An object with regularly repeating details such as a set of lines will give a different value for resolution. This is important as some convenient test objects are ruled lines in optics, and lattice fringes (sheets of atoms) in TEM. Near the resolution limit, images of two objects will overlap and this will reduce contrast. A precise measure of resolution requires a quantitative test to determine if the detail can be distinguished, and so there is a need for a quantitative definition of contrast. The contrast of a feature is generally defined as

$$C = |I_0 - I|/I_0 \quad (3.4)$$

where I_0 is the background intensity and I is the intensity at the detail [6, 10, 37]. It is often expressed as a percentage: $100 \times |I_0 - I|/I_0 \%$. In some cases, it may not be clear which is the "background" and which the "feature," so another definition is [38]

$$C = (I_{max} - I_{min})/I_{max} \quad (3.5)$$

In other cases, a logarithmic definition is more useful [6, 39]:

$$C = \log(I_0/I) \quad (3.6)$$

This means that great care is required in comparing results from different sources. The most complete method of defining resolution is to measure the ratio (contrast in the image)/(contrast in the object) as a function of detail separation. This is the *modulation transfer function* (MTF), also called the contrast transfer function (CTF), which describes how the modulation or contrast in the object is transferred to the image. Clearly, when MTF falls to some very small value, no object detail is reproduced in the image.

The MTF may be measured using test objects of known size and contrast, as when testing lithographic processes. A better method for TEM is to use a random object that contains details of all sizes and determine how these details are reproduced in the image. A Fourier transform of the image shows which modulations have been transferred. The resolution is then taken as the detail size at which the MTF first falls to some arbitrary value. The *Rayleigh criterion* for resolution is that the intensity between two bright spots should be no more than 80% of the peak value [1, 4]. This is derived from tests of human vision and corresponds to an MTF of $(1 - 0.8)/(1 - 0) = 0.2$. Contrast of 0.05 is needed in an image detail for it to be visible to the eye, so objects of intrinsic contrast >0.25 will be visibly resolved when MTF ≥ 0.2. In the TEM, MTF can be a complicated function of feature size [10, 25, 40]. For high resolution work, a plot of this function is more informative than any single number called resolution.

3.1.4.1 Limitations to Resolution

Four factors may limit the resolution of an image, and these are:

- The diffraction limit;
- Lens imperfections (*aberrations*);
- Noise;
- Detector resolution.

The diffraction limit depends on the wavelength of the radiation and the angular acceptance of the objective. The diffracted beams caused by a periodicity d come off at angles given by $\sin\phi = \lambda/d$ or greater. Therefore the lens must accept a semi-angle $\alpha \geq \phi$ for the periodicity to affect the image and the lens cannot resolve details smaller than $\lambda/\sin\alpha$. If oblique coherent illumination is considered, some rays can enter at an angle ($-\alpha$) to the optic axis, be diffracted by $\phi = 2\alpha$ to an angle ($+\alpha$), and still be accepted into the imaging system. Then by this *Abbe* or diffraction theory of imaging, the smallest resolvable detail on the object, d_{min}, will be $0.5\lambda/\sin\alpha$. More complete calculations give a very similar result for diffraction limited resolving power with any illumination, $0.6\lambda/\sin\alpha$ [1, 4, 5, 8]. The wavelength λ used in the formula above refers to its value in the space between specimen and objective. If, as is more common in optical microscopy, λ is taken to be the free space wavelength, the resolution d becomes $d = 0.6\lambda/\text{NA}$. NA, the *numerical aperture*, is ($n \sin \alpha$), where n is the refractive index of the medium in front of the objective. Table 3.2 shows the diffraction limit for typical high and low power objective lenses in the light microscope.

Lens aberrations of concern are chromatic aberration, where the focal length of the lens depends on wavelength and the five Seidel geometric aberrations:

1. Spherical aberration (axial);
2. Astigmatism (off axis and axial);
3. Coma (off axis);
4. Field curvature (off axis);
5. Distortion (off axis).

Generally optical lenses are very well corrected; they have been able to obtain resolutions very close to the theoretical diffraction limit for a century or more. However, full correction of chromatic aberration, spherical aberration, and field curvature to give a "Plan Apochromat" will be expensive. In contrast, lenses for electron microscopy are generally uncorrected except for axial astigmatism. Practical lenses that correct for spherical aberration have become available only very recently (see Section 6.2.4). These promise a jump in the resolving power of electron microscopes, but at the moment the very highest resolution in electron microscopes is of limited use in polymer studies. Spherical and chromatic aberration in the TEM is described in the following section.

Noise is not a problem in imaging with lenses unless radiation damage limits the number of electrons that can be used to form the image; this will be dealt with in Section 3.4.4.

Detector resolution: The usual plots of intensity versus position in the image, used to demonstrate the Rayleigh criterion for resolution, are smooth continuous curves. This would imply a detector of infinitely fine grain. Table 3.2 shows how the object resolution matches the total magnification M of the system and the resolution of the detector—in this case, the eye. The image resolution obtained (object resolution × magnification) is 200–300 μm. For visual observation the (virtual) image will be at the standard viewing distance of 25 cm from the eye. The detail therefore subtends an angle of 10^{-4} rad at the eye, near but comfortably within the eye's angular resolution of 3×10^{-5} rad. Much higher magnification would be empty, producing no further information.

When a digital detector is used it is easy to see that a pixel spacing the same as the separation of the spots will not allow them to be resolved. At least one pixel must be in between the spots, so that it can have a lower intensity and separate them. This is just one example of the Nyquist criterion in signal processing, that is, the sampling frequency must exceed twice the highest frequency in the original signal or information will be lost. If the projector lens had the same

TABLE 3.2. Diffraction-limited resolution d for objective lenses in the optical microscope, $\lambda = 0.5\,\mu m$

Objective magnification	NA	$d = 0.6\lambda/\text{NA}$ (μm)	Total system magnification with 10× eyepiece	Image resolution (μm)
10×	0.1	3	100	300
60×	0.75	0.4	600	240
100×	1.25	0.22	1000	220

power, 10×, as the eyepiece in Table 3.2, then the image resolution of 200–300 μm would require a pixel size of 100–150 μm or smaller. Much smaller pixels, say 30 μm, would give significant oversampling. An oversampled image may look smoother, but it might be better to change to a lower magnification projector lens, allowing a larger field to be viewed. On the other hand, some degree of oversampling is probably a good idea.

Figure 3.4 shows two Gaussian spots: in panels A and B they are just resolved, and the pixel size is very small. Going to pixels half the resolution limit still resolves two spots when they are centered on the pixel in Fig. 3.4C. But move the spots so that their peaks are at the boundary between pixels, and they are no longer resolved, Fig. 3.4D. Obviously, moving the spots makes no difference if the pixels are small, as seen in Fig. 3.4B.

3.1.4.2 Depth of Field

Once a resolution limit has been defined, the *depth of field* can be determined. The depth of field is the range of object distances where

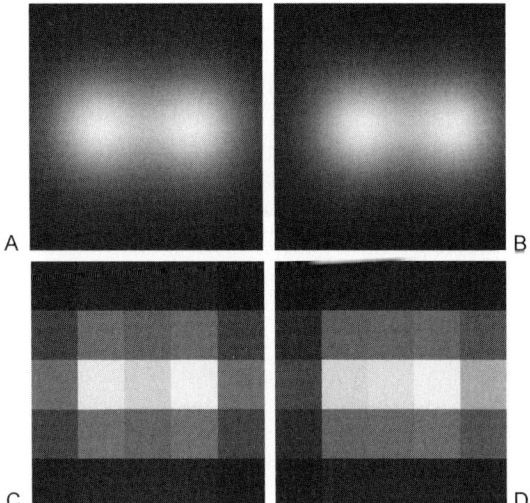

FIGURE 3.4. Two Gaussian spots separated to be near the Rayleigh resolution limit are shown with very high detector resolution in (A) and (B) and with pixels one half the spot separation in (C) and (D). When the spots are centered on the large pixels, they are still resolved (C); displaced to the pixel edge, they are not (D).

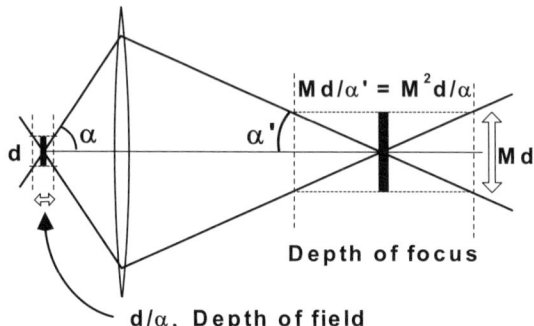

FIGURE 3.5. The depth of field depends on the resolution d and the angular acceptance of the lens, α. The object may be moved by d/α and a sharp image of size Md can still be formed.

defocus produces a spreading of a point image equal to the image resolution limit, Md. Figure 3.5 takes the whole imaging lens system to be a single thin lens. It shows that the depth of field is the resolution of the object, d, divided by the divergence angle α. *Depth of focus* is the range of image plane positions that produces the same spreading of the image, and is $M^2 \times$ (depth of field).

It can be seen in Table 3.3 that the depth of field at high magnifications in the optical microscope is very limited, less than the resolving power. The depth of field in the TEM is as small in absolute terms, but very much greater than the resolution and greater than a useful specimen thickness at that resolution. Thus, all the thickness of a TEM specimen is in focus at once, and the depth of field is never a problem.

3.1.4.3 Resolution in the TEM

The off-axis aberrations (off-axis astigmatism, coma, field curvature, and distortion) all vanish for rays parallel to the optical axis. All rays in the TEM are at such small angles to the optical axis that these aberrations are not usually important. The axial astigmatism can be corrected by adjustment of stigmators that cancel the residual non-circularity of the objective. This can be difficult, but assuming that it is done correctly, the remaining effective aberrations are chromatic and spherical.

TABLE 3.3. Depth of field

Objective lens	α (rad)	Resolution (μm)	System magnification	Depth of field (μm)	Depth of focus
Optical 10×	0.1	3	100	30	30 cm
Optical 100×	0.93	0.22	1,000	0.15	15 cm
TEM	0.0066	6×10^{-4}	100,000	0.1	1 km!

Chromatic Aberration

Chromatic aberration affects any image when the electrons that contribute to it do not all have the same energy. Modern microscopes have highly stable accelerating voltage, but there is an intrinsic energy spread of about 1 eV in electrons leaving a tungsten filament source. The chromatic limit to resolution due to this is only 0.2 nm at 100 keV, not important as a limit for polymer microscopy.

For polymers the major effect of chromatic aberration appears when a thick specimen causes the transmitted electrons to lose energy. Just how serious this is depends on the exact nature of the material as well as the specimen thickness. However, a largely carbon material of density 10^3 kg m^{-3} (specific gravity 1) will cause a 100 keV electron to lose about 400 eV/μm [10, 41, 42]. Thus a foil 100 nm thick causes 40 eV loss, for a chromatic limit to resolution of 6 nm. This chromatic aberration is the basis for the rule of thumb (often given for biological specimens) that details smaller than one tenth of the specimen thickness will not be resolved. Energy filtering (EFTEM [43]; see Section 6.5.3) will remove this effect and has been used on very thick stained biological specimens [44]. However, contrast in unstained materials may not always improve [31]. Since high energy electrons lose energy at a lower rate, and the chromatic aberration effect varies as $\Delta E/E$, going to higher accelerating voltage is also very effective in reducing this form of chromatic aberration.

Spherical Aberration

Spherical aberration occurs because rays passing through the outer portions of the lens are diverted too much and come to a focus short of the ideal focal plane (Fig. 3.6). It causes the image of a point at the ideal (Gaussian) focal plane to become a circle of radius $M\alpha^3 C_s$ corresponding to a resolution on the object of about $\alpha^3 C_s$. C_s is the spherical aberration coefficient, typically 1–2 mm for 100 kV lenses. This resolution limit is proportional to α^3, and the diffraction resolution limit is proportional to $1/\alpha$ (for small angles where sin $\alpha \approx \alpha$), so there is an optimum value of α that gives the best resolution. It is given by:

$$\alpha_{opt}^3 C_s = \lambda/\alpha_{opt} \quad \alpha_{opt} = (\lambda/C_s)^{1/4} \quad (3.7)$$

At this divergence, the resolution is

$$d_{min} = (C_s \lambda^3)^{1/4} \quad (3.8)$$

Focus Considerations

It is apparent from the ray diagram of Fig. 3.6 that there is a plane, B–B′, nearer to the lens than the geometric or Gaussian focus plane (A–A′), where the resolution is improved. It is called the "plane of least confusion" or Scherzer focus [4, 8, 10], and it is close to where the rays from the outermost parts of the lens intersect the axis. From the figure, the distance $\delta Z'$

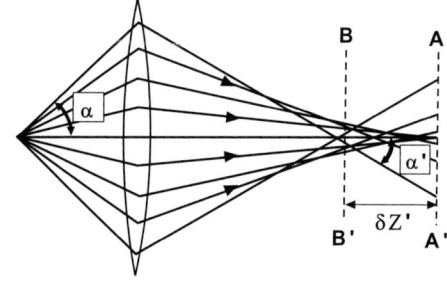

FIGURE 3.6. Spherical aberration causes rays at larger angles to the axis of the lens to come to a focus short of the ideal focal plane. The smallest image of a point source appears in the plane B–B′, at the Scherzer focus position.

between this plane and the Gaussian image plane is approximately the radius of the image disk in the Gaussian plane divided by α'. The radius is $M\alpha^3 C_s$ and $\alpha' = \alpha/M$, so

$$\delta Z' \approx C_s (\alpha M)^2 \quad (3.9)$$

Reducing the lens power, *underfocusing*, to move the image plane by $\delta Z'$ or the object plane by δZ (where $\delta Z = \delta Z'/M^2$) improves resolution. A detailed calculation gives:

Resolution: $d_{min} = 0.43(C_s \lambda^3)^{1/4}$ (3.10)

Optimum divergence: $\alpha_{opt} = 1.41(\lambda/C_s)^{1/4}$ (3.11)

At defocus: $\delta Z_{opt} = (C_s \lambda)^{1/2}$ (3.12)

(Defocus is conventionally given as a motion of the object, as one would focus an optical microscope.)

Comparing Scherzer focus to geometric focus, the resolution is improved by a factor of about 2, if the included divergence angle is increased by $\sqrt{2}$. Taking values of $\lambda = 0.004$ nm (for 100 keV electrons) and $C_s = 2$ mm, these become $\alpha_{opt} = 10$ mrad (0.5°) and $d_{min} = 0.3$ nm. This is not high resolution for the TEM but is much higher than normally obtainable in the TEM of polymers. For really high resolution in the TEM, using 200 keV reduces λ, and high performance objectives have lower C_s.

3.1.5 Phase Contrast and Lattice Imaging

Phase contrast means that scattered radiation has a phase shift applied to it, and this results in intensity changes that form an image. The basic reason for this can be seen in Fig. 3.7. Normally, scattered radiation is 90° out of phase with the unscattered radiation. The result of recombining them is a phase shift in the transmitted wave but no change in amplitude (Fig. 3.7A). This phase shift is normally described in terms of a refractive index not equal to 1. If the sample is homogenous and all the scatter is in the forward direction, there is no way to affect the phase of one wave and not the other. But if inhomogeneities make the scattered radiation appear at some angle, then a phase plate can put another 90° phase difference between the two beams, based on their position.

If the phases add, the scattered radiation is 180° out of phase, reducing the amplitude. Here a scattering object appears dark, similar to bright field, and this is called positive phase contrast (Fig 3.7B). Negative phase contrast makes the scattering object brighter than the background by canceling the original phase difference (Fig. 3.7C).

In the light microscope, the phase plates are usually designed to absorb some of the unscattered light, so that the effect of adding or subtracting a small scattered amplitude is greater. That is, they enhance phase contrast at the expense of image brightness.

In the TEM, the phase shift of scattered waves is a complicated function of defocus and

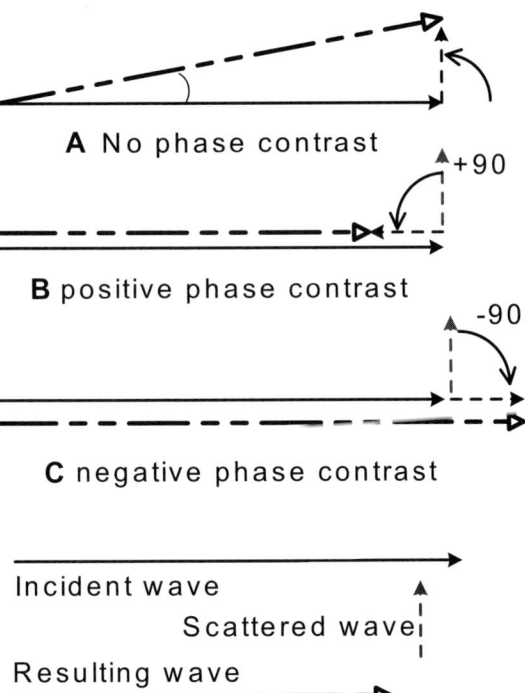

FIGURE 3.7. Waves involved in phase contrast are shown as vectors, with amplitude given by their length and phase by their direction. An extra phase shift makes the resulting amplitude depend on the scattering amplitude.

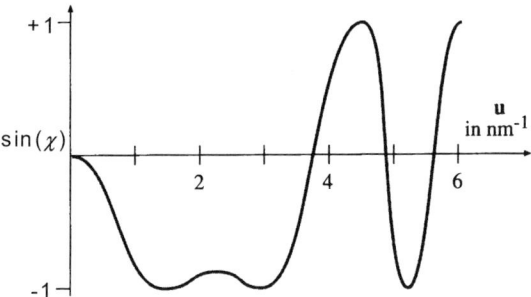

FIGURE 3.8. Plot showing the phase shift as a function of spatial frequency in a well-adjusted TEM at 200 kV, C_s 1 mm, Δf 58 nm. (Adapted from Williams and Carter [25], © (1996) Springer; used with permission.)

modeling HREM include chromatic aberration and beam divergence [45, 46].

The full theory of lattice imaging is quite difficult, but the basic principle is simple to explain. A plane wave, wavelength λ, falling on a three-dimensional repeating structure with spacing d, will produce diffracted waves at angles 2θ given by Bragg's law, $n\lambda = 2d \sin\theta$ (see Section 3.1.2). In Fig. 3.9, the structure is shown schematically as a row of objects, and the first diffracted wave ($n = 1$) is shown at angle $2\theta = \lambda/d$ (sin $\theta \sim \theta$ for the scattering angle. It is usually expressed as $\chi(u)$ where χ is the phase angle and u is the spatial frequency = (scattering angle/wavelength). Figure 3.8 is a plot of $\sin(\chi)$ as a function of u for a 200 keV beam, with a spherical aberration constant C_s of 1 mm and a defocus of −58 nm.

If $\sin(\chi)$ is near zero, there is no phase contrast. When $\sin(\chi)$ is near to ± 1, the phase shift is near $\pm 90°$ and there is strong negative or positive phase contrast. Because there is a broad band of good contrast in the region from 1 to 3.5 nm^{-1} in Fig. 3.8, this setting would be good for imaging specimen detail in the range 0.3–1 nm. If the specimen is crystalline, these details will include lattice spacings. The resolution limit of the image is normally taken to be the point just before the first zero in this function [25]. At optimum focus, this is close to the point-to-point resolution calculated in Section 3.1, at about $0.5(C_s\lambda^3)^{1/4}$.

The appearance of the image will change in a complicated manner with very small changes in focus. Numerical simulation of the expected image from model structures is a very important part of high resolution electron microscopy (HREM) and this requires accurate determination of all the operating parameters, [40, 45, 46]. Other factors besides $\sin(\chi)$ that affect image formation and must be taken into account in

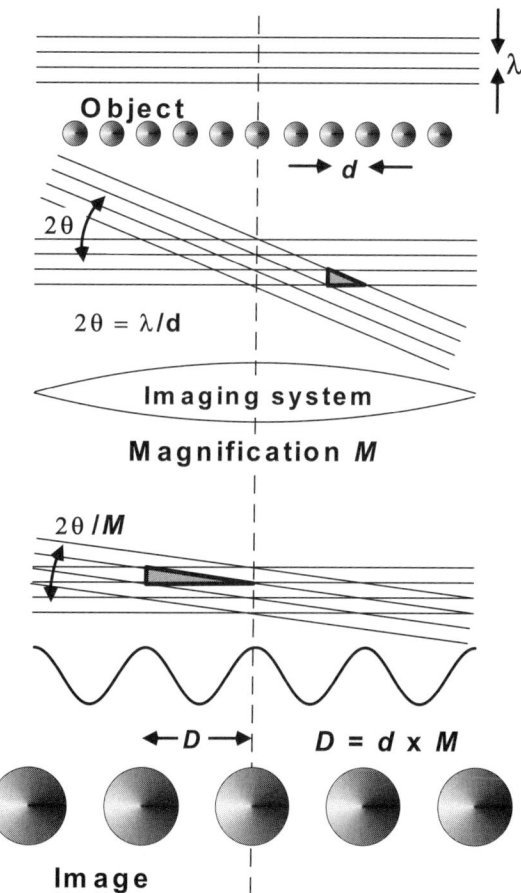

FIGURE 3.9. Schematic diagram of the process of resolution of lattice planes. The object structure, spacing d, causes a diffracted wave to form at $2\theta = \lambda/d$. The imaging lenses reduce the off-axis angle to $2\theta/M$ so that if the waves recombine, there is a lateral periodicity $D = Md$.

small angles). The interference pattern of this wave with the transmitted wave has a lateral repeat of d. To obtain lattice images, diffracted and transmitted waves are allowed to pass through the objective aperture and the imaging lenses, indicated as a single lens in the figure. This forms an image of the specimen with magnification M. As shown in Section 3.1, divergence angles α at the object are reduced to α/M at the image. Thus the diffracted waves recombine at an angle $2\theta/M$. Now the interference of these waves has a lateral periodicity $D = Md$. If the phase contrast is set up so that recombination produces intensity change, then a magnified image of the lattice spacing has been formed. The figure shows the sinusoidal intensity variation that might be produced if only this one diffracted beam reaches the image plane and also the schematic magnified image that would be formed from many beams.

For polymers, radiation sensitivity (see Section 2.6 and Section 3.4) will limit the number of electrons that can be used to form an image. This may limit the resolution because of the minimum signal to noise ratio required for a visible feature. Because the HREM image of a polymer must normally be recorded without viewing it, it is important to have the largest area possible in the image. Electronic detectors have a limited size (~3k × 4k pixels today) while digitally scanned film is equivalent to 13k × 27k pixels [47]. Thus film is still the preferred detector for these experiments, though this may change. As the spatial resolution of the film is 5–10 μm, the magnification is kept relatively low. Low contrast and noisy lattice fringes recorded will not be visible to the naked eye or even with a 10× magnifying lens. Laser light scattering from the image will show up areas with a periodic image [46].

Different polymers have a wide range of sensitivity to the electron beam. Aromatic, rigid rod and other high melting point polymers [48] will be more resistant to beam damage, needing a flux J_a of more than 100,000 electrons/nm^2 to be made amorphous. Low melting point flexible chain aliphatic polymers may withstand only a few hundred electrons/nm^2 (see Section 3.4.1 and Fig. 3.38) [47]. The quality of the images that can be obtained is as widely variable.

3.1.6 Illumination Systems

The illumination system must collect flux from the source and direct sufficient intensity onto the required field of view. As mentioned in Section 2.1.1, both light microscopes and TEMs use a system of two lenses to do this. Typical characteristics of sources were given in Table 2.2, which showed that the small size of the electron sources gives them a very high brightness. Unless rays are stopped by apertures, the brightness B (flux · unit area^{-1} · unit solid angle^{-1}) remains constant:

$$B = \frac{4i}{(\pi d\alpha)^2} \qquad (3.13)$$

where i is the beam current and d the spot size. Figure 3.3 shows that d is proportional to M and α is proportional to $1/M$ so that B is unchanged.

3.1.6.1 Light Microscopy

The illuminating system of the optical microscope is described in detail by Hartley [49]. At high powers a wide cone of rays is needed to fill the aperture of the objective lens. The first condenser lens is relatively weak and collects light for a strong substage (second) condenser that demagnifies the source giving a high divergence. To completely fill the aperture of an oil immersion objective, an immersion condenser will be required.

The illumination in optical microscopes has been the subject of much confusing discussion about whether the filament itself should be focused onto the specimen (critical illumination). With modern sources this gives an unevenly lit field of view. The Köhler system shown in Fig. 3.10 gives an evenly lit field and is the standard for the optical microscope. (At high resolution, the condenser lens should be *aplanatic*, that is, corrected for spherical aberration.) It uses the back focal plane of the field lens as the object that is imaged by the substage condenser into the specimen plane.

The *field diaphragm*, at the back focal plane, controls the area illuminated but has no effect on divergence. The aperture at the focal plane of the second (substage) condenser controls the

Imaging with Lenses

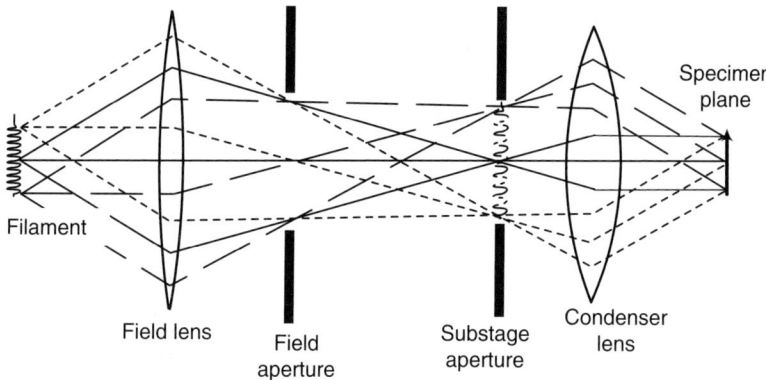

FIGURE 3.10. Köhler illumination; the field lens produces a magnified image of the filament at the substage aperture. This image is at the focal plane of the condenser lens. The condenser lens object plane is the back focal plane of the field lens. Each part of the specimen receives light from the whole filament.

divergence angle of the illuminated rays and has no effect on the area illuminated. When the system is properly adjusted in this way, the two apertures act completely independently. The field diaphragm is closed to match the illuminated area to the viewed area, reducing glare. The second aperture is usually set to fill 0.6–0.8 of the objective back focal plane to balance resolution and contrast.

The only disadvantage of this system is that the fixed demagnification of the condenser lens makes it impossible to illuminate a large field, as needed for a low-magnification image (objectives <10×). To do this, the substage condenser must be changed to a lower power. Most light microscopes have condensers with a top lens that can be removed for low power work. This allows a large field to be illuminated but breaks the Köhler system (Fig. 3.11). Under these conditions, the field aperture is used to control glare and adjust contrast; the substage aperture is normally left open as closing it will cause the edges of the image to dim (vignetting). Figure 3.11 shows the field lens unchanged, giving an image of the diffuser at the plane of the substage aperture. In some microscopes, a supplementary field lens should also be removed when changing to very low magnification.

This discussion describes normal bright-field condensers. Phase contrast and differential interference contrast (DIC) condensers have the required inserts that match specific

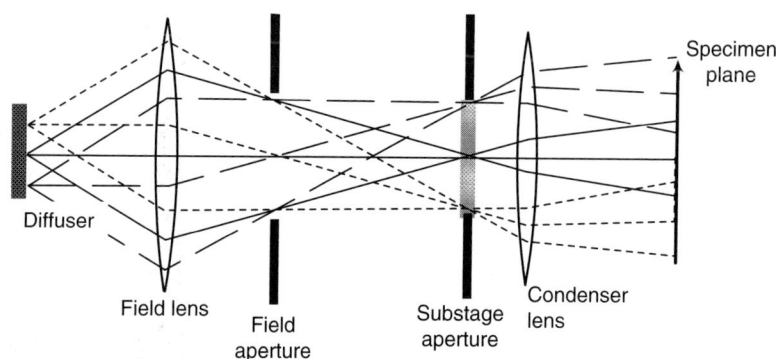

FIGURE 3.11. Light microscopy illumination for low magnification. The condenser lens is weak. Because a point on the image is illuminated by a restricted region of the filament, a diffuser (ground glass) is used to make the illumination more even.

objective lenses. For dark field in transmission, a condenser that gives annular illumination at angles above the acceptance angle of the objective is used.

3.1.6.2 Transmission Electron Microscopy

Transmission electron microscopy illumination has two important differences from light microscopy. One is practical; in the TEM the lens powers are continuously variable, but the apertures can be varied only in large steps. This is exactly the opposite of the optical case. The other is that the TEM most often needs a nearly parallel beam of illumination, rather than the strongly convergent beam needed to fill the objective in the optical microscope.

In conventional TEM, the first condenser, C_1, is strongly excited and forms a demagnified image of the source (not the filament, as the crossover of the electron gun is the effective source). Since beam divergence increases as the spot size falls, much of the flux is not accepted by rest of the system. The highest intensity at the object and the largest divergence angle are both obtained when the second condenser, C_2, produces a focused image of the source on the specimen. Under these conditions, the power of C_1 adjusts the spot size and the C_2 aperture controls the divergence.

Because the illumination is then very uneven, in normal operation the second condenser is defocused (usually underfocused) to "spread the beam." Now increasing the power of C_1 reduces the intensity and divergence of the illumination by reducing the size of the demagnified source. Altering the power of C_2 changes the area illuminated, and changing the diameter of the C_2 aperture does the same in large steps.

If the TEM is being used in scanning transmission electron microscopy (STEM) mode or for analysis using a small probe (and the source is not a field emission gun [FEG]), then the first part of the objective lens is used as a strong condenser lens to demagnify the source a lot and focus the illumination onto the specimen. The second condenser is turned off in this mode. Parallel illumination in this type of instrument is obtained by using C_2 to focus the source on the back focal plane of the upper objective lens. If the microscope has a FEG, its source is so small the first condenser may need to magnify, not demagnify it for conventional imaging.

At very low magnifications in the TEM, the first condenser lens must be substantially weakened, or turned off to get a large illuminated area. This is the equivalent of removing the top lens of the substage condenser in the optical microscope.

3.1.7 Polarized Light

Light is a transverse vibration. It consists of an electric field **E** and a magnetic field **B**. These are at right angles to each other and form the edges of a cube with **P**, the direction of propagation of the light ray. The electric field interacts strongly with materials, so it is used to describe the light. At any instant in time, the amplitude and direction of the electric field define a point on a plane perpendicular to the direction of the light ray. To an observer looking toward the source, this point would trace out a curve as the field varied. If the curve is simple and repetitive, the light is *polarized* and the form of the curve defines the state of polarization as shown in Fig. 3.12. If the curve is irregular and chaotic, the light is *unpolarized*.

When the electric field oscillates in amplitude but has a fixed direction, the curve traced out is a straight line and the field remains in one plane. This is called the plane of polarization, and the light is *linear* or *plane polarized*.

When the electric field is of constant amplitude but changes its direction, the point traces out a circle and the light is called *circularly polarized*. If both amplitude and direction change in a regular way, the curve traced out is an ellipse, and the light is *elliptically polarized*. This is the most general polarized state possible.

Any state of polarization can be considered as a combination of two perpendicularly plane polarized waves with different amplitudes and a specific phase difference. Adding two such waves can produce any polarization state. For a fixed result the two waves must be coherent, so that the phase relation between them remains the same.

Imaging with Lenses

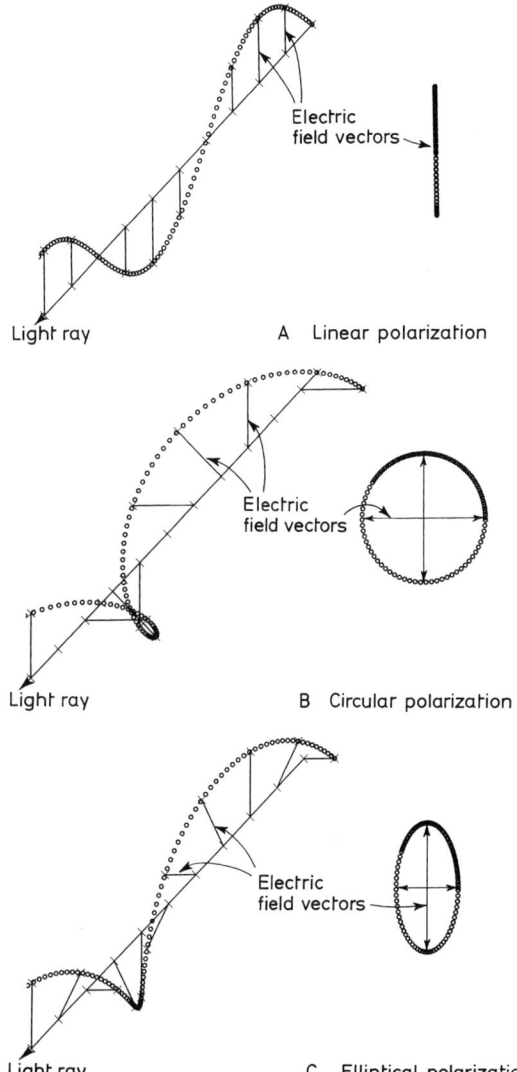

FIGURE 3.12. Schematic diagram of the electric field vector showing possible states of polarized light. With linear or plane polarized light (A), the electric field remains in a plane. In circularly polarized light (B), the electric field changes direction but has constant magnitude. In elliptically polarized light (C), the electric field changes in magnitude and direction, tracing out an ellipse.

3.1.7.1 Anisotropic Materials

Birefringent materials have a refractive index that depends on the direction of the electric field in the light [5]. They may be single crystals or oriented polymers, either amorphous or semicrystalline. Light waves passing through a birefringent material can be considered as divided into two plane polarized waves, polarized in the principal planes (perpendicular directions of extremal refractive index). If the different refractive indices are n_1 and $n_2 < n_1$ for the two plane polarized waves, there is a relative *retardation* $R = (n_1 - n_2)t$, and a relative phase shift $\delta = (2\pi/\lambda)(n_1 - n_2)t$ after passing through thickness t of the material. This will usually change the polarization state.

A complicating factor is that in anisotropic materials the electric displacement **D** is not generally parallel to the electric field **E**. **E** is perpendicular to the light ray (direction of energy propagation) but **D** is tangent to the wavefront. Therefore care must be taken when talking of "*the* direction of the light."

The optical properties of a birefringent material are shown by a surface called the *index ellipsoid* or *indicatrix*. The radial distance from the center to each point on this surface is proportional to the refractive index of light that has its electric displacement **D** in that radial direction. This is *not* the refractive index of light that is traveling in that direction. Cross sections of this surface are ellipses. The indicatrix of an isotropic material would be a sphere with circular cross sections.

In general the indicatrix is like a squashed (American) football. There are three principal refractive indices on perpendicular axes: the maximum, the minimum, and an intermediate value perpendicular to both of these (**A**, **C**, and **B** in Fig. 3.13). For example, a nylon crystal has $n(\mathbf{D}$ parallel to chains$) > n(\mathbf{D}$ parallel to hydrogen-bonded sheet$) > n(\mathbf{D}$ parallel to intersheets$)$. Such a material is *biaxial*. If two of these principal refractive indices are equal, the indicatrix is rotationally symmetric—an ellipsoid of revolution—and the material is *uniaxial*. For example, a polyethylene crystal has a larger refractive index for (**D** parallel to chains) and a smaller refractive index for any (**D** perpendicular to chains) (Fig. 3.14, left drawing).

A uniaxial material has one *optic axis* and a biaxial material has two. A birefringent material appears to be isotropic when a plane light wave passes through it along an optic axis. In terms of the indicatrix, the cross section in the

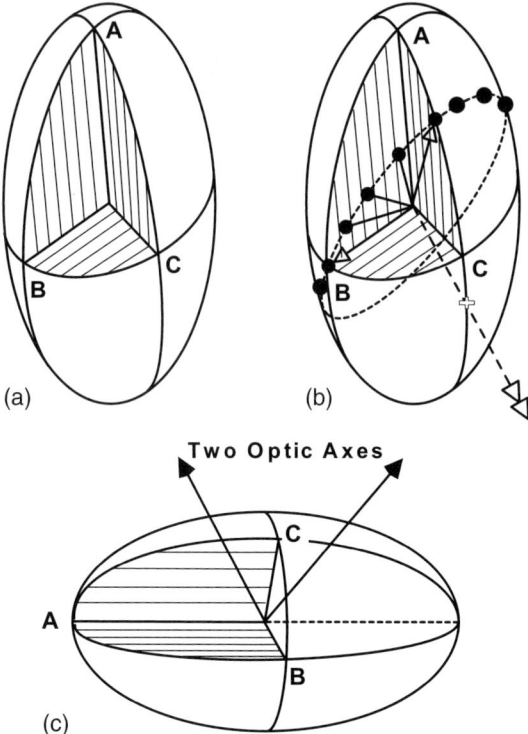

pairs of these points lie on two planes that also contain the intermediate value. These planes have the same principal radii—they are circular and the two optic axes are perpendicular to them (see Fig. 3.13B, C).

The discussion so far has been in terms of the optical properties of a substance. This may be different in kind, sign or magnitude from the optical properties of a sample. For example, a fiber of nylon where the chains are aligned but the crystals take up all orientations around the chain axis will be uniaxial. If a cross section of this fiber is taken as a sample it may appear isotropic in regular view because the light used to observe it passes down the optic axis. An unoriented amorphous sample will really be isotropic. A partially aligned sample will have a smaller birefringence than the material it is made from.

The sign of birefringence of a sample is given with reference to a distinct direction in the sample. For a fiber this is the fiber axis, for a spherulite it is the radius. Thus polyethylene, which has a positive birefringence as a material, forms positive fibers (chains along the fiber axis) but negative spherulites (chains perpendicular to the radial direction, Fig. 3.15).

FIGURE 3.13. (a) The indicatrix of a biaxial birefringent material. The principle refractive indices are A > B > C. The heavy dots in (b) outline the circular cross section. An optic axis is shown perpendicular to this plane. The two optic axes of the biaxial material are shown in (c). They lie in the plane A–C.

plane perpendicular to the optic axis is circular. Then the refractive index for light traveling along the axis does not depend on the plane of polarization. The birefringence of such a uniaxial material is $n(\mathbf{D}$ parallel to optic axis$) - n(\mathbf{D}$ perpendicular to optic axis$)$. If the unique principal refractive index is a maximum, the material has *positive birefringence* and if it is a minimum, the material has *negative birefringence* (Fig. 3.14B).

An ellipsoid of revolution must have circular cross sections perpendicular to the rotational symmetry axis, and so this is the optic axis. Biaxial materials are not so simple. Consider the elliptical cross section that contains the maximum and minimum values. At some (four) points it must have the same radial distance as the third intermediate principal value. The two

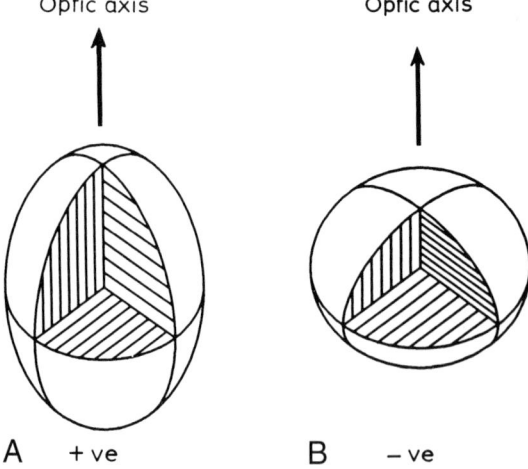

FIGURE 3.14. The indicatrix of uniaxial birefringent materials: (A) positive, (B) negative. For uniaxial materials, the optic axis is the axis of symmetry of the ellipsoid.

Imaging with Lenses

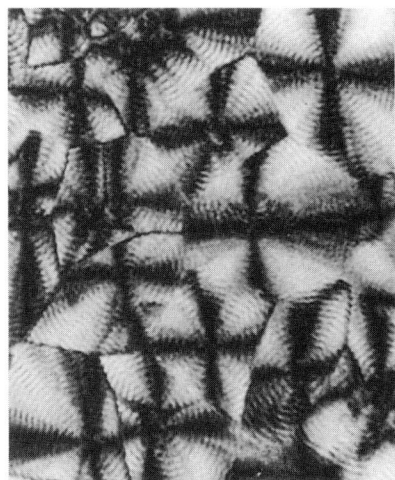 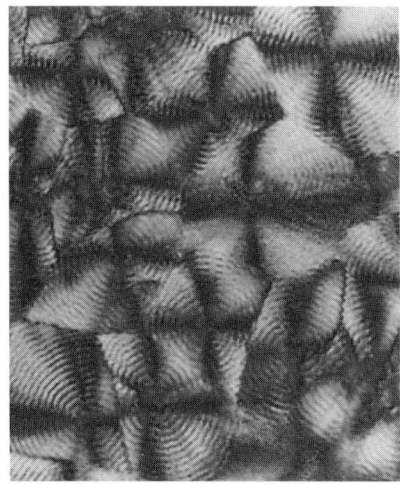

FIGURE 3.15. Two images of a thin melt cast film of high density polyethylene: the region is $200 \times 250\,\mu$m. The left hand image was taken in crossed polars. The radial "Maltese cross" is due to the extinction position. The spherulites in this material have dark circumferential bands. The crystals twist as they grow, and their orientation in these bands has the optic axis perpendicular to the specimen plane. The right hand image is the same area when a first order red plate is also used. The blue and yellow colors show that the spherulites are negative. (See color insert.)

3.1.7.2 Polarizing Microscopy

Polarizing microscopy involves the interaction of materials that have anisotropic optical properties with polarized light. The full theory of this interaction is complex. Simple results can be obtained in the polarizing microscope without much theory, but quantitative measurement requires basic understanding. General optics texts (see Section 3.1.1) cover the topic [1–5]; there are also specialist books at introductory [20] and higher levels [50, 51]. Of several texts that deal with polarizing microscopy [6, 7, 19, 52, 53], Hartshorne and Stuart [19, 54] contains a section on polymers, practical information, and also covers theoretical tools, whereas Hemsley [53] is entirely concerned with polymer applications.

If the specimen is isotropic, the field of view in crossed polars is ideally completely dark. A birefringent specimen will give an altered polarization state for light leaving the specimen. This will have some component transmitted by the analyzer, so it appears relatively bright. If one of the principal planes of the specimen coincides with the incident plane of polarization, there is no component in the other plane. Retardation cannot then affect the state of polarization. Thus the specimen will be dark or *extinguish* at four positions 90° apart when it is rotated. It will be brightest at 45° from extinction, when the two waves excited in the specimen have equal amplitude. This is the standard orientation for measuring retardation ($R = t\Delta n$).

There are three reasons why an anisotropic material may be dark between crossed polars:

1. The sample may be oriented so that it is being viewed down an optic axis and so appear isotropic.
2. The sample may be in an extinction position.
3. In monochromatic light, R can be an exact multiple of λ so that the phase change is $2m\pi$, where m is an integer. The incident plane polarized light is then reformed with no change.

In white light, this last condition can be met for some colors and not others, so the specimen becomes brightly colored. If the birefringent specimen is wedge shaped and viewed with white light in crossed polars, it will show a sequence of polarization colors.

The Michel–Levy chart of these polarization colors can be found in many texts, [7, 50, 54, 55], and Delly [56] has made a detailed comparison of the available versions. The chart can be used to estimate the retardation of a specimen within the range 200–1600 nm. For example, first order red is at 575 nm; third order green is at 1250 nm. At higher retardations the color is pale pink or pale green due to overlap of the orders. A material that is colored or has a birefringence dependent on the wavelength gives nonstandard colors that will be useless for determining retardation.

Retardation is measured more accurately with compensators. These are calibrated adjustable birefringent objects inserted to cancel (compensate) the effects of the specimen. To use them, the sample and the compensator are set at ±45° between crossed polars. The compensator is adjusted until the specimen appears dark, when the retardation of sample and compensator are equal and opposite, R(net) = 0. If the polarization color change on inserting a compensator shows an increase in retardation, then R(net) > R(specimen), and the compensator is adding its effect to that of the specimen. The specimen must be rotated by 90° so that the effects are subtracted as required.

Several types of compensator are available, covering different ranges of retardation. The simplest is the *quartz wedge* with range of R of 2000 nm. This is exactly what it says it is—a wedge of birefringent quartz. Some wedges are calibrated for retardation but most are used for rapid qualitative observations.

For large values, the *Berek* compensator is a disk of uniaxial material cut on the basal plane (normal to the optic axis) and mounted on a ring which can be tilted about a horizontal axis. When horizontal it has no retardation; when tilted, it can reach 2000 nm if it is quartz and more than 50,000 nm if it is calcite.

The *Elliptic* (or *Bräce–Köhler*) compensator is used for measuring the smallest retardations, with a maximum range of 20 or 50 nm and accuracy to 0.1 nm [57]. It has a birefringent plate that revolves in the horizontal plane. The *Senarmont* compensator has an intermediate range, to 150 nm. It uses a λ/4 plate and a rotating analyzer. For an explanation of how these work, see [19, 51, 52, 55].

Accurate measurement of retardation requires the use of monochromatic light (usually Hg green, 546 nm), but in monochromatic light R(net) = 0 cannot be distinguished from R(net) = λ or 2λ. White light is required first, as in white light (normally) only R(net) = 0 is black. If the birefringence changes significantly with λ, as it does in several polymers, the black band may not be at R(net) = 0 [19]. Using white light will then give the wrong answer. For example, the black fringe will "jump" one position every 6λ of retardation if the sample is poly(ethylene terephthalate) (PET) and the compensator is quartz [58]. One solution given by Hartshorne [58] is to prepare a wedge at the edge of the sample and follow the R(net) = 0 fringe inward to the thicker part of the sample as compensation is increased.

Two devices of fixed retardation are generally useful, and they are the *quarter wave plate*, R = 150 nm = λ/4 (yellow light), and the *first order red plate*, R = 575 nm = λ (yellow light). The λ/4 plate, when set at 45°, turns linearly polarized light into circularly polarized light. As has been mentioned, using crossed circular polarizer and analyzer means that there are no extinction positions. The first order red plate is also known as the *sensitive tint plate*. It is used to find the sign of the birefringence of objects of small retardation, ε (these have a gray color in crossed polars). With the first order plate in place, the net retardation is (575 + ε) nm, which is second order blue, or (575 − ε) nm, which is first order yellow. The eye is very sensitive to color changes in this range, so a very small value of ε has a visible effect.

Figures 3.15 and 3.16 show spherulites in crossed polars and with the sensitive tint plate. In crossed polars, the spherulites are bright except at the four perpendicular radial directions where the crystals are in the extinction position. High density polyethylene (see Fig. 3.15) produces spherulites that are unusually perfect, and also have circumferential dark bands. The bands are regions of apparent isotropy where the optic axis is perpendicular to the specimen plane. With the first order red plate placed so that its slow direction is at +45° (top right to bottom left of the image), the

Imaging by Scanning Electron Beam

FIGURE 3.16. Two images of a thin melt cast film of polycaprolactone. As with Fig. 3.15, the region is 200 × 250 μm and the left hand image was taken in crossed polars. The spherulites in this material are much less regular and some show colors under crossed polars, indicating a thicker film or a larger birefringence. The right hand image is the same area when a first order red plate is also used. The colors can still show that these spherulites are also negative. (See color insert.)

spherulites are clearly blue in the upper left and lower right quadrants and yellow in the other two. Blue means that the retardations of red plate and specimen are added, so that the larger refractive index ("slow") is perpendicular to the radius and the spherulite is negative.

Figure 3.16 shows similar images of polycaprolactone, which gives reasonably large but more realistically irregular spherulites. Note that some show colors in crossed polars. The colors that arise when the first order red plate is added do not mean anything unless the sample has only a small retardation and appears gray in crossed polars. The spherulites in this figure do not look at all radially symmetric, but a comparison of the two images shows that they are also negatively birefringent.

3.2 IMAGING BY SCANNING ELECTRON BEAM

3.2.1 Probe Formation

The basic purpose of the lenses in a scanning electron microscope (SEM) is to form a focused beam spot at the specimen. By simple geometry, the diameter of the focused spot would be (source diameter × demagnification). However, diffraction and lens aberrations will increase the focused spot size, just as they increase the resolved feature size in image formation by lenses (see Section 3.1.4). The tungsten filament or lanthanum hexaboride (LaB_6) source has a diameter of between 5 and 50 μm, and three condenser lenses are normally used. The effective source size from FEGs is only 5–15 nm (see Table 2.2), so only one stage of demagnification is required, but two or three lenses may be used to give the optical system flexibility. The divergence angle of the beam increases as the source is demagnified, so the electrons are at the largest off-axis angles within the final condenser. This is then the critical lens, as the objective lens is in a TEM (it is often called the objective lens of the SEM, as its control will be labeled "focus") and the limiting aperture for the system is within this lens.

The STEM is in principle optically the same as the SEM, but the TEM/STEM is a little different. Here a special objective lens can be highly excited, so that it acts as two lenses in STEM mode [10]. The part before the specimen acts as a final condenser and the remainder acts as a magnifying objective lens. This lens reduces the divergence of the transmitted electrons, allowing their transfer to the transmission detector, typically located below the fluorescent screen. In TEM mode the lens power is reduced and it acts as a normal objective. The following discussion of the "final condenser lens" will include the pre-specimen field of the TEM/STEM objective.

It is important to remember that image resolution depends on the interaction volume, which may be much greater than the probe diameter (see Section 2.3 and Section 3.2.2). Because brightness B is fixed by the electron source, the geometric (or Gaussian) focused spot size d_g can be given in terms of the probe current i using Eq. (3.14).

$$d_g = \frac{2}{\pi\alpha}\sqrt{\frac{i}{B}} \qquad (3.14)$$

α is the (semi) angle of convergence of the probe at the specimen. An estimate of the final probe diameter d is taken to be the sum in quadrature of the diameters given by all the limiting factors:

$$d^2 = \frac{4i}{B(\pi\alpha)^2} + \frac{1.83}{E\alpha^2} + \frac{C_s^2\alpha^6}{4} + \left(\frac{\Delta E C_c \alpha}{E}\right)^2 \qquad (3.15)$$

Here ΔE is the energy spread in the beam in eV, C_s the spherical aberration coefficient of the final condenser lens, and C_c its chromatic aberration coefficient.

The first, geometric, term is related to noise in the image through the probe current, i; the next two are the diffraction and spherical aberration limits to resolution that were discussed in Section 3.1.4.1 for the TEM. The last term is the probe size due to chromatic aberration. The theoretical minimum probe diameter d_{min} can be calculated by finding the value of α that minimizes this expression [10, 59]. This is a lower limit to the size of the resolved detail. Alternatively, the equation can be used to calculate the current obtainable in a probe of a given size.

If chromatic aberration is neglected and the beam current is very small, d_{min} approaches the calculated minimum size of the resolved detail in TEM, $(C_s\lambda^3)^{1/4}$. Very high resolution can then be obtained when C_s and λ are as small as possible. Realistically, chromatic aberration cannot be neglected, especially if the beam energy E is low. With a tungsten filament source, for which ΔE is ~2 eV, and a lens of large C_c, the chromatic term can completely dominate the probe diameter. High resolution at low voltage, say 1 kV, requires an electron source of lower ΔE, a lower value of C_c, or both [60–63].

Because C_s, C_c, and the focal length are generally of the same magnitude, high resolution requires a short focal length final condenser. Figure 3.17 shows three designs for this lens. (Designs involving aberration correction are discussed in Section 6.2.4.) Of these three, the "in-lens" design (Fig. 3.17B), where the specimen is completely immersed in the magnetic field of the lens, gives the smallest aberration coefficients. However, the sample cannot be more than a few millimeters in size, and there is no space near it for detectors. Special stages are limited, as in the TEM. The secondary electrons spiral up in the magnetic field of the lens and can be collected above the lens.

The snorkel design (Fig. 3.17C) is a good compromise, with low aberration and good specimen access. There is room for a detector below the lens as well as above it, and as these will give different images, the better one can be chosen for each specimen. Most high-resolution SEMs (HRSEMs) use a semi-in-lens design like this. It is possible, for example, to use such a lens for high resolution with short working distance at low voltage (1 kV) and then as a weaker lens with long working distance at 15 kV. This would give enough room to insert an x-ray detector, and for x-ray analysis the beam spot size is not important.

The final beam-limiting aperture was originally in the gap of the final lens, but only the design in Fig. 3.17A has sufficient room. The apertures shown are "virtual apertures" at an optically equivalent position, and their actual diameter must be divided by the demagnification of the final lens to give the effective aperture diameter D.

Probe sizes calculated on the basis of Eq. (3.15) are shown in Table 3.4. The columns compare a conventional SEM with one using

Imaging by Scanning Electron Beam

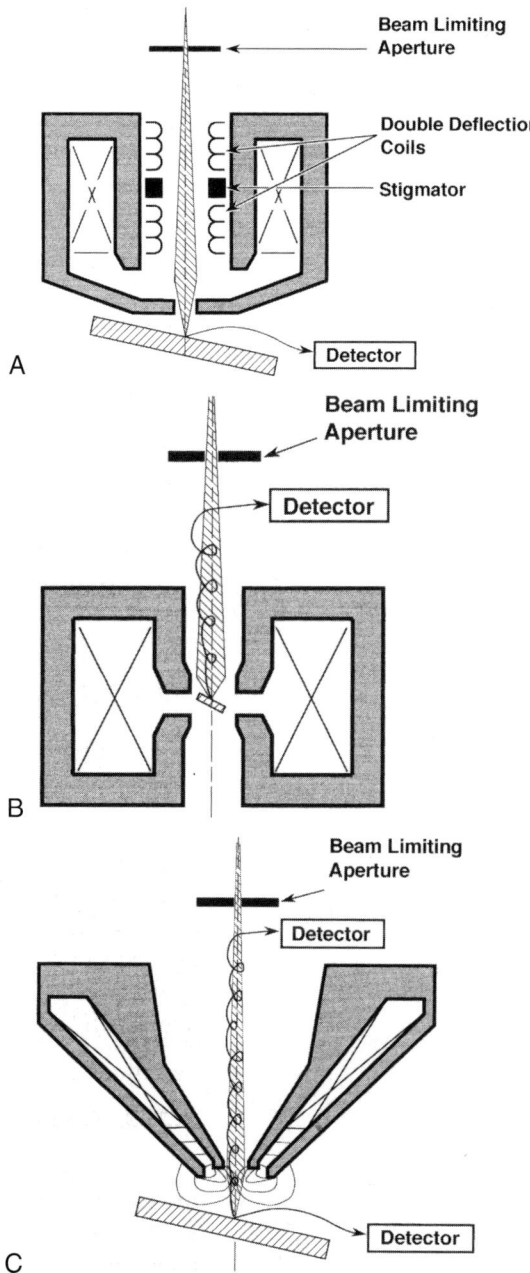

FIGURE 3.17. Three designs for the probe-forming lens in the SEM. (A) The conventional "pinhole" lens, with long working distance, but large aberrations. (B) The "in-lens" design, which has small lens aberrations, but restricted sample space. (C) A "snorkel" lens, a "semi-in-lens" design. Here, the magnetic field extends to the specimen giving small aberrations while allowing a large specimen. (Adapted from Goldstein et al. [38], © (2003) Springer; used with permission.)

the same lens but a cold field emission (CFE) source (a field emission scanning electron microscope; FESEM) and a HRSEM with the CFE source and a short focal length lens. When the probe current is set to 1 pA, a suitable value for recording an image, the FESEM maintains good resolution to 5 kV and the HRSEM all the way to 1 kV. The optimum probe divergence angles range from 1 to 10 mrad. More detailed calculations indicate that the final probe diameter may be as much as 50% less than this estimate [64]. Figure 5.49 is a direct comparison of images of a polymer blend obtained at 5 kV with a LaB_6 electron gun and with a FEG; the image from the FESEM contains more information.

When aberrations are dominant (e.g., for the conventional SEM at low beam voltage), the best divergence angle is small. Reduced aberrations in the HRSEM allow larger divergences. At 100 pA, which might be required for dynamic microscopy or analysis, the probe size is considerably degraded at low beam voltages. Here Eq. (3.15) fails for the CFE source, because it assumes that it is always possible to increase the current by increasing α. But the total current available is limited so the effective brightness drops as the current requirement is increased.

Table 3.4 uses "reasonable parameters" but individual instruments and adjustments may be considerably different [38, 65]. Note also that the optimization of α may not be fully possible as the divergence of the probe in the SEM is controlled by the working distance S and the effective final aperture diameter D; $\alpha = D/2S$. Both D and S have a restricted range of values in a given instrument. Manufacturers show what is to be expected from their SEMs for a given final aperture, accelerating voltage and working distance.

Typical values for D and S are 0.2 mm and 20 mm, giving $\alpha = 5$ mrad, close to the theoretical optimum. The rays passing through the final lens are also deviated by the scanning coils. This angle is small when the field of view is small at high magnifications, and at low magnifications, distortion is more important than spherical aberration. As when imaging with lenses, the depth of field is the resolution divided by the divergence angle α. At a resolution of 5 nm and $\alpha = 5$ mrad, the depth of field is 1 μm.

TABLE 3.4. Minimum probe size, d_{min}, at a beam current of 1 pA and 100 pA at different beam voltages

	Conventional SEM tungsten filament		Cold field emission gun		HRSEM	
ΔE (eV)	2		0.3		0.3	
C_s (mm)	20		20		2	
C_c (mm)	10		10		1	
Probe current	1 pA	100 pA	1 pA	100 pA	1 pA	100 pA
Beam voltage (kV)	d_{min} (nm), at α_{opt} (mrad)		d_{min} (nm), at α_{opt} (mrad)		d_{min} (nm), at α_{opt} (mrad)	
30	5.7 at 5.1	23 at 10	1.3 at 4.0	2.0	0.8 at 7	1.0
5	15 at 2.5	40 at 7	3.7 at 4.0	5.6	1.6 at 8.5	4.0
1	35 at 1.2	90 at 3.5	12 at 2.8	60	3.8 at 8.5	40

3.2.2 Probe-Specimen Interactions

High energy electrons entering a solid are scattered in two ways: they interact with the electrostatic field of the positively charged nuclei and with individual electrons. In the first case, *elastic scattering*, the electrons change direction but do not lose energy. The most probable scattering angle is small, a few degrees, but large changes of direction do occur, and after a number of collisions, some electrons can be turned around and leave the specimen. They have become backscattered. In the second case, the incident electron does lose energy, so it is *inelastic scattering*. Most often the energy transfer is small and the scattering angle is very small.

If a larger amount of energy is transferred to an inner shell electron, it leaves the atom producing an inner shell vacancy. The atom can de-excite either by emitting a characteristic x-ray, which is likely to leave the solid and be detectable, or by emitting an Auger electron, which will only escape if the target atom is very near the surface. These processes all occur statistically, and the interaction is modeled using Monte Carlo methods. By having a computer follow the random tracks of thousands of electrons, it is possible to predict not only the number of backscattered electrons, secondary electrons, and x-rays but also their spatial distribution—where they leave the surface—and for x-rays, the distribution of depths where they were produced [38, 61].

The inelastic scattering can be modeled as a continuous process causing an energy loss/unit distance (dE/ds) [66]. At 20 keV, this is about 1 eV/nm for polymers, density ~1 g/cm^3, and 7 eV/nm for silver, density ~10 g/cm^3 [38]. The energy loss rate increases less than linearly with (electron) density because some excitations are not available in heavy elements. For example, Ag Kα is >25 keV.

The higher energy loss rate at low electron energies means that the electron range falls. That is, there is less penetration of the incident beam and a smaller interaction volume. The range of electrons, R, can be approximated as $R = aE^n$, where E is the electron energy, and a and n are constants. According to the general empirical expression of Kanaya and Okayama [38, 67], for a carbon of density 1 g/cm^3, R in nm, E in keV, $a = 67$, $n = 1.4$ This gives R ~ 70 nm at 1 keV and 3 μm at 10 keV. Experimental data for carbon is fitted by $a = 60$, $n = 1.4$, under the same conditions [60], giving 60 nm at 1 keV and 1.5 μm at 10 keV, in approximate agreement with the general expression.

Figure 3.18 shows this dramatically using simulation of electron trajectories in the sample. The region containing the electron tracks is the interaction volume. This serves as a reminder that the interaction volume controls the resolution of backscattered electrons and x-ray analysis (see Section 2.3, Fig. 2.4 and Fig. 2.5). The spread is particularly severe in low density, low atomic number material such as polymers.

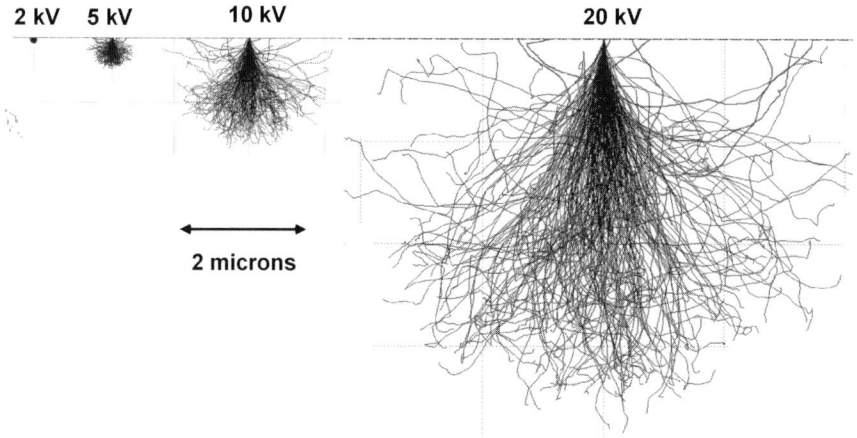

FIGURE 3.18. Simulation of the paths of 200 electrons entering a polymer (nylon-6) at different beam voltages. At 10 kV, backscattered electrons leave from an area 2 μm across. Below 2 kV, the interaction volume is too small to show on this scale. Drawn using CASINO [75].

3.2.2.1 Backscattered Electrons

The probability of elastic scattering varies strongly with atomic number, as Z^2, but the probability of backscattering, η, varies much less. For 20 keV electrons it is 0.06 for carbon and increases to 0.5 for silver [68] (shown in Fig. 3.21 of [38]). If the scattering is very efficient, in a short distance into the specimen, half the electrons will be going forward and half backward. Thus the backscattering coefficient saturates near 0.5; there is only a few percent difference between values for silver and uranium.

Backscattered electrons from such heavy elements typically do not travel far in the material before leaving it again. They have not lost much energy due to inelastic scattering, so their most probable energy is >90% of the incident beam energy [69]. For carbon, most backscattered electrons have penetrated quite deeply and the most probable energy is only half the beam energy. These backscattered electrons are produced in a significant fraction, perhaps the top one third, of the total interaction volume. They can therefore come from several micrometers deep in polymers when the beam voltage is 10 kV or more [38, 59] and leave the surface of the sample over a wide area.

There is generally little change in backscattering coefficient with beam voltage. This may be surprising, as higher energy electrons penetrate further into the specimen, but they are also elastically scattered less often. The result is that for high-Z elements their tracks have a similar shape, though on a different scale. The number of tracks leaving the specimen surface does not depend on this scaling. For carbon (and other low-Z elements), the pear-shaped interaction volume (see Fig. 2.4 and Fig. 3.18) does change shape, and the backscattering coefficient increases to 0.1 at 5 keV. Low beam voltage will always reduce both the depth and area from which the backscattered electrons come [60]. At very low voltage, the surface condition is important. However, at 20 keV a thin layer of gold on a sample of low atomic number will have little effect on the backscattering yield or resolution.

3.2.2.2 Secondary Electrons

The energy spectrum of electrons emitted by a specimen in an electron beam has two maxima. One is at high energy where most of the backscattered electrons are, and the other is at only a few eV. These are secondary electrons, which by definition include all electrons emitted at less than 50 eV. The number emitted divided by the number of incident electrons is the secondary emission coefficient δ. For 20 keV incident electrons in the SEM, the coefficient is close to

0.1 for all elements except carbon (0.05) and gold (0.2), although individual results do vary significantly [70] (shown in Fig 3.21 of [38]). Thus, a thin gold coating on a polymer specimen will increase the yield of secondaries without degrading their resolution. The general insensitivity to composition has been attributed to a contamination layer that exists on all specimens not cleaned and kept in ultrahigh vacuum (UHV) [38]. Certainly, the low energy of secondary electrons means that they are very sensitive to surface conditions.

Secondary electrons are divided into three main groups depending on their source (see Section 2.3.2.2). Those arising from scattering of the incident beam are called SE_1. SE_2 arise from the backscattered electrons as they leave the specimen and SE_3 from backscattered electrons when they strike the lens, the walls and other surfaces in the vacuum chamber. Here δ counts electrons leaving the specimen, so it includes SE_1 and SE_2 but not SE_3. In high-Z samples where the backscattering coefficient is large, secondaries derived from backscattered electrons can dominate the secondary signal.

The secondary emission coefficient δ increases as the beam energy falls from 20 keV, and this is to be expected. At low beam energies, more of the interactions occur near the surface, so more of the low energy electrons produced can escape. Eventually at very low beam energy the coefficient falls, as the incident electrons do not have the energy to do very much. [60, 71].

Figure 3.19 is a schematic of total electron yield as a function of beam energy. For SEM imaging, the important feature is that the total emitted electron coefficient exceeds one, $(\delta + \eta) > 1$, for a range of beam energies between E_1 and E_2. These crossover energies where $(d + \eta) = 1$ are points where the sample does not become charged in the beam and ideally does not need a conductive coating. The upper crossover energy, E_2, is a good choice for imaging, being in the range 0.5–2 keV for organic materials and 2–4 keV for inorganics [38, 72]. E_1 is too low at <200 eV for practical imaging at the moment, but E_2 has another important advantage: it is a stable point.

At a high incident beam energy, $E > E_2$, the number of electrons emitted from the specimen

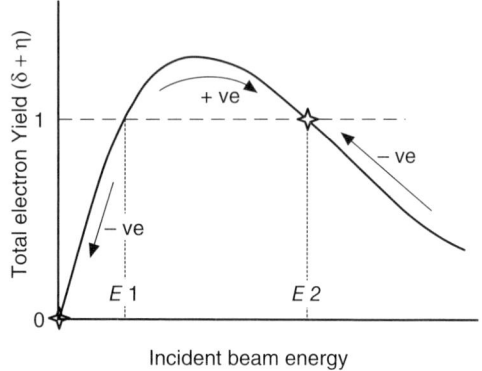

FIGURE 3.19. Total electron yield as a function of incident beam energy. When yield is >1, at $E_1 < E < E_2$, the sample becomes positively charged and the incident energy increases. at $E < E_1$ and $E > E_2$, the sample becomes negatively charged. The stars indicate stable points at 0, E_2.

is less than the number arriving. The specimen in the beam therefore becomes negatively charged, and this potential repels and slows down the electrons in the incident beam. As a result, the incident energy of the beam falls and continues to fall until it reaches E_2, when no more charging occurs.

Conversely, at low beam energy $E < E_2$, the yield is more than 1, so the specimen becomes positively charged. The charge attracts the incoming primary beam electrons, so that they arrive with more energy. This process continues until the electrons arrive with energy E_2, which is thus a stable point. The other crossover point E_1 is unstable. If the incident beam energy $E > E_1$, the specimen becomes positively charged until the incident energy reaches E_2. If $E < E_1$, the specimen becomes negatively charged until the beam is completely repelled. The other stable point is at $E = 0$, $(\delta + \eta) = 0$: no electrons in and no electrons out!

In practice it is found that operation at or near to E_2 with an insulating sample always reduces charging considerably, but for an entirely stable image it may still be necessary to apply a thin conductive coating. As both δ and η increase if the sample is tilted (see Section 3.2.3), the crossover energy will depend on sample tilt as well as the composition [60, 71, 73]. A tilted specimen surface releases more secondaries, and will have a higher value of E_2 than a specimen with no tilt. Exact matching at every point is thus not possible for any sample that shows topographic or atomic number contrast. The stability of E_2 will allow each region to become stable without exact matching, but only when there is a different surface charge at the different points. Large local charge differences may disrupt the beam or be unstable.

3.2.2.3 X-rays

When high energy electrons bombard a solid, characteristic x-rays are produced. Each requires inner shell ionization followed by x-ray emission. The ionization step requires the electron to have enough energy, $E > E_C$ where $E_C = 4.04$ keV for CaK x-rays, for example. The ratio E/E_C is the *overvoltage*, u. The probability of ionization goes as $\ln(u)/u$, zero at $u = 1$, maximum at $u = e$. Because electrons lose energy as they go through the sample, the maximum production of x-rays for a beam falling on a solid sample is at a higher energy, $u \sim 4$. The x-rays come from the region of the interaction volume where $E > E_C$. For $u \geq 4$, this will be a large fraction of the total interaction volume (see Fig. 2.4) and so the spatial resolution will not be good.

Inner shell ionization is rare, and the number of detectable x-rays per ionization, the *fluorescence yield*, is generally low, 0.1 for CaK. So even at the optimum beam voltage, x-ray production is not efficient, with only one electron in 10^4 or 10^5 producing a detectable characteristic x-ray [38, 74]. The efficiency falls for low atomic number elements. X-rays are thus too few for good image formation and are used mainly for elemental analysis.

Quantitative microanalysis is difficult in principle but largely automated for homogenous samples [38, 59, 69]. If the sample has layers of thickness comparable with the penetration depth, or particles of size comparable with the interaction volume, analysis can be difficult.

3.2.2.4 Interactions in the STEM

Thin specimens are used in the STEM so there is little lateral spreading of the beam. The interaction volume is limited to the probe diameter and the sample thickness. This is the basic difference between the SEM and the STEM. X-ray resolution can go from >1 μm in the SEM to <1 nm in the STEM [74]. As the interaction volume is so small, a very large incident beam flux is needed to produce a statistically significant x-ray analysis in the STEM. A bright source and a probe-forming lens with low C_s will be required to get enough current into the smallest spots. No polymer can withstand such irradiation without change, but the heavier elements being analyzed may stay in place even if the matrix is altered.

Backscattering from a thin sample will also have higher resolution and lower efficiency than that from a thick sample. However, scattering of all sorts is strongly peaked in the forward direction, so a very much larger signal produced by the same elastic scattering interactions appears below the specimen. The STEM thus uses forward rather than backward scattered electrons; these electrons pass through the specimen, and their interactions with it are the same as those in the TEM. The STEM can use electrons scattered to larger angles, and electrons that have lost significant amounts of energy, to form images. This is because there is no objective lens and therefore no spherical or chromatic aberration to consider for transmitted electrons.

For secondary electrons the STEM can act as a high-performance SEM. A very thin sample has no low resolution background due to secondaries produced by distant backscattered electrons, but secondaries can come from both surfaces. This means that topographic information from both sides of the specimen is superimposed on the image, making interpretation complicated. If the specimen is thicker, but still transmits the electrons in a spread beam, there will be a large low resolution signal from the back surface of the specimen. In this case, it would be better to mount the specimen on a solid substrate to make it nontransparent. Current interest in the STEM is largely as a very high resolution instrument, and it is not yet clear what the applications to polymers may be.

3.2.3 Image Formation in the SEM

Both secondary electron and backscattered electron emission depend strongly on the surface topography. If the beam falls on a tilted surface or onto a peak, more of the interaction volume is near to the surface, so more SE_2 and SE_3 secondaries will be produced as more backscattered electrons can escape. Figure 3.20 shows this using simulated electron trajectories in nylon-6 coated with 4 nm of gold [75]. There is a very high density of tracks near the tilted surface and near the vertical edge. The effect of a nearby edge becomes important when it is within the radius of the interaction volume. In Fig. 3.20, the beam voltage is 5 kV, and this radius is about 500 nm.

Clearly the scale of the effect will vary strongly with beam voltage. Figure 3.18 shows this by simulation using different voltages, and Fig. 3.21 shows the effect experimentally, using a silicon sample produced by microlithography that has sharp perpendicular edges [60]. At high SEM operating voltages there is a lot of glare near the edges, and they are ill-defined. At 30 kV the glare spreads from one edge to the other, making the whole raised structure brighter. As the incident beam voltage is reduced, this effect goes away and bright lines appear at the edges of the feature. At the same time, the detailed surface features become clearer; at the lowest beam voltage, thin surface layers become prominent.

The contrast due to the raised feature drops as the beam voltage falls. This is because at low voltage, most of the beam-specimen interactions are very near the surface, so that the low energy electrons produced have a good chance of escaping as secondaries. Tilting, or a nearby edge, will increase this chance, as shown in Fig. 3.20. However, at high voltage nearly all these electrons are produced deep in the sample. They normally have no chance of escaping and being detected, but a vertical edge allows many deep electrons to escape as secondaries, giving a very high contrast signal.

The behavior of the usual Everhart–Thornley (E-T) detector, as usually set up, has been described in Section 2.3.2.2 (see Fig. 2.6). Secondary electrons from the specimen are all collected, no matter what the orientation of the specimen surface. Backscattered electrons are directional. In Fig. 3.21C the image comes from E-T detector, and the detector is in the recommended position at the top of the image as viewed. The low voltage image, Fig. 3.21D, shows that while the topographic contrast is altered, the image retains the appearance of a shadowed view of a three dimensional object, which makes the SEM images so easily interpretable. The image is clearer, but light and dark patches on the structure are likely due to surface contamination, not to intrinsic specimen features.

In the SEM the brightness and contrast of each image are routinely optimized as an ele-

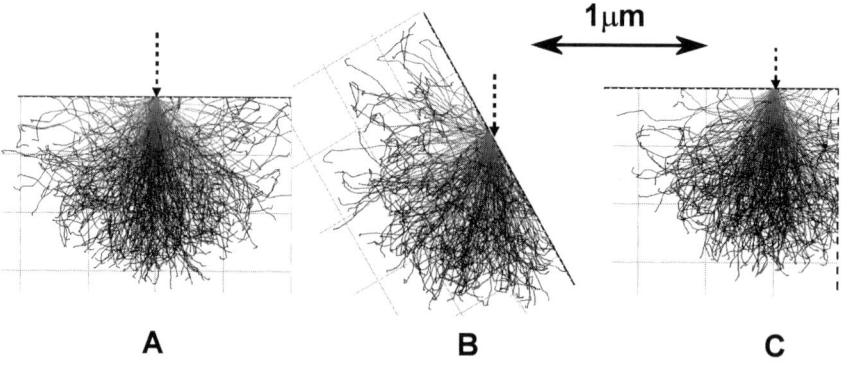

FIGURE 3.20. Simulated electron trajectories [75] for nylon-6 coated with 4 nm gold at 5 keV: (A) 0° tilt, (B) 60° tilt, (C) edge 350 nm from the point of incidence. The depth and width of this interaction volume is about 1 μm. A tilted surface produces more secondary emission because more of the interaction volume is near the surface. The tendency for electrons to scatter forward allows more backscattered electrons to escape. The trajectories are coded for energy, high energy being paler.

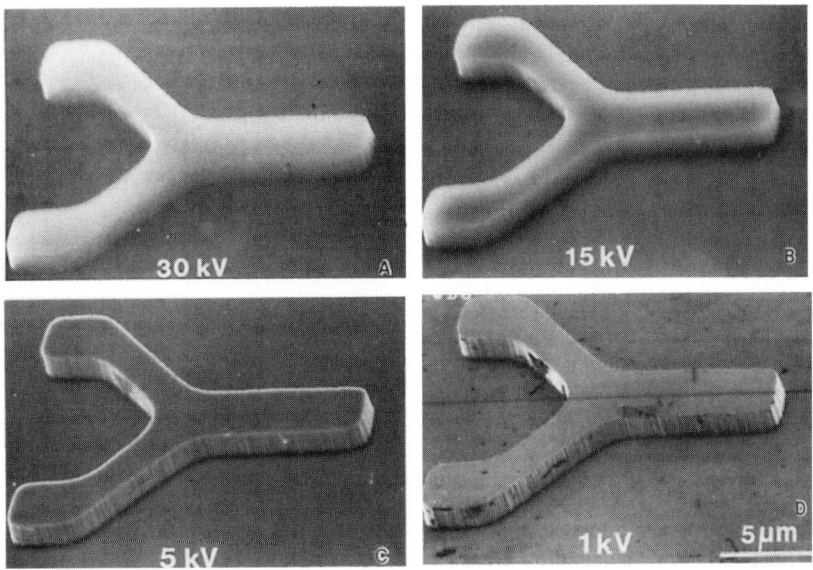

FIGURE 3.21. SEM images of a raised bar on silicon, tilted 45°, with the beam voltage (A) 30 kV, (B) 15 kV, (C) 5 kV, and (D) 1 kV. The excess emission due to the edges spreads right across the bar at 30 kV. At 5 kV and 1 kV, there is only a narrow and a very narrow bright line near the edge of the bar. (From Reimer [60]; used with permission.)

mentary adjustment. Typically a constant background signal is subtracted, and the amplification is increased to set the minimum and maximum signal levels to the dynamic range of the instrument display or storage system. This increases the contrast of any feature and the noise. The visibility limit of 0.05 contrast (described in Section 3.1.4) therefore does not apply to the original signal. If all the image structure is made visible, a detail must have one twentieth of the maximum image contrast. It is common practice to increase the contrast of the detail of interest even at the cost of lost information at high or low intensity levels. In this case visibility of a detail that has been resolved is controlled by noise.

For a perfect detector system, the noise is due to the random arrival of the detected particles, usually electrons or x-ray photons. It is $N^{1/2}$ if N particles arrive, so in a detector chain the noise depends on the link where the smallest number of particles is involved. Let this number be fN where N is the number of electrons striking the specimen. In x-ray imaging, the smallest number is the number of x-rays arriving at the detector, $f = 10^{-4}$ or less; in secondary electron imaging, $f \sim 1$.

Consider an image feature of intrinsic contrast C. The signal is CfN and the noise is $(fN)^{1/2}$ so the signal to noise ratio $k = C(fN)^{1/2}$. The probability p that a detail is visible through the noise is $k/(1 + k)$. The visibility limit is usually taken to be an 85% probability that the detail can be seen, so that $k \geq 5$. N depends on the area of the detail, the scan rate and the beam current. $Cf^{1/2}$ depends on the detail and the imaging mode. Take as an example a single pixel detail in a 512 × 512 image formed at television scan rates (25 or 30 frames s^{-1}) with a beam current of 100 pA: $N \approx 140$, and if $f = 1$, then for $k \geq 5$, $C \geq 0.4$. This limit is unaffected by an increase of contrast in image processing, as the noise is also increased. In the x-ray case, $f = 10^{-4}$ so even at $C = 1$, N must be 250,000 and the minimum size of a visible detail is about 45 × 45 pixels. This shows that while x-ray data may be collected at video scan rates, in practice any x-ray image is obtained by integrating the signal over a much longer time.

There are two ways to reduce noise in the image; increase the beam current or reduce the scan rate. Neither of these may be practical. At any resolution the beam current is limited by the brightness of the electron source (see Section 3.1.6) and the scan time is limited by practical concerns including the patience of the operator. For polymers there is also the real possibility that radiation damage of the specimen limits the total number of electrons that can be allowed to strike a given specimen area.

The effect of specimen charging in the beam may also depend on scan rate, with slow scans more sensitive than fast scans. In fast scanning mode, none of the image points have much time for the charge to leak away, so (with luck) all are at about the same potential. The uniform charge may subtly alter the image, but the effects are much less than at slow scan rates. In this case, there are large potential differences between different points on the image that were scanned at different times. The electric field caused by these potential differences can distort or destroy the image. Since a slow scan may produce a poor image due to charging, and a fast scan has too much noise, the answer is to make many fast scans and sum the images. This is very easy to do with digital image processing.

3.2.4 Low Voltage SEM

"Low voltage" for an SEM is not a very well defined concept. It used to mean operating with a beam voltage below 5 kV, whereas now it means a beam voltage of around 1 kV, and new instruments are extending this down to 0.1 kV. The drive to low voltage comes from a need to get the interaction volume down, so that all the signals have high resolution, and not just the relatively rare SE_1 electrons. Users of microscopes capable of high resolution over a wide range of beam voltages may often find that a low voltage in the range 0.5–3 kV gives the best image [62, 76]. The ability to image non-conducting specimens without the need for a metal coating (or at least with an extremely thin coating) is a very important bonus for low voltage operation in polymer microscopy.

Instrumental limitations originally made resolution at low voltages very poor. Two of these, discussed in detail in Section 3.2.1, are gun brightness and chromatic aberration. High brightness FEGs, both cold and thermal, are now much more reliable and consistent in their performance, and there are new lens designs with reduced chromatic aberration coefficient C_c. Other very important practical problems include contamination, stability, and the influence of stray fields, all generally worse at low beam voltages. General improvements in optical column and vacuum system design have reduced these effects.

Operation at low voltages has been described in detail [60] and in general texts on SEM [38, 59]. For minimum charging, the beam energy should be near to the crossover energy E_2 (see Section 3.2.2.2). This energy depends on the sample, and Table 3.5 shows a summary of data for many polymers from Butler et al. [61]. However, it is difficult to predict E_2 for a new material from this data.

Charging is a time-dependent phenomenon, and this provides a simple method for finding E_2 for a particular material and tilt. Select an area at low magnification, then switch to a higher magnification (say five times higher) for a few seconds, and return to the lower magnification. The more highly irradiated central area

TABLE 3.5. Experimental crossover energies E_2

Beam energy (keV)	0.9	1.0	1.1	1.2	1.3	1.4	1.5	1.6
Polymers	PS		PSulfone	N-6	HIPS	iPP	HDPE	PMMA
	SAN		PBT	N-66	PC		LDPE	
	PET		ABS	PE-PP	PE-VAc			
					EVOH	Kraton		

Source: Butler et al. [61].

that was the field of view at high magnification will still be charged more than the rest of the field of view. If it is brighter, then it is negatively charged and the beam voltage is too high, above E_2. If it is darker, the beam voltage is below E_2. Repeat until E_2 is found [77]. This is one of the nine methods for determining E_2 given in Reimer's text [60].

Early applications of low voltage SEM (LVSEM) to polymers were reviewed in 1995 [62]. Among many reports of the use of the LVSEM on polymers, Price and McCarthy [78] found the reduced interaction depth particularly valuable for imaging low-density polymer foams. Himmelfarb and Labat [79] compared images of polymer blends at low voltage and at higher voltages. The contrast was greater at higher voltages when one component of the blend was stained with a heavy metal, but otherwise visibility of small regions of either phase was better at low voltage. More recent practical examples are described in detail in Chapters 4 and 5.

3.2.4.1 Low Voltage SEM Detectors

When the beam voltage is very low, the detector systems have to be altered to accommodate the change. The secondary electrons, by definition, are emitted with the same low energy, and the standard Everhart–Thornley set with a grid voltage of +250 V accelerates them enough to make them excite the scintillator. This works well at 1 kV. But at very low beam voltages the electric field not only attracts the secondaries to the detector, and the backscattered electrons, but it can also affect the primary beam trajectory and thus distort the image.

Two similar detectors, one on each side of the sample chamber, will reduce the central field [80, 81] but it may be better to use a semi-in-lens design with a through-lens detector (see Fig. 3.17C). In this case the field that extracts the secondary electrons is the same one that is used to form the probe, so cannot disturb it.

A common detector for backscattered electrons at high energy is the "passive" scintillator with a large area and large collection angle. However, at <10 keV, the scintillator will not work well, as the electrons do not have enough energy to excite many photons. Similarly, for a solid state diode detector to work, the electrons must penetrate the gold coating, and to the buried p-n junction layer. Most designs that are efficient at 10–30 keV do not work below 2–5 keV. What then does work at low voltage? One system allows backscattered electrons to hit a target with high secondary yield (such as MgO) and detect the secondaries. That is, it increases the yield of SE_3 [60, 82]. If SE_1 and SE_2 are to be excluded from the signal, a grid at −50 V is placed around the specimen.

A channel plate electron multiplier detector also creates low energy electrons when a backscattered electron hits it, but the process occurs within the plate. A voltage of a few kilovolts across the plate accelerates electrons down the "channels"—many fine holes in the plate. Each time an electron strikes a wall of a channel, it emits more secondaries, producing a cascade [82, 83]. This is an efficient detector even at 100 eV. Some through-lens systems now have separate secondary and backscattered detectors at different positions in the column. They rely on the different energies of the two classes of electrons giving them different angular distributions after passing through the lens field [84].

X-ray microanalysis is still possible at low beam voltages, but it may not be easy. There are the advantages of better spatial resolution and reduced absorption correction, as the volume generating the x-rays is small and near the surface. The disadvantages are that the x-ray intensity produced is small, and not all elements can be analyzed. In principle, elements of higher atomic number can be identified using L- and M-shell x-rays, but these are much more complicated than the simple K-shell emissions.

The surface sensitivity may be useful sometimes, but it is more likely to be a problem; a thin conducting coat may be a significant part of the electron range, and without it the specimen may charge up. (Operation at voltages as low as E_2 is not likely for x-ray microanalysis.) Having a beam stationary on the sample for a long time to integrate the weak signals is a perfect way to produce contamination. Buildup of contamination can stop the electrons in a low voltage beam before they reach the specimen surface. Then the analysis does not relate to the specimen at all.

3.2.5 Variable Pressure SEM

The variable pressure SEM (VPSEM) operates with a gas pressure in the specimen chamber in the range 10–4000 Pa [85–87]. This makes it quite a different instrument [88]. As the normal column pressures (<1 mPa) are called "high vacuum," some authors call the VPSEM a "low-vacuum SEM." Low vacuum is a recognized term, but unfortunately its abbreviation cannot be distinguished from that of low voltage SEM. An important pressure for biological, polymer or mineral samples that must be observed wet and not frozen is the vapor pressure of water at room temperature, 3.2 kPa (25 torr or 0.46 psi). Specimens can be maintained at their normal water content in the specimen chamber when there is this pressure of water vapor, independent of the pressure of any other gases. A microscope that can operate at this pressure is called an environmental SEM, or ESEM [89] (ESEM is a trademark of FEI Inc. See Appendix VI.). Newbury [90] points out that the pressure limit for observing a wet specimen can be reduced to ~700 Pa by cooling to 2°C and suggests calling operation at <100 Pa VPSEM and >100 Pa ESEM. For imaging synthetic polymers, the lower pressures can be just as important, as ionization of the gas can prevent charging of insulators at high beam voltages.

Any increase of pressure in the gun chamber from its normal very low value will cause the electron gun to fail after a very short time. To protect it, the VPSEM contains some limiting apertures that restrict the flow of gas up the optical column from specimen to gun chamber. In the ESEM there may be as many as four apertures. Two are conventionally placed, separating the gun chamber from the column, and the column from the specimen chamber (the final aperture). Two more are placed just above the final aperture, dividing the vacuum system into five separate regions in all. Each of these has its own vacuum pumping system, and the result is a mechanically complex microscope with a sequence of pressures that the electrons experience as they move from gun to specimen: 10^{-5}, 10^{-4}, 10^{-2}, 10 and 1000 Pa.

The *gas path length* is the distance that the beam travels through the highest pressure region. It cannot be much more than the mean free path of the primary beam electrons in the gas or the focused probe will be lost as the electrons are scattered by gas atoms. Either the working distance—the separation of specimen and final aperture—is kept small, or (for EDS analysis) a tube can extend the high(er) vacuum region down toward the specimen. If the beam voltage is low or the gas pressure high, the restriction on working distance (or the effect on resolution) is more severe, as the electrons are more rapidly scattered. Beam voltages are typically kept at 10–20 keV. In the ESEM, the gas path length is 2–3 mm, while in the lower pressure VPSEM it is usually 10–15 mm.

The result of gas scattering is that part of the beam is unaffected, forming a fine probe, and the rest is scattered to hit the specimen over a very wide area, 20–200 µm in diameter. For imaging this gives a constant background, increasing noise and reducing contrast but otherwise not affecting the signal. However, part of the x-ray signal now comes from this wide area; the resolution will be degraded and spurious elements may appear in the spectrum [38, 90].

The gas in the specimen chamber has some effect on the primary beam electrons and the backscattered electrons, but a greater one on the low energy secondaries, as they have a very small mean free path. The secondaries ionize the gas atoms as they collide with them producing positive ions and more electrons. A positive voltage on a detector will make all the electrons drift rapidly toward it. If the attractive electric field is sufficiently large, (~100 V/mm), each drifting electron will be accelerated enough to gain the energy to cause more ionization at its next collision with a gas atom. As each collision can release more than one electron, the result is an amplification of the secondary electron current [86, 91]. Too high a field would cause avalanche and electrical discharge in the gas.

The detector system just described is analogous to a gas-filled detector used for x-rays. At low applied voltage it is like an ionization chamber, and when there is amplification, like a proportional counter. An x-ray proportional counter has a limited response rate, of about 10 kHz, but the x-ray systems are normally operated at pressures close to atmospheric, with local fields of up to 100 MV/m that give amplifications of several thousand times. These conditions can produce very high concentrations of positive ions by the central wire. The ions reduce the applied field, and the response rate is limited by the need for them to diffuse away. In the case of the VPSEM, the field is more uniform, and the pressure and the field are lower. All these factors tend to reduce local ion concentrations and thus increase the response frequency. However, it is still slower than the E-T detector (which cannot be used as it requires 10 kV on the scintillator, high enough to cause electrical breakdown of the gas).

The gas amplification will work best over only a limited range of pressures. Too high a chamber pressure makes the mean free path of electrons very small; a high field is then needed to accelerate them, and the high voltage required may not be practicable. Too low a pressure and the effect will be small because the electrons will have few collisions as they travel from specimen to detector. The exact form of the signal contrast is difficult to predict because of the complicated effects that local topography may have on the electric field. Nevertheless, the images look very like those of normal SEM secondary imaging.

At lower gas pressures, most VPSEM use a passive scintillator as the imaging detector. This is an efficient detector of backscattered electrons in the energy range used, and there are typically more backscattered than secondary electrons. Most VPSEM/ESEM images are taken at low magnification where the secondary electron resolution is that of the SE_2, that is, the same as for the backscattered electrons. The secondary electron image will therefore be very similar to the backscattered electron image.

Charging of the specimen is suppressed in the VPSEM, as the space above the specimen is full of ionized gas—charged particles of low energy. In particular, as insulating specimens normally charge negatively at these incident beam voltages, the positive gas ions are attracted to the specimen, neutralizing it. Local small differences in potential will be stably neutralized as the ions redistribute themselves. If conditions are far off, for example, if the gas pressure is too low, so there are insufficient ions formed, then there will be some negative charge on the surface. Similarly, if the pressure is high, there may be too many positive ions repelled to the surface [92] (see Fig. 5.25 of [38]). Typically charge balance requires a gas path length of about 10 mm at 30 Pa and 2 mm at 200 Pa, and these are typical operating conditions of the VPSEM and ESEM, respectively.

It is the combination of high surrounding gas pressure and no requirement for conductive coating that makes these instruments uniquely useful. Experiments that demand its use include the dynamic observation of reactions involving a solid and a gas or liquid, such as oxidation or other corrosion, as well as observation of wet specimens. However, if specimen charging is the only problem, a low voltage SEM may be a more practical solution.

The range of applications for the VPSEM/ ESEM is potentially very broad. Observations range from high temperature corrosion [93] and the melt processing of polymers [94] to wet delicate samples [95] and dental materials [96]. Oily as well as wet specimens can be imaged directly, and this is relevant to those involved in oil production [97] and to those planning to use polymeric materials to clean up oil spills [98]. Metal coating can improve the image in the VPSEM and may be applied to samples that are superficially dried but would still contaminate a high-vacuum system (see Fig. 5.87, Section 5.3).

3.3 IMAGING IN THE ATOMIC FORCE MICROSCOPE

3.3.1 Microscope Components

Figure 3.22 shows the essential elements of an atomic force microscope (AFM). These include a cantilever to support the tip, a system for detecting cantilever deflection with a feedback loop, and a scanner to control the relative

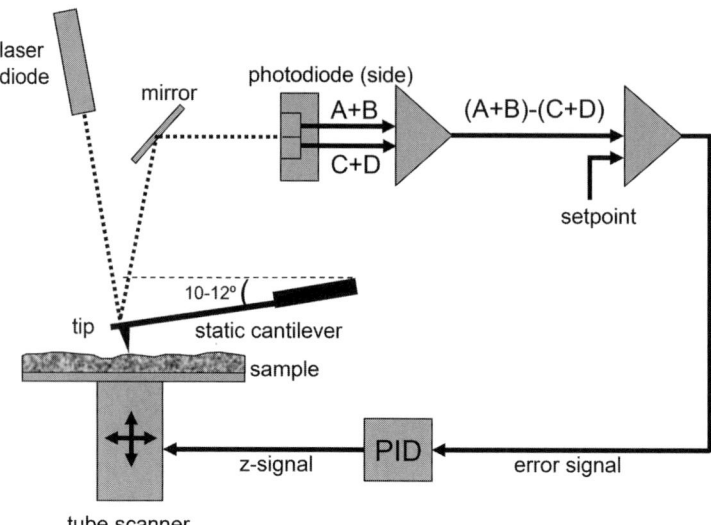

FIGURE 3.22. The principal components of an AFM. The instrument has the scanner attached to the sample and is operating in contact mode.

position of tip and specimen. The figure shows the scanner acting on the sample, and the signal in the feedback loop shows that the AFM is operating in contact mode.

3.3.1.1 Cantilever Parameters

For a given static deflection of the cantilever (as in contact mode) the applied force normal to the surface will depend on the bending stiffness or spring constant, k, given in units of N/m. The spring constant for a single-beam cantilever depends on the geometry of the lever (width, w, length, l, and thickness, t), and the material modulus, E.

$$k = \frac{Ewt^3}{4l^3} \qquad (3.16)$$

In contact mode AFM, k might range from 0.01 to 1.0 N/m; for intermittent contact mode AFM (IC-AFM), often called tapping mode AFM (TMAFM), it ranges from 3 N/m to 500 N/m.

For modes like IC-AFM where the cantilever is oscillating, there are two other properties that are important, in addition to the spring constant. One is the resonance frequency of the fundamental vibrational mode, ω_o. This depends on k and the equivalent mass, m^*.

$$\omega_o = \sqrt{\frac{k}{m^*}} = \sqrt{\frac{Ewt^3}{4l^3 m^*}} \qquad (3.17)$$

Generally, a high resonance frequency is desirable so that a number of oscillations takes place at each image point, even during rapid scanning.

The other cantilever property is the quality factor, Q, a measure of cantilever damping. This is inversely proportional to the damping factor. It is also the ratio of the resonance frequency ω_o to the full frequency width at half maximum of the resonance, defined in terms of energy.

$$Q = \frac{\omega_o}{\Delta \omega} \qquad (3.18)$$

In terms of amplitude, $\Delta \omega$ is the frequency range where the amplitude is greater than $1/\sqrt{2}$ of the value at resonance.

Silicon is a nearly ideal elastic material, so its intrinsic quality factor is very high. However, this quality factor is only realized if the device is operated in high vacuum, where Q may be 10,000 or more. Otherwise the damping effect of a surrounding medium reduces Q. In air, it is reduced to a few hundred. In water, it is reduced much more to 1–10.

The value of Q is important. For example, as IC-AFM operates by detecting the effect of the surface on near-resonant oscillations, a sharp resonance will be more changed by a small effect. Thus a high Q makes IC-AFM more sensitive. High Q cantilevers are more sensitive to attractive potentials near the surface than lower Q cantilevers. On the other hand, a high Q oscillator keeps on "ringing" and its amplitude is slow to respond to changes in its environment. This limits the speed at which the cantilever can be scanned. The Q defined by Eq. (3.18) is not strictly a cantilever property; it can be changed by active control of the oscillation [99–101]. In this case it may be increased when operating in a liquid [102] and reduced when operating in a vacuum, but more generally, it is adjusted to optimize the image formation. It may also be reduced to increase the speed of scanning [103].

3.3.1.2 Cantilever Deflection

Deflection detection schemes have been developed that use self-sensing piezoresistive cantilevers [104] or optical interferometry [105, 106]. But in nearly all practical systems cantilever deflection is measured with an optical lever system [107], as shown in Fig. 3.22. A small spot of laser light is reflected from the back of the cantilever and the reflected light is directed to an adjustable mirror. This in turn reflects the light onto a position-sensitive detector, a four-quadrant photo diode, shown schematically in Fig. 3.23.

Alignment is done with a static cantilever well away from any surface. The system is initially adjusted to maximize the total amount of light reaching the detector. This is done by centering the laser spot near the end of the cantilever and adjusting the alignment mirror. The laser spot is then centered on the detector by adjusting the postion of the photo diode detector vertically and horizontally. The cantilever is angled slightly (10° to 12°) off-parallel to the surface (see Fig. 3.22). This makes sure that stray light reflected from the surface does not reach the detector. In AFM operation, the normal deflection signal is then the output selected for further processing.

In contact mode, the value of this output is simply compared with a set point value. The difference, or error signal, is then conditioned using a proportional-integral-differential (PID) controller and sent to the Z-control.

3.3.1.3 Scanners

All scanned probe microscopes, including AFM, need to control the relative position of probe and specimen with extremely high precision, and they all use scanners made from piezoelectric ceramic material, usually lead zirconate titanate (PZT) to do this. Strictly speaking, it is the inverse piezoelectric effect that is used; an electric field is applied and the ceramic changes shape slightly. Two types of scanners used commercially are shown in Fig. 3.24. The original form was the tripod, with three independent piezo elements providing motion in the x, y and z directions. The most common today is the tube scanner, a single element that extends for z control and bends for x, y motion.

A typical tube scanner might have a sensitivity of 10 nm/V and a maximum voltage of ±500 V. This would give a range of motion of 10 μm. Voltage control to 1 mV would give positional control to 10 pm, and this 1 ppm precision would need 20 bit digital electronic control. Scanners of different range and sensitivity may be

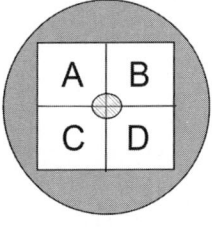

SUM Signal: A+B+C+D
Normal Deflection: (A+B) − (C+D)
Lateral Deflection : (A+C) − (B+D)

FIGURE 3.23. Schematic of the four-quadrant detector used in optical lever detection AFM.

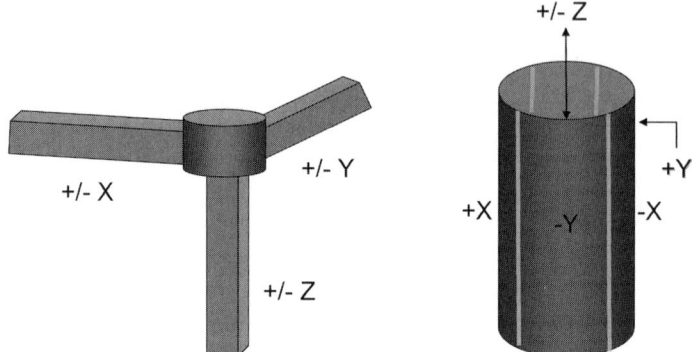

FIGURE 3.24. Two types of piezo scanners used in commercial AFMs: the tripod (left) and tube (right).

available for a given AFM, and then changing the scanner is equivalent operationally to changing the objective lens in an optical microscope.

Most commercial systems are sold with either tip scanning (tip attached to the scanner) or sample scanning (sample attached to scanner; sample cannot be too large). To get sensitivity in z and a wide range of motion in x, y, some designs use a "butted" approach with two tubes; an x, y tube glued to a separate z tube. Other designs split the motion control, scanning the sample in x, y and the tip in z.

Actuating the opposed pairs of electrodes on a tube scanner bends the tube so that a tip attached to the end moves in a circular arc. Thus motion in x or y involves motion in z. Ideally, in tripod scanner, the x, y and z motions are independent, but inaccuracy of construction or the use of a pivot point to amplify the motion in x and y couples the motions. Correction for this and other scanner errors is an important feature of AFM design, and image artifacts may result from uncorrected non-ideal behavior (see Section 3.3.7).

One additional piezoelectric element is the bimorph stack used to drive the cantilever oscillation in dynamic AFM. The stack is made from blocks of piezoelectric ceramic sintered with a central electrode. A voltage on this central electrode puts fields of opposite sign on the two halves (if the outside is grounded). One half expands, the other contracts, so the bimorph bends. Stacking gives large displacements, which are desirable for low voltage operation.

3.3.2 Probe-Specimen Interaction

3.3.2.1 Atomic Interaction

All atomic force microscopes operate by detecting, measuring or controlling the forces between the tip and surface [108–111]. The theory of AFM operation therefore begins with an understanding of the interaction forces between two solids at small distances [112]. The usual first step is to consider the interaction between two isolated atoms. A good approximation to the energy of this interaction is the Lennard–Jones potential $U(r)$, $U(r) = A/r^{12} - B/r^6$ and the force between the atoms is:

$$F(r) = -\frac{dU}{dr} = 12A/r^{13} - 6B/r^7 \quad (3.19)$$

Potential and force are shown in Fig. 3.25. Positive forces repel, and the first term dominates at short distances. The negative term is the van der Waals force that attracts and dominates at relatively large distances. For an AFM with an ideally sharp tip it might be better to consider the interaction of an atom with a plane. In this case, adding all the attractive forces from the more distant atoms makes the attractive force vary less strongly with separation, as $1/r^4$. But for a qualitative description of the interaction, the exact power law of attraction is not important. The essential form is a steep inner repulsive force region and a much larger region of weaker attractive force at greater distances.

The potential is at a minimum where the net force is zero, and this defines where contact

Imaging in the Atomic Force Microscope

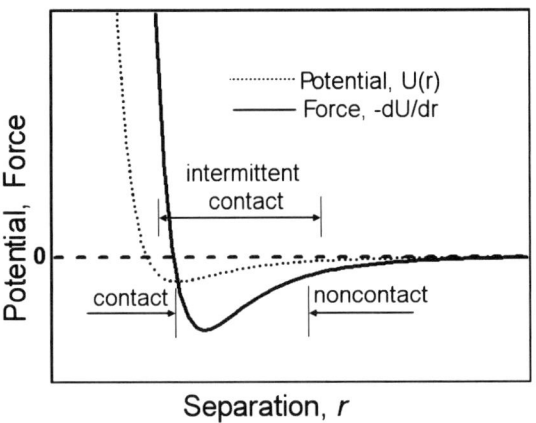

FIGURE 3.25. A Lennard–Jones potential between two atoms and the associated force between them. The separation regions for various AFM operating modes are indicated.

begins. Distances smaller than this are the regime of contact mode AFM, and well away from contact is the regime of noncontact AFM (Fig. 3.25). The IC-AFM or tapping mode spans both regimes.

A real tip containing many atoms is more complicated. "Contact" or zero net force for the whole tip will involve repulsive interaction of the atom or atomic cluster at the end of the tip and attractive interactions from atoms all around this. The local repulsive forces can be relatively large at zero net force if the attractive forces include electrostatic or capillary forces as well as van der Waals forces. A soft, weak sample could undergo significant local elastic or plastic deformation even at zero net force.

3.3.2.2 Cantilever Effects

The other complicating factor in the AFM is that the tip position is not under direct control. Instead it is on the end of a cantilever, and if the spring constant k of the cantilever is low, control of the tip position is weak. The effect is best thought of in terms of the force gradient $\frac{dF}{dr} = \frac{d^2U}{dr^2}$. If the cantilever base is slowly moved toward the specimen from far away, the tip approaches the surface and the force gradient due to the interaction increases, as in Fig. 3.25. Consider a small deflection of the cantilever dz towards the surface. The tip separation is reduced by dz, causing an increase in the attractive force, $(dF/dr)dz$. At the same time the deflection causes a restoring force kdz pulling the cantilever back in the other direction.

If $dF/dr > k$ the situation is unstable, as a small inward deflection causes the net attractive force to increase. The tip "jumps in" to contact [113, 114]. Figure 3.26 shows the jump-in and jump-out. There is hysteresis, even when the basic force interaction has no hysteresis.

The instability may be controlled by using a stiffer cantilever spring, where k is greater than the maximum value of dF/dr. However a stiff cantilever will lead to higher static forces in contact mode and thus more sample damage on soft materials. Active control of the cantilever can also be used to control the instability. For example, a magnetic coating on the cantilever and a nearby electromagnetic coil allows a variable force to be applied. A feedback loop can stabilize the cantilever deflection, effectively giving a cantilever with a variable spring constant [115].

If the cantilever base is kept at a fixed position and the tip is forced to move toward the sample surface, the restoring force keeps increasing, and the important factor is now not the force gradient but the sum of the forces.

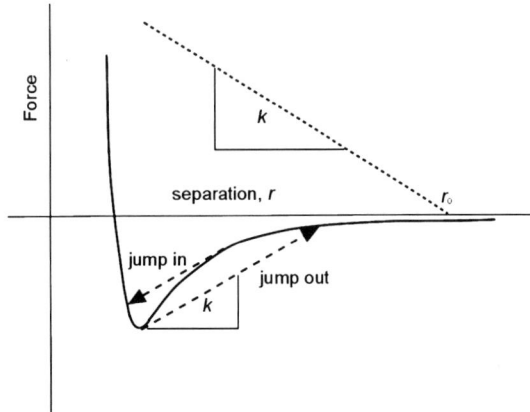

FIGURE 3.26. Force–separation curve, as in Fig. 3.25, with jump-in and jump-out shown for a cantilever of given spring constant. The dashed line is the restoring force for an oscillating cantilever, rest position r_o, which always exceeds the attractive force.

This case is relevant to an oscillating cantilever, as in IC-AFM. Figure 3.26 shows the case where the fixed position is at a far distance, and the restoring force always exceeds the attractive force, even though the spring constant of the cantilever is low. The oscillating tip does not jump in or out of contact.

3.3.2.3 Force-Separation Curves

The primary output of the AFM is the cantilever deflection signal, which can be calibrated to give the tip displacement and then the force if the spring constant k is known. The tip position is obtained by adding Δz, the motion of the z-piezo (either cantilever support motion or specimen motion), to the tip displacement. A measurement of tip deflection as a function of Δz with the x,y scanning turned off can thus be transformed into a force versus tip-sample separation curve. Figure 3.27A shows the cantilever deflection versus Δz plot for the simple case of an ideally rigid specimen. As this sample is ideally rigid, the tip cannot penetrate it and the motion of the scanner inward must be the same as the relative motion of tip and cantilever support. The curve in Fig. 3.27A thus contains its own internal calibration of deflection as displacement. A calibration of the spring constant is still required to convert the displacement to force.

The derived force versus tip separation curve is shown in Fig. 3.27B. Why is this not the same as the curves in Fig. 3.25 and 3.26? On the scale used, the forces in the noncontact regime before jump-in are very small, and since the surface is ideally rigid, the repulsive force curve is vertical and not merely steep. With a real specimen, the surface forces may be large enough to lead to deformation during the jump-in. At smaller separations, if the contact pressure in the repulsive regime exceeds the yield strength of the surface, then plastic deformation will occur. This is one of several possible types of force curves. They depend on many factors such as the nature of the interaction, surface forces, the deformability of the surface, and the medium between the tip and surface (air, liquid or vacuum) [116, 117]. Force curves can be used to study and verify fundamental theories of contact mechanics and adhesion on polymer surfaces [118].

3.3.3 Contact Mode AFM

The previous sections showed that for an ideally rigid substrate, the operation and interpretation of contact mode AFM is straightforward. The feedback set point (see Fig. 3.22) can be chosen to correspond to any net force that keeps the tip in contact with the surface (see Fig. 3.27). As long as the feedback system is sufficiently sensi-

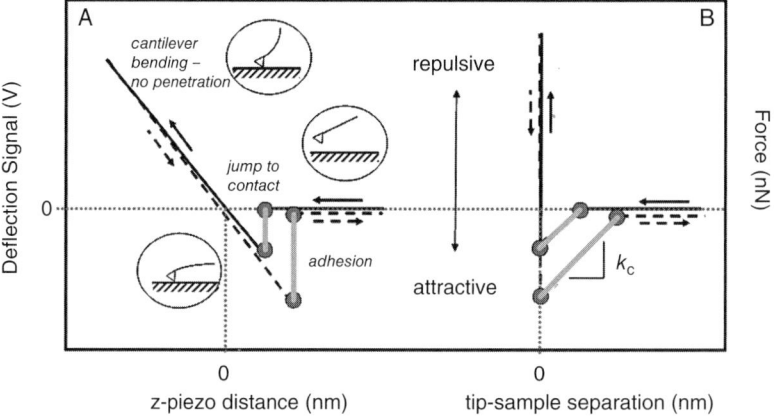

FIGURE 3.27. (A) Cantilever deflection versus z-piezo movement for a rigid surface. (B) The deflection signal has been converted to force and the displacement signal has been corrected for the cantilever compliance. The tip cannot penetrate the ideally rigid sample surface. The gray lines show the jump in and out; there is no data in these regions.

tive and responsive, keeping the force constant at the set point, the z-piezo position will trace out the sample surface topography.

For a real surface to appear rigid, the forces on it must be small. Using a cantilever with a small spring constant k keeps the forces small. A deflection signal in Fig. 3.27A corresponds to a smaller force in Fig. 3.27B when k is small. Contact mode cantilevers have $k < 1$ N/m, typically 0.2 N/m.

It would appear that small local forces can be ensured by adjusting the feedback set point to correspond to zero net force or to some point in the attractive region with net negative force. However, operation in the attractive region is usually too unstable a situation. If there is a sudden dip in the surface, the cantilever deflection increases to the point where the tip jumps out of contact before the feedback loop can correct it, and that is the end of that scan. As was pointed out above, zero net force can be a combination of strong repulsive forces at the contact point and attractive forces from the surrounding region.

If the surface is compliant, then the tip will penetrate the surface to some extent. Variations in compliance will appear as variations in surface height, with more rigid regions appearing to be higher. If the surface is weak, then the forces applied in scanning will cause permanent deformation and damage; in the worst case the image might have little relation to the true specimen structure.

What magnitude of force would be needed to cause damage to polymers? The answer depends on the resolution of the probe. Take a very simple model where the applied force is spread evenly over a contact area. A force of 1 nN will give a stress of 50 MPa when applied over an area of 5×10^{-18} m^2, a region ~2 nm across; 50 MPa is a reasonable value for the yield strength of a large sample of nylon or polystyrene. A surface region might be more mobile and weaker, or a small region away from defects could be much stronger than a bulk sample. Also, damage is more likely to be due to lateral forces than to the normal forces considered here. In any case, it is clear that very small forces are required for reliable atomic resolution in contact mode AFM.

3.3.3.1 Capillary and Electrostatic Forces

Most surfaces in air will have a contamination layer on the surface. Generally this is from water due to natural humidity and will be more pronounced on hydrophilic surfaces. The effect of this layer is to impart strong capillary forces that want to hold the tip against the surface. These forces can be significant and set a lower limit on the minimum force that can be achieved during imaging. They can be minimized by environmental control; by imaging in dry air or nitrogen or by imaging in water. Other contaminants may also be present on the surface of polymer materials as a result of processing (slip aids, oligomers, etc.) or from handling. Surface cleaning may be advised as long as the solvents do not alter morphology. Electrostatic forces can often hinder contact mode AFM analysis. Most polymers do not dissipate surface charge readily, and charge can accumulate in handling or when peeling apart blown film samples. The forces are long range and can attract or repel low spring constant cantilevers over large distances—millimeters not micrometers. Antistatic guns or α-particle emitters (Po^{210}) that are commercially available and commonly used in electron microscopy and microtomy may reduce this effect.

3.3.3.2 Lateral Forces and Frictional Force Microscopy

In contact mode AFM, where the normal force is the detected signal, there can also be a large contribution from lateral or shear forces. These can be significant enough so as to lead to deformation, abrasion and wear. These are generally not desired, however, this behavior has been used to probe the role of chain entanglements in polystyrene films as a function of molecular weight [119].

In *lateral* or *frictional force microscopy* (FFM) [120, 121], lateral deflection of the cantilever is the signal of interest. The single-beam cantilever is scanned in a direction perpendicular to its long axis to enhance the twisting motion. The "top to bottom" signal remains under feedback control to keep the normal force constant while the "right to left" signal is monitored (see Fig. 3.23).

The equivalent of the force-separation curve in FFM is called a *friction loop* [121]. It shows the lateral force and its hysteresis in a scan along a line on the specimen, forwards and then back again. Frictional force microscopy images have been used to generate friction coefficient maps of poly(methyl methacrylate)/polystyrene (PMMA/PS) blends [122, 123]. Interpreting friction images in terms of surface properties can be complex. The lateral force is sensitive to surface chemistry (adhesion) [124], orientation [125], humidity [126], elasticity [127], and dissipative processes or viscoelasticity [128]. Calibration of the lateral deflection in terms of frictional force is still a challenging problem [129]. Because the frictional force may depend on the vertical force, this must be known. Further, there is cross-talk between the two signals. The topographic signal is a force perpendicular to the local surface and will have a lateral component on a slope. Frictional force acting on the cantilever that is not perpendicular to its long axis can also influence the topography signal [130].

3.3.3.3 Force Modulation Imaging

A natural extension of contact mode imaging is to operate in a mode sensitive to material properties, *force modulation imaging* [131]. The setup for this is similar to contact mode, with three differences:

1. A small amplitude (1–5 nm), low-frequency (5–10 kHz) modulation in vertical position is applied to the cantilever support using a bimorph piezo stack.
2. The cantilever used is at least 10× stiffer than the usual contact mode cantilever.
3. The signal used to form the image is the amplitude of the cantilever motion at the modulation frequency.

As the setup is with the tip in contact with the surface and the applied modulation amplitude is small, the total force is always repulsive and the result is an oscillating force on the sample. The usual feedback loop (see Fig. 3.22) that keeps the mean deflection (or force) constant is in operation, so that the sample topography is tracked as usual during scanning. The oscillating cantilever displacement explores the force-displacement curve in a region close to the set point. As shown in Fig. 3.27A, a rigid surface does not allow the tip to penetrate, so all the displacement appears as a cantilever deflection. A soft region allows the tip to penetrate, so the cantilever deflection is reduced, as shown in Fig. 3.28.

The cantilever has to be reasonably stiff, or it would not penetrate the sample, and if it was too stiff, it would always penetrate. The cantilever and the sample are mechanically in series, and the best sensitivity is when the two are of similar stiffness. Stiff samples require stiff cantilevers.

The result is an image that reflects local stiffness variation across a surface. The image can be acquired at the nominal pixel resolution of the instrument, but the stress field extends into the material. The true resolution, controlled by interaction volume, may not be good, especially if the applied force is large. A very stiff cantile-

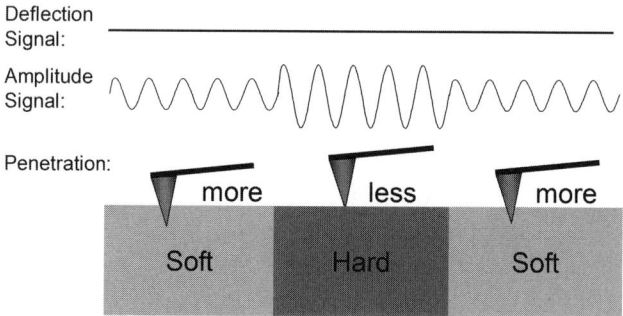

FIGURE 3.28. Schematic of the operation of force modulation microscopy. The dynamic response of the cantilever-bimorph drive amplitude is monitored while the cantilever deflection is held constant. The amplitude attenuation is related to the local surface stiffness.

ver and a large force of 1 mN was used to distinguish carbon fibers in a composite, with a resolution of about 60 nm [131]. Much better resolution has been obtained of rubbery phases in propylene/ethylene–propylene copolymers [132], isobutylene-based elastomer blends, and cross-linked rubbers [133].

Phase identification is relatively easy, especially when one phase is known to be much more compliant than the other. It is more difficult to obtain quantitative values for the elastic properties of the specimen. For this, the geometry of the tip and the surface must be known, and the effects of surface energy have to be considered [114, 134].

Force modulation imaging requires careful identification of the appropriate contact resonance. The frequency spectrum of the combined piezo stack and cantilever is generally contaminated with system resonances. It is important to identify the fundamental resonance that will attenuate on contact with the surface. In force modulation imaging, the tip is still subjected to lateral forces and this can cause problems.

3.3.3.4 Force Volume Imaging

This mode of operation has a name that could be confusing. There is no imaging of a volume of the sample, only the surface is investigated. The "volume" is of the image itself, which contains a deflection-distance curve (as in Fig. 3.27A) at every point [110, 116]. This gives a three dimensional data set with values of deflection—force when calibrated—at every value of x, y and z. The probe is kept at the fixed x, y position while the force curve is generated, so there is generally no lateral force problem. Each curve within the scan is set up to trigger at a specific cantilever deflection.

The data can be processed to give many different two dimensional images designed to show specific features. The downside is that data collection can be slow. Because of this, data is typically obtained over a coarse array of 64×64 or 128×128 positions. A 64×64 point array contains 4096 force curves. If each curve takes 1s to collect, the entire array will take over an hour.

An image can be reconstructed from the force curve array by taking a slice through the image volume at constant piezo distance. Multiple image slices may be formed by image reconstruction based on multivariate data (such as a spectrum) acquired at each point in the image (see Section 6.4). In force volume imaging in the AFM, the multivariate data is the force-distance curve.

Each force curve has an approaching (extending) and a retracting part, so it is possible to have two different force values at each Z position of the piezo. Usually one part is selected to form the image, the extending for stiffness or the retracting for adhesion. Force volume imaging has been used to generate relative adhesive force images of low friction additives on PET surfaces [135], phase separated domains in block copolymers at sub 50 nm resolution [136], and particle adhesion to gelatin surfaces [137]. Relative stiffness maps of heterogeneous polymer composites such as filled elastomeric antifouling coatings have been used to explain near-surface distribution of filler [138]. For soft materials, it has been shown that "zero" force height images can be reconstructed from force volume images that are more accurate than height images obtained from contact mode or IC-AFM [139].

3.3.4 Intermittent Contact AFM

The most useful imaging mode for polymer characterization is intermittent contact, IC-AFM, also called tapping mode AFM [140, 141]. In IC-AFM a cantilever with stiffness typically 5–50 N/m is driven into oscillation using a bimorph piezo driver at its base. The drive signal is of high frequency, at or near the resonant frequency of the cantilever, typically 50–400 kHz. The amplitude of motion at the base is small, typically <1 nm. In air, this slight motion is translated into a larger motion, 10–100 nm, at the end of the cantilever. This motion is much like gripping a long flexible stick and shaking it by small movements of the wrist. This would not work too well if the stick was immersed in water, because of viscous damping of the oscillation, and this is also true in IC-AFM. For operation in liquids, a cantilever with a magnetic coating at the tip can be driven directly by applying a time-varying current to a nearby coil [142]. This is effective even in the presence of strong damping.

The mean tip height is a little less than the amplitude of its oscillatory motion, so the tip comes into brief intermittent contact with the sample once every cycle. At 150 kHz, each oscillation takes 6.7 μs. For a 1 Hz line scan rate with 512 points per line, each location on the surface experiences roughly 300 of these "taps." The cantilever oscillation is affected by this interaction with the surface, and the signal for the feedback loop in regular IC-AFM is the amplitude of oscillation of the cantilever (Fig. 2.9 and Fig. 3.29). Because the cantilever is relatively stiff, and the mean distance from tip to sample surface is tens of nanometers, the mean static deflection of the cantilever is negligible.

Intermittent contact AFM has three great advantages. First of all, the cantilever has sufficient energy to penetrate and break out of the fluid layer on surfaces that are imaged in air (see Fig. 3.26) so that capillary forces are much less important. Second, the signal is at high frequency so a phase-sensitive detector can reject noise and measure small changes with great sensitivity. Third, the lateral forces are greatly reduced. The tip spends most of its time out of contact and its lateral motion during each brief period of contact is extremely small.

3.3.4.1 Modeling Cantilever Oscillation

The amplitude of the oscillation of the cantilever (with no influence from the specimen surface) depends on the cantilever material, its geometry, the damping characteristics of the medium and the magnitude of the drive signal. The material and geometry of the cantilever control the resonant frequency, ω_o (Eq. 3.17), and the material and its surrounding medium control the quality factor Q (Eq. 3.18).

The oscillation can be described as a driven, damped, harmonic oscillator [101], and the amplitude, A, and phase, ϕ, are given by the following equations, where a_d is the drive amplitude and ω is the frequency of oscillation:

$$A(\omega) = \frac{a_d \cdot Q \cdot \omega_o^2}{\sqrt{\omega_o^2 \omega^2 + Q^2 (\omega_o^2 - \omega^2)^2}} \quad (3.20)$$

$$\phi(\omega) = \arctan\left(\frac{\omega_o^2 \cdot \omega}{Q \cdot (\omega_o^2 - \omega^2)}\right) \quad (3.21)$$

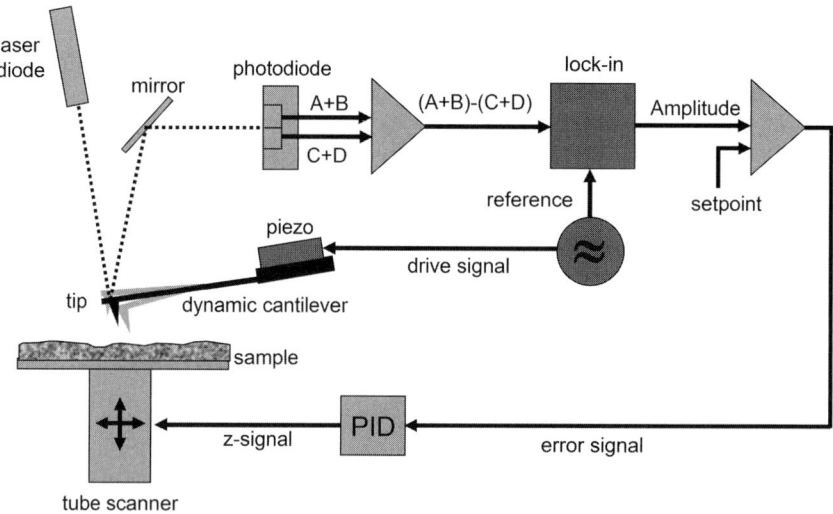

FIGURE 3.29. The principal components of an IC-AFM, with the scanner attached to the sample. Compared with the contact mode shown in Fig. 3.22, note the addition of the drive piezo at the base of the cantilever, the detection of average amplitude by the optical detection system, and the lock-in for determination of the phase shift.

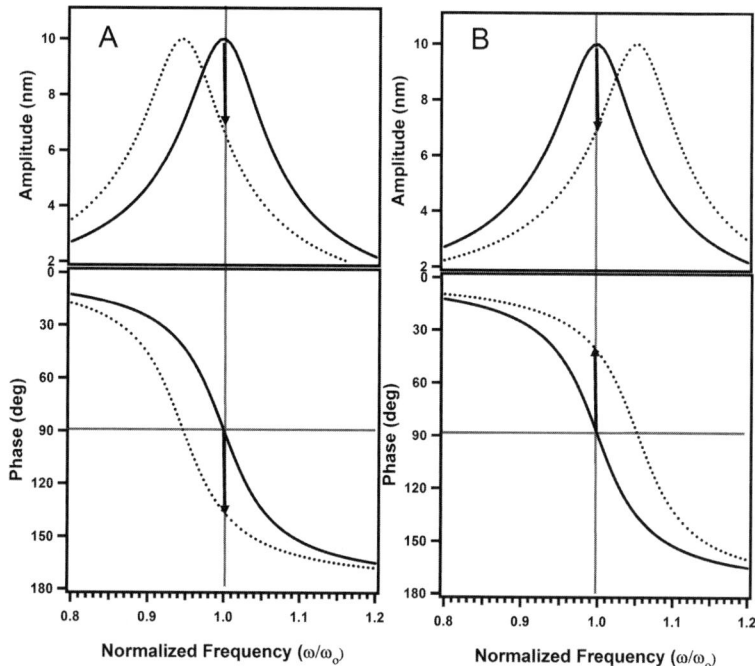

FIGURE 3.30. Amplitude and phase versus frequency for an idealized cantilever with $a_d = 1$ nm; $Q = 10$, and $\omega_o = 150$ kHz. Solid lines denote response for the cantilever driven at resonance according to a damped driven harmonic oscillator model. In (A), dashed lines represent the effect of an attractive force gradient—reduction in amplitude and decrease in phase. In (B), dashed lines represent the effect of a repulsive force gradient—reduction in amplitude and increase in the phase signal.

Note that at resonance, when drive frequency is equal to resonant frequency ($\omega = \omega_o$), the resonant amplitude, A_o, and phase, ϕ_o, are given by simple relations:

$$A_o = A(\omega_o) = a_d \cdot Q \quad (3.22)$$

$$\phi_o = \phi(\omega_o) = \frac{\pi}{2} \quad (3.23)$$

At resonance, the amplitude increases linearly with drive amplitude and quality factor.

The amplitude and phase derived from Eq. (3.20) and Eq. (3.21) are plotted in Fig. 3.30 using the following values: $a_d = 1$ nm, $Q = 10$ and $\omega_o = 150$ kHz. Experimental data of this type is commonly referred to as *frequency sweep data*. It is used to establish the initial conditions for IC-AFM. When obtained with the probe well away from the surface, it shows the resonant frequency of the system. This is the frequency with maximum amplitude, which coincides with the maximum slope of the phase signal.

The correct or "true" phase, φ, given by Eq. (3.20) is taken with reference to the driving signal. It increases with driving frequency, from 0° well below resonance to 180° well above resonance. It is 90° at resonance (Eq. 3.23). Some manufacturers have adopted different conventions for the instrument phase signal (still labeled "phase"). It may be $(180 - \varphi)°$ or $(\varphi - 90)°$ [143]. In the last case, the phase signal is zero at resonance, negative below and positive above resonance. This turns out to be a useful (although not rigorous) convention. A negative phase shift then indicates an attractive interaction and a positive phase shift indicates a repulsive interaction. In some cases the signal is plotted in an inverted manner, with lowest values at the top of the axes. This may make some conventions appear to be correct, but the phase signal may then "drop" on the plot, but increase numerically (as in Fig. 3.30) [143]. To summarize, "phase" is often set to be zero at resonance in the AFM, and the sign is confusing. It is best checked with a sweep such as in Fig. 3.30.

Figure 3.30 also shows schematically the effect of attractive and repulsive force gradients on the cantilever when it interacts with the surface. If the force is primarily attractive, the tip is delayed at the end of its swing, spending more time there, and the result is a reduction in the resonant frequency. Therefore, the response curve to the sweep of driving frequency is shifted to the left. Similarly, a repulsive force will repel the tip at the end of its swing; it will spend less time there, increasing the resonant frequency. The curve is displaced to the right in Fig. 3.30B.

The large changes shown by the dashed lines in Fig. 3.30 alter both the amplitude and the phase of the oscillation. The amplitude is reduced in both cases, while the phase change is of opposite sign. If the change in resonant frequency was very small, the effect would be controlled by the slope of the frequency sweep curve. Thus if the driving frequency is set exactly at the resonant frequency, the phase signal has a maximum sensitivity, but the amplitude is not sensitive at all. Amplitude sensitivity is obtained when the driving frequency is off-resonance.

3.3.4.2 Real Cantilever Oscillation

As the oscillating tip approaches the surface, it is affected by long range forces well before it comes into the range of intermittent contact. Within 10 μm of the surface, it is subjected to squeeze film damping. Motion of air near the surface is restricted and there is compression at each down-stroke of the lever. Restricted motion means more effective damping, and the effect is to reduce the Q of the system, broadening the resonance peak.

In the range 1 μm to 100 nm of the surface, the tip is subjected to electrostatic forces that can be either repulsive or attractive depending on the surface. Within 100 nm, the tip will experience capillary attractive forces. When operating in air, this is due to surface water. Although this is not the problem that it is in contact mode (see Section 3.3.3), these forces do affect the cantilever oscillation. Figure 3.31 shows the amplitude and phase signal (as $[90 - \varphi]°$) in a frequency sweep of a cantilever far from the surface, and then near to but not contacting the surface. Far from the surface, 200 μm above it, Q is 640 and the resonant frequency is

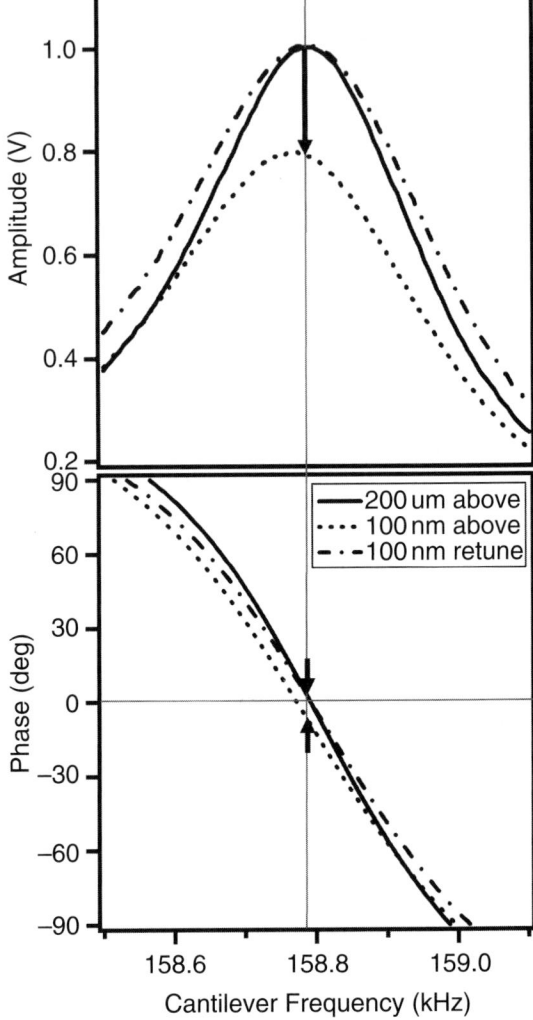

FIGURE 3.31. Frequency sweeps for a cantilever oscillating at 200 μm and 100 nm above a surface. The solid lines show the resonance at 200 μm above the surface. At 100 nm, dotted lines, the amplitude falls by 20%, while the resonance frequency and the phase are slightly reduced. The tip is retuned at 100 nm above the surface by increasing the drive voltage (dot-dashed line). Broadening of the amplitude peak and flattening of the phase response indicate a fall in Q.

158,790 Hz. At 100 nm above the surface the amplitude falls by about 20%. The frequency is shifted ~10 Hz lower and the phase signal drops. Comparing this with Fig. 3.30A shows that these effects are due to attractive forces near the surface.

Imaging in the Atomic Force Microscope

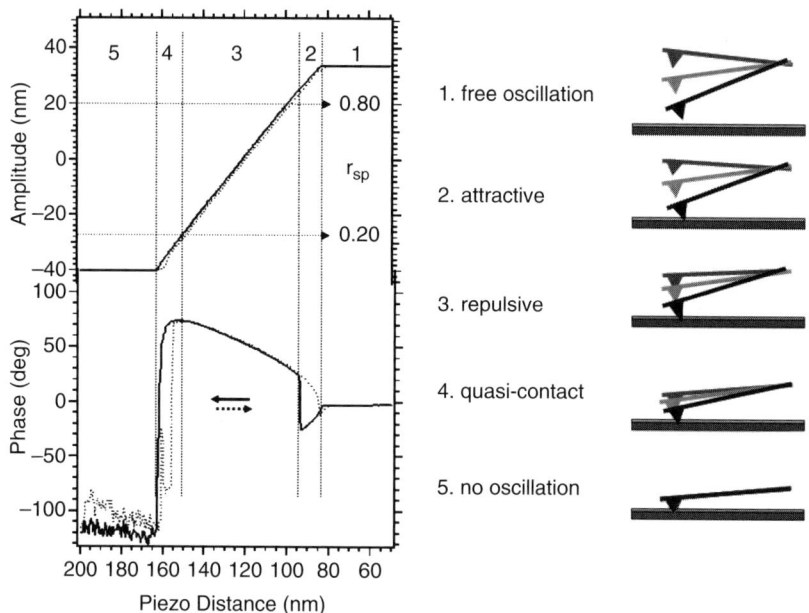

FIGURE 3.32. Amplitude and phase versus z position for a silicon cantilever tapping on a mica surface. The approach (solid) and retract (dotted) curves are shown. The amplitude falls linearly with displacement as there is no penetration into the rigid mica surface (2–3–4). The phase signal goes negative at small amplitude attenuation (2) and then abruptly changes to a positive value (3) when the forces become repulsive. Eventually, the vibration is fully damped (5). On retraction, there is hysteresis in both the regions 4 and 2. The amplitude scale has an arbitrary zero. For this curve, true zero amplitude occurs at an indicated −40 nm.

To control imaging conditions, it is good practice to tune the cantilever properties with the tip oscillating near the surface but not in intermittent contact. In early practice the cantilever was tuned slightly below resonance with the tip oscillating well above the surface. This was to anticipate a small shift in the peak resonance due to near-surface adhesive interactions as shown in Fig. 3.31.

Most commercial systems now permit the cantilever to be "retuned" while it is near the surface to establish the starting point conditions more precisely. The dot-dash line in Fig. 3.31 shows the effect of such retuning. The oscillation has been returned to the desired amplitude by slightly increasing the drive voltage on the piezo bimorph. The curve has been shifted on the plot to bring the resonant frequency back to the same point. This makes it clear that the resulting peak is broadened; the new Q value is 530.

During this retuning near the surface the vibration is still free of contact. From here the conditions for intermittent contact can be determined. A useful way to do this is by measuring amplitude and phase of the oscillation as the z-piezo distance is varied, extending the sample toward the tip and then retracting it away again. This curve is analogous to the deflection versus distance curve shown in Fig. 3.27 for contact mode operation. An example for a silicon tip and mica surface in air is shown in Fig. 3.32.

Figure 3.32 shows that the phase suddenly changes sign as the surface interaction changes from net adhesion in region 2 to net repulsion in region 3 (see Section 3.3.4.4). The amplitude of oscillation is attenuated further as the tip-sample distance is reduced until a situation of quasi-contact occurs (region 4). Now the tip spends more time in contact during each cycle than out of contact. Finally, the tip vibration is completely stopped with no detectable oscillation (region 5). The tip is always in contact with the surface and the situation is close to that of force modulation imaging (see Section 3.3.3.3).

Beyond this point the cantilever will have an increasing static deflection. As the cantilever is stiff, this will likely result in damage to the surface.

3.3.4.3 Imaging Conditions in IC-AFM

For imaging in IC-AFM (see Fig. 3.29), the amplitude of cantilever oscillation is used for feedback control, and the set point, A_{sp}, is less than the free oscillation amplitude, A_o. This mode of operation is also referred to as amplitude modulation (AM) AFM. A convenient way to standardize the description of tapping conditions for both stiff and compliant materials is to use A_o, A_{sp}, and A_{sp}/A_o [108]. This ratio is called the set-point ratio r_{sp}.

A_o and A_{sp} can be given either in terms of the voltage applied to the piezo stack used to drive the cantilever or as the real amplitude, the distance that the tip moves. Using real amplitude requires calibration, but it gives a better description of the imaging conditions. Different cantilevers and systems will have different amplitudes from the same drive voltage. Cantilevers with the same nominal spring constant will have some variation due to manufacturing tolerances. Any lack of rigidity in the mounting, giving some compliant contamination in the mechanical loop between the drive piezo and cantilever base, will also lead to a low amplitude response. Reporting A_o and r_{sp} as well as the nominal characteristics of the cantilever are recommended for a full description of imaging conditions.

A simple description of the operating condition used in IC-AFM mode would say that A_o is high, medium, or low, and r_{sp} is high, medium or low. This would mean an amplitude of >150 nm, 50–150 nm or <50 nm. In terms of the actual control parameter, drive voltage, this depends on the specific system used. In Figs. 3.31 to 3.33, the ranges would be >3 V, 1–3 V and <1 V. A high value of set-point ratio would be $r_{sp} > 0.8$, meaning a light tap that may be largely within the region of net attractive forces (regime 2; see Fig. 3.32). A medium value is $0.2 > r_{sp} < 0.8$, regime 3 of Fig. 3.32. A low value, $r_{sp} < 0.2$, corresponds with hard taps, moving into regime 4 of Fig. 3.32.

The imaging signal in IC-AFM may be the z-piezo position, or for phase imaging, the phase of the oscillation. In a heterogeneous sample the tip-surface interactions will generally be different for different components. This will alter the height at which a given set-point

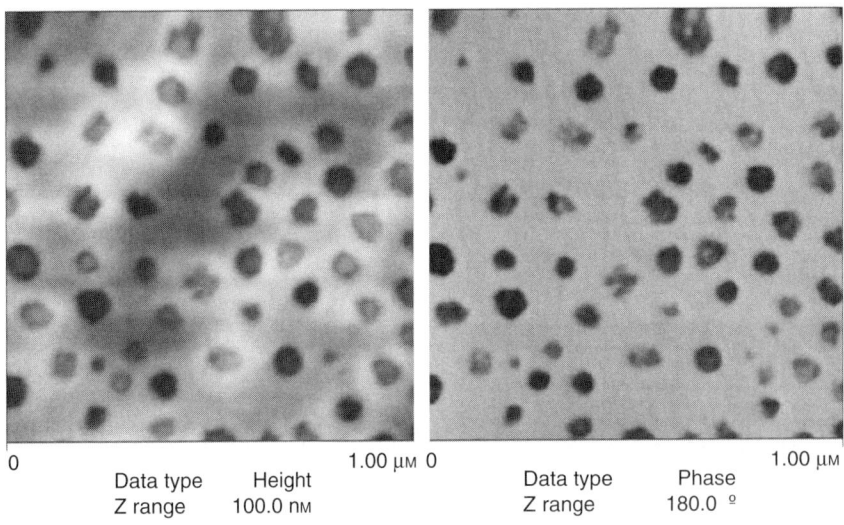

FIGURE 3.33. IC-AFM images of the free surface of a PMMA film containing 100 nm PBA latex particles. The height mode is on the left and the phase image is on the right. Free amplitude, $A_o = 38$ nm, $A_{sp} = 30$ nm, at a set-point ratio of 0.80.

Imaging in the Atomic Force Microscope

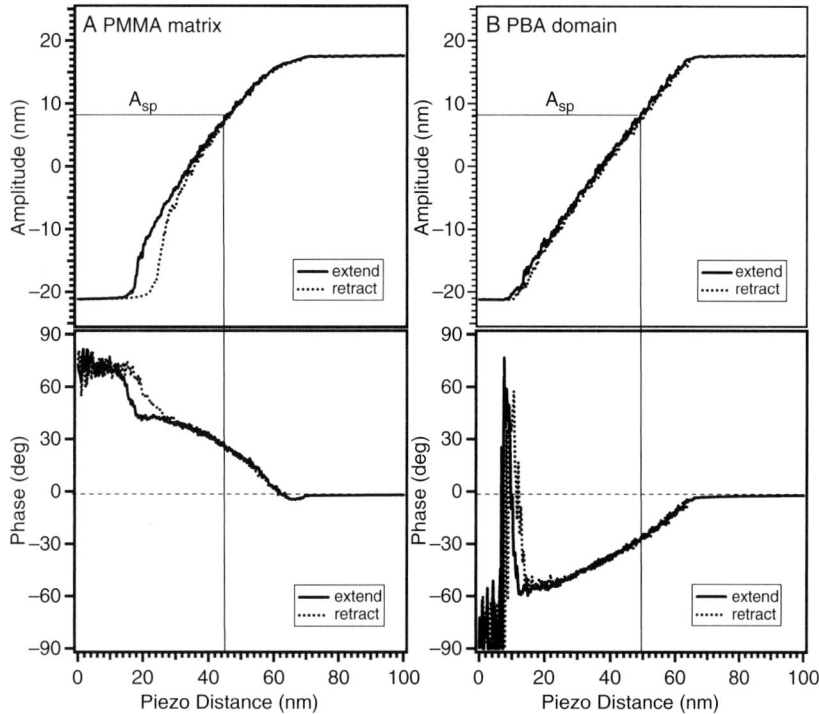

FIGURE 3.34. Amplitude and phase z-sweep data for the PMMA and crosslinked PBA domains imaged in Fig. 3.33. As before, the free amplitude, $A_o = 38$ nm. The imaging set point for Fig. 3.33, $A_{sp} = 30$ nm, is marked on the curves.

amplitude is obtained and will alter the phase at that set point. Compare mica in Fig. 3.32 with the two different polymers in Fig. 3.34. The phase response can be very sensitive to changes in the interaction and is not mixed with the topographical signal (except for large local slopes). This is why it is a very successful method for the imaging of polymer systems. The z-piezo response will contain both the surface topography and differences in composition.

Consider the height and phase images of a PMMA film containing well dispersed 100 nm crosslinked poly(butyl acrylate) (PBA) latex particles (see Fig. 3.33). The images are of the film's free surface. Both images show a light matrix with dark regions in the same locations. In the height image, dark regions correspond to depressions. The height image also has slower variations in intensity in the matrix which are absent in the phase image; these are topographic features of the free film surface.

Amplitude and phase z-sweep data obtained for the matrix and particle domains is shown in Fig. 3.34. These used the same free amplitude as the images, and the set-point amplitude used to form the images is marked on the figure. At the set point the amplitude attenuation is about 10 nm greater for the crosslinked PBA domain, and the phase has followed an attractive path with a $-20°$ phase shift. The PMMA matrix, however, has repulsive phase shift of about $+25°$. The softer PBA domains have a net attractive interaction to the tip, and the PMMA matrix is repulsive to the tip. Note that the z-sweep data would predict lower phase contrast at much lighter tapping (higher r_{sp}), even approaching no phase contrast.

The high contrast afforded by intermittent contact imaging for multicomponent polymer blends has enabled AFM to become a leading tool for morphology characterization, complementing TEM in many cases, but without the requirement of staining [141, 144–146].

3.3.4.4 Modeling IC-AFM

Interpretation of IC-AFM images is complicated by the fact that the tip-sample force is a nonlinear function of tip-sample separation. The tip-surface interactions in IC-AFM have been modeled extensively and have been recently reviewed [109, 143]. Two important conclusions have come from the modeling. First, the nonlinear interaction of the dynamic tip with the surface can lead to two stable oscillation states: one that follows a net attractive path and the other that follows a net repulsive path [147, 148]. A hint of this is seen in the phase versus frequency plot (see Fig. 3.32) where the cantilever initially oscillates along an adhesive path and then abruptly transitions to the repulsive path. Simulated amplitude and phase (z-sweep) curves can reproduce those determined experimentally. These have been interpreted in terms of force based interaction models that include the effect of capillary forces and adhesive forces when they are known or can be estimated. The transition between the bistable states depends on a number of factors including the cantilever Q, A_o, and r_{sp}, and the drive frequency as well as the surface properties [149]. In general high Q cantilevers or small A_o favor the net attractive path.

The second outcome of the modeling is interpretation of the tip-sample contact in terms of energy balance. The sine of the phase shift is related to energy dissipation or power lost in the tip-sample contact [150]. Dissipation images can be obtained by suitable instrument configuration to output a signal proportional to sine of the phase [151]. The effect of tip penetration and reduced tip-sample energy dissipation for elastomeric surfaces has also been experimentally determined for some systems [152]. Typical energy dissipation per tap is on the order of tens of eV. Typical bond strengths are on the order of a few eV and the contact area, tens of nanometers squared, contains many bonds. Therefore the tap is unlikely to disrupt bonding, and IC-AFM is generally nondestructive.

3.3.5 Noncontact AFM

Noncontact (NC) AFM for the study of polymer surfaces falls into two categories. In the first case, noncontact mode is used at low amplitude for high-resolution scanning of surfaces. Typically this experiment is carried out in high-vacuum or UHV environments. In UHV the cantilever has a high Q (>10,000) so it is more sensitive to weak attractive force gradients. The vibration amplitude is controlled to be a few nanometers or less, and a relatively stiff (~20 N/m) cantilever is used. Amplitude as a set-point signal for feedback control does not work well in this case, because the high Q makes for a slow response. Instead a frequency modulated (FM) detection system is used. Here the cantilever is allowed to oscillate at resonance and the change of resonant frequency due to the surface interaction is measured. The piezo distance is controlled to keep a constant frequency shift, and a separate feedback loop keeps the amplitude constant. This system can respond much more quickly to changes in force gradients.

This technique was first described in 1991 [153] and has since been reviewed [143, 154]. It has been used to give molecular resolution images of electrically conductive polymers [155] and PS-*b*-PMMA block copolymers [156] in UHV. Because the technique is truly noncontact, it does not have the complexity of dual interaction paths (either repulsive or attractive) that are inherent to amplitude modulation IC-AFM. This makes the technique suitable for the study of metal-polymer bonding under controlled ambient conditions [157].

The second application of noncontact AFM is in near-field detection of force gradients at 5–50 nm above the surface. This is the approach used in *electric force microscopy* (EFM) [158] and *magnetic force microscopy* (MFM) [159].

Usually in these modes the topography of the sample is first determined using IC-AFM. This topography is stored in memory and a second noncontact scan of the same area is made, using the stored height information to make the probe stay a fixed distance above the sample (a lift height of typically 5–50 nm). Feedback is disabled for this second scan and the direct amplitude or frequency shift of the cantilever due to near field force gradients is measured (Fig. 3.35). This second scan gives the EFM or MFM information.

This approach ensures the separation of topography from the effects of electric or magnetic fields. In EFM the field is generated using a bias voltage between a conductive tip and the sample surface. In MFM the static magnetic field of the sample interacts with a tip coated or fabricated with ferromagnetic material. The MFM is commonly used to evaluate magnetic bit integrity in magnetic recording tapes and drives. The EFM has been used to detect percolation thresholds of carbon black in polymer composites [160] and to image microphase domains in heterogeneous polymer blends based on dielectric constant [161].

Another interesting application of NC-AFM is in the detection of thermal events such as melting and glass transition temperature in polymers on heating. The technique uses FM detection with the tip fixed above the surface as the polymer temperature is ramped underneath [162]. The mechanism behind this technique is not well understood.

3.3.6 Practical Considerations for AFM Imaging

3.3.6.1 Feedback Loops

Feedback loops in AFM use proportional-integral-differential (PID) gain controls to condition the feedback signal. This type of control minimizes response time and overshoot when correctly set. The correct settings depend on the sensitivity and response of the system and so are not the same for the various imaging modes discussed. For instance, the proportional gain for NC mode may be a factor of 10 times that for IC operation. When adjusting the gain controls it is useful to monitor the line-by-line traces in oscilloscope mode. Proportional gain is increased until ringing (overshoot) is seen in the signal. The gain should then be reduced for imaging.

3.3.6.2 Scan Speed

Generally, contact mode imaging can be obtained at higher scan speed than IC-AFM mode. For the probe to track the surface topography in a reliable manner, its absolute speed over the surface cannot be too large. Therefore,

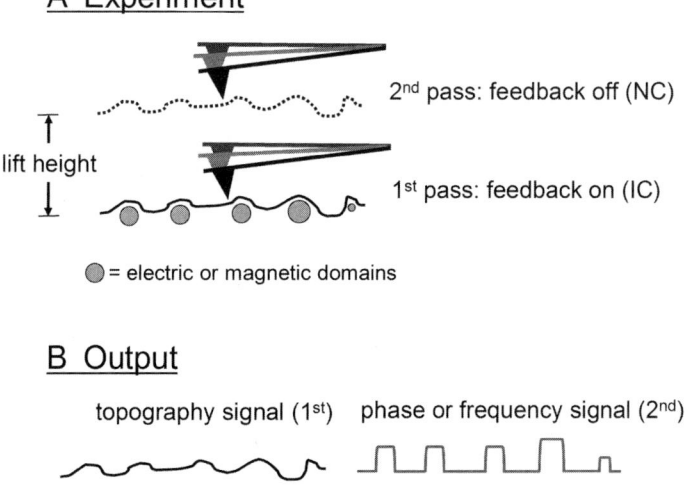

FIGURE 3.35. Dual pass technique used to collect signals due to force gradients 5–50 nm above the surface (A). The resulting output signal traces are shown in (B). Note that the phase or frequency may have a direction owing to the sign of the interaction (e.g., whether positive or negative bias is used in EFM).

scanning larger areas requires longer scan times. Collecting images with aspect ratios of 1:2 or 1:4 can save time when optimizing imaging parameters, so most instruments allow for non-square images. However, Fourier analysis will require collection of square pixel arrays. Higher-speed scanning instruments are discussed in Section 6.3.3.

3.3.6.3 Resolution

Good pixel resolution is required for image analysis. There must be sufficient pixels in a feature for it to be recognized as a feature instead of noise. If a particle size distribution is broad, then a single image may not be able to sample both small and large particles. This may require combining images of various size scales. Larger image formats with more pixels can reduce this problem, and scanning probe microscope (SPM) vendors are beginning to offer images at up to 5k × 5k pixels. Collecting images of this density will require long scan times. This will place extra demands on microscope stability over the long image acquisition. The advantage of having good statistics from a single image can outweigh the difficulties.

3.3.6.4 Presentation

Atomic force microscopy presents a challenge in data reporting. The topographic signal contains actual height information, and it is now common to record more than one channel of information. The three dimensional nature of the images means that they can be rendered in a variety of ways, height as intensity in a top view (see Fig. 3.33A) or in perspective view, with height encoded in pseudocolor and rendered with shading to enhance the appearance of features. The second channel, phase for example, can be shown in a separate frame (see Fig. 3.33B) or superimposed by color-coding on a topographic image. Compare this with the SEM, where the topographic image cannot be directly rendered as a height, and only low resolution elemental analysis information is regularly used as multichannel information. The information content in AFM is especially rich, but the layout and rendering required to demonstrate this in a specific image can take a lot of time and make comparison with different images from different sources more difficult.

3.3.7 Artifacts in SPM Imaging

All microscopes are subject to artifacts and scanning probe microscopy is no exception. It could be said that artifacts are more easily seen in SPM than when using other types of microscope. This is because it is usually possible to obtain some sort of image in SPM, even if it is incorrectly set up, while the TEM would show only a blur. Recognizing and perhaps removing artifacts is an important part of obtaining and interpreting images. Artifacts in SPM can arise from many sources [163]. These include:

- scanner motion
- tip geometry
- control electronics (e.g., improper feedback gains)
- noise (mechanical, acoustic, or electronic)
- drift (thermal or mechanical)
- signal detection (e.g., laser spillover in optical lever schemes [164])
- improper use of image processing (real time or postprocessed)
- sample preparation
- sample environment (e.g., humidity)
- tip-surface interaction (e.g., excessive electrostatic, adhesive, shear or compressive forces).

The most common artifacts are those that are related to scanner motion and tip geometry [165, 166].

3.3.7.1 Scanner Motion Artifacts

Scanning probe microscopes use piezoceramic elements to control the motion of the probe or sample on the nanometer scale (see Section 3.3.1.3). In *scanned tip* systems the cantilever/probe is attached to the scanning element. In *scanned sample* systems the sample is attached to the scanning element. Scanner-related artifacts arise from two sources:

1. Non-ideal behavior of the piezo elements in their response to applied voltage. There are

three common forms: *nonlinearity, hysteresis, and creep* [166–169].
2. Non-ideal motion of the scanner due to coupling of motion between the axes.

The sensitivity of a piezo scanning element is typically expressed in nm/V and is usually calibrated at the center of its range of motion. Nonlinearity means that the sensitivity changes as a function of applied voltage. Because linear ramp voltages are used in x and y to drive the probe over the sample, the effect is of a changing magnification over the field of view. This could easily go unnoticed on an irregular polymer specimen or a specimen whose features are not well known. The error becomes obvious when using a calibration or test sample that has a periodic structure. The features will appear unevenly spaced and linear features may appear curved. For open loop correction, there are conditioning algorithms (software) that adjust the applied voltage to linearize the response. This is usually accomplished during instrument calibration with suitable pitch and height standards. Systems with closed loop correction will achieve this in real time using feedback from an independent position sensing system (hardware).

If displacement versus an alternating voltage is plotted for a scanner with hysteresis, the result will be a loop. That is, the response of the scanner is different when the voltage is increasing and when it is decreasing. It can also depend on the rate of change of voltage. This hysteresis causes a variation in the lateral spacing of features that depends on the direction of scanning (either trace or retrace) or on the rate of scanning. In systems with a dual-channel oscilloscope, it is often useful to monitor the overlap of the trace and retrace signals to determine if hysteresis is present. Hysteresis in the z-piezo element would give rise to errors in measured step heights, particularly of large features where large voltage changes are necessary [169].

The response of a piezo to a large step change in voltage is not a step change in position. Most of the motion occurs very quickly, but then there is a slow continuing motion, called creep. This means that after a rapid change, the tip position will not be stable until the creep is complete. These changes may be due to moving up or down a large step and are particularly noticeable after zooming or offsetting to a new scan area. Systems without any closed loop correction to motion will usually need a few scans to "settle" the creep.

In tripod or tube scanners the z motion is coupled to x and y motion. Artifacts due to coupled motions include *bowing* and *Abbe offset errors*. Bowing results when the scanner moves in a circular arc when scanning in x and y. If a flat sample is being scanned, it extends further in the z direction at the edges of the scan, and appears to be dished. If the tip is being scanned in an arc over a stationary sample, the sample appears to bulge at the center. This error is predictable from the scanner geometry and can be corrected by plane fitting in real time or during post-processing.

The Abbe offset error is common to all metrology and arises when the motion and measurement axes are not collinear. If the sample is mounted on a tube scanner, as in Fig. 3.22, the sample height variation extends the total lever arm that includes the sample and scanner. A change in sample thickness changes the scanner length and causes a systematic error in the magnification [47, 170]. Considering the tube scanner as rigid and pivoting about its base, the scan area will be increased by the ratio of the sample thickness to tube length. The magnitude of this error can be quite large. For example, a tube scanner 25 mm long with a sample 5 mm thick would scan an area 20% larger than the end of the tube alone for the same arc. In real tube scanners the motion is better described as bending and is analytically more complex [171]. In polymer SPM, it is not uncommon to examine thin films or pieces of molded plaques or small microtome-compatible sample holders that will be of variable heights. For scanned sample systems multiple calibrations might be required to accommodate these situations, but for scanned tip systems this is not an issue.

3.3.7.2 Artifacts Due to Probe Geometry

Any finite-sized probe must cause an apparent increase in size of small protruding objects and an associated apparent reduction in size of pits.

Since all probe tips are finite, such artifacts are very common! If the surface features are all small or have a small slope, then only the very tip of the probe is important, whereas an abrupt step might interact with the probe well away from the tip.

The shape of the probe and its tip is vital for detailed interpretation of AFM or scanning tunneling microscope (STM) images. The nominal characteristics of commercially available tips make a good starting point [172]. Important characteristics of the tip include the opening half-angle, the aspect ratio, and tip radius. The overall shape (e.g., conical, pyramidal or parabolic) and the material of construction are important for such applications as nanoindentation where an accurate description of the probe shape is required. Detailed descriptions of analytical approximations to probe geometries can be found in the relevant ASTM standard [165].

All tip-related artifacts come from the *nonlinear geometric mixing* of the tip shape and surface shape. The closest or proximal point determines the tip height and this is not always the end of the tip. It can be the shank or even the cantilever itself. One of the first widely recognized artifacts in SPM imaging is the broadening that occurs when scanning features whose radii of curvature are smaller than that of the tip [173]. This is noted for scanning macromolecules like DNA on flat substrates [174]. Consider the simple case of a spherical tip with radius R_t scanning across a small fibril with a circular cross-section of radius R_s (Fig. 3.36). The measured width of the fibril will be broadened. For the spherical assumptions the measured width of the fibril will be $4(R_t \times R_s)^{1/2}$ [175]. The height will be measured accurately assuming no deformation. This describes the special case of a spherical tip and spherical surface feature. References to more general descriptions can be found elsewhere [176]. This is an important consideration in trying to measure the width of lamellar structures in semicrystalline polymers, block size in phase separated block copolymers, or even individual macromolecules deposited onto flat substrates.

Geometric mixing is clearly demonstrated when scanning undercut features such as large fibrils or steep-walled structures such as in patterned polymer thin films. In Fig. 3.37, the resulting image includes contributions from the feature but also the tip. The artificial sidewall on the left and right hand side of the large feature is a result of the edge of the feature scanning a partial profile of the tip (and masking the smaller feature underneath).

Mixing leads to a reduction in feature size when scanning nanoscale porous structures, nanoscale cracks and pits, or features that have downward excursions from the mean surface plane. Here the tip may not be able to fully penetrate into the feature before the tip contacts the other side. This can lead to images where the true feature size is reduced and/or situations where the bottom of the features is never contacted by the tip. Many examples of these cases can be found in [165].

Other geometrical considerations include axial symmetry of the tips and the angular tilt of the tip in the microscope. Tips with conical profiles have axial symmetry and present the same opening angle of the tip to the sample regardless of scan direction. This is not the case with tips that lack this symmetry such as faceted, pyramidal tips where the opening angle may be the half the angle between opposite faces, or half the angle between opposite edges. Another subtle but important consideration is the angle the tip makes relative to the surface. In most commer-

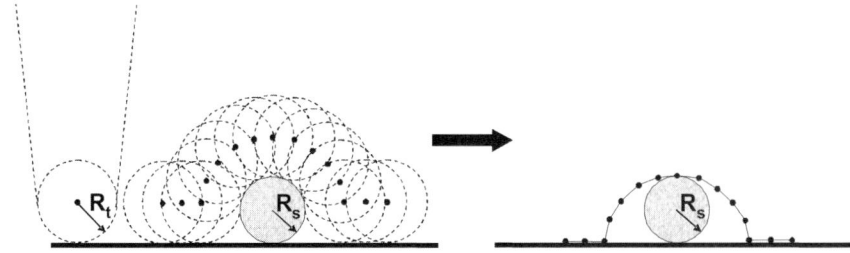

FIGURE 3.36. Broadening that occurs in the case of a spherical tip scanning a spherical object.

Imaging in the Atomic Force Microscope

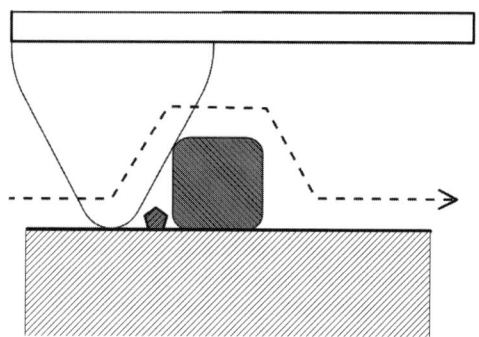

FIGURE 3.37. Regions of surface inaccessibility result from geometric mixing of tip shape and surface features.

cial SPM systems that use optical lever detection the tip is tilted off axis by 10° to 12°. This means that in one direction the actual opening angle will be increased by the tilt angle; in the other scan direction it will be reduced by this amount.

It is important to consider the role that geometric mixing may play when analyzing topographic images. It is possible that images may contain regions where the dimensions of structures of the true surface are not accurately represented or features may be missing [176–178]. The mathematical formalism that is used to describe the image formation is the process of morphological dilation. A common mistake is to describe this mixing as a convolution but this is not correct. A convolution implies that the real surface can be deconvolved from the image. Figure 3.37 shows that a particle near to a step does not affect the path of a conical probe tip. Because it has no effect, such a feature can never be recovered from the image by deconvolution or any other method.

Fortunately, there are mathematical approaches that can be used to recover a better estimate of the real surface profile and indicate which regions are subject to some uncertainty. In general these approaches use tip-shape reconstruction. Many researchers have suggested that using the tip broadening may provide information about the tip shape when scanning small objects of known, regular cross-sectional shapes (called "characterizers") such as gold colloids [179, 180], latex spheres [181], starburst polymers [182], and rod-shaped biomolecules [174]. This approach can be used to estimate the tip radius assuming spherical or parabolic profiles. Broader application of this method can include characterizers of any known shape or mixtures of characterizers designed to sample different regions of the tip. The process of reconstructing the tip shape is a mathematical erosion of the characterizer shape from the image of the characterizer, and the process produces an outer bound estimate of the tip shape [183].

More general approaches use the procedure of blind reconstruction [183–186]. Here the initial shape of the characterizer is not known in any great detail other than that the surface features will contact the tip. From the image and an initial guess of the tip shape, the tip is reconstructed by thresholding features in the image that sample various portions of the tip until a consistent tip shape is determined. This is an iterative process, and the resulting outer bound shape must be able to produce all features in the starting image. Characterizers of different sizes may be used in this approach—some that sample the apex only and larger ones that sample the tip shaft—and the resulting outer bound shape can be knitted together [187–189]. In either case, the outer bound reconstruction of the tip shape can be used to process an image using mathematical erosion to obtain an improved estimate of the real surface. In principle, it is possible to indicate which regions of the surface are completely restored and which may be subject to some uncertainty due to inaccessibility [186, 187].

3.3.7.3 Tip Defects, Wear, and Contamination

Real tips are not generally ideal in shape and may also contain defects as a result of the manufacturing process [190]. Tips can become worn in use or become contaminated by picking up debris during scanning. In both cases the local geometry of the tip is changed and so geometric mixing of a new or varying shape can occur. This can be a problem when trying to use a rigid characterizer for reconstruction of a tip that was used to scan soft materials like polymers [191]. A fairly common defect that occurs with tips is development of a double or multiple tips. Here multiple points of contact, usually from asperities in close physical proximity to the apex, can occur simultaneously and lead to "ghosting" of features in images. These may result from tip wear or contamination and can therefore occur during scanning. Good experimental practice dictates recording images on the same sample with different tips to verify that a consistent morphology can be obtained. When scanning hard materials such as metal oxides, metals and ceramics, tip wear can be anticipated especially if the tip modulus is much lower than that of the materials being scanned. If intermittent contact mode is used on such surfaces, then conditions favoring low contact forces should be used (lower amplitude and higher set-point ratios). Many filled polymer systems contain relatively hard fillers (clay, talc, silica, and carbon) that can lead to tip wear, but for polymer systems, tip contamination is generally the greater concern. Low modulus, rubbery polymers or impact modifier phases (either with T_g well below room temperature) can be adherent to the tip surface material (generally silicon dioxide). The high contact frequency imparted by intermittent contact helps to mitigate some of the softness problems with polymers as they effectively behave as though they are stiffer at high frequency [152, 192]. Nie has suggested the use of biaxially oriented polypropylene (BOPP) as a suitable sample to test for contamination in IC-AFM [193].

3.4 SPECIMEN DAMAGE IN THE MICROSCOPE

As was remarked in Section 2.6, the best way to avoid radiation effects is to avoid irradiating sensitive specimens, either by using light microscopy or AFM or making the specimens insensitive, for example by staining. This section is for those times when irradiation is unavoidable. Some photosensitive polymers are affected by all visible light, so that optical microscopy is not possible without chemically changing the specimen. Others are sensitive only to blue light, so the use of "safelight" colors permits optical observation without damage. However, these problems are confined to lithographic materials. The major concern here is with the effect of radiation on polymers in the electron microscope, for a high energy electron beam alters all organic materials to some extent. Electron micrographs of polymers cannot be interpreted accurately without considering the possibility that radiation damage has changed the image in some way.

3.4.1 Effect of Radiation on Polymers

Hughes [194] is a basic introduction to radiation chemistry, and the radiation chemistry of polymers is described in detail in several texts [195–199]. There are reviews of radiation damage of organic materials in the TEM [200–202]; reviews of polymer microscopy also deal with the subject [203, 204]. More recent summaries concentrate on biological or inorganic materials.

The electrons in the beam of the TEM or SEM interact with electrons in the specimen. They typically transfer tens of electronvolts of energy to an electron at the site of the interaction. X-rays or γ-rays act in much the same way, locally depositing enough energy to break many chemical bonds. High energy electrons can also interact with atomic nuclei in the sample, knocking them out of position. This is not an important damage mechanism in organic materials.

The energy transferred is enough to have several outer electrons leave their atoms. Bonds are broken, and free radicals form [195] with a high concentration of reactive species in a small volume. Most of these will recombine very rapidly, re-forming the original local chemical structure and dissipating the absorbed energy as heat. But some will form new structures, changing the chemistry. If the material is initially crystalline, defects form as the molecules change shape and eventually it must become amorphous.

When organic molecules are irradiated, specific bonds or types of bonds are likely to be disrupted. These are not always the least stable energetically; for example, in *n*-paraffins, the C–H bonds break much more often than the C–C bonds. Table 3.6 indicates which bonds break in typical polymers, with approximate G values for the specific products of the reaction. G(P) is the number of units of the product P formed for every 100 eV of radiation absorbed by the specimen. It depends on the target, its physical state, and its temperature. It may also depend on the type of radiation and the dose rate. If radiation breaks a bond in a polymer that is part of the main chain, it will undergo *degradation* or *scission* at a rate G(S) to pro-

TABLE 3.6. Radiation yield of various reactions and products for common polymers in terms of G value. A G value of 1 means that one such reaction occurs or one product is formed when 100 eV of radiation is absorbed

Polymer	Formula unit	G values			
		Scission	Crosslinking	Products	
				Hydrogen	Other
C–C main chain hydrocarbons					
Polyethylene	–CH_2–CH_2–	0.2	1.0	3.7	—
Polypropylene	–CH_2–CH(CH_3)–	0.2	0.16	2.8	—
Polyisobutylene	–CH_2–C(CH_3)$_2$–	4.0	—	2.1	2 CH_4
Other side groups					
PMMA	–CH_2–C(CH_3)– –(COOCH_3)	3.5	—	—	2.5 Ester group
Polystyrene	–CH_2–CH(C_6H_5)–	0.01	0.03	0.03	—
Poly(vinyl chloride)	–CH_2–CHCl–	—	0.1	—	13 HCl
PTFE	–CF_2–CF_2–	0.3	—	—	—
Other main chains					
Polyoxymethylene	–CH_2–O–	11	6	1.7	—
PDMS	–O–Si(CH_3)$_2$–	—	3	3	3 CH_4
Polybutadiene	–CH_2–CH=CH–CH_2–	4.0	—	—	–12 C=C

PMMA, poly(methyl methacrylate); PTFE, polytetrafluoroethylene; PDMS, poly(dimethyl siloxane).

ducts of lower molecular weight. If the bond that breaks is part of a side group of the polymer, a small fragment will be lost. This free valence may bond to another chain, so the two chains become joined by *crosslinking* at a rate G(X). The product has a higher molecular weight than the starting material.

Aromatic compounds are much less sensitive to radiation than aliphatic ones. A phenyl group in a compound can reduce the sensitivity of other chemical groups over 1 nm away [205], so even a few aromatic groups can make a material more radiation resistant. The presence of small amounts of oxygen can also have a large effect on the radiolytic yields, due to the formation of peroxides. In cryomicroscopy, water can be a significant source of oxygen.

G values generally rise at high temperature, are lower in the solid than the liquid state, and are lower still if the solid is crystalline. Consider the constraining effect of the surrounding molecules. In a crystal all the surroundings are fixed, and the excited atoms are held close to their original positions. They are therefore likely to re-form the original chemical structure, and the yield of changes will be lower. At high temperature the atoms are more separated and less likely to re-form bonds as before. This behavior is called the "cage effect," as the surroundings act as a cage for the excited species. The opposite effect can also occur. For example, a fully substituted carbon atom is likely to break one of its C–C bonds on irradiation. The other substituent groups get in the way of excited species and limit recombination.

Polymers are generally divided into two main groups, those that crosslink and those that degrade, for although many polymers do both, one of these processes generally dominates. For example, polyethers will degrade because the C–O–C linkage is easily broken by irradiation. The specimen in the electron microscope receives very large doses of radiation, so that degradation does not stop with a polymer of reduced molecular weight. It continues until the fragments are small enough to evaporate and leave the specimen. In an extreme case there will be a loss of all the material irradiated in a thin specimen, leaving a hole.

Continued crosslinking in the other group of polymers produces an infusible, insoluble, brittle solid of high carbon content. When both

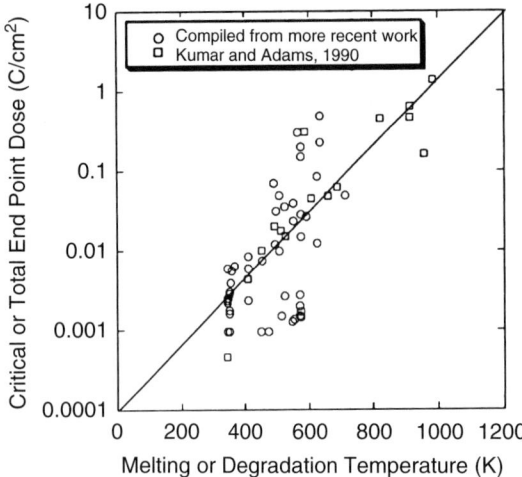

FIGURE 3.38. Plot of end-point dose (a measure of radiation stability) versus melting or degradation temperature (a measure of thermal stability). (Adapted from Martin et al. [47], © (2005) Wiley Interscience; used with permission.)

processes occur, there will be significant loss of mass, and the residue will be this crosslinked char. There may be some residue even from a sample supposed to degrade completely, because it changes chemically so much during irradiation. More generally, the results of radiation chemistry experiments at low doses, such as G values, cannot always be applied to electron microscopy. The most reliable predictor of stability in the electron beam for polymers is simply the melting point or degradation temperature as shown in Fig. 3.38 [47].

3.4.2 Radiation Doses and Specimen Heating

Radiation effects depend primarily on the energy absorbed. The *gray* (Gy) is the SI unit for absorbed energy, and 1 Gy is the absorption of 1 joule of ionizing radiation per kilogram of material. In terms of old units, 1 Gy = 100 rad. Absorbed doses may also be quoted in units of 100 eV absorbed energy per gram. This strange unit multiplied by the G value gives the concentration of radiation products in the sample. It is difficult to measure absorbed energy in the electron microscope, so the radiation dose is normally defined as an *incident dose* or electron flux, in Coulombs per square meter (Cm^{-2}), or electrons per square nm, ($e\,nm^{-2}$), where $1\,e\,nm^{-2} = 0.16\,C\,m^{-2}$.

The conversion factor between incident electron dose in Cm^{-2} (or $e\,nm^{-2}$) and absorbed dose in rad or Gy depends on the rate of energy loss of the incident electrons. For 100 keV electrons and a model material for organic polymers, calculations [10, 41, 42, 206, 207] give 400 to 450 $eV\mu m^{-1}$. If the specimen is very thin (<10 nm), a significant fraction of the energy transferred to it may be lost by the escape of secondary electrons. For 100 keV electrons, at 10 nm thickness nearly half the energy is lost and only 250 $eV\,\mu m^{-1}$ is absorbed [42].

Using 400 $eV\,\mu m^{-1}$ as the energy loss rate, $1\,Cm^{-2}$ deposits 400 J in 0.001 kg and is therefore equivalent to 400 kGy. Similarly, $1\,e\,nm^{-2}$ is equivalent to 64 kGy. Remembering that the lethal dose for humans is only 6 Gy (600 rad), these are extremely high doses. For high resolution TEM studies of metals, an incident flux of $1\,A\,cm^{-2}$ is common, which translates into $4\,GGy\,s^{-1}$.

It is important to realize that irradiation in the electron microscope, even at this dose rate, does not have to cause a large rise in the temperature of the specimen. This is because the illuminated area is a very small object, so it has a large surface area per unit volume. It can be efficiently cooled by thermal conduction into the rest of the specimen. Polymers are poor conductors, but most polymer studies use much lower beam intensities and the result is that the heating is normally 10 K or less.

In the ideal case of perfect thermal contact, the temperature rise is approximately proportional to the beam current. For a polymer film firmly mounted on a 200 mesh grid and irradiated with 100 keV electrons, the rise is 1–3 $K\,nA^{-1}$ [10, 42, 208, 209]. A spot 2 μm in diameter with the high beam current density of $1\,A\,cm^{-2}$ has a total current of 3×10^{-8} A and a temperature rise of 30–90 K. At the more usual beam current density, 10 or 100 times less than that used in the example above, the temperature rise is small. On the other hand, poor thermal contact with the support or a restricted thermal path, as when a thicker particle sits on a thin film support, can produce very high temperatures [25].

Heating effects are more difficult to calculate for scanning electron microscopes. The SEM produces smaller temperature increases than TEM because energy is deposited in a thin surface layer and can be conducted away into the depth of the sample [210–212]. Even with beam currents in the microamp range, it is not possible to melt the surface of polymer blocks with an SEM.

3.4.3 Effects of Radiation Damage on the Image

3.4.3.1 Mass Loss

In thin samples, low molecular weight fragments will rapidly diffuse to the surface and evaporate. Bubbles may form at high dose rates in thick specimens when volatile products are trapped, and this is often incorrectly taken to imply that the specimen temperature is high. In any case, the mass of the specimen decreases during observation. Polymers that degrade will lose a lot of mass, and those that crosslink will lose less.

In TEM the bright field transmitted intensity increases as mass is lost. This increase of intensity with dose has been measured [213, 214]. Calibration with films of known thickness allows the mass loss to be calculated if the film composition does not change too drastically. In the SEM the unirradiated bulk of the sample always acts to stabilize the form of the irradiated surface. Mass loss may cause a uniform depression of the sample surface or holes and cracks may appear [215].

More detailed changes in sample chemistry and loss of light elements during irradiation can be detected with electron energy loss spectroscopy [216–218]. Direct chemical analysis of thin polymer films with a large area irradiated with electrons has also been used [41, 194]. The chemical composition and the mass change for the first $0.1–1\,\text{C cm}^{-2}$ and then stabilize. The final composition contains more carbon and nitrogen than the initial material and less oxygen, hydrogen, and halides. The approach to a steady state makes sense, as groups susceptible to scission and removal are removed during irradiation leaving a more stable material behind.

In a favorable case, the final composition may not be much different from the starting composition. Polystyrene, of elemental composition $(CH)_x$, loses only 15–20% of its total mass before stabilizing at a composition near $(CH_{0.8})_x$. This allows useful quantitative measurement of mass thickness even after long periods of irradiation [219]. In contrast, poly(oxymethylene) ($-CH_2-O-$), retains only 15% of its mass, and the images after long irradiation are normally useless.

Mass loss destroys the sample to a greater or lesser extent, but it can be regarded positively as an etching process using the electron beam, and put to use. Two polymers in a blend or copolymer may have little contrast initially, but one phase loses more mass and the contrast increases or reverses [220]. Even in a homopolymer, the cracks and voids that appear can sometimes be related to the microstructure, making it more visible. For example, cracks form in molded poly(oxymethylene) samples during irradiation in the SEM [215], which follow spherulite radii and show up the oriented skin (e.g., see Fig. 4.34).

3.4.3.2 Loss of Crystallinity

The random insertion of chemical changes by irradiation ruins the regularity of the chains, and any existing crystallinity is destroyed. In some polymers this is a slow process, but most quickly become amorphous under normal viewing conditions in the TEM. The original crystalline electron diffraction pattern may contain spots, arcs or sharp rings depending on the perfection of orientation of the sample. Whatever the form of the original pattern, radiation damage will transform it into diffuse rings. The local molecular orientation may persist to very high doses; so may differences in density between regions originally crystalline and regions originally amorphous.

The dose required to change the diffraction pattern into diffuse rings, J_a, has often been used to determine the radiation sensitivity of materials in the microscope [10, 207, 221, 222]. If high resolution of the crystalline structure is required, this is an overestimate, and the decay of the relevant diffraction spots is more appropriate [223]. The change in the diffraction pattern takes place in two ways, seen most clearly when single crystals are used to give an initial sharp spot pattern. In some polymers, the sharp spots simply fall in intensity without changing their position or width and disappear into a diffuse background. In others, the spots spread out and shift their position as they become less intense (Fig. 3.39) [221, 224].

Poly(oxymethylene) belongs to the first class and polyethylene belongs to the second. Materials that degrade lose mass but are otherwise unchanged, so their diffraction patterns simply lose intensity. Crystals in materials that crosslink are distorted and so the diffraction spacings and peak breadth change. Loss of crystallinity causes all diffraction contrast features and lattice fringes in the TEM image to fade away. During irradiation, new contrast features—radiation artifacts—can appear temporarily and then fade with the rest. For example, polyethylene crystals become covered with fine lines or speckles in dark field and these can mask real features [203, 204].

3.4.3.3 Dimensional Changes

Large dimensional changes can be induced by radiation that distorts the object and changes the image permanently. The distortion may be almost instantaneous under the viewing conditions normally used, and the result is a stable and maybe misleading image. If the beam intensity is reduced the changes can be seen, but now recorded images are smeared and useless because the specimen is moving. Very much lower dose images catch the specimen before distortion starts. In some cases this is not worth the effort, for the distorted image can have high contrast features related to the original microstructure [225, 226].

The distortion of crystalline regions has been described in greatest detail for polyethylene. Large scale changes in the shape of thin films in the TEM and SEM have been related to changes in shape of lamellar crystals due to crosslinking [227, 228]. Similar effects are seen in isotactic polystyrene, natural rubber and

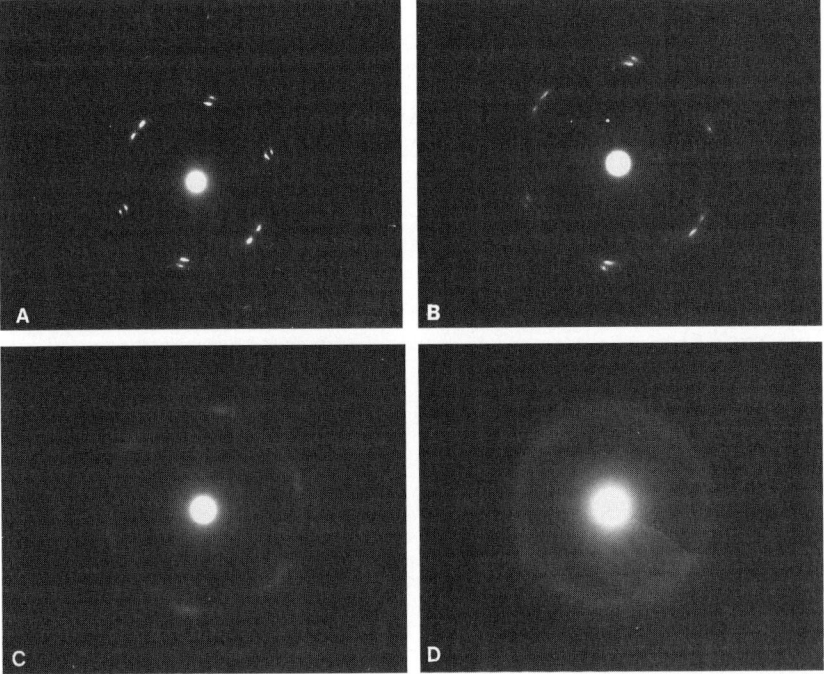

FIGURE 3.39. Sequence of electron diffraction patterns from a polyethylene crystal at 100 kV, showing how the sharp spots fade and spread so that the final result is a ring pattern. The crystals have become completely amorphous because of radiation damage. The doses are (A) 35–37 C m^{-2}; (B) 53–55.5 C m^{-2}; (C) 70.5–74 C m^{-2}; (D) 123–130 C m^{-2}.

nylon, all of them cross-linking materials [229–231]. When thick specimens are used in the SEM, the underlying material may prevent large-scale distortions. Changes are seen on a fine scale, less than the penetration depth of the electrons. In spherulitic polyethylene, this causes initially smooth surfaces to become full of fine structure-related detail [228, 232–234].

3.4.3.4 Radiation Damage in the SEM

It is not easy to predict how radiation damage varies with beam voltage in the SEM. Secondary emission is greater at low voltages, so fewer low energy incident beam electrons are needed to form a given image, and each deposits less energy in the specimen [76]. However, the electron range decreases as about $E^{1.5}$, so the energy deposited per unit mass (the radiation dose; see Section 3.4.2) is much the same or slightly greater at low voltages.

Radiation affects only a thin surface layer, but this is exactly the surface layer that provides all the information in the image. The unirradiated bulk of the sample serves to stabilize the damaged surface layer (see Section 2.6). The stabilization should be better if the damaged layer is thinner, and if mass is lost, the mass loss will be lower [60]. If the beam is stationary (or nearly so at high magnifications), the irradiated mass depends on the interaction volume ~ range^3 and the radiation dose will be much greater at low voltages and concentrated in a very small spot. Nevertheless, visible changes to the image are less at low beam voltages [76, 235].

If the sample is hydrated in the VPSEM/ESEM, then the irradiation of water may strongly increase the effects of radiation damage. This is due to the water acting as a source of small, highly mobile free radicals, which accelerates specimen degradation [236].

3.4.4 Noise Limited Resolution

As described in Section 3.2.3, when a finite number of electrons N are used to form an image, there will be a noise limit to resolution. Consider a radiation dose of J e nm^{-2} incident on a feature of size d nm, contrast C. If f is the fraction of the incident electrons that contribute to the image, there will be $N = d^2Jf$ electrons in the image of that feature. The statistical noise is $d(Jf)^{1/2}$ and the signal is $C(d^2Jf)$. If the signal-to-noise ratio k is required to be 5 or more, then $k = Cd(Jf)^{1/2}$, $d = k/C(Jf)^{1/2}$, $d \geq 5/C(Jf)^{1/2}$.

For high resolution J must be large, but if the feature is radiation sensitive, J cannot exceed the dose required to destroy it. J_a, the dose required to make the material amorphous, will be the upper limit for all crystallographic contrast features. As features shift and fade, calculation of optimum dose and the resolution obtained can only be a guide. An experimental approach, trying out a sequence of images at low and increasing doses, may be more useful. A study of polyethylene single crystals showed that the artifacts swamped the real structures at 300 e nm^{-2}, 60% of the dose that would be estimated from the decay of diffracted intensities [237].

Ideally, the best resolution will be obtained when $Cf^{1/2}$ is a maximum, and much of the calculation of resolution limits for radiation sensitive materials becomes a calculation of $Cf^{1/2}$, which depends on the imaging mode. For example, dark field gives a high C and low f. Estimates led to a resolution limit of 5nm for TEM at 100kV when the sample was PE crystal fragments 10nm thick [237]. Similar values have been obtained for biologically important molecules [238]. These calculations assume that all the electrons containing useful information form a single image. If m images of the same area are required, the number of electrons available for each falls to J/m. The resolution becomes $5m^{1/2}/C(Jf)^{1/2} = dm^{1/2}$ and the required magnification falls by the same factor. Long lines or fringes are more visible than random spots, so lattice fringes of much finer spacing can be imaged. For example, 0.22 nm fringes have been imaged in polyethylene at 200kV [239].

A similar calculation for scanning electron microscope images has the same basic equation, $d_{min} = 5/C(Jf)^{1/2}$. Differences between STEM and conventional TEM in the imaging of radiation-sensitive materials relate to the different values of C and f expected. These considerations are by no means simple [240] and depend on the type of object being imaged.

In principle any detector of high quantum efficiency is suitable for recording these noisy images. More important is that every electron passing through the sample should contribute to the image. In normal microscope operation, a large flux is used to focus the image and adjust the specimen position. This would destroy a sensitive sample and must be avoided. An image intensifier or high-speed CCD system allows a much lower intensity to be used without causing eye strain. The low-dose operation mode, available on most TEMs, uses the beam deflecting system in the TEM to focus and align on a nearby area and then switch the illumination back to record the image [241]. A very small irradiation may be used to check that an area of interest lies in the area to be imaged, or it may be recorded "blind." The latter approaches the ideal of using the maximum number of electrons to form the image, at the cost, perhaps, of many attempts before a useful micrograph is recorded.

References

1. F.A. Jenkins and H.E. White, *Fundamentals of Optics*, 4th ed. (McGraw-Hill, New York, 2001).
2. E. Hecht, *Optics*, 4th ed. (Addison Wesley, Reading, MA, 2001).
3. W.T. Welford, *Optics* (Oxford University Press, Oxford, 1988).
4. L.C. Martin, *The Theory of the Microscope* (Blackie, London, 1966).
5. M. Born and E. Wolf, *Principles of Optics*, 7th ed. (Pergamon, Oxford, 1999).
6. E.M. Slayter and H.S. Slayter, *Light and Electron Microscopy* (Cambridge University Press, Cambridge, 1992).
7. T.G. Rochow and P.A. Tucker, *Introduction to Microscopy by Means of Light, Electrons, X-Rays, or Acoustics*, 2nd ed. (Springer, New York, 1994).
8. S.G. Lipson and H. Lipson, *Optical Physics* (Cambridge University Press, Cambridge, 1969).
9. M. Spencer, *Fundamentals in Light Microscopy* (Cambridge University Press, Cambridge, 1982).
10. L. Reimer, *Transmission Electron Microscopy, Physics of Image Formation and Microanalysis*, 4th ed. (Springer, Berlin, 1997).
11. J.R. Waring, R. Lovell, G.R. Mitchell and A.H. Windle, *J. Mater. Sci.* **17** (1982) 1171.
12. I.G. Voigt-Martin, *Adv. Polym. Sci.* **67** (1985) 196.
13. Z. Zhang and X. Yang, *Polymer* **47** (2006) 5213.
14. J. Buerger, *Elementary Crystallography* (Wiley, New York, 1963).
15. W. Borchardt-Ott, *Crystallography*, 2nd ed. (Springer, New York, 1995).
16. D.E. Sands, *Introduction to Crystallography* (Dover, New York, 1994) [reprint].
17. C. Hammond, *The Basics of Crystallography and Diffraction* (IUCr Oxford University Press, Oxford, 1997).
18. C. Giacovazzo, Ed. *Fundamentals of Crystallography* 2nd ed. (Oxford University Press, Oxford, New York, 2002).
19. N.H. Hartshorne and A. Stuart, *Crystals and the Polarizing Microscope* (Elsevier, New York, 1970).
20. P. Gay, *An Introduction to Crystal Optics* (Longmans, London, 1967).
21. L.H. Schwartz and J.B. Cohen, *Diffraction from Materials* (Springer-Verlag, Berlin, 1987).
22. J.M. Schultz, *Diffraction for Materials Scientists* (Prentice-Hall, Englewood Cliffs, NJ, 1982).
23. J.M. Cowley, Ed. *Electron Diffraction Techniques* (Oxford University Press, Oxford, New York, 1992–1993).
24. J.C.H. Spence and J.M. Zuo, *Electron Microdiffraction* (Plenum Press, New York, 1992).
25. D.B. Williams and C.B. Carter, *Transmission Electron Microscopy: A Textbook for Materials Science* (Plenum Press, New York, 1996).
26. B. Wunderlich, *Macromolecular Physics*, Vol. 1 (Academic Press, New York, 1973).
27. P. Allen and M. Bevis, *Proc. Roy. Soc.* **A341** (1974) 75.
28. A.J. Lovinger and H.D. Keith, *Macromolecules* **12** (1979) 919.
29. D.L. Dorset, *Structural Electron Crystallography* (Springer, New York, 1995).
30. D.L. Dorset, *Rep. Prog. Phys.* **66** (2003) 305.
31. A. Du Chesne, *Macromol. Chem. Phys.* **200** (1999) 1813.
32. R.H. Geiss, G.B. Street, W. Volksen and J. Economy, *IBM J. Res. Develop.* **27** (1983) 321.
33. W. Claffey, K. Gardner, J. Blackwell, J. Lando and P.H. Geil, *Philos. Mag.* **30** (1974) 1223.
34. J.R. Fryer and D.L. Dorset, Eds. *Electron Crystallography of Organic Molecules* (Kluwer Academic Publishers, Dordrecht, Boston, 1991).

References

35. L.E. Alexander, *X-ray Diffraction Methods in Polymer Science* (Wiley-Interscience, New York, 1969).
36. M. Kakudo and N. Kasai, *X-ray Diffraction by Polymers* (Elsevier, Amsterdam, 1972).
37. R.D. Heidenreich, *Fundamentals of Transmission Microscopy* (Wiley, New York, 1964).
38. J.I. Goldstein, D.E. Newbury, D.C. Joy, C.E. Lyman, P. Echlin, E. Lifshin, L.C. Sawyer and J.R. Michael, *Scanning Electron Microscopy and X-ray Microanalysis*, 3rd ed. (Plenum, New York, 2003).
39. C.E. Hall, *Introduction to Electron Microscopy*, 2nd ed. (McGraw-Hill, New York, 1966).
40. P. Buseck, J. Cowley and L. Eyring, Eds. *High-Resolution Transmission Electron Microscopy and Associated Techniques* (Oxford University Press, Oxford, 1988).
41. K. Stenn and G.F. Bahr, *J. Ultrastruct. Res.* **31** (1970) 526.
42. D.T. Grubb, *J. Mater. Sci.* **9** (1974) 1715.
43. L. Reimer, Ed. *Energy-Filtering Transmission Electron Microscopy* (Springer, New York, 1995).
44. K.H. Kortje, U. Paulus, M. Ibsch and H. Rahmann, *J. Microsc.* **183** (1996) 89.
45. J.C.H. Spence, *High-Resolution Tranmission Electron Microscopy*, 3rd ed. (Oxford University Press, Oxford, 2003).
46. D.C. Martin and E.L. Thomas, *Polymer* **36** (1995) 1743.
47. D.C. Martin, J. Chen, J. Yang, L.F. Drummy and C. Kuebel, *J. Polym. Sci. B Polym. Phys.* **43** (2005) 1749.
48. S. Kumar and W.W. Adams, *Polymer* **31** (1990) 15.
49. W.G. Hartley, *Proc. Roy. Microsc. Soc.* **9** (1974) 167.
50. W.A. Shurcliff and S.S. Ballard, *Polarized Light* (Van Nostrand, New York, 1964).
51. G.N. Ramachandran and S. Ramaseshan, in *Handbuch der Physik*, edited by S. Flugge (Springer-Verlag, Berlin, 1961).
52. C.W. Mason, *Handbook of Chemical Microscopy* (Wiley, New York, 1983).
53. D.A. Hemsley, Ed. *Applied Polymer Light Microscopy* (Elsevier Applied Science, London, New York, 1989).
54. N.H. Hartshorne and A. Stuart, *Practical Optical Crystallography*, 2nd ed. (Elsevier, New York, 1969).
55. F.D. Bloss, *Optical Crystallography* (Mineral Society of America. 1999).
56. J.G. Delly, *The Michel-Lévy Interference Color Chart—Microscopy's Magical Color Key*. (2003) Available at http://www.modernmicroscopy.com. Accessed October 2006.
57. M.M. Swann and J.M. Mitchison, *J. Exp. Biol.* **27** (1950) 226.
58. N.H. Hartshorne, *Sci. Prog.* **50** (1962) 11.
59. L. Reimer and P.W. Hawkes, *Scanning Electron Microscopy, Physics of Image Formation and Microanalysis* (Springer-Verlag, Berlin, 1998).
60. L. Reimer, *Image Formation in Low-Voltage Scanning Electron Microscopy* (SPIE, Bellingham, WA, 1993).
61. J.H. Butler, D.C. Joy, G.F. Bradley and S.J. Krause, *Polymer* **36** (1995) 1781.
62. D.L. Vezie, E.L. Thomas and W.W. Adams, *Polymer* **36** (1995) 1761.
63. R. Gauvin, K. Robertson, P. Horny, A.M. Elwazri and S. Yue, *JOM* **58** (2006) 20.
64. J.E. Barth and P. Kruit, *Optik* **101** (1996) 101.
65. S.K. Chapman, *Scanning Microscopy* **13** (1999) 141.
66. H.A. Bethe, *Ann. Phys.* **5** (1930) 325.
67. K. Kanaya and S. Okayama, *J. Phys. D Appl. Phys.* **5** (1972) 43.
68. K.F.J. Heinrich, in *Proc. 4th Intl. Conf. on X-ray Optics and Microanal.*, edited by R. Castaing, P. Deschamps and J. Philbert (Hermann, Paris, 1966).
69. K.F.J. Heinrich and D.E. Newbury, *Electron Probe Quantitation* (Plenum Press, New York, 1991).
70. D.B. Wittry, in *X-ray Optics and Microanalysis, 4th Intl. Cong.*, edited by R. Castaing, P. Deschamps and J. Philibert (Herman, Paris, 1966).
71. D.C. Joy, *Scanning* **11** (1989) 1.
72. D.C. Joy and C.S. Joy, *Microsc. Microanal.* **4** (1998) 475.
73. J.B. Pawley, *J. Microsc.* **136** (1984) 45.
74. J.J. Hren, J.I. Goldstein and D.C. Joy, *Introduction to Analytical Microscopy* (Plenum, New York, 1979).
75. D. Drouin, CASINO. Available at http://www.gel.usherbrooke.ca/casino/. Accessed July 2006.
76. D.C. Joy and J.B. Pawley, *Ultramicroscopy* **47** (1992) 80.
77. J. Chang, S. Krause and R. Gorur, *Proc. XIIth Intl. Congr. Electr. Micr.*, edited by L.D. Peachey and D.B. Williams (San Francisco Press, Inc., 1990), p 1108.
78. C.W. Price and P.L. McCarthy, *Scanning* **10** (1988) 29.

79. P.B. Himelfarb and K.B. Labat, *Scanning* **12** (1990) 148.
80. B. Volbert and L. Reimer, *Scanning Electron Microscopy* **4**, (1980) 1.
81. M. Brunner and R. Schmid, *Scanning Microsc.* **1** (1987) 1501.
82. L. Reimer and B. Volbert, *Scanning* **2** (1979) 238.
83. M.T. Postek, W.J. Keery and N.V. Frederick, *Rev. Sci. Instrum.* **61** (1990) 1648.
84. H. Jaksch, M. Steigerwald, V. Drexel and H. Bihr, *Microsc. Microanal.* **9** (2003) 106.
85. G.D. Danilatos and R. Postle, *Scanning Electron Microsc.* **Part 1** (1982) 1.
86. G.D. Danilatos, *Adv. Electronics Electron Phys.* **71** (1988) 109.
87. V. Robinson, *J. Comp. Assisted Microsc.* **4** (1992) 247.
88. D.J. Stokes, *Philos. Trans. Roy. Soc. London A* **361** (2003) 2771.
89. G.D. Danilatos, *Microsc. Res. Techn.* **25** (1993) 354.
90. D.E. Newbury, *J. Res. Natl. Inst. Stand. Technol.* **107** (2002) 567.
91. R. Durkin and J.S. Shah, *J. Microsc.* **169** (1993) 33.
92. B.J. Griffin and C. Nockolds, in *Proc. 14th Intl. Conf. on Electron Microscopy*, edited by H. Calderon (Institute of Physics, London, 1998), p. 359.
93. R.A. Rapp, *Pure Appl. Chem.* **56** (1984) 1715.
94. K. Ramani, C.J. Hoyle and N.C. Parasnis, *ASME Mater. Div. Publ. MD* **46** (1993) 633.
95. D.J. Stokes, *Adv. Eng. Mater.* **3** (2001) 126.
96. N. Franz, M.O. Ahlers, A. Abdullah and H. Hohenberg, *J. Mater. Sci.* **41** (2006) 4561.
97. S. Mehta, in *Proc. SPE Annu. Tech. Conf. Exhib.* **Pi** (*Production Operations and Engineering pt* 2) (1991) 445.
98. H.-M. Choi and H.-J. Kwon, *Text. Res. J.* **63** (1993) 211.
99. B. Anczykowski, J.P. Cleveland, D. Kruger, V. Elings and H. Fuchs, *Appl. Phys. A* **66** (1998) S885.
100. T.R. Rodriguez and R. Garcia, *Appl. Phys. Lett.* **82** (2003) 4821.
101. A. Schirmeisen, B. Anczykowski and H. Fuchs, in *Applied Scanning Probe Methods*, edited by B. Bushan, H. Fuchs and S. Hosaka (Springer-Verlag, Berlin, 2003), p. 3.
102. A. Humphris, A. Round and M. Miles, *Surf. Sci.* **491** (2001) 468.
103. T. Sulcheck, R. Hsieh, J. Adams, G. Yaralioglu, S. Minne, C. Quate, J. Cleveland, A. Atalar and D. Adderton, *Appl. Phys. Lett.* **76** (2000) 1473.
104. M. Tortonese, R. Barrett and C. Quate, *Appl. Phys. Lett.* **62** (1993) 834.
105. R. Erlandsson, G. McClelland, C. Mate and S. Chiang, *J. Vac. Sci. Technol. A* **6** (1988) 266.
106. A. Ruf, M. Abraham, J. Diebel, W. Ehrfeld, P. Guenther, M. Lacher, K. Mayr and J. Reinhardt, *J. Vac. Sci. Technol. B* **15** (1997) 579.
107. S. Alexander, L. Hellemans, O. Marti, J. Schneir, V. Elings, P. Hansma, M. Longmire and J. Gurley, *J. Appl. Phys.* **65** (1989) 164.
108. S. Magonov and D. Reneker, *Annu. Rev. Mater. Sci.* **27** (1997) 175.
109. R. Colton, *J. Vac. Sci. Technol. B* **22** (2004) 1609.
110. D. Chernoff and S. Magonov, in *Comprehensive Desk Reference of Polymer Characterization and Analysis*, edited by R. Brady (Oxford University Press, New York, 2003).
111. G. Binnig, C. Gerber and C.F. Quate, *Phys. Rev. Lett.* **56** (1986) 930.
112. J. Israelachvili, *Surface and Intermolecular Forces, 2nd Edition: With Applications to Biological and Colloidal Systems* (Academic Press, New York, 1992).
113. J.B. Pethica and P. Sutton, *J. Vac. Sci. Technol. A* **6** (1988) 2400.
114. N.A. Burnham, R.J. Colton and H.M. Pollock, *J. Vac. Sci. Technol. A* **9** (1991) 2548.
115. S.P. Jarvis, H. Yamada, S.-I. Yamamoto and J.B. Pethica, *Nature* **384** (1996) 248.
116. W. Heinz and J. Hoh, in *Trends Biotechnol.* **17** (1999) 143.
117. H.-J. Butt, B. Cappella and M. Kappl, *Surf. Sci. Rep.* **59** (2005) 1.
118. K. Feldman, G. Hahner and N. Spencer, in *Microstructure and Microtribology of Polymer Surfaces*, edited by V. Tsukruk and K.J. Wahl (American Chemical Society, Washington, DC, 2000), p. 272.
119. G. Meyers, B. DeKoven and J. Seitz, *Langmuir* **8** (1992) 2330.
120. C. Mate, G. McClelland, R. Erlandsson and S. Chiang, *Phys. Rev. Lett.* **59** (1987) 1942.
121. R. Carpick and M. Salmeron, *Chem. Rev.* **97** (1997) 1163.
122. M. Paige, *Polymer* **44** (2003) 6345.
123. S. Breakspear, J. Smith, T. Nevell and J. Tsiboukllis, *Surf. Interface Anal.* **36** (2004) 1330.
124. K. Feldman, T. Tervoort, P. Smith and N. Spencer, *Langmuir* **14** (1998) 372.
125. G.J. Vancso and H. Schonherr, in *Microstructure and Microtribology of Polymer Surfaces*, edited

by V. Tsukruk and K.J. Wahl (American Chemical Society, Washington, DC, 2000), p. 317.
126. R. Piner and C. Mirkin, *Langmuir* **13** (1997) 6864.
127. M. Motomatsu, H.-Y. Nie, W. Mizutani and H. Tokumoto, *Thin Solid Films* **273** (1996) 304.
128. G. Haugstad, W. Gladfelter, E. Weberg, R. Weberg and R. Jones, *Tribol. Lett.* **1** (1995) 253.
129. R. Cain, S. Biggs and N. Page, *J. Colloid Interface Sci.* **227** (2000) 55.
130. R. Piner and R. Ruoff, *Rev. Sci. Instrum.* **73** (2002) 3392.
131. P. Maivald, H.-J. Butt, S. Gould, C. Prater, B. Drake and J. Gurley, *Nanotechnology* **2** (1991) 103.
132. B. Nysten, C. Meerman and E. Tomasetti, in *Microstructure and Microtribology of Polymer Surfaces*, edited by V. Tsukruk and K.J. Wahl (American Chemical Society, Washington, DC, 2000), p. 304.
133. A. Galuska, R. Polter and K. McElrath, *Surf. Interface Anal.* **25** (1997) 418.
134. A.L. Weisenhorn, M. Khorsandi, M. Kasas, V. Gotzos and H.-J. Butt, *Nanotechnology* **4** (1993) 106.
135. B. Beake, G. Leggett and P. Shipway, *Surf. Interface Anal.* **27** (1999) 1084.
136. H. Schonherr, C. Feng, N. Tomczak and G.J. Vancso, *Macromol. Symp.* **230** (2005) 149.
137. G. Willing, T. Ibrahim, F. Etzler and R. Neuman, *J. Colloid Interface Sci.* **226** (2000) 185.
138. F. Arce, R. Avci, I. Beech, K. Cooksey and B. Wigglesworth-Cooksey, *J. Chem. Phys.* **119** (2003) 1671.
139. Y. Jiao and T. Schaffer, *Langmuir* **20** (2004) 10038.
140. Q. Zhong, D. Inniss, K. Kjoller and V. Elings, *Surf. Sci. Lett.* **290** (1993) L688.
141. S. Magonov, in *Applied Scanning Probe Methods*, edited by B. Bushan, H. Fuchs and S. Hosaka (Springer-Verlag, Berlin, 2003), p. 207.
142. W. Han, S. Lindsay and T. Jing, *Appl. Phys. Lett.* **69** (1996) 4111.
143. R. Garcia and R. Perez, *Surf. Sci. Rep.* **47** (2002) 197.
144. G. Bar, Y. Thoman and M.-H. Wangbo, *Langmuir* **14** (1998) 1219.
145. A. Pfau, A. Janke and W. Heckman, *Surf. Interface Anal.* **27** (1999) 410.
146. G. Bar and G. Meyers, *MRS Bull.* **July** (2004) 464.
147. P. Gleyzes, P. Kuo and C. Boccara, *Appl. Phys. Lett.* **58** (1991) 2989.
148. R. Garcia and A.S. Paulo, *Phys. Rev. B* **60** (1999) 4961.
149. G. Haugstad and R.R. Jones, *Ultramicroscopy* **76** (1999) 77.
150. J.P. Cleveland, B. Anczykowski, A.E. Schmid and V.B. Elings, *Appl. Phys. Lett.* **72** (1998) 2613.
151. B. Anczykowski, B. Gotsmann, H. Fuchs, J.P. Cleveland and V.B. Elings, *Appl. Surf. Sci.* **140** (1999) 376.
152. G. Bar, M. Ganter, R. Brandsch, L. Delineau and M.-H. Whangbo, *Langmuir* **16** (2000) 5702.
153. T.R. Albrecht, P. Grutter, D. Home and D. Rugar, *J. Appl. Phys.* **69** (1991) 668.
154. F.J. Giessibl, *Rev. Mod. Phys.* **75** (2003) 949.
155. S. Tanaka, B. Grevin, P. Rannou, H. Suziki and S. Mashiko, *Thin Solid Films* **499** (2006) 168.
156. M. Brun, S. Decossas, F. Triozon, R. Rannou and B. Grevin, *Appl. Phys. Lett.* **87** (2005) 133101.
157. A. Schirmeisen, D. Weiner and H. Fuchs, *Surf. Sci.* **545** (2003) 155.
158. Y. Martin, D.A. Abraham and H.K. Wickramasinghe, *Appl. Phys. Lett.* **52** (1987) 1103.
159. Y. Martin and H.K. Wickramasinghe, *Appl. Phys. Lett.* **50** (1987) 1456.
160. N. Yerina and S. Magonov, *Rubber Chem. Technol.* **76** (2003) 846.
161. A.V. Krayev and R.V. Talroze, *Polymer* **45** (2004) 8195.
162. M. Meincken, L.J. Balk and R.D. Sanderson, *Macromol. Mater. Eng.* **286** (2001) 412.
163. P. West and N. Starostina, *Microsc. Today* **11** (2003) 20.
164. A. Mendez-Vilas, M.L. Gonzalez-Martin and M.J. Nuevo, *Ultramicroscopy* **92** (2002) 243.
165. ASTM E2382-04, Guide to Scanner and Tip Related Artifacts in Scanning Tunneling Microscopy and Atomic Force Microscopy, in *ASTM Annual Book of Standards*, Vol. 03.06 (2004).
166. R. Howland, in *Atomic Force Microscopy/Scanning Tunneling Microscopy*, edited by S.H. Cohen, M.T. Bray and M.L. Lightbody (Plenum Press, New York, 1994), p. 347.
167. S.M. Hues, C.F. Draper, K.P. Lee and R.J. Colton, *Rev. Sci. Instrum.* **65** (1994) 1561.
168. J. Fu, *Rev. Sci. Instrum.* **66** (1995) 3785.
169. J.E. Griffith and D.A. Grigg, *J. Appl. Phys.* **74** (1993) R83.
170. D. Snevity and J. Vancso, *Langmuir* **9** (1993) 2253.
171. X. Tian, N. Xi, Z. Dong and Y. Wang, *Ultramicroscopy* **105** (2005) 336.

172. ASTM E1813-96, Standard Practice for Measuring and Reporting Probe Tip Shape in Scanning Probe Microscopy, in *ASTM Annual Book of Standards*, Vol. 03.06 (2002).
173. K.L. Westra, A.W. Mitchell and D.J. Thomson, *J. Appl. Phys.* **74** (1993) 3608.
174. T. Thundat, X.Y. Zheng, S.L. Sharp, D.P. Allison, R.J. Warmack, D.C. Joy and T.L. Ferrell, *Scanning Microsc.* **6** (1992) 903.
175. J. Vesenka, M. Guthold, C. Tang, D. Keller, E. Delaine and C. Bustamante, *Ultramicroscopy* **42–44** (1992) 1243.
176. D. Keller, *Surf. Sci.* **235** (1991) 353.
177. G.S. Pingali and R.C. Jain, *Proc. SPIE Int. Soc. Opt. Eng.* **1823** (1992) 151.
178. U.D. Schwarz, H. Haefke, P. Reimann and H.-J. Guntherodt, *J. Microsc.* **173** (1994) 183.
179. J. Vesenka, S. Manne, R. Giberson, T. Marsh and E. Henderson, *Biophys. J.* **65** (1993) 992.
180. S. Xu and M.F. Arnsdorf, *J. Microsc.* **173** (1994) 199.
181. C. Odin and J.P. Aime, *Surf. Sci.* **317** (1994) 321.
182. M.J. Allen, N.V. Hud, M. Balooch, R.J. Tench, W.J. Siekhaus and R. Balhorn, *Ultramicroscopy* **42** (1992) 1095.
183. J.S. Villarrubia, in *Applied Scanning Probe Methods*, edited by B. Bushan, H. Fuchs and S. Hosaka (Springer-Verlag, Berlin, 2003), p. 147.
184. J.S. Villarrubia, *Surf. Sci.* **321** (1994) 287.
185. J.S. Villarrubia, *J. Vac. Sci. Technol. B* **14** (1996) 1518.
186. J.S. Villarrubia, *J. Res. Natl. Inst. Stand. Technol.* **102** (1997) 425.
187. S. Dongmo, M. Troyon, P. Vautrot, E. Delain and N. Bonnet, *J. Vac. Sci. Technol. B* **14** (1996) 1552.
188. S. Dongmo, J.S. Villarrubia, S.N. Jones, T.B. Renegar, M.T. Postek and J.F. Song, *AIP Conference Proc.* **449** (1998) 843.
189. L.S. Dongmo, J.S. Villarrubia, S.N. Jones, T.B. Renegar, M.T. Postek and J.F. Song, *Ultramicroscopy* **85** (2000) 141.
190. P. Grutter, W. Zimmermann-Edling and D. Brodbeck, *Appl. Phys. Lett.* **60** (1992) 2741.
191. H.Y. Nie, M.J. Walzak and N.S. McIntyre, *Rev. Sci. Instrum.* **73** (2002) 3831.
192. G. Bar, Y. Thoman, R. Brandsh and H.-J. Cantow, *Langmuir* **13** (1997).
193. H.-Y. Nie and N.S. McIntyre, *Langmuir* **17** (2001) 432.
194. G. Hughes, *Radiation Chemistry* (Clarendon Press, Oxford, 1973).
195. M. Dole., Ed. *The Radiation Chemistry of Macromolecules* (Academic Press, New York, 1973).
196. R.L. Clough, S.W. Shalaby and Eds. *Radiation Effects on Polymers*, ACS Symposium Series, 475 (ACS, Washington, DC, 1991).
197. A. Charlesby, *Atomic Radiation and Polymers* (Pergamon, Oxford, 1960).
198. A. Chapiro, *Radiation Chemistry of Polymeric Systems* (Interscience, New York, 1962).
199. S. Okamura, *Recent Trends in Radiation Polymer Chemistry* (Springer-Verlag, Berlin, 1993).
200. G.F. Bahr, F.B. Johnson and E. Zeitler, *Lab. Invest.* **14** (1965) 1115.
201. B.M. Siegel, D.R. Beaman and Eds. *Physical Aspects of Electron Microscopy and Microbeam Analysis* (Wiley, New York, 1975).
202. M.S. Isaacson, in *Principles and Techniques of Electron Microscopy*, edited by M. Hayat (Van-Nostrand Reinhold, New York, 1977).
203. D.T. Grubb, in *Developments in Crystalline Polymers*, edited by D.C. Bassett (Applied Science, London & New York, 1982).
204. E.L. Thomas, in *Structure of Crystalline Polymers*, edited by I.H. Hall (Applied Science, London, 1984).
205. P. Alexander and A. Charlesby, *Nature* **173** (1954) 578.
206. H. Kiho and P. Ingram, *Makromol. Chem.* **118** (1968) 45.
207. P.B. Hirsch, A. Howie, R.B. Nicholson, D.W. Pashley and M.J. Whelan, *Electron Microscopy of Thin Crystals* (Butterworths, London, 1965).
208. B. Gale and K.F. Hale, *Br. J. Appl. Phys.* **12** (1961) 115.
209. J. Ling, *Br. J. Appl. Phys.* **18** (1967) 991.
210. L.G. Pittaway, *Br. J. Appl. Phys.* **15** (1964) 967.
211. Y. Talmon and E.L. Thomas, *J. Microsc.* **111** (1977) 151.
212. H. Kohl, H. Rose and H. Schnabl, *Optik* **58** (1981) 11.
213. V.E. Cosslett. *Proc. Europ. Conf. on Electron Microscopy*, edited by A.L. Houwink and B.J. Spit (Nederlandse Vereniging voor Electronenmikroskopie, Delft, 1960).
214. R. Freemen and K.R. Leonard, *J. Microsc.* **122** (1981) 275.
215. J.W. Heavens, A. Keller, J.M. Pope and D.M. Rowell, *J. Mater. Sci.* **5** (1970) 53.
216. M.S. Isaacson, *Ultramicroscopy* **4** (1979) 193.
217. R.F. Egerton, *Ultramicroscopy* **5** (1980) 521.
218. R.F. Egerton, *J. Microsc.* **126** (1982) 96.

References

219. B.D. Lauterwasser and E.J. Kramer, *Philos. Mag.* **A39** (1979) 469.
220. D. Vesely, *Polym. Eng. Sci.* **36** (1996) 1586.
221. K. Kobayashi and K. Sakaoku, *Lab. Invest.* **14** (1965) 1097.
222. D.T. Grubb, *J. Phys. E Sci. Instrum.* **4** (1971) 222.
223. D.L. Dorset and F. Zemlin, *Ultramicroscopy* **17** (1985) 229.
224. D.T. Grubb and G.W. Groves, *Philos. Mag.* **24** (1971) 815.
225. F.P. Price, *J. Polym. Sci.* **37** (1959) 71.
226. J. Dlugosz and A. Keller, *J. Appl. Phys.* **39** (1968) 5776.
227. D.T. Grubb and A. Keller, *J. Mater. Sci.* **7** (1972) 822.
228. D.T. Grubb, A. Keller and G.W. Groves, *J. Mater. Sci.* **7** (1972) 131.
229. E.H. Andrews, *Proc. Roy. Soc.* **A270** (1962) 232.
230. C.G. Cannon and P.H. Harris, *J. Macromol. Sci. Phys.* **B3** (1969) 357.
231. J. Dlugosz, D.T. Grubb, A. Keller and M.B. Rhodes, *J. Mater. Sci.* **7** (1972) 142.
232. J.E. Breedon, J.F. Jackson, M.J. Marcinkowski and M.E. Taylor, *J. Mater. Sci.* **8** (1973) 1071.
233. D. Fotheringham and B. Paker, *J. Mater. Sci.* **11** (1976) 979.
234. S. Bandyopadhy and H.R. Brown, *Polymer* **19** (1978) 589.
235. Y. Chen and J.B. Pawley, in *Multidimensional Microscopy* edited by P.C. Cheng, T.H. Lin, W.L. Wu and J.L. Wu (Springer-Verlag, Berlin, 1994).
236. C.P. Royall, B.L. Thiel and A.M. Donald, *J. Microsc.* **204** (2001) 185.
237. E.L. Thomas and D.G. Ast, *Polymer* **15** (1974) 37.
238. R.M. Glaeser, *J. Ultrastruct. Res.* **36** (1976) 466.
239. L. Yin, *Polymer* **44** (2003) 6489.
240. D.L. Misell, *J. Phys D Appl. Phys.* **10** (1977) 1085.
241. R.C. Williams and H.W. Fischer, *J. Mol. Biol.* **52** (1970) 121.

Chapter 4
Specimen Preparation Methods

4.1	SIMPLE PREPARATION METHODS 132	
	4.1.1 Optical Preparations 132	
	4.1.2 SEM Preparations 133	
	4.1.3 TEM Preparations 133	
	4.1.3.1 Specimen Support Films 134	
	4.1.3.2 Dispersions 135	
	4.1.3.3 Single Crystal Formation 135	
	4.1.3.4 Crystalline Structures ... 137	
	4.1.3.5 Disintegration 137	
	4.1.3.6 Casting Thin Films from Solution or the Melt ... 137	
	4.1.3.7 Drawing Thin Films... 139	
	4.1.4 SPM Preparations 140	
4.2	POLISHING 142	
	4.2.1 Limiting Artifacts 142	
	4.2.2 Polishing Specimen Surfaces ... 143	
	4.2.2.1 Polishing Surfaces 144	
	4.2.2.2 Polishing Thin Sections 145	
4.3	MICROTOMY 146	
	4.3.1 Peelback of Fibers/Films for SEM 146	
	4.3.2 Microtomy for OM 147	
	4.3.2.1 Microtomes 147	
	4.3.2.2 Specimen Mounting ... 148	
	4.3.2.3 Specimen Embedding... 149	
	4.3.2.4 Microtomy 149	
	4.3.3 Microtomy for SEM 150	
	4.3.4 Microtomy for TEM and SPM 150	
	4.3.4.1 Specimen Drying 151	
	4.3.4.2 Embedding Media ... 151	
	4.3.4.3 Mounting Specimens and Block Trimming... 152	
	4.3.4.4 Ultramicrotomy 152	
	4.3.4.5 Literature Review 153	
	4.3.5 Cryomicrotomy for TEM and SPM 154	
	4.3.5.1 Literature Review 155	
	4.3.5.2 Summary 157	
	4.3.6 Microtomy for SPM 158	
	4.3.6.1 Literature Review 159	
	4.3.7 Limiting Artifacts in Microtomy 160	
4.4	STAINING 160	
	4.4.1 Introduction 160	
	4.4.1.1 Literature Review 161	
	4.4.2 Osmium Tetroxide 162	
	4.4.2.1 Preferential Absorption 163	
	4.4.2.2 Two Step Reactions ... 163	
	4.4.2.3 General Method 164	
	4.4.2.4 Inclusion Methods ... 164	
	4.4.2.5 Staining for SEM and SPM 165	
	4.4.3 Ruthenium Tetroxide 166	
	4.4.3.1 Literature Review 167	
	4.4.3.2 General Method and Discussion 169	
	4.4.3.3 Staining for SEM and STEM 170	
	4.4.3.4 Examples for TEM ... 170	
	4.4.4 Chlorosulfonic Acid and Uranyl Acetate 173	
	4.4.4.1 Literature Review 173	
	4.4.4.2 General Method and Examples 174	

Specimen Preparation Methods

- 4.4.5 **Phosphotungstic Acid** 175
  - 4.4.5.1 *Literature Review* 175
  - 4.4.5.2 *General Method and Examples* 176
- 4.4.6 **Ebonite** 177
  - 4.4.6.1 *General Method and Examples* 177
- 4.4.7 **Silver Sulfide** 178
- 4.4.8 **Mercuric Trifluoroacetate** ... 178
  - 4.4.8.1 *General Method* 179
- 4.4.9 **Iodine and Bromine** 179
- 4.4.10 **Summary** 179
- 4.5 ETCHING 181
  - 4.5.1 **Solvent and Chemical Etching** 181
    - 4.5.1.1 *Literature Review* 181
  - 4.5.2 **Acid Etching: Overview** 183
    - 4.5.2.1 *Literature Review* 183
  - 4.5.3 **Permanganate Etching** 184
    - 4.5.3.1 *Introduction* 184
    - 4.5.3.2 *General Method* 185
    - 4.5.3.3 *Literature Review* 185
    - 4.5.3.4 *Literature Review: AFM* 188
  - 4.5.4 **Plasma and Ion Etching** 188
    - 4.5.4.1 *Literature Review* 189
    - 4.5.4.2 *General Method and Examples* 191
    - 4.5.4.3 *Summary* 193
  - 4.5.5 **Focused Ion Beam Etching** ... 194
    - 4.5.5.1 *Literature Review and Examples* 194
  - 4.5.6 **Summary** 195
- 4.6 REPLICATION 195
  - 4.6.1 **Simple Replicas** 197
    - 4.6.1.1 *Replication for OM* ... 197
    - 4.6.1.2 *Replication for SEM* ... 197
    - 4.6.1.3 *Methods and Examples* 197
  - 4.6.2 **Replication for TEM** 198
    - 4.6.2.1 *Direct Replicas* 198
    - 4.6.2.2 *Two Stage Replicas* ... 199
    - 4.6.2.3 *Extraction Replicas* ... 201
- 4.7 CONDUCTIVE COATINGS 201
  - 4.7.1 **Coating Devices** 202
    - 4.7.1.1 *Vacuum Evaporators* ... 202
    - 4.7.1.2 *Sputter Coaters* 202
    - 4.7.1.3 *High Resolution Coating Devices* 202
  - 4.7.2 **Coatings for TEM** 203
    - 4.7.2.1 *Carbon Coatings* 203
    - 4.7.2.2 *Shadowing* 203
  - 4.7.3 **Coatings for SEM and STM** ... 203
    - 4.7.3.1 *Sputter or Evaporative Coating* 204
    - 4.7.3.2 *Ion Beam Sputter Coating* 204
    - 4.7.3.3 *Coatings for X-ray Microanalysis* 206
  - 4.7.4 **Artifacts** 207
    - 4.7.4.1 *Charging* 207
    - 4.7.4.2 *Beam Damage* 209
    - 4.7.4.3 *Low Voltage FESEM Imaging Artifacts* 210
  - 4.7.5 **Gold Decoration** 211
- 4.8 YIELDING AND FRACTURE ... 212
  - 4.8.1 **Fractography** 212
    - 4.8.1.1 *Fracture Types* 212
  - 4.8.2 **Fracture: Standard Physical Testing** 213
    - 4.8.2.1 *Fiber Fractography* ... 213
    - 4.8.2.2 *Fracture of Plastics* ... 214
    - 4.8.2.3 *Composite Fractures* ... 217
  - 4.8.3 **Crazing** 217
    - 4.8.3.1 *Preparation for TEM* ... 219
    - 4.8.3.2 *Deformation Methods* ... 221
  - 4.8.4 ***In Situ* Deformation** 221
    - 4.8.4.1 *In Situ Deformation in the SEM* 222
    - 4.8.4.2 *In Situ Deformation in the TEM* 223
    - 4.8.4.3 *In Situ Deformation in the AFM* 223
- 4.9 CRYOGENIC AND DRYING METHODS 226
  - 4.9.1 **Simple Freezing Methods** 226
  - 4.9.2 **Freeze Drying** 227
    - 4.9.2.1 *General Method and Examples* 227
    - 4.9.2.2 *Literature Review* 228
  - 4.9.3 **Critical Point Drying** 230
    - 4.9.3.1 *General Method and Examples* 230
  - 4.9.4 **Freeze Fracture-Etching** 231
    - 4.9.4.1 *Biological Method* ... 231
    - 4.9.4.2 *Literature Review* 231
  - 4.9.5 **Cryomicroscopy** 232
    - 4.9.5.1 *Cryo-SEM* 232
    - 4.9.5.2 *Cryo-TEM* 233
- **References** 234

4.1 SIMPLE PREPARATION METHODS

Specimen preparation ranges from direct and simple methods to complex, time consuming, and even frustrating ones. Fortunately, there are a number of simple methods that are quite adequate for some materials. For example, many particulate or fibrous materials may be handled by simple methods. This section covers a wide range of these simpler and generally more direct methods, which are described in broad subsections: optical microscopy (OM), scanning electron microscopy (SEM), transmission electron microscopy (TEM), and scanning probe microscopy (SPM) preparations. It must be emphasized that quick observation of most materials by a combination of a simple microscopy technique and direct preparation methods is often helpful in shedding light on the problem. This aids determination of the best approach to a solution. In many cases, there is no one *correct* approach, but there may well be approaches that can save time, if they are conducted early in the study. Complementary correlative techniques are at least useful, and at times essential, for problem solving, requiring several specimen preparation methods for various microscopy techniques. The more complex preparation methods are described in sections of this chapter.

Trade names of products used in specimen preparation are mentioned in the text and, unless otherwise stated, these are standard materials available from a variety of suppliers, some of which are noted in Appendix V. Specific microscopes are not mentioned, but some microscope vendors are listed in Appendix VI.

4.1.1 Optical Preparations

The single most important preparation instrument is the stereo binocular microscope. These instruments are inexpensive and readily available, permitting materials to be observed in either transmitted or reflected light, and the result often provides insights into the problem. Even rather large parts may be examined as part of the important first step in choosing the area of a sample to be analyzed.

Transparent specimens, generally less than $100\,\mu m$ thick, may be directly mounted onto standard glass microscope slides with coverslips, for transmitted light observation. For magnifications less than 100×, this preparation might be sufficient. For higher magnification examination, suitable mounting media are usually required in order to reduce surface reflections. Immersion oils of specific refractive indices, or special mounting media, such as Permount (available from Fisher Scientific, Waltham, MA), may be chosen to provide contrast between the material and the mountant, in the case of particles, whereas matching refractive index oils may be used with fibers and films to permit observation of internal structures. Fibers, particles, small strips of films, and membranes can be prepared in this manner for optical study. Simple reflected light microscopy of large and irregularly shaped specimens can be aided by pressing the underside of the specimen into modeling clay to provide an even top surface.

A direct method of preparing fluids is to use a cavity slide to permit a known fluid thickness to be examined optically. A crystal suspension may be examined in this way [1, 2]. Solutions or solid materials may be placed in a cavity slide or onto a slide with a coverslip (under an inert or dry atmosphere, if needed). Solid materials may be sliced with a razor blade or scalpel, placed on a microscope slide, covered with a coverslip, and heated using a hot plate or a more sophisticated hot stage for the microscope. After heating, pressure on the coverslip can provide a thin specimen for viewing by transmitted light in bright field, phase contrast, or polarized light microscopy (PLM).

Crystallization of polymers can be studied by polarized light microscopy using a hot stage with the microscope. This is very commonly done to correlate morphology changes at melting and other transition temperatures with complementary thermal analysis. Liquid crystalline monomers and polymers (see Section 5.6) are often studied using this technique. Sample preparation is straightforward, as small slices or particles of the polymer are placed between a slide and coverslip and the sample is heated and cooled while being examined for its crystal structure. For example, crystallization of polylactide diblock copolymers was studied by x-ray scattering, differential scanning calorimetry and PLM [3].

4.1.2 SEM Preparations

A major advantage of the SEM for surface observations is that sample preparation is generally simple. In the simplest case, the material to be examined, chosen carefully from a larger sample, is placed on double sided sticky tape on a specimen stub. In order to maintain electrical contact with the stub for conduction, the tape covers only part of the specimen stub so the sample is placed partly on the tape and partly on the stub. Conductive paints, such as silver or carbon suspensions, are used to attach and ground the specimen to the sample holder. Such paints can also be dabbed onto the tape or the base of the specimen to provide contact with the stub. Another common method is to make a glue solution by dissolving sticky tape in chloroform, placing a drop on the specimen stub, and allowing it to dry before use. Such simple preparation methods are very successful for particulate materials, fibers, films, membranes, and even rather large plastic parts. Dry material can be sprinkled onto the specimen stub, or using a stereo binocular microscope, small particles can be selected and moved using fine forceps. Care must be taken not to abuse the material during handling. Common difficulties are encountered when fine particles or fibrous material becomes embedded in the paints, glues, or tapes, or worse, when these materials wick onto the polymer surface of interest and are viewed and misinterpreted as morphologies associated with the specimen. Fine materials can be suspended in water or another fluid and a drop placed on a glass coverslip, attached to a SEM stub, and dried. Additional substrates for SEM (and for SPM see Section 4.1.4) include glass (generally glass coverslips), silicon, highly oriented pyrolytic graphite (HOPG), and freshly cleaved mica, all of which are flat and easy to prepare. Mica and HOPG are layered minerals that can be cleaved by attaching adhesive tape and peeling it apart to produce a fresh, clean, and atomically smooth surface.

Preparation is not necessarily complete as polymers generally require a conductive coating (see Section 4.7.3) to be applied for imaging. Another possibility is the use of low voltage or variable pressure SEM to avoid charging without such a coating (see Section 4.7.4). Dudler et al. [4] studied the conductive pathway created by ionic conducting polymer in polyolefin samples containing a percolating, antistatic stabilizer by low voltage SEM (LVSEM) and without use of a conducting coating. Samples were prepared by cutting the compound with a mechanical puncher and the film was glued with conductive carbon adhesive onto an electrically grounded sampler holder. Kugge et al. [5] compared the structure of coating layers prepared on nonabsorbent substrates, composed of mineral pigments, latex binders, thickeners, and dispersants, by both environmental and conventional mode SEM, the latter gold coated. No preparation was used for the former technique; similar features were viewed at low magnification.

Hearle et al. [6] described several specimen holders that are useful for simple preparations of fibers and fabrics where the sample is attached to the holder. Each SEM manufacturer and the various electron microscope (EM) supply companies provide different microscope stubs and modifications can be easily made.

4.1.3 TEM Preparations

Samples for TEM must fit onto a specimen support known as a grid or screen. This *grid* is a metal mesh screen, generally 2–3 mm in diameter, which fits into the specimen holder of the microscope. Grids come in a variety of mesh sizes and shapes. Sizes in general use range from 50 to 400 mesh, that is from 50 to 400 holes per inch. The grid mesh used is typically square; but for some materials, slotted screens, rectangles, hexagons, or single holes might be preferred. Grids are made of copper, for the most part, but beryllium, gold, polymer, and nickel grids are used for various applications, for their chemical resistance, and for x-ray analysis.

Microtomed sections, which contain polymer embedded in a resin, are usually directly supported on the grid. Very small sections or sections that break apart may require a support to hold them onto the grid. Particles, crystals, emulsions, and other fine materials are placed on an electron transparent support film on the TEM grid. The preparation of such support films will be described below.

Specimen preparation for TEM generally involves the formation of a thin film of the material less than 100 nm thick. For high resolution EM (HREM) imaging, the sample must be as thin as possible and be supported on a thin, stable substrate. The methods used for this preparation depend upon the nature of the polymer and its physical form. In the case of thick or bulk specimens, microtomy is generally used. In the case of solutions, powders, or particulates, simpler methods can provide a thin, dispersed form of the material. Three types of simple preparations will be described in this section: dispersion, disintegration, and film casting. The more complex methods such as microtomy, replication, etching, and staining will be described in other sections of this chapter.

4.1.3.1 Specimen Support Films

Plastic, carbon, and metal films (see Section 4.7) are used as specimen supports on TEM grids. There are two plastic support materials in general use: collodion (0.5% solution of nitrocellulose in amyl acetate) and formvar (0.25% polyvinyl formal in ethylene dichloride). These polymers are available as powders, solutions, or prepared films on TEM grids. (The suppliers provide information on handling for hazardous materials.) Formvar films, especially holey ones, are used as substrates for the formation of holey carbon films. Collodion is less commonly used as it is not as stable in the electron beam as formvar or carbon films.

Premixed solutions of formvar and collodion are recommended for high quality film supports as the powders take several days to dissolve and are less uniform. Collodion is generally film cast on a water surface. A large Petri dish filled with distilled water is allowed to settle and collodion is dropped onto the water surface and allowed to dry. The first film cast is used to clean the water surface. The films are placed onto the grids by any one of several methods. In one method, the grids are placed upside down on the film surface and are lifted up with a glass microscope slide placed on their surface and then scooped down into the water and up with the grids on the glass, under the film. The film is allowed to dry, and a razor is used to separate the film around the grids. In another method, the grids are placed on a mesh screen below the water surface, and the water is slowly removed allowing the film to settle on the grid surfaces. A sintered glass filter apparatus is useful in this case.

Formvar solutions are placed in any tall glass container that has a lid. Clean glass slides, pretreated with a detergent to aid release of the cast film, are dipped into the solution and then quickly lifted up above the solution, covered, and permitted to dry slowly in the solvent vapor. To free the film, the glass slide is scored around the perimeter with a razor, scalpel, or needle. The film may then be floated onto a water surface and the grids placed on it and picked up, as for collodion. Alternately, the slide can be scored into about 3 mm squares. Either form of film is slipped into the water by placing the slide at an angle to the surface and slowly immersing it beneath the surface. The small squares of film are picked up individually by placing the grids under the water surface and scooping them up. Coated grids are dried on filter paper and subsequently carbon coated to increase their beam stability. Breathing on the slide prior to immersion in the water bath aids release of the film from the glass.

Support films generally used for microscopy above 50,000× magnification are either perforated, holey, carbon coated plastic, or carbon or holey carbon films for highest resolution. In the case of holey plastic supports, the presence of moisture or some other immiscible liquid in the solution used for coating will provide holes in the films. A fuller discussion on this topic is found in the specimen preparation text by Goodhew [7]. In our experience, moisture from the air or from one's breath on the slide causes the formation of holes and, in fact, it may be more difficult to prepare continuous films. These perforated supports provide an area of the specimen supported by the film with other areas suspended over the holes.

4.1.3.2 Dispersions

Solutions or suspensions of polymers can be dispersed by several methods including atomizing or spraying, or the fluid might simply be allowed to spread over the support film on the TEM grid. The grids are placed on filter paper to remove the excess material. Preparation of polymer materials suspended in oils requires removal of the oil prior to application of a conductive coating. A droplet of the material in the oil is permitted to spread over the support film. The grids are placed on a fine mesh metal screen in a Petri dish containing a solvent at the level just below the screen. The solvent is replenished until the oily deposit is removed. Metal shadowing (see Section 4.7.2.2), evaporated at an angle of 30° to 60° to the specimen, is generally employed to add contrast to the material prior to examination.

Dilute solutions of some high molecular weight polymers can be dispersed directly onto carbon films to provide macromolecules collapsed into spherical shapes, which can be observed in the microscope. The size of the particle is related to the molecular weight. Hobbs [8] described a process for the examination of single molecules of rubbers or glassy polymers that involved the addition of a nonsolvent to aid collapse and the spraying of a fine mist onto a carbon coated TEM grid. Dissolving polyethylene in a 0.005% n-hexadecane solution was shown to be successful [9]. Dilute latex emulsions, with a glass transition above room temperature, can be atomized or sprayed using a fine pipette attached to a standard can of gas. Difficulties arise if the latex is not dilute, as clumps of particles are observed. With low glass transition latexes, cryogenic or chemical hardening methods must be used to ensure the mechanical stability of the polymer particles.

Powders are suspended in a fluid, mixed, and dropped onto a plastic or carbon coated support grid. If the powders do not disperse by this method, the powder can be rolled onto a plastic support film on a glass slide, cut, scored, floated off on a water surface, and picked up. Another variation is that powders may be mixed with the collodion or formvar and a film cast with the material held in the support film. This works best when the particles are larger than the film thickness. Crystallites can be powdered and deposited with no solvent onto holey carbon films on Au microscope grids and imaged by high resolution TEM at 400 kV [10] or higher as well.

4.1.3.3 Single Crystal Formation

Early TEM studies of polymer single crystals [11] provided much of the fundamental knowledge of polymers. Single crystals are readily formed by precipitation during cooling of dilute solutions of polymers, such as polyethylene (PE) from xylene. Polyethylene single crystals can be prepared for TEM by placing a drop of the crystal suspension in a solvent for the polymer on a carbon coated grid and shadowing to enhance contrast. Bassett and Keller [12] used this method to show the effect of temperature on PE morphology. An example of a TEM micrograph of PE single crystals is shown in Fig. 4.1A. Bassett et al. [13] used this technique to prepare polyoxymethylene in bromobenzene and poly(4-methylpentene-1) in xylene. Single crystals of linear PE were grown isothermally in xylene by self seeding [14]. Single crystals of poly(tetramethylene adipate) were prepared from dilute solutions in ethanol, n-propanol, n-butanol, and other solvents [15]. In a typical preparation, the solution was maintained at 80°C for 30 min and transferred to an oven at 2° above the clouding temperature and held 16 h before cooling. A drop of the turbid solution was placed on a carbon coated grid and shadowed. Two different forms of crystals, identified by electron diffraction, were produced depending on the solvent used.

Chang et al. [16] prepared single crystals of high density polyethylene (HDPE) by casting a xylene suspension on carbon coated glass slides and evaporating the solvent. Carbon platinum shadowed specimens (see Section 4.7.2.2) were studied to determine the nature of the

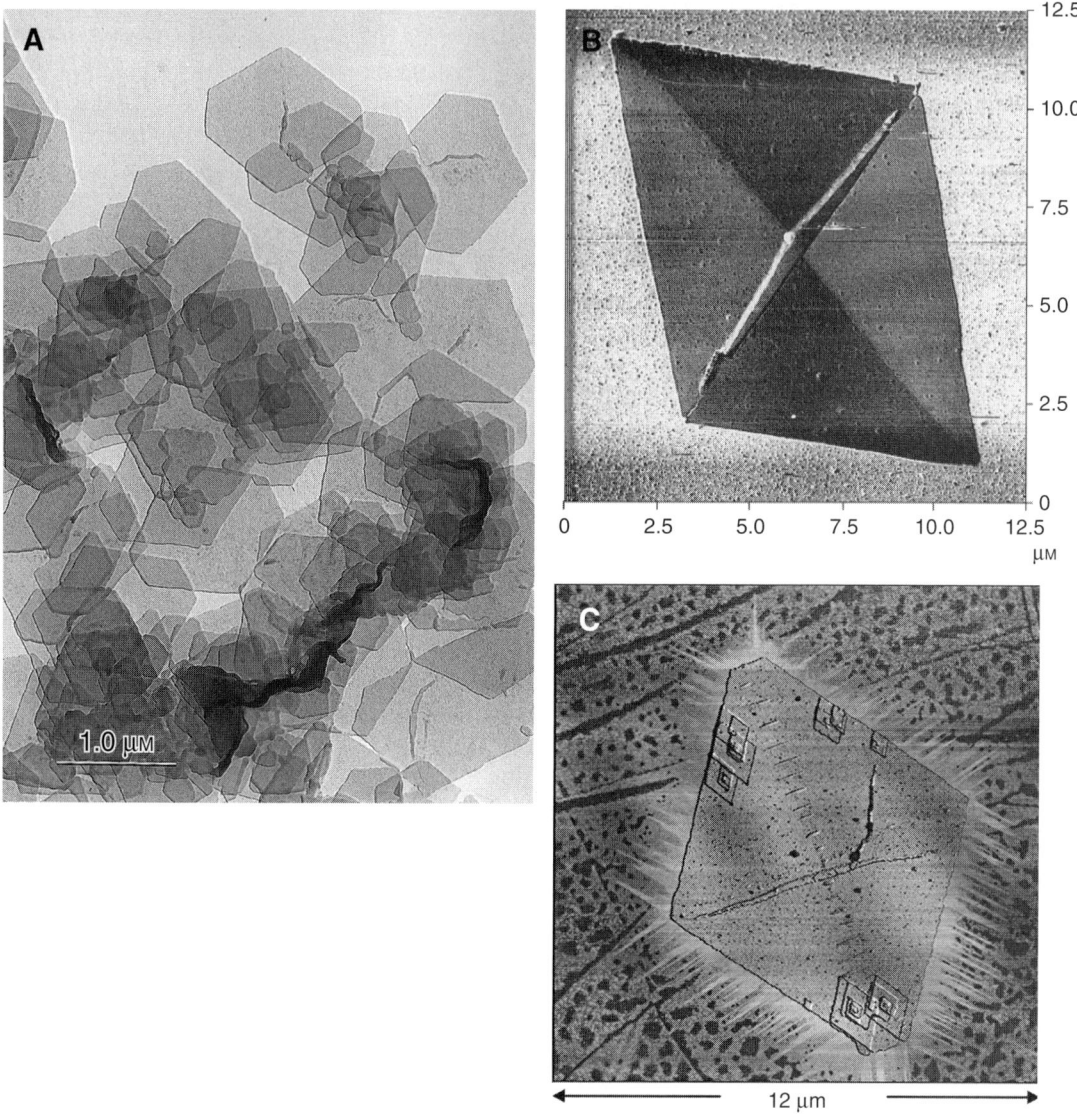

FIGURE 4.1. A TEM micrograph of polyethylene single crystals produced by self-seeding from a 0.05% solution of Sclair 2907 in xylene at 83°C (A) is a typical image produced from single crystals for many decades. It is interesting to compare it to a friction force micrograph (B) of a solution grown PE lamella deposited on mica [20, 21] and an AFM image of a single crystal of PE (C) deposited on graphite and annealed at 75°C [22]. (Figure 4.1B from Vansco and Schönherr [20], © (2000) American Chemical Society; used with permission. Figure 4.1C from Magonov and Yerina [22], © (2005) Springer; used by permission.)

chlorination reaction. Single crystals of poly(p-xylylene) have been observed directly in a high voltage TEM [17–19]. Crystals were obtained by dissolving Parylene-N film in chloronaphthalene by heating at 254°C and holding the 0.05% to 0.1% solution at 210°C overnight, followed by cooling to room temperature. Individual chains in an unknown form of the crystal have been imaged [17] and the structures analyzed [18]. Direct observations have been made of dislocations in poly(p-xylylene) [19].

As an example of SPM using a TEM preparation method, the friction force micrograph in Fig. 4.1B is of a solution grown PE lamella deposited on mica [20, 21]. In this case, the image was collected in air and the contrast is related to the orientation of the folds on the polymer surface, which are oriented along the crystal edge in each sector. Figure 4.1C is an example of atomic force microscopy (AFM) imaging of a single crystal of PE deposited on a flat substrate (graphite) and annealed at 75°C [22]. Significant research is being conducted on crystallization and annealing using SPM techniques.

4.1.3.4 Crystalline Structures

Studies have continued using HREM to provide molecular details. Kübel et al. [23] used low dose HREM to image the structure of regularly twisted poly(*m*-phenylene isophthalamide) (MPDI) strands formed by slow crystallization from solution. The MPDI solution was diluted to less than 0.1 wt.%, exposed to water, and the precipitated powder suspension in water was dropped onto a carbon coated copper TEM grid and water allowed to evaporate. Kübel and Martin [24] studied variations in poly(1,6-di(*N*-carbazolyl)-2,4-hexadiyne) (polyDCHD) nanocrystals prepared by precipitation from acetone solution as an aqueous dispersion. The acetone solution of the DCHD monomer was injected into vigorously stirred deionized water, resulting in a cloudy dispersion. Crystallization was followed by ultraviolet irradiation to start photopolymerization. Droplets of the dispersion were distributed on carbon coated mica, dried overnight, submerged into water, and the floating film collected with copper grids.

Dendrimers have been studied by low dose electron microscopy, using samples prepared by deposition of a dilute solution of the PAM family of molecules (based on a cyclic phenylacetylene backbone) in toluene onto amorphous carbon coated mica sheets [25]. Once the solvent is dried, the carbon film and crystallite layer were floated off the mica substrate and collected on copper grids; gold was evaporated on the samples as a calibration standard for electron microscopy.

4.1.3.5 Disintegration

The use of an ultrasonic bath to disintegrate cellulose into fibrils observable in the TEM was described in 1950 [26]. Even earlier papers were referenced, where either mechanical or ultrasonic vibrations were used to provide thin TEM specimens. Morehead [26] cut samples into short lengths and treated them in water in an ultrasonic bath for about 20 min. Hearle and Simmens [27] described the disintegration of fibers using a blender or an ultrasonic disintegrator.

More recently, high modulus fibers, such as aromatic polyamides, which are difficult to prepare by other methods, were fragmented into fibrillar fine structures [28] by high wattage ultrasonic irradiation in water. This is extremely useful for producing very finely divided material for transmission methods such as lattice imaging and electron diffraction. The major drawback is that the position of the sample in the original fiber is unknown. The choice of liquid used in the sonication preparation is also very important. The effect of the liquid depends on the boiling point, surface tension, and polymer interaction, but ethanol, water, or ethanol/water mixtures are the best choices. In addition, for best results, the materials must be mechanically broken down prior to sonication. Materials often require 30 min or more to fibrillate and the liquid must be kept cool. The sonicator can be pulsed on and off or the bath cooled directly.

4.1.3.6 Casting Thin Films from Solution or the Melt

Thin films can be cast from solutions of crystalline or noncrystalline polymers where the film thickness is controlled by the solution concentration. Polyethylene films have been cast from boiling dilute solutions in xylene [29], generally by dipping a glass slide into the solution, slowly drying in the vapors of the solvent. Roche et al. [30] prepared poly(butylene terephthalate) (PBT) in a 1% solution in hexafluoroisopropanol (HFIP) by depositing drops onto a glass slide placed on an incline in a beaker of the solvent to limit fast drying. Thicker regions on the bottom of the slide were used for optical study while the thinner upper regions were used in the TEM. Samples were used for scanning transmission electron microscopy (STEM) and electron diffraction in order to define the nature of the spherulites.

Vadimsky [31] described a useful method for the preparation of thin films from the melt or solution. Polymer is cast onto carbon coated mica or fractured NaCl crystal substrates or the substrate is dipped into a polymer solution. After solvent evaporation, the film is scored and removed from the glass by floating it onto a water surface. Geil [32, 33], in a variation of this method, deformed PE single crystals by deposition on a Mylar substrate and drawing it before carbon coating and TEM examination. Martin and Thomas [34] used this method as well as several others for high resolution imaging of ordered polymers. They reviewed the history of polymer HREM up until 1995, listing the polymers that have been examined and compared the polymers theoretically and experimentally.

Thin films may also be formed by casting onto a liquid surface to allow easy removal. A range of liquids have been used, such as glycerol, ortho-phosphoric acid, and mercury. Grubb and Keller [35, 36] placed a few drops of ortho-phosphoric acid or glycerol on a microscope slide, on a hot bench, and subsequently a drop of the PE slurry was allowed to fall on this hot surface. The polymer melted as the solvent evaporated, and the film was solidified by cooling. The film was floated onto a water surface and picked up on TEM grids. These self-supported thin films were found to be spherulitic by optical observation, and large areas of the film were thin enough for TEM [36]. Thin films were prepared for HREM and electron diffraction by surface tension spreading in the nematic melt onto hot phosphoric acid, at about 240°C [37]. After quenching, the polymer film was transferred to a water surface and picked up on copper grids and carbon coated to aid stability in the electron beam. In a study by Chu and Wilkes [38], films of poly(ethylene terephthalate) (PET) were cast from trifluoroacetic acid and the spherulitic films were studied by PLM, SEM, and light scattering.

Howell and Reneker [39] prepared thin polymer films from solutions spread on water for HREM studies; for example, poly(ether ether ketone) (PEEK) in *a*-chloronaphthalene, polystyrene (PS) in benzene, PE in hot xylene, were collected on carbon coated grids that had been thinned in an oxygen glow discharge plasma to produce holes or cracks in the carbon. The polymer films were then stained in ruthenium tetroxide (RuO_4) vapor for about 15 min (see Section 4.4.3) to add contrast to the molecular scale features observed by HREM. Highest resolution was achieved by focusing with a large objective aperture; the lattice planes in the particles of graphite were imaged to ensure highest resolution. Thin layers were also formed for HREM [39] by trapping polymer solutions between freshly cleaved sheets of mica separated by a wedge. The wedged mica sheets were placed in a solution of PE in boiling xylene, the wedge removed, and the mica sheets clamped together, withdrawn from the solution, cooled, and the solvent evaporated. The mica sheets were separated, shadowed with platinum/carbon, coated with evaporated carbon (see Section 4.7.2), floated off the mica, and picked up onto grids for HREM imaging.

Drummy et al. [40] used a simple method to prepare thin films, in the case of a semiconductor powder, by atomizing a toluene solution onto carbon coated mica for imaging using a recently developed, inexpensive, tabletop, low voltage electron microscope. Dilute solutions of polymer have also been spin-cast onto a single crystal silicon wafer, previously coated with a thin layer of Au [41]. PET was prepared by this method and by microtomy for examination by several complementary techniques, including electron energy loss spectroscopy (EELS) in STEM, for comparison of radiation damage rates. Windle and coworkers [42, 43] prepared thin films by shearing molten polymers on freshly cleaved rock salt with a sharp razor blade. The films, of thermotropic copolyesters (see Section 5.6), were quickly quenched, the rock salt dissolved, and the films annealed to assess crystallite growth.

Cast films exhibit a range of morphologies due to the effect of solvents, substrates, and orientation. The use of liquids provides a surface with little to no structure of its own compared with solid substrates. In the case of block copolymers, the choice of solvent is quite important to the final structure. Also, thin films may be annealed to reveal equilibrium morphologies that may not be seen under real processing conditions. Spherulites, a common textural structure observed in crystalline polymers, are formed in many industrial processes where the polymer is melted prior to forming the article of interest. Films produced by these industrial processes differ from films used in model studies as the former are usually not thin in the microscopic sense. In true thin films, the spherulitic texture is two dimensional, whereas in these thicker materials the spherulites are three dimensional.

4.1.3.7 Drawing Thin Films

Thin cast films may be drawn manually at room temperature or above. The film to be drawn is transferred to a glass slide coated with an inactive liquid, for example glycerol or silicone oil. Needles or blunted razor blades are pressed into the film and drawn apart. Small regions of drawn film suitable for TEM are produced, but there is no control over the draw ratio or the draw rate. Petermann and Gohil [44] developed a method for forming highly oriented, ultrathin films of crystalline thermoplastics under more controlled conditions. The first step is to make a small quantity of dilute solution (0.3% to 1%) of the polymer. A thin film of molten polymer is then made by placing a drop of the solution onto a hot glass slide and allowing the solvent to evaporate. If the polymer dissolves only at elevated temperatures, the finely divided suspension that forms on cooling the solution can be used, as the polymer redissolves as the drop heats up. The hot glass plate is kept at a temperature where quiescent crystallization is very slow. For HDPE, this is about 125°C. An enclosure may be used to limit temperature fluctuations. A glass rod coated with the polymer is touched to the melt film and slowly drawn away, at a few $cm\,s^{-1}$. The local orientation causes immediate crystallization, producing a solid drawn film that can be as thin as 20 nm. A schematic diagram of the drawing of a film is shown in Fig. 4.2 [44]. The thin drawn film is collected on a clean glass microscope slide, cut

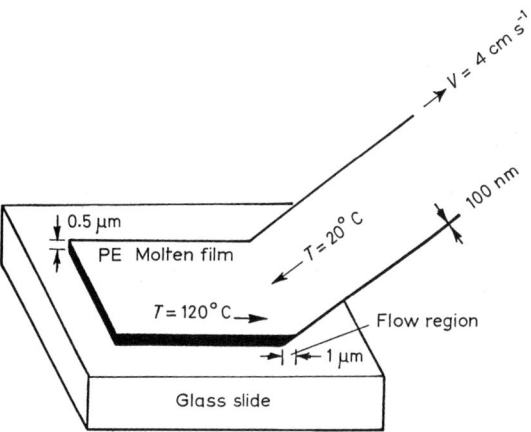

FIGURE 4.2. The method of producing a highly oriented melt drawn film is shown in this schematic. (From Petermann and Gohil [44]; used with permission.)

into small pieces, floated off onto a clean water surface, and mounted on grids for the TEM. The melt film remaining on the glass plate may be used directly in the OM after cooling. The drawing zone is limited to a region of about the same size as the melt film thickness, about 1 μm, so a draw rate of only $1\,cm\,s^{-1}$ corresponds with a very high strain rate, $10^4\,s^{-1}$. A longitudinal flow gradient of this magnitude is close to the conditions that are used in the industrial processes of high speed spinning.

This method has been used primarily for the study of HDPE (linear polyethylene), but it also works for other crystalline polyolefins. The oriented films that are formed are smooth surfaced and uniform in structure. They contain stacks of parallel lamellae on edge, and the lamellae extend through the entire film. This makes the interpretation of transmission images particularly simple, as there is no overlapping of adjacent structural regions. For the same reason, defocus imaging [45–47] is very straightforward and enhances the visibility of the crystals in the film (see Fig. 5.24). The HDPE specimens produced by this method are thin and uniform, with a well defined crystal orientation and arrangement. Preparation is comparatively easy and needs no special equipment beyond a hot plate. Therefore, these materials make excellent training aids for a microscopist who is learning how to take electron diffraction patterns or dark field micrographs from radiation sensitive polymers.

4.1.4 SPM Preparations

One of the most exciting advances in polymer microscopy during the past two decades is the development of SPM techniques and their application to the study of polymer morphology. Early investigators found very simple sample preparation methods could be used because the material could be imaged on a fairly large stage at room temperature with no need for a vacuum chamber. As the techniques have improved, especially for imaging soft biological and polymer materials (recently reviewed [48]), however, the crucial factor that specimen preparation plays in high resolution imaging has become apparent. As a result, traditional preparation methods, including simple dispersions, as noted above, continue to play a role, but cryoultramicrotomy, staining, and etching are also used for specimen preparation. The early history of sample preparations and the simpler methods will be described in this section, and other methods will be found in the appropriate sections to follow.

Scanning tunneling microscopy (STM) was first used to image polymers using many of the same preparation methods used for SEM and TEM. Atomic force microscopy (AFM), especially now using the tapping mode [49], and termed either Tapping Mode™ (trademark of Veeco Instruments, Woodbury, NY) AFM (TMAFM) or intermittent contact AFM (ICAFM), is more commonly used for polymers [50, 51] although not without the need for significant care in specimen preparation and image interpretation. Atomic force microscopy is used for imaging coarse and fine structures, the latter by high resolution surface imaging, and for compositional mapping, electric and magnetic properties [50]. Scanning probe microscopy techniques can be used in a variety of environments, including at elevated temperatures [52] and under various liquids. For example, a study of polyamide membranes was conducted by AFM, in a liquid cell under water, and compared with TEM of stained, embedded, and sectioned membranes [53]. As with more conventional techniques, the potential of artifacts is ever present and from new and possibly unexpected sources. A series of examples of SPM on organic materials systems, with comparison to other microscopy techniques, is found, for example, in Tsukruk and Reneker [54], Jandt [55], Magonov [50, 56], and Bar and Meyers [51].

Surfaces of many materials, such as coatings, paints, films, and so forth, may be observed without any preparation. Methods used for SEM and TEM may be applied for SPM preparations, including mounting on various substrates, dispersion, and casting of films by methods, such as spin casting, melting on the substrate, and melt drawing. Substrates that are most often used for SPM are glass, polycrystalline silicon, HOPG, freshly cleaved mica, and indium-tin-oxide (ITO) coated glass, many of which have been used for decades as they are all flat and easy to prepare. Early work in this field, reviewed by Leggett et al. [57], has shown that a number of features of HOPG surfaces, such as cracks and steps, can produce image features through thin polymer and biopolymer layers. Even worse, some graphite features and artifacts resemble molecules and are easily misinterpreted [57]. The recurrent theme, the need to conduct complementary imaging, especially when using relatively new imaging techniques, thus continues to be of utmost importance for the use of SPM techniques. Microtomy at room and cryogenic temperatures (see Section 4.3) is commonly used to form a flat surface from a bulk material. In addition, SPM can use the microtomed block faces without the need to prepare thin sections on substrates.

Coating with a conducting layer is usually required for STM, unless the sample is very thin, as is likely for Langmuir–Blodgett (LB) films. Albrecht et al. [58] imaged molecular scale orientation and morphology of LB films using both STM and AFM on samples prepared in three ways. Films were prepared by (a) raising a horizontally held graphite substrate from the bottom of the LB trough, (b) by lowering the horizontally held graphite to bring it into contact with the spread film, and (c) by dipping the graphite vertically into the water and then lifting it vertically out. Fibrillated liquid crystalline polymer (LCP) fragments were suspended in a water/ethanol mixture and dropped onto HOPG and thin peels and sections were also mounted on silicon and HOPG for field emission SEM (FESEM), STM [59, 60], and AFM [61], showing similar structures by these complementary techniques. Polystyrene films formed by spin coating solutions onto silicon were studied [62] after carbon coating for STM, which showed isolated molecules. Suzuki et al. [63] also studied spin cast films of polyimide precursors on ITO coated glass, followed by imidization at elevated temperatures. The STM images of gold coated films showed the details of the film surface compared with images of the substrate. Reneker et al. [64] studied PE samples prepared from solution as done for HREM [39] (see Section 4.1.3.6). Once the solution of PE was trapped between the mica sheets, the device was removed from solution, cooled, the solvent evaporated, and the sheets, separated and shadowed, were imaged by STM; complementary study showed the same lateral resolution by both techniques but much better vertical resolution by STM.

The feasibility of imaging polymer crystal samples was shown by Piner et al. [65] who conducted STM imaging of gold and chromium coated PE lamellae. They found that thick regions of crystals were more difficult to image than thinner regions on edges. Magonov et al. [66] used AFM to study cold extruded PE with atomic resolution and reported a fibrillar structure with highly oriented molecules was present. Fritsche et al. [67] studied polyethersulfone ultrafiltration membranes by drying the samples by a solvent exchange procedure for SEM, using isopropanol and hexane, and by observing the wet membranes directly by AFM. They were able to detect and measure pores by both methods although neither technique could resolve the surface pore structures, and the porosity seen in AFM and SEM differs, as expected, due to the drying and metal coating for SEM versus observation of the virgin membrane via AFM. In AFM, vertical profile measurements yield the pore diameters, however the accuracy of such measurements will depend on the sharpness of the tip.

Baldwin et al. [68] described a preparation method for AFM of thermally sensitive starch powder and fibrous samples by use of spin casting cyanoacrylate glue onto a substrate, and then sprinkling the powder or laying down the fibers onto the wet glue surface for comparison with SEM. In another study [69], coating formulations were prepared at 10 wt.% solids and cast on microscope slides by pulling the liquids across the surface with Kimwipes EX-L Wipers (Kimberly-Clark, Dallas TX). Other samples were spin coated on silicon wafers in a class 100 clean room and then cured. Atomic force microscopy characterization of these water based, nonstick, hydrophobic polymeric coatings was used to relate wettability to surface composition and adhesion. An electrospray method was used for conversion of isolated molecules of atactic polystyrene in solution to solid, single molecule particles for SEM and AFM [70]. Solvent, concentration, and spraying time are dependent on the polymer and particle size. Solutions were sprayed onto freshly cleaved graphite or mica, mounted with a small strip of double sided tape applied at one edge of the target; particle size distributions were analyzed from the SEM and AFM images.

Bar et al. [71] characterized the morphology of blends of poly(styrene)-*block*-poly(ethene-co-but-1-ene)-*block*-poly(styrene) with isotactic and atactic polypropylene block copolymers by ICAFM. Samples deposited from solution onto a glass substrate, dried, and annealed or quenched from the melt and by samples cut by an ultramicrotome were compared with earlier TEM results. The polymer film on the side of the film-glass interface was studied rather than the free surfaces of the polymer.

The annealing behavior of long chain alkane single crystals was observed by AFM in the intermittent contact mode, using a heating and cooling stage produced by Linkam Scientific Instruments (Surrey UK) that permits *in situ* evaluation [72]. Three preparation methods were tested on reference compounds, including a polysaccharide, which showed each had potential artifacts, but when used in concert the structure could be determined [73]. The preparation methods included drop deposition directly onto a mica surface; ultracentrifugation onto freshly cleaved mica sheets; and adsorption, by placing mica vertically in the sample solution, the latter being preferable. It has also been found that annealing is required to ensure solvent is removed even for ultrathin spin coated films [74]. Chemical contrasting in single polymer molecules has been studied [75] by staining a thin film on a silicon or mica substrate with hexacyanoferrate (HCF) anions acid solution bath for 3 min, rinsing, and drying the film for AFM. Atomic force microscopy studies of latexes was reported earlier [76], while a more recent study of poly(vinyl acetate) (PVAC) prepared with poly(vinyl alcohol) (PVOH) by applying 3 drops of latex on a glass slide cleaned with chromic acid, dried for 3 days in a desiccator, was imaged in the intermittent contact mode [77].

4.2 POLISHING

Mechanical polishing of surfaces, often conducted for metals, ceramics, and rocks, also finds use for polymers and polymer composites containing hard materials. The methods work remarkably well for complex composites that are hard to section, such as glass, carbon, or ceramic fiber composites, and tire cords and large specimens have been polished for SEM and x-ray microanalysis. Polished samples may be etched with solvents or plasmas for examination by SEM and for backscattered electron imaging (BSI). Variations in the polishing method are used to provide thick specimens with a polished surface for reflected light, SEM, or AFM, and thin specimens for transmitted light. Micrographs of polished composites are routinely used for quantitative image analysis to compare the size, shape, and distribution of the reinforcements with their mechanical and physical properties. There is little published on the polishing of polymers, although such methods have been used since the early 1970s in the Celanese laboratory [78].

A review [79] gives a description of the adaptation of metallographic and petrographic preparation techniques to the preparation of polymers for light microscopy, and Goodhew [7] provides a general text on preparations in materials science. Machined semicrystalline polymer and short glass fiber composite surfaces were shown by SEM to have artifacts produced by machining [80], which showed the need for polishing. Trempler [81, 82] reviewed the topic of the preparation of plastics for optical microscopy, covering the basic methods and the potential artifacts, which are incorporated below (Table 4.1). Anyone spending significant time polishing specimens is referred to that work. Two recently published chapters discuss polishing of polymers [83] and preparation of hard materials [84], relevant to some polymer composites. Finally, Samuels [85] has a book on metallographic polishing that provides comprehensive coverage of the topic but does not include polymer materials.

4.2.1 Limiting Artifacts

As with all specimen preparation methods, selection of the specimen is of utmost importance, especially when a small sample is to be prepared representing the material. Polishing can produce artifacts at each stage of the preparation (Table 4.1), especially cutting the specimen and final polishing. The initial cutting, generally with some type of saw, can cause the formation of microcracks that can be misinterpreted as voids or cracks in the specimen. In order to limit such cracking, the specimen should be rough cut several inches from the surface of interest and then cut more carefully. Bates and Wang [86] studied the effect of cutting moldings of nylon 66 using a band saw to see the effects on mechanical properties and surface morphologies, showing that high blade speed and a high number of teeth per unit length resulted in higher strength for glass fiber reinforced specimens. Conditions of low feed and blade speeds, however, appear best for unreinforced nylon 66; polishing the cut surface improves impact resistance.

Grinding should be done with gentle pressure on the mount and should never be done dry. Each grit should be used only until the scratches from the previous grit are removed.

TABLE 4.1. Defects caused during grinding and polishing of polymer specimens

Defect/cause of defect	Effect on the specimen	Suggestions for prevention
Damage to ground surface due to localized melting	Surface exhibits a matte, rough appearance without grinding marks	Use more coolant, use a metal disk substrate and/or apply less pressure
Grains of grinding abrasive embedded in the surface	Block shaped inclusions visible	Improve rinsing with more coolant, less applied pressure, better quality abrasive paper, rinse and clean ultrasonically
Grains of polishing abrasive embedded in the surface	Small inclusions visible under the microscope	Generally due to softening of the specimen surface. Ensure good cooling, low applied pressure, final polish only with water. Do not clean ultrasonically.
Grinding/polishing scratches	Straight and/or irregularly curved fine lines on the surface of the specimen	Use finest possible abrasives, clean carefully between each stage. Do not use ultrasonic. Finally, polish on very soft cloths.
Edge rounding	Differences in height between phases of differing hardness or between specimen and mountant	Prepolish on cloth as hard as possible and final polish only a short time with a soft cloth

Cleaning of the specimen is also important as larger grits or polishes on the specimen can contaminate the fine polishing cloth. A 30 s ultrasonic rinse in water followed by drying off with a can of Freon gas is adequate to remove the grit. However, the gas can should not be turned upside down as this will result in a frozen and ruined specimen polish. Cleaning the specimen surface for microanalysis is especially important as contaminants can result in artifacts, such as additional x-ray peaks. Finally, undercutting of the specimen, that is preferential polishing of the embedding resin compared with the specimen, or a soft portion of the specimen itself, is another problem that must be overcome when developing a polishing method. An interesting paper by Howell and Boyde [87], on specimens of human bone embedded in poly(methyl methacrylate) (PMMA), compared polishing and milling using a reflection confocal laser scanning microscopy (CLSM) and SEM. Confocal laser scanning microscopy provides topographic details showing that milling (after polishing) created less relief than polishing alone.

4.2.2 Polishing Specimen Surfaces

Standard metallographic methods [7, 82] are used for grinding and polishing. Large samples are precut with a saw to fit into 1 to 2 inch diameter metal or rubber molds. Small samples are mounted on a jig, such as a piece of cardboard, to hold them in the proper orientation, and then they are placed in the molds. Epoxy is mixed, poured into the mold, and cured to provide support. Curing can be done in a vacuum oven to ensure removal of trapped air in complete filling of voids. Epoxide resins and hardeners are acceptable for most polymer and composite materials. Fast cure resins are not recommended as the exothermic reaction creates large bubbles that hinder polishing. A water cooled diamond saw is used to precut cured specimens, and wet grinding is done with sandpapers ranging from 120 to 600 grit. As the specimen is ground on each of the successively finer papers, it is turned 90° so that the new scratches are clearly visible compared with the coarser scratches being replaced. Water flowing over the grinding papers keeps the specimen from heating up and also carries the debris away. In special cases, glycerine may be used as a lubricant. For hygroscopic materials, kerosene can be used as a lubricant with proper attention to safe handling. A hard cloth is used to prepolish and fine polish to 1 μm; soft cloths are used for final polishing. Polishing is conducted using graded alumina suspensions in water (1, 0.3, and 0.05 μm), chromium oxide slurries, or diamond paste suspended in oils. Polishing is done on a rotating wheel with a layer of fluid on the surface of the polishing cloth or using an automatic polisher. A benchtop ultrasonic cleaning unit is used with distilled water to remove the polishing media from the specimen, but care must be taken to keep the sample from heating. The quality of the polish can be checked in the optical microscope.

4.2.2.1 Polishing Surfaces

Figures 4.3 and 4.4 show two applications of polishing polymer materials. A carbon fiber composite, in an epoxy matrix, is shown in a reflected light micrograph (Fig. 4.3A) in which the small, round, white regions are fiber cross sections, whereas the longer white regions are fibers oriented parallel to the polished face. The black regions are voids in the original composite. At a higher magnification, another composite (Fig. 4.3B) exhibits uniform packing of the fibers in the polymer matrix. The flow pattern of the polymer is seen in a longitudinal polished section of a polymer extrudate (Fig. 4.4A), and reflected differential interference contrast (DIC) more clearly reveals the surface detail (Fig. 4.4B).

Specimens for OM or SEM can be prepared by polishing followed by etching, either chemically or with plasma or ion beams (see Section 4.5) to bring out relief or detail. Bartosiewicz and Mencik [88] polished nylon and polypropylene prior to etching to reveal the polymer structure. The specimen surface was wet ground with 320 and 400 grit silicon carbide paper and 600 grit aluminum oxide paper followed by polishing using a 15 μm aluminum oxide slurry on medium high nap felt cloth followed by use of 1 and 0.5 μm aluminum oxide slurries; final polishing was with a 0.05 μm chromium oxide slurry. In a study of glass fiber composites [89], samples cut perpendicular to the fiber direction

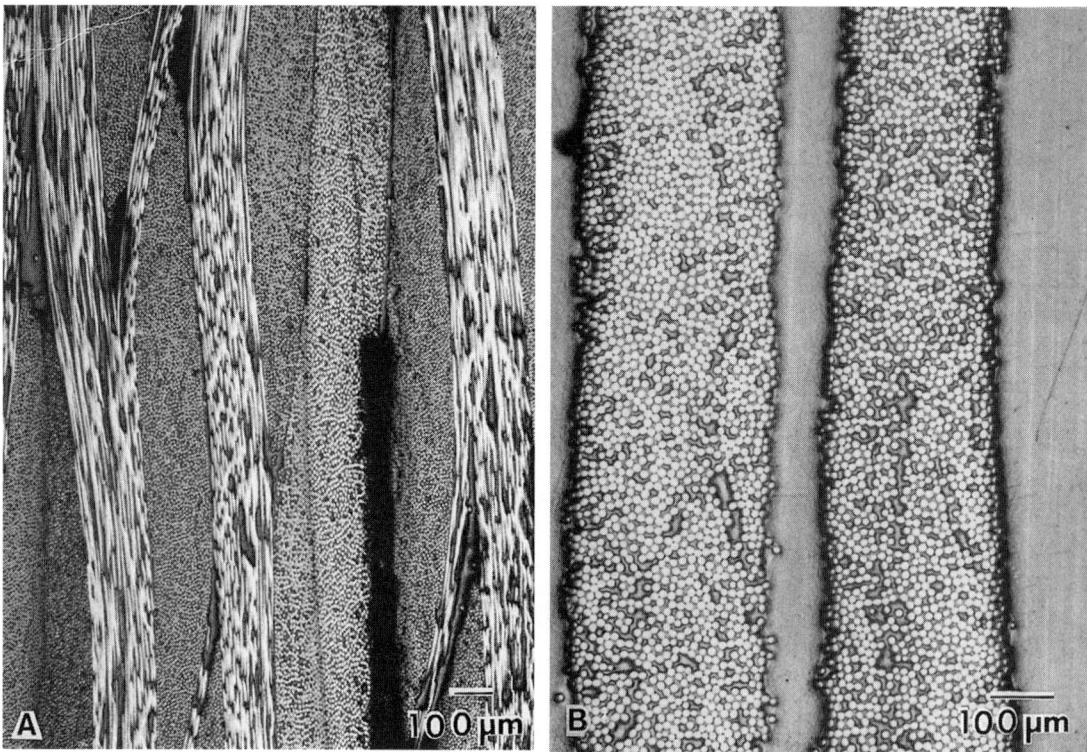

FIGURE 4.3. Reflected light micrographs of polished composite specimens show the carbon fibers (white) and their orientation within the polymer matrix. Black regions are voids. In (A), various layers are oriented normal to one another; in (B), two uniaxially oriented fiber tows are shown.

Polishing

FIGURE 4.4. Reflected light micrographs of polished longitudinal sections of a polymer extrudate show the flow pattern at low magnification (A). Differential interference contrast (DIC) of a similar specimen (B) shows this flow pattern in greater detail.

were mounted in cold mounting resin, ground, and diamond polished, followed by etching in hydrofluoric acid (40%) for 5-10s to reveal the glass fibers more clearly. Summerscales et al. [90] used polishing to prepare fiber reinforced composites to study the effect of processing on microstructure. Cut sections were polished on "wet-and-dry" diamond papers of 200, 320, 400, and 600 grades in turn until the surface appeared flat, followed by polishing using an automatic metallurgical machine with diamond pastes and a specimen load of 1 kg. Images were examined in bright field light microscopy and the microstructures analyzed in an automatic image analyzer.

Fiber orientation was studied for different thermoplastic polymers reinforced by short glass fibers, using a special specimen preparation treatment for reflected light microscopy [91]. Preparation includes embedding, cutting, and polishing of the specimens, which have hard, brittle fibers embedded in a flexible, soft, ductile matrix, at times with poor bonding. These authors use automated polishing with low pressure, ending with a 1 μm diamond polishing suspension, followed by etching the sample with oxygen ions in a vacuum at an acceleration voltage of 1 kV for 1 h. Although the fibers are above the surface, sputtering with platinum is used to remove this effect.

In a follow up paper, Mlekusch [92] makes quantitative measurements by image analysis to determine the fiber orientations. In an edited book on the microstructural characterization of fiber reinforced composites [93], polishing was used to determine yarn shape and for quantitative image analysis of the composite microstructure using methods found in the earlier editions of this text and by Hemsley [94].

4.2.2.2 Polishing Thin Sections

Thin sections (ca. 2–40 μm thick) provide ideal specimens for study using optical microscopy, especially polarized light microscopy to image the crystalline structure. However, some materials are too tough, brittle, or hard to be sectioned by microtomy. In geology, thin sections of rocks are commonly made by polishing techniques, and this method also works well for polymers that cannot be microtomed because they are too hard at room temperature. Samples are first cut, embedded in a support resin, ground, and wet polished on one face. The sample is glued, polished face down, to a well cleaned glass microscope slide with fast cure epoxy, cyanoacrylate glue, or Canadian Balsam [82], being sure that the sample is not heated during this process. The cured sample is cut, ground, and polished on the opposite side until it is thin enough to transmit light. The thin sectioning method has an added problem in that the thinner the specimen, the better the optical image but also the greater likelihood of the specimen pulling out of the mount. The suggestion here is caution; polish for short times, such as 30–60 s, clean and examine the specimen before proceeding.

Applications of the polished thin sectioning method [79] include extrudate cross sections, glass fiber reinforced polycarbonate, and rubber toughened nylon. The fine structure of thermotropic liquid crystalline polymers (see Section 5.6.2) has been observed in thin specimens prepared by the polishing method [95]. This method was required as microtomy was not possible for these tough polymers. Gammon and Ray [96] provide an in-depth instruction on the metallographic preparation of ultrathin sections of fiber reinforced polymeric materials. Examples used are a thermoset; glass fiber-filled Ultem 2300; and a polyamide prepreg (carbon fiber mixed with resin is termed a prepreg). A description is provided of alignment and embedding of multiple specimens, grinding and then polishing the surfaces using alumina slurries or automated polishing. A wide range of transmitted optical microscopy techniques are used for specimen evaluation.

4.3 MICROTOMY

Generally, *microtomy* refers to the preparation of thin slices of material by sectioning for observation in an optical microscope by transmitted light. Microtomed sections are cut with steel or glass knives to about 1 to 40 µm thickness. *Ultramicrotomy* methods involve the preparation of ultrathin sections of material for observation in an electron microscope. Ultramicrotome sections are cut with glass or diamond knives to a thickness ca. 30–100 nm. If imaging is to be done via many techniques, the TEM preparation method can be utilized to prepare thin sections for OM, TEM, and AFM, and the flat block face is used for SEM and/or SPM.

Ultramicrotomy is very commonly used in the preparation of biological and polymer specimens for electron microscopy. Polymers must be carefully embedded to provide sections with visible fine structure that represents the original material. Stains (see Section 4.4) are used to enhance contrast by increasing the electron density in specific structures in the material. Some polymers are too soft to be sectioned at room temperature, and these must be hardened either chemically or by cooling below room temperature during microtomy. This latter method, *cryomicrotomy*, or *cryoultramicrotomy*, is an increasingly popular specimen preparation method for many materials for imaging by TEM and SPM.

4.3.1 Peelback of Fibers/Films for SEM

Synthetic fibers were among the earliest polymers produced for commercial applications. Morphology studies were undertaken in order to understand structure-property relations and to improve properties. During the 1950s, the SEM was not available, and although microtomy was already being used (e.g., by Scott and Fergerson [97]), little could be seen by direct observation of thin sections in the TEM, resulting in a need for other preparation methods. One of these methods is the peelback method, or longitudinal splitting of fibers [98]. The early peelback technique was used to provide surfaces for replication or thin films, both for the TEM, whereas the method is now used to prepare fibers and films for SEM or SPM study. Scott [98] described the peelback method as a split section or cleavage technique by what he called orientation and cleavage plane splitting. The idea is to open a fiber (or film) with minimal disruption. Orientation splitting involves cutting of a fiber with a razor blade, at an oblique angle, halfway into the fiber, followed by a cut along or parallel to the fiber axis. This second cut below the first one is literally peeled back with forceps to provide a thin section aligned with the longitudinal axis. In cleavage plane splitting, a razor is used to make an oblique cut halfway through the fiber and then the polymer is peeled back along the fiber axis with forceps. The split works its way along the fiber to the outer surface where it then continues along the surface skin. Additional splitting of the same

Microtomy

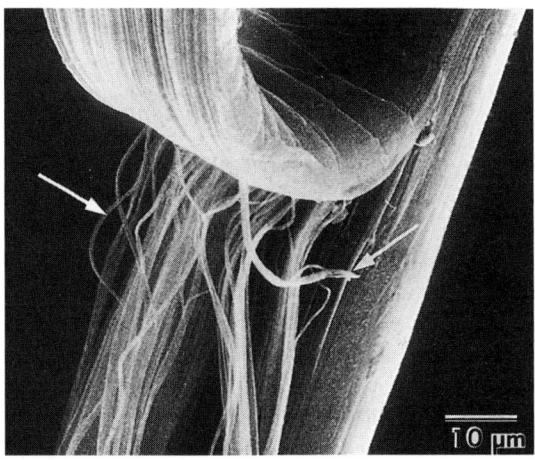

FIGURE 4.5. Scanning electron microscopy micrograph of a polymer fiber prepared by the peelback method reveals the internal fibrillar texture (arrow).

fiber provides layers aligned along the molecular fiber axis. An example of a peeled back PET fiber, showing a fibrillar texture, is seen in an SEM image in Fig. 4.5.

4.3.2 Microtomy for OM

Specimens for transmitted light microscopy must be transparent; as the practical resolution of the optical microscope is about 0.5–1 μm, specimens about 1–40 μm thick must be prepared. Very simple handheld devices using a razor blade may be sufficient or complex preparations involving several steps prior to microtomy may be required. For example, alignment of fibers for sectioning involves mounting the fibers in a specific orientation, placing them in a capsule or mold, adding appropriate embedding resins, curing, and trimming before sectioning. Much of the preparation is common to both optical and TEM sectioning so the reader should be aware that both sections of this chapter should be read for a complete discussion. Because fibers are among the most difficult specimens to cut, they are used as an example in the following sections.

4.3.2.1 Microtomes

Sectioning is accomplished using several types of manual and mechanical techniques. Handheld and homemade forms of sectioning devices with razor blades are commonly found in any textile fiber laboratory. Stoves' book *Fibre Microscopy* [99] describes several of these methods, including a handheld metal plate, the Hardy microtome, and other mechanical sectioning devices. Such instruments for cutting sections for optical microscopy were in use as early as the 18th century [100]. A collection of different devices for fibers, yarns, and composites has also been shown by Annis and Quigley [101].

The simplest and most useful of these devices for fibers, films, yarns, and fabrics is a modified metal plate, as shown in Fig. 4.6. Sections cut using such a plate tend to be ragged around the edges, nonuniform in thickness, and often damaged and distorted. The plate is 3 × 1 inch in size, with slots from the edge to the center, ending in small holes, about 1 mm across. The specimen is combined with a colored filler yarn and forced into the slot. A single edged razor blade is used to cut both top and bottom surfaces, the plate is placed on a glass microscope slide, and a drop of immersion oil and a

FIGURE 4.6. A metal plate used to provide rapid cross sections of fibers, yarns, and fabrics. The plate is the same size as a 1 × 3 inch glass microscope slide.

coverslip are placed on the top surface before viewing in transmitted light. Rotary and sledge microtomes are used with steel knives to provide sections for observation in the optical microscope. These mechanical systems provide consistent thin sections in the range 5–40 µm thickness. Thinner sections are cut using glass knives that provide good sections 1–4 µm thick.

4.3.2.2 Specimen Mounting

Specimens for microtomy must either be cut to fit in the chuck of the microtome and be self-supporting or they must be embedded in a supporting medium. Pieces of molded or extruded plastics are trimmed to fit the chuck using small saws and razor blades. In the case of films and fibers, however, the samples must be embedded in a resin for both support and orientation. The steps for mounting specimens will be described here for OM, and a more complete description of mounting and embedding media for TEM and SPM will be found in Section 4.3.4.

One method for mounting samples for sectioning is as follows.

1. A glass microscope slide is put on top of a piece of graph paper to aid alignment. Fibers are placed lengthwise on the glass slide and secured at the ends with double sided tape.
2. Fibers can be consolidated by tying them at intervals with a cord or yarn, preferably of a different color than the specimen.
3. Samples are supported with cardboard frames, cut in a C or oval donut shape, placed under the fibers, and glued to the frame with Elmer's white glue or Eastman 910 fast drying glue.
4. Fibers are mounted for cross sections as shown in Fig. 4.7A. For longitudinal

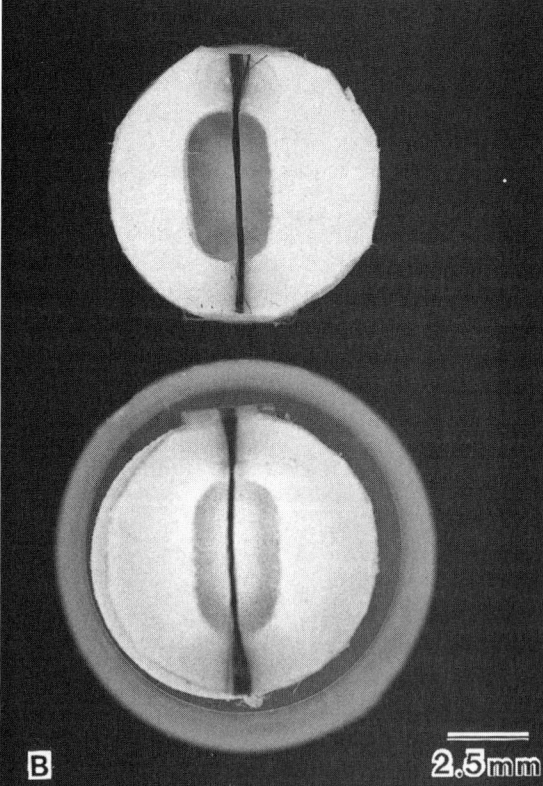

FIGURE 4.7. Fibers are shown mounted on a C-shaped frame (A), punched out of light cardboard, in order to orient them for cross sections. For longitudinal sections, fibers are mounted perpendicular to the direction shown but on two C-frames or they are mounted on a frame that fits into an embedding capsule (B).

sections, two frames can be glued together and slipped under the fibers so they are parallel to the short axis of the frame, or they are mounted across a donut shaped frame (Fig. 4.7B, top) and placed in the top of the capsule, face down (Fig. 4.7B, bottom).

5. The Elmer's glue is hardened in a 60°C to 70°C oven for about 15 min and the frame placed in a capsule or mold.

4.3.2.3 Specimen Embedding

Embedding media are chosen, mixed, added to the capsule, and allowed to cure, according to the details supplied by the manufacturer. One or two embedding media are in general use while a wider range is on the shelf, in case they are needed. In this way, different types of materials may be embedded at once, and the person performing the microtomy can become familiar with the sectioning characteristics of those specific resins. Paraffin, long used for embedding biological and polymer materials, will not be described in any detail as its use is not reliable: cutting is poor at room temperature, it is messy to use, and maintaining the hot paraffin is time consuming. There are specific studies, such as embedding of experimental bone implant specimens, using polyethylene particles, that still use paraffin for embedding [102]. In another study [103], failure of Navy coating systems were studied by SEM/energy dispersive x-ray spectroscopy (EDS) of the cross sections and chemical depth profiling of weathered high-solids aliphatic poly(ester-urethane) military coatings, prepared by immersion in liquid N_2, and embedding in histological wax. SEM imaging of the sections showed the topcoat-primer interface and the EDS maps depicted the spatial distribution of the elements present.

Room temperature curing epoxies work well for optical sectioning of most dry polymer materials. Wet materials are more compatible with water soluble resins as materials need not be totally dehydrated prior to embedding. Processing is rapid, and high quality sections can be obtained using a glass knife. Where optical studies will be followed by TEM, the EM embedding media are best used for both studies.

Media used for TEM will be described in the next section.

A variety of molds are available to encapsulate samples for embedding. BEEM type capsules are useful for small pieces of materials that can be placed in the tip, which then provides pretrimmed specimen blocks. Gelatin capsules are rounded in shape and thus must be trimmed, but they have the advantage that they can be removed by soaking in water. Other shapes are good for specific sample forms. Flat embedding molds are excellent for films and membranes, especially when orientation of the specimen is important. A range of these capsules and molds should be kept in the laboratory.

4.3.2.4 Microtomy

For trimming, the bulk specimen, or the cured block, is placed in a vise and cut with a jeweler's saw. Fine trimming with a razor blade can be done under a stereomicroscope to provide a square or trapezoidal block face shape (Fig. 4.8). The trimmed sample is secured in the chuck of the microtome and sectioned to the desired thickness. A common problem is that sections tend to curl due to distortions caused by the sectioning process. Curled sections can be flattened by manually uncoiling them under a stereomicroscope using a small soft brush or forceps and placing them in immersion oil between a glass slide and a coverslip. The

FIGURE 4.8. The photograph shows the trapezoid shaped face of a trimmed block ready to be sectioned.

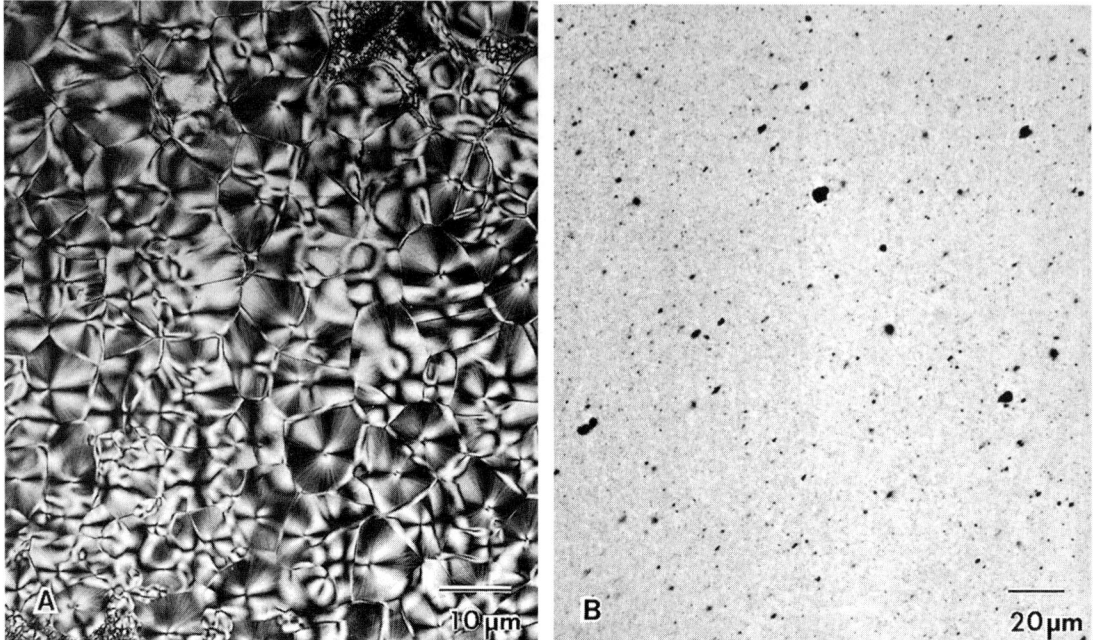

FIGURE 4.9. Two examples of microtomed sections viewed in the optical microscope are shown: (A) a section of a nylon pellet, in polarized light, reveals a coarse spherulitic texture; and (B) a section from a black, molded nylon part, in a bright field micrograph, shows the size and distribution of the carbon black filler.

sections can also be relaxed by heating with an infrared lamp or by flotation on water, or hot glycerol, if they are not sensitive to heating.

Typical polymer materials studied in optical thin section include extrudates or molded parts, such as semicrystalline polyoxymethylene; multiphase polymers, such as rubber toughened nylon; filled polymers, such as carbon black filled nylon; fibers, such as polyester and rayon; and films that are too thick to directly transmit light. An example shown here is nylon, imaged in polarized light (Fig. 4.9A), revealing its spherulitic texture, and imaged in bright field (Fig. 4.9B), revealing dispersed carbon black particles.

4.3.3 Microtomy for SEM

The SEM is also used to study the surfaces of polymer materials, such as plastics, fibers, yarns, membranes, films, and composites. The outer surfaces can be studied or bulk SEM studies can be performed on samples that have been fractured (see Section 4.8) or sectioned. The flat surface remaining after sectioning can be used as a sample for SEM, particularly for x-ray microanalysis. A conductive coating (see Section 4.7.3) is generally applied to the surfaces of interest for SEM or carbon coating for microanalysis. Hearle et al. [6] described several special holders that are modifications of the metal plate method. Specimen holders that have holes for packing materials such as fibers, yarns, or fabrics for razor blade cutting are useful methods that involve the insertion of a looped thread through the hole and the placement of the material in the loop. This material is then pulled through the hole and sectioned with a razor blade. Modifications to the stubs on any SEM can be made readily in a machine shop. Many of the methods used to provide cross sections for the SEM are modifications of optical techniques.

4.3.4 Microtomy for TEM and SPM

Ultramicrotomy is an obvious example of a preparation method developed by biologists and

Microtomy

now in general use for polymers. In the early 1950s, Porter and Blum [104], Sjostrand [105], and Haanstra [100] were constructing microtomes to provide sections, about 0.1 μm thick (or less), for TEM study. Today, ultramicrotomes are available from many companies that permit ultrathin sections of polymers to be obtained on a routine basis. Microtomy permits the preparation of polymers for observation of the actual structure in a bulk material, which is not possible by methods such as thin film casting or surface replication. The first step in the process is to determine the temperature to use for sectioning. If the glass transition of the polymer is below room temperature, then cryoultramicrotomy is needed. Examples of polymers that can be sectioned at room temperature are polycarbonate (PC), PMMA, HDPE, epoxies, nylons, and rigid polyurethanes; and prestained and thus hardened high impact polystyrene (HIPS), polypropylene (PP), acrylonitrile-butadiene-styrene (ABS), and PC/ABS. Polymers that may need to be sectioned at cryotemperatures are PP, PE, rubber, polytetrafluoroethylene (PTFE), poly(vinyl chloride) (PVC), ABS, HIPS, latexes, paints, silicones, and so forth.

A general method will be described for the preparation and sectioning of polymers, pointing out those areas of difficulty and citing some of the excellent references available. The aim of the preparation is sections that truly represent the original bulk structure and that are thin, flat, and not deformed, have contrast, and can be used for TEM images that can be easily interpreted. The resulting block face is the sample of interest for SPM imaging. The steps involved in sample preparation for ultramicrotomy include (1) fixation/staining (if needed), (2) drying (if needed), (3) specimen mounting (see Section 4.3.2), (4) embedding and curing, (5) trimming and sectioning, and (6) carbon, metal, or polymer coatings. Staining is often required in order to increase the electron scattering of polymers selectively and, thus, aid contrast and resolution of details. Staining is performed either on small pieces of the material before embedding or by poststaining the sections themselves, or both. This topic is so vast and important to the polymer microscopist that it is treated separately (see Section 4.4).

4.3.4.1 Specimen Drying

Most polymer materials are dry or can be air dried without structural change, whereas others, such as some membranes and emulsions, are wet and require drying prior to embedding. Air drying may destroy or distort structures due to the deleterious effects of surface tension. A method that has been used in the case of wet, porous membranes [106] is to cut strips of the membrane in a jar containing water and to transfer the strips directly into the embedding resin in a disposable weighing pan. This sample is placed in vacuum for several hours at room temperature to remove the water, which is replaced by resin. After vacuum impregnation, the membrane samples are placed in flat embedding molds in fresh resin. Other methods of drying might include critical point drying, freeze drying, or chemical drying (see Section 4.9).

4.3.4.2 Embedding Media

There are many embedding media available, almost too many for the novice to choose one over another for a given specimen. Glauert [107, 108] are excellent references that include descriptions of the various media and the reasons for their use. There are three media in general use: epoxy, polyester, and acrylic resins. The ease of sectioning and stability in the high vacuum and the electron beam are important, with the epoxy resins the most stable. A hardness match of the polymer and embedding media is also important to good sectioning. Variations can be made in the recipe for each medium in order to change its hardness. Availability and good results have been the general experience with the epoxies and they are used in most applications except where the greater potential infiltration of low viscosity resins is required. Epoxies can be reactive toward some surfaces and thus distort morphology. In these cases, the surface may be coated with a thin, conformal metal (such as chromium) prior to embedding. The embedding media should be handled using the safety information supplied and cured using the times and temperatures recommended by the supplier.

Recently, microwave techniques have been used for more rapid embedding [109–113], and recipes have been developed for a reduction in curing from 48 h to 15 min or less at a power of 700 W in a 2540 MHz microwave oven [110]. Only laboratory microwave ovens should be used, and these with precautions taken, such as the use in a hood [108]. Embedding media can also be cured or polymerized by UV irradiation instead of by heat [108]; special chambers are available or can be constructed. This method has the advantage of being used at lower temperatures, but this is not needed for epoxy resins, which are cured at room temperature. A specialized method of embedding, generally of geological or biological specimens, for determination of carbon content or distribution, is to use sulfur embedding, although this has many challenges as to safe use of this brittle material [114].

4.3.4.3 Mounting Specimens and Block Trimming

Specimen preparation methods, trimming blocks, and sectioning were discussed (see Section 4.3.2) and are covered in detail for biological materials by Reid [115]. A factor to consider for TEM sectioning is the orientation of the specimen in the trimmed block, which is important to section quality and structural interpretation, especially of anisotropic specimens. Longitudinal sections of fibers and films should generally be aligned so that the long axis of the material is *not* parallel or perpendicular to the knife edge. The specimen should be oriented at an oblique angle to the bottom of the trapezoid or perpendicular to it for best sectioning. A simple method has been described for orienting silk and other flexible fibers for transverse or longitudinal sections [116]. Prior to embedding in epoxy resin, the silk is wound around a notched support made from polyester film. After curing, the film can be peeled away to reveal oriented silk or other fiber threads that can be cut with a microtome. A glass knife was used for initial cutting, followed by use of a diamond knife. Compression molding thin fibers and free standing polymer films into a transparent thermoplastic that is easy to microtome is a viable procedure to obtain cross sectional specimens [117]. While the experimental conditions may have to be modified for each system, the procedure is easy and rapid for generating samples that are ready for microtomy. This procedure has been successfully applied to obtain thin sections and block faces of encapsulated poly(imide) films suitable for TEM and AFM, respectively.

Trimming a block to a specific shape and size is more critical for TEM and SPM than for OM. The cured specimen block is cut close to the specimen plane with a jeweler's saw or a razor blade. The pretrimmed block in a microtome chuck is placed in a holder built to fit in the stage of a stereomicroscope. The top surface is cut or faced and then the sides of the block are cut at an oblique angle to provide a trapezoidal shaped block face. Where possible, the epoxy is all trimmed away; where this is not possible, a minimum sized block face with little epoxy is prepared. Smaller block faces are better for TEM. The bottom of the section, the part that will be cut by the knife first, should be the longer side of the trapezoid. Block faces on the order of 250 μm or less help to minimize chatter artifacts that are readily imaged by SPM.

4.3.4.4 Ultramicrotomy

Glass knives are used to face the block, up to the specimen and for final trimming, either in the ultramicrotome itself or in a microtome made for that purpose. Ultrathin sections are cut with either a glass or a diamond knife. Glass knives have the advantage of being inexpensive, easy to make, and are sharper than razor blades, however they do not remain sharp. Glass knives provide good sections of samples that are softer than glass, such as unfilled nylon, polyoxymethylene (POM), and PE. In the case of very hard materials that might crack a diamond knife, such as polymer composites, or very high modulus materials, it is better to use an expendable glass knife. Diamond knives are sharp, can be resharpened, and can cut very thin sections, but they are very expensive and can chip or crack. Old diamond knives are quite useful for cutting hard materials that require diamond but might chip or crack a good knife. In practice, proficiency should be demonstrated with a glass knife prior to attempting diamond

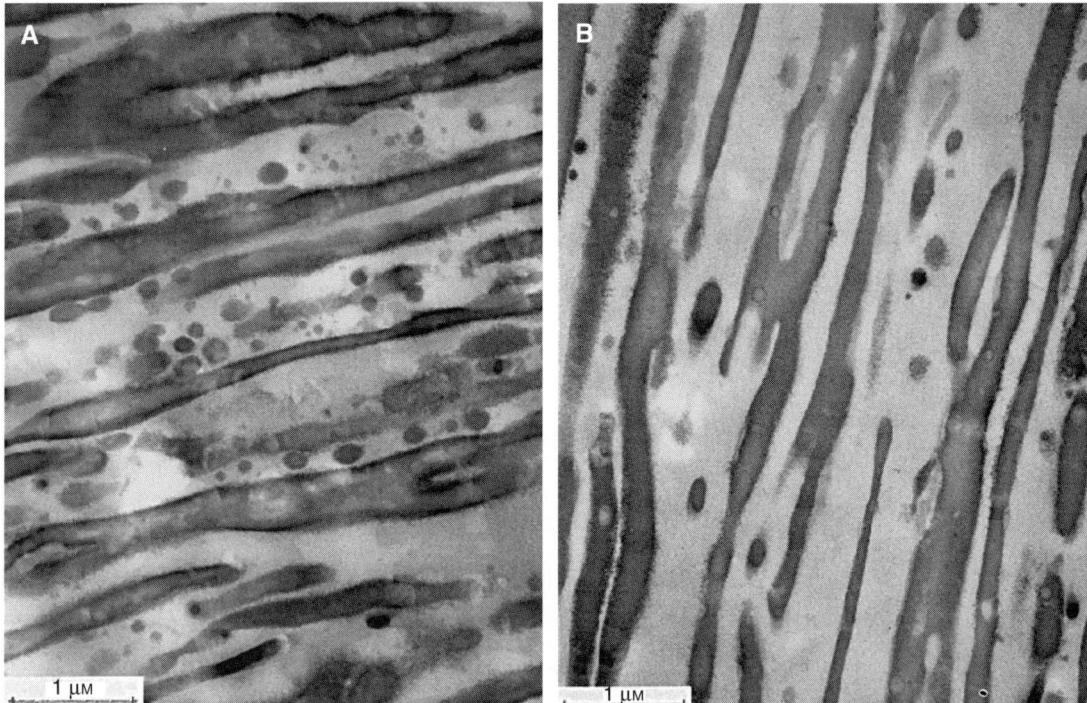

FIGURE 4.10. Ultrathin sections of polypropylene and polyethylene sectioned at room temperature using a 35° angle diamond knife (A) and sectioned at room temperature using an oscillating diamond knife (B). (From Vastenhout and Gnagi [119]; used with permission of Cambridge University Press, courtesy of the Microscopy Society of America.)

knife sectioning. Finally, the nature of the trough, and the trough fluid used to float the sections so they can be picked up on grids, must be considered. Diamond knives are supplied with a trough that is simply filled with water during room temperature sectioning. Glass knives are made as they are used and various sticky tapes may be used to produce troughs.

Developments have been made to reduce the compression of ultrathin sections by use of an oscillating diamond knife [118]. A prototype of this oscillating diamond knife was evaluated and tested to prepare ultrathin sections for comparison with the same sample prepared with a 35° angle diamond knife [119], as shown in Fig. 4.10 of a blend of polypropylene and polyethylene. A block face from such sectioning should be flatter and be more useful for AFM imaging as well.

4.3.4.5 Literature Review

Plummer [120] provided an excellent review in his "reflections" on the use of microtomy for materials science specimen preparation, with a table of material classes, references, and comments. The review includes examples of polymer composites, blends, membranes, elastomers, and coatings. Kink bands and shear deformation were shown in polybenzobisoxazole (PBO) fibers microtomed and studied by TEM [121]. Fibers treated to cause compressive deformation were taped to sheet polycarbonate and coated with spray acrylic to fix them to the substrate. Small sections (<5mm) attached to the polycarbonate were cut, trimmed, and microtomed to a thickness of 40–80nm with a new area of a diamond knife and picked up on 400 mesh grids. Ericson and Lindberg [122] showed that when the sample holder of the ultramicrotome is instrumented with a force transducer, it is possible to measure the very small sectioning force and calculate the energy dissipated. The method was demonstrated on two amorphous polymers, PMMA and epoxy. Potential applications are research on nanoscale fracture, characterization of molecular anisotropy, and development of the ultramicrotome. Pellets of PET were microtomed at room temperature, about 50 nm thick, and covered by a holey carbon film for EELS analysis [123]. Nanofoams [124] were prepared for microtomy by partial submerging of the films in capsules filled with embedding epoxy, with the ends of the films held between two clean glass slides above the submerged portion of the films while the epoxy dried. The epoxy was a support and the film was sectioned with no embedded epoxy to prevent filling of the pores. Sections were stained for TEM study. Polyethylene or any other polymer film, coextruded films, fiber or molded sample can be cryo-faced with a glass or diamond knife, followed by staining in ruthenium tetroxide vapors, and cutting thin sections <100 nm using a diamond knife at ambient temperature [125]. This method was initially developed for the analysis of domain morphology of stained polyolefin blends by LVSEM.

A method developed for biological tissue might find use for polymers to examine materials in an x-ray microscope and by complementary TEM, SEM, and SPM [126]. Glycol methacrylate (GMA), a water soluble plastic, was used as an embedding medium; blocks were sectioned with glass knives and the flattened sections were placed in a square silicon wafer frame. Near surface plastic deformation under surface scratches was shown in injection molded isotactic polypropylene blends with more than 20% ethylene-propylene-diene monomer (EPDM) rubber modifier (thermoplastic polyolefin, or TPO) [127] by polarized light microscopy and x-ray photography of thin sections, and TEM of ultrathin sections, after staining the block with ruthenium tetroxide. A thin layer of gold was evaporated to mark the surface, the blocks were embedded in low viscosity epoxy resin, and ultrathin sections cut with a diamond knife. Results obtained suggest that a strong interfacial adhesion exists between the rubber phase and polypropylene. Plummer et al. [128] studied the microdeformation in heterogeneous polymers by TEM, SEM, and AFM imaging. Notched specimens were used by inserting a wedge between the crack faces, embedding, staining, and sectioning around the crack tip.

Ionomer powders were sectioned for imaging by first compression molding them and then sectioning at room temperature, collecting the sections with a wet method by floating them onto deionized water in the knife boat [129]. Imaging of the ethylene- and styrene-based ionomers was conducted by bright field and high angle annular dark field scanning transmission electron microscopy (BF- and HAADF-STEM).

4.3.5 Cryomicrotomy for TEM and SPM

Cryomicrotomy and cryoultramicrotomy are sectioning methods performed at low temperatures to produce thin or ultrathin sections, respectively, of polymers too soft for room temperature sectioning. The methods are not routine, but they have gained in popularity as a direct result of the ease of use of new commercial instruments and the need for flat block faces for SPM as well as for traditional OM and TEM study. Sections for OM, especially for hot stage work, may be cut about 2.5–5 μm thick using the same sample as for TEM. Polymers that have a glass transition temperature below room temperature are soft and can be hardened by cooling. Sectioning in the −20°C to −40°C range is fairly straightforward as liquids may still be used to remove sections off the knife. At lower temperatures (e.g., −120°C) sectioning is more difficult as a dry knife must be used.

The following hints and tips are from a polymer microscopist [130]: Use a diamond knife with an angle of 35° for polymer materials; section thickness of about 90 nm is best to start, with change to lower temperature if there is severe compression and reduction of thickness to 60–70 nm once sections are uniform; slow cutting speed, below 1 mm/s, is preferred for polymer specimens.

Dry sections are difficult to pick up without damage. Ionizing antistatic devices are used to inhibit sections flying away due to static charge. Special instrumentation is required to control the temperature of the knife and the specimen accurately, as the section thickness is affected by thermal expansion and contraction. The cryomicrotome must also limit frost buildup on the specimen, as frost is a problem in sectioning at low temperatures. Zierold [131] provides an excellent discussion of the optimal conditions for the preparation of ultrathin cryosections of biological materials. The mean ice crystal size was shown to be critical, and the section quality was seen to increase as the ice crystal size decreased. The theoretical and practical applications of low temperature techniques, including cryoultramicrotomy, are found in a comprehensive text by Echlin [132].

Complementary imaging by TEM and AFM is useful for better understanding of the morphology of materials, especially blends and copolymers. Intermittent Contact mode AFM images of the block face of HIPS, obtained using cryomicrotomy at −110°C, are shown in the height image (Fig. 4.11A) and the phase image (Fig. 4.11B) [133]. Comparison with a TEM image (Fig. 4.11C) of a thin section obtained by room temperature microtomy of a block face exposed to OsO_4 vapor shows similar microstructure [133].

4.3.5.1 Literature Review

Chappius and Robblee [134] used solid carbon dioxide and alcohol as a coolant while many others (e.g., [135, 136]) used liquid nitrogen to cut about 50 nm thick sections of rubber. Andrews et al. [136] examined sections of several semicrystalline polymers produced by cryoultramicrotomy at −150°C using a modified microtome. Ethylene glycol was used to wet the knife edge, but no fluid was used to pick up the sections. Dlugosz and Keller [137] produced sections of spherulitic PE by cryoultramicrotomy and studied the beam induced band structures. Early results with cryomicrotomes were described [138] for polyurethane elastomer, a blend of crystalline and noncrystalline polymers, which showed spherulitic textures after sectioning at about −70°C. Injection molded PP was also sectioned at about −70°C, whereas PTFE was sectioned at much lower temperatures. Extruded styrene-butadiene-styrene (SBS) copolymer was prepared by cryosectioning with a diamond knife in liquid air at −85°C to −115°C, followed by osmium tetroxide (OsO_4) vapor staining for 1 h [139], revealing the alternating sequence of the PS and polybutadiene lamellae. Odell et al. [140] prepared extruded triblock copolymer by first chemically hardening the polybutadiene, with OsO_4, followed by cryoultramicrotomy to produce 30 nm thick sections: parallel PS rods were observed in the SBS copolymer. Ultramicrotomy and selective staining with OsO_4 was also used in the preparation of a binary blend of PP and thermoplastic rubber [141].

Fridman et al. [142] sectioned polyester-based polyurethanes using glass knives with a 45° cutting angle and propanol at −90°C as the trough fluid for floating off the cryosections. The trough was filled with water and allowed to freeze, forming a shelf onto which a drop of propanol or isopentane was placed. The alcohol lubricates the knife edge permitting sections to be brushed onto a TEM grid. The structure and morphology of segmented polyurethanes was also shown by TEM observation of samples prepared by cryoultramicrotomy [143].

Transmission electron microscopy was used to examine the morphology of extruded thermoplastic starch/poly(ethylene-vinyl alcohol) (EVOH) blends [144]. The samples were sectioned in a cryoultramicrotome at a sample temperature setting of −20°C to 15°C and a knife temperature from −20°C to 0°C. Changes in temperature were made based on the blend, with lower temperatures needed as the EVOH level increased. As with many other materials, the T_g of starch is a strong function of the relative humidity. Samples were collected from the knife with an eyelash tool and deposited onto a droplet of ethanol on a TEM grid; the grids were blotted with filter paper to remove any excess ethanol. Some sections were exposed to iodine vapor for staining the starch. The TEM was also used to examine the morphology of nanocomposites composed of an organoclay and PP by cryoultramicrotomy using a diamond knife of samples at −90°C, collecting the sections on copper grids and imaging at 120 kV [145].

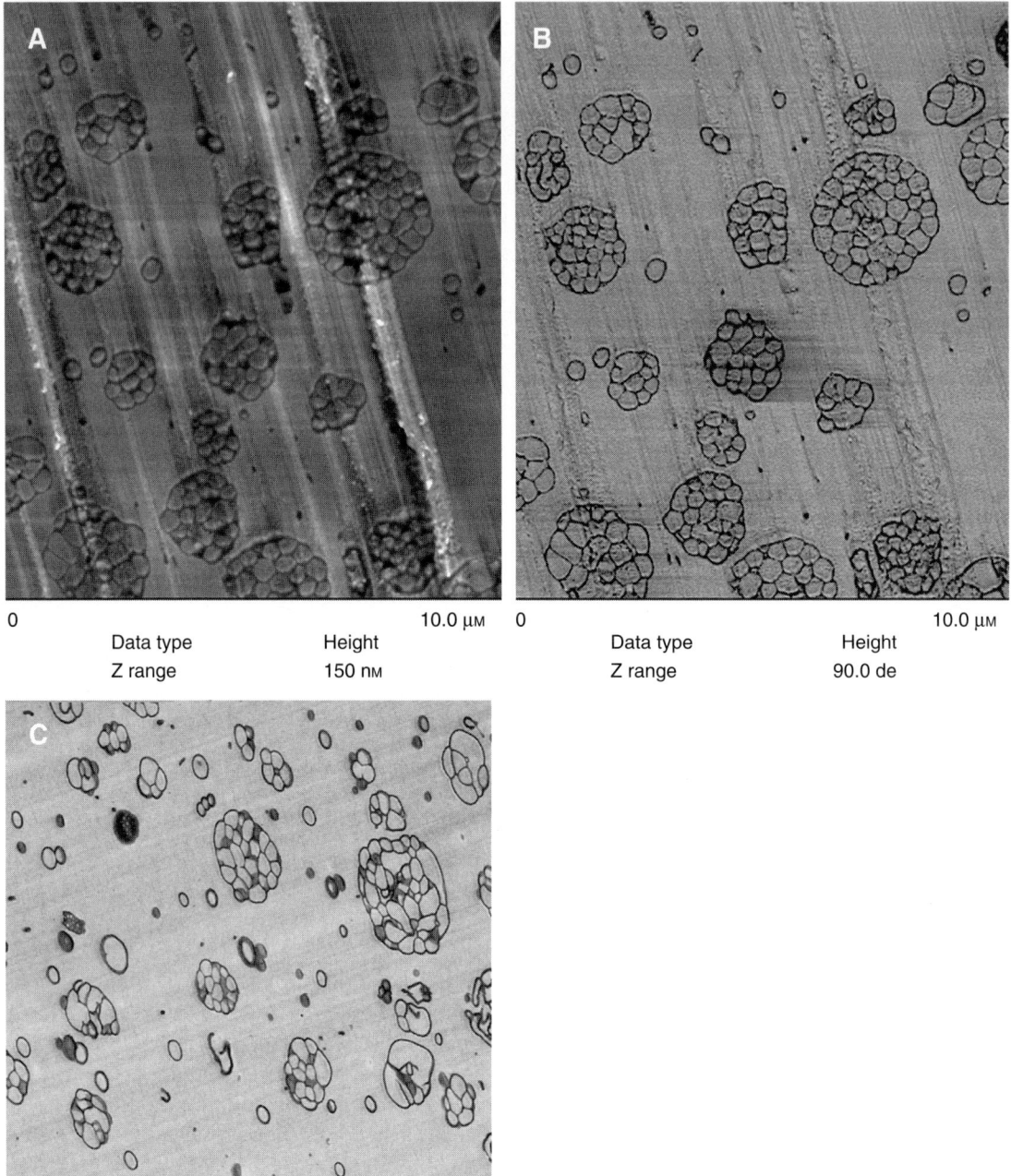

FIGURE 4.11. Intermittent contact mode AFM images of the block face of high impact polystyrene (HIPS), obtained using cryomicrotomy at −110°C. Note that the surface imperfections due to knife marks are obvious in the height image (A), whereas the phase image (B) provides contrast due to the material with less visible artifacts [133]. Comparison with a TEM image (C) of a thin section obtained by room temperature microtomy of a block face exposed to OsO_4 vapor shows detail of the HIPS structure and evidence of some compression in the section. (From Li et al. [133], © (2004) Taylor & Francis; used with permission.)

The application of fracture mechanics to polymer systems is an emerging field [146]. The investigation of a ductile/brittle temperature transition based on Izod tests in toughened nylon blends was conducted by characterizing both fracture toughness and yield strength and the morphology of the cryoultramicrotomed blend by TEM. Deformed regions from the center of test specimens were sectioned at −90°C to about 90 nm thick using a diamond knife and the sections stained in either phosphotungstic acid solution or ruthenium tetroxide vapor. Energy filtering transmission electron microscopy (EFTEM) was applied to investigate interfaces between a polymer and an adhesive. Poly(butylene terephthalate) (PBT) sheets laminated with an epoxy adhesive were heat aged at 180°C in air for >9 h, then were cut into small pieces and sectioned to 100 nm, cutting parallel to the interface with a diamond knife at −80°C. Evaluation using EELS and EFTEM showed the heat aging decreases the adhesion strength drastically due to formation of a weak boundary interface. Microfibrillar reinforced polymer composites were also studied by means of TEM of ultrathin sections prepared at −60°C [147] and compared with SEM of cryogenic fracture surfaces. Sections for TEM were either stained with ruthenium tetroxide vapor for 10 min or the bulk sample was stained overnight prior to sectioning. Honeker and Thomas [148] assessed the orientation of a poly(styrene-*block*-isoprene-*block*-styrene) (SIS) triblock copolymer with a PS cylinder morphology that was globally oriented via roll casting. Transmission electron microscopy was conducted on sections that were microtomed at temperatures of −90°C (knife) and −110°C (sample) of about 70 nm thick and stained with OsO_4 vapors for 1–2 h and carbon coated. Transmission electron microscopy of sections viewed parallel and perpendicular to the cylinder axis provided real space information on the deformed morphology. Microphase separation was characterized in ABA triblock copolymers containing polyhedral oligomeric silsesquioxane pendant groups by cryoultramicrotomy at −80°C, followed by transfer of sections into a glass jar for staining with ruthenium tetroxide vapor [149].

A polycarbonate/styrene-*co*-acrylonitrile blend was examined as a representative of blends with all components well under T_g at room temperature by low voltage TEM and AFM [150, 151]. Ultrathin sections were prepared using a diamond knife with edge angle of 45° at room temperature or low temperature cutting at −130°C with a knife edge angle of 35°, held at −60°C. Cut surfaces and the surfaces of ultrathin sections on freshly cleaved mica were observed by AFM to have rough sample relief. A rougher surface relief was found on the ultrathin sections than on the remaining cut surfaces. The surface relief was assumed to occur in the course of the cutting when extensive shearing occurs. Correspondence of the surface relief with the phase structure proves the influence of different mechanical behavior of individual blend components on the resulting morphology. Modified cryomicrotomy methods for viewing morphologies of latexes containing rubbery polymers was reported [152]. In the first method, a drop of latex is frozen in liquid nitrogen, sectioned with a diamond knife, and vapor stained with OsO_4 for TEM. A chemical fixation method is described for imaging the morphology of such rubbery latex particles by adding glutaraldehyde to the latex, followed by OsO_4. The sample is then dehydrated in ethanol, epoxy resin added, and the sample cured, ultramicrotomed, and imaged with TEM.

4.3.5.2 Summary

Cryoultramicrotomy is not a routine method of specimen preparation for polymer specimens, however, new instruments are now available that permit better control of the operating variables. Cryosectioning has several advantages: specimen embedding is not required, which limits the potential of chemical reaction; soft polymers can be sectioned, which may not be possible at room temperature; and hardening is not performed by a chemical reaction. Disadvantages are: it is time consuming; special equipment is required to control knife and specimen temperature; static charge affects picking up of sections; and frost buildup limits the method. This method is used more often in polymer microscopy than in the past, as it is valuable for materials science studies.

4.3.6 Microtomy for SPM

Some SPM techniques require a very flat specimen, which is best produced by microtomy. The quality of information from an ultramicrotomed surface provides a specimen with no smearing and minimum surface roughness. Microtome manufacturers, such as Leica, have produced special AFM sample holders to aid preparation of specimen faces free from artifacts. These holders allow clamping of the specimen, preparation of the block face, and then direct examination in the AFM without releasing the specimen from the insert. A new sample holder that allows combined microtomy for AFM and TEM was described [153] that has a small central part that is used directly for the AFM measurements after the TEM ultrathin sections are made (Fig. 4.12A). Figure 4.12B, C shows an AFM phase image and TEM images of a syndiotactic polypropylene/ethylene butylene copolymer blend prepared by precipitating

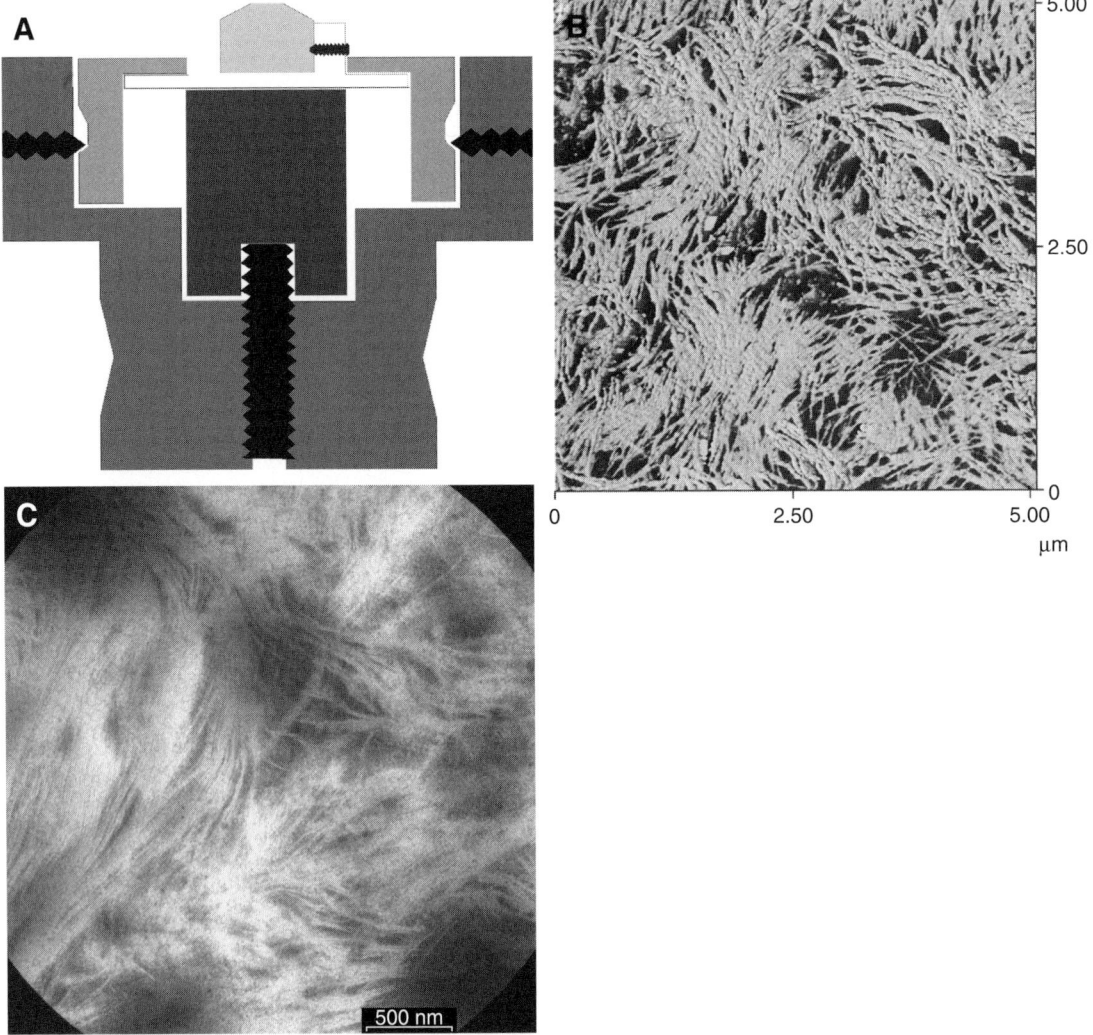

FIGURE 4.12. A sample holder for combined AFM and TEM microtomy is shown (A) with a central inner part that can be used for AFM. Micrographs taken of an s-PP/PEB blend in a phase AFM image (B) is compared with a TEM image of a RuO_4 stained section from the same sample (C). (From Thomann et al. [153], © (1999) Blackwell Publisher; used with permission.)

the respective polymers from a hot xylene solution, drying, annealing, and crystallizing, followed by cryomicrotomy using the sample holder described here.

A paper that provides details on this method is one in which heterophasic PP copolymers were prepared by ultramicrotomy at cryogenic temperatures of epoxy embedded samples to supply a smooth surface for AFM [154]. A small handsaw was employed to cut the sample, and sample trimming was performed with a water cooled grinding device. A minimum of force was applied to avoid cracks, stress whitening, heating with melting of low crystalline regions, and orientation of structures within the samples. The samples were embedded in epoxy and mounted in the cryomicrotome. The temperature applied during cutting was below the T_g or other transitions of the sample. The effect of cutting speed has not been thoroughly studied, however, a low cutting speed (below 0.5 mm/s) gave more uniform section thickness and the surface was better preserved than if a higher cutting speed was employed.

4.3.6.1 Literature Review

Bar et al. [71] characterized the morphology of blends of poly(styrene)-*block*-poly(ethene-*co*-but-1-ene)-*block*-poly(styrene) with isotactic and atactic polypropylene block copolymers by ICAFM. Samples deposited from solution onto a glass substrate, dried and annealed or quenched from the melt, and samples cut by an ultramicrotome were compared with earlier TEM results. The polymer film on the side of the film-glass interface was studied rather than the free surfaces of the polymer. Li et al. [155] observed the crystallization process directly under ICAFM phase imaging of a thin film of about 300 nm, prepared by spin coating the polymer solution onto a silicon chip. Transmission electron microscopy and AFM were used to study multiphase polymers, using cryomicrotomy [156]. Chlorine containing polyethylene was characterized by AFM after melting the film, in an uncovered pan under nitrogen in a DSC instrument, holding at elevated temperature and cooling [157]. The film was cryoultramicrotomed at −75°C, and etched for 30 min with a 2:1:0.07 sulfuric acid/*o*-phosphoric acid/potassium permanganate solution to remove the surface marks caused by microtomy. Comparative AFM and TEM study of nylon 6 rubber blends was conducted using microtomy, which did not affect the morphology of the rubber particles [158]. Samples were stained by either ruthenium tetroxide vapor or phosphotungstic acid and cryoultramicrotomed parallel and perpendicular to the flow direction with a diamond knife at −45°C. Surfaces were analyzed by AFM using phase contrast imaging and sections were imaged by TEM. Digital analysis was used to measure particle sizes. Self-assembled diblock copolymers were also ultramicrotomed, at −50°C, followed by immersion in methanol for 20 min to extract the diblock copolymer for imaging by AFM [159]. Multilayered PC/PET composites were also prepared for AFM by cryoultramicrotomy and compared with similar sections stained, using ruthenium tetroxide vapor for several hours at room temperature, to selectively stain the less dense PC component for study by TEM [160]. Freshly broken glass knives were used for cryoultramicrotomy of ethylene/1-octene copolymers and their blends with high density polyethylene to form 40 μm thin films [161]. The films, between two freshly cleaved mica sheets, were then melted and cooled using a special miniaturized heating device.

The ultra-low-angle microtomy (ULAM) technique has been developed to form a cross sectional, ultra-low-angle taper through polymeric materials such as coatings and paints [162]. Ultra-low-angle microtomy employs a conventional rotary microtome in combination with high precision, angled sectioning blocks to fabricate ultra-low-angle tapers. Scanning electron microscopy and AFM were used to investigate the morphology and topography of the polymer surface. X-ray photoelectron spectroscopy (XPS) or time-of-flight secondary ion mass spectrometry (ToF-SIMS) was used for compositional depth profiling or "buried" interface analysis. Apparently, correctly mounted polymeric samples, sectioned with a sharp microtome knife, display little perturbation or smearing of the resulting polymeric surface.

TABLE 4.2. Causes and potential remedy of sectioning problems

Problem	Potential cause	Potential remedy
Striations or lines in the section	Nicks in knife edge; hard particles or contaminants	Change knife or knife edge used for cutting
Section thickness is not uniform	Knife angle too low; block hard; temperature not correct for specimen	Adjust knife angle; choose correct temperature; retrim to improve shape
Sections compressed or wrinkled	Knife not sharp or clean; knife angle not correct; sectioning speed too fast	Exchange knife or change cutting zone; cool block; thin sections require slow cutting speed
Sections do not form ribbons	Specimen block may be too cold; inclination angle too large; too thick; or knife blunt	Increase temperature of block, expose it to light; adjust knife angle; select thinner sections; exchange knife or cutting zone
Curved ribbons	Edge of block not parallel; alignment incorrect with knife edge	Trim block edges to be parallel; align block to knife edge
Chatter marks, generally at right angles to the sections	Knife clearance angle incorrect; material too hard for knife used; vibration during cryosectioning	Adjust knife angle; use softer embedding media; for hard specimens choose greater angle; if cryosectioning, freeze on specimen disk
Cryosectioning: sections smear	Specimen may not be cold enough so they thaw	Select lower temperature and wait until cold enough
Cryosectioning: sections crack	Specimen too cold	Select higher temperature
Sections not flattened	Specimen may not be cold enough or the surface too large	Select lower temperature; trim specimen or increase section thickness

4.3.7 Limiting Artifacts in Microtomy

Sectioning problems, collection, and troubleshooting are all described by Reid [115]. For example, artifacts have been observed in ultramicrotomy of liquid crystal polymers, when compared with x-ray diffraction, due to compression that modified the structure of the polymer [163]. The oscillating diamond knife, discussed earlier, is also intended to reduce compression of ultrathin sections [118]. The list in Table 4.2 is not complete but is intended to provide insight into the cause and potential remedy of sectioning problems [164].

4.4 STAINING

4.4.1 Introduction

Image contrast in TEM is the result of variations in electron density among the structures present. Unfortunately, most polymers in common with biological materials are composed of low atomic number elements, and thus they exhibit little variation in electron density. In addition, the production of very thin specimens, for example, by microtomy (see Section 4.3), is difficult. Transmission electron microscopy micrographs of multiphase polymers often do not provide enough contrast to image the phases clearly. The primary methods that have proved useful in contrast enhancement are staining, generally by the addition of heavy atoms to specific structures. Staining involves the incorporation of electron dense atoms into the polymer, in order to increase the density and thus enhance contrast for TEM, or to add color for light microscopy. The term *staining* refers to either the chemical or the physical incorporation of the heavy atom. This section provides the methods and references for staining specific polymers. Several authors [45, 165, 166] have provided reviews that include some of the stains actively used for polymers, many of which have been adapted from biological methods [167–169].

Many of the stains applied to polymers are positive stains. In positive staining, the region of interest is stained dark by either a chemical interaction or by selective physical absorption. Chemical reactions are preferred as stains that are only physically absorbed (such as iodine) may be removed in the vacuum of an electron microscope. In microstructural work, staining may occur chemically after the staining agent penetrates some regions because of a higher diffusion rate. In negative staining, the shape of small particles mounted on smooth substrates is shown by staining the regions surrounding the particles rather than the particles themselves. Such staining methods are often applied to latex or emulsion materials.

Development of high resolution TEM, SEM, and SPM techniques has raised questions about the need for staining. Correa and Hage [170] used energy filtering transmission electron microscopy (EFTEM) for studying the polymer morphology of unstained samples to produce conventional bright field unscattered images and inelastic filtered images of high impact PS ultramicrotomed samples compared with samples stained with osmium tetroxide. Correa et al. [171] compared backscattered imaging and chemical element mapping with energy dispersive spectroscopy with EFTEM of unstained chlorine-modified polycarbonate, in which the contrast was achieved by filtering out electron inelastic scattering from the bright field images. Review of the literature continues to show staining as an important preparation method for microscopy study of polymers. Most of the chemicals used as stains are hazardous and/or toxic; specific methods of handling are left to the researcher to carefully review the material safety data sheet (MSDS) supplied with these chemicals.

4.4.1.1 Literature Review

Dyes and dye mixtures are used for identification of textile fibers, and their distinctive cross sectional shapes aid identification by light microscopy [172]. Water soluble dyes have been used with PS films imaged by energy loss spectroscopy imaging (ESI-TEM) [173]. The two step procedure includes dye adsorption in latex particles, followed by polymer plasticization with a suitable solvent. Sudan Black B, usually used for lipids, was found to be an excellent stain for water-loaded polyurethane grafts when diluted with ethanol [174]. The staining procedure was coupled with a computerized image analysis system to evaluate void sizes and structure.

Boylston and Rollins [175] reviewed the use of osmium tetroxide, iodine, phosphotungstic acid, and uranium salts, stains that were initially used by biologists. Staining of textile fibers with high atomic number elements has been employed since the mid-1950s. Maertens et al. [176] used OsO_4 and silver nitrate to stain cellulose, and Hess et al. [177] deposited iodine in natural cellulose fibers. Phosphotungstic acid was used as a negative stain [178] for cellulose. Iodine was described [175] as a stain for nylon 6 fibers; phosphotungstic acid was proposed as a stain for nylon, polypropylene, and Terylene; and OsO_4 was advocated as a stain for polyester and polybutadiene. Hagege et al. [179] described an interesting method of inclusion of stainable unsaturated polymers within the fibrillar framework of cellulosic fibers. Walters and Keyte [180] first observed dispersed particles in blends of rubber polymers by phase contrast optical microscopy. Marsh et al. [181] studied elastomer blends by both optical phase contrast and TEM. Electron microscopy was applied to study blends of natural rubber, styrene-butadiene rubber (SBR), *cis*-polybutadiene (PB), and chlorobutyl rubber [182]. It became obvious that both hardening of the rubber and staining were necessary for producing sections with contrast for TEM.

Today, the most common methods of observing multiphase polymers are by phase contrast OM of thin sections, TEM of stained ultrathin sections, SEM of etched or fractured surfaces, and SPM of microtomed or etched surfaces. Osmium and ruthenium tetroxide are the most commonly used stains for observation of the dispersed phases in multiphase blends, whereas other stains have more limited application. Detailed fine structure of polymers is also made visible by staining. For example, chlorosulfonic acid staining enhances the lamellar texture of PE [183]. There are cases where a stain has been associated with a specific functional group of polymers. A specific stain for nylon [184] showed the sizes of the macrofibrils and microfibrils. Fibers were immersed in 10% aqueous solution of $SnCl_2$ for 10 min at 100°C, rinsed, placed in NH_4OH solution to convert the tin chloride to insoluble SnO, and then embedded for ultrathin sectioning.

Staining of polymers can be conducted either before or after sectioning. The sample is cut into small blocks, about 1–3 mm across, and immersed in the stain solution or exposed to the vapor. Materials can be embedded and the blocks faced and then stained, especially when the stain diffuses into the polymer slowly. This method permits the sectioning and collection of the near surface material, which is the most thoroughly stained. If sections can be cut prior to staining, then they are stained either in the vapor, immersed in the solution, or placed on the surface of a stain droplet.

4.4.2 Osmium Tetroxide

Forty years after its first application, the method of OsO_4 staining is still widely and successfully applied to unsaturated rubbers and latexes, toughened epoxy adhesives, and many other polymers for imaging by TEM, SEM, and now by SPM as well, although phase imaging is generally used to enhance contrast in the latter. The staining and hardening of rubber phases with OsO_4 was introduced by Andrews and Stubbs [135] and Andrews [185], who stained unsaturated synthetic rubbers, and then further developed by Kato [186–188], to show the morphology of rubber modified plastics and unsaturated latex particles, which flatten and aggregate upon drying. The polybutadiene in ABS polymers is not apparent in unstained cross sections in the TEM, but staining results in contrast enhancement due to increased density of the unsaturated phase.

Osmium tetroxide reacts with the carbon-carbon double bonds in unsaturated rubber phases enhancing the contrast in TEM by the increased electron scattering of the heavy metal in the rubber compared with the unstained matrix. The reaction is very important as it both fixes and stains the polymer (Scheme 4.1). This fixation, as it is termed in biology, is a chemical cross linking, or bridging, of the rubber, which causes hardening and increased density. Staining is also known to take place by selective absorption in both semicrystalline and amorphous polymers. The reaction is slow, often taking days to weeks, when staining a block of material in an aqueous solution. Solvents are often added in order to increase diffusion of the stain. The high vapor pressure of OsO_4 is beneficial, making vapor staining of sections viable; however, this vapor pressure, combined with the toxicity of the stain and its low exposure limit [108], makes it very dangerous to use, and appropriate care must be taken to handle this material in a hood with good ventilation (see suppliers for MSDS and safe handling of all stains).

Some examples of OsO_4 staining are worthwhile to discuss as they describe staining methods for specific polymers. Vapor staining with osmium is a general method for thin, melt crystallized polymer films. OsO_4 staining and ultrathin cryosectioning [140] of SBS copolymers showed structures that contain the styrene phase in cylindrical form arranged in a regular hexagonal macrolattice. A combination of prestaining, cooling with liquid air to harden the

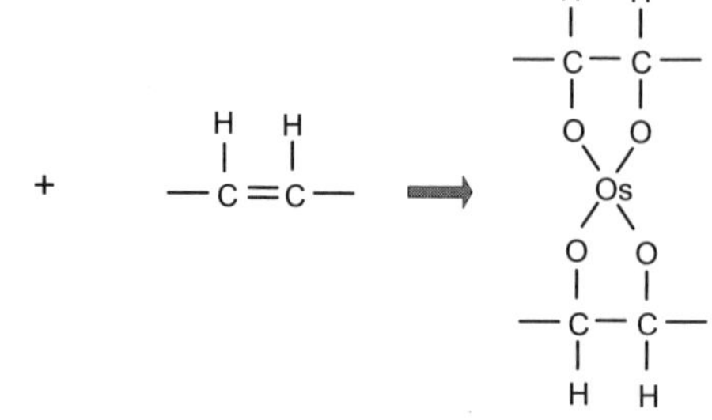

SCHEME 4.1.

polymer, and subsequent vapor staining of the sections with OsO_4 (24 h) revealed the now classical structure. Early work by Molair and Keskkula [189], Kato [188], and Matsuo et al. [190] showed staining of HIPS.

4.4.2.1 Preferential Absorption

Preferential absorption of OsO_4 has been shown [191] to reveal spherulites in semicrystalline PET. Contrast in ABS/polycarbonate blends was shown by selective absorption as the styrene-acrylonitrile copolymer (SAN) polymer contains the osmium stained rubber particles whereas the polycarbonate was not stained [192]. Niimoni et al. [193] found that there is often enough phase contrast in stained copolymers, which have different degrees of unsaturation or functional groups like –OH, –O–, or –NH_2, as they each vary in reactivity with the stain. Fridman and Thomas [194] used OsO_4 to reveal the structure of crystalline polyurethanes in which the unsaturated, hard segment was preferentially stained.

An example of the staining of semicrystalline PET is shown in Fig. 4.13. Unoriented polyester chip was melt crystallized at 235°C for 2–3 h and subsequently stained by immersion in 4% aqueous OsO_4. The amorphous regions in the spherulites appear to have enhanced electron density, due to the stain. The polarized light micrograph (Fig. 4.13A) shows the overall texture of the spherulites, whereas the higher magnification TEM micrograph of an ultrathin section (Fig. 4.13B) shows the detailed spherulitic texture where the amorphous regions exhibit enhanced electron density.

4.4.2.2 Two Step Reactions

Several two step reactions have extended the range of OsO_4 staining to materials that cannot be stained directly. Riew and Smith [195] exposed rubber modified epoxy resins to OsO_4 dissolved in tetrahydrofuran (THF), which diffuses into the epoxy, speeding the reaction with the rubber. Aqueous formaldehyde was used in the staining of polyamides [196]. Thin films of

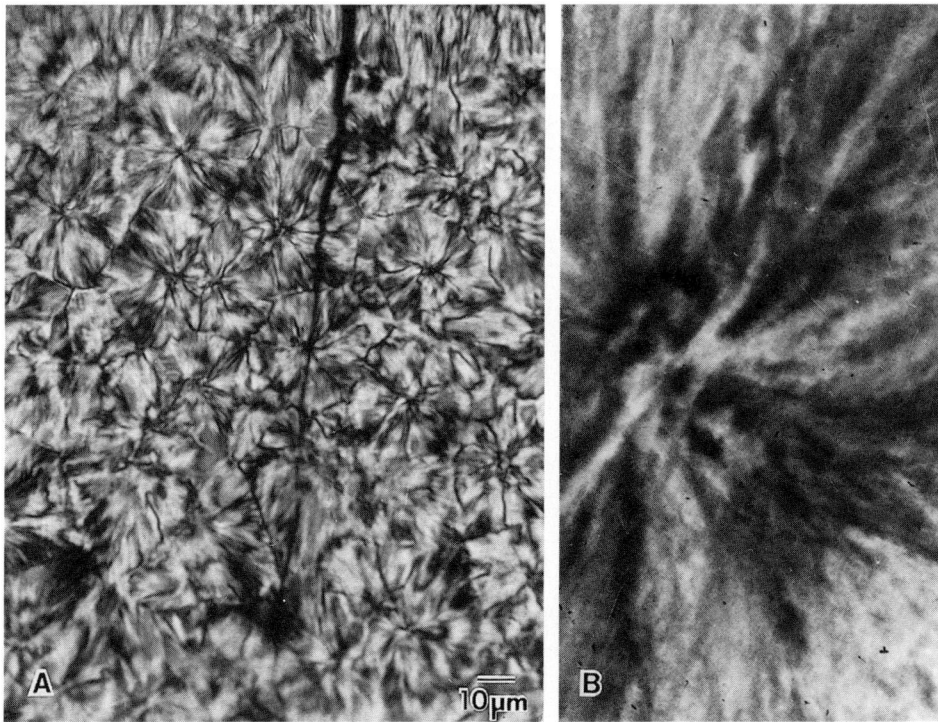

FIGURE 4.13. Unoriented, melt crystallized polyester stained for 7 days by immersion in 4% osmium tetroxide exhibits a spherulitic texture in polarized light (A) and TEM (B).

melt crystallized polyamide samples were placed in a mixture of equal parts of 30% aqueous formaldehyde and 1% OsO_4 for 3 days. The osmium tetroxide is reduced by the formaldehyde and reacts selectively, resulting in better definition of the polyamide structure. Kanig [197] developed a stain for butyl acrylate rubber by treatment with hydrazine or hydroxylamine and poststaining with OsO_4, which works for polymers containing acid and ester groups. The best known staining methods for polyesters are two stage procedures; one of which is based on the reduction of the ester bonds by borane, followed by oxidation with OsO_4 or RuO_4. Bulk samples of the polyesters were trimmed and exposed at 30°C to 35°C for 0.5 to 10h to the vapor phase of a borane-dimethyl sulfide complex; this was followed by several hours at room temperature in an OsO_4 solution, washing, drying, and microtoming [198]. Hutchins [199] used solvent-assisted osmium staining to characterize butylacrylate and ethylene propylenediene in a SAN matrix. Faced microtomy specimens were soaked in 1,7-octadiene, to introduce C=C sites and then stained in 1% aqueous OsO_4 at 60°C for microtomy and TEM.

Vesely [200] provided a review of the microstructural characterization of polymer blends by a wide range of microscopy techniques, as has been done in this text, and they introduced another two step method for STEM imaging. This method involves the sectioning of blends, such as PMMA/PS/PC, irradiation of the blend in the electron beam, using 200 keV electrons to cause changes in the chemistry prior to staining with OsO_4. In this case, the PMMA is imaged as a white phase, PS as a gray phase, and the PC as a black phase clearly differentiating the various phases as was also shown for a blend of PVC/SAN. The advantage of the method is to enhance the contrast and show miscibility at the boundaries of the particles, and the obvious disadvantage is the multiple steps required as seen in Scheme 4.2.

4.4.2.3 General Method

Osmium tetroxide is available in small ampoules either as crystals, ready to dissolve in water, or as

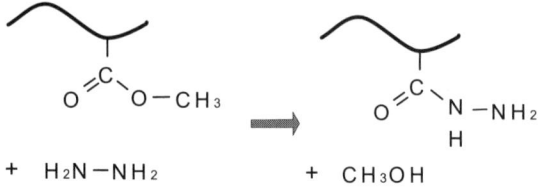

SCHEME 4.2.

premixed solutions. The ampoules are generally prescored and require scoring to cut open and pour into glass containers. Staining prior to embedding can enhance contrast and harden the material. Penetration is rather poor, and days to weeks may be needed to stain a specimen by immersion in a 1% to 2% aqueous solution. Embedded and faced specimen blocks can also be stained by immersion. Crystals can be placed in the bottom of a tube and the specimens or sections on grids placed above them for vapor staining. To speed the reaction, the tube used for vapor staining is placed in a beaker of water on a hot plate. Vapor staining at 50°C requires about 8h for a bulk specimen or 1–2h for sections. Some authors have reported vapor staining in 20min for thin films of SBS triblock copolymer cast from solution [201]. Diluted emulsion or latex particles are dropped onto coated grids and stained over 1% aqueous OsO_4 in a closed vessel for about 30min. Stain times are dependent upon the form of the specimen, the mechanism of reaction, the degree of unsaturation, and the temperature used.

4.4.2.4 Inclusion Methods

Polymers containing unsaturated rubber and semicrystalline polymers are often effectively stained using OsO_4. What about materials that do not show such differential staining? Two examples will be described where reactive (unsaturated) materials are included into the polymer to provide reaction sites. Inclusion of a stainable unsaturated polymer was shown for cellulosics [179] and synthetic fibers [202]. The initial work focused on improvement of the properties of cellulosics by inclusion of an elastomer between the microfibrils. OsO_4 staining revealed that a lamellar sheet structure was present. Marfels and Kassenbeck [202] used a similar method with polyester and nylon fibers.

Staining

An outline of the isoprene inclusion method is as follows:

1. Treat the fibers overnight in a 1% solution of benzoyl peroxide catalyst, in freshly distilled isoprene.
2. Rinse the fibers in fresh isoprene.
3. Suspend the fibers in a metal autoclave filled with 5–20 ml distilled isoprene. Seal and heat to 90°C for 6–8 h at about 2–3 kg/cm^2 pressure.
4. Dry and embed the fibers for ultrathin sectioning.
5. Stain sections in OsO_4 vapor for 1 h at 50°C.

Application of this method to polyester fibers [203] is shown in the TEM micrographs in Fig. 4.14 of longitudinal sections of a PET fiber before (A) and after (B) inclusion with isoprene and staining with OsO_4. Dense, stained isoprene regions are observed in teardrop shaped voids adjacent to delustrant particles. This method is useful for observation of void sizes and shapes, which relate to factors such as dyeability.

The second reactive inclusion method was developed [204] for microporous membranes. Stretched polypropylene, Celgard 2500 (trademark, Celgard LLC, Charlotte NC), shows little fine structure after ultrathin sectioning and examination in the TEM (Fig. 4.15A), although SEM study clearly reveals a surface pore structure. In order to enhance contrast, the membrane was treated with an unsaturated surfactant followed by OsO_4 staining and ultrathin sectioning. A 1% to 2% solution of polyoxyethylene allyl ether, Brij 97 (available from ICI Americas Inc., Bridgewater NJ), in 50/50 methanol/water solution was used to treat the membrane for about 10 min. Strips of the treated membrane were placed in a test tube containing OsO_4 crystals, the tube was sealed with a cork and placed in a beaker of water at 50°C for about 8 h in a well ventilated hood. Figure 4.15B shows that the pores in the membrane have been coated with the surfactant and stained, revealing the microporous structure. Higher magnification micrographs of similar membranes are shown in Section 5.2. This method has general utility for revealing the structure of porous polymer materials.

4.4.2.5 Staining for SEM and SPM

Rubber particle morphology has also been shown [205] by removal of the rubber from impact polystyrene (IPS) and ABS polymers and examination in the SEM. The method

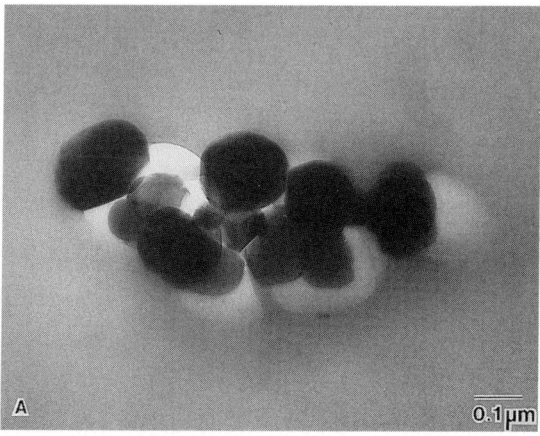 

FIGURE 4.14. Transmission electron microscopy micrographs of longitudinally sectioned PET fibers taken before (A) and after (B) isoprene inclusion and staining reveal major differences. The untreated fiber exhibits no structural detail and aggregated particles and holes within the aggregates. After treatment, dense regions of stained isoprene are observed adjacent to the particle aggregates, confirming that these regions were originally holes in the fiber, about 10 nm wide.

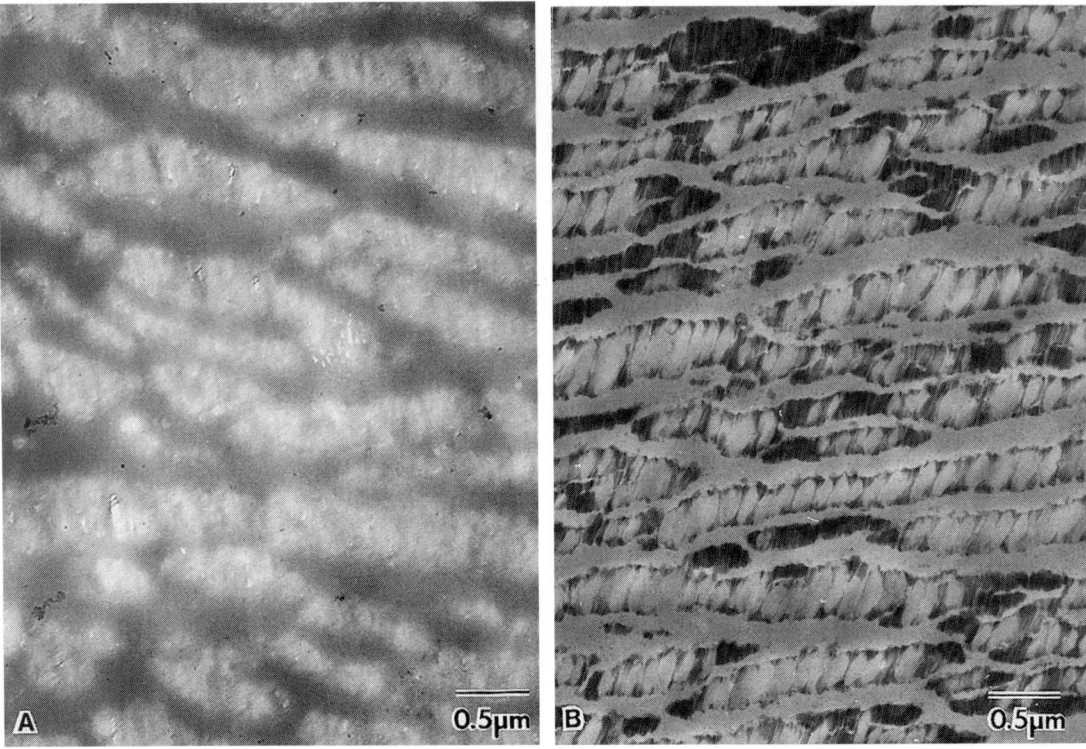

FIGURE 4.15. Transmission electron microscopy micrographs of a sectioned microporous membrane show little detail of the structure due to the low electron scattering of the polymer (A). Treatment with an unsaturated surfactant followed by osmium staining results in sections with enhanced contrast, which permits assessment of the microporous structure (B).

involved reaction of a small piece of polymer with 1% OsO_4 solution in cyclohexane on a steam bath and separation of the PS from the hardened particles by washing with isopropanol. Comparison with other methods shows this to be interesting for the study of particle size, shape, and deformation. Some caution is in order, however, as comparison of diameter measurements made by electron microscopy of stained diblock copolymers of polybutadiene spheres in a PS matrix and small angle neutron scattering have shown some discrepancy [206]. Scanning electron microscopy imaging of unvulcanized natural rubber/high density polyethylene (HDPE) blends was investigated by quick quenching samples after mixing, cutting surfaces using a cryomicrotome set at −140°C, and then OsO_4 vapor staining for 15 min followed by carbon coating [207].

Cryosectioning and staining is often used for preparation of specimens for AFM. In an early study of cross-linkable epoxy thermoplastics with 5% grafted rubber concentrate, contact AFM was used to image sections before (Fig. 4.16A) and after (Fig. 4.16B) OsO_4 vapor staining [208]. The study showed that the OsO_4 reacts, swells, and hardens the rubber, which allowed imaging. Without staining, the rubber was compressed during scanning making it difficult to conclusively confirm that rubber was there. The development of phase contrast imaging in ICAFM now allows for stain-free contrast in these impact modified systems.

4.4.3 Ruthenium Tetroxide

Ruthenium tetroxide is known to be a stronger oxidizing agent than osmium tetroxide and supposedly superior for staining rubber [209]. This chemical oxidizes aromatic rings yielding either mono- or dicarboxylic acids [210]. Although osmium tetroxide was the more predominant

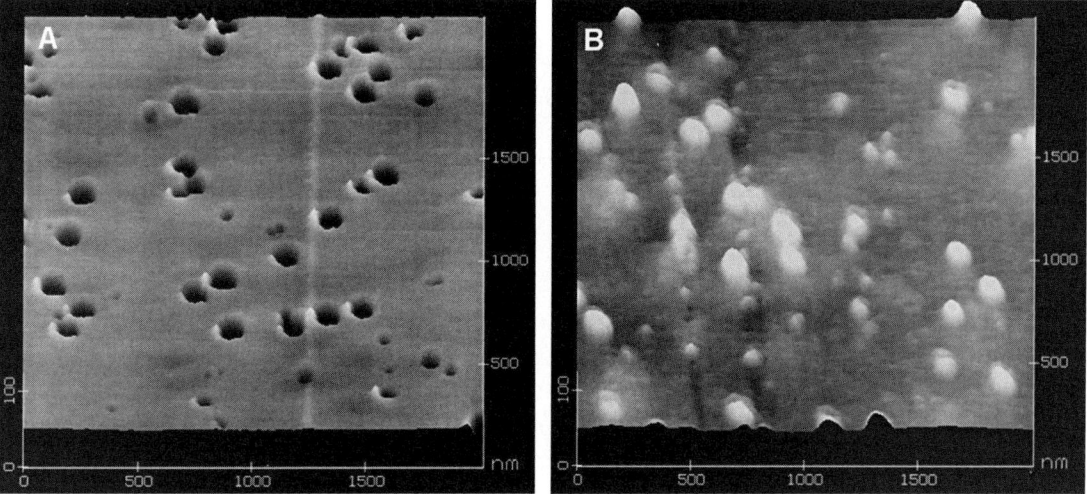

FIGURE 4.16. Contact AFM of cryosections before (A) and after (B) OsO_4 vapor staining. The OsO_4 reacts, swells, and hardens the rubber, which confirmed the presence of residual rubber phase. (From Meyers et al. [208, 602]; used with permission.)

stain in the 1990s, RuO_4 appears to have become more popular for polymers, likely due to its ability to stain latex and resin materials by cross linking ester groups. Aqueous solutions (1%) or vapor staining can be conducted for ether, alcohol, aromatic or amine moieties. Combined use of OsO_4 and RuO_4 reveals interesting details in some multiphase polymers.

Two groups independently introduced the use of RuO_4 as a staining agent. Vitali and Montani [209] showed the staining of latex and resin materials. Polybutadiene latexes, treated with OsO_4 and RuO_4 vapor, for comparison, showed no significant differences. Acrylonitrile-butadiene-styrene resins were treated by both reagents, and the TEM thin sections showed they both reacted with the unsaturated rubber. Treatment of a saturated acrylonitrile-styrene-acrylate (ASA) resin resulted in no staining with OsO_4, whereas the saturated rubber phase was hardened and stained by 1% RuO_4 treatment. The reaction is attributed to a cross linking mechanism on the ester groups of the acrylate phase. Trent et al. [211–213] introduced RuO_4 stain for the electron microscopy of polystyrene/poly(methyl methacrylate) (PS/PMMA), ABS, and nylon 11. Films were cast from toluene onto glass slides and sections were picked up on TEM grids. Staining was conducted by mounting the grids on a glass slide suspended over a 0.3% aqueous solution of RuO_4 (30 min) at room temperature.

4.4.3.1 Literature Review

RuO_4 stains polymers containing ether, alcohol, aromatic or amine moieties [212], and no reaction was observed for PMMA, PVC, and PAN, whereas HDPE, isotactic PP, and atactic PP were said to be lightly stained. Thin films cast from 1% solutions were stained in RuO_4 vapor for 120 min. High impact polystyrene was cast from toluene; ABS was cast from ethyl acetate; nylon 11 was cast from 50/50 phenol/formic acid; and PS was cast from toluene. The combined use of OsO_4 and RuO_4 staining revealed interesting detail in ABS. The spherulitic texture of nylon 11 was observed after a thin film was exposed to RuO_4 vapor. Morel and Grubb [214] conducted staining experiments on amorphous and spherulitic films of isotactic PS by vapor staining from a 1% solution for 5 min, revealing the lamellar morphology of melt crystallized PS with the added benefit of stability against radiation damage. In contrast with OsO_4, RuO_4 appears to stain surface layers and react more with the lamellar surfaces than the amorphous regions. The method has also been used to stain crazes [214, 215]. Combination of OsO_4 and RuO_4 was used to image crazes and the rubbery phase in modified PMMA [216]; these authors recently reviewed the preparation and imaging of rubber modified amorphous thermoplastic polymers.

Frochling and Pijpers [217] used 0.5% aqueous solution of RuO_4 for 30 min and 2% aqueous solution of OsO_4 for 10 min to stain the amorphous rubber phase in impact modified polyamide containing poly(propylene glycol). The OsO_4 staining did not reveal contrast but the RuO_4 clearly stained the polyether. Ohlsson and Tornell [218] used RuO_4 vapor to stain blends of PP and SBS. The solid polymers were cooled in liquid nitrogen, cut in a microtome, and vapor stained for 3 to 24 h for backscattered electron imaging in an SEM and for sectioning for TEM. Selective etching with xylene for 24 h dissolved the thermoplastic elastomer for comparison by SEM. Ruthenium in hypochlorite was used as a stain for TEM of ternary blends of PP, EPDM, and HDPE prepared by two different methods [219]. Samples were trimmed for microtomy, treated for 16 h with a 2% solution of ruthenium trichloride (see below), sectioned, and examined by TEM. This method worked very well and showed the lamellar structure of the PP matrix and the dispersed HDPE phase as the amorphous regions were stained.

Hobbs et al. [220, 221] studied the morphology and toughening mechanisms of blends of poly(butylene terephthalate) (PBT) and bisphenol A (BPA) polycarbonate using a combination of staining and etching to prepare samples for SEM and TEM. The impact modifier reacted with OsO_4, as it oxidizes double bonds; immersion in a 1% solution in hexane for 30 min was sufficient. The PC absorbed RuO_4 and the PBT did not. Copper grids with thin sections on them were glued to a glass slide and suspended above the RuO_4 solution in a stoppered bottle for 30 min. The RuO_4 solution was prepared by the first method described in the next section [222]. Scanning electron microscopy was conducted on samples crystallized in a hot stage and on cross sections faced with an ultramicrotome after etching the PC with diethylene triamine (DETA), which had little effect on the PBT. This thorough work [220, 221] showed that the various combinations of stains and etching were effective in revealing blend morphology. The PBT was seen to be the continuous phase, with the core–shell impact modifier isolated in islands in the PC. The toughening mechanisms were determined by correlation of morphology and physical properties.

The study of core-shell morphologies is very important to the field of emulsion polymers and also to toughened polymer blends. Shaffer et al. [223–225] are well known for their work in this field and have developed many methods for analyzing morphology. Methods used include negative staining with phosphotungstic acid [223] (see Section 4.4.5), OsO_4 [223] (see Section 4.4.2) and RuO_4 [224]. A few drops of the latex mixture are combined with a few drops of 2% aqueous solution of uranyl acetate, which acts as a negative stain. A drop of this mixture is deposited on a formvar-coated stainless steel grid. The grid is then exposed to RuO_4 vapors to differentiate the phases in the core and shell in SAN, HIPS, ABS, ASA, and nylon 11; HDPE and PP are only lightly stained by this method.

Injection molded syndiotactic polystyrene (s-PS) was studied by polarized light microscopy (2 μm thick sections) and TEM to evaluate morphology as a function of mold temperature [226]. Various etching and staining procedures were used for TEM, including chlorosulfonation and post-treatment with 1% aqueous uranyl acetate; etching with potassium permanganate in sulfuric acid; etching with concentrated nitric acid; and staining with RuO_4, both commercial and fresh solutions. The latter was prepared by reaction of ruthenium trichloride hydrate with 5 ml of 5.5% sodium hypochlorite aqueous solution at room temperature. Pieces were stained both prior to microtomy and after by exposure to RuO_4 vapor. Staining was the most effective method for increasing contrast and showing the skin-core effect and shish kebab morphologies. A modified *in situ* method of formation of RuO_4, by oxidation of ruthenium dioxide in a saturated aqueous solution of sodium periodate, was used to vapor stain and reveal the spherulites in polyoxymethylene by a two stage method of prestaining the sample and then the sections for TEM [227]. Loo et al. [228] directly imaged PE crystallites within block copolymer microdomains using RuO_4 staining of ultrathin sections. The morphology of semicrystalline block copolymers was observed by TEM of stained sections; a thick slice was cut in a cryomicrotome using a glass knife at −110°C, RuO_4 stained for 3h at room temperature and then microtomed with a diamond knife for TEM [229].

High resolution EM was used to characterize the displacement fields near edge dislocations in ordered polymers [230]. Samples of ABC triblock copolymer polystyrene-*block*-poly(ethylene-*co*-butylene)-*block*-poly(Me methacrylate) (SEBM) were microtomed and stained with RuO_4. This analysis makes it possible to predict and explain the variation in tilt of different lattice planes in the vicinity of dislocations in isotropic solids, anisotropic crystals, and liquid crystals in terms of their elasticity constants. The low drawability of syndiotactic polypropylene (sPP) was studied by TEM of samples embedded in epoxy, trimmed, stained at 60°C for 3h by RuO_4 vapor, and ultrathin sectioned [231]. The same method was used to investigate the mechanism of isothermal crystallization from the melt for sPP [232]. Melt grown PBT crystals were prepared based on an optimized staining method at 25°C for 8h and then sectioned, revealing formation of fringed micellar crystal nuclei in the early stage and folded chain fringed micellar crystals in the later stage of isothermal crystallization at 40°C from the melt [233]. These authors prepared samples by staining before and after sectioning and at various temperatures and found that higher temperature staining resulted in structural change not observed at the lower temperatures.

A study of PET by Haubruge et al. [234] was conducted in the vapor phase, on both spin coated thin films and sections exposed at room temperature to vapors of the freshly made solution for 5 min to 1 h. The 13% active chlorine aqueous sodium hypochlorite used is stronger than shown in Scheme 4.3. A study of micro- and nanostructured surface morphology on electrospun polymer fibers was conducted on

$$2NaIO_4 + RuO_2 \xrightarrow[1°C]{H_2O \text{ at}} 2NaIO_3 + RuO_4$$

4g   0.6g   100ml

SCHEME 4.3.

fibers stained before embedding [235]. Chou et al. [236] used transmission EELS to study the effect of RuO_4 staining on PS, showing there is alteration of aromatic character. Samples on grids were stained in vapor from a solution of 0.5% RuO_4 in water. Imaging and selected area electron diffraction show that a layer of RuO_2 nanocrystals, about 2–5 nm in size, forms on the surface of bulk PS specimens exposed to RuO_4, independent of the chemical nature of the specimen, limiting resolution at nanometer length scales. This group [237] also uses phase contrast TEM imaging as an alternative to staining for study of morphology in nanoscale objects such as globular polymer particles. Specific details of the moieties stained by ruthenium tetroxide continue to be a potential source of artifacts, and care must be taken in image interpretation of complex materials. Debolt and Robertson [238] used TEM of a RuO_4 stained ternary blend of nylon 66 and polystyrene in a polypropylene matrix with and without compatibilization by an ionomer resin (for nylon 66) and a styrene-*block*-ethylene-*co*-butylene-*block*-styrene (SEBS) copolymer (for polystyrene). Samples were prepared by diamond knife cryoultramicrotomy at −120°C with a cutting speed of 0.4 mm/s, followed by exposure of the sections to RuO_4 vapors for 30 min. This resulted in staining the PS dark, the nylon medium gray, and the PP light gray with the ionomer remaining unstained.

4.4.3.2 General Method and Discussion

The general method for ruthenium tetroxide staining is to stain sections over a *fresh* 1% solution for about 5–30 min. This is a problem as RuO_4 solutions are quite unstable, though Trent et al. [212] have frozen solutions in sealed glass containers for periods up to 6 months and report that ruthenium tetroxide can be prepared by oxidation of hydrated ruthenium dioxide using sodium periodate (available from

Morton Thiokol, Inc., Alfa Products, Chicago IL). Literature with currently available solutions should be used to determine stability. The reaction, shown in Scheme 4.3 [239], is complete in 3–4 h [222].

Trent [222] and Montezinos et al. [240] reported this instability and suggested alternative reactions for preparation. Montezinos et al. [240] discussed the preparation of RuO_4 by oxidizing ruthenium dioxide or trichloride with sodium metaperiodate and extracting the tetroxide with chloroform. However, it seems this is not simple or fast, whereas oxidation of ruthenium trichloride with sodium hypochlorite [241] is a viable alternative (Scheme 4.4). This latter, one step procedure was applied [240] to PE films and blends of PE and PP with elastomers. Treatment times of 30 min to 3 h were followed by drying and embedding in Spurr epoxy resin for ultramicrotomy. Care must be taken as ruthenium tetroxide is volatile and toxic although little is known regarding health hazards. The use of MSDS supplied with these chemicals is recommended. The expected reaction of unsaturated chains with ruthenium tetroxide is given by Scheme 4.5.

4.4.3.3 Staining for SEM and STEM

Brown and Butler [125] developed a modified method of formation of RuO_4 for the analysis of domain morphology of stained polyolefin blends. Sodium hypochlorite solution (1 ml of 10 w/v %) is added to 0.02 g of $RuCl_3·3H_2$) in a 5 ml glass vial, mixed, and capped for vapor staining. Blocks of the sample were cut in a cryomicrotome and stained in RuO_4 vapor for 2.5 h for LVSEM imaging, and sections were collected for TEM. The stained block was recut to remove the overstained skin. Scanning transmission electron microscopy was conducted on

$$8NaClO + 2RuCl_3 \cdot 3H_2O \xrightarrow{H_2O}$$
10 ml 5% aq   0.2 g   100 ml

$$\longrightarrow 8NaCl + 2RuO_4 + 3Cl_2 + 3H_2O$$

SCHEME 4.4.

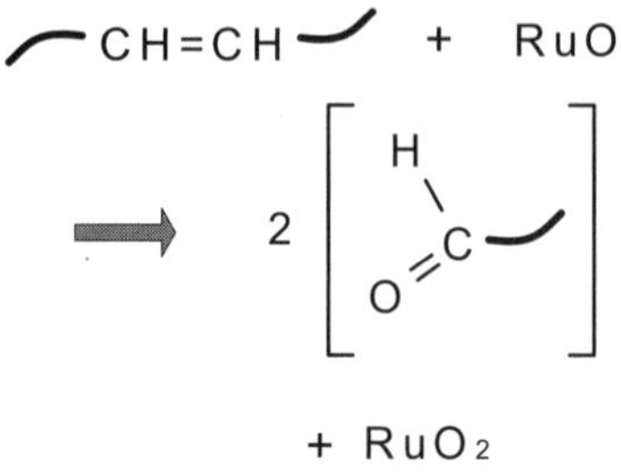

SCHEME 4.5.

blends of PP and cycloolefin copolymers (COC), stained using the RuO_4 preparation just described, for 100 min. Scanning transmission electron microscopy micrographs showed light PP matrix with dark COC particles that compared well with SEM images of fractured samples [242]. A practical method for high resolution imaging by LVSEM was described using RuO_4 stained films of core-shell latexes embedded in a PMMA matrix [243]. Three preparation methods were used: staining of a diamond knife smoothed block face, directly imaged; ultrathin sections (50 nm thick) from a stained specimen using a diamond knife for TEM imaging, picked up on copper TEM grids; and the same stained sections placed on a thin metal film on an SEM stub. The use of a stained thin section on a thin metal film of high conductivity (e.g., Cu or Al) and/or low atomic number (e.g., Al or Ti) eliminated the need for a conductive coating for LVSEM.

4.4.3.4 Examples for TEM

Microwave methods have been used for fixation and staining of biological samples [112, 113, 244] and for polymers by Wood [244] who showed the benefit of microwave stimulated heavy metal staining with RuO_4 of conventional and ultrahigh molecular weight polyethylene, segmented block copolymers, ethylene-propylene copolymers, and PS. The staining agent used was RuO_4 vapor generated from ruthenium trichloride [240] in an industrial microwave oven installed in a chemical fume hood. A staining vessel with sections on grids suspended over the solution was transferred to an oven operated at 17% power for 10 to 30 min. Changes in staining were attributed to heating the aqueous mixture, enhanced reaction rate, and improved interactions of the water and polymers. Wood [244] showed interfacial staining at the boundary of polyethylene domains dispersed in HIPS was enhanced relative to room temperature staining (Fig. 4.17). The polyethylene-rich regions also contain small unstained domains of cross-linked PE as well as two other ethylene copolymers (Fig. 4.17A). A comparison of TEM images of occluded PS particles in the HIPS prepared using microwave (Fig. 4.17B) versus conventional staining (Fig. 4.17C) shows greater contrast at the HIPS/rubber interface, which may contain some grafted polymer, in the case of microwave staining. The effect of two different stains in a pigmented Noryl blend of poly(phenylene oxide) (PPO) and HIPS [245] is shown: the pigment particles are best seen in an unstained section (Fig. 4.17D); OsO_4 reveals the butadiene-rich structure in HIPS (Fig. 4.17E), and regions rich in PS are seen best by RuO_4 (Fig. 4.17F).

Transmission electron microscopy images of latex particles prepared by staining with OsO_4, RuO_4, and phosphotungstic acid [225] are found

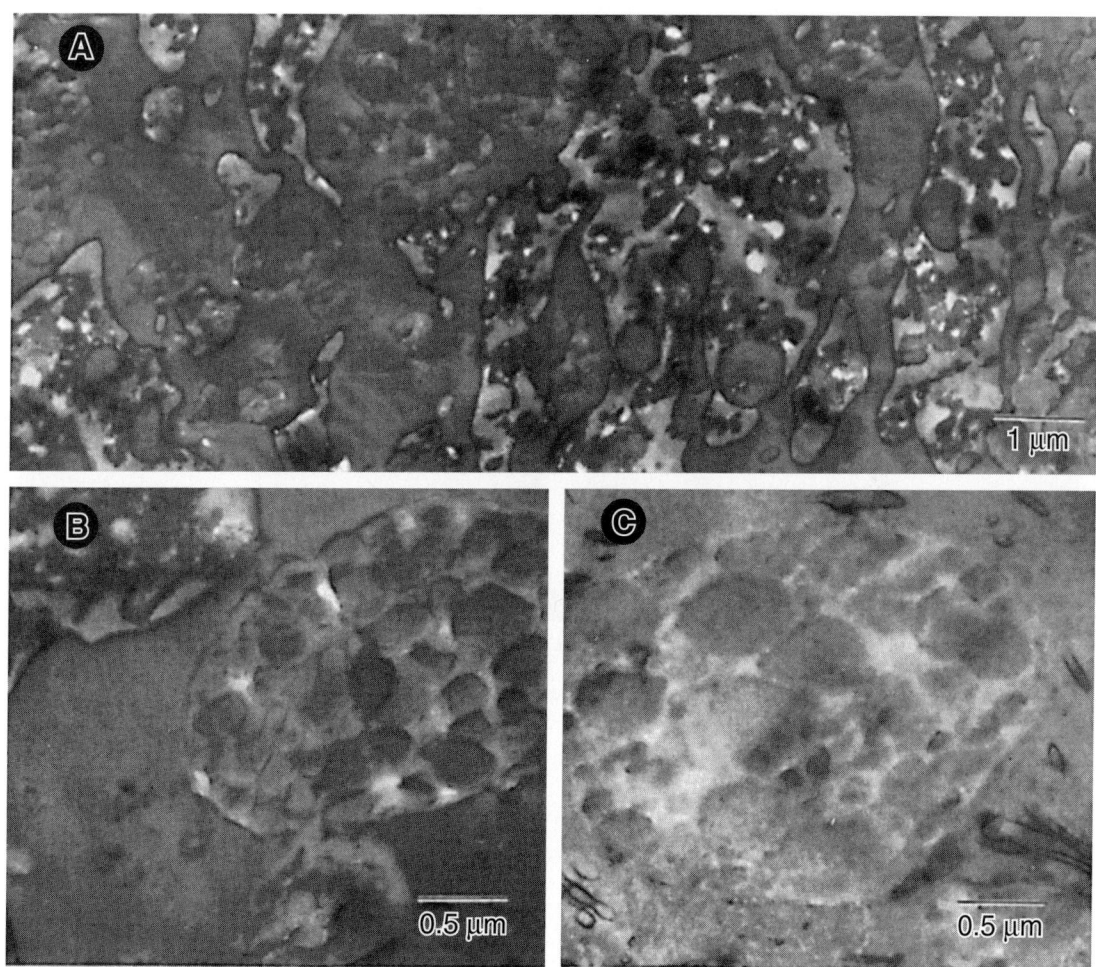

FIGURE 4.17. Transmission electron microscopy images [244] of microtomed section of HIPS/PE/ethylene copolymer blend stained with RuO_4 in a microwave oven, showing selective staining at the HIPS/PE interface (A). Higher magnification of HIPS region of material (B); HIPS prepared by conventional RuO_4 staining (C); rubber phase appears white.

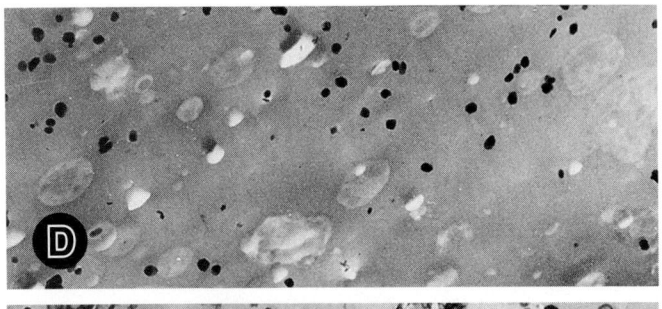
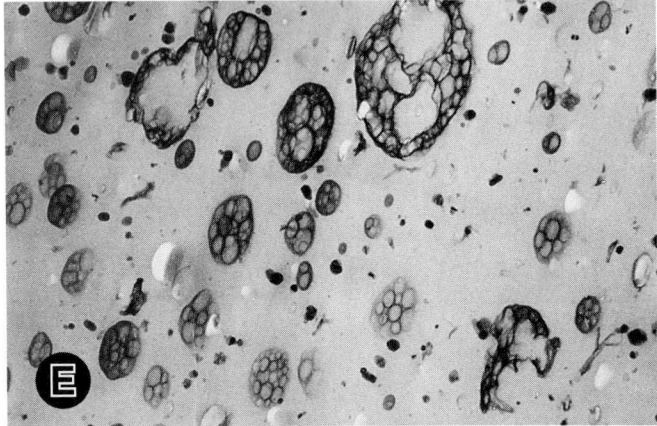
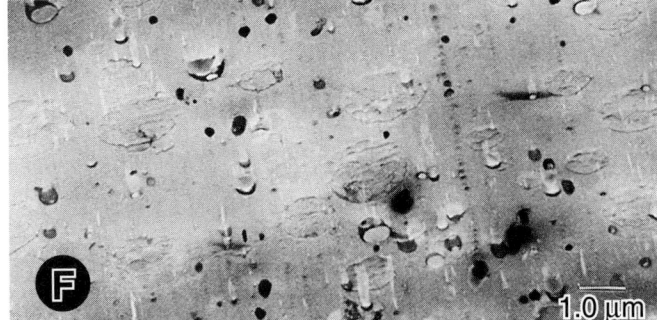
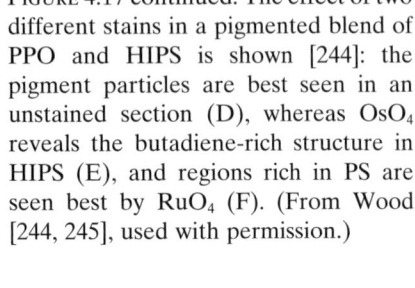

FIGURE 4.17 continued. The effect of two different stains in a pigmented blend of PPO and HIPS is shown [244]: the pigment particles are best seen in an unstained section (D), whereas OsO$_4$ reveals the butadiene-rich structure in HIPS (E), and regions rich in PS are seen best by RuO$_4$ (F). (From Wood [244, 245], used with permission.)

in Fig. 4.18. A latex with a core of poly(butylacrylate-butadiene) (PBA-PB) and PMMA shell was diluted in distilled water and stained by adding three drops of 2% aqueous OsO$_4$ to the diluted dispersion. This dispersion was then diluted in 2% aqueous phosphotungstic acid (PTA) and a drop placed on a carbon coated formvar stainless steel TEM grid. The grid was then placed on a glass slide and stained in RuO$_4$ vapor for 10–20 min and air dried. Figure 4.18A shows latex particles with dark phases, the core of PBA-PB stained with OsO$_4$ and RuO$_4$, and negative staining due to the PTA [225]. "Lumps" of PMMA can be seen on the core. Figure 4.18B is of a core-shell latex of PS with a PBA shell, prepared by the same method without the OsO$_4$. The RuO$_4$ stained the PS core, and this is seen against a lighter PBA incomplete shell [225].

In summary, RuO$_4$ is an oxidizing agent that appears somewhat similar to OsO$_4$ in the staining of unsaturated phases. Some saturated polymers may be stained with this reagent by vapor phase reaction of sections for short times (30 min). Although the reactive moieties appear to include ethers, alcohols, aromatics, and

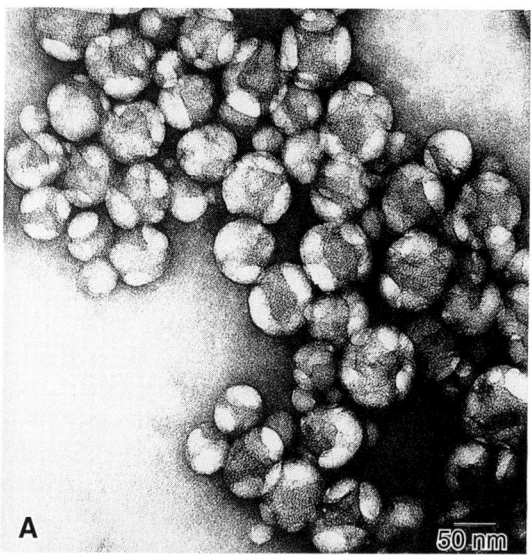 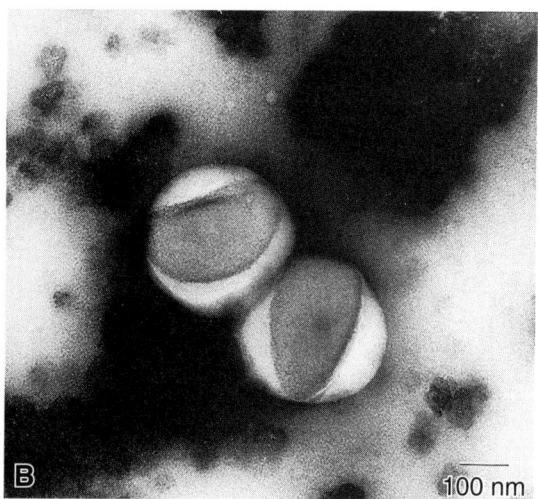

FIGURE 4.18. Transmission electron microscopy images of latex particles prepared by staining with OsO$_4$, RuO$_4$, and phosphotungstic acid [225]. Transmission electron microscopy image (A) shows latex particles with dark phases, the core of PBA-PB stained with OsO$_4$ and RuO$_4$, and negative staining due to the PTA. "Lumps" of PMMA can be seen on the core. Transmission electron microscopy image (B) shows a core-shell latex of PS with a PBA shell, prepared by the same method without the OsO$_4$; the RuO$_4$ stained the PS core, and this is seen against a lighter PBA incomplete shell. (From O.L. Shaffer, unpublished [225].)

amines, it must be remembered that RuO$_4$ is known to oxidize aromatic rings and cleave double bonds or rings rather than bonding to them.

4.4.4 Chlorosulfonic Acid and Uranyl Acetate

An important staining technique was developed by Kanig [246] for the enhanced contrast of PE, a material that has been a model compound for fundamental polymer studies. Polyethylene crystals cannot be sectioned, nor are they stable in the electron beam, due to radiation damage. The chlorosulfonation procedure cross links, stabilizes, and stains the amorphous material in crystalline polyolefins, permitting ultrathin sectioning and stable EM observation. Chlorosulfonic acid diffuses selectively into the amorphous material in the semicrystalline polymer, increasing the density of the amorphous zone compared with the crystalline material. The treatment stains the surfaces of the lamellae primarily due to incorporation of chlorine and sulfur. Treatment with a salt solution results in a reaction with the polar groups, and metal ions are deposited resulting in increased electron density. Poststaining with uranyl acetate intensifies and stabilizes the contrast due to the high electron density of the uranyl group. Lamellar structures are revealed in PE that are now known to be typical of semicrystalline polymers.

4.4.4.1 Literature Review

Chlorosulfonic acid was used to show the lamellar structure in both linear and branched PE and to study the effects of drawing and annealing [197, 247, 248]. Dlugosz et al. [249] examined cryosections of drawn, rolled and annealed, bulk oriented PE that clearly showed a lamellar texture. Lamellae were shown to have two preferred orientations and to be arranged in stacks perpendicular to the draw direction. Experiments using chlorosulfonation on other olefins and polyesters were shown to result in their dissolving in the acid rather than staining. Hodge and Bassett [250] prepared and observed the lamellar texture in bulk PE documenting an evaluation of the staining method, mechanism, application, and limitations using PE in the chain extended form. The reaction time was shown to vary depending on the diffusion channels open. Voigt-Martin et al. [251] prepared chlorosulfonic acid stained cryosections [252] of melt crystallized linear PE and compared the morphology with TEM of replicas and light scattering. Results showed lamellar crystallites in the entire molecular weight range studied. Kanig [253] described the lamellar crystallization of PP from the melt using this method. Staining was also applied (120°C) to the crystallization of PE [253, 254]. The chlorosulfonation method has been applied to bulk polymers and high modulus fibers [255–257]. Smook et al. [257] studied the fracture process of ultrahigh strength PE fibers and used the method in a unique application to show the nature of the kink bands present. Highly oriented ultrahigh molecular weight PE fibers were shown to be preferentially attacked at these kink bands when exposed to the acid for 45 min at 80°C.

Schaper et al. [258] developed structure-property relations for surface grown PE fibers before and after zone drawing by studying ultrathin sections stained with chlorosulfonic acid. Computer processing was used to reveal detail in HREM imaging, which showed the microfibrillar superstructure. Additionally, shish kebab structures were shown by three methods, chlorosulfonic acid staining, gold decoration (see Section 4.7.5). and after permanganate etching (Section 4.5.3). Scanning electron microscopy showed fiber buckling and kinking, which is another effect of the highly oriented fibrillar structure [258].

4.4.4.2 General Method and Examples

A general method for staining PE with chlorosulfonic acid is as follows:

1. Treat the sample with chlorosulfonic acid for 6–9 h at 60°C.
2. Wash the stained sample in concentrated sulfuric acid and then in water.
3. Dry and embed the sample in epoxy resin.
4. Cut ultrathin sections with a diamond knife.
5. Poststain the sections in 0.5% to 1% aqueous uranyl acetate (3 h) [108].

The stained lamellar texture is shown (Fig. 4.19) to result from the chlorosulfonation of linear PE crystallized isothermally from the melt [183]. Although large parts of the section show no detail, some regions do contain the parallel dark lines shown in this figure. These lines are a few tens of nanometers apart, and they appear and disappear as the section is tilted in the microscope. The interlamellar surfaces are electron dense, showing that they are stained. It should be remembered that uranyl acetate alone can also be used for staining polymers, although it is generally used as a secondary or tertiary stain. Roberts [259] used a 5% solution of uranyl acetate in isobutanol saturated water for aiding stain intensity in biological tissue. As with all the stains mentioned, care must be taken in handling and disposal of uranyl acetate due to its radioactivity and toxicity.

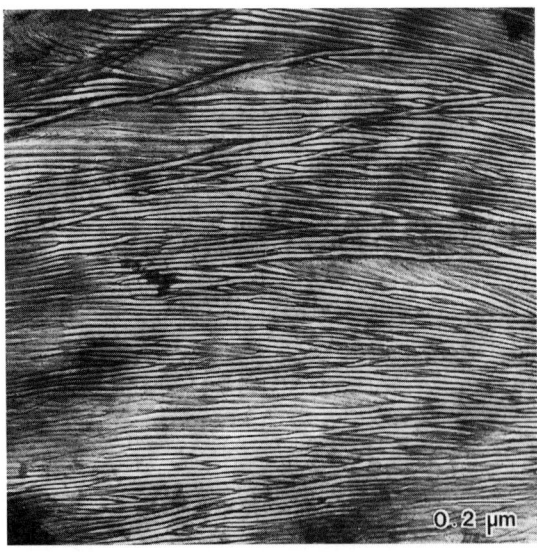

FIGURE 4.19. Transmission electron microscopy micrograph of a chlorosulfonic acid stained linear polyethylene crystallized isothermally from the melt reveals the electron dense interlamellar surfaces typical of polyethylene.

Ultrahigh molecular weight PE (UHMWPE), used as an implant material in artificial joints, was examined by in an early study done by SEM, AFM, and TEM [260] using chlorosulfonic acid to attack the crystal/amorphous interface leaving behind atoms that react to heavy metal stains. Pieces of polymer were trimmed, exposed to the chlorosulfonic acid for 1 h, and cryoultramicrotomed at −90°C using a diamond knife to cut sections about 90 nm thick, stained overnight in PTA. Permanganate etching was used to remove the amorphous material, and carbon/platinum replicas were prepared for TEM. Samples for AFM and SEM were prepared by trimming with a razor blade, fracturing after immersion in liquid nitrogen, and cleaning with trichloroethylene. The block face resulting from cryomicrotomy at −170°C was also used for AFM imaging. Chemically etched surfaces required less preparation for AFM imaging than cryomicrotomy. Ultrahigh molecular weight PE was also evaluated for wear by TEM [261] using the general method of staining with chlorosulfonic acid and poststaining in 2% uranyl acetate for 2.5–3 h. A representative TEM image is shown in Fig. 4.20, of chemically cross linked compression molded UHMWPE showing short and numerous lamellae.

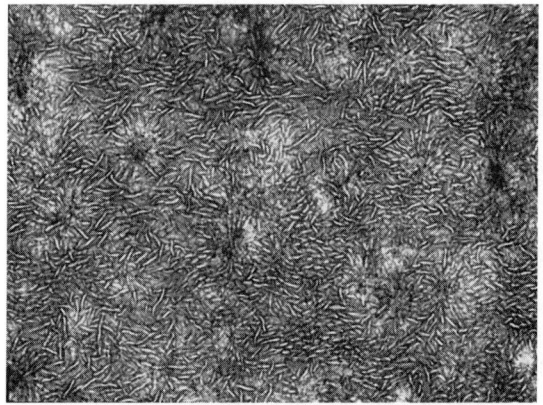

FIGURE 4.20. A representative TEM image of a chemically cross linked UHMWPE, used as an implant material in artificial joints, prepared by staining with chlorosulfonic acid and poststaining in 2% uranyl acetate [261], shows numerous lamellae. (From Kurtz [261]; used with permission.)

4.4.5 Phosphotungstic Acid

Phosphotungstic acid was first used for biological staining of structures about 1945. Hayat [262] described that early work and what is known about the mechanism of staining. Phosphotungstic acid is an anionic stain with a high molecular weight (3313.5 g/mol), which imparts high density to the stained material, generally acting as a negative stain. There is no agreement, apparently, among biologists as to the interaction of this stain with organic materials, although it is known to stain proteins. Two interpretations are the formation of a complex in aqueous solution and ionic precipitation. In any case, the specificity of staining, at least in biological tissue, appears related to the pH of the solution, due to the fact that the PTA molecule is unstable and degrades when the pH is higher than 1.5.

4.4.5.1 Literature Review

Phosphotungstic acid staining was used to show the fine structure in nylon 6 fibers [263] by soaking the fibers in 9% to 11% aqueous salt solutions and staining with 1.5% and 4.8% PTA. Longitudinal periodicities were shown for the stained fibers by TEM, but the lamellae showed a change in size, depending on the concentration. At 4.8% PTA, the lamellae were about 7 nm. These authors reported [264] that 9% to 11% HCl treatment bound 15% of the weight of PTA in the unoriented or amorphous regions of nylon. Transmission electron microscopy and electron diffraction showed a range of periodicities in nylon from the ordered, oriented, and unoriented regions. Spit [265, 266] showed detailed spherulitic structures in solvent cast nylon 6 and nylon 6,6 films cast from formic acid onto water and stained with 2% PTA. According to Boylston and Rollins [175], PTA was used to reveal the fine structure of polyesters [267, 268]; polyoxymethylene was also stained with PTA [269].

The staining of polypropylene was not as straightforward as that of the polyamides and polyesters. Hock [270] developed a method for staining melt crystallized PP and showed it to be composed of spherulites containing lamellae. The melt crystallized polymer was boiled for 4 h in 70% HNO_3 at 120°C, and then chips of the oxidized polymer were reacted in 5% aqueous PTA for 3 days at room temperature. Functional groups that polyamides have in common with proteins might be the reason why they can be stained with PTA. However, the direct staining of olefins by PTA would not be expected. In the case of oxidized PP, staining apparently occurred where the folded chains had been cut by the acid and the $-NO_2$ and $-COOH$ groups were attached to the remaining short molecules. Staining clearly revealed the interlamellar regions adjacent to the unstained crystalline lamellae, and lamellar thicknesses were measured as a function of the crystallization temperature.

4.4.5.2 General Method and Examples

Phosphotungstic acid is known to react with monomer epoxy resins, which extract the stain [167], precluding its use prior to epoxy embedding. Therefore, PTA stained material is usually either embedded in glycol methacrylate or polyester resins, or sections are poststained in cured epoxies. Phosphotungstic acid penetration is slow and about $100 \mu m$ penetration into a block of material can be expected. Sections on grids are immersed into the solution or placed on a droplet. Pretreatment in absolute ethanol increases exposure to the stain. Martinez-Salazar and Cannon [271] reported staining of nylon 6 and nylon 6,6 using 2% PTA and 2% benzyl alcohol; thin films on a specimen grid were floated on a drop of the solution for 10 min and then washed in water several times.

Phosphotungstic acid reacts with surface functional groups such as hydroxyl, carboxyl, and amines [272] as a positive stain. Shaffer et al. [273] used PTA for negative and positive staining to enhance the contrast in TEM imaging of latexes, such as poly(butyl acrylate) and poly(ethyl acrylate). One drop of latex was added to 1 ml *fresh* 2% PTA to stain deformable or low glass transition temperature material. A drop of the stained latex was placed on a carbon-formvar coated, stainless steel grid. After removing excess fluid with filter paper, the specimen was placed in a TEM cold stage and frozen. Figure 4.21 shows TEM micrographs of a latex, observed using a cold stage, with and without staining [274]. The unstained latex shows (Fig. 4.21A) discrete particles that are somewhat aggregated, whereas after PTA staining the particles could be imaged more

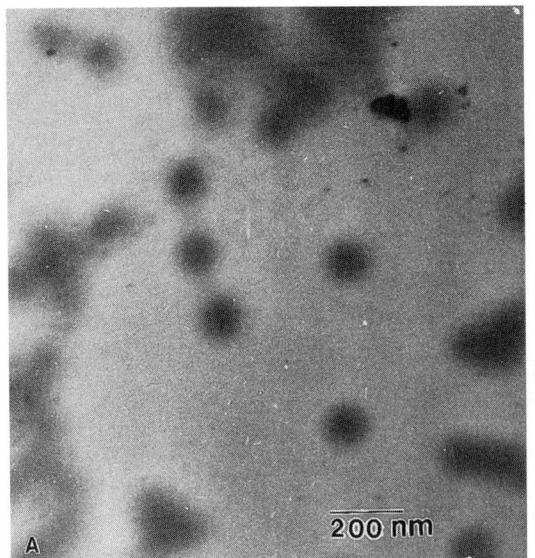
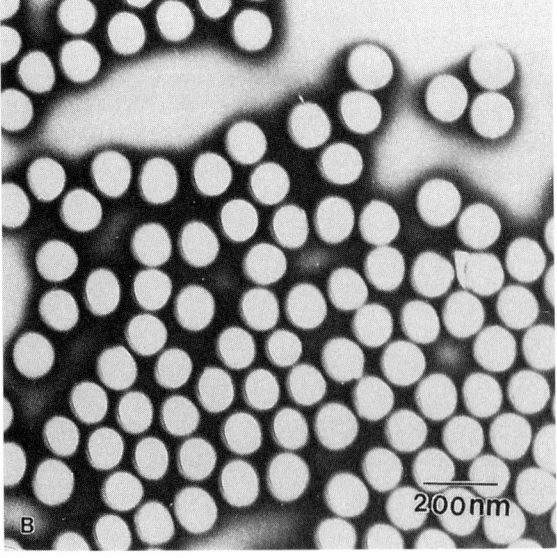

FIGURE 4.21. Transmission electron microscopy micrographs of latex particles are shown: images taken in a microscope with a cold stage (A) and after both staining with PTA and using the cold stage (B). (From Shaffer [274]; unpublished.)

clearly (Fig. 4.21B) [225]. Core-shell latex particles, such as electron dense PS, can be observed by this method. Sections can also be stained; for example, poly(vinyl acetate) sections were stained with 5% PTA for 30 min [273]. In a more recent work [275], TEM imaging of several different polymers using negative staining with PTA was demonstrated in water and organic solvents (e.g., dimethylformamide (DMF), dimethylsulfoxide (DMSO), and tetrahydrofuran (THF)). Polymer particle size, size distribution, and shape seen in negative staining were shown to correlate with those of unstained materials. Polymer solutions examined included block copolymers, microgels, and polymer brushes. Solutions of uranyl acetate and PTA 2% were used in water and several solvents. A series of nylon 6 blends with maleated rubbers were analyzed for morphology and fracture analysis by cryogenic microtomy at −50°C and then treated with a 2% solution of PTA for 30 min at room temperature [276]. Rubber particle sizes were determined with a semiautomated digital analysis technique. TEM images (see Fig. 4.18) of latex particles prepared by staining with OsO_4, RuO_4 and phosphotungstic acid [225] were described earlier (see Section 4.4.3.4).

4.4.6 Ebonite

The study of phase size and compatibility requires that the different phases be observed, distinct from one another, and that there is minimum distortion in the polymers. However, multiphase polymers often cannot be stained or sectioned uniformly. There are composite structures that are combinations of soft rubbers, coatings, and oriented fibers that cannot be stained with a single staining agent and the sections may be deformed or distorted, limiting observation, measurement, and interpretation. The ebonite method developed by Smith and Andries [277] is used to stain and uniformly harden composite polymers.

4.4.6.1 General Method and Examples

The recommended method [277] uses molten sulfur to transform rubber to ebonite that can be polished for surface examination. A control method was to cryosection with subsequent OsO_4 staining. Blends of styrene-butadiene rubber (SBR) with chlorobutyl rubber (CB) or cis-PB, prepared by the two methods revealed that the overall structures were the same, but the ebonite method was the simpler of the two. With ebonite, both phases stained, but the SBR appeared to have more contrast. With newer and easier to use cryomicrotomes, it may be that the ebonite method has less current utility. It is certainly not as popular as OsO_4 even though the materials are less toxic to use. Three diblock copolymers of 1,4-polybutadiene and cis-1,4-polyisoprene and blends of these copolymers with the corresponding homopolymers were prepared by the ebonite method [278], providing useful observations of the dispersed phases.

A modified ebonite method [279] was used to study the interfaces associated with polymer tire cords. Tire cords composed of PET, rayon or nylon fibers are generally bonded to rubber with a resorcinol-formaldehyde-latex (RFL) adhesive. The nature of the interfaces is of major interest. OsO_4 may be used to stain and harden the RFL, but the soft rubber is not affected by this treatment, and, in fact, it forms a barrier to stain penetration. The ebonite reaction hardens the rubber and hardens and stains the RFL while maintaining the geometrical integrity of the composite.

The reaction medium consists of molten sulfur/accelerator (N,N-dicyclohexyl-2-benzothiazolylsulfenamide)/zinc stearate in the weight ratio 90/5/5. Small pieces of the cord are cut from the tire carefully trimming some of the rubber but leaving a thin, undisturbed surface layer. Eight hours are required for the reaction at 120°C. Samples are removed, scraped off, and placed in a 120°C oven to remove the excess ebonite. The treated cords are embedded and sectioned to 50–60 nm thick with a diamond knife. The fiber-RFL-rubber interfaces are all observed in the tire cord cross section (Fig. 4.22). The ebonite method can be used routinely for polymer blends or composite specimens, such as tire cords, where hardening, penetration, and staining are required. Although sulfur is a poison that must be used

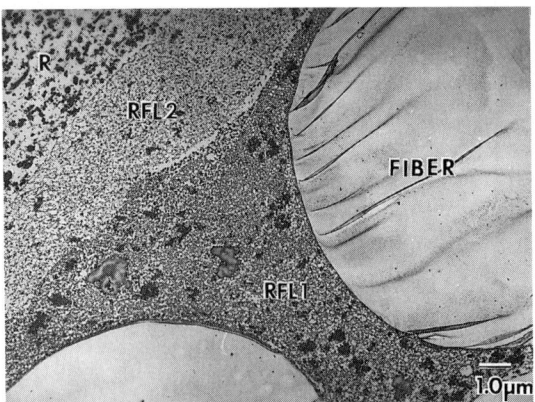

FIGURE 4.22. Transmission electron microscopy micrographs of sectioned industrial tire cords are prepared by the ebonite method to enhance the contrast of the various structures and to harden the adhesive and the rubber for sectioning. Fiber cross sections, two adhesive layers (RFL), and the rubber (R) are shown by TEM.

in a hood, it is easier to use and less toxic than OsO_4 and less time consuming. The resolution possible with the ebonite method is quite good, enabling the imaging of the fine structure of the latex adhesive.

4.4.7 Silver Sulfide

Microporous structures in biological and polymer materials are often difficult to observe, even in the TEM, as there is limited differential electron scattering and thus limited contrast between the embedding resin and the specimen. The electron contrast can be increased by the addition of a high atomic number element, such as silver, atomic number 47, to porous regions. This was shown earlier for osmium tetroxide (see Section 4.4.2.3). The silver sulfide method has been described for use in both natural and synthetic fibers. Sotton [280] studied wool, regenerated cellulose, poly(vinyl alcohol), acrylics, and polyesters. Hydrogen sulfide was liquefied and the samples treated in an autoclave under pressure prior to treatment with a silver nitrate (0.1 N) solution, which precipitated silver sulfide in the pores. Silver nitrate can be used with water, acetone, or ethylene glycol as solvents.

Hagege et al. [281] used a similar silver sulfide insertion technique in the study of aramid fibers. The fibers were treated according to the method described above and also by a method [179] using isoprene but excluding the staining. Sections examined in the TEM showed details of the void microstructure. The Leeds group (e.g., [282, 283]) used the silver sulfide method to prepare new aramid fibers for TEM in order to evaluate the nature of the microvoids and to determine their influence on the tensile and compressive behavior of the fibers:

Polymers that may be stained include esters and aromatic polyamides.

1. Treat the specimen with gaseous hydrogen sulfide at a pressure of 1,380 kPa at 20°C for 16 h; wash in alcohol (see suppliers for safe handling methods).
2. Immerse the specimen in a 5% aqueous solution of silver nitrate, at 20°C for 3–4 h [284].
3. Embed the specimen in a low viscosity resin (e.g., Spurr resin).
4. Ultrathin section with a diamond knife.

4.4.8 Mercuric Trifluoroacetate

The structural elucidation of multiphase polymers is a continuing problem as, at times, the commonly used staining reagents are not satisfactory. Mercuric trifluoroacetate has been described for the staining of several polymers [285]: PS, poly(2,6-dimethyl-1,4-phenylene oxide), and saturated styrene-butadiene-styrene block copolymers. The reaction of mercuric trifluoroacetate is said to take place by electrophilic substitution and is activated by electron donating groups, such as oxygen, which are attached to the ring. Apparently, bisphenol A polycarbonate/polyethylene blends can also be stained with this reagent but must be diluted with water and used for shorter times (20 min) to prevent crumbling of the sample [285]. This staining reagent is very toxic and it must be handled in a hood, with proper protective clothing, although little is known about this specific substance. The slow rate of diffusion limits penetration of the reagent into the polymer. Long exposure times are limited by acid attack, which results in swelling and disruption of the specimen surface [286]. In addition, the major disadvantage of the method is that there is no hardening of the polymers, making room temperature ultramicrotomy difficult.

4.4.8.1 General Method

Trimmed and faced blocks are stained by immersion in a 10% solution of HgO in trifluoroacetic acid for 10–60 min. Samples are washed in a dilute solution of trifluoroacetic acid followed by distilled water. Poly(phenylene oxide) appears to have a higher mercury uptake than PS in a bonded laminate of the two films [8]. Blends made by coextrusion of SBS block copolymers and PPO show dispersed particles in a matrix with PPO taking up more stain. In summary, mercuric trifluoroacetate staining has been shown for several polymers where the dispersed phase particles are differentiated by this stain. The method has limited application.

4.4.9 Iodine and Bromine

Early treatment of materials with iodine revealed structures in nylon 6 filaments [287, 288] and the longitudinal periodicity of nylon was shown by the differential absorption of the iodine in the crystalline and noncrystalline regions. Hess and Mahl [177] treated poly(vinyl alcohol) fibers with 5% iodine, which revealed a long period (16 nm) that relates to the absorption of iodine in the noncrystalline or amorphous regions. Hess et al. [289] detected structures of the order tens of nanometers in cellulose treated with 12% iodine. Iodine dissipates upon standing in air and is known to vaporize in the vacuum of the electron microscope so it is no longer used for staining of PE and has been replaced by the chlorosulfonic acid method. The structure of PET fibers was described [290] using the iodine sorption method. Oil-based finishes on fiber or yarn surfaces, such as PET, have been stained with iodine [291] by placing the yarn overnight in an airtight glass chamber containing iodine crystals. The treated yarns were assessed by using low magnification (250×) on an ordinary research microscope and using a stereo binocular microscope for finish uniformity highlighted by the adsorption of iodine. Overall, there appears to be simple physical adsorption of the iodine, preferentially in the amorphous regions of semicrystalline polymers. There are few current applications of the iodine staining or sorption method found in the literature, and most are using optical microscopy.

The surface morphology of bromine treated Kevlar fiber was shown by SEM to have a rougher texture than the untreated fiber, and the tensile strength decreases with the increase of bromine treating time [292]. The fibers were immersed in saturated bromine water, taken out and neutralized, and then rinsed and dried. Polymer stabilized liquid crystals or anisotropic gels were studied using three stains for comparison [293]. In order to observe a cross section through the network, it was embedded in a resin and then sectioned using an ultramicrotome. Reagents used for network staining were OsO_4 vapor for up to 48 h; chlorosulfonic acid vapor staining for 18 h; and the successful method — bromine vapor for 1 h or 24 h. Thus, there is no one stain useful for all polymers, and even iodine and bromine have some utility.

4.4.10 Summary

Staining of polymers is an important part of sample preparation for microscopy as it provides the enhanced contrast required to image the structures. There are a few staining techniques that work for a range of polymers and others with limited applicability. For many polymers, there is no proven stain, and thus preparative treatments must be found by experimentation. This section is summarized in Tables 4.3 to 4.5 listing those polymers that have been shown to be stained by the various reagents described in this section. In addition, several stains that are applicable for only a few

TABLE 4.3. Polymer functional groups and stains

Polymers	Stains
Unsaturated hydrocarbons, alcohols, ethers, amines	Osmium tetroxide
Acids or esters	(a) Hydrazine (b) Osmium tetroxide
Unsaturated rubber (resorcinol-formaldehyde-latex)	Ebonite
Saturated hydrocarbons (PE and PP)	Chlorosulfonic acid/ uranyl acetate
Amides, esters, and PP	Phosphotungstic acid
Ethers, alcohols, aromatics, amines, rubber, bisphenol A, and styrene	Ruthenium tetroxide
Esters, aromatic polyamides	Silver sulfide
Acids, esters	Uranyl acetate

TABLE 4.4. Specific functional groups, examples, and stains

Functional group	Examples	Stains
–CH–CH–	Saturated hydrocarbons (PE, PP) (HDPE)	Chlorosulfonic acid Phosphotungstic acid Ruthenium tetroxide
–C=C–	Unsaturated hydrocarbons (polybutadiene, rubber)	Osmium tetroxide Ebonite Ruthenium tetroxide
–OH, –COH	Alcohols, aldehydes (polyvinyl alcohol)	Osmium tetroxide Ruthenium tetroxide Silver sulfide
–O–	Ethers	Osmium tetroxide Ruthenium tetroxide
–NH$_2$	Amines	Osmium tetroxide Ruthenium tetroxide
–COOH	Acids	Hydrazine, then osmium tetroxide
–COOR	Esters (butyl acrylate) (polyesters) (ethylene–vinyl acetate)	Hydrazine, then osmium tetroxide Phosphotungstic acid Silver sulfide Methanolic NaOH
–CONH$_2$ –CONH–	Amides (nylon)	Phosphotungstic acid Tin chloride
Aromatics	Aromatics Aromatic polyamides Polyphenylene oxide	Ruthenium tetroxide Silver sulfide Mercury trifluoroacetate
Bisphenol A-based epoxies	Epoxy resin	Ruthenium tetroxide

Etching

TABLE 4.5. Osmium tetroxide staining

(a) Multiphase polymers stained by OsO_4

Acrylonitrile-butadiene-styrene

Acrylonitrile-styrene-acrylate

Styrene-butadiene-styrene

High impact polystyrene

Impact poly(vinyl chloride)

Copolymers of 1,4-polybutadiene and *cis*-1,4-polyisoprene

Blends with unsaturated rubber, isoprene, or isoprene included in fibers, e.g., PET or nylon

Polyoxyethylene allyl included in membranes

(b) Polymers requiring pretreatment prior to OsO_4

Polymer	Pretreatment
Acids, esters	Hydrazine
Epoxy thermosets	Tetrahydrofuran
Ethylene-vinyl acetate copolymers	Alkaline saponification
Chlorinated PE	Bicyclic amine
Polyesters	Allyl amine

polymers are also listed here, even though they were not fully described.

4.5 ETCHING

Etching is another preparation method that potentially enhances the information available by microscopy while at the same time providing opportunity to introduce many artifacts. Samples are etched and then generally used directly by AFM, or replicas made for TEM, or a conductive coating is applied to the etched surface for imaging by SEM.

There are two general categories of etching: chemical attack or bombardment with charged particles as in plasma and ion beam etching. Chemical attack can be further divided into several categories, including dissolution, which implies the removal of whole molecules of a material as it dissolves acids and other chemicals. Solvent extraction with xylene was used for the study of PE [294]; however, dissolution is not recommended due to the artifacts that can result from swelling or reprecipitation. In chemical etching, there is a chemical attack and removal of fragments from the specimen. Acid treatment that selectively oxidizes one phase present in a multiphase material, aiding contrast between the various phases, is a form of chemical etching. A special case of acid etching, with permanganate, is still the most widely used etchant and will be discussed in detail. Charged species activated by high voltages (ion etching) or in a radiofrequency plasma [295] are used to modify the chemistry of a surface or etch away surface material differentially. Plasmas and ion beams are employed to bombard or sputter and remove surface atoms and molecules. Just as specimen preparation methods first used for biological materials are used for polymers (e.g., microtomy), ion and plasma etching were first used for metals and ceramics and are now applied to polymers. Focused ion beam (FIB) etching, well established for the semiconductor industry, is now finding use for polymers and will be discussed. Etching will be considered in five sections: (1) solvent and chemical etching, (2) acid etching, (3) etching with potassium permanganate, (4) plasma and ion etching, and (5) focused ion beam etching.

4.5.1 Solvent and Chemical Etching

Solvent and chemical etching is a complementary technique in the determination of microstructure. Chemical etching has been arbitrarily divided into two sections for the purpose of this discussion, and acid etching will be dealt with separately. Early studies involved etching to reveal the interior of polymers for replica formation (see Section 4.6).

4.5.1.1 Literature Review

Peck and Kaye [296] immersed cellulose acetate specimens in acetone, at −50°C, and then flooded the surfaces with cold absolute alcohol, followed by replication, which showed the skin, orientation, voids, and pigment. Reding and Walter [29] etched PE with hot carbon tetrachloride (high density PE), benzene (low density PE), or toluene, which removed the amorphous material. Bailey [297] used a rapid xylene etch to reveal spherulites in PE and PP, while Li and Kargin [298] etched with benzene. Ohlsson and Tornell [218] used RuO_4 vapor to stain blends of PP and SBS for TEM and also did selective etching with xylene for 24 h to dissolve the thermoplastic elastomer for comparison by SEM. More recently, hot xylene was used to remove PE from blends with LCPs to reveal the dimensions and shapes of the LCP particles [299]. Samples were punched from molded circular disks, freeze polished (see Section 4.2), and treated in three steps at successively higher temperatures (105°C, 120°C, 130°C) for 1 week each to release layers of LCP free of PE (revealed by IR). Scanning electron microscopy showed the LCP lamellar structure and correlated the morphology with permeability measurements.

Isopropanol vapor was used to dissolve the matrix in polymer blends [300]. Williams and Hudson [301] etched microtomed blocks of HIPS so that the rubber particles protruded from the matrix. Later, Kesskula and Traylor [205] removed rubber particles from HIPS and ABS polymers by dissolving the matrix in a cyclohexane solution of osmium tetroxide and extracting the dispersed phase for SEM. Amine etching was used to reveal the structure of PET as early as 1959 [98] when fibers were etched with n-propylamine for replica formation. Methylamine was also used [302], although the selectivity of the reagent was questioned. Tucker and Murray [303] etched PET filaments with 42% aqueous solutions of n-propylamine at 30°C. Warwicker [304] showed disk forms in PET swollen in dichloroacetic acid. If aminolysis is to be used to study PET, then the excellent review of the reagents, reactions, and earliest studies by Sweet and Bell [305] should be evaluated. These authors used primary amines to degrade PET selectively by removing the less ordered regions. In some cases, the aminolysis was shown to degrade and crystallize PET, which limits the use of the method.

An etching technique was developed to study the structure of crystalline polymers, such as nylon 6, nylon 6,6, and PP [88]. Aromatic and chlorinated hydrocarbons were used to etch surfaces of ground and polished plastic parts, which removed the surface detail with a series of polishing media down to a 0.05 μm alumina slurry. Choice of specific etchant was made so that it would be a poor solvent for the plastic so as to suppress swelling and dissolution of the crystalline phase. Results were evaluated using reflected light and Nomarski differential interference contrast optical microscopy.

Another etching method was developed for the observation of the internal crystalline structure of water-soluble poly(ethylene oxide) using sodium ethoxide in ethanol. Bu et al. [306] found the optimal etching condition is to use 21 wt.% NaC_2H_5 in ethanol for 10 min at ca. 298 K with frequent agitation. The sample must be washed in absolute ethanol for 10 min and then dried prior to making one stage replicas by shadowing with heavy metal, backing with carbon, and then removing the polymer with water. Blends of thermoplastic polyurethane elastomer and ABS resin were observed by SEM for the morphology of the blends after etching with methyl ethyl ketone (MEK) [307]. Samples showed variations in structures depending on the etch time with 3 h adequate to etch these blends, whereas serious artifacts were observed at 4 h. This points to the need for carefully controlled experiments.

Hobbs et al. [220, 308] determined the morphology and deformation behavior for toughened blends of PBT, PC, and PPE using complementary staining and TEM as well as etching for SEM. Samples for SEM were faced with an ultramicrotome and etched by brief immersion in diethylene triamine, a selective etchant for PC. This method was said to be preferable to solvent or plasma etchants that attack both components to various degrees and can obscure fine details. Zeronian et al. [309, 310] studied the surface modification of polyester by alkali-metal hydroxides and amines to change the properties of polyester fabrics, such as drape and hydrophilicity. Weight loss occurs rapidly with time in methanolic sodium hydroxide at 21°C, whereas it is much slower for treatment with 10% aqueous sodium hydroxide at 60°C. The reaction is thought to occur at the surface of PET fibers with chain scission products removed into the solution. In the SEM, surface pitting is obvious as is a decrease in the fiber diameter, with larger pits being observed for fibers with titanium dioxide delustrant or other inert material [309, 310]. The reader interested in the use of these techniques to modify the properties of the polyester fibers is referred to a review of this topic [309]. The morphology of PMMA, poly(ethylene oxide) (PEO) blend, and the grafted copolymer poly(methyl methacrylate-g-ethylene oxide) was observed by SEM after etching the PEO phase out of the blend [311]. The PEO phase of the copolymer was stained with OsO_4 and examined using a backscattering image; the blend was also etched with methanol for 4h, the time chosen based on achieving constant weight.

Wallheinke et al. [312] studied the location of different compatibilizers at the interface in blends of thermoplastic polyurethane (TPU) and polypropylene (PP) by SEM and TEM. Selective dissolution of the TPU matrix in dimethylformamide with 1% dibutylamine resulted in a solution of the TPU with PP/EC particles that was filtered, dried, and the membrane with particles imaged by low voltage SEM. In the uncompatibilized blend, the surfaces of the PP particles are very smooth. Therefore, the surface coverage by the compatibilizers can be detected easily. Compared with TEM, using SEM on separated particles has the advantage of a fast and easy preparation without microtoming and staining.

Shahin and Olley [313] studied poly-3-hydroxy butyrate (PHB) that forms large banded spherulites by SEM and TEM by etching techniques, acknowledging the need for a reagent that gives a true and easily interpreted view of the morphology. They found potassium permanganate did not work well with PHB, although it did with other polymers (see Section 4.5.3), but potassium hydroxide in ethanol, propanol, or methanol did work at room temperature for about 1 h. Samples were coated with gold for SEM and direct replicas made for TEM using tantalum/tungsten alloy at an angle of 30°.

Hu et al. [314] studied the solid state structure and oxygen transport properties of smectic poly(diethylene glycol 4,4'-bibenzoate) (PDEGBB) by AFM. Samples were sandwiched between glass cover slides, melted, pressed to spread the sample, cooled, and one slide was removed and the sample was vacuum dried and remelted. Free surfaces were etched with 40 wt.% aqueous methylamine solution for 48h and then examined by AFM, providing insight into the structure related to oxygen permeability.

4.5.2 Acid Etching: Overview

Oxidizing acids have been used to etch chemically resistant semicrystalline polymers, such as polyolefins, as the acid preferentially diffuses into and attacks the amorphous regions. After slight etching, the crystalline lamellae remain proud of the surface, and details of the morphology may be seen in the SEM after metal coating or in the TEM after shadowing and replication. Large scale features, such as spherulites, may be visible by reflected light microscopy. Permanganate acid etching is the major etchant used today to study polymer materials, and thus it will be described separately after the historical overview below.

4.5.2.1 Literature Review

The general historical trend has been to start with very severe etchants and move to weaker ones, which more reliably show fine scale structures. The progression of acids with time has been as follows: nitric, chromic, permanganate, permanganate/sulfuric, permanganate/sulfuric/phosphoric acids. Palmer and Cobbold [315] first etched bulk samples of melt crystallized PE with fuming nitric acid (95%) at 80°C and observed the lamellar morphology, and Hock [316] used boiling nitric acid (70%) to reveal the microstructure of PP. Hock [317] saw spherulites in the oxidized polymer and, upon sonication, also observed lamellar fragments. Kusumoto and Haga [318] treated nylon 6,6 with 18% nitric acid at 60°C and showed it was more easily oxidized than PE.

A major problem with nitric acid etching is that it is too strong, etching not only the amorphous surface regions but also the bulk. Chromic acid etching was employed to alleviate this problem. Armond and Atkinson [319] treated PP with fuming nitric acid and then with chromic acid to reveal the cracks and fractures of bulk annealed PP. Bucknall et al. [320] studied polymer blends after etching with chromic and phosphoric acids. Bucknall and Drinkwater [321] developed a method for ABS blends by fracturing in liquid nitrogen and etching, at 40°C for 5 min, in concentrated aqueous chromic acid solution. Briggs et al. [322] used chromic acid etching on blends with polyolefins, and Bubeck and Baker [323] applied this method to etching PE. Epoxy resins have also been etched with chromic acid [324] and sulfuric acid [325]. Sheng et al. [326] used chromic acid etching of three propylene polymers and conducted contact angle measurement, x-ray photoelectron spectroscopy (XPS), attenuated total reflection infrared (ATR-IR), SEM, and an adhesion test, showing that etching with such an oxidative acid has a significant effect on the surface roughness.

To understand the mechanism of common surface defect formation in the injection molding process, Edwards et al. [327] investigated the effect of processing parameters on the morphological features of a polycarbonate/acrylonitrile-styrene-acrylate rubber modified thermoplastic blend using acid and alkaline etching and SEM characterization. Chromic acid etching revealed that the process had a large influence on the morphology of the thermoplastic, particularly on the surface skin layer. The degree of molecular orientation on the surfaces of molded parts had a significant effect on the efficiency of both the acid and alkaline etching techniques. Thus, methods have not been in general use during the past decade, whereas potassium permanganate is still used in many laboratories.

4.5.3 Permanganate Etching

4.5.3.1 Introduction

Permanganic acid, a weaker acid than nitric acid, has been found to selectively remove the amorphous regions of polyolefins and many other crystalline polymers to reveal their internal lamellar organization. It was developed and has been applied by Olley and Bassett [328–330] and others [251, 252, 331] as a complementary method to chlorosulfonation. The original permanganate etching method involved the use of 7% potassium permanganate ($KMnO_4$) in concentrated sulfuric acid; treatment for 15 min at 60°C, or for 15–60 min at room temperature [330]. This method has been used to reveal lamellar detail in surfaces of PE, isotactic PP, and isotactic poly(4-methylpentene-1). Addition of ortho-phosphoric acid to the reagent decreases the presence of artifacts in drawn linear PE and in blends. Figure 4.23 [332, 333] shows the types of structures observed by permanganate etching followed by replication and TEM. Naylor and Phillips [332] developed a method, used in the production of the micrograph in Fig. 4.23, with 2% w/w $KMnO_4$ in concentrated sulfuric acid.

FIGURE 4.23. Portion of a banded spherulite of linear polyethylene, also showing a spherulite boundary, as revealed by etching for 1 h in 2% w/w $KMnO_4$ in concentrated sulfuric acid, as described by Naylor and Phillips [332]. The micrograph is of a chromium shadowed carbon replica studied in the TEM where each disk-like object in the bands is composed of lamellae nucleated at that point by a screw dislocation. Scale bar is 2.0 μm. (From Phillips and Philpot [333]; used with permission.)

This method shows details of the structure with minimum artifacts. These authors also adapted the permanganate acid etching method for SEM evaluation by a study of etching as a function of time, temperature, and concentration for a series of polyethylenes. They identified artifacts that limit study but that can be minimized by treatment in an ultrasonic bath. Bassett and Olley [334] studied the lamellar morphology of isotactic PP, which involves treatment of a glass knife microtomed specimen with 0.7% w/v solution of $KMnO_4$ in 2:1 concentrated sulfuric acid/dry phosphoric acid for 15 min. Nomarski differential interference contrast optics were used to judge the etched surfaces. Two stage replicas were prepared for TEM by metal shadowing at the first stage.

4.5.3.2 General Method

It has been known for more than 20 years that a wide array of polymers can be etched using potassium permanganate [331], although care must be taken to limit the effect of artifacts. The list includes linear and branched PE, PP, PS, poly(4-methylpentene-1), poly(butene-1), PVF_2, PEEK, PET, and various copolymers such as EPDM terpolymers, and LCPs. The past decade has seen refinement of the method by many of the same researchers:

1. 1% w/v $KMnO_4$ dissolved by continuous stirring in an acidic mixture of 10 volumes concentrated sulfuric acid, 4 volumes 85% ortho-phosphoric acid, 1 volume water, respectively, with the sample shaken for 2 h. (10:4:1 sulfuric acid, orthophosphoric acid, water.)
2. Recovery of the etched specimens is typically by adding a small quantity of hydrogen peroxide to the chilled acid mixture.

The standard two stage replica process was used with cellulose acetate, shadowed with tantalum/tungsten, followed by deposition of a carbon film and extraction of the replica (see Section 4.6.2).

4.5.3.3 Literature Review

As predicted in the past two editions, there has continued to be further development and utilization of the permanganate method for a variety of polymers. It is important to understand that the correct "recipe" for etching is very dependent on the specific composition of the polymer, its history and morphology. Care must be taken to identify the appropriate reference and then do controlled experiments on your own materials. Key references that discuss specimen preparation methods developed during the past 20 years will be summarized here, but they only represent some of the important studies that have been conducted.

Bashir et al. [335] for instance conducted a comparative study of preparation methods for microscopy of a melt processed extrudate of PE using $KMnO_4$ and chlorosulfonic acid treatment. They found the permanganic acid etch required 2 h at 60°C compared with 15 min for spherulitic PE, whereas the chlorosulfonic treatment took 1 day at 70°C. The etched samples were replicated and compared with thin sections of the chlorosulfonic acid treated samples by TEM and both showed interlocked shish kebab structures. A modification of the procedure was developed for PEEK that revealed the spherulitic structure [336]; this required 2% w/v solution of $KMnO_4$ in a mixture of 4:1 by volume of ortho-phosphoric acid/water, with no sulfuric acid as it is a solvent for the polymer. Samples were etched at room temperature for 50 min, then etching was stopped by adding the reagent to twice the volume of hydrogen peroxide solution followed by washing and replication.

Lemmon et al. [337] appear to be among the first to use $KMnO_4$ to etch thermotropic liquid crystalline polymers (TLCPs) (see Section 5.6). Thin films formed from the melt on glass substrates were etched for a few minutes in a solution of potassium permanganate in sulfuric and phosphoric acids. Discrete entities corresponding with the "non-periodic layer crystallites" are observed in the SEM of an etched film of a very low molecular weight polymer. Ford et al. [338] also claim to be among the first to report on the application of permanganic etching to TLCPs. In this work, the etchant was prepared by dissolving potassium permanganate (10 mg to 1 ml) in a 2:1:1 mixture of ortho-phosphoric acid, sulfuric acid, and water for 45 min at ambient temperature, followed by washing as mentioned earlier in this section. These authors dried the specimens and coated them with gold for examination by OM and SEM. The use of gold coating implies that the work was not conducted at very high resolution. Bedford and Windle [339] studied shear-induced textures in thermotropic LCPs also using the method of Bassett and coworkers [329]. Hanna et al. [340] from the same laboratory further discussed the dimensions of crystallites in the TLCPs, using two specimen preparation methods. A melt shearing process was used to prepare thin films for TEM (see Section 5.6) and samples for SEM were prepared by treating thin films on glass substrates with 2 wt.% $KMnO_4$ in ortho-phosphoric acid, followed by washing and metal coating. Anwer and Windle [341] used the same method for study of TLCPs after magnetic alignment, and fracturing for imaging in the SEM. The etching revealed crystallites that lie normal to the local chain axis. Hudson and Lovinger [342] used a similar technique [329] to investigate the morphology of a TLCP using 0.5% $KMnO_4$ with a solvent ratio of 2:2:1 of water, phosphoric acid, and sulfuric acid for about 2 h at 30°C. They used a two stage replica process to prepare samples for TEM, which showed the fine fibrillar structure of these polymers.

Sutton and Vaughan [343] described a method of preparation using permanganic etching of specimens that are difficult to handle for geometrical reasons. Polypyrrole p-toluenesulfonate films embedded in acrylic or epoxy resins were etched with a number of different reagents and etching rates. Procedures were developed to etch the polymer and also to attack the embedding media. The method has the advantage for handling thin films and fibers, but caution must be used when using a range of etchants in the interpretation of the resulting images. In a later study [344], SEM was used to assess transverse fracture, and TEM was conducted after permanganic etching of microtomed cross sections [343]. Polyethylene samples, crystallized from the melt both by quenching and by isothermal crystallization at each of five different temperatures, have been characterized by small angle x-ray scattering (SAXS) to monitor the lamellar separation [345]. The same materials were prepared for TEM using fixation with chlorosulfonic acid, followed by sectioning, and permanganic etching, for 2 h in an ultrasonic, followed by replication. Comparison with SAXS spacings, showed too low a value for lamellar separation when samples were fixed for insufficient time in chlorosulfonic acid, and too high a value for lamellar separation when samples were shadowed at too low an angle, after permanganic etching. The time for chlorosulfonic acid treatment is dependent on the material and temperature; the shadowing angle used for replication should be as high as possible to get correct measurements.

Hosier et al. [346] studied PE blends and the effect of morphology on electrical breakdown strength by treating microtomed samples as shown above. Samples for SEM were gold coated and shadowed carbon replicas were made for TEM. White and Bassett [347] studied isotactic polypropylene using this same $KMnO_4$ mixture, for 2 h at room temperature, followed by making two stage carbon replicas showing the row structures. Two step etching methods have been developed for binary polymer blends of linear low density polyethylene (LLDPE) with high density polyethylene (HDPE), and for blends of atactic and syndiotactic PS [348]. For both cases, two different etchants have been identified for the component neat polymers; sequential etching has been employed to reveal the distribution of the component polymers within the blend. The polyolefin blend component was etched in 1 wt.% $KMnO_4$ in sulfuric acid for 2 h and examined by SEM, AFM, and TEM. Morgan et al. [349] studied co-crystallization in polyethylene blends using two methods of sample preparation for TEM, permanganic etching/replication [345] and chlorosulfonic fixation/sectioning [197], and found the latter method better for revealing detailed morphological features between large lamellae.

Sue et al. [350] studied the morphologies of various diglycidyl ethers of 4,4′-dihydroxy-α-methylstilbene–based liquid crystalline epoxy (LCE) formulations used as matrices for high performance composites by polishing, to 0.3 μm finish, followed by etching with $KMnO_4$ in a mixture of 1:2:2 volumes of sulfuric acid, ortho-phosphoric acid, and water for 75 min at 60°C. They evaluated the composites by reflected light microscopy, micro-Raman spectroscopy, TEM of replicas, and selected area electron diffraction.

Shahin et al. [351] refined the permanganic reagents (2h) to characterize a special feature of polyethylene spherulites, namely the S-profile of dominant lamellae seen as growing toward the observer. Direct and two stage replicas were used for TEM. The morphology of woven polypropylene cloth compactions has been examined by electron microscopy after etching as above for 1 h [352]. Samples were examined by SEM after gold coating, and two stage replicas were produced for TEM. With increasing temperature, the interior structure of the fibers undergoes progressively greater melting and recrystallization in the form of shish kebab structures while the volume of melt surrounding the fibers increases. Jones and Lesser [353] used the method described above [345] to etch aliphatic polyketones using an ultrasonic bath for 2h. They washed the etched material in three consecutive baths of (1) 30% aqueous hydrogen peroxide, (2) distilled water, and (3) acetone, and made C/platinum/palladium replicas for TEM.

The internal morphology of high modulus PE fibers, both melt and gel-spun and PP fibers, have also been revealed by permanganic etching [354]. Fibers were embedded between two sheets of polystyrene-ethylene-propylene block copolymer, Kraton (trademark, Kraton Polymers US LLC) cut with a diamond knife to produce transverse sections, and then etched for SEM examination. Longitudinal sections were made after embedding fibers in the copolymer, etching, and making two stage replicas for TEM. The "dry etchant" mixture for transverse sections was 1 wt.% $KMnO_4$ in a mixture of 2:1 of sulfuric acid and ortho-phosphoric acid (prepared by boiling off the water) for 75 min. The "wet etchant" was as in the method above, for 2h for longitudinal sections and 45 min for transverse sections. Hosier and Bassett [355] used a similar mixture to this wet etchant for monodisperse n-alkanes, quenching the etch using a 1:4 mixture of hydrogen peroxide in a 2:7 mixture of sulfuric acid in water, precooled over dry ice for at least 10 min. Methanol was used to remove the etchant before replicas of tungsten/tantalum metal were evaporated at 35° to the horizontal, followed by vertical carbon coating. Highly drawn sheets of PE [356] were embedded in a block copolymer as above, microtomed at −70°C with a glass knife, and then etched at room temperature for 1h in the mixture as in the method above. Standard two stage carbon replicas were studied by TEM, and gold coating was used for SEM. Etching revealed that in transverse sections of the unannealed specimens, the legacy of the banded spherulitic morphology is seen at draw ratio 10× but appears to have been overwritten by subsequent developments at higher draw ratios. However, after annealing, the specimens show recrystallized regions that follow the pattern of the original banded spherulites drawn affinely.

The spherulitic and lamellar morphologies of melt crystallized isotactic polypropylene (iPP) studied earlier [357] have been investigated by SEM for various crystallization temperatures and times [358]. The optimum etchant formulation found was 3 wt.% $KMnO_4$ with 2:1 sulfuric and phosphoric acids for 7h at room temperature. Samples were washed with water, hydrogen peroxide, and acetone sonication, sputtered with a thin layer of gold, and examined by SEM. The morphologies after etching revealed the microstructure of iPP. Shahin and Olley [313] set out to etch poly(3-hydroxy butyrate) (PHB) and poly(oxymethylene) (POM) with permanganic acid but found only the latter was amenable to that etchant while the PHB was stained by potassium hydroxide. Poly(oxymethylene) was etched for 1h with a reagent similar to that used for PEEK [336]. Etched samples were imaged by SEM after gold coating and TEM of replicas. Shahin [359] also etched two highly oriented poly(p-phenylene terephthalamide) (PPT) fibers, Kevlar and Twaron. The fibers were stretched over a copolymer, buried in a thin layer of molten trans-polyisoprene at 60°C, etched for 1h in 1% $KMnO_4$ in 4:1 orthophosphoric acid and water [336]. Etched fibers were investigated under a

Nomarski differential interference contrast microscope in the extinction position between crossed polars, which showed bright and dark interference fringes and a periodic pleat structure for the fibers. Weng et al. [360] studied the changes in the melting behavior with the radial distance in iPP using permanganate etching [351] for 0.5 h. Standard two stage process was used to form a replica for TEM. The wear mechanisms of untreated and gamma irradiated UHMWPE for total joint replacement was also etched [351] for study by TEM and SEM [361]. The study showed a relation between the microstructure and wear leading to insight into the wear mechanisms.

4.5.3.4 Literature Review: AFM

Schoenherr et al. [362] studied the spherulitic morphology of iPP by a combination of optical and atomic force microscopes. Thin films were prepared *in situ* by using a hot stage and then $KMnO_4$ etched; AFM revealed details of the structure such as mother/daughter lamellae, cross hatching, and thickness and orientation of the lamellae. Figure 4.24A is an AFM image of a rapidly quenched iPP spherulite [362]. Hedrites, superstructures that consist of more or less parallel, centrally connected lamellar crystals with a polygonal appearance, have been imaged by AFM in the height and deflection modes after 1% permanganic etching showing their three dimensional structure [363]. Atomic force microscopy images showing views observed from the y axis and the z axis near the center of an iPP hedrite crystallized at 140°C is seen in the deflection images in Fig. 4.24B and Fig. 4.24C, respectively [363].

Permanganate etching was used to prepare polyethylene spherulites for both TEM and AFM [364]. Both techniques showed the band spacing and its relationship with molecular weight. Atomic force microscopy imaging of permanganically etched, metallocene catalyzed, high density PE has been done by melt crystallizing the samples in a hot stage, cooling to room temperature, and etching for 3 h in 1% w/v $KMnO_4$ in 2:1 sulfuric and dry orthophosphoric acids. Markey et al. followed developments [365] with the optical microscope and observations done directly using ICAFM, revealed three dimensional information without the need for replicas and staining. A mathematical model of the etching process was developed to demonstrate that band asymmetry can stem from differential etching and to fully simulate the entire resultant relief. The model not only explains the slope asymmetry but also gives a good simulation of band height and its radial variation, etching depth, and apparent band period. Further refinement of the modeling work on other semicrystalline polymers amenable to permanganic etching is also a possibility.

Propylene based random copolymers with co-units of ethylene, 1-butene, 1-hexene, and 1-octene were studied after rapid and isothermal crystallization by AFM of etched film surfaces and ultramicrotomed plaques. To enhance contrast, the surfaces were permanganate etched by the method described above, using 1% w/v $KMnO_4$ for 10 min. Comparison of the structures were made by OM, AFM, and SAXS [366].

4.5.4 Plasma and Ion Etching

Etching is commonly used today to thin, or ion mill, metals and ceramics for TEM and to produce SEM specimens. Barber [367] applied ion bombardment to prepare thin foils of nonmetals for TEM. The theory and practice of sputtering metal specimens has been described [368–374]. Artifact or cone formation has been commonly observed in metals [370, 375] and a roughening of the surface and the production of conical protrusions has been observed [376, 377] as have "ripples" and dune structures [378, 379]. Barnet and Norr [380] used an oxidizing plasma to etch embedded carbon fibers for SEM by a method that has been applied to polymer fibers. With polymers, artifacts may be formed, and the factors affecting the etching process must be understood before etching studies can be successful. Of all the methods of specimen preparation, etching is the most prone to such artifacts and thus image interpretation is very difficult. Etching preparations are useful for comparison with structures formed during other specimen preparation processes, especially

Etching

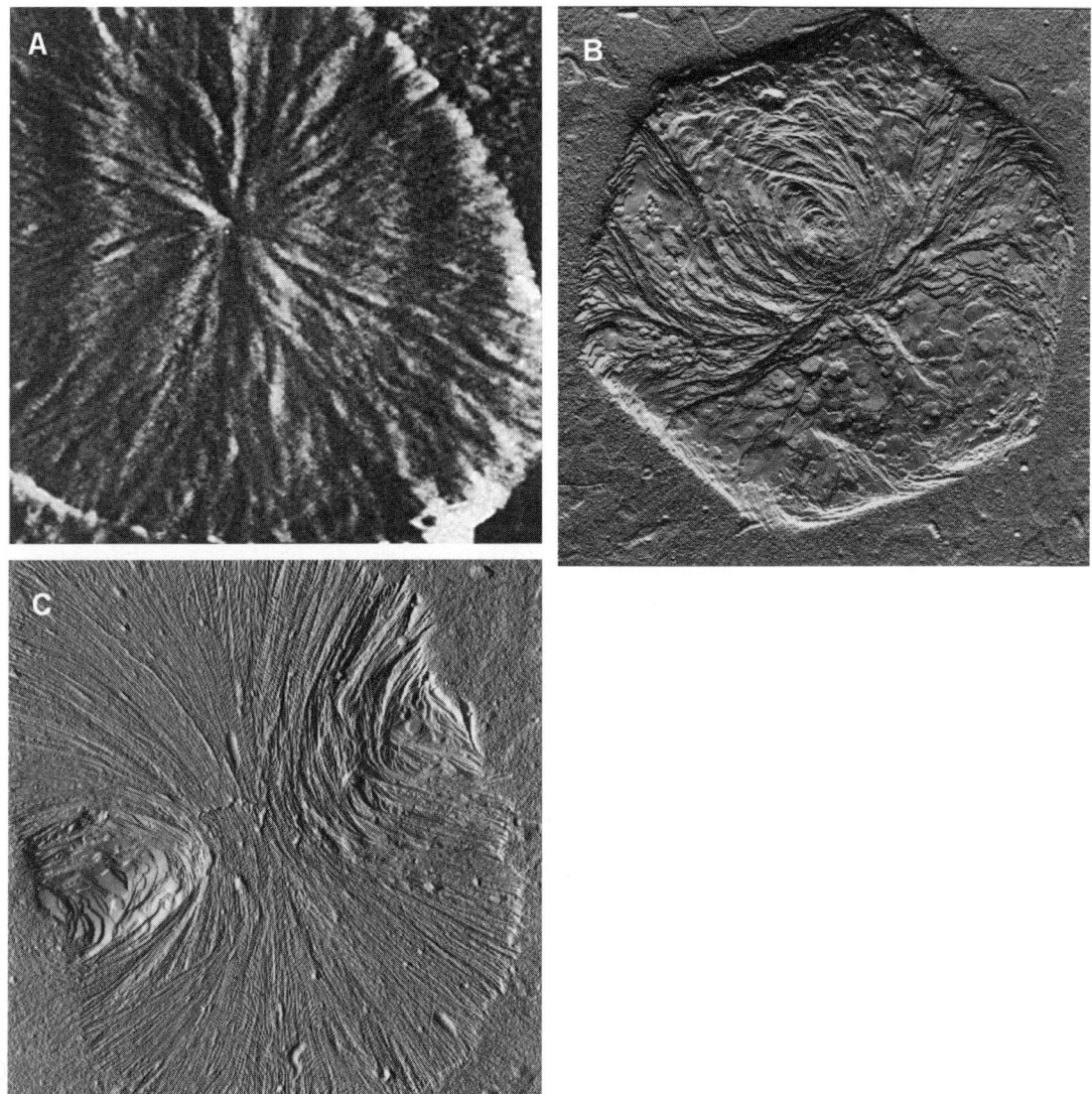

FIGURE 4.24. Atomic force microscopy of a rapidly quenched spherulite of iPP formed on a hot stage, quenched with KMnO$_4$ (A). Atomic force microscopy images of iPP show deflection image views observed from the y axis (B) and the z axis (C) near the center of a hedrite crystallized at 140°C. (Figure 4.24A from Vancso et al. [362], © (1993) Springer; used with permission. Figure 4.24B, C from Vancso et al. [363], © (2000) Wiley-Interscience; used with permission.)

microtomy. Such complementary studies are essential to the determination of the true polymer structure.

4.5.4.1 Literature Review

Etching has been used to reveal structures in polymer fibers [381, 382], polymer blends [383, 384], and in bulk polymers, such as PET [385] and PE [386]. Periodic structures were seen to develop on argon plasma etched, oriented materials, such as nylon and PET, and fine scale structures were observed for etched polyamide [387], PET [27, 388], and aramid fibers [387], which showed ripple structures transverse to the fiber axis. Goodhew [389–392] discussed the formation of large scale structures (0.1–1 μm) perpendicular to the drawn fiber axis upon ion etching carbon fibers; such structures were not formed on glass fibers which were used as an unoriented, amorphous control. Optimization of the ion etching method was achieved [393] by first acid etching the carbon fibers to remove the surface striations. In general, authors using controls and optimum conditions for both plasma and ion etching are able to relate the structures observed to the original crystalline structure of the material.

Blakey and Alfy [394] used oxygen plasma etching (5–30 min) to reveal the nature of delustrant particles in polymer fibers. Friedrich et al. [395] applied plasma etching to HDPE where differential etching resulted from greater etching of the amorphous regions compared with the crystalline regions, and Friedrich et al. [396] used oxygen plasma etching to reveal the nature of polymers blended with glass fibers. Poly(butylene terephthalate) and poly(vinyl acetate) with glass fibers were ion etched (5 kV, 5–30 min) [8] to degrade polymer blocks selectively after sectioning. Kojima and Satake [397] ion etched PP, HDPE, and propylene-ethylene block copolymers and revealed lateral structures resulting from preferential etching of the amorphous regions. Woods and Ward [398] studied the oxygen gas plasma treatment of high modulus PE fibers in order to determine possible mechanisms that result in improvements in fiber-resin adhesion. The mechanisms are surface oxidation, cross linking, and surface etching that could be related to improvement in interlaminar shear strength of the composites. Scanning electron microscopy of the plasma treated fibers showed the presence of micro surface cracks after relatively short plasma treatment times. A comparison of ion beam and radiofrequency plasma etching for biological ultrastructure is quite useful [399]. Clearly, artifacts may be formed during the etching process, and such effects must be correctly interpreted for the method to have any utility. Mijovic and Koutsky [400] and Hemsley [401] have reviewed polymer etching.

Mechanical strength of cold plasma treated PET fibers were shown to degrade (SEM) after cold plasma treatment [402]. Single fibers were treated with oxygen and a mixture of oxygen and tetrafluoroethylene in a cold plasma reactor for 30, 100, and 200 s. The single fibers were then tested in tensile mode and the mechanical strength was analyzed by using the Weibull distribution function. Plasma treatment of fracture surfaces from particulate calcium carbonate/polypropylene composites has been shown to expose filler particles in the fracture surface for easier SEM image interpretation than for composites etched with permanganate acid [403]. Fracture surfaces were prepared by breaking test specimens in a Charpy impact tester after cooling in dry ice and alcohol followed by treatment in cold oxygen plasma for between 3 and 15 min.

A low voltage, high resolution SEM was used to examine the morphology of poly(styrene)-poly(butadiene) diblock copolymer films exposed by using a nonselective fluorine based reactive ion etching (RIE) technique [404]. By controlling the depth of the RIE etch, surface layers of poly(butadiene) were exposed to assess the microstructures as a function of depth with a 4 nm lateral resolution and a 10 nm depth resolution. Experiments suggest this method could be used to supplement microtomy of block copolymers on substrates. A series of PE films [405] with different degrees of crystallinity were treated with a radiofrequency tetrafluoromethane (CF_4) gas plasma (48–49 W, 0.06–0.07 mbar, and continuous vs. pulsed treatment). The etching behavior and surface chemical and structural changes of the PE films were studied by weight measurements, x-ray photoelectron spectroscopy (XPS), static and dynamic water contact angle measurements, SEM, and AFM. With increasing crystallinity (14% to 59%) of PE, a significant and almost linear decrease of the etching rate was found. Gas plasma etching of poly(ethylene oxide)/poly(butylene terephthalate) (PEO/PBT) segmented block copolymer films with a radiofrequency carbon dioxide (CO_2) or with argon (Ar) plasma showed preferential etching on surface structure, topography, chemistry, and wettability, as studied by SEM, AFM, and x-ray photoelectron spectroscopy [406]. The wettability increased after plasma treatment inducing enhanced cell adhesion and/or growth compared with untreated biomaterials, suggesting that plasma treated PEO/PBT copolymers have a high potential as scaffolds for bone tissue regeneration. The ozone etching of a poly(styrene)/poly(isoprene) (PS/PI) block copolymer was studied by AFM as these materials have potential for nanotechnology applications [407]. Etching by ozone showed no degradation of polymer at low doses, but extended ozone treatment resulted in more obvious but nonselective degradation.

The effect of plasma treatment on surface characteristics of PET films was investigated using helium and oxygenated-helium atmospheric plasmas [408]. Sample exposure to plasma was conducted in a closed ventilation test cell inside the main plasma chamber with variable exposure times. The percent weight loss of the samples showed an initial increase followed by decrease with extended exposure time, indicating a combined mechanism of etching and redeposition. AFM revealed increased surface roughness, as well as evidence of redeposition of etched volatiles. There are industrial applications that require deposition of thin metallic films to provide functional properties to a polymer surface, such as wear protection, electromagnetic shielding, and for decorative purposes such as painting automobiles. Kupfer and Wolf [409] applied plasma and ion beam assisted metallization to a variety of high temperature polymer materials to control the adhesion of these coatings and investigated the surface detail produced by SEM and AFM. The plasma etch attacks the polymers leaving behind mineral particles, which provide mechanical adhesion sites for the metal layers.

Buck and Fuhrmann [410] modified the spin casting film procedure by dropping the solution of a diblock copolymer onto an already rotating substrate, precleaned with an air plasma at 100 W for 1 min. They used air plasma for selective etching of the PMMA microphase in the PS/PMMA film, at a power of 40 W in steps of 5 s, keeping the sample in vacuum for 15 min after each etch period. Blend morphology and clay dispersion has been successfully studied by AFM of samples polished and physically and chemically etched, the latter by immersion in formic acid, for 2 h at room temperature, to dissolve the nylon phase in nylon/polypropylene/clay systems [411]. The polished surface was physically etched by argon ion bombardment at 500 eV [412].

4.5.4.2 General Method and Examples

There is no one general method that is applicable for either ion or plasma etching polymers. Etch times are generally short, on the order of 5–30 min, as it is important not to leave a residue on the sample surface. Rotation of the specimen and the target and cooling of the ion guns are factors that minimize potential artifact formation. In addition, the smoother the original surface texture, the fewer artifacts are expected. In our experience, plasma etching is much less complicated and there is less chance of artifacts than in ion etching.

Controlled etching experiments were designed, keeping in mind that etching could provide some insights into polymer fine structure *if* great care was taken. Etching, as conducted in a low temperature radio frequency (RF) plasma asher (LTA), is generally used to oxidize organic matter and thus such devices provide a nondirectional plasma that minimizes artifact formation. Oxygen and argon were used in these experiments, at 162°C, from 5 to 30 min. Temperatures below the crystalline melting temperature should be used. Etching time is affected by the surface area of the specimen. Ion etching was conducted using the Ion Tech microsputter gun (see Section 4.5.4) with two water cooled ion guns, at 7 kV, for 10–30 min. The guns are directed at the region of interest as the specimens are rotated. A major benefit of the system used was the diffusion pumped vacuum system that permitted slow etching with a fast pump rate. This results in reduced ash deposition and limits the formation of coarse structures that result from restructuring of the material.

The etching study was conducted to complement ultrathin sectioning of high modulus oriented fibers [413]. Representative results of the plasma etching experiment are shown in the secondary electron images (SEI) (Fig. 4.25). Glass fibers and amorphous polyester film were used as controls. Oriented PET and aramid

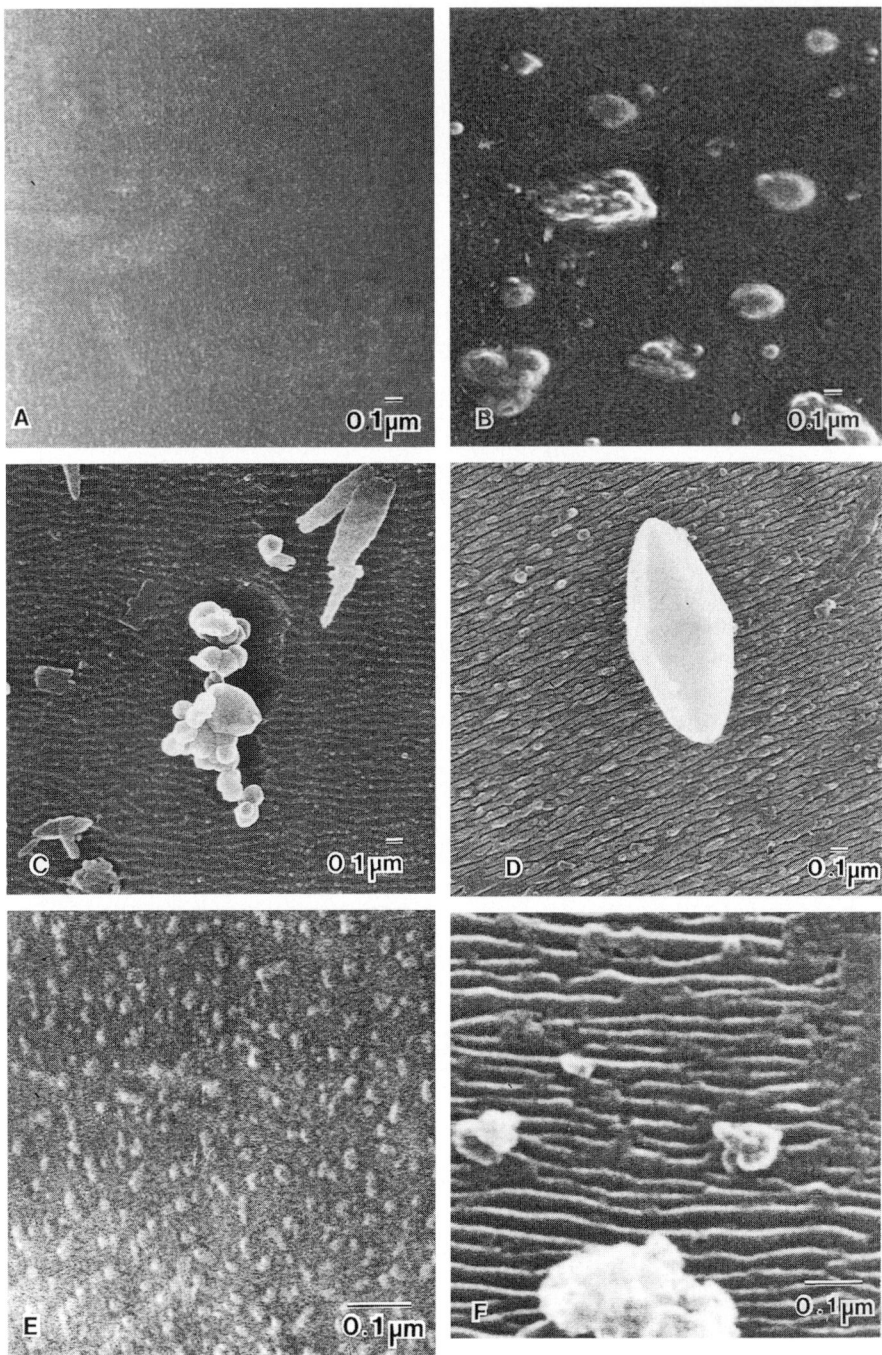

FIGURE 4.25. Secondary electron images show the result of plasma etching with argon: (A) a glass fiber surface shows no ordered details; (B) amorphous PET film with particles but no order; (C) an oriented crystalline PET fiber surface with lateral striations; and (D) an aramid surface also with lateral striations. The effect of plasma etching with oxygen is shown for (E) a glass fiber surface, which reveals no detail, and (F) an aramid fiber, which exhibits a lateral striated texture.

fibers were etched for 15 min in argon in order to evaluate the effect of the treatment on oriented crystalline and liquid crystalline materials, respectively. Glass fiber surfaces (Fig. 4.25A) show a very fine, disordered texture, whereas amorphous PET shows that delustrant particles are present, but no detailed surface textures are revealed (Fig. 4.25B). Poly(ethylene terephthalate) fibers (Fig. 4.25C) show delustrant particles, and a striated texture is observed normal to the fiber axis as it is in the etched aramid (Fig. 4.25D) where a large particle of unknown origin is seen. Oxygen etching is shown for a glass fiber surface (Fig. 4.25E) and an aramid fiber (Fig. 4.25F) for comparison with argon etching. The glass fiber surface is mottled, but it has no structure, whereas the aramid fiber has a striated texture normal to the fiber axis. This striated texture, observed in semicrystalline and liquid crystalline materials [414], is similar for both argon and oxygen etching. An aramid fiber etched with the Ion Tech ion beam microsputter gun is shown in a representative secondary electrons (SE) image in Fig. 4.26. The structures observed are similar to those formed by plasma etching but there is a tendency to some directionality in the structures depending upon sample orientation with respect to the ion guns.

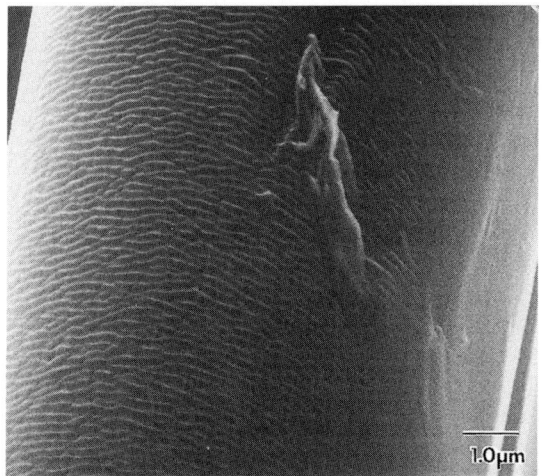

FIGURE 4.26. Secondary electron imaging shows the effect of argon ion etching on an aramid fiber surface where lateral striations are observed normal to the fiber axis.

4.5.4.3 Summary

Factors known to limit artifacts in etched polymers are as follows: plasma etching appears less directional than ion etching; argon generally results in cleaner surfaces than oxygen; slow etching and fast pump rates provide cleaner surfaces; specimen rotation and cooling limit heating effects; ion gun cooling and low current also limit heating effects; and low angle of incidence of two ion guns minimizes artifacts.

In most etched, oriented semicrystalline polymers, striated, or lateral, structures are observed normal to the draw direction. These structures, termed *ripples*, *striations*, or *corrugations*, appear to fall into two size ranges: 5–20 nm and 100 nm or greater. The fine textures often correspond with the SAXS fold period (the lamellar crystal periodicity); these are most likely related to the material microstructure. The larger textures are more likely to be artifacts of the method. Experiments have shown that the striations are observed in oriented, semicrystalline materials, strongly suggesting that there is preferential etching of the amorphous regions resulting in a texture associated with the original microstructure of the polymer. There is no consensus on interpretation of the striations or just what these textures imply regarding liquid crystalline polymers. These examples of the etching method were used specifically because the textures are obvious and have been observed in many laboratories: as yet, there is a range of opinion regarding their interpretation.

Different specimen types yield a range of results upon ion or plasma etching. Multiphase polymers generally etch differentially, enhancing the contrast. Melt crystallized polymers can be etched to reveal the spherulites. Surface protuberances and particulate fillers can and do form cones or ridges when etched. Oriented semicrystalline polymers, on the other hand, appear to be the most controversial with respect to the resulting surface textures. Clearly, in such cases the specimen should be prepared by other methods for comparison, and control experiments are essential. There are problems in the industrial laboratory that can be solved, in part, by microscopy of surfaces prepared by etching techniques: however, these are far fewer than those addressed by other specimen preparation methods.

4.5.5 Focused Ion Beam Etching

The past decade has seen the growing development of focused ion beams for the preparation of semiconductor specimens for TEM evaluation; for example, as compiled by Anderson and Walck [415], in an introductory book on the techniques and practice by Giannuzzi and Stevie [416], and in a paper describing the relation between ion beams and material interactions and FIB specimen preparation [417]. Only recently has the method been used in other materials sciences, such as metallurgy, and very little has been published on polymers. It is worthwhile to explore work done in this exciting area, especially in conjunction with TEM, AFM, and SEM imaging.

FIB is a physical process that permits preparation of surfaces with a finely focused ion beam (<10 nm probe size). The liquid-metal ion source generally generates positive gallium ions with a typical energy between 10 and 50 keV. The samples are coated with a conductive protective metal layer prior to etching or milling to dissipate the charge. In much the same way as an electron beam scans a sample in the SEM, the ion beam scans and gradually etches the surfaces away. Control of the beam current and energy, the scanning speed and time are used to produce flat surfaces. Samples for FIB preparation for TEM are generally polished or sectioned to form a thin film less than $100\,\mu m$ thick and then FIB removes the remaining material, leaving a thin film less than 100 nm thick. Alternatively, the FIB can mill two adjoining craters into the surface of a monolithic specimen where the wall between the craters can be thinned to 100 nm. The thinned wall becomes a TEM foil that can be cut out and removed using a lift-out technique. In the case of AFM or SEM, only one side of the specimen needs to be etched. As with all specimen preparation methods, there is the potential for artifacts to form, in this case due to the high energy ion beam, especially with polymer materials. Polymer changes can include, but are not limited to, chain scission and/or cross linking, chain shrinkage, changes to the surface chemistry and crystallinity, among many others (see Chapter 3). It can be confusing, but should be noted, that FIB is used for both sample preparation and for imaging, such as in an SEM; a dual-beam FIB has electron and ion beam columns on one instrument.

4.5.5.1 Literature Review and Examples

White et al. [418] used focused ion beam/liftout to prepare block copolymer films spin cast on silicon substrates for TEM and AFM. This is a particularly difficult cross section to prepare as the silicon is hard and cannot be microtomed. The samples were sputter coated with gold/palladium to protect the surface from damage and then cross sectioned in less than 1 h to 20 nm thickness using a finely focused Ga+ ion beam operating at 30 kV. The thin transparent samples were lifted out and placed on a carbon coated copper mesh grid for TEM, as has been described [415]. Although there was some minor damage at the free surface, this method was apparently an improvement over KOH etching, which destroyed the interface. Loos et al. [419] used FIB to prepare cross sections of polymer solar cells deposited on glass, in about 3 h, another specimen that cannot be prepared by microtomy due to the soft nature of the polymer and hard glass. Applying the lift-out method, no pretreatment of the sample was required; the electron transparent membrane was removed from the bulk and analyzed by TEM. The technique for preparation of wafer cross sections and applications in electron beam lithography of PMMA resists was also shown by the FIB method for cross sectional SEM [420]. Another complex composite specimen, a nanowire polymer based nanocomposite for a prototype thermoelectric device, was also prepared by FIB for SEM characterization of the cross sections [421]. A FIB/SEM system was also used for imaging polymer films by ion and electron beams. This work involved imaging of PET polymer films rather than specimen preparation and thus uses much lower beam currents with a claim of 5 to 7 nm resolution. Atomic force microscopy has also been used to characterize materials with nanoscale filler particles prepared by FIB milling after mechanical polishing [422].

Virgilio et al. [423] reviewed FIB preparation of polymers for study by AFM and described a method of preparation and compared ternary blends by AFM and SEM. The method was used to prepare and analyze model interphase thicknesses in HDPE/PS/PMMA ternary polymer blends as this blend exhibits a dispersed phase composed of a well segregated PMMA core and a PS shell. The samples were cryomicrotomed at −160°C, and plasma coated with gold/palladium under pulse for 20 min (to dissipate the heat). FIB preparation was conducted using a finely focused 30 kV Ga+ ion beam to remove a layer of the blend sample. Specimens were fixed on a metallic support using silver glue or graphite tape for AFM. Samples for SEM were cryomicrotomed, the PS was selectively dissolved with cyclohexane (at room temperature for 1 day), and then plasma coated. Intermittent contact mode AFM images of the composite droplets were used to measure their mean diameter and PS shell thickness. The work showed that the three polymer components have different ion beam etching rates, which results in topologic contrast between the phases of the blends when viewed by ICAFM. In this case, PMMA has the highest etching rate, whereas PS has the lowest and HDPE is intermediate. This high level of contrast between the phases allows for clear identification of the PS interphase. Comparison was made with SEM measurements to check on the dispersed phase size data. The benefit of FIB preparation is to limit the artifacts of microtomy, such as deformation of the polymers and debonding at the interphase, and also to limit the need for extraction with solvents and staining to improve contrast for SEM or TEM.

As with any specimen preparation method, care must be taken to use complementary preparation methods and microscopy techniques as high energy ion bombardment induces physical and chemical modifications on polymers [423]. Other potential disadvantages include electrostatic discharge, radiation damage, surface ripples, and damage from heating effects [424], but all methods of formation of thin films have some associated artifacts. Some general artifacts arising from FIB milling have been discussed for metal matrix composites, some of which are relevant to polymers [425].

4.5.6 Summary

Table 4.6 includes functional groups and polymers and their respective etchants. Chemical etching, such as with solvents and acids, and ion and plasma etching are conducted in order to

TABLE 4.6. Etchants for polymers and multiphase polymers

(a) Polymers

Polymer	Etchant
PE	Hot carbon tetrachloride, benzene, or toluene
PE, PP	Xylene or benzene
Melt crystallized PE	95% fuming nitric acid (80°C)
PE, isotactic PP	1% permanganate/in mixture of 10:4:1 sulfuric acid, 85% ortho-phosphoric acid, and water, for 15 min at 60°C or 1–2 h at room temperature
PET	42% n-propylamine, 1 h, room temperature, or o-chlorophenol or methylamine
Nylon 6, or 6,6	Aromatic and chlorinated hydrocarbons
Cellulose acetate	Acetone at −50°C, then cold ethanol
Polycarbonate	Triethylamine, chloroform vapor
Polyoxymethylene	Iodobenzene and HFIP

TABLE 4.6. (*Continued*)

(b) Multiphase polymers

Polymer	Etchant
Matrix in HIPS	Cyclohexane in osmium tetroxide
HIPS	100 ml sulfuric acid, 30 ml phosphoric acid, 30 ml water, 5 g chromic acid
ABS–rubber phase	10 M chromic acid, 5 min, 40°C
ABS–rubber phase slower on SAN matrix	Sulfuric, chromic acids and water for 5 min at 70°C
PU/ABS or PU/SAN (polyester based)	2% aqueous osmium tetroxide for 24 h, liquid nitrogen fracture, then use the following:
SAN	Methyl ethyl ketone, 4 h
Polyurethane	Tetrahydrofuran vapor, 1 h or dimethylformamide
PMMA in SAN/ PMMA	Chain scission in electron beam
Poly(vinyl methyl ether)	Chain scission in electron beam

reveal selectively structures in polymers that may not be observed directly. In all these methods, interpretation of the structures formed can be more difficult than specimen preparation. Accordingly, the etching methods are best used to complement other methods, such as microtomy, fractography, and staining. Controls and complementary microscopy are essential to ensure that the experimentalist is not led astray imaging artifacts, hills, and valleys, or missing fine structure, lost in the wash baths or reprecipitated using any etchants. Controls are essential to any experiment of this type, but, with care, the structures of semicrystalline polymers and polymer blends may be observed.

4.6 REPLICATION

Replication is one of the oldest methods used for the production of thin TEM specimens and it is also used for specimen preparation for optical microscopy and SEM. The procedure was first introduced by Bradley [426, 427], and it is well documented in texts on specimen preparation [428, 429]. Replicas have the surface characteristics or topography of the original specimen without the need to directly image beam sensitive materials. Reflected light microscopy of polymer surfaces is often difficult because glare from the surface limits visible detail, although metal coatings reduce the glare. Replicas have been used for SEM samples too large (e.g., orthopedic implants), too volatile (e.g., light oils on sample surfaces), or beam sensitive materials. High resolution secondary electron imaging and FESEM, which has matched the resolution available by TEM replication, and the increased use of SPM techniques all have resulted in limiting the need for replicas, especially in the industrial research laboratory.

Replica methods most often used for TEM imaging include single or direct replicas, double or two stage replicas, and extraction replicas. Direct replicas have the highest resolution but are the most difficult to prepare. Double or two stage replicas are easiest to prepare, but it is quite time consuming. Extraction replicas provide a thin layer of the specimen attached to the replica, permitting analytical study. Application of conductive coatings and shadowing, an integral aspect of the replication process, will be described in the next section. Specimens for replication are often pretreated in order to reveal the internal or bulk structures by such methods as etching with solvents, chemicals, ions, or plasmas (see Section 4.5). The overriding disadvantages of replication are that they are time consuming to prepare, and the interpretation of the resulting images is difficult, at best. Woodward and Zasadzinski [430] studied the thermodynamic limitations on the resolu-

Replication

tion obtainable with metal replicas of Pt/Pd, Pt, and Pd by scanning tunneling microscopy (STM) showing the grain sizes and structure.

4.6.1 Simple Replicas

4.6.1.1 Replication for OM

Hemsley [401] described two methods for preparing replicas: application of a 5% PS solution in benzene (or xylene) or a solution of gelatin in water. These are dried, stripped from the surface, and metal shadowed or coated to enhance the detail. A silastic replica method is used for replication of internal surfaces where either surface topography or critical size measurements are required. Dow Corning Silastic, such as RTV silicone rubber, with Silastic E curing agent (Dow Corning, Midland, MI), can be used as the replicating medium by dropping or brushing it onto the specimen. If the part has a fine opening, the part itself can be attached to the stopper of a vacuum flask and a vacuum applied while adding the Silastic mixture. The replica is allowed to cure overnight and the Silastic is carefully removed and metal coated to enhance surface detail. Examination is generally by reflected optical microscopy.

4.6.1.2 Replication for SEM

Several methods have been specifically developed for the preparation of replicas for the SEM. Peck [431] developed a method to determine the nature and location of volatile and nonvolatile smoke deposits on cigarette filters. Eastman 910 adhesive, methyl-2-cyanoacrylate monomer liquid, was heated and vaporized adjacent to the specimen as the *in situ* replicating material. Eastman 910 apparently works well because it polymerizes rapidly in a water environment. However, as a result of the reaction in water, best results are obtained when producing the replica in a dry nitrogen atmosphere.

Oliver and Mason [432] used replica methods to assess the effect of surface roughness on the spreading of liquids and to measure contact angles. For stationary studies, small beads of PMMA were melted and the molten droplets spread and solidified on the surfaces. Dynamic studies involved polymer melts mounted on a remotely controlled hot stage stub in the SEM and the experiments were video recorded. Plastic and silicone replicas for the SEM were compared in a study of large specimens (orthopedic implants made of HDPE) which could not be destroyed [433]. Negative replicas were made using collodion (2% to 4%), in amyl acetate or a silicone replication medium, Xantopren Blue (a silicone substance manufactured by Unitek for dental impressions, Monrovia CA).

4.6.1.3 Methods and Examples

The replication methods described have a basic limitation in that they are all negative impressions of the specimen, which is a major factor in image interpretation. Wood [434] developed a method to provide positive replicas of polymer fiber surfaces. Polymer fibers often have surface coatings, such as finishes in textile fibers, that aid handling but may be volatile in the SEM. However, reflected light microscopy, at high magnifications, reveals only a portion of the curved surface at any one level of focus. Therefore, the depth of focus of the SEM is required.

The general replica method for SEM of fibers or other polymers is as follows:

1. Place the specimen onto and into thick tape to permit removal after replication.
2. Pour a mixture of a silicone and a cure agent over the fibers, permit it to cure, and peel it off. The replicating mixture can be Dow Corning RTV 3112 Encapsulant and Catalyst F (fast cure) [434], or Dow Corning RTV silicone rubber with Silastic curing agent, or Xantopren (Unitek) [433].
3. Clean the specimen in an ultrasonic bath.
4. Drop a low viscosity embedding resin, such as Epotek 301 (Epoxy Technology Inc., Billerica, MA) into the molded silicone (negative) replica. Turn the mold over onto an SEM stub and allow to cure overnight.
5. Remove the silicone from the resin replica, and metal coat the positive replica for the SEM.

An SEM image of such a fiber replica is shown in Fig. 4.27. The rough fiber surface texture is associated with the finish. Gordon [435] reviewed the artifacts associated with

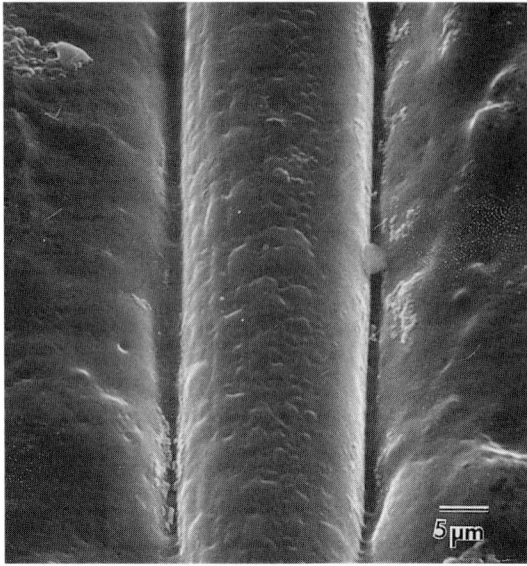

FIGURE 4.27. Scanning electron microscopy of a fiber surface replica shows details of the surface finish, which is not possible in cases where the finish vaporizes in the vacuum of the SEM. (From Wood [434]; unpublished.)

silicone elastomer surface replicas, especially bubbling, and suggests that these methods must be used with great care.

4.6.2 Replication for TEM

4.6.2.1 Direct Replicas

Direct or single stage replicas have the best possible resolution, are the fastest, but, unfortunately, are the most difficult to prepare. The method involves the deposition of the replicating media and its removal or dissolution of the polymer. Materials used for direct replication include polymers, evaporated carbon films, or metal oxides. Carbon is widely used for replication, especially if the polymer specimen to be replicated can be readily removed or dissolved.

Carbon replicas [427] are formed by the evaporation of a thin layer of carbon in a vacuum evaporator. Metal shadowing, at an angle of 20° to 45° to the specimen surface is performed while the specimen is in the evaporator. The highest resolution direct replica material is carbon/platinum (C/Pt). After evaporation, the thin replica is stripped from the substrate by etching, dissolving, or some other method. In some cases, the carbon replica film can be scored and floated off onto a water surface and then picked up onto TEM grids. Carbon replicas do not usually just simply float off the substrate, and treatment is required for their removal. Evaporation of a wetting release agent, such as Victawet, prior to carbon evaporation aids replica stripping although it can affect resolution.

The removal of direct carbon replicas is dependent upon the polymer. Boiling xylene vapor was used to remove drawn PE from replicas [436] in work on drawn polymer morphology. A direct carbon replica method for a PBT impact fracture surface was described by evaporation of platinum at 20° and PBT removal in hexafluoroisopropanol (HFIP) [437]. Latex film coalescence in poly(vinyl acrylate) homopolymer and vinyl acrylic copolymer latexes was studied using direct replicas [438]. As the latex films have a low glass transition temperature, they were cooled by liquid nitrogen to about −150°C in the vacuum evaporator and shadowed with Pt/Pd at 45° followed by deposition of a carbon support film at 90° to the specimen surface. The latex films were dissolved in methyl acetate/methanol. Transmission electron microscopy micrographs of the latex films show the difference between films aged for various times (see Section 5.5.2).

Polymers are used to aid removal of thin replicas when the specimen cannot be dissolved. Polymers that are used include poly(acrylic acid) (PAA) [439, 440], gelatin, formvar, and collodion. In the former work, C/Pt replicas of drawn PE were backed by a 5% aqueous solution of PAA dried and stripped off the specimen and the PAA was dissolved in water. The advantage of using PAA or gelatin is that the replicas are left floating on the water surface and they can be readily picked up on a TEM grid. A variety of methods [441] are used to wash away the backing plastic. Carbon replicas, with the plastic film, are placed on the TEM grid, on filter paper in a dish, and a volatile solvent is introduced onto the paper. The addition of solvent into the covered dish is often successful in dissolving the plastic while the carbon replica settles onto the grid. Alternatively, the solvent can be allowed to drip over the grid on a mesh screen, or a sophisticated extraction apparatus can be used to permit the solvent vapor to extract the plastic material.

Hosier and Bassett [355] used a permanganate etch for monodisperse n-alkanes, using methanol to remove the etchant before replicas of tungsten/tantalum metal were evaporated at 35° to the horizontal, followed by vertical carbon coating, directly on the etched grids. Shahin and Olley [313] studied poly-3-hydroxy butyrate (PHB), which forms large banded spherulites, by SEM and TEM using etching methods, followed by replication. Direct replicas were made for TEM using tantalum/tungsten alloy at an angle of 30° for PHB [313] and at an angle of 35° for PE [442].

An example of direct, or single stage, carbon replicas is shown in Fig. 4.28 of a grooved polycarbonate disk [443]. The polycarbonate disk was shadowed with Au/Pd at a shallow angle (ca. 30°) and then a thin carbon layer was deposited in a vacuum evaporator. The disk was dissolved using methylene chloride and the replicas were placed on copper grids for TEM evaluation. In this particular case, the replicas were made by shadowing from opposite directions, normal to the grooves to illustrate the asymmetry of the groove structure (Fig. 4.28), as one side has a steeper slope than the other. The TEM study was complementary (see Section 5.3.2) to other studies of thin sectioned substrates, cut perpendicular to the grooves, viewed in TEM. Scanning electron microscopy of an ion beam sputtering (IBS) coated (see Section 4.7.3) disk showed details, and these were compared with FESEM images of uncoated disks.

4.6.2.2 Two Stage Replicas

Two stage, or double, replicas provide positive impressions of the specimen surface, although they require more time to prepare. The steps involved in formation of a two stage replica are as follows: form the replica; strip the replica from the specimen; shadow cast with metal in a vacuum evaporator; carbon coat in a vacuum

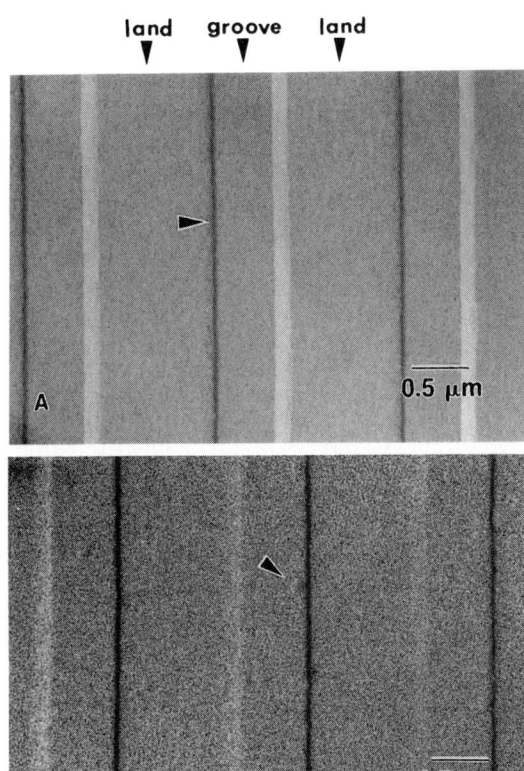

FIGURE 4.28. Transmission electron microscopy micrographs of replicas from a polycarbonate disk, made by shadowing from opposite directions normal to the grooves to illustrate the asymmetry of the grooves. In the top photo, the shadowing direction is radially inward, whereas for the bottom photo the shadowing direction is radially outward [443]. The inner edge (arrowhead in top photo) was found to have a much steeper slope than the outer edge (arrowhead in bottom photo). (From Baro et al. [443]; used with permission.)

evaporator; place the replica on grids and dissolve or extract the plastic.

Replicating films or solutions may be used to form a first stage replica. The replica is stripped from the specimen, using tweezers or double sided sticky tape. Breathing on the replica is another method to get it to release, generally followed by floating it off onto a water surface. In some cases, a second thick layer of the same plastic is applied in order to provide a backing. Unfortunately, the thicker films aid stripping at the expense of longer drying times. Examples of replicating plastics (along with their solvents) are cellulose acetate (acetone), gelatin (water or dilute sodium hydroxide), acrylic resin (acetone) and poly(vinyl alcohol) (water). Solutions of 2% formvar in chloroform or dioxane and 1% to 4% collodion in amyl acetate are appropriate for rough surfaces where tapes and films are not easily removed. Washing and extractions are similar to those described above for one stage replicas. A major problem can be the incomplete dissolution of the plastic replica, which interferes with TEM imaging.

Replicas of fibers and yarns are difficult to strip off due to their size and shape. Fibers can be prepared for replication by semiembedding them first in a resin or gelatin [444–446]. The fibers are replicated with metal and carbon and then stripped. Scott [98] used the peelback method to prepare fibers for internal or bulk study by coating the peeled fibers with 10 nm of chromium at a 30° angle to the specimen. Fibers were placed, chromium side down, onto a 3% solution of PAA in water and allowed to dry before removing the fiber and leaving the replica on the PAA. The replica was coated with PS in CCl_4, the PAA removed with water and the styrene with CCl_4.

A commonly employed method for the preparation of positive, two stage replicas, uses PAA as the first stage replica:

1. Place the fiber or film on a slide and shadow with a metal (such as chromium) at 30° to the specimen.
2. Place 2–3 drops of 3% PAA in water on the shadowed specimen and allow to dry overnight, or place the specimen onto a drop of the PAA [447].
3. Peel the PAA/Cr replica from the specimen and turn the PAA side down on the glass for carbon evaporation onto the Cr.
4. Float the replica onto distilled water for 4–6h to dissolve the PAA and then pick up on TEM grids.

In the modified method [447], the specimen is peeled away leaving the Cr face up on the PAA ready for carbon evaporation. After carbon coating, the entire slide is placed in water to surround the PAA, leaving the Cr/C replica to float. Hudson and Lovinger [342] used a concentrated aqueous solution of PAA, cast on the polymer, dried, detached, shadowed with Pt, and coated with carbon. Olley et al. [448] made an impression of etched PE fibers in softened cellulose acetate, shadowed with tantalum/tungsten, followed by deposition of carbon and extraction of the replica. Figure 4.29 is a TEM micrograph of a replicated experimental, stretched polypropylene film made by this method with chromium as the shadowing metal [332, 333]. The replica shows the elliptically shaped pores formed by stretching the PP. Unstretched lamellae are seen in rows between the rows of pores. The replica method is very time consuming as the drying and dissolution of the PAA extend the preparation over several days.

Polyethylene samples, crystallized from the melt both by quenching and by isothermal crystallization, have been prepared by several specimen preparation methods for TEM, including permanganic etching followed by replication [345]. Replicas were made by mounting the samples, etched side up on a microscope slide, lightly shadowing with Pt/Pd at about 40°, and then coating with carbon at normal incidence using carbon rope for evaporation. A modification of the method listed

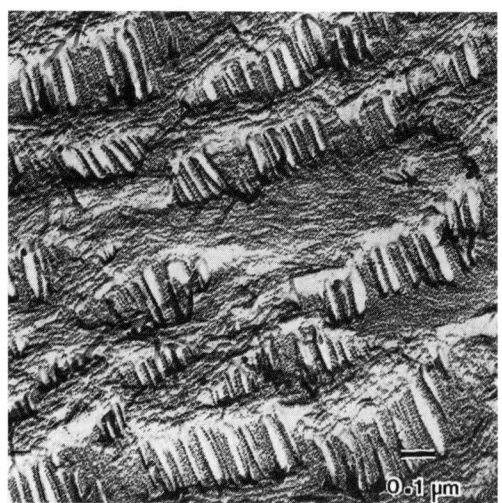

FIGURE 4.29. Transmission electron microscopy micrograph of a two stage replica made of the surface of an experimental stretched polypropylene film. The replica was shadowed with chromium at an angle of 30°. Slit-like voids are seen that are formed in rows or channels separated by oriented fibrils.

above was used, by placing drops of PAA on the etched surface, drying in a desiccator for 2–3 days, and then floating off the PAA coated replica onto water, and picking up the replicas on TEM grids. Etched surfaces of binary polymer blends have been used directly for AFM and for TEM by preparation of two stage Pt/C replicas [348]. Standard two stage shadowed carbon replicas are often still used for TEM after etching preparation, using cellulose acetate moistened with acetone [347, 350, 360] or PVA [350], or with Pt/Pd, carbon, or the standard PAA method [349, 351].

In summary, there are many plastic materials available to prepare two stage replicas, although these methods are quite time consuming. The plastics that dry in less than 30 min (e.g., formvar and collodion) are the most useful as many replicas may be made of the surfaces under study. Although the PAA method is the most commonly used, it has the decided disadvantage of taking more than 24 h to dry and taking additional hours to dissolve after replication. Thick replicating tapes also take longer and do not reproduce topography as accurately as thinner tapes. If the rapid methods cannot be used, plan on a week long effort and use PAA.

4.6.2.3 Extraction Replicas

Extraction or detachment replicas provide a replica of the surface along with some thin fragments of the actual specimen. The advantage is that the surface topography is recorded and analytical experiments, such as electron diffraction, may also be conducted. The method is complementary to microtomy or ultrasonication. Methods for such replication are similar to those described above for single and double stage replicas. The samples are either small, such as powders or crystals, or they are pretreated to encourage fibrillation or splitting. Much work on single crystals has been conducted by this method where the *in situ* replicas, the crystals themselves, are present in the final preparation. Similar methods can be used for bulk crystallized polyethylene.

Bassett and Keller [449] studied the *in situ* shape of PE crystals by extraction replication. A PE solution was evaporated directly onto viscous poly(vinyl alcohol) (PVOH) in water, and the PVOH was allowed to dry. The crystals were shadowed and carbon coated, the PVOH was dissolved, and the crystals were picked up on carbon coated support grids. Solution grown PE single crystals on Mylar were elongated and replicated for TEM [32, 33] and the crystals examined directly. The crystals were removed by being stripped with PAA, metal shadowed, backed with carbon, and the PAA dissolved. Petermann and Gleiter [450] deformed carbon coated grids by 5% in a miniature tensile device, forming cracks about 1–5 μm wide, prior to the evaporation of a xylene suspension of PE crystals onto the grids. The samples were stretched a second time to deform the areas where the original cracks were formed so that the fibrils extending across the broken carbon film could be imaged.

Structural studies were conducted on PE and PTFE fibrils extracted by coating with C/Pt and stripping with a backing layer. In this case [439, 440], the backing layer was 5% PAA in water. After dissolution of the PAA, fibrils were left for TEM observation. Thin shreds or layers of rubber were torn off blends of natural rubber, polyisoprene, SBR, and neoprene using gelatin for the extraction replica [451].

Martin and Thomas [34] reviewed work on HREM of polymers and described a range of sample preparation methods to provide thin sections, including detachment replication. For fibers, the method takes advantage of their relatively weak lateral bonding. Fibers are embedded into a collodion substrate and the bulk of material removed leaving thin fragments that are suspended on a holey carbon grid by dissolving away the collodion with amyl acetate.

4.7 CONDUCTIVE COATINGS

Polymers are typically nonconducting specimens that collect rather than dissipate the electrons from the electron beam resulting in a range of spurious image details referred to as *charging* effects. Such samples are generally metal coated to provide an electrically conductive layer, to suppress surface charges, to minimize radiation damage, to reduce the effects of heating, and to increase electron emission. Carbon coating is used for certain TEM preparations to support the sample and is also used to coat samples for microanalysis. Microscopy techniques used for nonconductive polymers without a conductive coating include LVSEM (generally with a FEG) or variable pressure SEM, the latter at some loss of resolution. Charging effects and other artifacts will be discussed in some detail as they affect image interpretation.

4.7.1 Coating Devices

There are a range of devices available for the deposition of conductive coatings. The device commonly in use for preparation of TEM replicas and carbon support films (ca. 1–50 nm thick) is the vacuum evaporator. A range of sputter coaters are used for coating SEM specimens, to provide a conductive layer about 1–10 nm thick, which emits secondary electrons producing SEM images. The range of evaporative and sputter coating methods will be reviewed here; more detail is found elsewhere [452].

4.7.1.1 Vacuum Evaporators

Vacuum evaporators have been in use for several decades for the thermal evaporation of materials, such as metals, onto a specimen to provide a conductive layer and dissipate charge during electron microscopy. Typically, a 12 inch diameter bell jar is fitted onto a vacuum system that includes a rotary pump and diffusion pump. Electrodes are fitted onto the baseplate of the evaporator and connected to a transformer. These electrodes are used for attachment of the metals and the carbon rods for evaporation. High melting metals, such as tungsten, tantalum, and iridium, must be heated in a refractory crucible, whereas the lower melting noble metals may be placed in such a container or draped over a refractory wire. Thin films are formed by direct line of sight from the metal to the specimen, and there is a high potential for heating effects that must be managed with apertures and/or cold stages. An important accessory is a liquid nitrogen trap, which is fitted above the diffusion pump, and a filter on the rotary pump, to trap oil vapors and keep the vacuum clean, which is essential for clean carbon coating. Good practice also dictates cleaning the boats and wires to remove hydrocarbon contamination prior to evaporation.

The time needed for vacuum evaporation depends upon the state and speed of the vacuum system, but the actual coating time is usually about 2 min after 30 min pumping to obtain an appropriate vacuum. Preparation of metal wires for evaporation is straightforward. The length of the wire, its thickness, and the distance to the specimen determine the coating thickness. Metal wires are typically wrapped around a V-shaped filament of tungsten wire attached to the electrodes in the vacuum chamber and evaporated from the heated tungsten filament. The filaments can be purchased or bent manually from tungsten wire. Specimen rotation devices and shutters are available that limit specimen heating. Potential disadvantages of evaporative coatings include nonuniformity, heat damage to the specimen, oil contamination, and a grain size resolvable by modern instruments.

4.7.1.2 Sputter Coaters

In sputter coating [452], the metal (gold/palladium or platinum) on a target is dislodged by inert ions and directed onto the specimen. In simple units, the small chamber is evacuated with a rotary vacuum pump, the system is flushed several times with pure argon, and the argon flow is controlled during the sputter coating as argon ions bombard the metal target and the metal is deposited on the specimen. Diode systems have generally been replaced by triode systems, as the sample is the anode in the diode systems, which causes deleterious sample heating. A major problem with sputter coaters is heating of the specimen. Robards et al. [453] sputtered gold onto frozen specimens using a permanent magnet to confine the plasma and, thus, limited heating effects.

4.7.1.3 High Resolution Coating Devices

There are a variety of other coating methods, including electron beam (E beam), plasma magnetron coating, and Penning sputtering and saddle field ion beam sputtering. The methods use more expensive equipment and require more time and effort than the techniques described, but they result in finer grained coatings that are needed for high resolution SE imaging and for TEM. Ion beam sputtering will be described, as an example of these systems, in the section on coatings for SEM.

4.7.2 Coatings for TEM

Replication is a preparation using conductive coatings for TEM. Carbon is used as a support in two stage replicas, or as a direct replica, but it does not impart any electron contrast. The specimen is generally metal shadowed at an oblique angle, in order to highlight the surface topography, enhancing the electron contrast, followed by evaporation of a carbon support film. Carbon support films are used in place of plastic supports for high resolution imaging. Polymer sections often require a light (1–5 nm) carbon coating in order to limit charging and protect the specimen from heat and beam instabilities. This coating helps the situation, although beam damage and radiation damage still occur.

4.7.2.1 Carbon Coatings

Carbon films are used as specimen supports for replicas and as coatings for ultrathin sections. Carbon evaporation is difficult to perform as high current is needed, and the carbon rod often moves or breaks. Carbon rods are prepared by sharpening one end to a short cylinder about 4 mm long. The other rod is flattened at an oblique angle and the rods are set so that the cylinder faces the specimen while making contact with the flat rod; such rods are available from many suppliers. One method of carbon coating [454] is rapid application of current. In our experience, best results are obtained by applying the current smoothly, in about 20–30 s, trying not to shock the rods into losing contact. A shutter is used between the specimen and the carbon rods to limit heat exposure.

Carbon support films are evaporated directly onto freshly cleaved mica, NaCl, plastic coated grids, or plastic coatings on glass slides. There is difficulty in stripping the carbon film from the support. With mica and NaCl, the films are floated off onto water and picked up on TEM grids. The carbon is left on plastic support films unless very high resolution is required. Carbon films about 1–3 nm thick enhance beam stability of ultrathin sections, whereas about 10–20 nm may be required for replicas. A drop of diffusion pump oil is placed on a glass slide on top of a piece of white filter paper to aid thickness determination. The color of the carbon film can be used to judge the thickness; a shiny light gray color indicates the coating is too thick. Quartz crystal monitors are available for film thickness measurements.

4.7.2.2 Shadowing

Shadowing [429] is used to enhance specimen contrast by increasing the electron scattering. Replicas are metal shadowed prior to carbon coating, in order to accentuate surface topography by the application of a heavy metal in front of raised areas while no metal is deposited on the other side. Shadowing at known angles, generally 20° to 45°, is useful for the measurement of particle heights. Fine detail is accentuated by smaller shadowing angles and larger angles are used for larger structures. Metals generally used for shadowing include Au/Pd (60/40), Pt/Pd (80/20), gold, tungsten oxide, chromium, platinum, C/Pt, or platinum/iridium/carbon. Gold is very popular but it has larger grain sizes, whereas platinum is the most difficult metal to evaporate but provides the finest grained shadow. Chromium is quite coarse and useful for very low magnification studies. Aluminum readily oxidizes but is used for optical microscopy.

4.7.3 Coatings for SEM and STM

Scanning electron microscope and STM samples are generally coated to provide an electrically conductive layer, to suppress surface charges, to minimize radiation damage, and to increase electron emission. The coating is intended to be a thin, continuous *in situ* replica of the specimen surface. The thickness and texture of the coating must be minimized for smaller textures and higher resolution. Thickness can be judged in the same way as noted above for carbon – a drop of diffusion pump oil is placed on a glass slide on top of a piece of white filter paper to aid thickness determination. Gold or Au/Pd coatings should be pale in color; shiny coatings are much too thick. Coating techniques for SEM have been reviewed [452] although it must be noted that very low voltage SEM and variable pressure microscopes do not require conductive coatings.

4.7.3.1 Sputter or Evaporative Coating

There are several types of coating devices to choose from for the preparation of specimens for SEM, including sputter coaters and vacuum evaporators, described above. Sputter coaters have some definite advantages: short preparation time (5–10 min), multiple specimens, and uniform coatings on rough textured specimens. Special stages fitted onto the baseplate of an evaporator provide for sample rotation for thorough coating. Overall, finer grained and thinner conductive films are obtained with the vacuum evaporator, but it is not recommended for heat sensitive or rough textured specimens. When magnifications under 5,000× are required, the sputter coater is the preparation method of choice. As the texture of the specimen decreases in size and the required resolution increases, there are definite differences observed in the method of application of conductive coatings, metal type, and thickness.

There are no resolvable variations among different metal coatings sputtered to thicknesses of 20 nm or more when examined in microscopes that only resolve 10–15 nm. However, this is certainly not the case in the FESEM where resolutions can be less than 4 nm. The nature of the metal target is important as the various metals result in variations in fine structure. Thick gold coatings tend to be granular, cracked, and nonuniform and they may be resolved in the SEM, whereas Au/Pd and platinum are less likely to be resolved. Braten [455] showed that diode sputtered gold gave a granular and cracked appearance to the surface and recommended vacuum evaporation of Au/Pd. For highest resolution dedicated SEM imaging, a thickness of <5 nm of sputtered platinum is recommended [456–459].

A study of various coating devices was conducted several years ago to identify the best conductive coating for high resolution SEI taken using a TEM (with a scanning attachment) with a resolution of 3 nm [458]. The results of that study were then applied to FESEM imaging with potentially similar resolution. A stretched polypropylene membrane was used in this study as there are fine and coarse textures and the film is extremely beam and heat sensitive. Coatings were made of similar thickness (<5 nm) and metal (Au/Pd) in a range of devices. SE images taken in a late 1970s vintage dedicated SEM show that diode sputtering results in a cracked metal surface (Fig. 4.30A), whereas ion beam sputtering results in no cracking (Fig. 4.30B). SE images taken in an analytical electron microscope (AEM) show that evaporative metal coatings result in a granular texture (Fig. 4.30C) compared with an ion beam sputtered sample (Fig. 4.30D).

A mid-1980s dedicated SEM, with 4 nm resolution, was found to resolve even thin platinum sputter coatings that had not been resolved earlier. Contrary to earlier assumptions, these coatings were even resolved at low magnifications (5,000×). What is implied by the statement that coatings are "resolved" is simply that a granular texture not associated with the specimen is observed. This texture is an *artifact* if it is misinterpreted as being related to the specimen. Figure 4.31 is a high resolution SE image of an ion beam sputtered membrane where the fine texture is resolved.

4.7.3.2 Ion Beam Sputter Coating

The saddle field ion gun was first used for traditional ion thinning of materials, such as metals [460]. The ion beam sputtering (IBS) method uses a fine, collimated, and water cooled ion source that sputters metal from a target onto the specimen surface. Argon is directed into the ion gun, and the specimen is neither heated nor bombarded by the plasma as is generally the case in ordinary sputter coating. The device is

Conductive Coatings

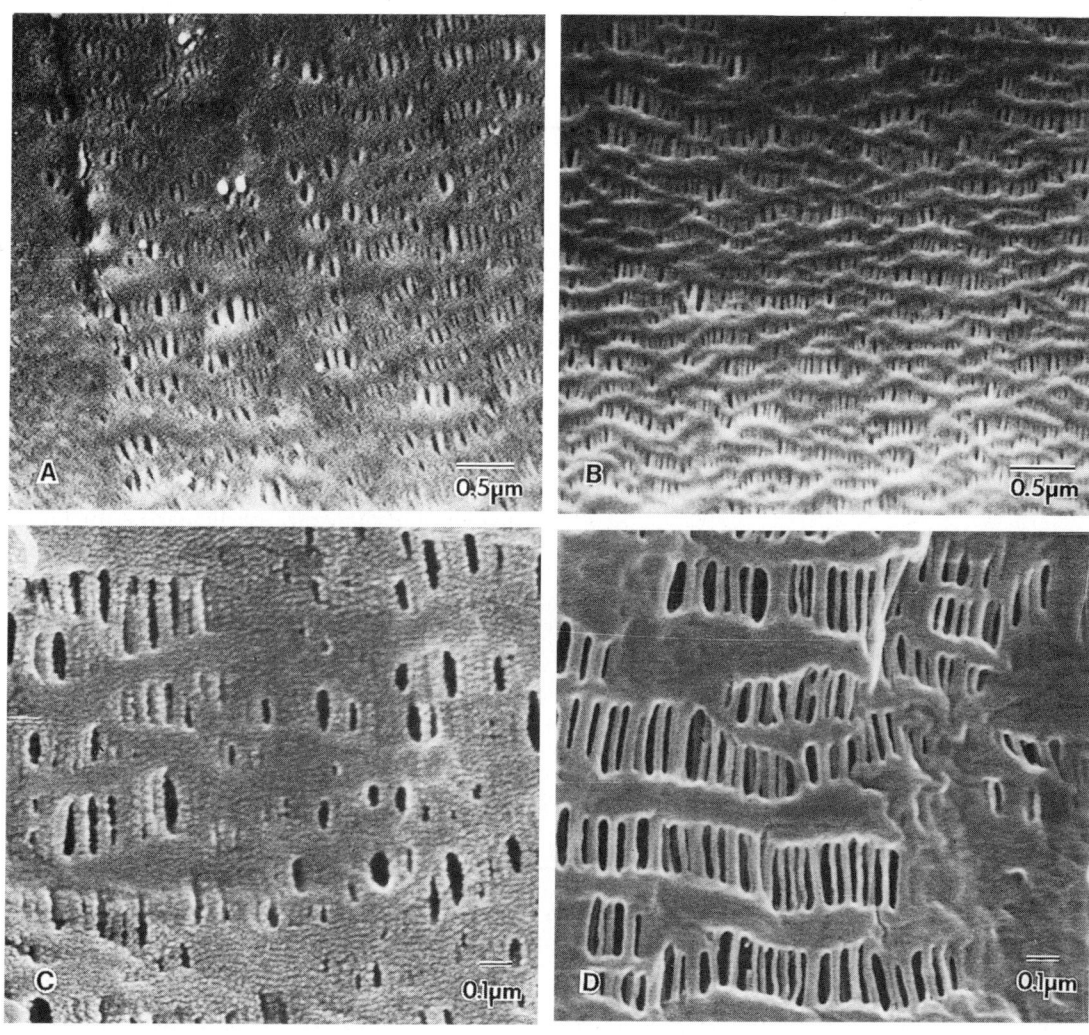

FIGURE 4.30. Micrographs of a commercial stretched polypropylene membrane (Celgard 2400) prepared for microscopy by different metal coating devices are shown by SEI in a dedicated SEM (A, B) and in an AEM (C, D): (A) Au/Pd sputter coating, (B) ion beam sputtered with Au (same magnification as in [A]), (C) Au/Pd evaporation and (D) ion beam sputtering with Au (same magnification as [C]).

used on a vacuum station with a diffusion pump and a liquid nitrogen trap to limit contamination, and pure argon gas is used to backfill the chamber. The best approach to a clean vacuum is use of a turbomolecular pump on both the coating device and the SEM. The geometry is such that the ion source impinges on the rotating target at an angle of about 35° to 45°. The specimen is placed at a 90° angle to the target, out of range of the ion source (Fig. 4.32). In a more complete description of this technique [461], the metal grain sizes were shown to be <1 nm by TEM cross sections.

Clay and Peace [462] compared evaporative, sputtering, and ion beam sputter coatings and showed that IBS produces a fine grained, uniform film of gold. Similar results can be expected for coatings prepared by Penning sputtering [452, 463] or by electron beam coating [464]. Unfortunately, these devices have some drawbacks that limit their use. Specimens generally must be fairly flat for coating. In addition, the apparatus

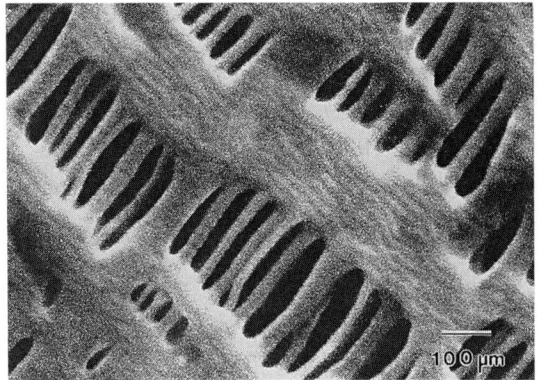

FIGURE 4.31. Field emission SEM image of Celgard 2500 taken at 5 kV with ca. 3 nm Pt coated by ion beam sputtering.

is more expensive and the coating times are longer than by any of the other systems. Magnetron sputtering [465] has also been used to produce fine grain metal coatings with 1–2 nm of Pt deposited on a cold substrate, but it has been found to have aggregated metal particles [452]. Organic specimens have also been treated with osmium vapor to prevent charging [452], although one attempt at such coating on a PP membrane resulted in a very poor nonuniform coating. Porous polyetherimide (PEI) membranes were examined by FESEM with no coating, and coating using magnetron sputtering, ion beam, and Penning sputter coating and compared with electron beam evaporation [466]. Polyetherimide is a very stable membrane and was shown in this study to be free of artifacts when prepared by use of a neutral particle source rather than one using charged particles. This study is an excellent example of the need for careful sample preparation by complementary methods.

Several factors were found important to the production of high resolution coatings with ion beam sputtering, including: slow deposition rate, clean environment, and the fact that the specimen is not exposed to high energy electrons [452, 458, 464]. Overall, the system provides thinner films with smaller grain sizes [467, 468] compared with most other coating methods. Ion beam sputtering with high purity argon gas is the coating method of choice for highest resolution with targets of platinum, tantalum, tungsten, or chromium, preferably <1 nm thick [452] and <4 nm of gold/palladium or platinum for routine SEM. In summary, compromises must be made in the choice of conductive coating methods, depending on resolution. Issues to consider are capital cost, coating deposition time, ease of use, and the resolution required for observation.

4.7.3.3 Coatings for X-ray Microanalysis

Chemical imaging and x-ray microanalysis modes in the SEM require that samples be conductive, and yet the information of interest cannot be masked by a heavy metal coating nor should it contribute to the x-ray spectrum. Polymer specimens are particularly difficult to analyze using energy dispersive spectrometers (EDS) or wavelength dispersive spectrometers (WDS) for the same reasons they are difficult to image in the SEM. The issue is that at the beam energies used in the SEM, it is possible to break, rearrange, or disrupt the chemical bonds, with resulting mass loss, and differential mass loss of some elements, making microanalysis difficult or impossible. A simple assessment of the x-ray count rate as a function of time will reveal the change of the chemistry and should be done prior to any judgment being made as to

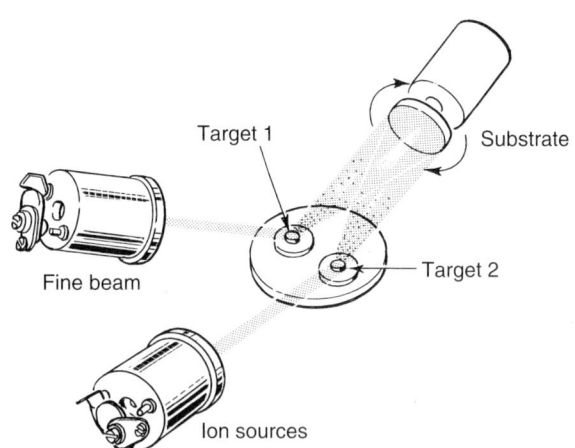

FIGURE 4.32. The geometry in the ion beam sputtering unit is shown. Two water-cooled ion guns form a cone of ions/atoms of argon that is focused on the metal target. The sputtered metal is then focused onto the rotating specimen.

Conductive Coatings

the composition of the specimen; methods have been developed for such analyses [469]. Carbon coatings are applied to such specimens in order to dissipate the charge, although carbon does not provide much electron emission, and often metal coatings must be applied for imaging. Carbon coatings are also used to provide a continuous coating prior to metal coating. Production of carbon coatings has been described above (see Section 4.7.2.1).

4.7.4 Artifacts

It is appropriate to discuss the potential "imaging effects," or *artifacts*, in specimen preparation. For this discussion, the term *artifacts* will be reserved for those textures that are an effect of the preparation method and are incorrectly attributed to the specimen. The polymer microscopist has to deal with such preparation effects daily and, thus, should be fully aware of both the causes and the appearance of the resulting structures. How can the microscopist know when the structures observed are an adverse preparation effect or a real structure of the material? Experience is the best teacher, but there are some effects of metal coating, charging, and beam damage that are generally known to result in artifacts with polymer materials.

4.7.4.1 Charging

Polymers are generally insulators, and poor conductive coatings will cause a charge buildup in the specimen. Charging is a reversible process associated with a high negative charge on the specimen that can cause bright spots in the image. Other effects, such as the specimen moving in the beam, image shift, poor signal output, and "snowy" images are all due to charging. Charging may be decreased by carbon coating prior to metal coating, thicker metal coatings, better contact between the specimen holder and the specimen, and lower accelerating voltage SEM operation. The top of the specimen surface may be connected to the specimen stub by silver paint or a carbon paste in the case of x-ray specimens. Typical charging effects are shown in the micrographs in Fig. 4.33.

Charging continues to cause image effects even with the use of field emission electron guns and low voltages. As the specimen discharges, there is unstable imaging. At high beam energies, 20 keV for example, the total electron yield is less than 1. The specimen emits fewer electrons than it receives and charges negatively. At lower beam energies, the incident electrons do not penetrate so far into the specimen. Electrons, particularly low energy

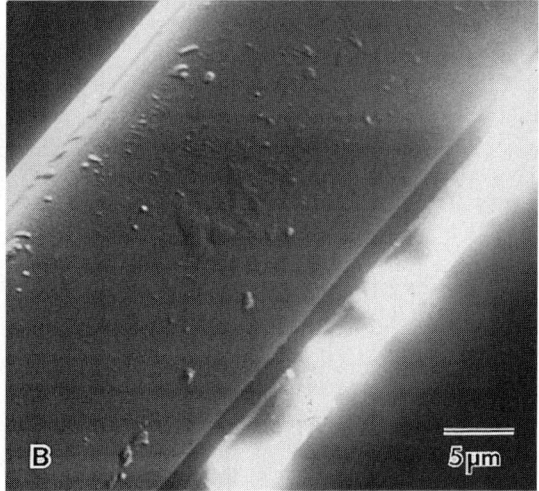

FIGURE 4.33. Scanning electron microscopy micrographs show a few of the spurious imaging effects that result from specimen charging. The banding shown in (A) and what appears as movement of the specimens are also due to charge buildup. The edges of the fiber are bright due to a bright edge effect (B).

secondary electrons, can escape the specimen more easily and the electron yield increases. An area that has charged negatively has a lower effective incident beam energy, a higher electron yield, and thus appears bright. At some low incident energy $E2$, the total electron yield becomes unity and there is a dynamic charge balance. If the beam voltage is reduced to give an incident energy of $E2$, an uncoated nonconducting polymer sample may, in principle, be imaged with no charging effects (see Section 3.2.2.2).

Joy [470–472] has provided careful experimental studies on the effect of imaging at low voltages in the SEM and the control of charging relating to many materials, including polymers. Calculated and measured values of $E2$ for a range of polymers are in the range 0.9–1.6 keV [472, 473] (see Table 4.7) and for polymers the $E2$ energy is dependent on their chemistry. The sample can be tilted to increase $E2$ to a higher voltage. He also developed a simple test for determining $E2$ for a specimen during viewing. When viewing the specimen at some energy $E > E2$, the charge of the specimen will be negative as more electrons are received by the specimen than it emits. When viewing at $E < E2$, there will be a positive charge. The SEM should be set at the lowest usable operating voltage and the following test should be done to determine where the setting should be ($E2$) for imaging:

1. Scan the area at low magnification.
2. Go to higher magnification and count to 5.
3. Return to the original magnification.
4. If the area in the center of the image is bright, there is negative charging and the beam energy, $E > E2$.
5. If the area in the center of the image is dark, there is positive charging and the beam energy, $E < E2$.

If the sample is charging positively, then increase the beam voltage before operating. If there is negative charging, then the beam voltage is too high; the specimen may be tilted and then tested again.

The beam voltage that gives charge balance depends on the exact nature of the specimen, and charging will produce some effects on the image even at low voltage. The charges build up slowly, so low beam current and rapid scanning gives less charging, but more noise. A digital framestore gives a convenient method for integrating over many rapid scans to remove this noise. Heating the specimen to 50°C to 200°C or putting it on a high atomic number conducting substrate may reduce charging by increasing leakage currents, but neither is as reliable as a light coating of metal.

Butler et al. [470] showed unexpected and unusual charging effects using a field emission LVSEM with an immersion type objective lens. For strongly insulating samples such as polymers, they showed that even working at $E2$, some charging effects were observed. Charging has been observed as a function of magnification; at high magnifications, the sample might charge positively, whereas at lower magnifications, the sample might charge negatively. Charging was also observed to be worse when the beam is exactly focused on the specimen, whereas slightly out of focus the charging is less. Positive and negative charging can also be observed within the same image. These authors [470] have termed these effects *dynamic charging* and have further explained this effect [474]. In our laboratory [475], after several years imaging in a field emission SEM, the practice to limit charging effects is to lightly coat the specimens with <3 nm Pt by the ion beam sputtering technique (see Section 4.7.3) and to image at 5 kV. The addition of a metal coating increases the apparent $E2$, and the charging effects and artifacts are limited by this

TABLE 4.7. Measured values of $E2$ for a range of polymers

Source	Polymer	Measured E2 (keV)
Ticona	Celcon	1.2
Ticona	PBT	1.1
Dow	Nylon 6	1.2
Exxon	HDPE	1.5
Exxon	EVOH	1.4
Exxon	LDPE	1.5
Exxon	PC	1.3
Exxon	PS	0.9
Goodyear	PET	0.9
Rohm & Haas	PMMA	1.6

Source: From [470].

Conductive Coatings

preparation. This method makes it easier and faster to focus and to collect images, either by traditional methods or by digital scanning, at a range of magnifications than by using lower voltages and no coating.

4.7.4.2 Beam Damage

The effect of beam induced damage must be distinguished from that of conductive coatings as cracking and buckling of the surface may be observed in both cases. All electron beam instruments can induce damage, including SEM and TEM instruments. A recent study of such damage to thin specimens in a TEM included hole formation and the accumulation of material within the irradiated area [476]. Examples of irreversible beam damage effects are shown in the SEM images in Figs 4.34 and 4.35. Beam damage causes different changes in different polymers. For instance, PMMA is very beam sensitive with changes thought to be caused by chain scission, whereas polycarbonate is thought to crosslink in the beam and is, thus, less sensitive [477]. Radiation damage in POM is rapid and is due to chain scission [478]. In the case of resolved metal coatings, the morphological effects are observed all over the specimen, and there is no additional charging in these cracked areas. In cracked or buckled textures resulting from beam damage, the textures and charging increase with time, and fresh areas do not exhibit these textures. Resolution of metal grain textures is more likely by FESEM or HRSEM and should not be misinterpreted.

A good example of a cracked, beam damaged specimen is the surface of a POM molding, shown by SEM in Fig. 4.34. The molded surface has a production related scratch mark and several particulate "pock" marks (Fig. 4.34A). Seconds after taking this micrograph, the region has the appearance shown in Fig. 4.34C. The diagonal lines could be misinterpreted as being cracks from a thick metal coating or as being due to the surface morphology of the specimen. However, reducing the magnification reveals the pattern in the picture frame (rectangle) that was exposed to the electron beam for the higher magnification micrograph (Fig. 4.34B). Another example of the effect of beam damage that

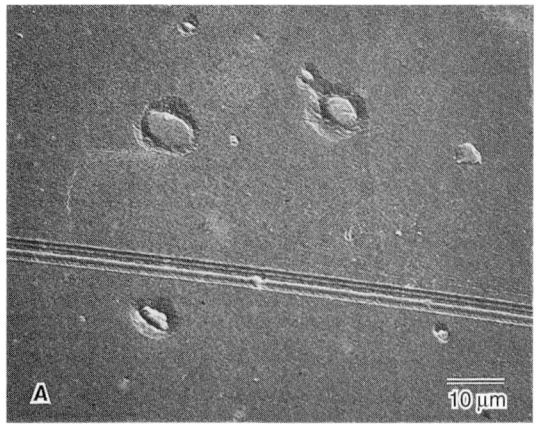

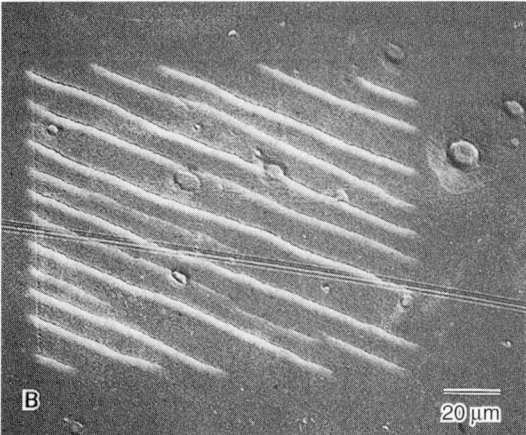

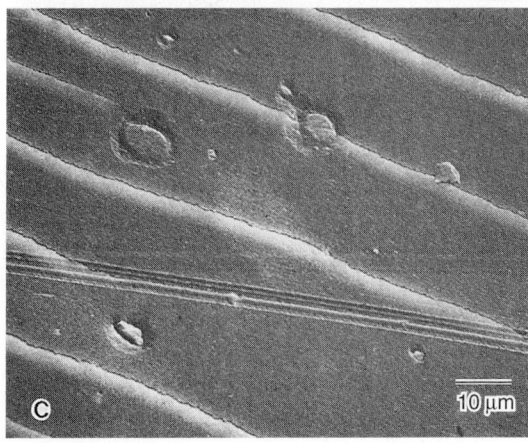

FIGURE 4.34. The effect of beam damage is shown here: (A) SEM of the surface of molded polyoxymethylene taken first, whereas (B) and (C) were taken afterward. The diagonal lines were caused by the electron beam, but they could be misinterpreted, perhaps as a cracked metal coating or a specimen defect, if only (C) is examined.

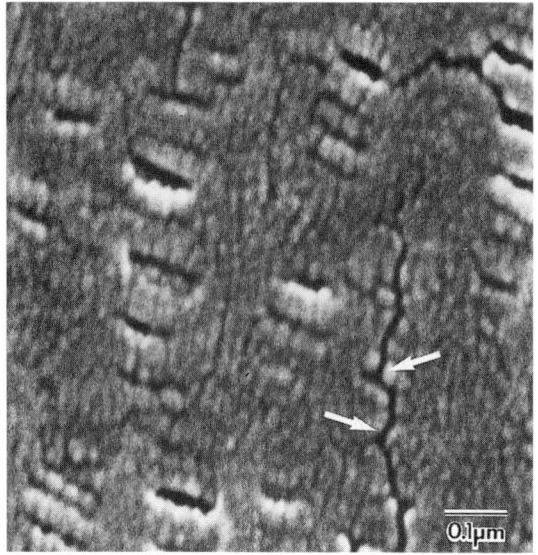

FIGURE 4.35. Even thick metal coatings do not always protect the specimen from beam damage. An SE image of a microporous membrane surface exhibits a granular, or pebbly, texture with fine cracks due to resolution of the evaporated Au/Pd coating.

4.7.4.3 Low Voltage FESEM Imaging Artifacts

The higher brightness of FEGs allows operation at lower beam voltages, and under these conditions there are new challenges in imaging and interpretation. A major effect of low voltage imaging is simply that it is more sensitive to surface detail. This can be an advantage of low voltage operation, but it means that the images are much more sensitive to surface contamination. An excellent example is shown in Fig. 4.36A of the surface of a grooved polycarbonate optical disk, imaged using FESEM at 1 kV. Several effects can be noted. First, the picture frame contrast, from focusing on the central region at higher magnification, is very obvious, and clearly the SEM was contaminated as shown by the dark region. The picture frame is also darker at the left side of the image due to the rastering of the beam. Figure 4.36B shows a mottled, contaminated surface even at 3 kV. A fine, linear pattern that runs horizontally is also observed in the images, especially in Fig. 4.36C [475], which is due to field emission noise, suggesting the microscope and gun settings were not optimized. Again, use of fine metal coatings and 3–5 kV limits the observation of contamination and field emission noise, which can detract from resolving specimen details.

might be mistaken for too thick a metal coating is shown in Fig. 4.35 of a membrane surface prepared for SEM by evaporation of Au/Pd. The metal granularity has been described earlier: however, the large cracks are due to beam damage.

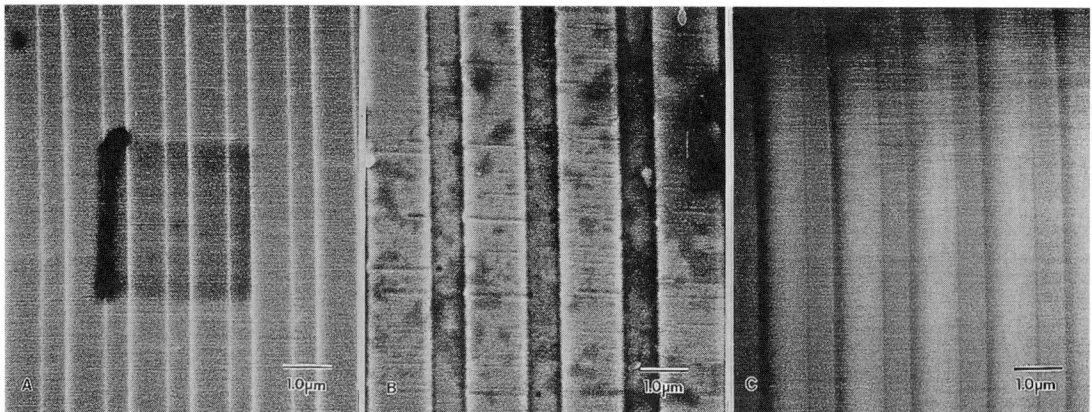

FIGURE 4.36. Field emission SEM image of a polycarbonate grooved disk at 1 kV: (A) clearly shows contamination in the "picture frame" caused by focusing at higher magnification; (B) shows mottled surface contamination; and (C) shows a horizontal pattern due to emission noise from the gun. (From Jamieson, unpublished [475, 480].)

Conductive Coatings

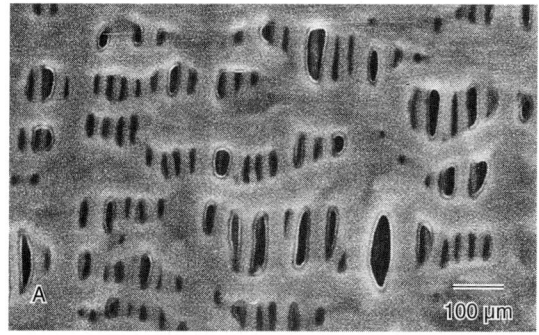
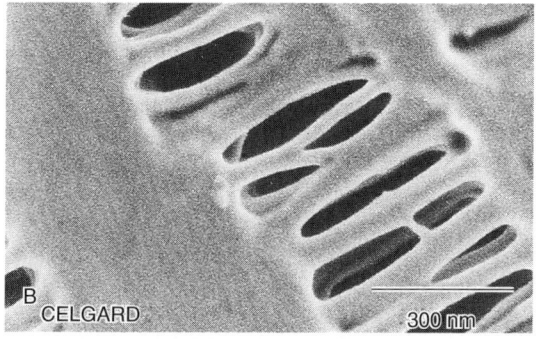
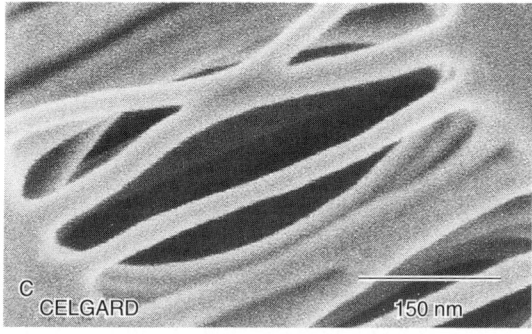

FIGURE 4.37. Field emission SEM of a Celgard microporous membrane shows (A) that the pores are filling in due to contamination, (B) fibrils are joining together and one pore region is filled, and (C) two fibrils are formed into an X pattern due to time in the electron beam. (From M. Jamieson, unpublished [480].)

Combined low voltage and field emission SEM have been used to study microporous membranes [479, 480] shown earlier in this chapter. Figure 4.37 shows some of the potential artifacts that have been imaged and are examples for consideration when imaging at very high magnifications and thus irradiating very small sample volumes [480]. In Fig. 4.37A, there is a thin layer within each of the pores in the membrane that accumulated over time in the microscope and was not present in the first images of this area; this is a contamination effect. Figure 4.37B, C shows less obvious artifacts. The fibrils that separate the pores in this membrane are not usually filled in nor are they collapsed on one another, except during longer times in the microscope. This might not be obvious at higher voltages but at 1 kV the surface detail is very clear and does not represent the original membrane (see Fig. 4.31). These are just some of the artifacts that are possible using lower voltages. Boyes [481] more recently studied LVSEM and the rate and dose dependent specimen damage that may increase at low voltages. Higher resolutions are possible, but the interpretation of the resulting images remains one of the key issues in these imaging techniques.

4.7.5 Gold Decoration

Gold decoration, described by Bassett [482], is a method of highlighting very fine surface steps using a very light coating of gold, about 0.3–1 nm, evaporated onto the specimen surface followed by carbon evaporation. The metal nucleates along the edges of the steps on crystal specimens. Thin materials are examined directly with the gold/carbon film. The carbon film with the gold may be stripped from thicker specimens. Polymer thin films and single crystals have been explored using the gold decoration technique. Studies using gold decoration have been performed in several laboratories, including thin solution cast films of nylon [483], stirred PE solutions [484], spherulites of block copolymers [485], and the internal structure of PE fibers [486]. The width of the active zone, a region of strain softened material at the craze-bulk interface, was measured as a function of temperature in PS using gold decoration [487]. This method has limited utility due to widespread use of SPM techniques.

4.8 YIELDING AND FRACTURE

4.8.1 Fractography

Microscopy is used to study the plastic flow and fracture of a range of biological, ceramic, and polymer materials. The most common technique used for polymers is the SEM of fracture surfaces. In many cases, specimens are frozen in liquid nitrogen prior to fracture; this is addressed in the next section. Traditional fractography studies are conducted by optical microscopy techniques, but as most fracture surfaces are very rough, the superior depth of field of the SEM makes it the better choice even at low magnifications. Optical microscopy is still used for some studies; for example, internal flaws (cracks, crazes, and shear bands) can be seen in transparent materials and so can stress whitening, which occurs in the plastic zone of some polymers. The fine structure of fracture surfaces used to be studied by TEM of replicas, but the modern SEM has sufficient resolution for this purpose. Transmission electron microscopy, especially at high voltage, is used for the study of local plastic deformation in shear zones or crazes, which are the precursors or early stages of fracture in glassy and semicrystalline polymers. Crazes have been studied by both OM and SEM, but TEM is best suited for resolution of the fine structure. The best approach to preparing fracture specimens is to use an easy and reproducible physical test method, such as in a tensile or impact machine at standard machine settings, so that the deformation is reproducible. Atomic force microscopy is also used to study fracture in which the surface of the material requires study at high resolution but not under vacuum. Special stages are now manufactured for most microscopes for *in situ* observation of deformation by fracture and temperature.

4.8.1.1 Fracture Types

The bulk fracture surface may come from one of three general sources: deformation of the sample in a standard mechanical testing device, such as a tensile tester (e.g., Instron, Norwood MA), an impact testing machine (e.g., Charpy or Izod), or a tear tester; deformation by fracture using *in situ* testing devices generally during observation in the SEM; yielding or fracture of a real article or product during service or testing.

Extensive discussions of polymer fractography, including crazing, are found in basic texts (e.g., by Kausch [488, 489] and Liebowitz [490]). In this section, discussion will focus on the techniques required to prepare fractured or deformed specimens for microscope observation. It is quite obvious that the preparation methods all involve damage processes.

In the first case above, mechanical deformation is part of the sample preparation, where the sample is fractured by testing used to obtain mechanical data. The purpose of conducting microscopy on the fractured sample is normally to determine the mode and propagation of failure and to correlate the mechanical data and sample microstructure. Scanning electron microscopy of the fracture surface provides these observations for comparison with standards of similar materials [491]. The fracture could simply be used to provide access to the bulk polymer for microstructural investigation. In this case, it must be remembered that the fracture surface is *not* a random section through the material, but one where the fracture required the least energy. Bhowmick et al. [492] studied tensile, tear, abrasion, and flexing failure modes of a nitrile rubber vulcanizate. Tear testing showed that the addition of particulate fillers caused strengthening. In the tensile case, a flaw-initiated fracture showed typical fracture morphology. The dispersed phase morphology and adhesion of the polymers was shown to relate to the region studied in the SEM. Observation of a fracture surface must be conducted carefully, relating the morphology to the nature of the fracture. For example, different regions of impact fracture surfaces in poly(butylene terephthalate) show variations in dispersed phase morphology [8].

In the second case, the deformation method will not be by a standard test, but suitable samples will be designed to fit the microscope. The *in situ* deformation of polymers is almost always conducted in the SEM at low magnification and is most often used for fibers and fabrics. Variable pressure SEM and AFMs are also

Yielding and Fracture

used for such studies. Other microdeformation methods are used to prepare thin films for TEM observation, to model fiber structure, or to investigate crazing. Because the mechanical deformation technique is controlled by microscopy, it is appropriate to describe it in this chapter.

In the third case, the failure and fracture may occur outside the laboratory, and microscopy is often used in a forensic manner to determine the failure mode or to aid product development. Microanalysis may be used to pinpoint a specific locus of failure and cause of the failure. Here, the special features of specimen preparation relate to the nature of the work, and optical photography may be required to document the relation of the specific SEM samples to the original object.

4.8.2 Fracture: Standard Physical Testing

Fractured bulk polymers and composites require only coating with a conductive layer (see Section 4.7.3) before observation in the SEM, although some composite fracture surfaces are so rough as to make deposition of a thin continuous conductive film very difficult. High resolution is rarely required in these materials, so the common solution is to use a thicker coating, and often carbon is evaporated followed by metal coating. Fibers, particularly textile fibers and thin films, have such a small cross sectional area that the main difficulty is in handling the broken sample.

Standard strain induced in an external tensile test device using semithin sections (ca. 500 nm) has been conducted prior to microscopic analysis of styrene/butadiene block copolymers by high voltage EM [493]. The sections, cut at cryogenic temperatures (−120°C), were strained and then exposed to OsO_4 vapor to selectively stain the polybutadiene phase, and the bulk morphology was investigated by AFM in the intermittent contact mode on the block face. Michler et al. [494] provided an overview of different micromechanical deformation processes leading to an enhancement of toughness in heterophase polymers. Morphology was studied by TEM of thin, stained sections and SEM of samples after deformation in a tensile tester. In order to obtain planar and flat surfaces, the deformed miniaturized tensile bars were epoxy embedded and microtomed down to the middle of the bar with a metal blade, which eliminated the influence of the surface layers. A special tensile stage was also used in a high voltage TEM or for the AFM (see Section 4.8.4). These complementary studies provide an improved understanding of the mechanism of particle toughening of semicrystalline polymers [493–495].

4.8.2.1 Fiber Fractography

Hearle and Cross [496] broke thermoplastic fibers at normal rates of extension in an Instron and examined them in the SEM. They developed a special stub for mounting the fiber ends. A step was cut from the circular stub and an elliptical hole cut in the remaining stub and a screw placed in the cut step. This provides a space in the center of the stub for fibers to be mounted on double sided sticky tape. The screw is used to attach the cut step portion to the remainder of the stub.

Studies have been conducted by breaking polymer fibers on devices, such as an Instron, and then selecting fiber ends for mounting and SEM examination. This method is time consuming, but many more well characterized fibers can be evaluated by breaking fibers in the microscope. Thus, the method of choice is the one where controlled failure occurs outside the microscope. Fibers fractured either by standard testing or by deformation in the SEM are prepared by the following method. The fibers are placed onto double sided sticky tape on a rectangular piece of a carbon stub and attached to another stub (Fig. 4.38). Mounting fibers is best done under a stereo binocular microscope. Only a few fibers can be studied by this method in a reasonable time; however, morphological analysis of the fracture aids determination of the nature and cause of failure.

Scanning electron microscopy images of a matched pair of tensile failed PET fibers are shown in Fig. 4.39. A classical slow fracture zone, or *mirror*, is seen adjacent to the locus of failure. A typical ridged or *hackle* morphology is exhibited as the crack propagates and

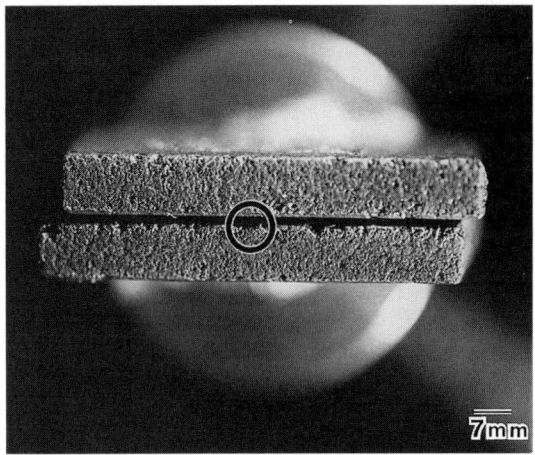

FIGURE 4.38. Photograph of matched fiber ends mounted between two carbon blocks to insert in the SEM for imaging of the fracture surface and x-ray microanalysis of the locus of failure.

accelerates away from the failure locus. In this study, an inorganic residue from the polymer process was shown to be the cause of failure [497]. The value of such a fracture investigation is that the flaws causing failure can be determined, and this information can be used to modify the process and improve mechanical properties. There are many good reviews of polymer fracture processes [488, 490, 498, 499].

4.8.2.2 Fracture of Plastics

The fracture process and morphology of thermoplastics, glassy thermoplastics, and thermosetting resins can all be studied by fracture methods. In one example, two samples of a nylon polymer exhibited very different elongation properties, although they were processed similarly. Scanning electron microscopy study showed that the specimen with poor elongation exhibited brittle fracture morphology (Fig. 4.40A, B), due to a contaminant, whereas the specimen with higher elongation properties exhibited ductile fracture morphology (Fig. 4.40C–F) more consistent with the polymer material.

Multiphase polymers are observed by fractography for evaluation of the rubber or the dispersed phase size, shape, and morphology. Bucknall [500] described the control of the structure during blend manufacture and the resulting effects on the properties including microscopy characterization. Reed [499] described the impact performance of polymers. Evaluation of fractured materials resulting from various physical property tests is conducted in the SEM where the fracture mechanism and the cause of failure are related to the properties. A detailed study [501] of impact toughened polyamides showed mechanisms of failure. Ductile fracture surfaces of water conditioned nylon 6 showed extensive yielding and shear bands. Stress whitened regions

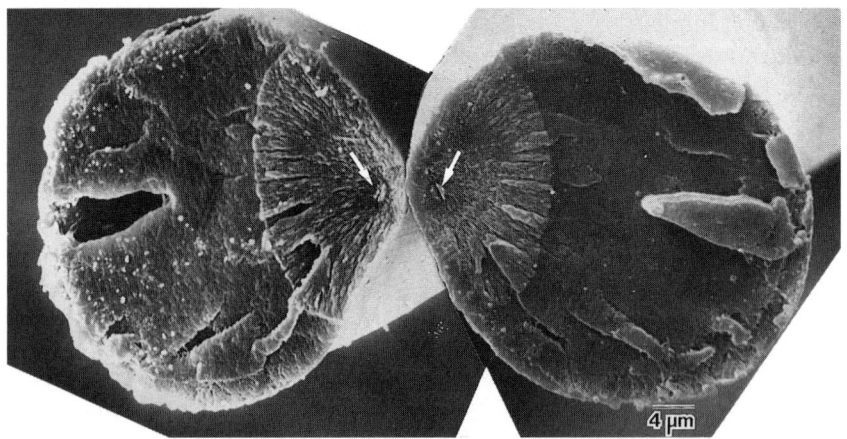

FIGURE 4.39. Scanning electron microscopy images of a matched fractured PET fiber show a defect at the locus of failure (arrows). The region surrounding the locus of failure, the mirror, is the slow fracture zone. As the fracture accelerates across the fiber, ridges (or hackles) are formed in the outer fracture surface.

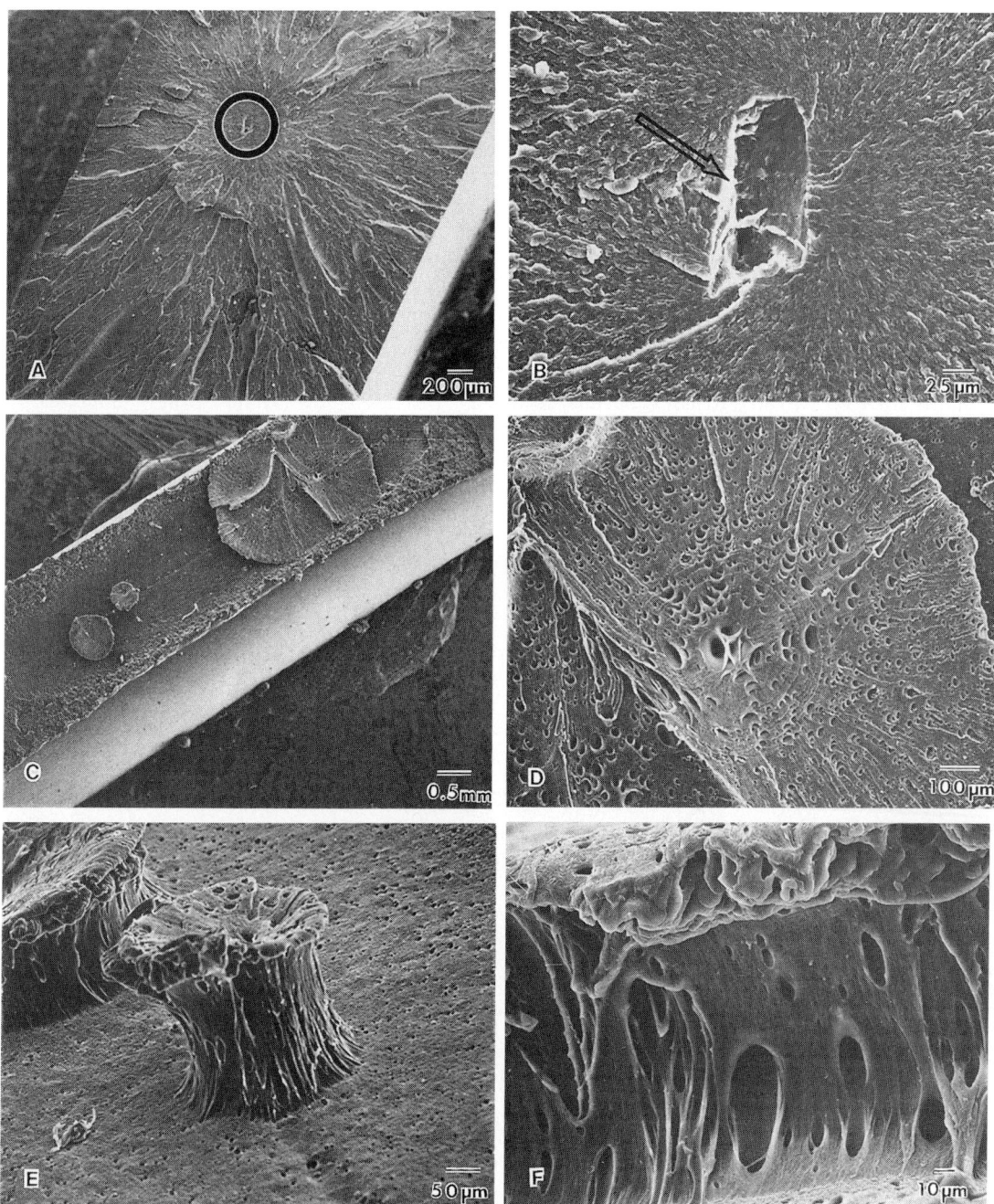

FIGURE 4.40. The SEM was used to compare the fracture morphology of two semicrystalline specimens that were processed similarly and yet produced different elongation properties. The nylon specimen with lower properties (A, B) has a brittle fracture morphology with a rectangular shaped particle at the locus of failure (B). The specimen with higher elongation properties (C–F) exhibited ductile failure.

beneath the ductile fracture surfaces were cryogenically dissected both transverse and longitudinally with respect to the ductile crack growth direction, and cavitation was observed around the rubbery particles. Fracture mechanics investigation of toughened nylons [146] addressed their inherent fracture toughness. Blends were produced and the crack growth resistance tested to characterize the fracture toughness. Morphology studies of the deformed regions from the center of the test blends were made using TEM of cryoultramicrotomed sections using diamond knives and compared with sections taken away from the crack growth area to assess particle size and dispersion.

Representative SEM images (Fig. 4.41) show a range of different multiphase polymers in notched Izod impact fractured specimens. A polymer with large, nonuniform, dispersed phase particles, not well adhered to the matrix, is shown in Fig. 4.41A. A much finer dispersed phase is shown in Fig. 4.41B with both particles and holes from particle pullouts. Smaller particles are not as obvious in Fig. 4.41C, although the dispersed phase accounts for 15% of the specimen. Finally, the SEM image in Fig. 4.41D does not reveal the elastomer, so the size and distribution of the dispersed phase must be provided by some other microscopy technique.

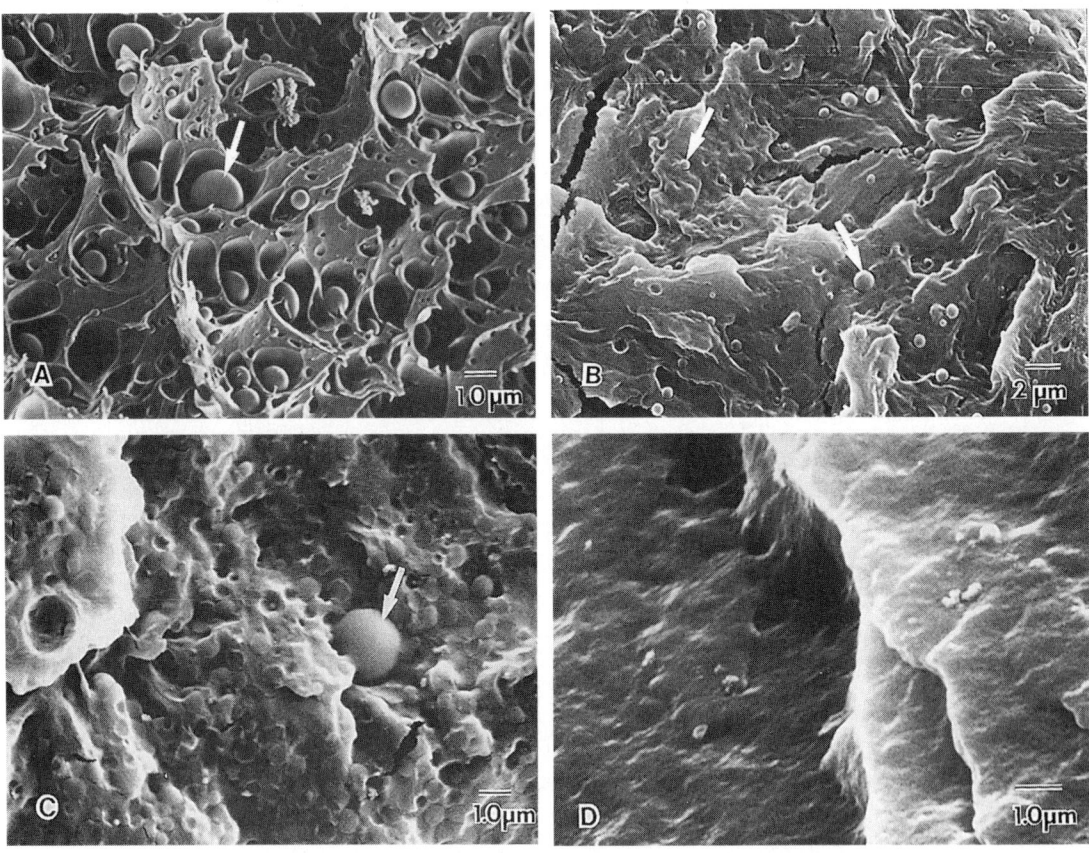

FIGURE 4.41. Scanning electron microscopy images of representative fractured, dispersed phase polymer specimens: (A) has obvious dispersed phase particles in a ductile matrix that have poor adhesion; (B) has much finer dispersed phase particles; (C) has particles and holes where particles pulled out and are on the other fracture face (the adhesion is much better in this case compared with [A]); (D) has barely distinguishable particles, although more than 20% elastomer is present.

4.8.2.3 Composite Fractures

Materials science studies relating structure to the mechanical behavior of fiber reinforced composites are important as these materials find application in both structural and nonstructural applications. The fracture behavior of composites is affected by many variables, including the nature of the fiber and matrix, the fiber-matrix bond, fiber orientation, stacking angle and sequence, void level, loading, and the environment. As a result of the breadth of the variables, Mulville and Wolock [502] stated that generalizations relating to fracture behavior are difficult to make. Some possible damage modes in composites, however, are matrix cracking, fiber-matrix interfacial bond failure, fiber breakage, void growth, and delamination. Scanning electron microscopy examination of the failure zone may reveal all of these failure modes with no clear evidence of the initiation site. However, the SEM does provide some insights into the nature of the failed surface as shown in Fig. 4.42 of two such composite fracture surfaces. The clean surface of the glass fibers, shown by SEM (Fig. 4.42A, B), results from poor bonding of the fibers to the matrix and failure at the fiber-matrix interface. In another case (Fig. 4.42C, D), the fibers have resin on their surfaces, and the failure appears to have taken place within the matrix rather than at the fiber-resin interface.

The identification, characterization, and quantification of fracture modes of graphite fiber resin composites in a specific deformation (tensile testing) mode have been determined using the SEM [503] where three different modes of fracture were identified for off-axis fiber-resin composites. Kline and Chang [504] investigated the fracture surface features associated with various failure tests, including tension, compression, and tension fatigue. Flexural fatigue loading of graphite epoxy composites results in matrix cleavage, hackle formation, and wear failure features [505].

Polymer matrices are also commonly reinforced with mineral fillers or fibers, such as calcium carbonate, talc, wollastonite, clay, and mica [506], and more recently fine additives are used to manufacture nanocomposites. Scanning electron microscopy images of fracture surfaces show the wetting behavior or adhesion of the filler by the polymer matrix or additives used to improve their dispersion. Figure 4.43 shows secondary electron image (A) and backscattered electron image (B) micrographs of a mineral filler in a matrix of a commercial polymer. Scanning electron microscopy does not reveal the nature of the filler in the matrix, whereas backscattered electron imaging (BEI) does reveal the mineral filler due to atomic number contrast. Backscattered electron imaging is important in the observation of mixed polymer and inorganic materials, enhancing the contrast between these materials. Kubat and Stromvall [507] studied the reinforcement of polypropylene and polyamide 6 with mineral fibers. Polymer fibers are also used to increase the fracture toughness of polymer composites.

4.8.3 Crazing

Crazing is the first stage of fracture in many glassy polymers and also in blends and semi-crystalline polymers where there is a glassy matrix. Crazing is a localized tensile yielding process that produces thin sheets of deformed, *crazed*, material with the sheets perpendicular to the principal stress axis. Within a craze, the continuous phase is void and there are fine (ca. 10 nm) fibrils of oriented polymer parallel to the stress axis that extend across the craze. Craze material might have an average density of only 20% of that of the bulk polymer (i.e., if 80% of the craze is void), but the craze can still support a significant load because of the oriented fibrils. The load bearing capacity of crazes can be seen directly in some tensile test samples of transparent polymers where crazes visible to the eye extend right across the specimen while it retains most of its strength [489].

Eventually, crazes break down to form cracks, and when the cracks grow to critical size the sample fails. Although crazes lead to failure in this way they can be useful, because if many crazes are produced before failure occurs, energy is absorbed by the material as local yielding takes place. The impact strength of modified blends is due to the large number of crazes formed. Even when there is no macroscopic indication of crazing, microscopy may

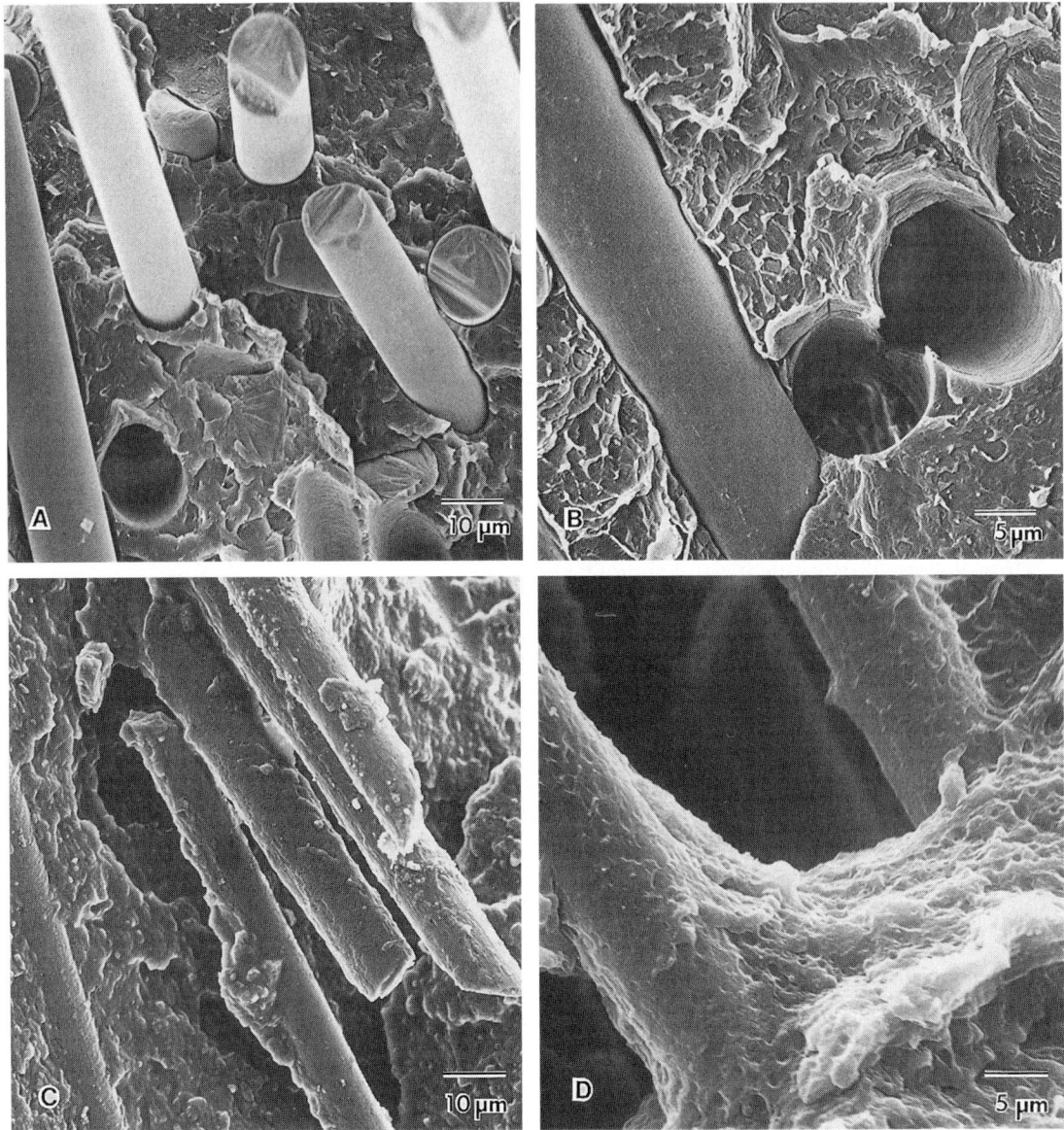

FIGURE 4.42. Scanning electron microscopy images of two glass filled composites with different adhesion properties are shown: (A) is a composite with poor adhesion, as shown by the clean glass fiber surfaces resulting from failure at the fiber-resin interface; (B) has holes where fibers with poor adhesion pulled out of the matrix; (C) shows failure in the matrix and thus there is resin on the fiber surfaces; also seen magnified in (D).

show local yielding, by crazing or shear deformation, taking place just in advance of a crack tip. This ductile zone controls the fracture toughness of the polymer just as it does in metals. Examples of polymers showing this effect are HIPS that has undergone fatigue cracking [508], polycarbonate [509], and lightly crosslinked epoxy resins. Kramer [510] described the formation and breakdown of fibrils quite thoroughly. The TEM is required to see the fine structure of crazes and their relation to second phase particles in blends. Crazing

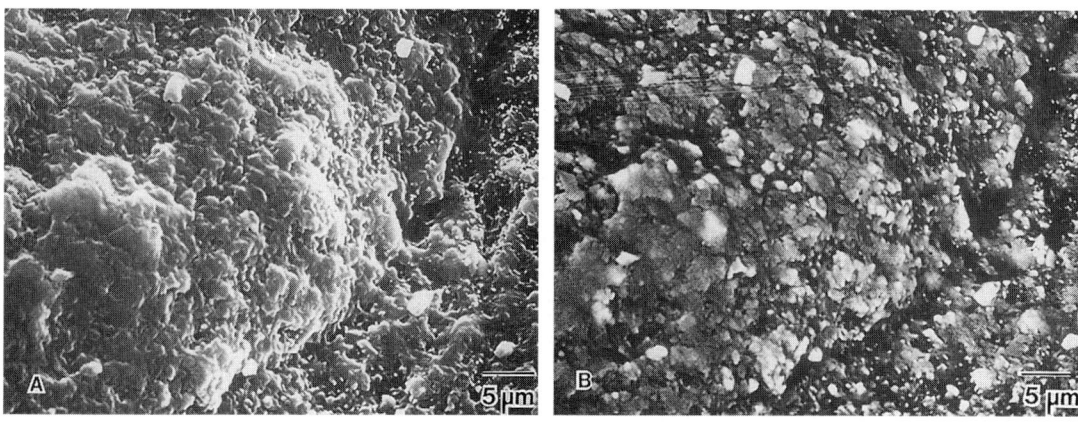

FIGURE 4.43. Secondary electron image of a mineral filled polymer composite (A) does not reveal the nature of the dispersed filler particles. Atomic number contrast in the backscattered electron image (B) clearly shows the mineral filler (brightness increases with atomic number).

and fracture have been reviewed by Kramer [511] and Kausch [488, 489]. Kausch et al. [498] have continued to study the deformation and fracture behavior of polymers as ways to improve them using microscopy methods. Microdeformation mechanisms were studied for amorphous methyl methacrylate glutarimide copolymers and amorphous semiaromatic polyamides by straining thin films that showed mechanisms involving scission crazing at low temperatures, crazing, and formation of deformation zones in a large temperature range and disentanglement crazing at elevated temperatures.

Real-time cryodeformation of PP and impact modified PP has been conducted in the TEM as a function of temperature using a commercially available cooling/straining holder in conjunction with a copper deformation cartridge [512]. The low temperature cooling stage permits studies of the ductile-brittle transition when the transition is between 23°C and −170°C. Thin cryosections were used without staining to assess the morphology. At room temperature, both samples deform by shear yielding, whereas below the ductile-brittle transition crazing was observed and recorded in real time using a CCD camera. A method for the study of microdeformation mechanisms of polymeric materials using tensile straining stages in the TEM allows the *in situ* observation of morphological changes while tensile strain is applied to a polymer material as a function of temperature [513].

With the advent of nanotechnology, nanomechanical properties have become important. Thin polystyrene films embedded with multiwalled carbon nanotubes (MWNTs) grafted with PS chains were prepared via solution casting for study using AFM, TEM, and SEM [514]. Thin films were bonded on copper grids, mounted in a strain jig, and stretched under an optical microscope to observe the growth of local deformation zones, or crazes. The films were then examined under an AFM and by FESEM to calculate the local nanomechanical information. Scanning transmission electron microscopy was used to observe the microstructure. Percolated network of entangled nanotubes was observed to be well dispersed in the PS thin films, and the films demonstrated strikingly different mechanical properties compared with the pristine PS film. The MWNT/PS films were very tough showing no microfracture at large strains beyond 20%. Complementary microscopy was beneficial in assessing the enhancement of properties due to the addition of MWNTs to the polymer matrix.

4.8.3.1 Preparation for TEM

Preparation of crazed polymers for TEM is quite difficult. First, the whole specimen must be stressed to failure, resulting in crazes that are weak and full of voids. Worse yet, the craze structure is unstable in the absence of applied stress. Relaxation of the oriented fibrils and of the elastic deformation of the bulk material tends to close up the craze, destroying its original structure. Sectioning a bulk sample containing crazes is therefore very likely to destroy the information that is being sought. If thin sections or solvent cast thin films are deformed, to avoid the problem of sectioning crazed material, the craze structure may not be the same as that in the bulk material, as the surfaces of a thin film have more freedom. One can imagine a thin film necking down to an even thinner film on drawing, but this could not happen if the same material was inside a thick specimen, without creating large voids. In the terms of mechanical testing, the thin films are deformed in plane stress, whereas inside the bulk such sheets of material are deformed in plane strain.

Microscopists have either attempted to stabilize the crazes in bulk samples by infiltration with supporting materials before sectioning or they have used thin films in the TEM and a range of techniques including diffraction to try to ensure that the thin film structures are representative. Kambour [515, 516] stabilized crazes in polycarbonate by impregnation with silver nitrate for 4–10 days. Selected area electron diffraction showed that the material deposited was metallic silver. Later, Kambour and Holik [517] used liquid sulfur to impregnate and reinforce crazes in PPO. The crazes were formed in ethanol and then were filled with formamide at 125°C for 1 h to keep the voids open. Specimens were transferred to liquid sulfur, at the same temperature, for 24 h. Carbon coating the sections supported the structure and permitted sublimation of the sulfur. This method is useful in cases where the material is not changed during treatment. Kambour [518] outlined the requirements for a craze "infusant" material: (1) the craze should be completely filled with a liquid, below the temperature where the craze loses its strength; (2) this material should be a solid at the temperature for microtomy; and (3) this material should have higher electron density than the polymer, if it remains in the craze.

Kramer et al. [519] developed a TEM method to observe and measure deformation and fracture mechanisms at planar interfaces that should have general applicability. The samples studied were PS and PVP homopolymers as an immiscible polymer pair and a PS-PVP block copolymer as a compatibilizer to reinforce the interface between the PS and PVP. Polystyrene and PVP were fabricated by compression molding. A thin film of the block copolymer was spin cast from toluene solutions on the PVP slab. The coated PVP molding was joined to the PS molding, annealed, and allowed to cool to permit diffusion at the interface. The sandwich was cut in slabs with a diamond saw for fracture toughness measurements and then cut into smaller pieces, embedded in epoxy resin, and cured at room temperature for 2 h, then microtomed with a glass knife perpendicular to the interfacial plane to obtain thin films, ca. 0.5–1.0 μm at a knife angle of 45° and clearance angle of 9°. The films were placed on ductile Cu grids with 1 mm squares previously coated with the block copolymer, so that the interface was aligned perpendicular to the straining direction. The film was bonded to the grid by exposure to vapor of the solvent and then dried at 50°C for 12 h in vacuum. The grid was strained in tension at a constant strain rate of $4 \times 10^{-4} s^{-1}$ with a servo controlled motor drive at room temperature and then exposed to iodine vapor at room temperature for 6–12 h to stain the PVP phase prior to TEM observation. Kramer et al. [520] also developed a technique for the study of crazing in glassy block copolymers. They prepared thin films ca. 0.5 μm thick by spin casting diblock and triblock copolymers of PS and PVP from benzene onto a rock salt substrate. The dry films were exposed to benzene vapor for 24 h and the glassy films were floated off the rock salt onto a water bath surface where they were picked up on a grid coated with a thin film of PS. The film adhered to the grid after a brief exposure to the solvent. An SEM beam was used to "burn" a thin slit in the material, and cracks 50–100 μm long by 10 μm wide were introduced in the center of each grid square. Grids were deformed as described above and examined by OM to locate areas of interest for TEM study and measurement of craze fibril extension ratios [521].

4.8.3.2 Deformation Methods

Even if it can be assumed that crazes formed in thin films are representative of bulk behavior, there has been a problem that the mechanical deformation itself has been poorly controlled. Deformation by handheld tweezers or by exposing strained films to solvent is a poorly controlled experiment (e.g., [522]) which cannot give well defined mechanical histories. This has made it difficult to relate the observed structure to mechanical properties. Lauterwasser and Kramer [521] solved this problem by bonding solvent cast films of PS to a ductile Cu grid (1 mm mesh) and drawing the grid in a micrometer screw driven tensile stage to a well defined strain. As the deformation of the grid is completely plastic, it can be removed from the tensile stage, and individual 1 mm squares of interest can be cut out for study in the TEM without relaxing the load on the polymer film. A quantitative analysis of craze shape and mass thickness contrast within the craze permitted derivation of the stress profile existing along a PS craze. This was extended to many other polymers, relating the mean density of craze material to entanglement density in the polymer glass and to toughness [523] without a basic change of preparation technique. Clearly, if a rubber modified blend contains particles larger than the film thickness (about $1\,\mu m$), the film cannot be representative of the bulk. For the study of craze tips and craze growth, a double tilting stage was used to obtain stereo pairs of the craze tip [524].

The microstructure observed for thick films shows fibrils, about 4–10 nm in diameter for PS, in agreement with SAXS measurements on the crazes in the bulk polymer. Very thin films of PS (100 nm) show modification in the craze structure as there is no plastic restraint normal to the film [525]. Deformation zones have also been studied in polycarbonate, polystyrene-acrylonitrile, and other polymers [526]. Crazes in thermosets can be studied in thin films spun onto NaCl substrates, which can be washed away when the film has been cured. Mass thickness measurements are difficult to make in radiation sensitive materials; that is why most TEM work has been done on PS and least on PMMA. After developing the techniques described above for TEM, Donald and Kramer [526] applied similar methods in optical microscopy to study radiation sensitive materials and the kinetics and growth of deformation zones. Thin films were strained on grids *in situ* in a reflecting OM. Change of interference color, which depends on the film thickness, was a very sensitive method for observing film deformation.

A study of a PS–PB block copolymer showed variation in craze behavior as a result of rubber particles added to modify the otherwise brittle, glassy polymers. Such copolymers were studied under the high strain of physical laboratory testing where the polybutadiene in the copolymer was stained with osmium tetroxide prior to microtomy [527]. The brittle behavior of the glassy polymers was shown by TEM and STEM to be modified by the rubber particles, which provide toughening by control of the craze behavior. In a study of the craze behavior in isotactic PS [215], films of PS were drawn from dichlorobenzene solution and cast onto glass microscope slides, followed by various treatments including isothermal crystallization. Transmission electron microscope specimens were produced, as described earlier [521], strained, and examined by OM and TEM. Crazes in amorphous isotactic PS were shown to be similar to crazes formed in atactic PS. Clearly, crazes are important to the understanding of the fracture behavior of both isotropic and oriented glassy polymers and also to the understanding of fracture toughness in toughened polymers. Transmission electron microscope images of a craze are shown in Fig. 4.44.

4.8.4 *In Situ* Deformation

Deformation experiments can be conducted within the microscope in order to assess the nature of the change in structure as a function of the specific deformation process. In this section, deformation will include not just fracture but also heating stages in various microscopes (cryostages will be considered in Section 4.9). Hot stages and deformation stages are commonly applied to optical microscopy

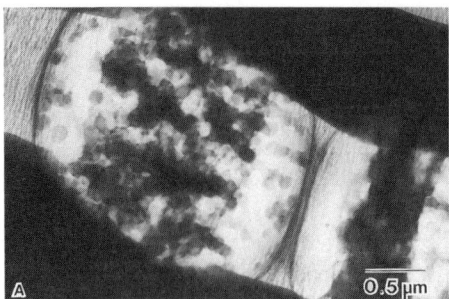 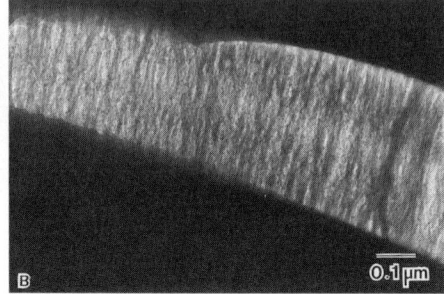

FIGURE 4.44. Typical TEM micrographs of crazed thin films showing a deformed rubber particle in HIPS (A) and craze fibrils in unmodified PS (B).

experiments, especially to assess the nature of the melting and crystallization behavior, phase structures and phase changes as a function of temperature, and the morphology of semicrystalline polymers. A relatively new high pressure hot stage has been described [528]. The chamber of the SEM is quite large and it can accommodate a variety of stages, including hot, cold, and wet stages, tensile and straining stages. *In situ* deformation of fibers and plastics is performed in the SEM and the chemical composition of the specimen can also be monitored by x-ray microanalysis techniques. Although there is little space in the specimen area of the TEM, stages can be accommodated. The types of structure information that are monitored in such *in situ* experiments are the crystal or phase structure and the formation of crazes, often using electron diffraction and dark field imaging. Micromechanical investigations in the AFM benefit from the nanometer resolution without the need for coating, staining, or etching and any radiation damage.

4.8.4.1 In Situ Deformation in the SEM

A major problem in deformation experiments with polymers in the SEM is that a conductive coating is normally required for such imaging. Where such a coating cannot be applied, due to the formation of new surfaces during deformation, other methods are applied to increase conduction or decrease charge buildup in the specimen. Charge neutralizer devices use low energy ion sources to neutralize the negative surface charge by irradiation with a flux of positively charged ions [529]. McKee and Beattie [530] described a stage for the SEM where they deformed fibers, yarns, and fabrics after spraying with either an antistatic spray or a commercial surfactant. Charge neutralization and antistatic sprays have in common usefulness only at low magnifications.

The SEM permits the observation of *in situ* deformation of small specimens, such as fibers and films, using specially built tensile and straining stages [531]. Fibers and woven and nonwoven fabrics have been deformed under tension [6] and have aided understanding of the failure mechanism. Experiments of this type have deformed the specimens from both ends so that the central region remains stationary [532, 533]. A common method to limit specimen charging during *in situ* deformation studies is to use low accelerating voltages, generally less than 1 kV, and/or to use a variable pressure or environmental SEM.

An example of a study conducted using a tensile stage in the SEM is the evaluation of the ductile failure of poly(vinyl chloride) [534] in which stamped dumbbell shaped pieces of polymer from 1 mm thick sheets were extended to a neck in an Instron tester and then strained in the SEM. Low accelerating voltage was used for imaging of the uncoated specimens. These experiments showed that, after neck formation, fracture occurs by crack propagation from a flaw or cavity within the surface craze. The *in situ* deformation of amorphous polymers by shear deformation and craze growth has been observed in optical microscope studies by Donald and Kramer [509]. Thin films of various polymers and polymer blends were prepared on copper grids that were strained in air on a strain frame held in an optical microscope. The films were precracked in an electron microscope by a method more fully described by Lauterwasser and Kramer [521].

Large strain deformation and failure behavior of mixed biopolymer gels have been investigated via *in situ* environmental SEM to explore the changes in the structure of the material while allowing the sample to stay hydrated as it was subjected to tensile strain [535]. The "dogbone" shaped samples were placed in a specially designed tensometer that fitted inside the specimen chamber. It was possible, therefore, not only to measure the mechanical properties of the hydrated material but also to observe any morphological changes occurring as it was being stretched.

4.8.4.2 In Situ Deformation in the TEM

Microdeformation and failure in polymers has also been studied by *in situ* tensile tests in a TEM. Straining stages were used to study a rubber toughened thermoset epoxy or thermoplastic material while tensile strain was applied over a temperature range of 165°C to 500°C [513]. Thin sections about 100 nm thick were microtomed using a diamond knife at room temperature or at cryogenic temperature, and picked up onto a special grid for the tensile stage after the grids were dipped into a glue solution to improve adhesion. For single edge notch tensile testing, one of the grid edges along with a section nearby is scored with a razor blade to concentrate the deformation of the grid and allow crack initiation.

Understanding of the mechanisms in rubber modified polymers have benefited from methods used by Michler et al. [493–495] for the *in situ* deformation of rubber modified amorphous polymers and butadiene-styrene block copolymers. The techniques used were microscopic investigations of deformed samples, including *in situ* deformation of thin sections by TEM and AFM. Deformation tests in the SEM included investigation of the samples using special tensile devices at different temperatures (from −150°C to 200°C) in an SEM or ESEM. Deformation tests are conducted on semithin or ultrathin sections, which can be deformed in tensile devices and examined by TEM or high voltage TEM. Issues with *in situ* deformation studies is the comparison of these microdeformation results with those measured in bulk tests; the character of the deformation might be the same but the absolute values are generally not [495]. A general problem for all *in situ* electron microscope tests of polymers is radiation damage that causes changes in the fundamental chemistry, which could affect the deformation process.

4.8.4.3 In Situ Deformation in the AFM

Direct study of surface structures is possible in the AFM without coating, staining, etching, or using a vacuum; with AFM there is no radiation damage. Atomic force microscopy can also be performed to study the mechanical and adhesive properties of materials. Hild et al. [536] used SPM to visualize the surface of hard elastic polypropylene (HEPP) film, both unstrained and after *in situ* stretching perpendicular to the extrusion direction, using a SPM with a home-built stretching device. Michler [495] used mapping of the phase of the cantilever oscillation in the intermittent contact mode and said it is the preferred technique to detect local variations in the composition, adhesion, and viscoelasticity. A tensile stage in the AFM aided the study of block copolymers, especially when combined with TEM and SEM, providing a view of the different micromechanical deformation processes leading to enhanced understanding of toughness in heterophase polymers [494, 495].

Rubbery materials are easily penetrated by an AFM probe, and the penetration depth can be evaluated as a relative measure of local stiffness and to know the depth of imaging. Bhushan [537] reviewed the use of indentation in conjunction with AFM to determine mechanical properties such as hardness and Young's modulus of elasticity on micro to picoscales. In nanoindentation, which is often applied for probing mechanical properties of polymer materials, with or without an AFM, stiff cantilevers and diamond tips are used.

Nanoindentation includes indenting of a sample site at a particular force and afterward imaging to reveal the shape and dimensions of the indent. Mechanical properties of surfaces and coatings can be examined by scratching. Nanoindentation properties were investigated for compression molded UHMWPE, used for total knee replacement prosthesis, by AFM and FESEM [538]. The cause and generation of wear debris is strongly related to the deformation mechanism of the material under applied stress, which in turn influences the tribological performance of the prosthesis and thus its life span. The elastic modulus and hardness of untreated and treated UHMWPE was measured using nanoindentation at various penetration depths. The core layers were studied using FESEM of block faces permanganate etched after cryoultramicrotomy at −150°C. Atomic force microscopy was used to assess the effect of surface preparation on the sample's average surface roughness. The accuracy of mechanical measurements of polymer modulus via AFM indentation requires careful calibrations (cantilever spring constant, optical lever sensitivity) and application of appropriate contact mechanics models [539].

Quantitative wear analysis was conducted using a new method to measure the volume of locally restricted topographic changes on surfaces due to wear with ultrahigh precision [540]. A stand-alone AFM was used in combination with a special sample holder so that the same section of the sample surface can be imaged before and after tribological stressing in a conventional tribometer. Measurements of wear events due to sliding motion on filled polymers were shown. A combined nanoindenter and AFM have been used to quantitatively study the scratch and mar resistance of polymer coatings with the ability to rank coatings under specific testing conditions according to a micro mar resistance (MMR) index [541].

Scanning probe microscope techniques can be used in a variety of environments, including at elevated temperatures [52] and under various liquids. Use of a hot stage with AFM has been shown to provide information on the organization of semicrystalline polymers at the nanometer scale and its evolution in the course of crystallization. Schoenherr et al. [362] studied the spherulitic morphology of isotactic polypropylene by a combination of optical and atomic force microscopes in which the thin films were prepared *in situ* by using a hot stage. Ivanov et al. [542] reviewed this topic, providing examples of AFM studies of homopolymers and polymer blends crystallized in the bulk, in thin films and in solution, including solution grown single crystals of polyethylene, melt crystallized poly(ethylene terephthalate), poly(trimethylene terephthalate), syndiotactic PS, isotactic and syndiotactic polypropylene. High temperature AFM with a hot stage was used for *in situ* monitoring of melt crystallization of poly(ethylene terephthalate) at 233°C [543]. The evolution of the lamellar structure was compared with the results observed by SAXS permitting the best choice of a structural model. A custom designed system for AFM using an elastic contrast mechanism was used to raise the specimen temperature between the glass transitions of the constituent polymers, enhancing the phase contrast mechanism and optimizing imaging of the sample components. This technique was demonstrated using films of a series of diblock copolymers in commercial and custom built heating stages by Fasolka et al. [544].

The process of melting in PEO was followed in real time at elevated temperatures by AFM using a simple hot stage apparatus [545]. A small k-type disk thermocouple was used to measure temperature. Atomic force microscope imaging of the morphology above the onset of melting revealed the dynamics of a complex melting process. Atomic force microscope deflection images of PEO spherulites grown at 55°C for 4 days are shown in the figures taken at room temperature (Fig. 4.45A), at 69°C (Fig. 4.45B), after 15 min (Fig. 4.45C), and after 34 min (Fig. 4.45D). Melting in the z direction was assessed using the indentation depth of the tip to understand the influence of the pressure exerted on the surface by the AFM tips. Such a hot stage makes *in situ* observations by AFM complementary to optical microscopy.

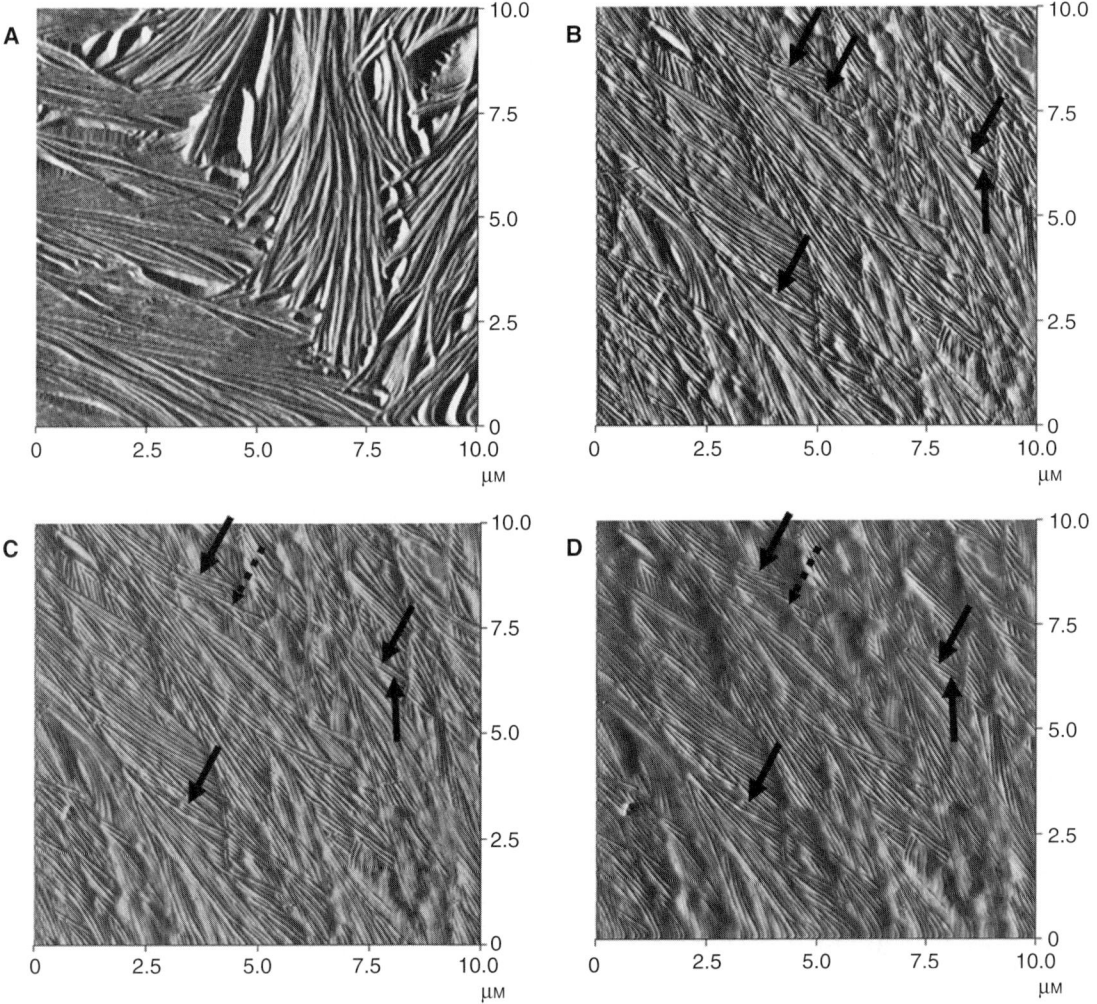

FIGURE 4.45. Atomic force microscopy deflection images of PEO spherulites grown at 55°C for 4 days shown (A) at room temperature and (B) at 69°C; (C) after 15 min and (D) after 34 min; imaging was conducted using a simple hot stage apparatus. (From Beekmans et al. [545], © (2002) Elsevier; used with permission.)

Different gas atmospheres may find utility for *in situ* studies in which the atmosphere can cause differences to occur, such as avoiding oxidation by use of inert atmosphere, or changes in structure based on air humidity [56]. For example, a study of polyamide membranes was conducted by AFM, in a liquid cell under water, and compared with TEM of stained, embedded, and sectioned membranes [53]. The nucleation and growth of poly(ε-caprolactone) (PCL) on PTFE was studied by *in situ* AFM using a hot stage under a nitrogen atmosphere [546]. Samples were prepared for AFM by casting from a PCL/xylene solution onto glass cover slide previously rubbed with PTFE. High resolution real-space information was observed for individual lamellae during the nucleation process at elevated temperature. Figure 4.46 shows AFM deflection images of PCL as observed near the center of a hedrite crystallized at 140°C, from the y axis (Fig. 4.46A) and the z axis (Fig. 4.46B).

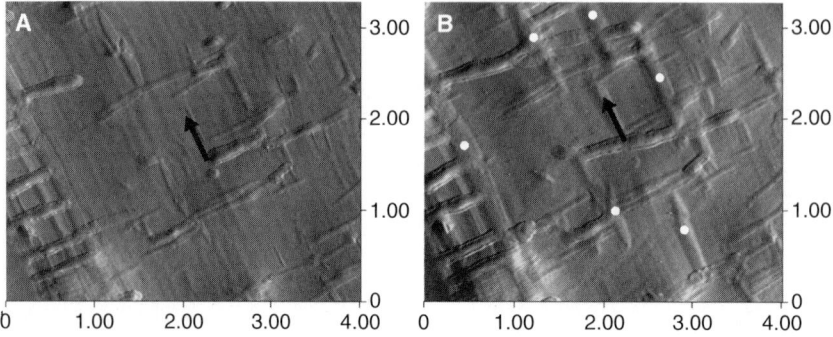

FIGURE 4.46. Atomic force microscopy deflection images of PCL as observed near the center of a hedrite crystallized at 140°C, from the y axis (A) and the z axis (B). (From Beekmans et al. [546], © (2004) American Chemical Society; used with permission.)

4.9 CRYOGENIC AND DRYING METHODS

Generally, the methods described here involve special drying of the specimen, so it maintains its original microstructure, or some kind of freezing technique, including the use of special cryostages for SEM and TEM. Materials as diverse as emulsions, latexes, paints, polymer blends, wet membranes, and ductile polymers often require these special methods to provide appropriate specimens for electron microscopy. In the case of emulsions, latexes, some adhesives, and wet membranes, the specimen of interest is wet, generally with water, and must be dried prior to electron microscopy. The deleterious effects of air drying result from the stress of surface tension forces. The methods used by biologists [132, 547–549] to avoid this problem are (1) the replacement of water with an organic solvent of lower surface tension, (2) freeze drying, or (3) critical point drying. Some latexes can be "fixed" in their original shape by chemical or physical treatment that makes the particles sufficiently rigid to withstand the surface tension forces [550]. Treatments that have been used are bromine [551], osmium tetroxide [187], and high energy irradiation [551]. Unfortunately, many polymers, especially those that are chemically saturated, such as vinyl acetate or acrylates, are unaffected by such treatments. Sectioning films of latexes is also possible, but the individual particles are often not observed by this method. In cases where the specimen is too soft or ductile for routine fractography, low temperature methods, such as freeze fracture and freeze etching, can provide an appropriate specimen for study. Special TEM preparation methods for such polymers, by both direct imaging techniques and replication methods, have been described [552].

4.9.1 Simple Freezing Methods

There are some simple freezing methods that provide adequate preparation for some polymers. Manual methods of freeze fracture are often useful in providing specimens for study in the SEM. An example of a "freeze shattering" method was described by Stoffer and Bone [553] for comparison with microtomy results. Polymers immersed in liquid nitrogen were mechanically shattered with a hammer, mounted, vacuum pumped, and sputter coated for observation. However, fine structural details are not conclusive when specimens are prepared by such methods. A better approach is to use a sharp blade to initiate the fracture, forming a stress fractured specimen. Multiphase polymers and blends often have such high impact properties that they do not fracture during room temperature impact testing. Prefreezing in liquid nitrogen, after notching, has been used successfully to fracture such polymers in an impact tester for property and structure evaluation. This method is far superior to shattering with any mechanical device that deforms the specimen nonreproducibly. Polymers that are ductile at room temperature and that smear upon fracturing may also be frozen and fractured in this way.

Fibers, composites, and molded and extruded polymers are often frozen in liquid nitrogen and then cryofractured, with the best method involving fracture while still in the liquid nitrogen. As an example of simple fracturing [554], films made by microlayer coextrusion of PEO with a $CaCO_3$ filled polyolefin were cryofractured in liquid nitrogen to form a cross section and then washed with ethanol to remove the PEO. Samples were coated with 9 nm of gold and examined by SEM to reveal the microlayer structure as part of the development of a tough, breathable film. Blends of styrene/ethylene-butylene/styrene rubbers were prepared for SEM study and comparison of their morphology and fracture properties by cryogenically fracturing them in liquid nitrogen, etching in toluene for 48 h to extract the rubber, and sputter coating with Au/Pd [555].

4.9.2 Freeze Drying

Of the three methods that are used to diminish the surface tension effects resulting from air drying, the replacement of water with an organic solvent, such as ethanol or amyl acetate [556], is certainly the simplest. However, controlled experiments must be conducted to ensure that the organic solvent has no detrimental effect on the polymer. The second method used to avoid surface tension effects, *freeze drying*, enjoyed great popularity during the 1970s. The method was introduced for biological specimens in 1946 [557] and has been developed and promoted over the years for biological [547, 549, 558–560] and polymer [550] specimens. Freeze drying permits the sublimation of water as a solid to the gas phase, thus avoiding surface tension effects. The fundamentals of freeze drying have been reviewed by Rowe [561, 562] and Echlin [132, 549].

Freeze drying involves rapid freezing, sublimation of the frozen water into water vapor, and application of a conductive coating to add stability to the material. The actual freeze drying takes place in a vacuum evaporation unit set up with a special specimen freezing device and a cold trap for trapping of the water molecules during the procedure. There are several points involved in freeze drying any material [563] including (1) rapid freezing of a thin specimen layer, (2) minimum ice crystal formation, and (3) slow but complete sublimation of the ice. Rather than describing those techniques used in biological studies [132, 547, 549, 560], a practical application of this method for polymers will be described in the next section.

4.9.2.1 General Method and Examples

Walter and Bryant [564] described a method for freeze drying latex specimens in a homemade vacuum system rather than in a commercially available device (as was typical of the state of the art at that time). Later, a freeze drying/image analysis method using commercially available equipment was described [563]. Important details of that method included specimen preparation, placement onto a TEM (or SEM) grid, the hardware for the experiment, and the metal coating.

A general method for the preparation of latex for TEM study has been used on film forming latex, important in coatings and adhesives with a glass transition below room temperature. Solutions must be very dilute, so as to obtain a monolayer of uniformly frozen particles. Two methods generally used to transfer latex particles to the TEM grid are spraying and placement of a microdroplet onto a plastic coated grid. Freezing must be rapid, so that it occurs before air drying. Direct spraying or dropping onto a frozen grid is difficult, at best, and often the specimen freezes in the air above the grid. A simple prefreezing table (adapted from Walter [565]) was constructed (Fig. 4.47A) to permit good sampling and rapid freezing. The steel or aluminum table is placed in a Styrofoam vat filled with liquid nitrogen up to the level of the bottom surface. Drops of solution are placed onto a room temperature grid, and then the specimen holder is quickly placed on the precooled table, so that freezing occurs prior to any air drying. The cover limits air access to the specimens and the possibility of frost formation.

An Edwards evaporator with freeze fracture accessory was used in this experiment, although any commercial freeze etch device can be used. The system must have provision for pumping liquid nitrogen into the specimen holder and temperature sensing and controlling devices.

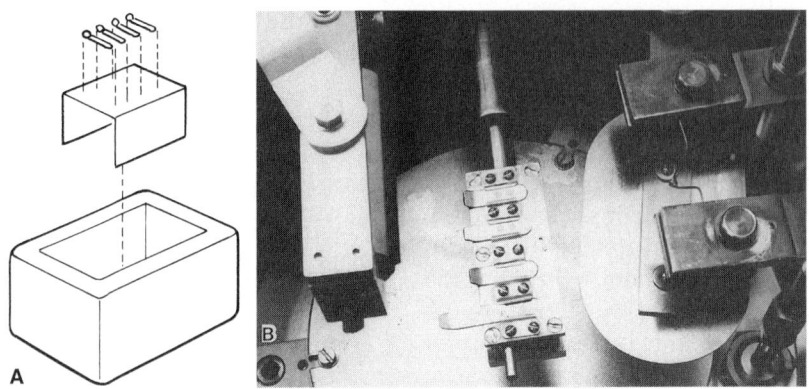

FIGURE 4.47. (A) A Styrofoam cooler, used as a prefreezing chamber, is shown in this sketch with an aluminum specimen table and small, rectangular holders for the TEM grids. The cooler lid (not shown) is important for limiting frost formation on the specimen. (B) A photograph of the specimen holders (with lips for easy accessibility) is shown on the cold stage in the vacuum evaporator.

The specimens are quickly transferred from the prefreezing table to the precooled specimen stage in the evaporator at its lowest possible temperature (−150°C). Transfer takes place through the port in the stainless steel collar, below the bell jar, under a flow of dry nitrogen, which is used to limit ice formation on the frozen specimen surface. Handling of the grids is simplified by placing them on the special lipped specimen holders (Fig. 4.47B), which permit the rapid transfer (i.e., less than 30s) of several specimens into the vacuum chamber. Samples are freeze dried at −60°C to −80°C, depending on the specific material, for about 8 h.

Freeze dried specimens are generally shadowed (at about −150°C) in the vacuum chamber. Shadowing provides a metal cap that is the shape of the frozen, undistorted particle and limits distortion during examination in the electron microscope [566]. Replication of the latex particles, or macromolecules, can also be done if the specimen is expected to change at room temperature. For SEM specimens, the same procedure is used, but the specimens are dried onto small glass coverslips that are attached to the stub with silver paint.

Results of the experiment described are shown in Fig. 4.48. A monodisperse latex of known particle size (Fig. 4.48A) was used both as a control and for calibration of the particle size distribution measurement [567]. A film forming latex is shown after both air drying (Fig. 4.48B) and freeze drying (Fig. 4.48C). Clearly, the flat, film forming, air dried particles are three dimensional after freeze drying.

Thus, the steps in the general specimen preparation are as follows:

1. Dilute the latex to about 1:1000 with water or until only a slightly milky blue cast is seen in the clear solution.
2. Place microdroplets of diluted latex on formvar coated grids that are on rectangular, lipped specimen holders. Quickly place holders on the precooled (liquid nitrogen) table and cover.
3. Transfer specimen grids on the holders to the precooled stage in the vacuum evaporator under flowing nitrogen.
4. Freeze dry specimens at −60°C to −80°C (about 8 h).
5. Lower the temperature to −150°C, and shadow samples with a heavy metal.

4.9.2.2 Literature Review

Various composites are made with microfibrils of wood and cellulose in polymer matrixes. A novel technique to produce cellulose microfibrils (defined as a bundle of nanosized fibrils making up a diameter of up to 1 μm) involves a combination of severe shearing followed by high impact crushing under liquid nitrogen [568]. Fibers in water were immersed in liquid nitrogen to freeze the water, and high impact grinding was per-

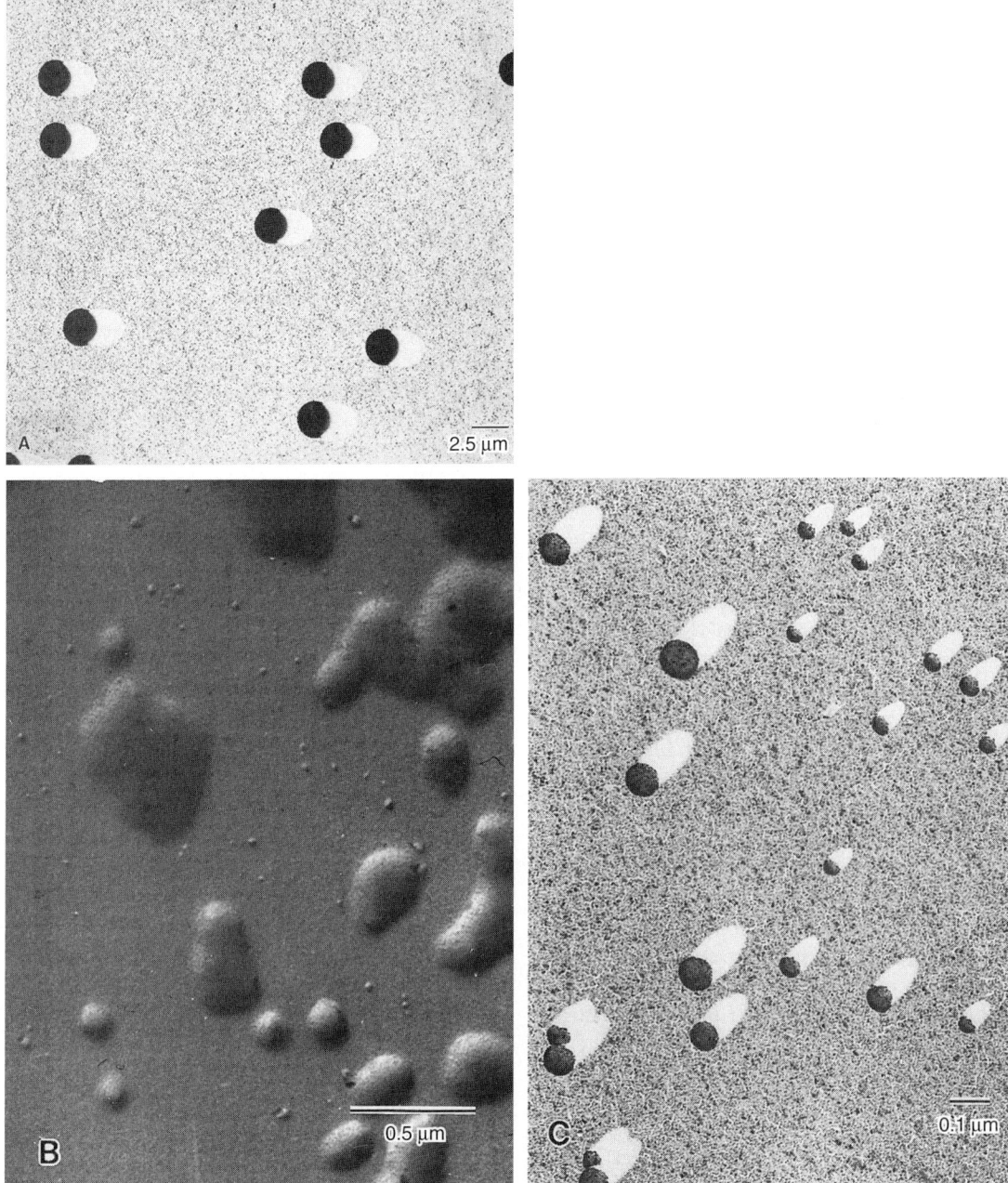

FIGURE 4.48. Transmission electron microscopy micrographs of freeze dried polystyrene latex (A) used as a control for the experiment shows three dimensional particles with no deformation, whereas an air dried film forming latex (B) shows flat regions that have no shadow. The same latex as in (B) after freeze drying is clearly three dimensional, based on the shadows present (C).

formed with a cast iron mortar and pestle under the surface to form individual fibrils that were subsequently freeze dried at −50°C and filtered. The fibers were characterized using SEM, TEM, AFM, and high resolution optical microscopy. The microfibrils have the potential to produce composites with high strength and stiffness for high performance applications. The microfibrils in water were compounded with polylactic acid polymer to form a biocomposite. Laser confocal microscopy showed that the microfibrils were well dispersed in the polymer matrix.

4.9.3 Critical Point Drying

The drying preparation method in greatest use among biologists today is critical point drying (CPD). The method was first described by Anderson [569–571] and then by Hayat and Zirkin [572]. A clear explanation of the method and applications is given by Anderson [571, 573] and Cohen [574], and a later review by Cohen [575] is recommended to the interested reader.

In ordinary drying, the liquid in a specimen evaporates, and the resulting surface (interfacial) tension can distort the structure. In critical point drying [574], heating a specimen in a fluid above the critical temperature to above the critical pressure permits the specimen to pass through the *critical point* (that temperature and pressure where the densities of the liquid and vapor phases are the same and they coexist and thus there is no surface tension). By definition, a gas cannot condense to a liquid at any pressure above the critical temperature. The *critical pressure* is the minimum pressure required to condense a liquid from the gas phase at just below the critical temperature. Thus, CPD allows the specimen and fluid to be taken directly to a gas phase without experiencing any surface tension effects and resulting distortion. At above the critical temperature, the gas is bled off leaving the specimen dried for study. Unfortunately, critical point drying procedures are potentially hazardous because of the high liquid and gas pressures. If only occasional use of this method is required, then use of an outside service lab is worth considering.

Critical point drying is conducted using transitional fluids that go from liquid to gas through the critical point. The critical temperature (more than 300°C) and pressure (above 21 MPa) of water are much too high for it to be used. Unfortunately, this requires the removal of water and its replacement by a transitional liquid. Water is removed and replaced by dehydration fluids, which are replaced by the selected transitional fluid. A typical transition sequence is as follows: a graded water/ethanol (or water/acetone) series for 10–20 min each at room temperature; 100% ethanol (or acetone) for about 15–30 min; transitional fluid, usually Freon (13 or 16) or CO_2.

Graded series are combinations of fluids that are used to gradually replace the water with the dehydrating fluid, such as water/ethanol: 90/10, 75/25, 50/50, 25/75, 0/100. Freon TF (134) is useful as an intermediate fluid, as it does not have to be fully flushed out of the system when used prior to drying with CO_2. Critical constants for Freon 13 (CCl_3), Freon 116 (CF_3–CF_3), and carbon dioxide (CO_2) are in the range 20°C to 40°C and 3.75–7.5 MPa. The CPD preparation is conducted in a pressure vessel to control both the temperature and the pressure.

4.9.3.1 General Method and Examples

One example [468] of this method will be briefly described. A wet polymer membrane composed of polybenzimidazole (PBI) becomes brittle and distorted upon air drying, due to surface tension effects. Samples prepared for the SEM by immersion in liquid nitrogen and then hand fractured are distorted, and often ductile fracture is observed. Standard critical point drying from carbon dioxide, following dehydration into ethanol, yields a membrane that fractures with no evidence of ductility. Most exciting is the fine structure visible in these fractures. Comparison of SEM micrographs of a membrane prepared by freeze fracturing (Fig. 4.49A) and a membrane dried by CPD (Fig. 4.49B) shows large differences in the structure. The freeze fracture method reveals a ductile, distorted structure, whereas the CPD fracture clearly reveals a monolayer of granular particles in the outer surface, or skin layer, and a more open substructure of similar granular particles. These results are consistent with a more complete microstructural study [106].

Finally, for critical point drying of polymers, a suitable series of fluids and conditions must be chosen that will not damage the specimen.

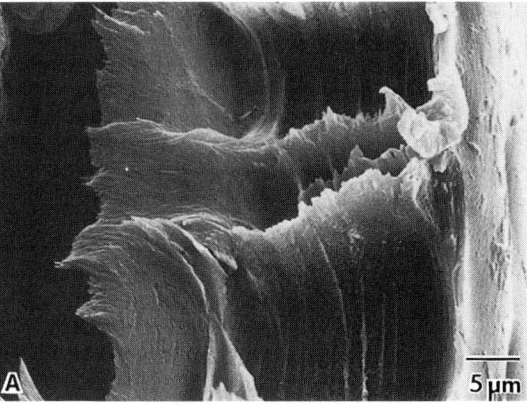 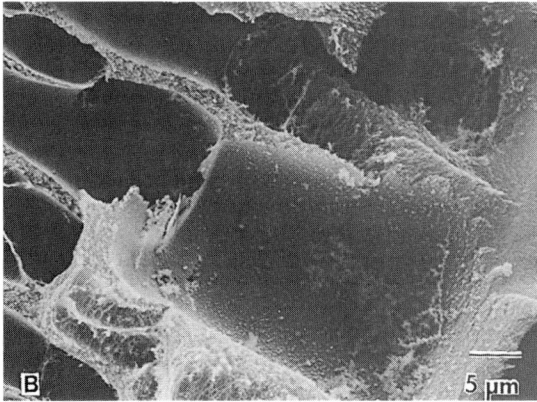

FIGURE 4.49. Scanning electron microscopy images show a comparison between fracturing of a wet membrane after immersion in liquid nitrogen (A) and after critical point drying (B). Fracturing after freezing results in a deformed ductile failure, whereas the fracture after critical point drying shows no deleterious effects of surface tension and the result is a brittle fracture with excellent detail of the internal morphology.

Carbon dioxide dissolves in epoxies and in PS. Apparently, long term high pressure exposure and then a sudden release can turn these polymers into popcorn. Microscopic comparison of a material prepared by a variety of different preparation methods is the best way to uncover any possible artifacts caused by the specimen preparation method.

4.9.4 Freeze Fracture-Etching

Freeze fracture and freeze etching are distinctly biological techniques that have also been used by the polymer microscopist. Freeze fracture means fast freezing of a specimen followed by fracture with a cold knife in a vacuum chamber to reveal the internal structure of a bulk specimen. Freeze etching is somewhat of a misnomer as this process is the surface freeze drying of the freshly fractured specimen or the sublimation of ice from the frozen surface. Typical conditions are to sublime the surface ice at about −100°C for 1–5 min at a good vacuum. This method has the advantage of exposing the underlying true surface features, by removing about 20 nm of ice, for replication and TEM evaluation or for direct SEM observation of the surface structures.

4.9.4.1 Biological Method

An excellent review of freeze fracturing is found in a chapter by McNutt [576]. The method has been described for the preparation of biological membranes. An older, improved version of the method was described by Steere [577] and Moor et al. [578], and a review of the method and application to membranes was described by Branton [579]. The method involves cementing a 1–2 mm piece of the material onto a copper disk with gum arabic dissolved in 20% glycerin and then transferring the material to liquid Freon 22 (chlorodifluoromethane). The specimen is fractured with a cold knife in a vacuum evaporator, and it may be etched prior to replication. A replica can be cast on the surface of the hydrated material at low temperature in a vacuum [576] for examination in the TEM; or the replica, or the shadow cast specimen surface, can be examined directly in the SEM.

4.9.4.2 Literature Review

Several polymer studies have been reported where the specimens were prepared by freeze fracture techniques. A modification of the freeze fracture method was used by Singleton et al. [580] in the preparation of plasticized PVC. The sample was notched, cooled, and fractured and then immediately replicated with platinum/carbon. Replicas were stripped after warming to room temperature. The authors noted that the preparation was not highly reproducible, perhaps due to nonuniform cooling of large specimens. The refolding of PEO chains in block copolymers was studied [581] using the freeze fracture replica method. Specimens in the form of viscous gels were quenched from −25°C to −160°C by immersion into liquid Freon 22 cooled with liquid nitrogen. The styrene and ethylene oxide block copolymer specimens were fractured and platinum/carbon replicas cast in a vacuum evaporator.

Freeze fracture has been used to study the structure of colloidal particles in water-oil mixtures stabilized by polymer emulsifiers. Microemulsions consisting of water, toluene, and graft copolymer composed of a polystyrene backbone and a PEO graft were deposited onto a small gold plate, quenched in liquid nitrogen in equilibrium with its own solid phase [582]. Replicas of the fractured surfaces were washed with tetrahydrofuran, which showed the micellar structure of the copolymers. Classical microemulsions have been studied [583], and micellar aggregates of copolymers have been shown [584, 585]. Polymer latexes have been prepared using similar methods by Sleytr and Robards [586]. The emphasis in this review was on the plastic deformation observed in the freeze fracture method and in ultrathin frozen sections.

Stokes et al. [587] conducted a comparative study between three modes of cryo-SEM (high vacuum, low voltage, and low vacuum) using ice cream as a model system; although this is not a traditional polymer specimen, the study appears worthwhile to note here. Samples were frozen in liquid nitrogen and then rapidly placed in a cryo-SEM specimen holder for transfer to nitrogen slush and then to the cryo-prep chamber with the cold stage at −150°C. Samples were freeze fractured and lightly etched at −90°C for 45 s and some coated with Au/Pd under pure argon using a cold magnetron sputter coater. Samples were transferred to the microscope chamber, under high vacuum, with the cryostage initially at −150°C. Specimens were investigated with and without Au/Pd coating and for a range of temperatures from <−110°C to −90°C. Standard high vacuum imaging of coated specimens gave the best results for fully frozen ice. Low voltages, such as 1 kV, could be used for imaging uncoated specimens at high vacuum, although charging was an issue, and contrast was poor. Low vacuum, involving small partial pressures of nitrogen gas, was suited to *in situ* sublimation work.

4.9.5 Cryomicroscopy

Cryo-SEM and cryo-TEM are the methods that have been developed for study of polymers at low temperatures, and these will be discussed in this section. Low temperature stages for AFM are just becoming available. There is some published work on PDES (polydiethylsiloxane) materials obtained in an AFM cooled to −50°C using thermoelectric cooling [588].

4.9.5.1 Cryo-SEM

Cryo-SEM was used to examine fracture surfaces of polymer-modified asphalts and compared with results from observation of the samples in an ESEM, which can be used with wet or dry specimens although with limited resolution [589]. The cryo-SEM fractured samples were coated with 4 nm of Pt at −165°C and transferred to the microscope; the contrast of the images was good and the information available significantly better than by ESEM. The morphology of these blends was also studied using confocal laser scanning microscopy (CSLM), which provided good contrast as well. The fracture morphology of bitumen/polymer blends was characterized using cryo-FESEM at −160°C, which was said to extend observations made by ESEM to much better resolution [590]. Special modules were used to precool the sample, transfer it to the cryo-prep module, coat the sample with 4 nm Pt at −160°C using ultrapure argon, and then transfer to the cold stage of the FESEM. Characterization of the distribution of the polymer nanoparticles aids understanding of the crack

4.9.5.2 Cryo-TEM

Cold stage microscopy of colloidal suspensions, microemulsions, and liquids is possible by fast freezing and examination of the thin, frozen specimen in a TEM. Talmon et al. [591] developed a rather interesting technique in which a thin sample is trapped between two polyimide films. The liquid layer is about 100 nm thick, whereas the films are about 40 nm thick. Film selection is quite important as the polyimide is more radiation resistant than traditional support films. The films are formed by dipping glass microscope slides into an 0.75% solution of the prepolymer in N-methylpyrrolidone and xylene. The slide is dried for 10 min at 90°C and then cured at 300°C for 3.5 h. The polyimide-coated slides are dipped into 12% hydrofluoric acid for 5–19 s and then floated off onto water and picked up onto grids. A drop of the specimen is placed onto one film-coated grid and covered with another film coated grid prior to immersion into liquid nitrogen. A transfer module was designed in this study. Examples were shown for benzene-plasticized polystyrene latex and a surfactant. Falls et al. [592] used this fast freeze cold stage in a study of hexagonal ice. Knowledge of this structure is essential to understand the morphology of frozen fluids.

Egelhaaf et al. [593] designed a new controlled environment vitrification system for cryo-TEM for application to surfactant solutions in order to reduce the lag-time from blotting to vitrification and allow rapid transfer of the sample to the TEM. Libera [594] imaged unstained polymers by plunging frozen hydrated samples into liquid nitrogen, cutting sections by cryoultramicrotomy at −175°C, and cryo-transferring sections to a TEM equipped with a cryostage. High angle annular dark field (HAADF) STEM images of the materials showed their structure, and low loss spectra were also collected to map the spatial distribution of oil, water, silicone, and alkane in one drop of the sample.

Talmon and his group have continued to work in the area, advancing the development of digital cryogenic TEM as an advanced tool for direct imaging of complex fluids [595] and using this new method for applications to surfactants, polymer-surfactant solutions, and microemulsions. The particular significance of digital imaging is that photographic film, and the artifacts it brings when condensed on a cold specimen, is avoided. A commercial surfactant mixture in isopropanol and water, at a concentration of 0.1 wt.% of active surfactant, was prepared in a controlled environment vitrification system (CEVS) at controlled temperature and humidity to avoid loss of volatiles [595, 596]. A drop of the solution was placed on a holey carbon coated TEM grid, blotted with filter paper to form a thin liquid film, and plunged into liquid ethane at its freezing temperature (−183°C). The specimens are stored under liquid nitrogen and observed by TEM at 120 kV using a cryospecimen holder maintained below −175°C. Specimens are examined by low dose imaging to minimize beam damage, and images are recorded digitally at up to 175,000×. This method permits imaging of the inner structural details of thread-like micellar systems. Other studies using this method include water soluble, hydrophobically modified polymers, used as viscosity modifiers [597]; block copolymer micelles containing solubilized liquid crystals [598]; block copolymer vesicles in nonaqueous systems [599]; and imaging the osmotic limit of polyelectrolyte brushes [600].

Cryo-TEM was used in a study of the nano and microparticles formed by complexation of PDAC [poly(diallyldimethylammonium chloride)] [601] and SDS (sodium dodecyl sulfate). The nature of the complexes was revealed by direct imaging cryo-TEM, showing nanometric details of the complexes, as shown in the images in Fig. 4.50 [601]. The nanostructure of the complexes strongly suggests they are made of a hexagonal liquid crystalline phase, which was further supported by SAXS. Overall, the changes in the specimen preparation methods used have increased the research possible in this exciting and interesting area.

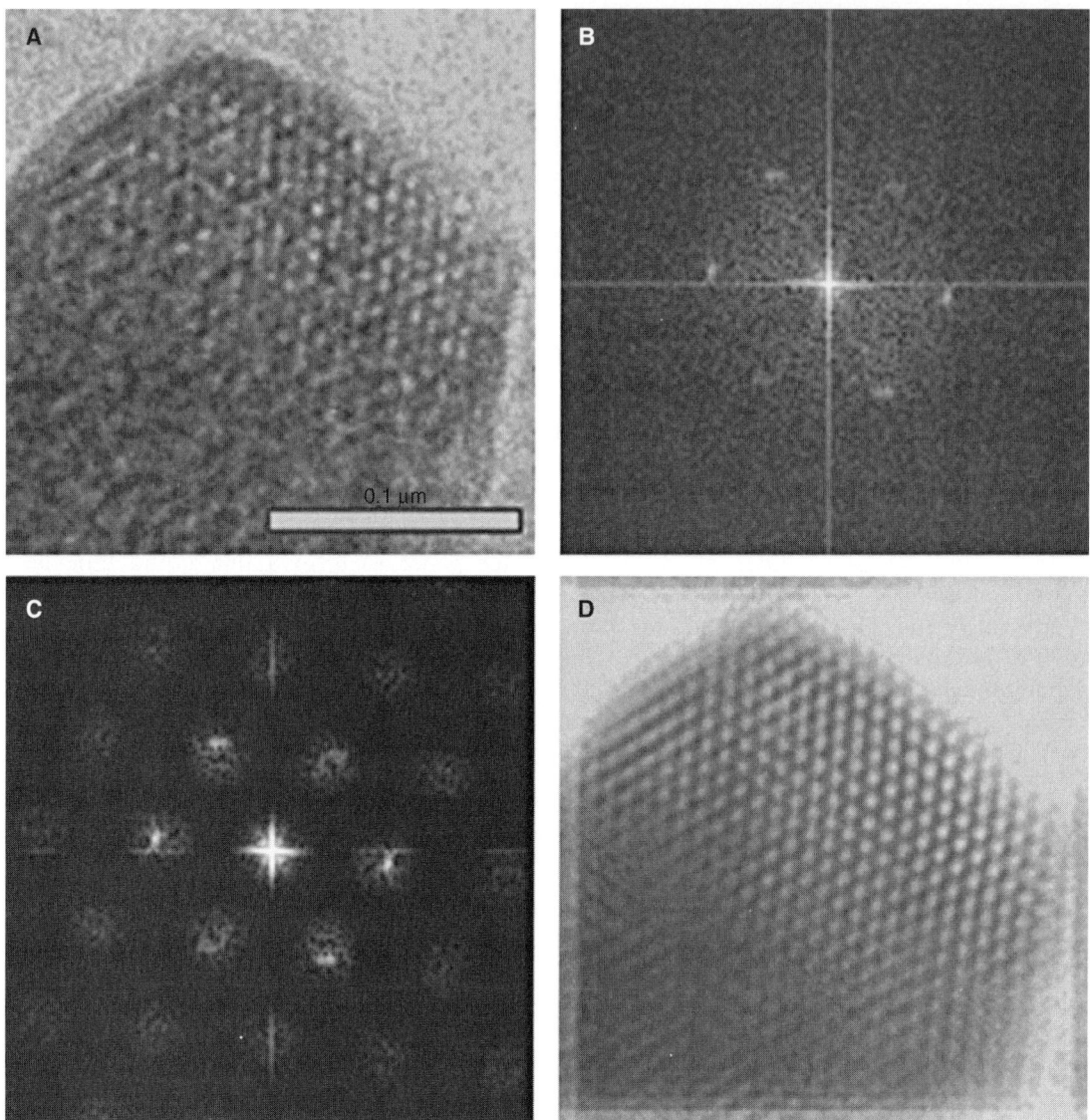

FIGURE 4.50. Digital filtering of a cryo-TEM image and complex aggregate are shown: (A) the original, unprocessed digital image; (B) fast-Fourier transform (FFT) of image (A); (C) filtered FFT of the transform of image (B); (D) filtered image of original micrograph. (From Talmon et al. [601], © (2004) American Chemical Society; used with permission.)

References

1. D.C. Bassett, F.C. Frank and A. Keller, *Philos. Mag.* **8** (1963) 1739.
2. S. Mitsuhashi and A. Keller, *Polymer* **2** (1961) 2.
3. I.W. Hamley, V. Castelletto, R.V. Castillo, A.J. Mueller, C.M. Martin, E. Pollet and P. Dubois, *Macromolecules* **38** (2005) 463.
4. V. Dudler, M.C. Grob and D. Merian, *Polym. Degrad. Stab.* **68** (2000) 373.
5. C. Kugge, V.S.J. Craig and J. Daicic, *Colloids Surf. A* **238** (2004) 1.
6. J.W.S. Hearle, J.T. Sparrow and P.M. Cross, *The Use of the Scanning Electron Microscope* (Pergamon, Oxford, 1972).
7. P.J. Goodhew, in *Practical Methods in Electron Microscopy*, Vol. 1, edited by A.M. Glauert

(North Holland-American Elsevier, Amsterdam, 1973).
8. S.Y. Hobbs, *J. Macromol. Sci. Rev. Macromol. Chem.* **C19** (1980) 221.
9. M. Furuta, *J. Polym. Sci. Polym. Phys. Edn.* **14** (1976) 479.
10. V.M. Castano, A. Alvarez-Castillo, G. Vazquez-Polo, D. Acosta and V. Gonzalez, *Microsc. Res. Tech.* **40** (1998) 41.
11. P.H. Geil, *Polymer Single Crystals* (Interscience, New York, 1963).
12. D.C. Bassett and A. Keller, *Philos. Mag.* **8** (1962) 1533.
13. D.C. Bassett, F.R. Dammont and R. Salovey, *Polymer* **5** (1964) 579.
14. D.J. Blundell and A. Keller, *J. Macromol. Sci. Phys.* **B2** (1968) 337.
15. R. Minke and J. Blackwell, *J. Macromol. Sci. Phys.* **B18** (1980) 233.
16. B.H. Chang, A. Siegmann and A. Hiltner, *J. Polym. Sci. Polym. Phys. Edn.* **22** (1984) 255.
17. M. Tsuji, S. Isoda, M. Ohara, A. Kawaguchi and K.I. Katayama, *Polymer* **23** (1982) 1568.
18. S. Isoda, M. Tsuji, M. Ohara, A. Kawaguchi and K.I. Katayama, *Polymer* **24** (1983) 1155.
19. S. Isoda, M. Tsuji, M. Ohara, A. Kawaguchi and K.I. Katayama, *Makromol. Chem. Rapid Commun.* **4** (1983) 141.
20. G.J. Vancso and H. Schönherr, in *ACS Symposium Series: Microstructure and Microtribology of Polymer Surfaces*, edited by W. Tsukruk and K.J. Wahl (American Chemical Society, New York, 1999), p. 317. Springer, Berlin.
21. G.J. Vancso, H. Hillborg and H. Schonherr, *Chemical Composition of Polymer Surfaces Imaged by Atomic Force Microscopy and Complementary Approaches*, 182 Ed. (Springer, Berlin, 2005).
22. S. Magonov and N.A. Yerina, in *Microscopy for Nanotechnology*, edited by N. Yao and Z.L. Wang (Kluwer Academic Press, New York, 2005), p. 113.
23. C. Kubel, D.P. Lawrence and D.C. Martin, *Macromolecules* **34** (2001) 9053.
24. C. Kubel and D.C. Martin, *Philos. Mag. A* **81** (2001) 1651.
25. C.J. Buchko, P.M. Wilson, Z. Xu, J. Zhang, J.S. Moore and D.C. Martin, *Polymer* **36** (1995) 1817.
26. F.F. Morehead, *Text. Res. J.* **8** (1950) 549.
27. J.W.S. Hearle and S.C. Simmens, *Polymer* **14** (1973) 273.
28. M.G. Dobb, D.J. Johnson and B.P. Saville, *J. Polym. Sci. Polym. Symp.* **58** (1977) 237.
29. F.P. Reding and E.R. Walter, *J. Polym. Sci.* **38** (1959) 141.
30. E.J. Roche, R.S. Stein and E.L. Thomas, *J. Polym. Sci. Polym. Phys. Edn.* **18** (1980) 1145.
31. R.G. Vadimsky, in *Methods of Experimental Physics*, edited by R.A. Fava (Academic Press, New York, 1980), p. 185.
32. P.H. Geil, *J. Polym. Sci.* **A2** (1964) 3813.
33. H. Kiho, A. Peterlin and P. Geil, *J. Polym. Sci.* **B3** (1965) 157.
34. D.C. Martin and E.L. Thomas, *Polymer* **36** (1995) 1743.
35. D.T. Grubb and A. Keller, *J. Mater. Sci.* **7** (1972) 822.
36. D.T. Grubb, A. Keller and G.W. Groves, *J. Mater. Sci.* **7** (1972) 131.
37. H.-T. Jung, S.D. Hudson and R.W. Lenz, *Macromolecules* **31** (1998) 637.
38. C.-m. Chu and G.L. Wilkes, *J. Macromol. Sci. Phys.* **B10** (1974) 231.
39. B.F. Howell and D.H. Reneker, *Mat. Res. Soc. Symp. Proc.* **115** (1988) 155.
40. L.F. Drummy, J. Yang and D.C. Martin, *Ultramicroscopy* **99** (2004) 247.
41. E.G. Rightor, A.P. Hitchcock, H. Ade, R.D. Leapman, S.G. Urquhart, A.P. Smith, G. Mitchell, D. Fischer, H.J. Shin and T. Warwick, *J. Phys. Chem. B* **101** (1997) 1950.
42. A.M. Donald and A.H. Windle, *Colloid Polym. Sci.* **261** (1983) 793.
43. R.J. Spontak and A.H. Windle, *J. Polym. Sci. B Polym. Phys.* **30** (1992) 61.
44. J. Petermann and R.M. Gohil, *J. Mater. Sci.* **14** (1979) 2260.
45. E.L. Thomas, in *Structure of Crystalline Polymers*, edited by I.H. Hall (Elsevier-Applied Science, London, 1984), p. 79.
46. J. Petermann and H. Gleiter, *Philos. Mag.* **28** (1973) 1279.
47. M.J. Miles and J. Petermann, *J. Macromol. Sci. Phys.* **B16** (1979) 1.
48. K. El Kirat, I. Burton, V. Dupres and Y.F. Dufrene, *J. Microsc.* **218** (2005) 199.
49. Q. Zhong, D. Inniss, K. Kjoller and V.B. Elings, *Surf. Sci.* **290** (1993) 688.
50. S.N. Magonov, in *Encyclopedia of Analytical Chemistry*, edited by R.A. Meyers (John Wiley & Sons Ltd, Chichester, 2000), p. 7432.
51. G.K. Bar and G.F. Meyers, *MRS Bull.* **29** (2004) 464.
52. S.N. Magonov and N.A. Yerina, *Langmuir* **19** (2003) 500.
53. V. Freger, J. Gilron and S. Belfer, *J. Membrane Sci.* **209** (2002) 283.

54. V.V. Tsukruk and D.H. Reneker, *Polymer* **36** (1995) 1791.
55. K.D. Jandt, *Mater. Sci. Eng. R* **21** (1998) 221.
56. S. Magonov, in *Applied Scanning Probe Methods*, edited by P. Avouris, D. Klitzing, H. Sakaki and R. Wiesndanger (Springer, Berlin, Heidelberg, 2004), p. 207.
57. G.J. Leggett, M.C. Davies, D.E. Jackson, C.J. Roberts and S.J.B. Tendler, *Trends Polym. Sci.* **1** (1993) 115.
58. T.R. Albrecht, M.M. Dovek, C.A. Lang, P. Grutter, C.F. Quate, S.W.J. Kuan, C.W. Frank and R.F.W. Pease, *J. Appl. Phys.* **64** (1988) 1178.
59. I.H. Musselman, P.E. Russell, R.T. Chen, M.G. Jamieson and L.C. Sawyer. *Proc. XIIth International Congress for Electron Microscopy*, Seattle, edited by W. Bailey (San Francisco Press, 1990), p. 866.
60. L.C. Sawyer, R.T. Chen, M.G. Jamieson, I.H. Musselman and P.E. Russell, *J. Mater. Sci.* **28** (1993) 225.
61. I.H. Musselman and P.E. Russell, *Microbeam Anal.* **26** (1991) 377.
62. T.G. Stange, R. Mathews, D.F. Evans and W.A. Hendrickson, *Langmuir* **8** (1992) 920.
63. M. Suzuki, T. Maruno, F. Yamamoto and K. Nagai, *J. Vac. Sci. Technol.* **A8** (1990) 631.
64. D.H. Reneker, J. Schneir, B. Howell and H. Harary, *Polym. Commun.* **31** (1990) 167.
65. R. Piner, R. Reifenberger, C. Martin, E.L. Thomas and R.P. Aparian, *J. Polym. Sci. C Polym. Lett.* **28** (1990) 399.
66. S.N. Magonov, K. Qvamstrom, V. Elings and H.J. Cantow, *Polym. Bull.* **25** (1991) 689.
67. A.K. Fritsche, A.R. Arevalo, A.F. Connolly, M.D. Moore, V. Elings and C.M. Wu, *J. Appl. Poly. Sci.* **45** (1992) 1945.
68. P.M. Baldwin, R.A. Frazier, J. Adler, T.O. Glasbey, M.P. Keane, C.J. Roberts, S.J.B. Tendler, M.C. Daview and C.D. Melia, *J. Microsc.* **184** (1996) 75.
69. D.L. Schmidt, B.M. DeKoven, C.E. Coburn, G.E. Potter, G.F. Meyers and D.A. Fischer, *Langmuir* **12** (1996) 518.
70. R. Festag, S.D. Alexandratos, K.D. Cook, D.C. Joy, B. Annis and B. Wunderlich, *Macromolecules* **30** (1997) 6238.
71. G. Bar, Y. Thomann and M.-H. Whangbo, *Langmuir* **14** (1998) 1219.
72. A.K. Winkel, J.K. Hobbs and M.J. Miles. *Polymer: International Symposium on Semicrystalline Polymers in Memory of Professor Andrew Keller at the ACS National Meeting, 23–26 Aug. 1999*, (Elsevier BV, New Orleans, LA, 2000), p. 8791.
73. E. Balnois and K.J. Wilkinson, *Colloids Surf. A* **207** (2002) 229.
74. G. Goldbeck-Wood, V.N. Bliznyuk, V. Burlakov, H.E. Assender, G.A.D. Briggs, Y. Tsukahara, K.L. Anderson and A.H. Windle, *Macromolecules* **35** (2002) 5283.
75. A. Kiriy, G. Gorodyska, S. Minko, C. Tsitsilianis, W. Jaeger and M. Stamm, *J. Am. Chem. Soc.* **125** (2003) 11202.
76. O.L. Shaffer, personal communication.
77. B.M. Budhlall, O.L. Shaffer, E.D. Sudol, V.L. Dimonie and M.S. El-Aasser, *Langmuir* **19** (2003) 9968.
78. M.A. Sieminski and L.C. Sawyer, unpublished data.
79. A.S. Holik, R.P. Kambour, D.G. Fink and S.Y. Hobbs, *Microstructural Science* (Elsevier-North Holland, Amsterdam, 1979).
80. B. Haworth, C.S. Hindle, G.J. Sandilands and J.R. White, *Plast. Rubber Proc. Appl.* **2** (1982) 59.
81. J. Trempler, *Praktische Metallographie* **40** (2003) 481.
82. J. Trempler, *Praktische Metallographie* **38** (2001) 231.
83. J.I. Goldstein, D.E. Newbury, D.C. Joy, C.E. Lyman, P. Echlin, E. Lifshin, L.C. Sawyer and J. Michael, in *Scanning Electron Microscopy and X-ray Microanalysis,* 3rd ed. (Kluwer Academic/Plenum/Springer, New York, 2003), p. 565.
84. J.I. Goldstein, D.E. Newbury, D.C. Joy, C.E. Lyman, P. Echlin, E. Lifshin, L.C. Sawyer and J. Michael, in *Scanning Electron Microscopy and X-ray Microanalysis,* 3rd ed. (Kluwer Academic/Plenum/Springer, New York, 2003), p. 538.
85. L.E. Samuels, *Metallographic Polishing by Mechanical Methods,* 4th ed. (ASM International, Materials Park OH, 2003).
86. P.J. Bates and C.Y. Wang, *Polym. Eng. Sci.* **43** (2003) 759.
87. P.G.T. Howell and A. Boyde, *Scanning J. Scanning Microsc.* **21** (1999) 361.
88. L. Bartosiewicz and Z. Mencik, *J. Polym. Sci. Polym. Phys. Edn.* **12** (1974) 1163.
89. F.J. Guild and B. Ralph, *J. Mater. Sci.* **14** (1979) 2555.
90. J. Summerscales, D. Green and F.J. Guild, *J. Microsc.* **169** (1993) 173.

91. B. Mlekusch, E.A. Lehner and W. Geymayer, *Compos. Sci. Technol.* **59** (1999) 543.
92. B. Mlekusch, *Compos. Sci. Technol.* **59** (1999) 547.
93. J. Summerscales, Ed., *Microstructural Characterisation of Fibre-Reinforced Composites* (CRC Press, Boca Raton, 1998).
94. D.A. Hemsley, Ed. *Applied Polymer Light Microscopy* (Elsevier Applied Science, London, New York, 1989).
95. L.C. Sawyer and M. Jaffe, *J. Mater. Sci.* **21** (1986) 1897.
96. L.M. Gammon and D.J. Ray, *Microstruct. Sci.* **25** (1998) 457.
97. R.G. Scott and A.W. Fergerson, *Text. Res. J.* **26** (1956) 284.
98. R.G. Scott, *ASTM Spec. Tech. Publ. No.* **257** (1959) 121.
99. J.L. Stoves, *Fibre Microscopy* (Van Nostrand, Princeton, New Jersey, 1958).
100. H.B. Haanstra, *Philips Tech. Rev.* **17** (1955) 178.
101. P.A. Annis and T.W. Quigley, Jr., in *Microstructural Characterisation of Fibre-Reinforced Composites*, edited by J. Summerscales (CRC Press, Boca Raton, 1998), p. 17.
102. O. Rahbek, S. Kold, S. Overgaard and K. Soballe, *J. Microsc.* **218** (2005) 225.
103. L.T. Keene, G.P. Halada and C.R. Clayton, *Prog. Organic Coatings* **52** (2005) 173.
104. K.R. Porter and J. Blum, *Anat. Record* **117** (1953) 685.
105. F.S. Sjostrand, *Experientia* **9** (1953) 114.
106. L.C. Sawyer and R.S. Jones, *J. Membrane Sci.* **20** (1984) 147.
107. A.M. Glauert, *Fixation, Dehydration and Embedding of Biological Specimens*, in *Practical Methods in Electron Microscopy*, Vol. 3, edited by A.M. Glauert (North Holland-American Elsevier, Amsterdam, 1975).
108. A.M. Glauert and P.R. Lewis, *Biological Specimen Preparation for Transmission Electron Microscopy* (Princeton University Press, 1998).
109. B.L. Giammara. *Proc. 43rd Ann. EMSA* edited by W. Bailey (San Francisco Press, SF, 1985), p. 706.
110. B.L. Giammara and J. Hanker, *Stain Tech.* **61** (1986) 51.
111. B.L. Giammara, *Scanning* **14** (1992) II.
112. G.R. Logan and A.M. Dvorak, *Methods of Microwave Fixation for Microscopy* (1993).
113. L.P. Kok and M.E. Boon, *J. Microsc.* **158** (1990) 291.
114. R.C. Hugo and S.L. Cady, *Microsc. Today* **9** (2004) 28.
115. N. Reid, *Ultramicrotomy*, in *Practical Methods in Electron Microscopy*, Vol. 3, Pt. 2, edited by A.M. Glauert (North Holland-American Elsevier, Amsterdam, 1975).
116. J.E. Trancik, J.T. Czernuszka, C. Merriman and C. Viney, *J. Microsc.* **203** (2001) 235.
117. A. Taubert, J.H. Ferris and K.I. Winey, *Microsc. Today* **11** (2003) 36.
118. D. Studer and H. Gnaegi, *J. Microsc.* **197** (Pt 1) (2000) 94.
119. J.S.J. Vastenhout, H. Gnagi, *Microsc. Microanal.* **8** (2002) 324.
120. H.K. Plummer, Jr., *Microsc. Microanal.* **3** (1997) 239.
121. C.C. Chau, J. Blackson and J. Im, *Polymer* **36** (1995) 2511.
122. M.L. Ericson and H. Lindberg, *Polymer* **38** (1997) 4485.
123. K. Varlot, J.M. Martin, C. Quet and Y. Kihn, *Ultramicroscopy* **68** (1997) 123.
124. J.S. Fodor, R.M. Briber, T.P. Russell, K.R. Carter, J.L. Hedrick and R.D. Miller, *J. Polym. Sci. B Polym. Phys.* **35** (1997) 1067.
125. G.M. Brown and J.H. Butler, *Polymer* **38** (1997) 3937.
126. B.W. Loo, Jr., I.M. Sauerwald, A.P. Hitchcock and S.S. Rothman, *J. Microsc.* **204** (2001) 69.
127. H. Tang and D.C. Martin, *J. Mater. Sci.* **38** (2003) 803.
128. C.J.G. Plummer, P. Beguelin, C. Grein, R. Gensler, L. Dupuits, C. Gaillard, P. Stadelmann, H.-H. Rausch and J.-A.E. Manson, *Macromol. Symp.* **213** (2004) 97.
129. B.P. Kirkmeyer, R.C. Puetter, A. Yahil, K.I. Winey, *J. Polym. Sci. B Polym. Phys.* **41** (2003) 319.
130. R. Vastenhout, personal communication, 2006.
131. K. Zierold, *Ultramicroscopy* **14** (1984) 201.
132. P. Echlin, *Low Temperature Microscopy and Analysis* (Plenum Press, New York, 1992).
133. J. Li, W. Liang, G.F. Meyers and W.A. Heeschen, *Polym. News* **29** (2004) 335.
134. M.M. Chappius and L.S. Robblee, *Rubber World* **136** (1957) 391.
135. E.H. Andrews and J.M. Stubbs, *J. R. Microsc. Soc.* **82** (1964) 221.
136. E.H. Andrews, M.W. Bennett and A. Markham, *J. Polym. Sci.* **A2** (1967) 1235.
137. J. Dlugosz and A. Keller, *J. Appl. Phys.* **39** (1968) 5776.

138. A.J. Cobbold and A.E. Mendelson, *Sci. Tools* **18** (1971) 1.
139. J. Dlugosz, M.J. Folkes and A. Keller, *J. Polym. Sci. Polym. Phys. Edn.* **11** (1973) 929.
140. J.A. Odell, J. Dlugosz and A. Keller, *J. Polym. Sci. Polym. Phys. Edn.* **14** (1976) 861.
141. A. Ghijels, N. Groesbeek and C.W. Yip, *Polymer* **23** (1982) 1913.
142. I.D. Fridman, E.L. Thomas, L.J. Lee and C.W. Macosko, *Polymer* **21** (1980) 393.
143. C.H.Y. Chen, R.M. Briber, E.L. Thomas, M. Xu and W.J. MacKnight, *Polymer* **24** (1983) 1333.
144. S. Simmons and E.L. Thomas, *Polymer* **39** (1998) 5587.
145. A.B. Morgan and J.D. Harris, *Polymer* **44** (2003) 2313.
146. D.D. Huang, B.A. Wood and E.A. Flexman, *Adv. Mater.* **10** (1998) 1207.
147. K. Friedrich, E. Ueda, H. Kamo, M. Evstatiev, B. Krasteva and S. Fakirov, *J. Mater. Sci.* **37** (2002) 4299.
148. C.C. Honeker and E.L. Thomas, *Macromolecules* **33** (2000) 9407.
149. J. Pyun, K. Matyjaszewski, J. Wu, G.-M. Kim, S.B. Chun and P.T. Mather, *Polymer* **44** (2003) 2739.
150. F. Lednicky, J. Hromadkova and Z. Pientka, *Polymer* **42** (2001) 4329.
151. F. Lednicky, Z. Pientka and J. Hromadkova, *J. Macromol. Sci. Phys.* **42B** (2003) 1039.
152. N. Subramaniam, J. Simpson, M.J. Monteiro, O. Shaffer, C.M. Fellows and R.G. Gilbert, *Microsc. Res. Tech.* **63** (2004) 111.
153. Y. Thomann, R. Thomann, G. Bar, M. Ganter, B. Machutta and R. Mulhaupt, *J. Microsc.* **195** (1999) 161.
154. B.S. Tanem, T. Kamfjord, M. Augestad, T.B. Lovgren and M. Lundquist, *Polymer* **44** (2003) 4283.
155. L. Li, C.-M. Chan, J.-X. Li, K.-M. Ng, K.-L. Yeung and L.-T. Weng, *Macromolecules* **32** (1999) 8240.
156. A. Pfau, A. Janke and W. Heckmann, *Surf. Interface Anal.* **27** (1999) 410.
157. C.H. Stephens, H. Yang, M. Islam, S.P. Chum, S.J. Rowan, A. Hiltner and E. Baer, *J. Polym. Sci. B Polym. Phys.* **41** (2003) 2062.
158. E. Radovanovic, E. Carone, Jr. and M.C. Goncalves, *Polym. Testing* **23** (2004) 231.
159. A.W. Fahmi and M. Stamm, *Langmuir* **21** (2005) 1062.
160. R. Adhikari, W. Lebek, R. Godehardt, S. Henning, G.H. Michler, E. Baer and A. Hiltner, *Polym. Adv. Technol.* **16** (2005) 95.
161. R. Adhikari, R. Godehardt, W. Lebek, S. Frangov, G. H. Michler, H.-J. Radusch, F. J. Baltá Calleja, *Polym. Adv. Technol.* **16** (2005) 156.
162. S.J. Hinder, C. Lowe, J.T. Maxted and J.F. Watts, *J. Mater. Sci.* **40** (2005) 285.
163. V. Tournier-Lasserve, A. Boudet and P. Sopena, *Ultramicroscopy* **58** (1995) 123.
164. Leica, in *Troubleshooting in Microtomy* (Leica Microsystems Nussioch GmbH, 2005), p. 1.
165. S.Y. Hobbs, in *Plastics Polymer Science and Technology*, edited by M.D. Bayal (Wiley-Interscience, New York, 1982), p. 239.
166. D.T. Grubb, in *Developments Crystalline Polymers*, edited by D.C. Bassett (Applied Science, London, 1982), p. 1.
167. M.A. Hayat, *Positive Staining for Electron Microscopy* (Van Nostrand Reinhold, New York, 1975).
168. M.A. Hayat, *Principles and Techniques of Electron Microscopy-Biological Applications,* 3rd ed. (CRC Press, Boca Raton, 1989).
169. P.R. Lewis and D.P. Knight, Eds. *Staining Methods for Sectioned Material* (North Holland-American Elsevier, Amsterdam, 1977).
170. C.A. Correa and E. Hage, Jr., *Polymer* **40** (1999) 2171.
171. C.A. Correa, B.C. Bonse, C.R. Chinaglia, E. Hage Jr. and L.A. Pessan, *Polym. Testing* **23** (2004) 775.
172. E.K. Boylston, *Microsc. Microanal.* **8** (2002) 192.
173. M. Braga, C.A.P. Leite and F. Galembeck, *Langmuir* **19** (2003) 7580.
174. G. Soldani, P. Losi, C. Milioni and A. Raffi, *J. Microsc.* **206** (2002) 139.
175. E.K. Boylston and M.L. Rollins, *Microscope* **19** (1971) 255.
176. C. Maertens, G. Raes and G. Vanderrneerssche. *Proc. 1st Eur. Reg. Congr. on Electron Microscopy*, Stockholm, 1956, p. 292.
177. K. Hess and M. Mahl, *Naturwissenschaften* **41** (1954) 86.
178. K. Muhlethaler, *Z. Schweiz. Forst.* **30** (1960) 55.
179. R. Hagege, P. Kassenbeck, D. Meirnoun and A. Parisot, *Text. Res. J.* **39** (1969) 1015.
180. M.H. Walters and D.N. Keyte, *Rubber Chem. Technol.* **38** (1965) 62.
181. P.A. Marsh, A. Voet and L.D. Price, *Rubber Chem. Technol.* **39** (1966) 359.
182. W.M. Hess, C.E. Scott and J.E. Callan, *Rubber Chem. Technol.* **41** (1968) 344.

183. D.T. Grubb and A. Keller, *J. Polym. Sci. Polym. Phys. Edn.* **18** (1980) 207.
184. A.C. Reimschuessel and D.C. Prevorsek, *J. Polym. Sci. Polym. Phys. Edn.* **14** (1976) 485.
185. E.H. Andrews, *Proc. R. Soc. Lond. A* **227** (1964) 562.
186. K. Kato, *J. Electron Microsc.* **14** (1965) 220.
187. K. Kato, *J. Polym. Sci. Polym. Lett. Edn.* **4** (1966) 35.
188. K. Kato, *Polym. Eng. Sci.* **7** (1967) 38.
189. G.E. Molair and H. Keskkula, *J. Polym. Sci.* **A14** (1966) 1595.
190. M. Matsuo, J. Sagae and H. Asai, *Polymer* **10** (1967) 79.
191. N.C. Watkins and D. Hansen, *Text. Res. J.* **38** (1968) 388.
192. D. Stefan and H.L. Williams, *J. Appl. Polym. Sci.* **18** (1974) 1451.
193. M. Niimoni, T. Katsuta and T. Kotani, *J. Appl. Polym. Sci.* **19** (1975) 2919.
194. I.D. Fridman and E.L. Thomas, *Polymer* **21** (1980) 388.
195. C.K. Riew and R.W. Smith, *J. Polym. Sci.* (1971) 2739.
196. G. Weber, D. Kuntze and W. Stix, *Colloid Polym. Sci.* **260** (1982) 956.
197. G. Kanig, *Proc. Colloid Polym. Sci.* **57** (1975) 176.
198. D.M. Huong, M. Drechsler, H.J. Cantow and M. Moller, *Macromolecules* **26** (1993) 864.
199. G.A. Hutchins, *Proc. 51st Ann. MSA* (1993) 900.
200. D. Vesely, *Polymer Engineering and Science, Joint Meeting of POLYBLENDS '95 and RETEC on Polymer Alloys and Blends, 19–20 Oct. 1995* **36** (1996) 1586.
201. G. Kim and M. Libera, *Macromolecules* **31** (1998) 2569.
202. H. Marfels and P. Kassenbeck, *Chemie. Text. Ind.* **27/79** (1977) 788.
203. M. Jamieson, unpublished.
204. T. Sarada, L.C. Sawyer and M. Ostler, *J. Membrane Sci.* **15** (1983) 97.
205. H. Keskkula and P.A. Traylor, *Polymer* **19** (1978) 465.
206. C.V. Berney, R.E. Cohen and F.S. Bates, *Polymer* **23** (1982) 1222.
207. P. Laokijcharoen and A.Y. Coran, *Rubber Chem. Technol.* **71** (1998) 966.
208. G.F. Meyers, B.M. DeKoven, M.T. Dineen, A. Strandjord, P.J. O'Connor, T. Hu, Y.-H. Chiao, H. Pollock and A. Hammiche, *Polymer Preprints (American Chemical Society, Division of Polymer Chemistry)* **39** (1998) 1222.
209. R. Vitali and E. Montani, *Polymer* **21** (1980) 1220.
210. U.A. Spitzer and D.G. Lee, *J. Org. Chem.* **39** (1974) 2468.
211. J.S. Trent, J.I. Scheinbeim and P.R. Couchman, *J. Polym. Sci. Polym. Lett. Edn.* **19** (1981) 315.
212. J.S. Trent, P.R. Couchman and J.I. Scheinbeim, *Polym. Sci. Technol.* **22** (1983) 205.
213. J.S. Trent, J.I. Scheinbeim and P.R. Couchman, *Macromolecules* **16** (1983) 589.
214. D.E. Morel and D.T. Grubb, *Polym. Commun.* **25** (1984) 68.
215. D.E. Morel and D.T. Grubb, *Polymer* **25** (1984) 417.
216. W. Heckmann, G.E. McKee and F. Ramsteiner, in *Mechanical Properties of Polymers Based on Nanostructure and Morphology*, edited by G.H. Michler and F.J. Balta-Calleja (CRC Press, Boca Raton, 2005), p. 435.
217. P.E. Frochling and A.J. Pijpers, *J. Polym. Sci. B Polym. Phys.* **25** (1987) 947.
218. B. Ohlsson and B. Tornell, *J. Appl. Polym. Sci.* **41** (1990) 1189.
219. Y. Tervorrt-Engelen and J.v. Gisbergen, *Polym. Commum.* **32** (1991) 261.
220. S.Y. Hobbs, M.E.J. Dekkers and V.H. Watkins, *J. Mater. Sci.* **23** (1988) 1219.
221. M.E.J. Dekkers, S.Y. Hobbs and V.H. Watkins, *J. Mater. Sci.* **23** (1988) 1225.
222. J.S. Trent, *Macromolecules* **17** (1984) 2930.
223. O.L. Shaffer, M.S. El-Aasser and J.W. Vanderhoff. *Proc. 45th EMSA*, edited by W. Bailey (San Francisco Press, 1987), p. 502.
224. I. Segal, O.L. Shaffer, V.L. Dimonie and M.S. ElAasser. *Proc. 51st MSA*, edited by W. Bailey (San Francisco Press, 1993), p. 882.
225. O.L. Shaffer, unpublished micrographs.
226. L.C. Lopez, R.C. Cieslinski, C.L. Putzig and R.D. Wesson, *Polymer* **36** (1995) 2331.
227. J.X. Li, J.N. Ness and W.L. Cheung, *J. Appl. Polym. Sci.* **59** (1996) 1733.
228. Y.-L. Loo, R.A. Register and D.H. Adamson, *J. Polym. Sci. B Polym. Phys.* **38** (2000) 2564.
229. S. Hong, A.A. Bushelman, W.J. MacKnight, S.P. Gido, D.J. Lohse and L.J. Fetters, *Polymer* **42** (2001) 5909.
230. L.F. Drummy, I. Voigt-Martin and D.C. Martin, *Macromolecules* **34** (2001) 7416.
231. J. Harasawa, H. Uehara, T. Yamanobe, T. Komoto and M. Terano, *J. Mol. Struct.* **610** (2002) 133.
232. J.I. Harasawa, H. Uehara, T. Yamanobe and T. Komoto, *J. Electron Microsc.* **51** (2002) 157.

233. N. Manabe, Y. Yokota, H. Minami, Y. Uegomori and T. Komoto, *J. Electron Microsc.* **51** (2002) 11.
234. H.G. Haubruge, A.M. Jonas and R. Legras, *Polymer* **44** (2003) 3229.
235. S. Megelski, J.S. Stephens, J.F. Rabolt and D. Bruce Chase, *Macromolecules* **35** (2002) 8456.
236. T.M. Chou, P. Prayoonthong, A. Aitouchen and M. Libera, *Polymer* **43** (2002) 2085.
237. T.-M. Chou, M. Libera and M. Gauthier, *Polymer* **44** (2003) 3037.
238. M.A. DeBolt and R.E. Robertson, *Polym. Eng. Sci.* **46** (2006) 385.
239. P.J. Beynon, P.M. Collins, D. Gardiner and W.G. Overend, *Carbohydr. Res.* **6** (1968) 431.
240. D. Montezinos, B.G. Wells and J.L. Burns, *J. Polym. Sci. Polym. Phys. Edn.* **23** (1985) 421.
241. S. Wolfe, S.K. Hasan and J.R. Campbell, *Chem. Commun.* (1970) 1420.
242. M. Slouf, J. Kolarik and L. Fambri, *J. Appl. Polym. Sci.* **91** (2004) 253.
243. C. Gaillard, P.A. Stadelmann, C.J.G. Plummer and G. Fuchs, *Scanning* **26** (2004) 122.
244. B.A. Wood. *Proc. 51st MSA*, edited by W. Bailey (San Francisco Press, 1993), p. 898.
245. B.A. Wood, in *Advances in Polymer Blends and Alloys Technology*, Vol. 3, edited by K. Finlayson (Technomic Publishing, Lancaster, 1992), p. 24.
246. G. Kanig, *Kolloid Z. Z. Polym.* **251** (1973) 782.
247. G. Kanig, *Kunststoffe* **64** (1974) 470.
248. D.T. Grubb, J. Dlugosz and A. Keller, *J. Mater. Sci.* **10** (1975) 1826.
249. J. Dlugosz, G.V. Fraser, D.T. Grubb, A. Keller, J.A. Odell and P.L. Goggen, *Polymer* **17** (1976) 471.
250. A.M. Hodge and D.C. Bassett, *J. Mater. Sci.* **12** (1977) 2065.
251. I.G. Voigt-Martin, E.W. Fischer and L. Mandelkern, *J. Polym. Sci. Polym. Phys. Edn.* **18** (1980) 2347.
252. G. Strobl, H. Schneider and I.G. Voigt-Martin, *J. Polym. Sci. Polym. Phys. Edn.* **18** (1980) 1361.
253. G. Kanig, *J. Crystal Growth* **48** (1980) 303.
254. G. Kanig, *Colloid Polym. Sci.* **261** (1983) 373.
255. D.T. Grubb, in *Developments Electron Microscopy and Analysis*, edited by D.L. Missell (Institute of Physics, Bristol, 1977), p. 399.
256. J.A. Odell, D.T. Grubb and A. Keller, *Polymer* **19** (1978) 617.
257. J. Smook, W. Hamersma and A.J. Pennings, *J. Mater. Sci.* **19** (1984) 1359.
258. A. Schaper, D. Zenke, E. Schulz, R. Hirte and M. Taege, *Phys. Status Solidi* **116** (1989) 179.
259. I.M. Roberts, *J. Microsc.* **207** (2002) 97.
260. P.H. Vallotton, Denn, M. M., Wood, B. A. and Salmeron, M. B., *J. Biomater. Sci. Polym. Edn.* **6** (1994) 609.
261. S.M. Kurtz, *The UHMWPE Handbook: Principles and Clinical Applications in Total Joint Replacement* (Academic Press, New York, 2004).
262. M.A. Hayat, *Positive Staining for Electron Microsopy* (Van Nostrand Reinhold, New York, 1975), p. 47.
263. K. Hess, E. Gutter and H. Mahl, *Naturwissenschaften* **46** (1959) 70.
264. K. Hess, E. Gutter and H. Mahl, *Kolloid Z.* **168** (1960) 37.
265. B.J. Spit. *Proc. Fifth Int. Congr. for Electron Microscopy* Philadelphia, 1962, edited by S.S. Breese Jr. (Academic Press, NewYork, 1962) p. BB7.
266. B.J. Spit, *Faserforsch. Textiltech.* **18** (1967) 161.
267. E. Belavtseva, *Vysokomol. Soed.* **5** (1963) 1847.
268. E. Belavtseva and K. Gumargalieva, *Zadodsk Lab.* **29** (1966) 966.
269. A. Peterlin, P. Ingram and H. Kiho, *Makromol. Chern.* **86** (1965) 294.
270. C.W. Hock, *J. Polym. Sci.* **A2** (1967) 471.
271. J. Martinez-Salazar and C.G. Cannon, *J. Mater. Sci. Lett.* **3** (1984) 693.
272. C.W. Pease, *J. Ultrastruct. Res.* **15** (1966) 555.
273. O.L. Shaffer, M.S. El-Aasser and J.W. Vanderhoff. *Proc. 41st Annu. Mtg. EMSA*, edited by W. Bailey (San Francisco Press, Phoenix, 1983), p. 30.
274. O.L. Shaffer, personal communication.
275. J.R. Harris, C. Roos, R. Djalali, O. Rheingans, M. Maskos and M. Schmidt, *Micron* **30** (1999) 289.
276. O. Okada, Keskkula, H., Paul, D.R., *J. Polym. Sci. B Polym. Phys.* **42** (2004) 1739.
277. R.W. Smith and J.C. Andries, *Rubber Chem. Technol.* **47** (1974) 64.
278. R.E. Cohen and A.R. Ramos, *Office of Naval Research, Task No. NR 356–646, Technical Report No. 2, July 6*, in Vol. 4 (1978).
279. G. Gillberg, L.C. Sawyer and A.L. Promislow, *J. Appl. Polym. Sci.* **28** (1983) 3723.
280. M. Sotton, *C. R. Acad. Sci. Paris* **270B** (1970) 1261.
281. R. Hagege, M. Jarrin and M. Sotton, *J. Microsc.* **115** (1979) 65.

282. M.G. Dobb, C.R. Park and R.M. Robson, *J. Mater. Sci.* **27** (1992) 3876.
283. M.G. Dobb, D.J. Johnson, A. Majeld and B.P. Saville, *Polymer* **20** (1979) 1289.
284. M. Sotton and A.M. Vialard, *Text. Res. J.* **41** (1971) 834.
285. S.Y. Hobbs, V.H. Watkins and R.R. Russell, *J. Polym. Sci. Polym. Phys. Edn.* **18** (1980) 393.
286. S.Y. Hobbs, personal communication (1985).
287. K. Hess and H. Kiessig, *Naturwissenschaften* **31** (1943) 171.
288. K. Hess and H. Kiessig, *Z. Phys. Chern.* **A193** (1944) 196.
289. K. Hess, H. Mahl and E. Gutter, *Kolloid Z.* **155** (1957) 1.
290. J. Gacen, J. Maillo and J. Bordas, *Bull. Sci.* **ITF6** (1977) 167.
291. G. Gillberg, A. Kravas and J. Langley, *J. Microsc.* **138** (1985) RP1.
292. J.-S. Lin, *Eur. Polym. J.* **38** (2002) 79.
293. M. Brittin, G.R. Mitchell and A.S. Vaughan, *J. Mater. Sci.* **36** (2001) 4911.
294. M.M. Winram, D.T. Grubb and A. Keller, *J. Mater. Sci.* **13** (1978) 791.
295. A. Garton, P.Z. Sturgeon, D.J. Carlsson and D.M. Wills, *J. Mater. Sci.* **13** (1978) 2205.
296. V. Peck and W. Kaye, *Text. Res. J.* **4** (1954) 295.
297. G.W. Bailey, *J. Polym. Sci.* **62** (1962) 241.
298. L.S. Li and V.A. Kargin, *Vysokomol. Soed.* **3** (1961) 1102.
299. G. Flodberg, M.S. Hedenqvist and U. Gedde, *Polym. Eng. Sci.* **43** (2003) 1044.
300. H. Keskkula and P.A. Traylor, *J. Appl. Polym. Sci.* **11** (1967) 2361.
301. R.J. Williams and R.W.A. Hudson, *Polymer* **8** (1967) 643.
302. G. Farrow, D.A.S. Ravens and I.M. Ward, *Polymer* **3** (1962) 17.
303. P. Tucker and R. Murray. *Proc. 33rd Annu. Mtg EMSA*, edited by W. Bailey (San Francisco Press, 1975), p. 82.
304. J.O. Warwicker, *J. Appl. Polym. Sci.* **22** (1978) 869.
305. G.E. Sweet and J.P. Bell, *J. Polym. Sci. Polym Phys. Edn.* **16** (1978) 1935.
306. H.S. Bu, S.Z.D. Cheng and B. Wunderlich, *Polymer* **29** (1988) 1603.
307. R.d.J. Santos, J.C. Bruno, M.T.M.B. Silva and R.C.R. Nunes, *Polym. Testing* **12** (1993) 393.
308. M.E.J. Dekkers, S.Y. Hobbs and V.H. Watkins, *Polymer* **32** (1991) 2150.
309. S.H. Zeronian and M.J. Collins, *Textile Inst.* **20** (1989) 1.
310. M.J. Collins, S.H. Zeronian and M. Semmelmeyer, *J. Appl. Poly. Sci.* **42** (1991) 2149.
311. M.C.V. Amorim and C.M.F. Oliveira, *Polym. Testing* **15** (1996) 517.
312. K. Wallheinke, W. Heckmann, P. Poetschke and H. Stutz, *Polym. Testing* **17** (1998) 247.
313. M.M. Shahin and R.H. Olley, *J. Polym. Sci. B Polym. Phys.* **40** (2002) 124.
314. Y.S. Hu, D.A. Schiraldi, A. Hiltner and E. Baer, *Macromolecules* **36** (2003) 3606.
315. R.P. Palmer and A.J. Cobbold, *Makromol. Chem.* **74** (1964) 174.
316. C.W. Hock, *Polym. Lett.* **3** (1965) 573.
317. C.W. Hock, *J. Polym. Sci.* **4** (1966) 227.
318. N. Kusumoto and Y. Haga, *Rep. Prog. Polym. Phys. Japan* **XV** (1972) 583.
319. V.J. Armond and J.R. Atkinson, *J. Mater. Sci.* **4** (1969) 509.
320. C.B. Bucknall, I.C. Drinkwater and W.E. Keast, *Polymer* **13** (1972) 115.
321. C.B. Bucknall and I.C. Drinkwater, *Polymer* **15** (1974) 254.
322. D. Briggs, D.M. Brewis and M.B. Kovieczo, *J. Mater. Sci.* **11** (1976) 1270.
323. R.A. Bubeck and H.M. Baker, *Polymer* **23** (1982) 1680.
324. K. Selby and M.O.W. Richardson, *J. Mater. Sci.* **11** (1976) 786.
325. D.J. Boll, R.M. Jensen, L. Cordner and W.D. Bascom, *J. Compos. Mater.* **24** (1990) 208.
326. E. Sheng, I. Sutherland, D.M. Brewis and R.J. Heath, *J. Adhes. Sci. Technol.* **9** (1995) 47.
327. S.A. Edwards, Choudhury, N. Roy, Provatas, M., *J. Appl. Polym. Sci.* **87** (2003) 774.
328. R.H. Olley, A.M. Hodge and D.C. Bassett, *J. Polym. Sci. Polym. Phys. Ed.* **17** (1979) 627.
329. R.H. Olley and D.C. Bassett, *Polymer* **23** (1982) 1707.
330. D.C. Bassett, *Principles of Polymer Morphology* (Cambridge University Press, Cambridge, 1981).
331. D.C. Bassett, in *Developments Crystalline Polymers*, edited by D.C. Bassett (Elsevier Applied Science, London, 1988), p. 67.
332. K.L. Naylor and P.J. Phillips, *J. Polym. Sci. Polym. Phys. Edn.* **21** (1983) 2011.
333. P.J. Phillips and R.J. Philpot, *Polym. Commun.* **27** (1986) 307.
334. D.C. Bassett and R.H. Olley, *Polymer* **25** (1984) 935.
335. Z. Bashir, M.J. Hill and A. Keller, *J. Mater. Sci. Lett.* **5** (1986) 876.

336. R.H. Olley, D.C. Bassett and D.J. Blundell, *Polymer* **27** (1986) 344.
337. T.J. Lemmon, S. Hanna and A.H. Windle, *Polym. Commun.* **30** (1989) 2.
338. J.R. Ford, D.C. Bassett, G.R. Mitchell and T.G. Ryan, *Mol. Cryst. Liq. Cryst.* **180B** (1990) 233.
339. S.E. Bedford and A.H. Windle, *Polymer* **31** (1990) 616.
340. S. Hanna, T.J. Lemmon, R.J. Spontak and A.H. Windle, *Polymer* **33** (1992) 3.
341. A. Anwer and A.H. Windle, *Polymer* **34** (1993) 3347.
342. S.D. Hudson and A.J. Lovinger, *Polymer* **34** (1993) 1123.
343. S.J. Sutton and A.S. Vaughan, *J. Mater. Sci.* **28** (1993) 4962.
344. S.J. Sutton and A.S. Vaughan, *J. Polym. Sci. B Polym. Phys.* **34** (1996) 837.
345. M. Patrick, V. Bennett and M.J. Hill, *Polymer* **37** (1996) 5335.
346. I.L. Hosier, A.S. Vaughan and S.G. Swingler, *J. Mater. Sci.* **32** (1997) 4523.
347. H.M. White and D.C. Bassett, *Polymer* **39** (1998) 3211.
348. M.S. Bischel, J.M. Schultz and K.M. Kit, *Polymer* **39** (1998) 2123.
349. R.L. Morgan, M.J. Hill and P.J. Barham, *Polymer* **40** (1999) 337.
350. H.-J. Sue, J.D. Earls, R.E.J. Hefner, M.I. Villarreal, E.I. Garcia-Meitin, P.C. Yang, C.M. Cheatham and C.J.G. Plummer, *Polymer* **39** (1998) 4707.
351. M.M. Shahin, Olley, R. H., Blissett, M. J., *J. Polym. Sci. B Polym. Phys.* **37** (1999) 2279.
352. J. Teckoe, R.H. Olley, D.C. Bassett, P.J. Hine and I.M. Ward, *J. Mater. Sci.* **34** (1999) 2065.
353. N.A. Jones and A.J. Lesser, *J. Polym. Sci. B Polym. Phys.* **37** (1999) 3246.
354. M.I. Abo El-Maaty, R.H. Olley and D.C. Bassett, *J. Mater. Sci.* **34** (1999) 1975.
355. I.L. Hosier and D.C. Bassett, *Polymer* **41** (2000) 8801.
356. T. Amornsakchai, D.C. Bassett, R.H. Olley, A.P. Unwin and I.M. Ward, *Polymer* **42** (2001) 4117.
357. R.H. Olley and D.C. Bassett, *Polymer* **30** (1989) 399.
358. J. Park, K. Eom, O. Kwon and S. Woo, *Microsc. Microanal.* **7** (2001) 276.
359. M.M. Shahin, *J. Appl. Polym. Sci.* **90** (2003) 360.
360. J. Weng, R.H. Olley, D.C. Bassett and P. Jaaskelainen, *J. Polym. Sci. B Polym. Phys.* **41** (2003) 2342.
361. J. Zhou and K. Komvopoulos, *Trans. ASME J. Tribol.* **127** (2005) 273.
362. H. Schoenherr, D. Snetivy and G.J. Vancso, *Polym. Bull. (Berlin)* **30** (1993) 567.
363. D. Trifonova-Van Haeringen, J. Varga, C.W. Ehrenstein and G.J. Vancso, *J. Polym. Sci. B Polym. Phys.* **38** (2000) 672.
364. J.J. Janimak, L. Markey and G.C. Stevens, *Polymer* **42** (2001) 4675.
365. L. Markey, J.J. Janimak and G.C. Stevens, *Polymer* **42** (2001) 6221.
366. I.L. Hosier, R.G. Alamo and J.S. Lin, *Polymer* **45** (2004) 3441.
367. D.J. Barber, *J. Mater. Sci.* **5** (1970) 1.
368. H.Z. Fetz, *Physik* **119** (1942) 590.
369. G.J. Wehner, *J. Appl. Phys.* **25** (1954) 270.
370. G.J. Wehner, *J. Appl. Phys.* **30** (1959) 1762.
371. E. Jakopic. *Proc. Eur. Reg. Conf. on Electron Microscopy*, 1960, edited by A.L. Houwink and B.J. Spit (Nederlandse Vereniging Noor Eletronenmik-roskopie, Delf, 1960) p. 559.
372. J. D. E. Harrison, N.S. Levy, J.P.J. III and H.M. Effron, *J. Appl. Phys.* **39** (1968) 3742.
373. M.J. Nobes, J.S. Collingon and G. Carter, *J. Mater. Sci.* **4** (1969) 730.
374. G. Carter, J.S. Collingon and M.J. Nobes, *J. Mater. Sci.* **6** (1971) 115.
375. R.S. Dhariwal and R.K. Fitch, *J. Mater. Sci.* **12** (1977) 1225.
376. A.D.G. Stewart and M.W. Thompson, *J. Mater. Sci.* **4** (1969) 56.
377. I.H. Wilson and M.W. Kidd, *J. Mater. Sci.* **6** (1971) 1362.
378. J.P. Ducommun, M. Cantagrel and M. Moulin, *J. Mater. Sci.* **10** (1975) 52.
379. I.S.T. Tsong and D.J. Barber, *J. Mater. Sci.* **7** (1977) 687.
380. F.R. Barnet and M.K. Norr, *Carbon* **11** (1973) 281.
381. B.J. Spit. *Proc. Eur. Reg. Conf. on Electron Microscopy*, 1960, edited by A.L. Houwink and B.J. Spit (Nederlandse Vereniging Noor Eletronenmik-roskopie, Delf, 1960) p. 564.
382. B.J. Spit, *Polymer* **4** (1962) 109.
383. J. Dlugosz, *Proc. Fifth Int. Congr. for Electron Microscopy*, Philadelphia, 1962, edited by S.S. Breese Jr. (Academic Press, NewYork, 1962) p. BB11.
384. L. Moscou, *Proc. Fifth Int. Congr. for Electron Microscopy*, Philadelphia, 1962, edited by S.S. Breese Jr. (Academic Press, NewYork, 1962) p. BB5.
385. A. Keller, *Proc. Fifth Int. Congr. for Electron Microscopy*, Philadelphia, 1962, edited by S.S.

Breese Jr. (Academic Press, NewYork, 1962) p. BB3.
386. J.E. Breedon, J.F. Jackson and M.J. Marcinkowski, *J. Mater. Sci.* **8** (1973) 1071.
387. G. Carter, A.E. Hill and M.J. Nobes, *Vacuum* **29** (1979) 213.
388. M.R. Padhye, N.V. Bhat and P.K. Mittal, *Text. Res. J.* **46** (1976) 502.
389. P.J. Goodhew, *J. Phys. E Sci. Instrum.* **4** (1971) 392.
390. P.J. Goodhew, *Nature* **235** (1972) 437.
391. P.J. Goodhew. *Proc. Fifth Eur. Congr. on Electron Microscopy*, 1972, p. 300.
392. P.J. Goodhew, *J. Mater. Sci.* **8** (1973) 581.
393. S.B. Warner, D.R. Uhlmann and L.H. Peebles, *J. Mater. Sci.* **10** (1975) 758.
394. P.R. Blakey and M.O. Alfy, *J. Text. Inst.* (1978) 38.
395. J. Friedrich, J. Gahde and M. Pohl, *Acta Polym.* **31** (1981) 310.
396. J. Friedrich, J. Gahde and M. Pohl, *Acta Polym.* **33** (1982) 209.
397. M. Kojima and H. Satake, *J. Polym. Sci. Polym. Phys. Edn.* **20** (1982) 2153.
398. D.W. Woods and I.M. Ward, *Surf. Interface Anal.* **20** (1993) 385.
399. R.W. Linton, M.E. Farmer, P. Ingram, J.R. Somner and J.D. Shelburne, *J. Microsc.* **134** (1984) 101.
400. J.S. Mijovic and J.A. Koutsky, *Polym. Plast. Technol. Eng.* **9** (1977) 139.
401. D. Hemsley, in *Developments in Polymer Characterization*, edited by J.V. Dawkins (Applied Science, London, 1978), p. 245.
402. D. Ferrante, S. Iannace and T. Monetta, *J. Mater. Sci.* **34** (1999) 175.
403. V. Khunova, J. Hurst, I. Janigova and V. Smatko, *Polym. Testing* **18** (1999) 501.
404. C. Harrison, M. Park, P.M. Chaikin, R.A. Register, D.H. Adamson and N. Yao, *Polymer* **39** (1998) 2733.
405. M.B. Olde Reikerink, J.G.A. Terlingen, G.H.M. Engbers and J. Feijen, *Langmuir* **15** (1999) 4847.
406. M.B. Olde Riekerink, M. B. Claase, G.H.M. Engbers, D.W. Grijpma and J. Feijen, *J. Biomed. Mater. Res. A* **65A** (2003) 417.
407. S. Collins, I.W. Hamley and T. Mykhaylyk, *Polymer* **44** (2003) 2403.
408. Y.J. Hwang, S. Matthews, M. McCord and M. Bourham, *J. Electrochem. Soc.* **151** (2004) 495.
409. H. Kupfer and G.K. Wolf, *Nuclear Instruments & Methods in Physics Research, Section B (Beam Interactions with Materials and Atoms), 10th International Conference on Radiation Effects in Insulators, 18–23 July 1999* **166–167** (2000) 722.
410. E. Buck and J. Fuhrmann, *Macromolecules* **34** (2001) 2172.
411. W.S. Chow, Z.A.M. Ishak and J. Karger-Kocsis, *J. Polym. Sci. B Polym. Phys.* **43** (2005) 1198.
412. J. Karger-Kocsis, O. Gryshchuk and S. Schmitt, *J. Mater. Sci.* **38** (2003) 413.
413. L.C. Sawyer, *J. Polym. Sci. Polym. Lett. Edn.* **22** (1984) 347.
414. L.S. Li, L.F. Allard and W.C. Bigelow, *J. Macromol. Sci. Phys.* **B22** (1983) 269.
415. R.M. Anderson and S.D. Walck, Eds., *Specimen Preparation for Transmission Electron Microscopy of Materials IV (Symposium held 2 April 1997, in San Francisco, California)*, in: *Mater. Res. Soc. Symp. Proc.* (1997) 480.
416. L.A. Giannuzzi and F.A. Stevie, *Introduction to Focused Ion Beams: Instrumentation, Theory, Techniques and Practice* (Springer, New York, 2005).
417. B.I. Prenitzer, C.A. Urbanik-Shannon, L.A. Giannuzzi, S.R. Brown, R.B. Irwin, T.L. Shofner and F.A. Stevie, *Microsc. Microanal.* **9** (2003) 216.
418. H. White, Y. Pu, M. Rafailovich, J. Sokolov, A.H. King, L.A. Giannuzzi, C. Urbanik-Shannon, B.W. Kempshall, A. Eisenberg, S.A. Schwarz and Y.M. Strzhemechny, *Polymer* **42** (2001) 1613.
419. J. Loos, J.K.J. van Duren, F. Morrissey and R.A.J. Janssen, *Polymer* **43** (2002) 7493.
420. W. Hu, T. Orlova and G.H. Bernstein, *J. Vac. Sci. Technol. B* **20** (2002) 3085.
421. A.R. Abramson, W.C. Kim, S.T. Huxtable, H. Yan, Y. Wu, A. Majumdar, C.-L. Tien and P. Yang, *J. Microelectromech. Systems* **13** (2004) 505.
422. D. Vogel, J. Keller and B. Michel, *Proc. SPIE* **5392** (2004) 148.
423. N. Virgilio, B.D. Favis, M.-F. Pepin, P. Desjardins and G. L'Esperance, *Macromolecules* **38** (2005) 2368.
424. L.D. Madsen, L. Weaver and S.N. Jacobsen, *Microsc. Res. Tech.* **36** (1997) 354.
425. P. Gasser, U.E. Klotz, F.A. Khalid and O. Beffort, *Microsc. Microanal.* **10** (2004) 311.
426. D.E. Bradley, *J. Appl. Phys.* **27** (1956) 1399.
427. D.E. Bradley, in *Techniques for Electron Microscopy*, edited by D.H. Kay (Blackwell, Oxford, 1965), p. 96.

428. P.J. Goodhew, *Specimen Preparation in Materials Science*, in *Practical Methods In Electron Microscopy*, Vol. 11, edited by A.M. Glauert (North Holland, Amsterdam, 1980).
429. J.H.M. Willison and A.J. Rowe, *Replica, Shadowing and Freeze-etching Techniques* in *Practical Methods in Electron Microscopy*, Vol. 8, edited by A.M. Glauert (North Holland, Amsterdam, 1980).
430. J.T. Woodward IV and J.A. Zasadzinski, *J. Microsc.* **184** (1996) 157.
431. V.G. Peck, *Appl. Polym. Symp.* **16** (1971) 19.
432. J.F. Oliver and S.G. Mason, *J. Colloid Interface Sci.* **60** (1977) 480.
433. P. Robbins and J. Pugh, *Wear* **50** (1978) 95.
434. E. Wood, personal communication 1980.
435. K.D. Gordon, *J. Microsc.* **134** (1984) 183.
436. K. Sakaoku and A. Peterlin, *J. Macromol. Sci. Phys.* **B1** (1967) 103.
437. S.Y. Hobbs and C.F. Pratt, *J. Appl. Polym. Sci.* **19** (1975) 1701.
438. B.R. Vijayendran, T. Bone and L.C. Sawyer, *J. Dispersion Sci. Technol.* **3** (1982) 81.
439. A. Peterlin and K. Sakaoku, *J. Appl. Phys.* **38** (1967) 4152.
440. K. O'Leary and P.H. Geil, *J. Appl. Phys.* **38** (1967) 4169.
441. P.J. Goodhew, *Specimen Preparation in Meterials Science*, in *Practical Methods in Electron Microscopy*, edited by A.M. Glauert (North Holland-American Elsevier, Amsterdam, 1973), p. 144.
442. M.M. Shahin, R.H. Olley and M.J. Blissett, *J. Polym. Sci. B Polym. Phys.* **37** (1999) 2279.
443. A.M. Baro, L. Vazquez, A. Bartolame, J. Gomez, N. Garcia, H.A. Goldberg, L.C. Sawyer, R.T. Chen, R.S. Kohn and R. Reifenberger, *J. Mater. Sci.* **24** (1989) 1739.
444. J. Dlugosz, *Proc. 1st Eur. Reg. Congr. on Electron Microscopy*, Stockholm, 1956, p. 283.
445. N. Ramanthan, J. Sikorski and H.J. Woods. *Proc. Int. Conf. Electron Microscopy*, 1954, p. 482.
446. M.H. Walters and D.N. Keyte, *Trans. Inst. Rubber Ind.* **39** (1962) 40.
447. P.H. Geil, *J. Macromol. Sci. Phys.* **B12** (1976) 173.
448. R.H. Olley, D.C. Bassett, P.J. Hine and I.M. Ward, *J. Mater. Sci.* **28** (1993) 1107.
449. D.C. Bassett and A. Keller, *Philos. Mag.* **6** (1961) 345.
450. J. Petermann and H. Gleiter, *J. Polym. Sci. Polym. Phys. Edn.* **10** (1972) 2333.
451. H. Kiyek and T.G.F. Schoon, *Rubber Chem. Technol.* **40** (1967) 1238.
452. J.I. Goldstein, D.E. Newbury, D.C. Joy, C.E. Lyman, P. Echlin, E. Lifshin, L.C. Sawyer and J. Michael, in *Scanning Electron Microscopy and X-ray Microanalysis*, 3rd ed. (Kluwer Academic/Plenum/Springer, New York, 2003), p. 647.
453. A.W. Robards, A.J. Wilson and P. Crosby, *J. Microsc.* **124** (1981) 143.
454. P.J. Goodhew, *Specimen Preparation in Materials Science*, in *Practical Methods in Electron Microscopy*, edited by A.M. Glauert (North Holland-American Elsevier, Amsterdam, 1973), p. 140.
455. T. Braten, *J. Microsc.* **113** (1978) 53.
456. I.M. Watt, *Proc. 9th Int. Electron Microscopy Congr.*, edited by W. Bailey (San Francisco Press, Toronto, 1978), p. 94.
457. P. Echlin and G. Kaye, *Scanning Electron Microsc.* **11** (1978) 109.
458. L.C. Sawyer, unpublished data.
459. H.S. Slayter, *Scanning Electron Microsc.* **13** (1980) 171.
460. J. Franks, P.R. Stuart and R.B. Withers, *Thin Solid Films* **60** (1979) 231.
461. J. Franks, C.S. Clay and G.W. Peace, *Scanning Electron Microsc.* **13** (1980) 155.
462. C.S. Clay and G.W. Peace, *J. Microsc.* **123** (1981) 25.
463. K.R. Peters, *Scanning Electron Microsc.* **13** (1980) 143.
464. P. Echlin, *Scanning Electron Microsc.* **14** (1981) 79.
465. T. Muller, P. Walther, C. Scheidegger, R. Reichelt, S. Muller and R. Guggenheim, *Scanning Microsc.* **4** (1990) 863.
466. M. Schossig-Tiedemann and D. Paul, *J. Membrane Sci.* **187** (2001) 85.
467. B.H. Kemmenoe and G.R. Bullock, *J. Microsc.* **132** (1983) 153.
468. L.C. Sawyer and R. Brozynski, unpublished data.
469. J.I. Goldstein, D.E. Newbury, D.C. Joy, C.E. Lyman, P. Echlin, E. Lifshin, L.C. Sawyer and J. Michael, *Scanning Electron Microscopy and X-ray Microanalysis*, 3rd ed. (Kluwer Academic/Plenum/Springer, New York, 2003).
470. J.H. Butler, D.C. Joy and G.F. Bradley, *Proc. 51st MSA*, edited by W. Bailey (San Francisco Press, 1993), p. 870.
471. D.C. Joy, *Scanning* **11** (1989) 1.
472. D.C. Joy, *Notes from the course on SEM, Lehigh University*, in Vol. 4 (unpublished, Lehigh Course).

473. D.C. Joy and C.S. Joy, *Microsc. Microanal.* **4** (1999) 475.
474. D.C. Joy and C.S. Joy, *Microsc. Microanal.* **1** (1995) 109.
475. M. Jamieson, unpublished data.
476. R.F. Egerton, F. Wang and P.A. Crozier, *Microsc. Microanal.* **12** (2006) 65.
477. A. Colebrooke and A.H. Windle, in *Scanning Electron Microscopy: Systems and Applications*, edited by W.C. Nixon (Institute of Physics, Bristol, 1973), p. 132.
478. D.T. Grubb and G.W. Groves, *Philos. Mag.* **24** (1971) 190.
479. L.C. Sawyer and M. Jamieson, *Proc. 47th EMSA*, edited by W. Bailey (San Francisco Press, 1989), p. 334.
480. M. Jamieson and L.C. Sawyer, unpublished data.
481. E.D. Boyes, *Microsc. Microanal.* **6** (2000) 307.
482. G.A. Bassett, *Philos. Mag.* **3** (1958) 1042.
483. B.J. Spit, *J. Macromol. Sci. Phys.* **B2** (1968) 45.
484. D. Krueger and G.S. Yeh, *J. Macromol. Sci. Phys.* **B6** (1972) 431.
485. M. Kojima and J.H. Magill, *J. Macromol. Sci. Phys.* **B15** (1978) 63.
486. K. Shimamura, *J. Macromol. Sci. Phys.* **B16** (1979) 213.
487. P. Miller and E.J. Kramer, *J. Mater. Sci.* **26** (1991) 1459.
488. H.H. Kausch, *Polymer Fracture* (Springer-Verlag, Berlin, 1978).
489. H.H. Kausch., Ed. *Crazing in Polymers, Adv. Polym. Sci. Ser. 52/3* (Springer-Verlag, Berlin, 1983).
490. H. Liebowitz., Ed. *Fracture of Non-Metals and Composites* (Academic Press, New York, 1972).
491. L. Engel, H. Klingele, G.W. Ehrentstein and H. Scherper, *An Atlas of Polymer Damage* (Prentice Hall, Englewood Cliffs, NJ, 1981).
492. A.K. Bhowmick, S. Basu and S.K. De, *Rubber Chem. Technol.* **53** (1980) 321.
493. R. Adhikari, G.H. Michler, E. Ivan'kova, R. Godehardt, W. Lebek and K. Knoll, *Macromol. Chem. Phys.* **204** (2003) 488.
494. G.H. Michler, R. Adhikari, S. Henning, *Macromol. Symp.* **214** (2004) 47.
495. G.H. Michler, *J. Macromol. Sci. Phys.* **B40** (2001) 277.
496. J.W.S. Hearle and P.M. Cross, *J. Mater. Sci.* **5** (1970) 507.
497. L.C. Sawyer and M. Jamieson, unpublished data.
498. H.H. Kausch, J.-L. Halary and C.J.G. Plummer, *Macromol. Symp.* **214** (2004) 17.
499. P.E. Reed, in *Developments in Polymer Fracture*, edited by E.H. Andrews (Applied Science, London, 1979), p. 121.
500. C.B. Bucknall, *Toughened Plastics* (Applied Science, London, 1977).
501. F. Speroni, E. Castoldi, P. Fabbri and T. Casiraghi, *J. Mater. Sci.* **24** (1989) 2165.
502. D.R. Mulville and I. Wolock, in *Polymer Fracture*, edited by E.H. Andrews (Applied Science, London, 1979), p. 263.
503. J.H. Sinclair and C.C. Chamis, *34th Annu. Tech. Conf. 1979, Reinforced Plastics/Composites*, (SPI, Washington DC, 1979), p. 1.
504. R.A. Kline and F.H. Chang, *J. Compos. Mater.* **14** (1980) 315.
505. R. Richard-Frandsen and Y. Naerheim, *J. Compos. Mater.* **17** (1983) 105.
506. J.E. Theberge, *Polym. Plast. Technol. Eng.* **16** (1981) 41.
507. J. Kubat and H.E. Stromvall, *Plast. Rubber Process.* (1980) 45.
508. J.A. Manson and R.W. Hertzberg, *J. Polym. Sci. Polym. Phys. Edn.* **11** (1973) 2483.
509. A.M. Donald and E.J. Kramer, *J. Mater. Sci.* **17** (1982) 1871.
510. E.J. Kramer, *Polym. Eng. Sci.* **24** (1984) 761.
511. E.J. Kramer, in *Developments in Polymer Fracture*, edited by E.H. Andrews (Chapman and Hall, London, 1979).
512. R.C. Cieslinski, H.C. Silvis and D.J. Murray, *Polymer* **36** (1995) 1827.
513. H. Tang, R. Cieslinski, N.E. Verghese and N. Jivraj, *ANTEC 2004—Annual Technical Conference Proceedings, May 16–20 2004*, (SPE Brookfield CN, Chicago, IL., 2004), p. 4018.
514. C.-C. Hsiao, T.S. Lin, L.Y. Cheng, C.-C.M. Ma and A.C.-M. Yang, *Macromolecules* **38** (2005) 4811.
515. R.P. Kambour, *Polymer* **5** (1964) 143.
516. R.P. Kambour, *J. Appl. Polym. Sci. Appl. Polym. Symp.* **7** (1968) 215.
517. R.P. Kambour and A.S. Holik, *J. Polym. Sci.* **A2** (1969) 1393.
518. R.P. Kambour, *J. Polym. Sci.* **D7** (1973) 1.
519. J. Washiyama, C. Creton and E.J. Kramer, *Macromolecules* **25** (1992) 4751.
520. C. Creton, E.J. Kramer and G. Hadziioannou, *Colloid Polym. Sci.* **270** (1992) 399.
521. B.D. Lauterwasser and E.J. Kramer, *Philos. Mag.* **A39** (1979) 469.
522. E.L. Thomas and S.J. Israel, *J. Mater. Sci.* **10** (1975) 1603.

523. A.M. Donald and E.J. Kramer, *J. Polym. Sci. Polym. Phys. Edn.* **20** (1982) 899.
524. A.M. Donald and E.J. Kramer, *Philos. Mag.* **A43** (1981) 857.
525. A.M. Donald, T. Chan and E.J. Kramer, *J. Mater. Sci.* **16** (1981) 669.
526. A.M. Donald and E.J. Kramer, *Polymer* **23** (1982) 1183.
527. A.S. Argon, R.E. Cohen, B.Z. Jang and J.B.V. Sande, *J. Polym. Sci. Polym. Phys. Edn.* **19** (1981) 253.
528. Y. Maeda and M. Koizumi, *Rev. Sci. Instrum.* **67** (1996) 2030.
529. J.R. White and E.L. Thomas, *Rubber Chem. Technol.* **57** (1984) 457.
530. A.N. McKee and C.L.B. Ill, *Text. Res. J.* **40** (1970) 1006.
531. R.M. Minnini and M. Jamieson, unpublished data.
532. S.Y. Hobbs, *Rev. Sci. Instrum.* **53** (1982) 1097.
533. R.H. Hoel and D.J. Dingley, *J. Mater. Sci.* **17** (1982) 2990.
534. K. Smith, M.G. Hall and J.N. Hay, *J. Polym. Sci. Polym. Lett. Edn.* **14** (1976) 751.
535. R. Rizzieri, F.S. Baker and A.M. Donald, *Rev. Sci. Instrum.* **74** (2003) 4423.
536. S. Hild, W. Gutmannsbauer, R. Luthi, J. Fuhrmann and H.-J. Guentherodt, *J. Polym. Sci. B Polym. Phys.* **34** (1996) 1953.
537. B. Bhushan, in *Applied Scanning Probe Methods I*, edited by B. Bhushan, H. Fuchs and S. Hosaka (Springer, Berlin, Heidelberg, 2004), p. 171.
538. S.P. Ho, L. Riester, M. Drews, T. Boland and M. LaBerge, *Proceedings of the Institution of Mechanical Engineers, Part H (Journal of Engineering in Medicine)* **217** (2003) 357.
539. M.R. VanLandingham, J.S. Villarrubia, W.F. Guthrie and G.F. Meyers, *Macromol. Symp.* **167** (2001) 15.
540. J. Schofer and E. Santner, *Wear* **222** (1998) 74.
541. W. Shen, L. Mi and B. Jiang, *Tribol. Int.* **39** (2006) 146.
542. D.A. Ivanov and S.I. Magonov, in *Polymer Crystallization. Observations, Concepts and Interpretations (Lecture Notes in Physics, Vol. 606)*, edited by J.-V. Sommer and G. Reiter (Springer-Verlag, New York, 2003), p. 98.
543. D.A. Ivanov, Z. Amalou and S.N. Magonov, *Macromolecules* **34** (2001) 8944.
544. M.J. Fasolka, A.M. Mayes and S.N. Magonov, *Ultramicroscopy* **90** (2001) 21.
545. L.G.M. Beekmans, D.W. van der Meer and G.J. Vancso, *Polymer* **43** (2002) 1887.
546. L.G.M. Beekmans, R. Vallee and G.J. Vancso, *Macromolecules* **35** (2002) 9383.
547. M.A. Hayat, *Principles and Techniques of Electron Microscopy: Biological Applications* (Van Nostrand Reinhold, New York, 1977).
548. P. Echlin, *Microsc. Microanal.* **7** (2001) 211.
549. J.I. Goldstein, D.E. Newbury, D.C. Joy, C.E. Lyman, P. Echlin, E. Lifshin, L.C. Sawyer and J. Michael, in *Scanning Electron Microscopy and X-ray Microanalysis*, 3rd ed. (Kluwer Academic/Plenum/Springer, New York, 2003), p. 621.
550. E.B. Bradford and J.W. Vanderhoff, *J. Colloid Sci.* **17** (1962) 668.
551. E.B. Bradford and J.W. Vanderhoff, *J. Colloid Sci.* **14** (1959) 543.
552. L. Bachmann and Y. Talmon, *Ultramicroscopy* **14** (1984) 211.
553. J.O. Stoffer and T. Bone, *J. Dispersion Sci. Technol.* **1** (1980) 393.
554. C. Mueller, V. Topolkaraev, D. Soerens, A. Hiltner and E. Baer, *J. Appl. Polym. Sci.* **78** (2000) 816.
555. S. Jose, S. Thomas, E. Lievana and J. Karger-Kocsis, *J. Appl. Polym. Sci.* **95** (2005) 1376.
556. A.K. Kleinschmidt, D. Lang, D. Jacherts and R.K. Zahn, *Biochim. Biophys. Acta* **61** (1962) 857.
557. R.W.G. Wyckoff, *Science* **104** (1946) 36.
558. R.C. Williams, *Biochim. Biophys. Acta* **9** (1952) 237.
559. R.C. Williams, *Exp. Cell Res.* **4** (1953) 188.
560. A.W. Robards and U.B. Sleytr, Low Temperation methods in biological electron microscopy, in *Practical Methods in Electron Microscopy*, Vol. 10, edited by A.M. Glauert (Elsevier, Amsterdam, 1985).
561. T.W.G. Rowe, *Ann. N. Y. Acad. Sci.* **85** (1960) 641.
562. T.W.G. Rowe, in *Current Trends in Cryo Biology*, edited by A.V. Smith (Plenum Press, New York, 1970).
563. L.C. Sawyer, B. Strassle and D.J. Palatini. *Proc. 37th Annu. Mtg EMSA*, San Antonio, edited by W. Bailey (San Francisco Press, 1979), p. 620.
564. E.R. Walter and G.H. Bryant. *Proc. 35th Annu. Mtg EMSA*, edited by W. Bailey (San Francisco Press, Boston, 1977), p. 314.
565. E.R. Walter, private communication.
566. S.A. McDonald, C.A. Daniels and J.A. Davidson, *J. Colloid Interface Sci.* **59** (1977) 342.

567. J.W. Vanderhoff, *J. Macromol. Sci. Chem.* **A7** (1973) 677.
568. A. Chakraborty, M. Sain and M. Kortschot, *Holzforschung* **59** (2005) 102.
569. T.F. Anderson, *Trans. N. Y. Acad. Sci.* **13** (1951) 130.
570. T.F. Anderson, *C. R. Prem. Congr. Int. Microsc. Electron.*, Paris, 1953, p. 567.
571. T.F. Anderson, *Proc. 3rd Int. Conf. on Electron Microscopy*, London 1954, Royal Microscopy Society, London, 1956, p. 122.
572. M.A. Hayat and B.R. Zirkin, in *Principles and Techniques of Electron Microscopy: Biological Applications*, edited by M.A. Hayat (Van Nostrand Reinhold, New York, 1973), p. 297.
573. T.F. Anderson, *Physical Techniques in Biological Research*, 2nd ed. (Academic Press, New York, 1966).
574. A.L. Cohen, in *Scanning Electron Microscopy*, edited by M.A. Hayat (Van Nostrand, New York, 1974), p. 44.
575. A.L. Cohen, *Scanning Electron Microsc.* **10** (1977) 525.
576. N.S. McNutt, Ed. *Dynamic Aspects of Cell Surface Organization* (Elsevier-North Holland Biomedical Press, Amsterdam, 1977).
577. R.L. Steere, *J. Biophys. Cytol.* **7** (1957) 167.
578. H. Moor, K. Muhlethaler, H. Waldner and A. Frey-Wyssling, *J. Biophys. Cytol.* **10** (1961) 1.
579. D. Branton, *Proc. Natl. Acad. Sci. U. S. A.* **55** (1966) 1048.
580. C.J. Singleton, T. Stephenson, J. Isner, P.H. Geil and E.A. Collins, *J. Macromol. Sci. Phys.* **B14** (1977) 29.
581. M. Gervais and B. Gallot, *Makromol. Chem.* **180** (1979) 2041.
582. F. Candau, J. Boutillier, F. Tripier and J.C. Wittmann, *Polymer* **20** (1979) 1221.
583. J. Biais, M. Mercier, P. Lalanne, B. Clin, A.M. Bellocq and B. Lemanceau, *C. R. Acad. Sci. Paris* **285** (1977) 213.
584. C. Price and D. Woods, *Eur. Polym. J.* **9** (1973) 827.
585. A. Rameau, P. Marie, F. Tripier and Y. Gallot, *C. R. Acad. Sci. Paris* **286** (1978) 277.
586. U.B. Sleytr and A.W. Robards, *J. Microsc.* **110** (1977) 1.
587. D.J. Stokes, J.-Y. Mugnier and C.J. Clarke, *J. Microsc.* **213** (2004) 198.
588. Y.K. Godovsky, V.S. Papkov and S.N. Magonov, *Macromolecules* **34** (2001) 976.
589. L. Champion, J.F. Gerard, J.P. Planche, D. Martin and D. Anderson, *J. Mater. Sci.* **36** (2001) 451.
590. L. Champion-Lapalu, A. Wilson, G. Fuchs, D. Martin and J.-P. Planche, *Energy Fuels* **16** (2002) 143.
591. Y. Talmon, H.T. Davis, L.E. Scriven and E.L. Thomas, *Rev. Sci. Instrum.* **50** (1979) 698.
592. A.H. Falls, S.T. Wellinghoff, Y. Talmon and E.L. Thomas, *J. Mater. Sci.* **18** (1983) 2752.
593. S.U. Egelhaaf, P. Schurtenberger and M. Muller, *J. Microsc.* **200** (2000) 128.
594. M.R. Libera, PMSE, Preprints, *224th ACS National Meeting, Boston, MA, United States, August 18–22, 2002* (ACS, Washington DC, 2002) 177.
595. D. Danino, A. Bernheim-Groswassen and Y. Talmon, *Colloids Surf. A* **183–185** (2001) 113.
596. Y. Talmon, in *Modern Characterization Methods of Surfactant Systems*, edited by B.P. Binks (Marcel Dekker, New York, 1999), p. 147.
597. S. Nilsson, M. Goldraich, B. Lindman and Y. Talmon, *Langmuir* **16** (2000) 6825.
598. I.W. Hamley, V. Castelleto, J. Fundin, Z. Yang, M. Crothers, D. Attwood and Y. Talmon, *Colloid Polym. Sci.* **282** (2004) 514.
599. E. Kesselman, Y. Talmon, J. Bang, S. Abbas, Z. Li and T.P. Lodge, *Macromolecules* **38** (2005) 6779.
600. A. Wittemann, M. Drechsler, Y. Talmon and M. Ballauff, *J. Am. Chem. Soc.* **127** (2005) 9688.
601. G. Nizri, S. Magdassi, J. Schmidt, Y. Cohen and Y. Talmon, *Langmuir* **20** (2004) 4380.
602. G.F. Meyers, B.M. DeKoven, M.T. Dineen, A. Strandjord, P.J. O'Connor, T. Hu, Y.H. Chiao, H. Pollock and A. Hammiche, Polymer Preprints, *216th ACS National Meeting, Boston, August 23–27* (American Chemical Society, Washington D.C., 1998) 1419.

Chapter 5
Applications of Microscopy to Polymers

5.1 FIBERS 250
  5.1.1 Introduction 250
  5.1.2 Textile Fibers 251
    5.1.2.1 Optical Microscopy of Textile Fibers 251
    5.1.2.2 SEM of Textile Fibers 253
    5.1.2.3 Fiber Fractography 254
    5.1.2.4 SEM of Woven and Nonwoven Fabrics 258
    5.1.2.5 TEM of Fibers 259
  5.1.3 Problem Solving Applications 260
    5.1.3.1 Characterization of Textile Fibers 262
    5.1.3.2 Metal Loaded Fibers ... 265
    5.1.3.3 Contamination 267
  5.1.4 Industrial Fibers 267
    5.1.4.1 Tire Cords 267
  5.1.5 High Performance Fibers 270
    5.1.5.1 Introduction 270
    5.1.5.2 High Modulus PE Fibers 270
    5.1.5.3 High Performance Fibers 272
    5.1.5.4 Examples of Fiber Studies 273
    5.1.5.5 Spider Silk Fibers 275
5.2 FILMS AND MEMBRANES 276
  5.2.1 Introduction 276
  5.2.2 Model Studies 278
    5.2.2.1 Introduction 278
    5.2.2.2 Literature Review 278
    5.2.2.3 Semicrystalline Films ... 280
    5.2.2.4 Amorphous Films 281
  5.2.3 Industrial Films 282
    5.2.3.1 Optical Microscopy ... 283
    5.2.3.2 Electron Microscopy 284
    5.2.3.3 Scanning Probe Microscopies 286
    5.2.3.4 SEM and SPM Examples 287
    5.2.3.5 Electronic Films and Devices 288
    5.2.3.6 Multilayered Films 292
  5.2.4 Flat Film Membranes 294
    5.2.4.1 Introduction 294
    5.2.4.2 Literature Review 295
    5.2.4.3 SEM: Surface and Bulk 298
    5.2.4.4 Reverse Osmosis Membranes 300
    5.2.4.5 Microporous Membranes 303
  5.2.5 Hollow Fiber Membranes 305
5.3 ENGINEERING RESINS AND PLASTICS 308
  5.3.1 Introduction 308
    5.3.1.1 Resins and Plastics 308
    5.3.1.2 Characterization 309
  5.3.2 Process-Structure Considerations 311
    5.3.2.1 Extrusion and Molding 311
    5.3.2.2 Spherulitic Structures ... 312
    5.3.2.3 Skin-Core Structures ... 315
  5.3.3 Single Phase Polymers 316
    5.3.3.1 Amorphous Polymers 316

5.3.3.2 Semicrystalline
　　　　　　 Polymers 318
　5.3.4 **Multiphase Polymers** 321
　　　5.3.4.1 Introduction 321
　　　5.3.4.2 Toughened Resins 323
　　　5.3.4.3 Processing of
　　　　　　 Multiphase Polymers ... 326
　　　5.3.4.4 Toughened Thermoset
　　　　　　 Resins 326
　　　5.3.4.5 Impact Modified
　　　　　　 Thermoplastics 328
　　　5.3.4.6 Block, Graft and
　　　　　　 Random Copolymers ... 337
　　　5.3.4.7 Polyurethanes 345
　　　5.3.4.8 Biodegradable
　　　　　　 Polymers 347
　5.3.5 **Failure or Competitive
　　　　Analysis** 349
5.4 **COMPOSITES** 354
　5.4.1 **Introduction** 354
　5.4.2 **Literature Review** 355
　　　5.4.2.1 Composites in
　　　　　　 General 355
　　　5.4.2.2 Contact
　　　　　　 Microradiography 356
　　　5.4.2.3 Carbon and Graphite
　　　　　　 Fiber Composites 356
　　　5.4.2.4 Hybrid Composites 357
　5.4.3 **Composite Characterization** ... 357
　　　5.4.3.1 OM Characterization ... 357
　　　5.4.3.2 SEM of Composites 359
　　　5.4.3.3 Problem Solving
　　　　　　 Application 362
　5.4.4 **Carbon and Graphite Fiber
　　　　Composites** 365
　5.4.5 **Particle Filled Composites** 366
　　　5.4.5.1 Introduction 366
　　　5.4.5.2 Carbon Black Filled
　　　　　　 Rubber 368
　　　5.4.5.3 Examples of Particle
　　　　　　 Filled Composites 369
　5.4.6 **Nanocomposites** 370
　　　5.4.6.1 Introduction 370
　　　5.4.6.2 Literature Review of
　　　　　　 Clay Nanocomposites... 373
　　　5.4.6.3 Literature Review of
　　　　　　 Carbon Nanotube
　　　　　　 Composites 375
　　　5.4.6.4 Problem Solving
　　　　　　 Applications 376

5.5 **EMULSIONS, COATINGS AND
　　ADHESIVES** 380
　5.5.1 **Introduction** 380
　5.5.2 **Emulsions and Latexes** 381
　　　5.5.2.1 General Literature
　　　　　　 Review 381
　　　5.5.2.2 Literature Review:
　　　　　　 SEM 382
　　　5.5.2.3 OM, SEM and TEM
　　　　　　 Characterization 383
　5.5.3 **Particle Size Measurements** ... 385
　5.5.4 **Adhesives and Adhesion** 386
　　　5.5.4.1 Literature Review 386
　　　5.5.4.2 Problem Solving
　　　　　　 Example 387
　5.5.5 **Wettability and Coatings** 388
　　　5.5.5.1 Introduction 388
　　　5.5.5.2 AFM
　　　　　　 Characterization 389
　　　5.5.5.3 Problem Solving
　　　　　　 Examples 389
　　　5.5.5.4 Cryo-TEM
　　　　　　 Characterization 394
5.6 **HIGH PERFORMANCE
　　POLYMERS** 398
　5.6.1 **Introduction** 398
　　　5.6.1.1 Introduction to Liquid
　　　　　　 Crystalline Polymers ... 398
　　　5.6.1.2 Chemistry of LCPs 399
　　　5.6.1.3 Microscopy of LCPs ... 400
　5.6.2 **Microstructure of LCPs** 400
　　　5.6.2.1 Optical Textures 400
　　　5.6.2.2 Banded Structures in
　　　　　　 OM and TEM 402
　5.6.3 **Molded Parts and
　　　　Extrudates** 403
　　　5.6.3.1 Structure of Unfilled
　　　　　　 Moldings and
　　　　　　 Extrudates 403
　　　5.6.3.2 Structure Models of
　　　　　　 Unfilled Moldings and
　　　　　　 Extrudates 406
　　　5.6.3.3 Structure of Filled
　　　　　　 Moldings 407
　　　5.6.3.4 Blends with LCPs 408
　5.6.4 **High Modulus Fibers** 409
　　　5.6.4.1 Aromatic Polyamides... 409
　　　5.6.4.2 Rigid Rod Polymers ... 411
　　　5.6.4.3 Aromatic
　　　　　　 Copolyesters 412

5.6.5 Structure-Property Relations in LCPs	412
5.6.5.1 Microstructure of LCPs	413
5.6.5.2 LCP Structure Model	417
References	418

5.1 FIBERS

5.1.1 Introduction

Characterization of the microstructure of polymer fibers can provide insights into the fundamental structures present and into the relationship between structure and properties important for applications. Morphological characterization provides information to help understand the effects of process history on mechanical and physical properties. Microscopy techniques are used to observe features such as fiber shape, diameter, structure (crystal size, voids, etc.), molecular orientation, size and distribution of additives, structure of yarn and fabric assemblages, and failure mechanisms. These features are directly related to specific mechanical and thermal properties. Emphasis in this section is on assessment of the structure of polymer fibers as it relates to solving problems or evaluating the effect of process modifications. Fibers prepared from liquid crystalline polymers require special methods and interpretation that will be described later (see Section 5.6). It must be emphasized that any study of polymer fibers will be incomplete if only microscopy techniques are applied. X-ray scattering (e.g., [1]), thermal analysis (differential scanning calorimetry (DSC), thermogravimetric analysis (TGA), heat shrinkage), and spectroscopy (IR, Raman, and photoelectron spectroscopy (XPS)) are among the many techniques that complement microscopy investigations (see Section 7.4).

The polymers used in fibers are linear, so the molecules are a few nanometers across and several hundred nanometers long. In unoriented materials, the molecules are coiled and folded into loose isotropic spheres. When a fiber is oriented, by drawing for example, the molecular chains become aligned parallel to the fiber axis (uniaxial fiber orientation), and the stiffness and strength improve. In most fibers, the molecules are still coiled and folded, although they are oriented. Only in ultrahigh modulus fibers (see Section 5.1.5) or in fibers formed from liquid crystalline precursors (see Section 5.6.4) are the molecules highly elongated and extended parallel to the fiber axis.

Natural and synthetic textile fibers were among the earliest materials studied by electron microscopy. Guthrie [2] and Stoves [3] described the techniques and applications of fiber microscopy to industrial practice. Somewhat later, evidence was provided for an oriented microfibrillar texture in polymer fibers [4]. X-ray diffraction suggested an arrangement of fine structures about 50 nm long and 5 nm wide in semicrystalline fibers [5, 6]. Peterlin [7, 8] observed the formation of fibrils and microfibrils by the deformation and transformation of spherulites using various microscopy techniques.

A basic element of semicrystalline fibers is the *microfibril*. Microfibrils may be bundled into fibrils, about several hundred nanometers thick. A mechanically weak boundary between the fibrils results in fibrillation during deformation. Barham and Keller [9] and Prevorsek et al. [10] discussed the microfibrillar model, and the latter authors summarized the effects of fiber structure on textile properties. Microfibrils are known to exist in most fibers and are also known to be present in drawn single crystals, such as single crystal polyethylene (PE) mats, crazes, melt extrudates, and solid state extrudates. In addition, it is known [10] that larger structures, macrofibrils, are composed of microfibrils and that crystallites, disordered domains and partially extended noncrystalline molecules, are present in fibers. Fiber structure and properties for nylon 6 and poly(ethylene terephthalate) (PET) fibers were further elaborated [11–14] using both electron microscopy and small angle scattering. A major point of these studies was evidence supporting the strong lateral interactions between the microfibrils.

Reviews of specimen preparation methods for fiber microscopy and instrumental techniques applied to fibers were published during the early 1970s [15–17] and also more recently [18]. This section contains applications of microscopy to the understanding of fiber microstructures used in the industrial laboratory for modification of fiber formation processes to improve specific mechanical properties or for problem solving. Textiles are fibrous materials made from fibers, such as filaments, yarn, cords, ropes, fabrics, nets, carpets, and rugs. These materials are used in many industrial applications, including clothing, protective clothing, geotextiles, construction, transportation, medical, consumer products, and aerospace and in many types of composites. Protective clothing has become more important in recent years for chemical and biological protection, for fire fighting, law enforcement, and for medical personnel. Fibers have been produced from a wide range of polymers (see Appendix III). A handy listing of common textile fibers is found in the *Textile World Manmade Fiber Chart*, issued by Textile World and available for order online [19]. This comprehensive chart lists the various fiber names, types, optical micrographs of cross sections and longitudinal views, mechanical properties, chemical reactivity, and so forth, of about 50 branded fibers.

5.1.2 Textile Fibers

5.1.2.1 Optical Microscopy of Textile Fibers

The optical microscope is used to study various fiber features, such as (1) size, (2) cross section (shape), (3) uniformity, (4) molecular orientation, and (5) distribution of fillers. Specimen preparation methods include direct observation and sectioning. Fibers are embedded prior to sectioning by microtomy (see Section 4.3) or polishing (see Section 4.2) methods. Video and image processing can be used for documentation of the structure of the fibers. Figure 5.1 contains optical micrographs showing longitudinal views of typical PET textile fibers (Fig. 5.1A) and a fabric composed of Orlon fibers (Fig. 5.1B). A range of fiber cross section sizes and shapes are shown in the optical micrographs in Fig. 5.2. A drawn PE fiber seen in a cross section (Fig. 5.3) exhibits a fine spherulitic texture. *Birefringence*, the difference between the refractive index parallel and perpendicular

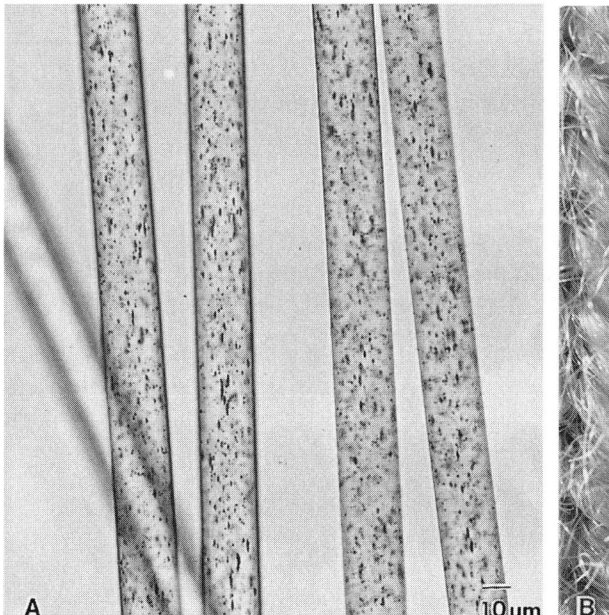

FIGURE 5.1. Transmitted optical micrograph of a polyester textile fiber (A) shows cylindrical fibers containing dense pigment particles. A low magnification optical view (B) shows a fabric woven with Orlon fiber containing yarns.

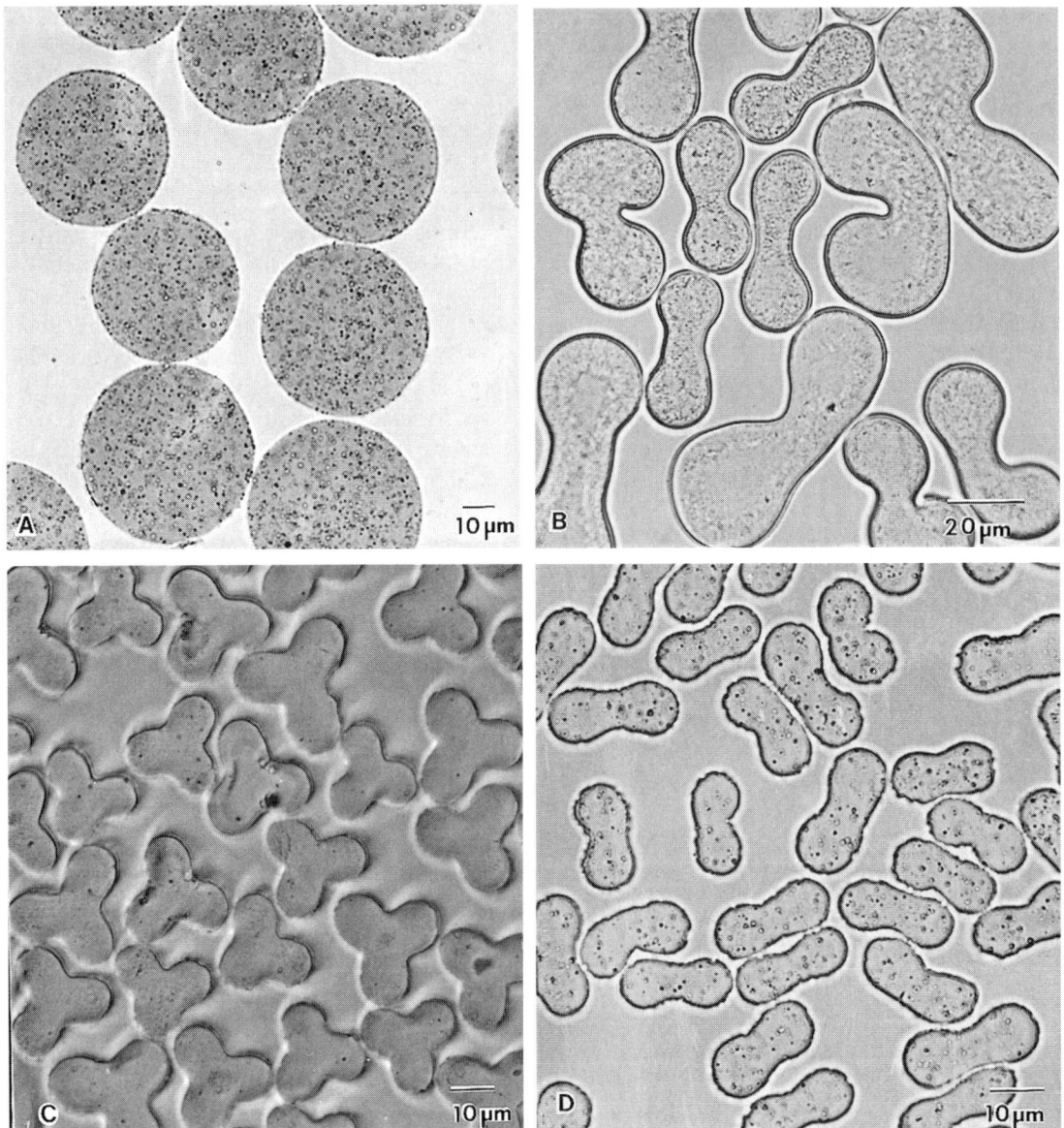

FIGURE 5.2. Cross sectional views of representative fibers show the fiber shapes and dense pigment particles. The fibers are (A) round polyamide, (B) irregularly shaped polyacrylonitrile, (C) trilobal shaped Orlon, and (D) dogbone shaped Orlon fiber sections.

to the fiber axis, is an important quantitative measure of molecular orientation [20]. Birefringence is measured by either the Becke line (immersion) method or a compensator method [3] (see Section 2.2.4 and Section 3.1.7). The Becke line method [21] measures the surface birefringence, whereas compensator methods measure the average birefringence of the fiber [22]. Combination of these methods provides a useful measure of the differential birefringence, a skin-core effect, if it is present.

When fibers are observed in the 45° position between crossed polarizers (polars), the change in thickness from the center to the fiber edge produces a series of polarization bands or fringes. An example of these fringes is shown in a PET fiber in Fig. 5.4. These fringes can be used to determine the retardation of the fiber, and the birefringence equals retardation/thickness. If the fiber is round, its thickness is the same as its width. For low birefringence fibers, measurement of the retardation is straightforward, as few of these bands must be counted. However, for higher birefringence fibers, there are many bands present, which are difficult to count. Additionally, the zero-order fringe must be identified for measurement of the birefringence. It may be difficult to know which fringe is correct if the dispersion of the birefringence of the fiber is different to that of the compensator. A useful trick is to cut a wedge at the end of the fiber and count the number of fringes along the wedge, which is the number of full orders of path difference [23]. The additional partial order is measured with a compensator [24] (see Section 3.1.7).

Birefringence provides a measure of the local orientation of a material (i.e., the mean orientation of monomer units). The relation between orientation and birefringence was known from early studies of polystyrene filaments, which described both the theory and measurement [23, 25]. They showed that the orientation was greater at the surface than in the core. Mechanical properties, such as tensile strength and elongation at break, have been shown to

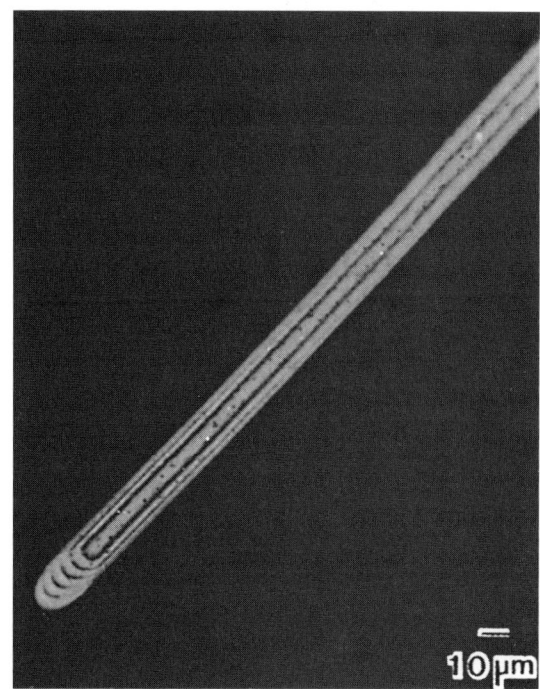

FIGURE 5.4. A polyester fiber in polarized light, aligned at 45° to the crossed polarizers. The dark bands are fringes that reflect the high order birefringence. In the orthogonal position, the same fiber would exhibit extinction and thus appear black. (See color insert.)

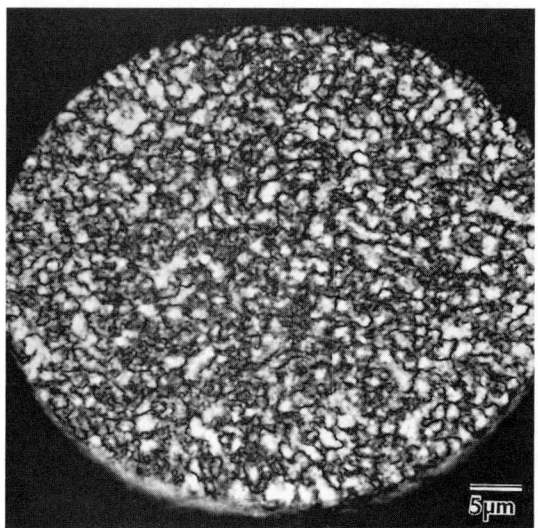

FIGURE 5.3. A drawn polyethylene fiber cross section reveals a fine, spherulitic texture in polarized light.

increase with orientation. Other techniques, such as the measurement of shrinkage on heating, give information on the molecular orientation on a larger scale and provide a measure of the entanglement length. The combination of birefringence measurements and other techniques has been demonstrated (e.g., in studies of thermal shrinkage [26] and of the effect of heat setting on the mechanical properties of PET fibers [27]). Birefringence measurements are effective in providing a structural parameter that may be used to relate process variables to mechanical properties.

5.1.2.2 SEM of Textile Fibers

The scanning electron microscope (SEM) has proved to be a very useful instrument for the assessment of fiber morphology. The three dimensional images produced clearly show surface features, such as the presence of surface modifications, finish applications, wear, and the nature and cause of fiber failure. The great depth of field, simple specimen preparation, and high resolution have resulted in the SEM providing a major contribution to the study of textile fibers. Textile fibers generally have a surface finish applied after or during spinning to aid handling of the fibers for production of yarns and fabrics and to provide special properties, such as flame retardancy. Fiber finishes have been observed by SEM of fiber surfaces [28] and in cross section by using x-ray techniques in the SEM [29]. Preparations are simple (see Section 4.1.2) and are generally followed by the application of a conductive coating (see Section 4.7.3). Unevenness and lack of uniformity of fiber coatings seen directly in the SEM have been correlated with spectroscopic and wettability studies [30].

Scanning electron microscopy micrographs taken in the normal mode do not always permit effective observation of features on fiber surfaces. Display modes such as deflection or Y modulation and imaging modes such as backscattered electron imaging (BEI) can provide clearer contrast, as shown in Figs 5.5 and 5.6. Heat aging of polymer fibers can cause cyclic trimers and oligomers to diffuse to the fiber surface. Scanning electron microscopy (Fig. 5.5A) of heat aged octalobal fibers shows oligomers of crystalline appearance on the surface although deflection modulation provides a clearer view of the surface detail (Fig. 5.5B). Solid state BEI detectors can be set to emphasize either topographic contrast ("topo") or atomic number contrast ("compo"), which suppresses the surface detail in the image. A BEI "compo" micrograph (Fig. 5.6A) shows some bright surface detail, which has higher mean atomic number than the fiber, and the appearance of surface cracks, which is the metal coating. This detail is not as obvious in the "topo" (Fig. 5.6B) mode.

5.1.2.3 Fiber Fractography

Textile fiber fractography was initially developed at UMIST (University of Manchester Institute of Science and Technology), especially by Hearle. Fiber fractography and the classes of fracture were reviewed by Hearle et al. [16, 31, 32], and these classes are shown in Table 5.1 with examples and early references. The mechanism of fiber failure can be determined by fractography studies (see Section 4.8) in the SEM. Fibers broken during a standard physical test, such as

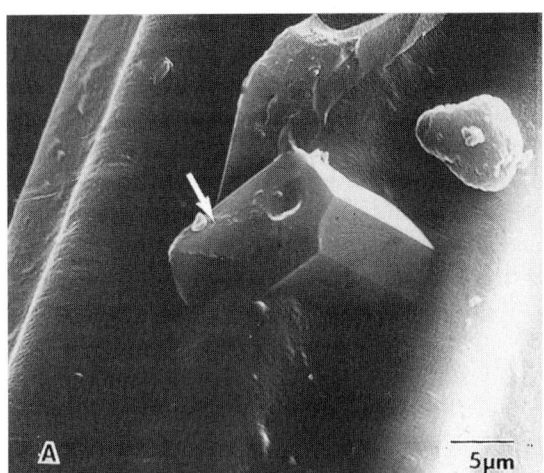 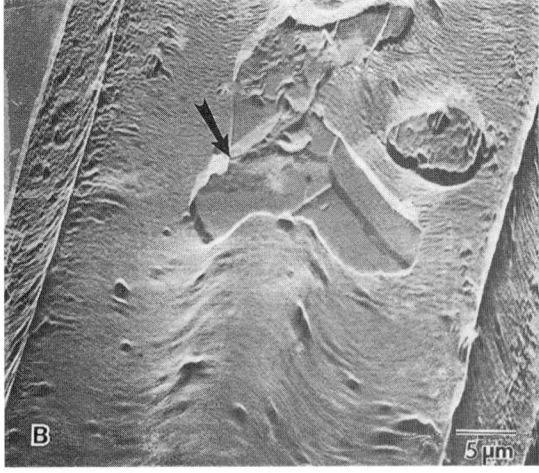

FIGURE 5.5. Heat aging polyester fibers draws oligomers to the fiber surface. Crystalline oligomer particles are shown (arrow) in an SEM of an octalobal fiber (A). The curved surface of the fiber does not exhibit much detail other than the oligomers. The micrograph in (B) is of a similar region of this fiber in the Y modulation mode, which accentuates the three dimensional surface material.

Fibers

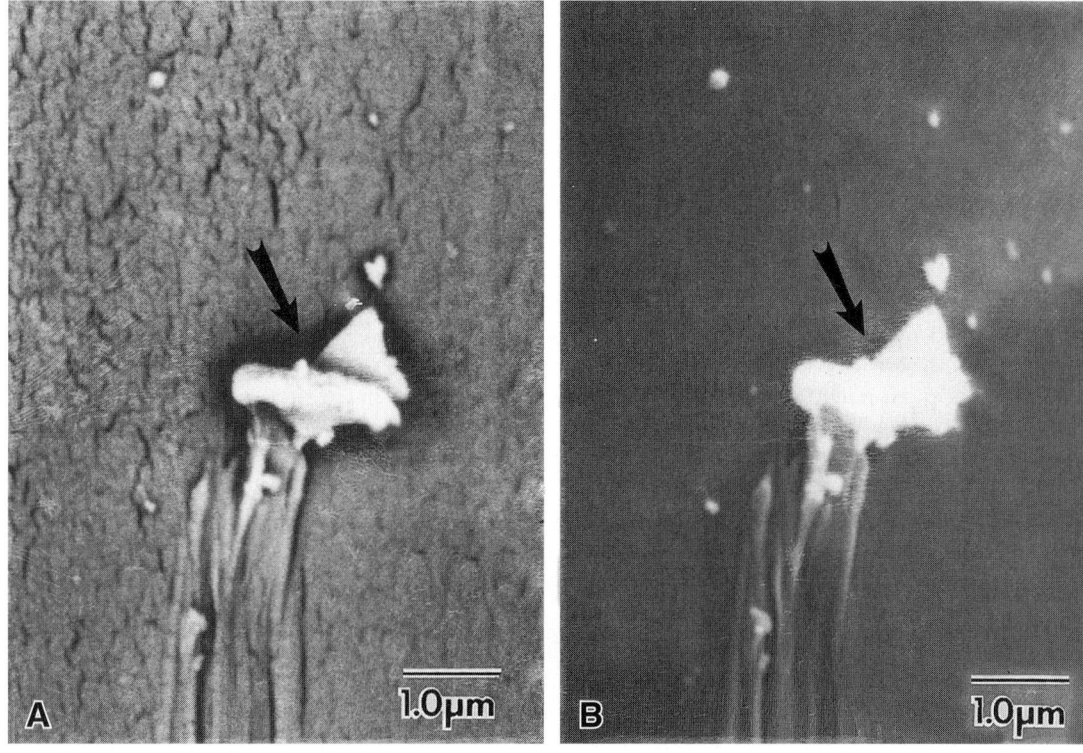

FIGURE 5.6. Fiber surfaces are shown here taken in BEI modes. A BEI "compo" micrograph (A) shows some bright surface detail, which has higher mean atomic number than the fiber, and the appearance of surface cracks, which is the metal coating. This detail is not as obvious in the "topo" (B) mode.

tensile testing in an Instron, are examined in the SEM for the nature of the failure mechanism and to identify the locus and cause of failure. Analytical microscopy (x-ray microanalysis) can be conducted in order to determine the chemical composition of any defects causing failure.

Aromatic polyamides (aramids) split longitudinally due to fibrillation, whereas nylon shows plastic deformation under the same conditions [33]. Gupta [34] showed skin-core effects in polyester fibers. Hearle and Wong [35] studied the fatigue properties of nylon 6,6 and PET.

TABLE 5.1. Fiber fracture

Fracture type	Form of fracture	Polymer types	Ref.
Brittle fracture	Tensile failure; brittle fracture	Elastomers, high modulus fibers (e.g., aramids)	[31]
Ductile crack	Crack, draw, "V" notch formation, catastrophic failure	Nylon, PET, acrylics	[38]
Axial fiber splitting	Split along length; tensile fracture, torsional fracture	Cotton, some acrylics	[39]
Fatigue splitting	Cracks along fiber initiate at surface and break with long, thin tails	Nylon, PET, acrylics, aromatic polyamides	[35, 36, 40, 41]
Lateral failure	Failure normal to the fiber axis	Rayon, acrylics	[16]
Kink band	Compression inside of bend or by flexing at 45° to the axis; regions of reorientation due to shear forces	PET, nylon, aramids	[16]

Fatigue is an important property as it is related to repeated loadings that are typical in general use. Oudet and Bunsell [36] reported on the loading criteria for the fatigue failure of polyamide fibers. Smook et al. [37] studied the fracture process of ultrahigh strength PE fibers by examination in the SEM. Fracture was shown to be initiated at surface kink bands leading to formation of a fibrillated fracture surface. Extensive study of fracture under various conditions showed a diameter dependence of the tensile strength, which is consistent with the Griffith relation. Application of fractography to crosslinked fibers showed a change in fracture morphology from a fibrillar to a brittle mechanism.

Fracture mechanics considerations, summarized by Kausch [42, 43], permit determination of the effect of defects on the fracture stress, or tensile strength, of bulk polymers and polymer fibers. It is important that such detailed study be performed on the original or *primary* fracture surfaces, which are the only surfaces that relate to the tensile stress. This is generally done by examining both failure surfaces to ensure that they are a matched pair. An example of matched, primary fracture surfaces was shown earlier (see Fig. 4.39). In that case, a definite defect site was observed in the brittle fracture surface. Such assessment provides the structure-property information needed to modify the process and produce fibers with higher strengths.

Another example of typical fracture morphology, with a defect at the locus of failure near the fiber surface, is shown in Fig. 5.7. Once fracture is initiated, it slowly propagates radially away from the failure locus and continues into the adjacent polymer, providing a smooth surface, nearly perpendicular to the tensile stress, termed a "mirror." As the crack propagates away from the flaw site, it accelerates causing crack branching. The region where the acceleration begins to occur is seen to have fine ridges, or "mist," present whose size is dependent on the fiber microstructure. Crack propagation accelerates quickly through the fiber causing catastrophic failure. This "fast fracture" morphology is generally characterized by ridges, or "hackles," at an angle to the path of the original failure.

The fracture morphology in tensile fatigue is quite different as shown in the micrographs of a polyamide and a polyester fiber in Fig. 5.8 [44]. The fibrillar nature of a polyamide fiber results in a break with a tail, as seen in the broken ends (Fig. 5.8A, B). The typical fatigue failure has a crack initiated at the fiber surface with radial penetration into the fiber. The crack then runs along the fiber, and finally the fiber

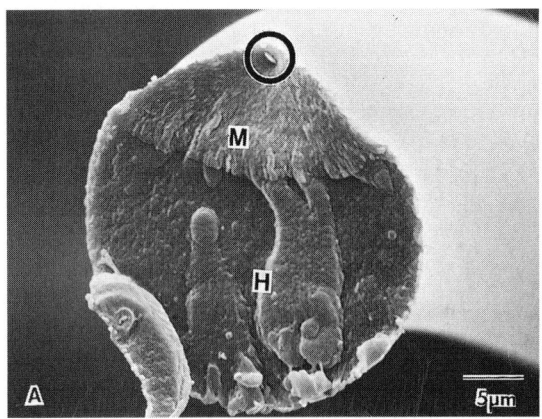
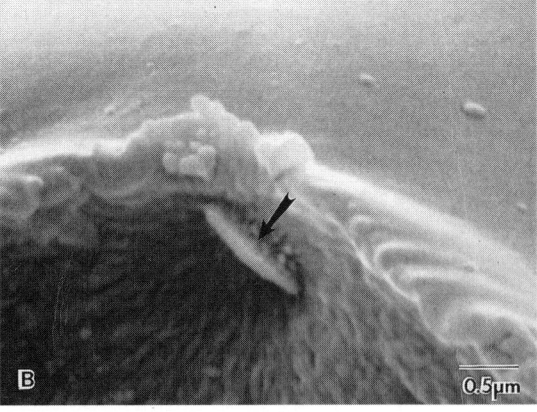

FIGURE 5.7. Scanning electron microscopy of fractured fibers reveals the flaws causing failure. A low magnification micrograph (A) shows the flat mirror (*M*) region, adjacent to the locus of failure, which is the region of slow crack growth. The fast crack growth region, or hackle (*H*), has large ridges. A higher magnification micrograph (B) reveals the flaw causing failure (arrow) in more detail.

Fibers 257

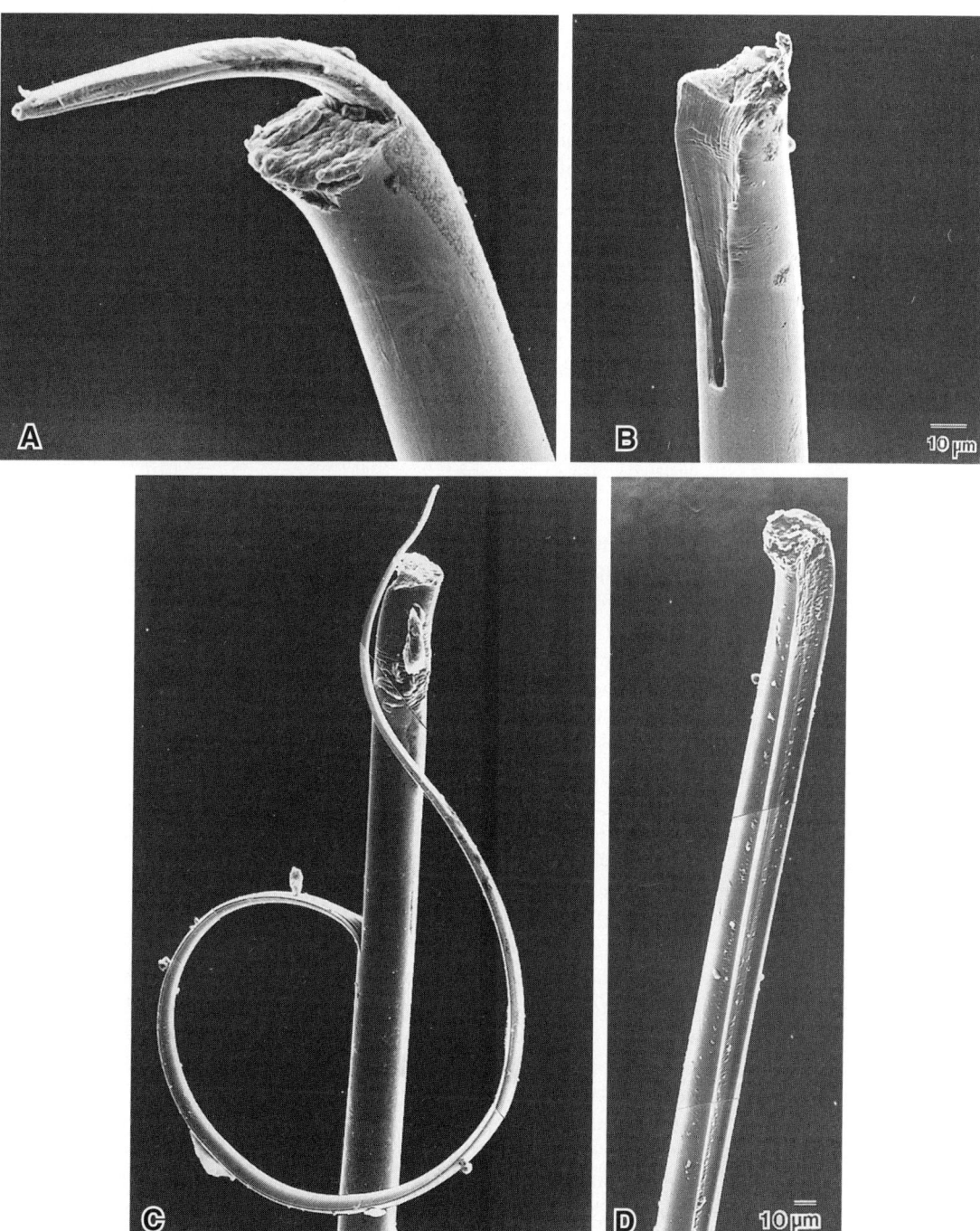

FIGURE 5.8. Typical fatigue failure of a nylon 6,6 fiber is shown by SEM (A and B) of a fiber tested at 0–55% of nominal breaking load, with a lifetime of 1.3×10^5 cycles at 50 Hz. Fiber failure is initiated at the surface resulting in a tail on one side (A) and long furrow on the other side (B). Fatigue failure of a polyester fiber, at 0–70% of nominal breaking load, with a lifetime of 3×10^5 cycles at 50 Hz, shows a somewhat different morphology. Fiber failure is initiated at the surface, but it continues along the surface resulting in a very long tail, on the one side (C), with splitting down that side as well, and a long shallow furrow on the other side (D). (From A.R. Bunsell [44]; unpublished.)

fails once the cross section cannot support the applied load. In simple tensile failure, the two fracture surfaces are mirror images, whereas after fatigue testing, complementary fiber ends are quite different. Fatigue fracture of a polyester fiber appears somewhat different in morphology (Fig. 5.8C, D). Although failure appears to be initiated at the fiber surface, the penetration is shallower. The crack extends along the near surface resulting in a very long tail on one fiber end. Tensile fatigue studies of synthetic polymer fibers have been discussed by Bunsell and coworkers [36, 41].

5.1.2.4 SEM of Woven and Nonwoven Fabrics

Yarns and fabrics are assemblages of fibers that have commercial application in the textile industry. The fabrics include those formed by weaving and also nonwoven fabrics. The geometry of the fabric, as well as the chemical composition of the polymer, influences mechanical properties and applications. The SEM is useful for evaluating: (1) construction, (2) coverage, (3) uniformity, (4) surface structure, and (5) effects of wear. The SEM provides a useful tool for the evaluation of fibers and fabrics during the manufacturing process. Fiber formation has been studied utilizing a bicomponent electrospinning approach [45] with results shown for polyvinyl chloride fibers using field emission SEM (FESEM) to image the morphology in the backscattered mode after sputter coating the fibers with 5 nm of Pt/Au to reduce charging.

The SEM was used to show yarns, composed of fibers twisted together, *woven* into a fabric (Fig. 5.9), and can be compared to an optical microscopy (OM) view of a similar fabric (see Fig. 5.1B). Important features of the fabric, "hand" and "coverage," relate to the yarn geometry, which can make a thick and comfortable fabric or a thin, uniform, and open structure. Tilted side views can be used to count the number and the length of protruding surface hairs, which are known to affect the feel or hand of the fabric and its mechanical properties [46, 47]. The complex subject of fabric wear

FIGURE 5.9. Scanning electron microscopy of a typical woven fabric shows twisted fibers in this fiber assemblage. The fibers overlap the space between the yarns providing coverage, or "cover," an important fabric parameter.

is particularly suited to SEM study, as the three dimensional structure and surface texture can be related to individual fiber failure [48, 49]. Kirkwood [48] described SEM studies that showed the wear of cotton used in military clothing into long fibrils and nylon into shorter, thicker fibrils. The mechanism of attrition is quite complicated and involves friction, surface cutting, and fiber rupture. The effect of both testing and wear can be evaluated using the SEM. Nylon 6,6, among other fibers, has been tested for wear by measuring the depth of surface damage in the SEM after sliding the fabrics on rough hard surfaces [49].

Nonwoven fabrics are another form of fiber assemblage, but they are less regular and uniform in structure than woven fabrics. A qualitative evaluation of nonwoven samples has been shown using a special stain (Pylam Products Co., Tempe, AZ) to identify fibers including polyester, rayon, wood pulp, and polypropylene as well as various blends and bonding points and binder distribution [50]. A spray spun nonwoven is shown in side view (Fig. 5.10A) on an SEM stub, and a surface view (Fig. 5.10B) shows the fabric has a range

Fibers

FIGURE 5.10. A spray spun nonwoven fabric is shown in the SE images (A and B). A side view of the fabric is shown (A) on an SEM stub, and a face view (B) shows there is a range of fiber diameters and shapes present with no specific pattern or arrangement. A calendered nonwoven surface is shown by SE images (C and D). The spots on the surface are regions of local melting that hold the fabric together (C). At higher magnifications (D), the fibers are seen to range from round to deformed shapes.

of fiber diameters and shapes with many deformed fiber cross sections. The coverage, size, and distribution of the open space and the individual fiber diameters can be important depending on the application. A more uniform, calendared nonwoven fabric is shown to have been spot calendared (Fig. 5.10C, D). The fabric has only been flattened out by the calendaring process in local regions, whereas other regions are three dimensional. Many large, deformed fibers are observed of varying diameter. A self-assembled honeycomb of polyurethane nanofibers has recently been shown by SEM imaging [51]; these fibers may be useful for drug delivery devices, protective clothing, and filters.

5.1.2.5 TEM of Fibers

Transmission electron microscopy (TEM) techniques are very important for the elucidation of details of fiber microstructure. The types of detailed structures that can be determined by TEM are:

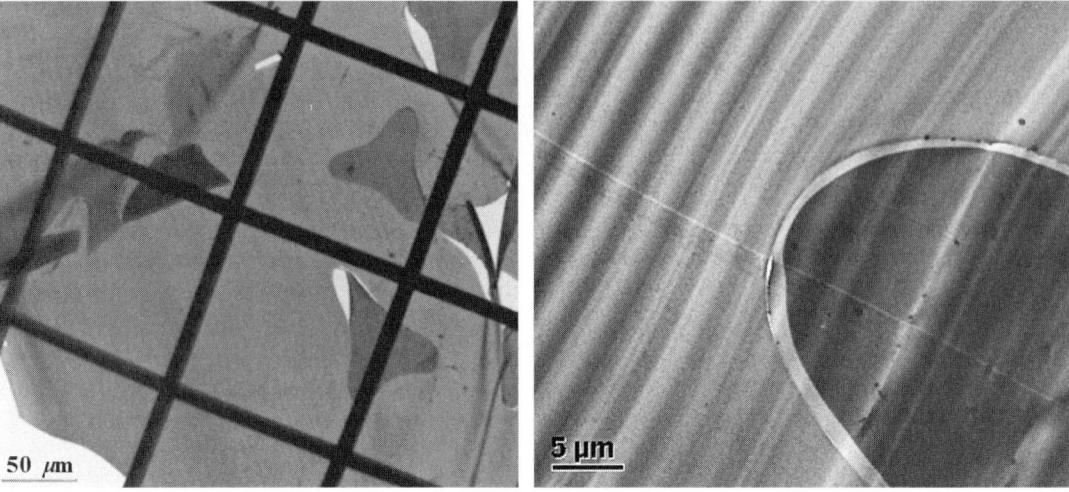

FIGURE 5.11. Transmission electron microscopy was used to image the sheath of trilobal carpet fibers by Wood [52], at 200 kV. The overview of the fiber sections (A) shows the trilobal shape and (B) shows the sheath at the fiber surface. (From Wood [52]; unpublished.)

1. polymer structure;
2. void size, shape, and distribution;
3. size, shape, and distribution of fillers;
4. local crystallinity;
5. crystallite sizes.

Early TEM studies were by replica methods [20], as in a study of replicated and etched fiber surfaces [16]. Such studies are now conducted by SEM of external and bulk structures and by ultrathin sectioning for TEM. Microstructural studies generally require complementary optical and SEM study to understand the arrangement of the fine structural details within the macrostructure. Transmission electron microscopy was used to image the sheath of trilobal carpet fibers by Wood [52], by embedding the fibers in a two-part epoxy, curing overnight at 60°C, and cryoultrathin sectioning with a diamond knife. Sections were accumulated in cold ethanol, transferred to water and to the surface of 1% phosphotungstic acid overnight, rinsed, and observed at 200 kV. Figure 5.11A, B shows the overview of the fiber sections and the sheath at the fiber surface, respectively.

A fairly simple example defining the structure in an experimental fiber is described to show that just one microscopy technique does not generally provide the complete structural picture. Three different techniques/methods are shown in Fig. 5.12. Scanning electron microscopy of the fractured fiber shows an overall view of the bulk structure and the fiber shape (Fig. 5.12A) and the internal porosity (Fig. 5.12B), although the size, shape, and distribution of the voids are not defined. Scanning electron microscopy of the outer fiber surface (Fig. 5.12C) shows that voids reach the fiber surface. Transmission electron microscopy micrographs of ultrathin cross sections (Fig. 5.12D, E) clearly provide a description of the void sizes and their local distribution. A pore gradient is observed, with smaller voids at the surface and coarser voids within the fiber. The smaller voids are located within a micrometer sized band around the fiber periphery. Complementary microscopy has been shown to describe the experimental fiber microstructure. The void sizes and distribution are parameters that are both affected by process modifications and relate to the end use properties.

5.1.3 Problem Solving Applications

Characterization of fiber microstructure normally requires several microscopy techniques to fully understand the details and solve

Fibers

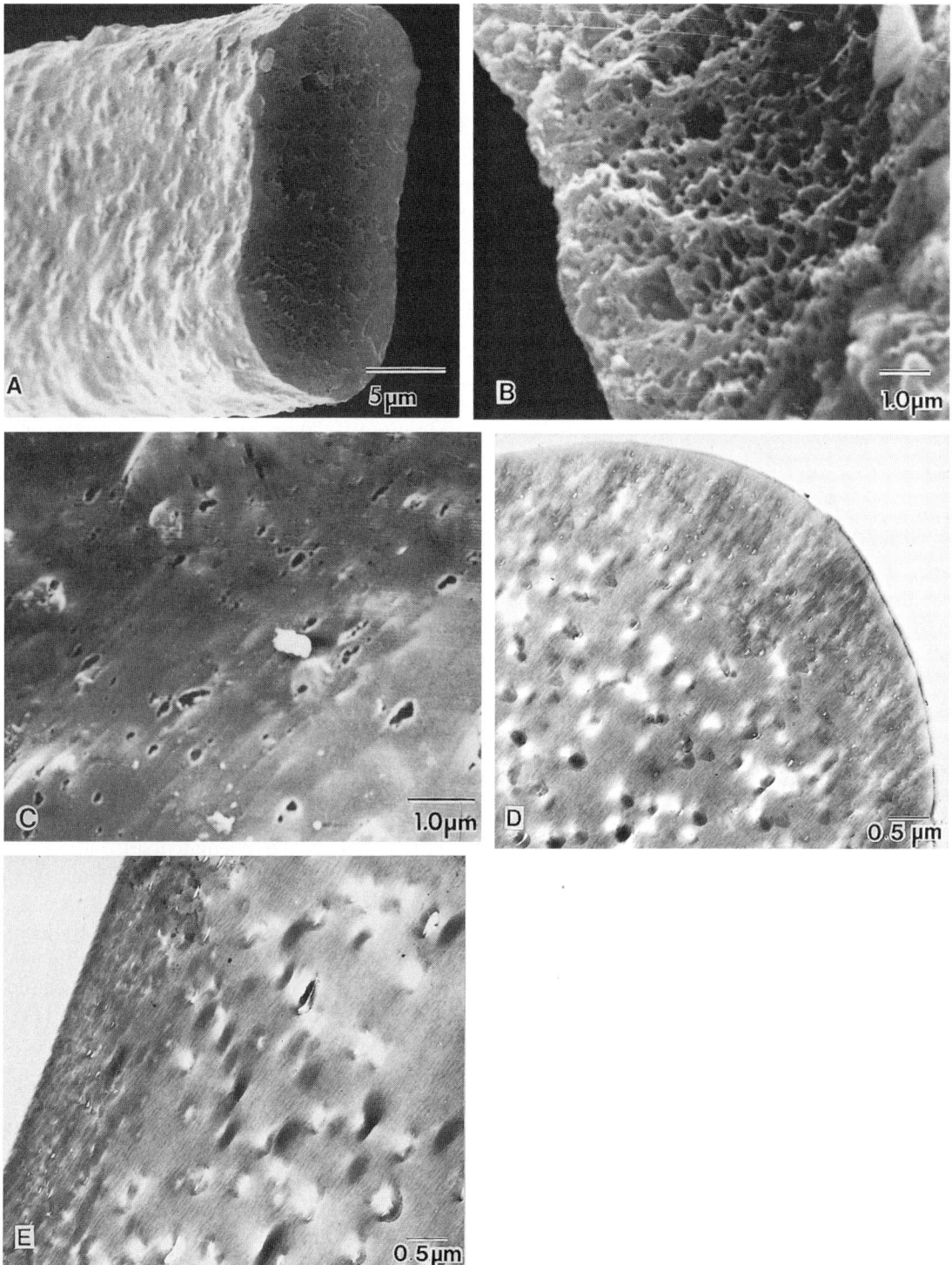

FIGURE 5.12. A microporous fiber is shown in the SEM and TEM micrographs. The fiber is seen to have voids in the fracture views (A and B), but the voids are not clearly defined. The outer fiber surface (C) also has voids, and particles are observed within some of them. Transmission electron microscopy micrographs of the fiber cross sections (D and E) reveal voids (white regions) that are smaller in size in an outer micrometer sized band than those in the central portion of the fiber.

problems. An optical cross section of a fiber may have a dogbone shape (Fig. 5.2D), and yet this image does not reveal much about the internal fiber structure. On the other hand, a fracture surface of a fiber may reveal the presence of internal detail when viewed in the SEM (Fig. 5.12) and yet not provide a complete picture of the structure. Clearly, complementary microscopy techniques and nonmicroscopy techniques must be applied to solving structural problems. Specific problem solving examples are described here that are representative of the wide range of studies conducted and documented in the many journals that publish polymer research.

5.1.3.1 Characterization of Textile Fibers

Delustrant and pigment particles are often used to provide modifications in the visual appearance of textile fibers and fabrics. For example, titanium dioxide particles are commonly added to polymers before fiber spinning to change the fiber luster. Additions to a polymer require monitoring the fiber formation process to assess the effect of the particles, for quality control of particle size and distribution and for failure analysis. Questions relate to the effect of particle size on mechanical properties and on the polymer structure. A wide range of practical problems can occur in any polymer process, and these problems are easier to solve when the fiber microstructure is fully known. Poor surface texture, fiber breakage during spinning, and nonuniformity in visual appearance are a few of the problems that may occur in any industrial plant producing textile fibers. Microscopic analysis is helpful in most of these cases where comparison can be made between standard controls and problem fibers. Rather than dealing with any one of these specific problems, the example in this section is a description of a typical, microstructural characterization of a high speed spun polyester fiber containing titanium dioxide.

Direct observation of a fiber in the optical microscope or study of a fiber cross section provides information relating to the size and distribution of added particles. A longitudinal view of a PET fiber in an OM micrograph (see Fig. 5.1A) reveals dense particles. The distribution of the particles can be seen, but their size and their relationship within the fiber are not clear at this magnification. Fiber cross sections (see Fig. 5.2) show the particle distribution more clearly. Only a small, thin section about $5\mu m$ thick is observed optically, and a great number of such sections must be examined to define a statistical distribution.

Electron microscopy techniques provide more resolvable detail than optical microscopy. Scanning electron microscopy of a PET fiber surface (Fig. 5.13A) shows splits and delustrant particles. The particles were determined by x-ray microanalysis to contain titanium. A surface discontinuity is shown by SEM (Fig. 5.13B) and by high resolution secondary electron imaging (SEI) (Fig. 5.13C) taken in an AEM (a TEM with a scanning attachment). The SEI image clearly shows particles protruding from holes in the fiber surface, although the image shows limited depth of field by this technique, due in part to the short working distance. At higher magnifications (Fig. 5.13D, E), particles about $0.1\mu m$ in diameter are seen adjacent to the fibrillar polymer texture. The microfibrils are about 50nm wide and are oriented parallel to the fiber axis. These structures have now been resolved and identified by high resolution SEM [53].

Cross sections of particulate loaded fibers are required for accurate particle size determination and for studying the effect of the particles on the microstructure. Optical cross sections of PET fibers (Fig. 5.14A, inset) have more particles present as the sections are significantly thicker (ca. $5\mu m$) than TEM sections (Fig. 5.14), which only show a few aggregates of dense particles. Skin-core textures resulting from high speed spinning [54] may also be seen in some sections. Particles and the adjacent holes are observed in the section. An important question is whether the holes are caused by the process or are a result of the sectioning method. Care must be taken in interpretation, and thus follow-up studies were conducted to determine the origin of the voids.

Further structural study can be conducted by simple peeling methods for SEM and by staining methods for TEM. The peeling of a segment

Fibers

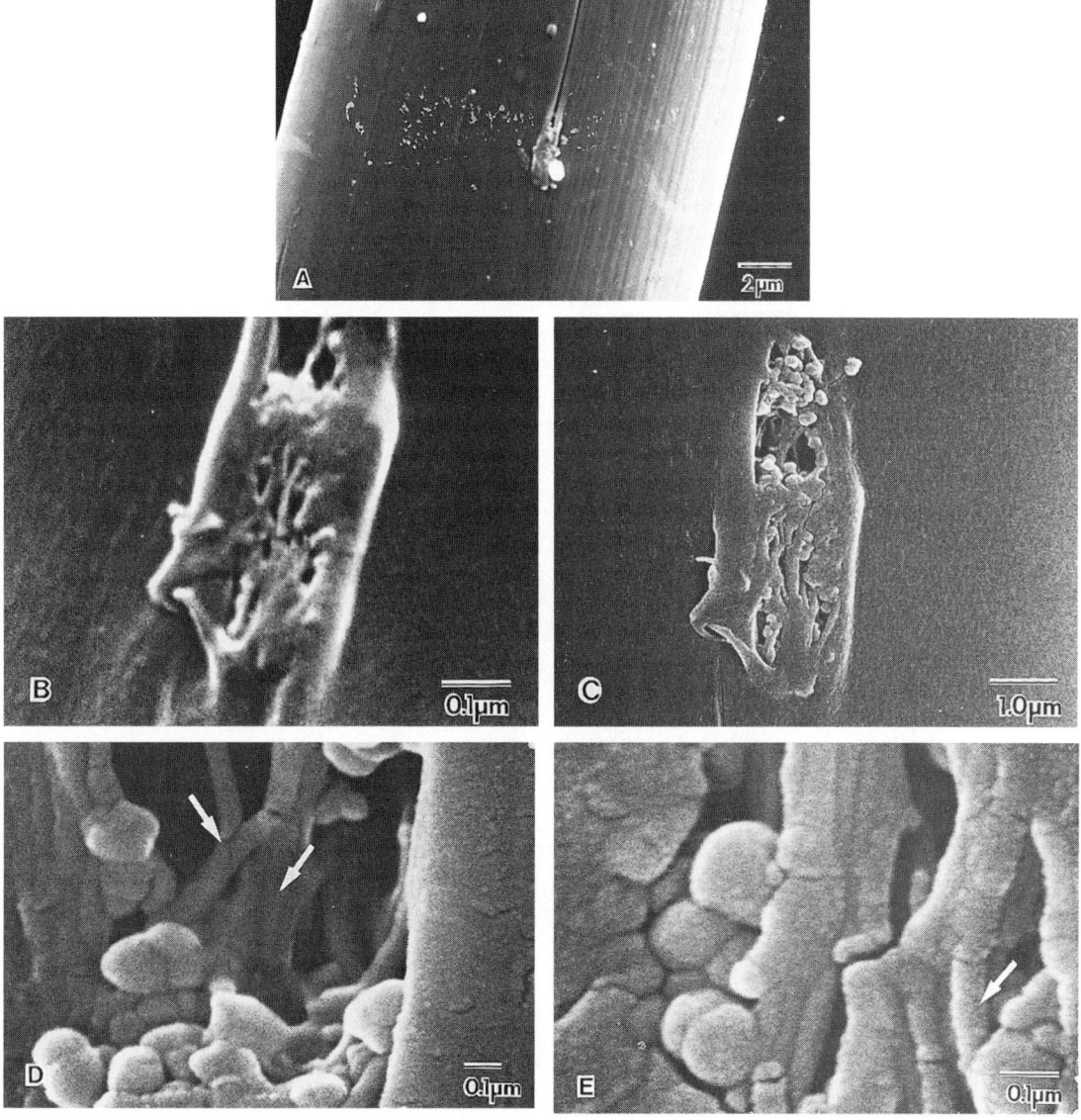

FIGURE 5.13. Fibers containing titanium dioxide are commonly employed for various textile applications. Scanning electron microscopy of a fiber surface, without finish (A), shows a surface split, likely caused by such a particle at or near the surface. A similar region at higher magnification is shown in both SEM (B) and high resolution SEI (C) micrographs. The defect region is not very clear in image (B) taken in a dedicated SEM, but the higher resolution image provides interesting detail (D and E). Particles, voids, and microfibrils (arrows) are observed.

of a fiber to reveal the internal structure, as first developed by Scott [20], has since been used to show the microfibrillar structure of nylon 11 and 12 [55] and of PE [56, 57]. The highly fibrillar structure that develops in PE fibers on drawing was correlated with the increasing crystalline orientation, as observed by x-ray diffraction, and with increased tensile strength and modulus. Poly(ethylene terephthalate) fibers peeled back to reveal their internal structure (see Fig. 4.5) show the microfibrillar texture and the titanium dioxide particles in

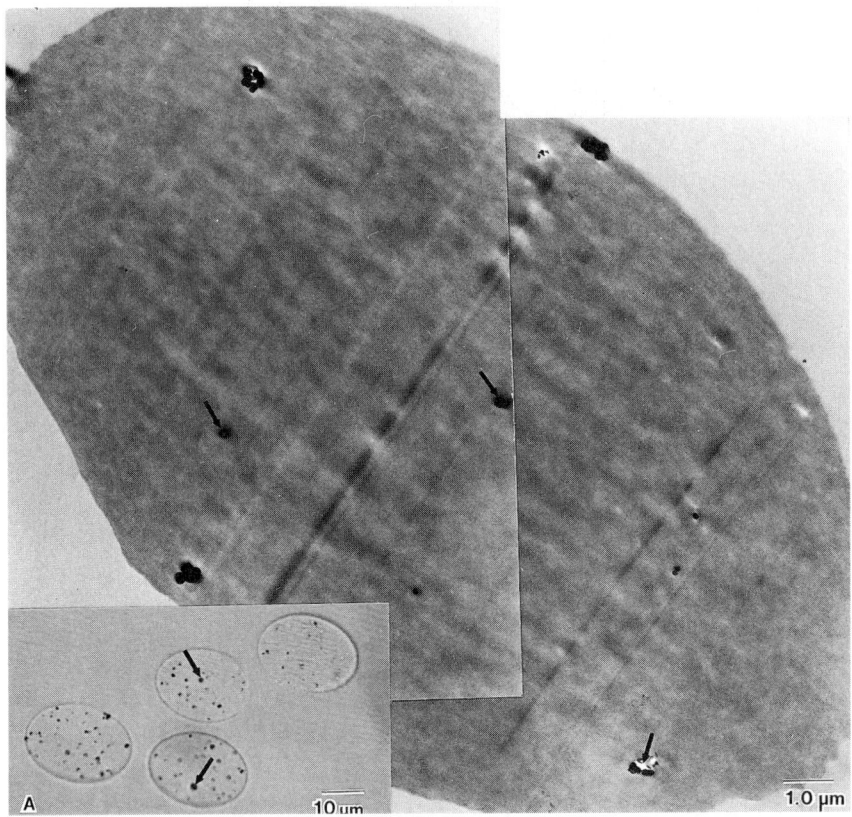

FIGURE 5.14. Transmission electron microscopy micrographs of polyester fiber cross sections reveal the size of the pigment particles. The optical inset (A) shows fibers with dense particles in greater amounts than seen by TEM due to the difference in section thickness. The dense particle aggregates are titanium dioxide particles. The fiber exhibits no major structural detail. The dense lines are knife marks produced during sectioning.

depressions within the fiber, shown in more detail in Fig. 5.15. As with the TEM sections, this morphology could be due to the deformation during peeling, although this study suggests that there are holes formed adjacent to the particles during the spinning process.

Another approach to the characterization of fiber microstructure is the isoprene inclusion method (see Section 4.4.2), applied to the study of PET fibers [58] and aramid fibers [59] for the purpose of showing their radial microporous and fibrillar texture. Any holes or voids are filled by inclusion of isoprene in the fiber, which is then stained by the reaction with osmium tetroxide. Longitudinal sections of high speed spun PET are shown in the TEM micrographs in Fig. 4.14A and Fig. 4.14B, before and after the reaction, respectively. After isoprene inclusion and staining, these regions (Fig. 5.16) are electron dense and elongated parallel to the fiber axis, with a fine pattern of elongated, dense regions, also aligned parallel with the fiber axis, which suggests there is an ordered arrangement of voids about 10 nm wide. Furthermore, voids are more prevalent near the outer fiber surface than within the fiber. The staining method has confirmed that voids are present in high speed spun PET fibers.

Characterization of the microstructure of high speed spun polyester fibers has been demonstrated using combined SEM of bulk peeled fibers and fiber surfaces, OM of thin sections, and TEM of sections both stained and unstained. The polyester fiber microstructure has been

Fibers

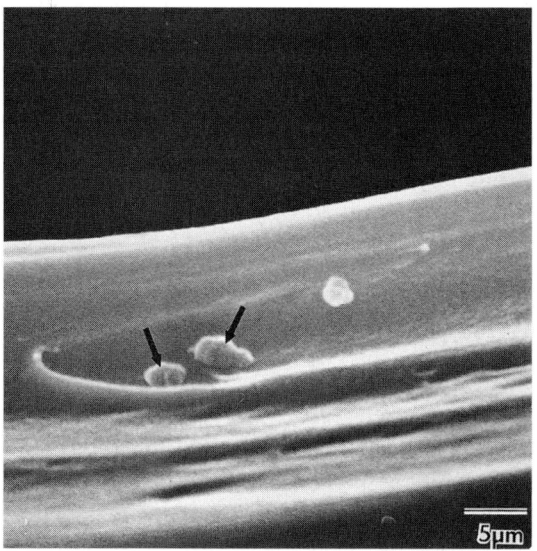

FIGURE 5.15. A PET fiber peeled back to reveal the internal textures was seen by SEM (see Fig. 4.5) and a more detailed view shows particles in furrows, and voids are observed adjacent to the particles.

shown to be microfibrillar. It contains microvoids, elongated parallel to the fiber axis, and has elongated voids adjacent to the particle aggregates, also aligned parallel to the fiber axis.

The effect of the titanium dioxide delustrant particles is to produce voids adjacent to the particle aggregates that elongate on drawing. This morphology suggests a high degree of orientation with enhanced mechanical properties [57]. Study of PET fibers by x-ray scattering, infrared, and birefringence [14] shows a microfibrillar structure, with microfibrils loosely held together in fibrillar units at least an order of magnitude larger in size. The microfibrillar structure shown here for PET fibers is similar to the structure shown for nylon stained with tin chloride [13].

5.1.3.2 Metal Loaded Fibers

Microscopy techniques can be used to evaluate the size and distribution of particles added to polymer fibers, such as metals that modify the physical, mechanical, or electrical properties. In general, ultrathin sections are examined in either scanning transmission electron microscopy (STEM) or TEM modes to reveal the particles within the polymer. Energy (EDS) and wavelength dispersive x-ray spectroscopy (WDS) methods are used to map for various elements in order to establish the relation between the particle morphology and chemical

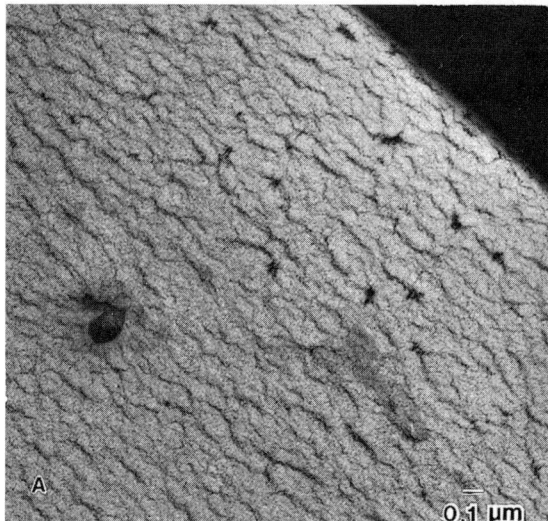 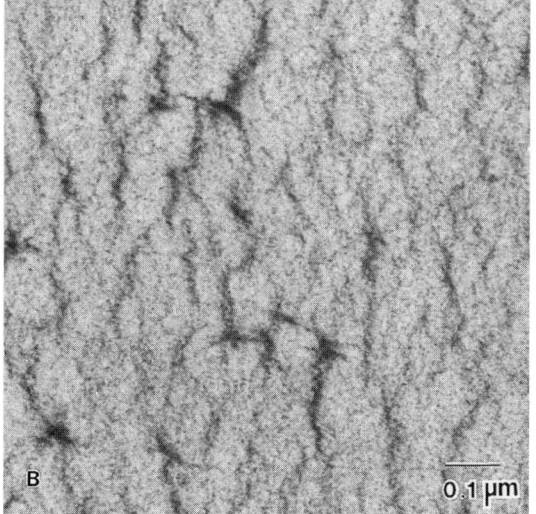

FIGURE 5.16. Transmission electron microscopy micrographs of longitudinally sectioned PET fibers taken after isoprene inclusion and staining reveal fine dense elongated regions, aligned with the fiber axis, representing voids about 10 nm wide. There is a higher void density near the fiber periphery than in the core, as shown in cross section (A) and at higher magnification (B). A circumferential, skin-core arrangement of voids is observed.

composition. A specimen preparation method for x-ray analysis in the SEM is to use a trimmed block face, which remains after cutting thin sections, or to study a thick section.

As an example of such a study, a particle loaded polymer fiber cross section, is shown by TEM, BEI, and in an x-ray map, providing a complementary assessment of the size, chemical nature, and location of the particles. A dense band around the fiber periphery is seen by TEM with dense particles ranging from about 5 to 20 nm in diameter (Fig. 5.17A, B). The BEI

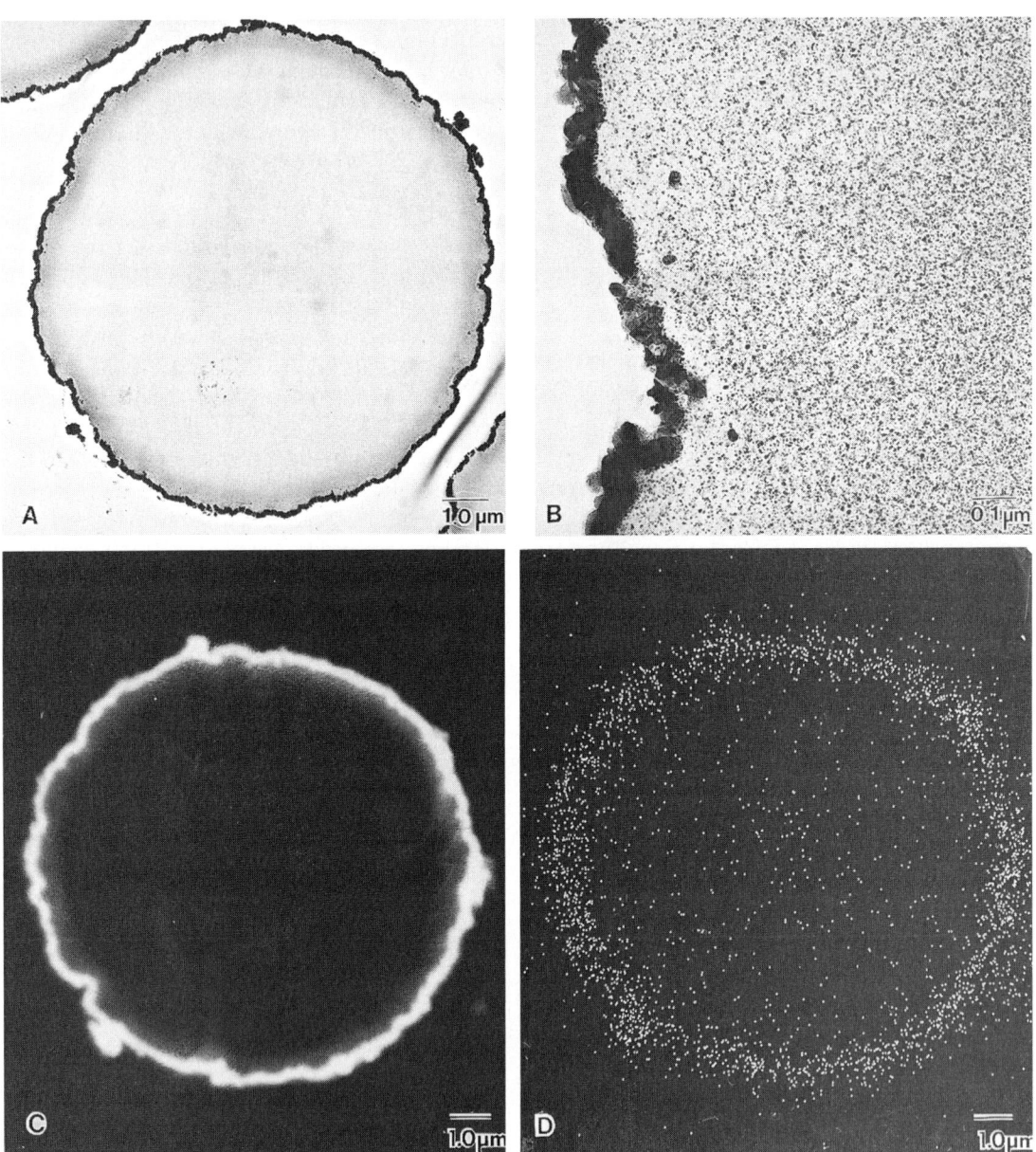

FIGURE 5.17. A metal particle loaded fiber is shown by complementary microscopy techniques. Sharp detail is observed by TEM (A and B), which shows a dense band around the fiber periphery and fine dense particles adjacent to and within the band. The chemical composition of the particles is seen to differ from the polymer in the BEI "compo" image (C), and x-ray spectra and mapping (D) show the distribution of the metal.

Fibers

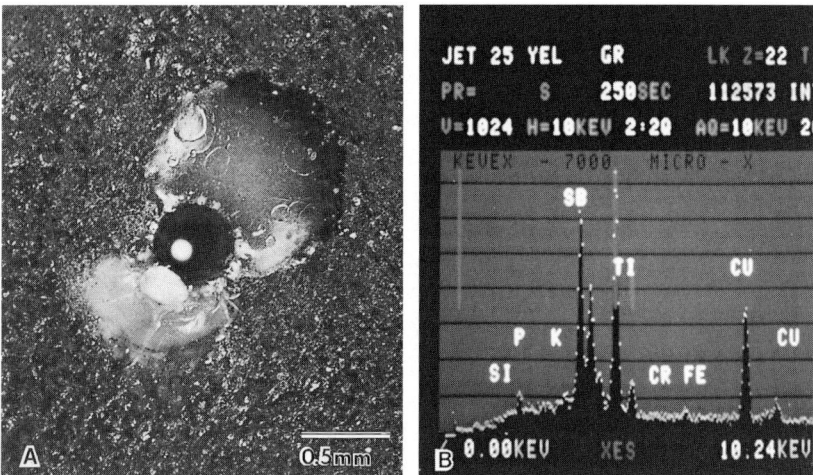

FIGURE 5.18. Combination of optical microscopy and EDS analysis permits the identification of contaminants plugging a spinneret. Rust colored material present on the spinneret (A) was scraped off to give the EDS spectra (B). The plugging material contains silicon, phosphorus, antimony, titanium, chromium, and iron. The copper is due to the specimen holder. (See color insert.)

image (Fig. 5.17C) shows that the dense outer layer contains higher atomic number material than the fiber. A WDS x-ray map (Fig. 5.17D) shows the elemental distribution. Thus, microscopy techniques provide a complete assessment of the particle size, chemical composition, and location of particles added to polymer fibers to modify properties.

5.1.3.3 Contamination

Contamination studies are often required to understand process related problems and to investigate the cause of property deficiencies. In many cases, the problem is not the polymer itself but the addition of some unknown material during handling or processing. Determination of the nature of such contaminants is a serious problem for the microscopist. Spinnerets used in fiber spinning can become clogged, stopping the process because of polymer plugging or contamination. The optical micrograph (Fig. 5.18A) shows the surface of a spinneret with dark, rust colored material as well as white polymer (see color inset). This material was scraped off the spinneret and dispersed on a carbon stub for EDS. The EDS spectra (Fig. 5.18B) of the rust colored material showed that antimony (Sb), silicon (Si), titanium (Ti), and iron (Fe) were present. This showed inorganic contaminants to be the cause of clogging of the jet.

5.1.4 Industrial Fibers

5.1.4.1 Tire Cords

Major industrial applications of fibers are in the production of tire cords and belts. Some tire cords are composed of yarns of polymer fiber, twisted together into cords coated with adhesives. Wilfong and Zimmerman [60] discussed tire yarn property criteria and the polymer and fiber structural factors that control the properties. Polyester, rayon, and nylon are common tire cord yarns. The tire cord composite is a complex system of multiple interfaces that must be well bonded to provide high strength. Characterization of tire cords by microscopy is used to understand the microstructure and apply that knowledge to solving problems. Some of the problems encountered relate to: (1) application of fiber finishes, (2) poor strength or strength reduction, (3) poor adhesion of fibers to adhesives, (4) fiber degradation, and (5) poor adhesion of the coated cord to the rubber.

Finishes must be applied uniformly to the fiber for protection, ease of handling, and compatibility with the adhesive to limit strength reduction. The adhesive is applied to the cord to enhance cord integrity, protect the fibers from the rubber, and to join the cord and rubber into a well ntegrated material. Essentially, there are several interfaces of interest: fiber-finish, fiber-adhesive, and adhesive-rubber. The adhesives are generally resorcinol-formaldehyde-latex (RFL) dips [61]. An optical cross section (Fig. 5.19A) provides an overview of an RFL coated tire cord. The outer surface of the coated cord (Fig. 5.19B) and a surface close up is shown by SEM (Fig. 5.19C) and BEI (Fig. 5.19D). The normal SEM image does not provide much detail, except that a filmy coating is present, whereas the BEI image provides detail on the disposition of the RFL. Regions that are brighter are likely of higher atomic number than the background. The adhesive coating appears to be "puddled" into the interstices between the fibers more than on the outer fiber surfaces, which would be expected

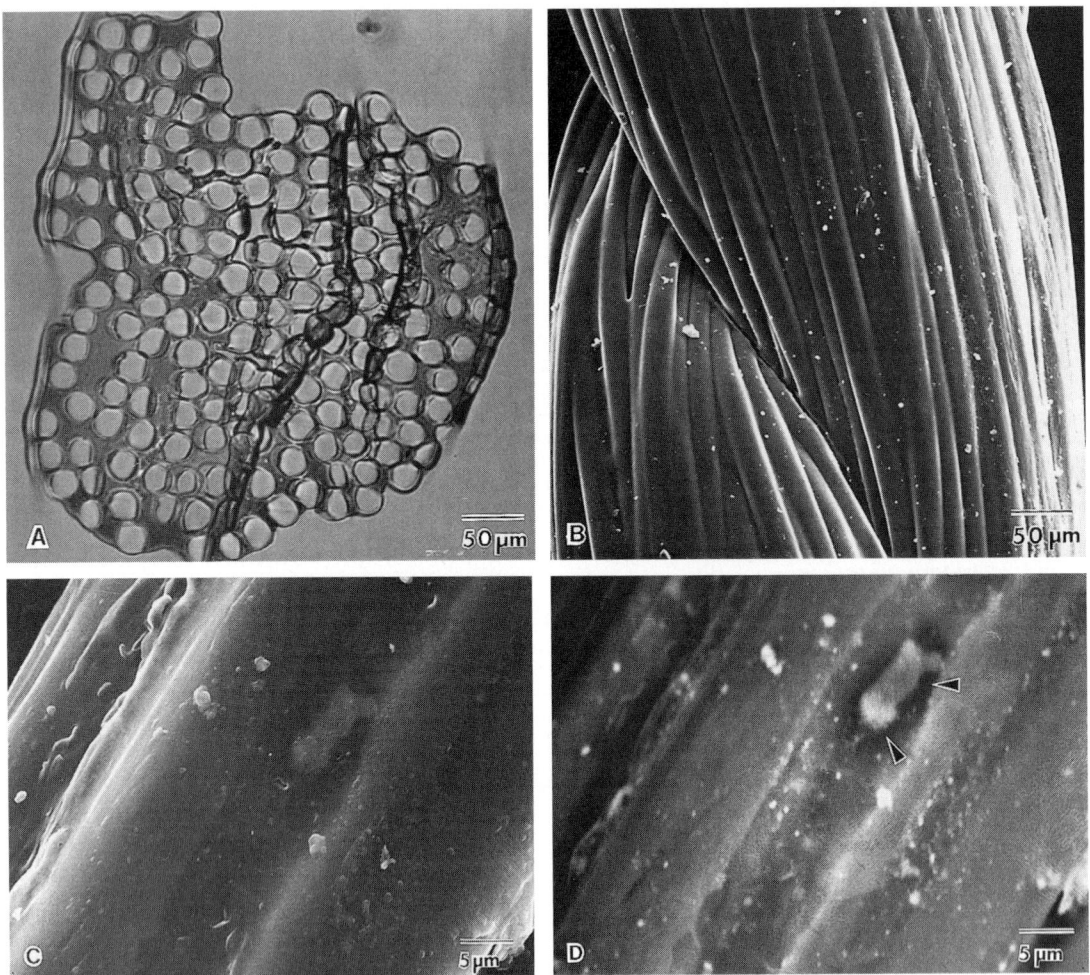

FIGURE 5.19. Tire cords are shown by several complementary microscopic methods. A cross section of an adhesive coated yarn is shown by optical microscopy (A) (see color insert). The surface of an RFL coated cord is shown by SEM (B). Comparative normal SEM (C) and BEI (D) images show the filmy adhesive coating. The BEI image shows that the surface has higher atomic number particles (arrowheads) and detailed surface structures that suggest that the adhesive is found in higher concentration in the interstitial regions, between fibers, than on the fiber surfaces in this experimental cord.

Fibers

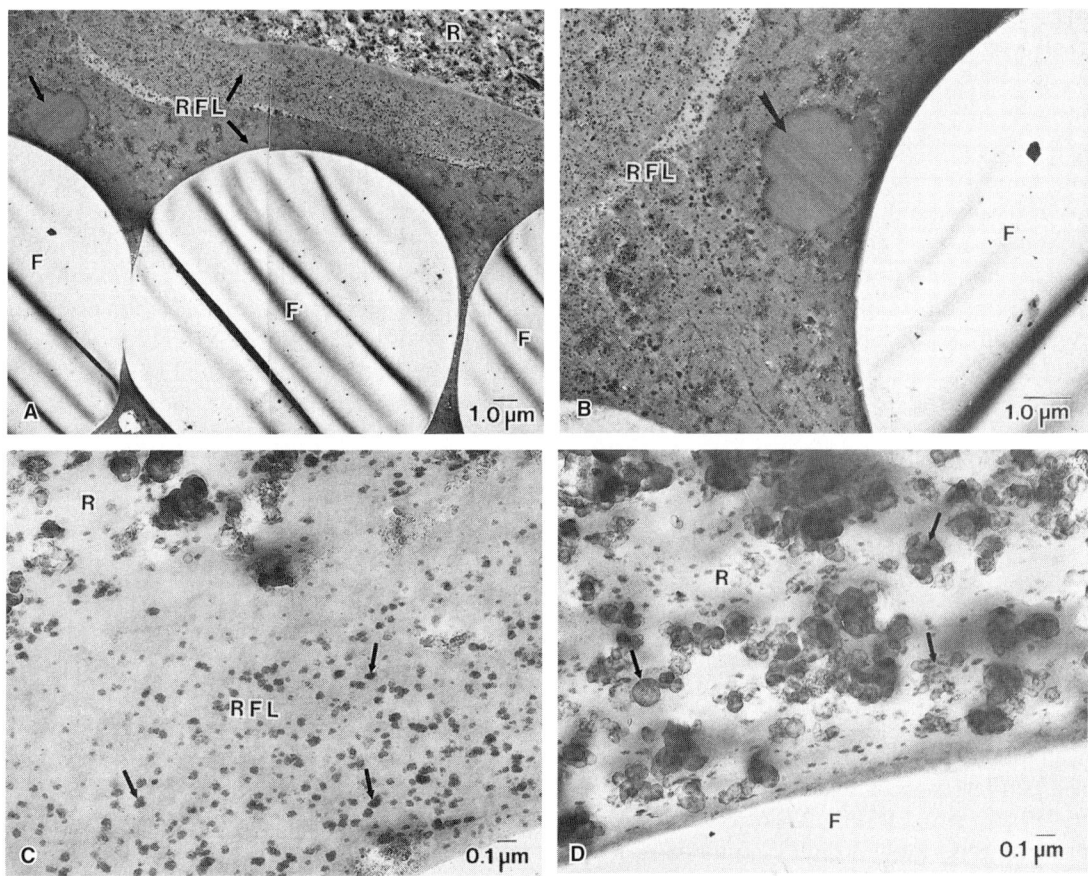

FIGURE 5.20. Ebonite treated tire cords are shown in TEM micrographs of ultrathin cross sections. The montage (A) provides an overview of the double dip RFL coating the polyester fibers (F). The rubber (R) is seen coating the RFL. Note that the dark lines in the fibers are wrinkles due to the sectioning. At higher magnification (B), a phase separated, globular region is seen within the first RFL dip. Good adhesion is exhibited between the fiber and the RFL. A micrograph showing the fiber-RFL and the RFL-rubber interfaces (C) shows that dense specks are present but not near the rubber interface. A view of the rubber (D) shows that carbon black aggregates are present (arrows) as well as other dense particles. Adhesion is seen to be good at the RFL-rubber boundary.

to result in incomplete adhesion to the rubber.

The combination of materials in a tire cord is a nightmare for the person performing the microtomy as each material has a different hardness and would be expected to pull apart during the cutting procedure. The ebonite diamond knife sectioning preparation (see Section 4.4.6) permits uniform hardening of the soft latex in the RFL and the rubber as was shown in an overview of the structure in the TEM (see Fig. 4.22). Detailed TEM images of cross sectioned tire cord specimens are shown in Fig. 5.20. An overview (Fig. 5.20A) shows that this cord has a two dip coating on the fiber surface with interfaces showing good integrity; thus there is good adhesion between the fiber and the RFL and between the RFL and the rubber. Specific morphological details of the two RFL layers are seen (Fig. 5.20B) that relate to the chemistry and behavior of the system. The fiber surface has a uniform RFL coating, containing dense specks that are depleted at the rubber interface (top) (Fig. 5.20C). A view of the rubber in a region with only a thin layer of adhesive (Fig. 5.20D) shows small, dense specks and larger carbon black particles within the rubber phase. Carbon black is a common filler used to reinforce rubbers and enhance strength. The ebonite method has been applied to optimize finish and adhesive application on tire cords and to better understand fiber-rubber adhesion, its influence on strength, and strength loss. Tire cords exhibiting poor strength, strength loss, and/or poor adhesion of the fiber to the adhesive have been observed, and fibers with good mechanical properties generally exhibit good adhesion overall.

5.1.5 High Performance Fibers

5.1.5.1 Introduction

Applications requiring superior mechanical properties combined with light weight have benefited from the use of polymers in fiber form. High performance organic fibers can be used to produce high tensile modulus, high strength, and high energy absorption structures of much less weight than the equivalent parts made from structural metals or ceramics. High modulus fibers made from PE, aramids, or thermotropic liquid crystalline polymers (TLCPs) are finding applications in cables, protective fabrics (bulletproof vests), and composites for automotive, marine, and aerospace use. Biopolymers such as wood pulp fiber and spider silk are studied for understanding structure-property relations and for their own applications.

Good reviews of these high performance polymer fibers include descriptions of their structure, morphology, production, and properties and their use in composites (e.g., [62–68]). Generally, the fibers have a highly oriented and largely extended chain structure. This structure of closely packed, rod-like molecules with weak intermolecular interaction produces a very anisotropic material that is weak in shear and compression. The tensile modulus along the fiber direction is largely controlled by the cross sectional area of the molecular chains, the chain stiffness (helical chains are soft, all-trans chains are stiff), and the perfection of chain orientation and extension. A high performance fiber may have the chain orientation distributed over a narrow range, so that the mean misorientation of molecular chains from the fiber axis is only one or two degrees. In diffraction, the fiber pattern then looks just like what would be obtained by rotating a single crystal about the chain axis. The mechanical properties of the chains are so anisotropic that even this small misorientation may reduce the axial modulus of the fiber by 30% or 50% from its ideal value, but it may still be as much as 100 times greater than the transverse modulus. Composite concepts, such as the aggregate model of Ward [69], describe how the fiber mechanical properties are derived from the fiber structure. Fibers made of aramids or TLCPs are quantitatively modeled as an aggregation of molecules with a range of orientation, but this simple model does not work for PE.

5.1.5.2 High Modulus PE Fibers

High modulus fibers produced from conventional polymers may be made from a range of flexible chain linear polymers using a number of processing routes. For PE, these include solid-state extrusion, die drawing, zone drawing, and gel spinning [62, 70]. Compaction processes using PE fibers to make composites have been explored using microscopy techniques as well (e.g., [71, 72]). Commercial high modulus PE fibers, such as Dyneema (trademark of DSM, Netherlands) or Spectra (trademark of Honeywell, Morristown NJ), are produced from extremely high molecular weight polymer by forming a gel and spinning it. High molecular weight gives high strength along with high modulus, and gel spinning gives the highest production rates—still very slow compared with those for conventional fibers. The polymer is first dissolved or highly swollen in solvent at elevated temperature, and the solution is wet spun into a quenching bath to form a gel filament. The gel can be dried by solvent extraction and then hot drawn to very high extension ratios. Alternatively, heating, drying by evaporation, and drawing can be done simultaneously.

Fiber strength is affected by the presence of defects such as chain ends and their arrangement [54, 62, 63, 70]. Any flaw or cluster of defects that can act as a stress concentrator may reduce the strength. Therefore, the highest strength fibers tend to have a close to homogeneous morphology, where the "amorphous" material is very highly oriented. Most high performance fibers are microfibrillar, with morphology similar to the structure shown in Fig. 4.5. A natural model for the mechanical properties of the fiber is that of a fiber composite, with the more crystalline microfibrils the reinforcing, load bearing elements. The high alignment and extreme anisotropy of these fibers makes peeling or transverse splitting easy, so that internal surfaces are made accessible. These must be preferred fracture surfaces, planes of weakness, so they may not be completely representative of the whole fiber.

It is easy to analyze these internal surfaces by observing fibrillar features by SEM, TEM replica, or SPM. If the preferred fracture planes are randomly distributed, there will be a wide range of feature sizes on the surface. A single microscopic technique will have some limit to resolution and to its field of view. Features approaching the size of the field of view must be rare in an image, so any single imaging technique can only show a limited range of sizes of object features. The most common will be a size above the resolution limit. Care must be taken to assess if there is a true hierarchy of structures or if this is due to sampling of a random continuum. As so often in microscopy, other techniques or supporting information from a totally different method is required for certainty in interpretation.

Light microscopy and low voltage SEM (LVSEM) have been used to study the crystallization behavior and morphology of PE based single polymer composites [65]. In this study, ultrahigh molecular weight PE (UHMWPE) fibers were studied by thermal analysis, TEM, and LVSEM at 1 kV. Samples were prepared by fixing both ends of four fibers on a glass slide and hot pressing a high density polyethylene (HDPE) film onto the fibers, annealing them in a hot stage at 138°C for 5 min and melting the HDPE film and not the UHMWPE fibers. These methods yielded insights on the interface morphology and the transcrystalline layer. High modulus PE fibers, made by melt and gel spinning, were examined after permanganic etching as were fibers after treatment by high temperature compaction [66]. Transverse sections of the fibers were obtained by embedding them between two sheets of polystyrene-(ethylene-propylene) block copolymer, Kraton (trademark of Kraton Polymers US LLC, Houston TX) and cutting with a diamond knife for SEM. Replicas for TEM were obtained using a two stage technique, with cellulose acetate on the etched surface and then coating with Ta/W at an angle of about 35° to the horizontal, followed by carbon coating vertically onto the replica. Transmission electron microscopy was conducted after solvent extraction.

Transmission electron microscopy of stained fiber samples is another technique, but preparation of these samples is much more difficult. Schaper et al. [73] conducted an extensive TEM study of surface grown, high modulus PE fibers. They used various preparations including chlorosulfonation (see Section 4.4.4) and ultramicrotomy. With computer image processing to improve feature visibility, they could clearly distinguish fibrils micrometers long and about 5–20 nm in width. They also saw defects, such as voids and kink bands similar to those seen in LCP fibers (see Section 5.6.4). The width of the fibril core did not vary with draw ratio, a result consistent with their model of formation of the extended chain fibrils. Calculations of tensile modulus based on the observed structure agreed well with the results of mechanical testing. This work, among many reviewed [73], supports the idea that there are microfibrils in this material that are the load bearing elements. A fibril width of 10–20 nm agrees with earlier work [74, 75] that used crystallographic dark field (DF) contrast in the TEM on specially prepared thin samples of surface grown material. The crystal lengths in the fiber direction were shown to follow a most probable distribution; that is, a mean length could be determined, but there was no preferred length.

Atomic force microscopy (AFM) (e.g., [76–79]) has been used to image the external and internal fracture surface of high modulus PE and other fibers with high degrees of molecular orientation. Maganov et al. [76, 77] studied cold extruded PE surfaces prepared with an ultramicrotome cutting along the extrusion direction. This controlled fracture gives the generally flat surface required for high resolution AFM. The rod-shaped samples were embedded in epoxy for microtomy, to produce a flat cut surface. The image size ranged from 700 × 700 nm to the atomic scale. At the lower magnifications, microfibrillar structures were seen, with fibrils of 20–90 nm diameter and indefinite length. The authors claim to see the molecular chain structure at higher magnifications, and the images are quite impressive, apparently showing local details of the molecular arrangement. However, true atomic or molecular resolution of the undisturbed sample is exceptionally difficult to obtain in AFM, especially when using soft materials like polymers. Atomic force microscopy instruments using intermittent contact AFM (ICAFM), allow imaging at lower applied forces (see Section 3.3.4). Atomic force microscopy requires a generally smooth surface, and if a fracture surface is used, artifacts may appear. Annis et al. [79] imaged fracture surfaces of extended chain crystals of PE grown at high pressure by both AFM and TEM of replicas (see Section 4.6.2). The images were very similar, with morphological details and features of the growing crystalline lamellae clearly shown. The AFM showed some sharp edges as rounded and could not image sharp protrusions. This is because the tip used to form the image is rounded, and the image is a convolution of the sample shape and the tip shape.

5.1.5.3 High Performance Fibers

There is a broad range of high performance fibers in general use today in addition to PE, which range from wood pulp fibers, used in composites, to protective fabrics used for firefighters, bulletproof vests, and so forth. The polymers range from cellulose, to higher performance fibers composed of PE (see previous section), aramids, polybenzobisoxazole (PBO), M5 (another rigid-rod-like PBO), and LCP fibers [80]. Aramids for instance exhibit better creep resistance and temperature resistance than PE fibers and are thus used in more demanding applications, but some of the newer fibers are lighter weight and more useful for protective applications. Much of the driving force of these newer developments is the military and other defense organizations that require protective materials that are flexible, lightweight, and bullet or fire proof. Assessment of these fibers benefits from the use of a range of microscopy techniques.

Study of compacted PET filaments has been conducted [68, 81] using a combination of SEM of the woven PET cloth and SEM of peel fracture surfaces, showing the evidence of recrystallization of the melted matrix phase and reporting on thermal analysis studies. Study of high modulus PET yarns for reinforcing applications, especially high modulus, low shrink (HMLS) yarns, have benefited from study by SEM, TEM, and optical microscopy [82]. Fibers were sliced longitudinally with a scalpel and peeled back for SEM after sputter coating with 5 nm Pt. Embedded and thin sectioned (ca. 100 nm thick) samples were examined by TEM, and polarized light microscopy (PLM) was used to examine fibers in an immersion oil. Birefringence was also measured in PLM. One result of the study was to show cold draw processing yields a sheath/fibrillar core microstructure in each fiber of the yarn.

A study has been conducted of the compressive deformation behavior of thermally crosslinkable poly(p-1,2-dihydrocyclobuta phenylene terephthalamide) (PPXTA) fibers [83]. The morphology of the failure zones was examined by SEM of Au/Pd coated samples at 2 kV, which clearly show the kink zones, and by dark field TEM at 400 kV of samples microtomed to less than 0.1 μm thickness. Compressive failure of the fibers changed from kink-dominated failure to brittle rupture with increased heat treatment temperature, evidently as the result of crosslinking or of chain degradation. A study of the nature of kink band formation in high performance fibers, in this case the compressive behavior of carbon and polymer fibers (DuPont Kevlar, Wilmington, DE poly(p-phenylene benzobisthiazole) [PBZT], and PBO), was measured using a microscale compression apparatus in an optical microscope [80]. With increasing compressive strain, kink band formation was observed, and the number of kink bands per unit length (referred to as kink band density) was determined. By extrapolating to zero kink band density, the critical compressive strain was obtained for the compression of PBO and PBZT fibers and was calculated to be a 0.42 to 0.57 µm diameter fibril, and not the smaller diameter microfibrils.

5.1.5.4 Examples of Fiber Studies

Atomic force microscopy has been used extensively during the past decade, but direct imaging of fibers is difficult as they are not flat unless they are embedded and a flat surface is produced by microtomy. The high sensitivity of phase contrast to material properties is shown here by example using a wood pulp fiber image (Chernoff and Magonov [84]). The height image (Fig. 5.21A) shows the topographic features of the fiber, whereas the phase image shows the different components present (Fig. 5.21B), notably the cellulose microfibrils and the amorphous lignin patches. Atomic force microscopy with chemically modified cantilever tips (chemical force microscopy) was used to study the adhesion forces on cellulose model surfaces and bleached softwood kraft pulp fibers in aqueous media [85]. Fibers were suspended in water and a drop placed on a glass cover slide, dried, and used for AFM imaging. Comparison between the cellulose model surfaces and cellulosic fibers in this experiment revealed that surface roughness does not affect adhesion strongly. XPS and fourier transform infrared (FTIR) spectroscopy revealed that both substrate surfaces have homogeneous chemical composition suggesting that chemical force microscopy can be used for the chemical characterization of cellulose surfaces at a nanolevel.

Nanofibers were imaged by low dose, high resolution electron microscopy (LD-HREM) to analyze the fiber structure of regularly twisted poly(m-phenylene isophthalamide) (MPDI) strands formed by slow crystallization from solution [86]. The MPDI chains were found to aggregate into regular assemblies exhibiting uniform twisting at several different length scales. The polymer backbone forms a flattened helical structure organized into twisted bundles, which promotes good lateral packing but leads to an open core running down the helical axis. The aromatic polyamide (sold

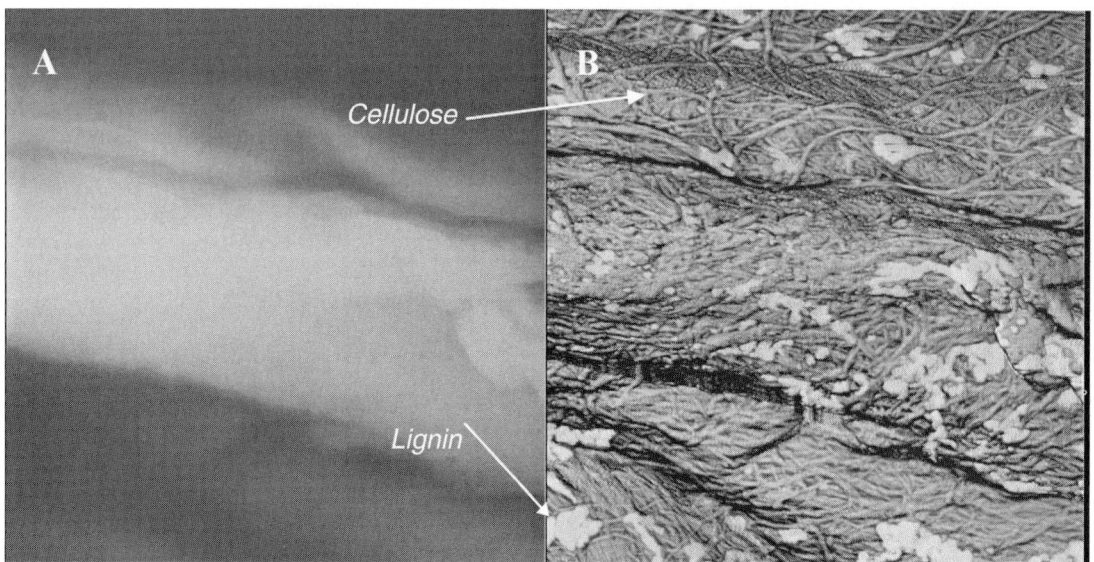

FIGURE 5.21. Atomic force microscopy height image (A) shows the topographic features of the fiber, whereas the phase image shows the different components present in the material (B), notably the cellulose microfibrils and the amorphous lignin patches [84]. Scale is 3 µm on a side. (From Chernoff and Maganov [84], © (2003) American Chemical Society; used with permission.)

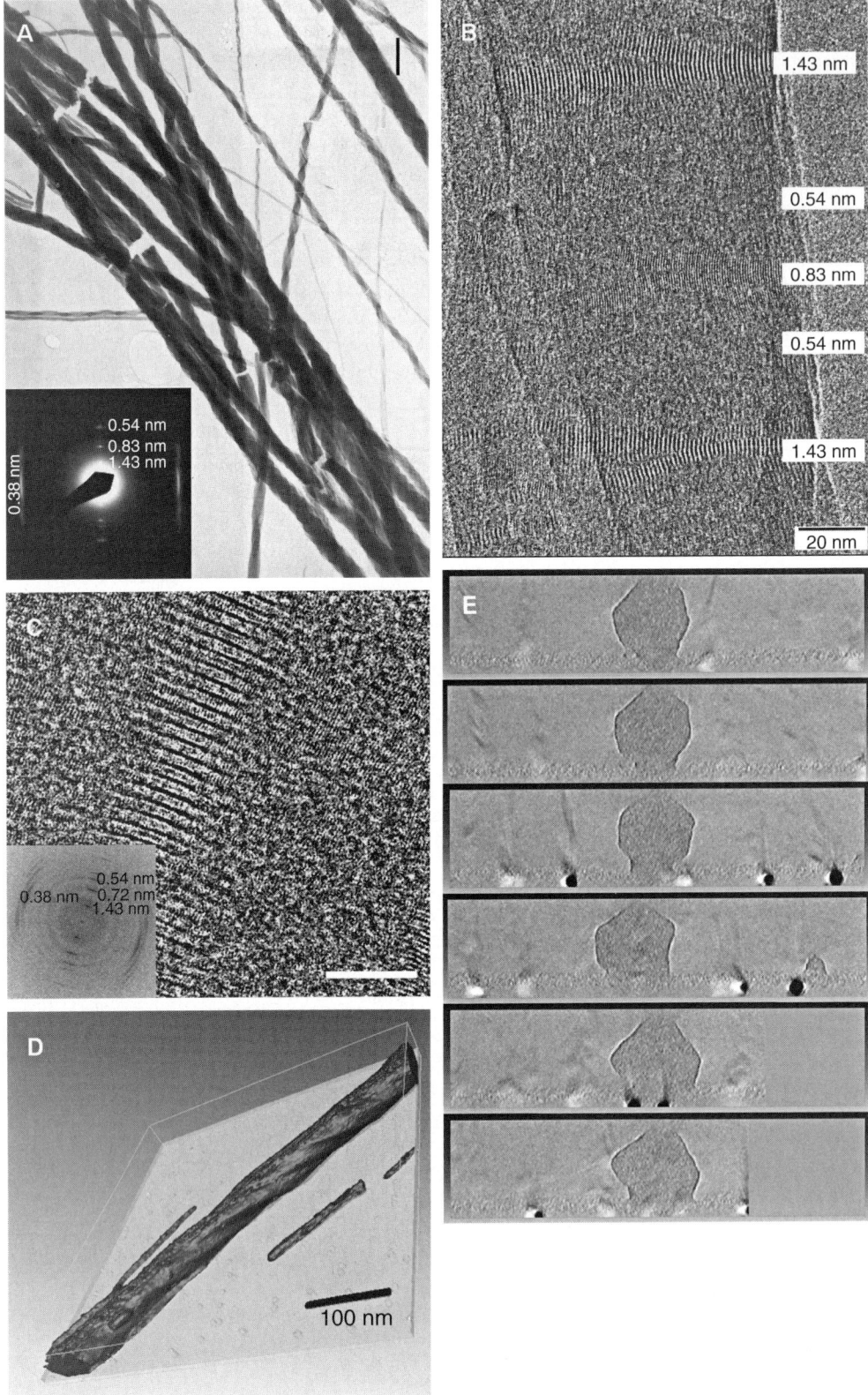

commercially as Nomex dry spun fibers, trademark of DuPont) was obtained as a solution, allowed to precipitate, and suspended on a carbon coated TEM grid, and the water was allowed to evaporate. Transmission electron microscopy bright field imaging of the twisted fibers was conducted at 200 kV; HREM was conducted at 400 kV. A TEM bright field image in Fig. 5.22A shows the uniformly twisted MPDI fibers, one to eight strands wrapped around each other forming a periodic structure like a rope; an inset selected area electron diffraction (SAED) pattern of one of the fibers indicates the highly ordered structure. Low dose HREM of the twisted fibers revealed different sets of lattice fingers in the digital fast Fourier transformations (FFTs) (Fig. 5.22B). Low dose HREM image of a small area of a twisted fiber exhibiting the 1.43 nm (100) and 0.54 nm (210) reflections is shown in Fig. 5.22C with the inset FFT of the image exhibiting a lattice resolution of 0.28 nm. The three dimensional reconstruction of the fiber by electron tomography [87] shows the overall shape of the fiber, and the cross sections indicating the strands are in close contact throughout their length (Fig. 5.22D, E). The images reveal the full structure of these complex nanofibers at molecular resolution. A full review of HREM of ordered polymers has been published by Martin et al. [88].

5.1.5.5 Spider Silk Fibers

Spider silk is of interest due to the unique combination of high tenacity and high elongation and toughness that polymer scientists may learn from in the development of synthetic fibers. Spiders synthesize silk, a polypeptide, and spin fibers using liquid crystalline solution spinning with a general structure of substituted nylon 2 (e.g., [89]). The morphology of spider silk is essentially round in cross section. Images of silk fibers by light microscopy and SEM as well as birefringence measurements are commonly performed on these fibers as with other polymer fibers (e.g., [90]). The unique ribbon morphology of some spiders has also been studied by bright field TEM, AFM, low dose TEM, and electron diffraction (e.g., [91]). An SEM stub with double sticky carbon tape was used to attach a web surface, and the samples were sputtered with Au/Pd. The silk ribbons were captured by passing a glass slide above a web for direct imaging by AFM and after floating off on water after brief immersion in very dilute HF for mounting on carbon coated TEM grids. The mechanism of silk processing has been studied by Kaplan (e.g., [92]) who used PEO to form a blend for study. These studies included SEM of blends and fractured fibers and AFM. Nanoscale fibers of natural silks were produced and observed by OM, SEM, TEM, and x-ray diffraction [93]. Collected fibers were coated with a very thin layer of evaporated carbon for TEM, and silk fibers were collected on silicon wafers and then coated for low voltage, high resolution SEM imaging.

It is not surprising based on the significant number of studies in the current literature that many researchers interested in high performance fibers, such as LCPs, are also studying spider silk fibers. Gould et al. [94] is an example of such work

FIGURE 5.22. Transmission electron microscopy a bright field image (A) of uniformly twisted MPDI fibers of one to eight strands wrapped around each other forming a periodic structure like a rope; an inset SAED pattern of one of the fibers indicates the highly ordered structure (scale bar represents 500 nm). Low dose, high resolution electron microscopy (LD-HREM) (B) of the structure of the twisted fibers revealed different sets of lattice fringes in the digital fast Fourier transformations (FFTs). LD-HREM image (C) of a small area of a twisted fiber exhibiting the 1.43 nm (100) and 0.54 nm (210) reflections with the inset FFT of the image exhibiting a lattice resolution of 0.28 nm (space bar corresponds with 10 nm). Nanofibers of MPDI were imaged by LD-HREM [86]. The three dimensional reconstruction of the fiber by electron tomography [87] shows the overall shape of the fiber (D) and the cross sections indicating the strands are in close contact throughout their length (E) (See color insert). (From Kübel et al. [86], © (2001) American Chemical Society, and Kübel [87]; used with permission.)

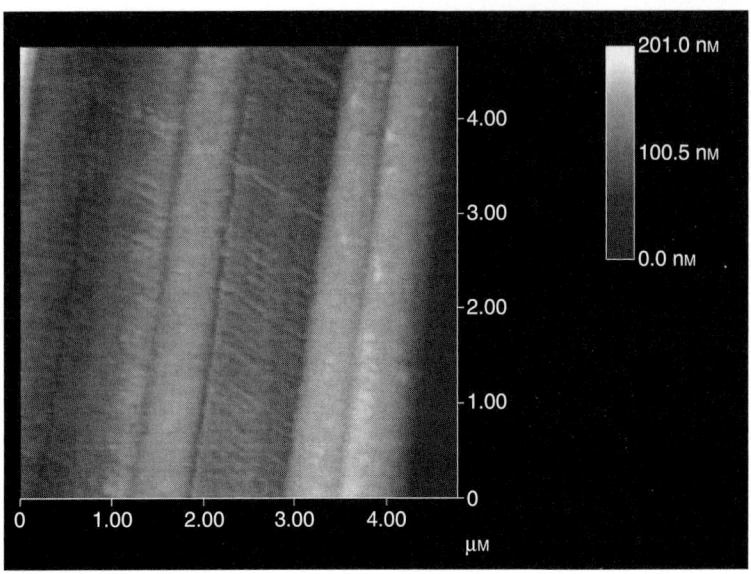

FIGURE 5.23. Large scale AFM image of the surface of unstretched threads shows fibers and fibrils. Image dimension is 4.8 × 4.8 μm. (From Gould et al. [94], © (1999) Elsevier; used with permission.) (See color insert.)

in which AFM is used to explore the surfaces of silk threads, showing both unordered and highly ordered regions. Atomic force microscopy images show that with increasing strain, both mean fiber and fibril diameters decrease and that fibrils align themselves more closely with the thread axis. The observation of fibers and fibrils within the cobweb threads has implications for current models of the secondary and tertiary structure and organization of spider silk. A large scale AFM image of the surface of unstretched threads showing fibers and fibrils in spider silk fibers, Fig. 5.23, appears remarkably similar to images of LCP fibers (see Section 5.6) [94].

5.2 FILMS AND MEMBRANES

5.2.1 Introduction

Films and membranes include a formidable array of materials that are widely used in a range of industrial applications. Films find application as coatings, packaging materials such as food wraps, and dielectric thin films used in electronic devices such as multichip modules, printed circuit boards, and active matrix liquid crystal displays. Newer applications include the use of LCP films in medical, chemical, electronic, and packaging applications and fuel cells, fluid distribution systems, and in other components. The varied permeability of polymer films permit their use in a wide range of applications including bottles, laminates, gas tanks, and liners for storage tanks. Films in the form of membranes are used for separations, controlled release, coatings and packaging barriers and contact lenses. Structural studies of films fall into two major categories: model film studies and the study of commercial films. Commercial films for this section include both free standing films and films on substrates. Discussion of latex films, coatings, and adhesives is found in Section 5.5.

Model studies are generally conducted in university research laboratories where the thin, flat structure of a melt cast or drawn film provides an ideal specimen for study. Such thin films can be examined *directly* or indirectly by TEM as well as by optical microscopy, SEM, or SPM. Specimen preparation is minimal, and interpretation of the image is made easier for several reasons:

1. Structures in very thin films are clearer because they do not overlap in the transmission image.

2. Model cast films are made with no possibility of damage or deformation during specimen preparation by microtomy or fracture.
3. Structures in model semicrystalline films may be larger and better defined than in commercial materials because of lower nucleation densities or lower quenching rates.

If the TEM is used, interpretation always requires care, even for this "easier" case. Problems of radiation damage of the polymer [95] (see Section 3.4) cannot be eliminated by using thin specimens.

Model film studies are more appropriate and easier to relate to commercial films in cases where the orientation is low or primarily uniaxial, as biaxial orientation is difficult to mimic on a small scale. Films with large second phase particles also cannot be studied by this method as the true structure cannot be reproduced in a thin film. Nevertheless, many of the basic concepts used in describing microstructure and its development in films, particularly semicrystalline films, have been derived from model film studies. Model studies have helped considerably in understanding blown PE films. The x-ray diffraction patterns from these films could not be interpreted unambiguously, and their microstructure was unclear until Keller and Machin [96] produced the same structures in model drawn films. They described the structure and showed how it was formed during crystallization from an oriented melt. This helped in many ways as, for example, the effect of molecular weight distribution on film properties could be understood by its effect on the structure formation. The orientation in films is measured by optical birefringence techniques, and such studies have shown the effect of resin molecular weight in stretching of parisons in the orientation of bottles [97].

Thin films may be divided into three classes, depending on their molecular orientation. Real films can be biaxially oriented, with the molecular chain axis most likely to lie along the direction of film extrusion (the "machine direction") and more likely to lie in the plane of the film than perpendicular to it. However, some films have little or no orientation, and their microstructure is similar to the bulk polymer, with differences possibly due to the presence of the film surfaces or the substrates. These allow rapid quenching and may increase nucleation of crystals. Other films are strongly stretched or drawn as part of the extrusion process, and molecular alignment along the machine direction dominates the properties. These films are fundamentally similar to fibers, although the geometry is different. The third class of films may be prepared by biaxial drawing, for example, to make the molecules lie in the plane of the film with no strong machine direction effect. These films have good properties in all directions in the sheet, and include for example, PET (Mylar; trademark of DuPont) and polyimide (Kapton; trademark of Kapton Polymers US LLC). Mylar is a strong polyester film that is used in consumer markets in food wrap, magnetic audio and video tape, capacitor dielectrics, packaging, and batteries. Kapton is a high performance polyimide film noted for durability and performance in extreme temperature environments and is used, for example, in miniaturized electronic components, high speed locomotive motors, and airbag seat sensors.

Membranes can be thought of as special types of films that provide specific end use characteristics. Membrane technology has replaced some conventional techniques for separation, concentration, or purification [98]. Applications include desalination, dialysis, blood oxygenators, controlled-release drug delivery systems, and gas separation. Processing of polymer films and membranes is well known to affect the morphology, which in turn affects the physical and mechanical properties. As is true for all films, membrane separation properties are based on both the chemical composition and the structure resulting from the process. Membranes are produced in two major forms, as flat films and as porous hollow fibers, both of which will be discussed.

A wide range of polymer chemical compositions is used in both films and membrane materials. A listing of commonly known polymers, including those found in films and membranes, is found in Appendix IV. The focus of this section is on a description of model film studies, industrial film applications, and flat film and

5.2.2 Model Studies

5.2.2.1 Introduction

Model studies of soluble semicrystalline polymers may be made quite simply by allowing a drop of a dilute solution of the polymer to fall onto a substrate, such as a glass microscope slide, and allowing the solvent to evaporate. The thickness of the resulting "thin cast film" may be controlled by changing the solution concentration, and films about 20–100 nm thick are used directly for TEM. A hot solution or substrate may be required to control film thickness. If the film adheres to the glass too strongly to be removed for mounting on a TEM grid without damage, an inert liquid or a water soluble material, such as NaCl, may be used as a substrate. For simple optical microscopy, the films are made thicker and not usually removed from the glass substrate.

Quiescently crystallized films may be used to determine the crystal morphology and, for example, its dependence on crystallization rate or polymer molecular weight. Drawing the films, while on a liquid or other deformable substrate, has been used to follow the spherulite to microfibrillar transition and so is clearly relevant to fiber drawing. Other industrial processes, such as fiber spinning, film extrusion, and blowing and injection molding, involve melting and deformation of the molten material and often crystallization under a high rate of extensional flow. Model films made under these conditions can be used to help understand the structure-property relations that result from this large class of industrial processes.

5.2.2.2 Literature Review

Strained and unstrained films of natural rubber were examined by Andrews [99, 100] who showed that the spherulitic morphology of the unstrained films changes to a fibrillar morphology with crystalline units on the order of 6–25 nm wide. This is consistent with the work of Scott [20] who showed that strain-induced crystallinity results in fibrillar structures parallel to the strain direction. A further development of this idea by Keller and Machin [96] showed the morphology of melt extruded PE sheets to be an arrangement of lamellar crystals, normal to the stress direction, arranged in fibrillar units parallel to that stress. The concept of model film studies providing information related to commercially produced materials was described in this study.

Electron microscopy of drawn polypropylene (PP) films extended the range of these microfibrillar observations. Sakaoku and Peterlin [101] prepared thin films by evaporation on a water surface and then transferred the films onto Mylar. They then drew the Mylar films in a simple tensile deformation frame and the thin PP film was simultaneously deformed by the same amount. The PP films were etched by ion bombardment or with acid, for replication. Dark field imaging and electron diffraction showed microfibrils about 20 nm wide that had formed in micronecks in the original crystal lamellae. The stress field at the neck has a negative hydrostatic component (trying to increase the volume of the material), and in thin films longitudinal voids form with microfibrils bridging the gap, similar to the structures later found to exist in crazes (see Section 4.8.3). The microfibrillar morphology of drawn materials, shown in the well known Peterlin model [7], has broad implications for the structures of both fibers and films. Tarin and Thomas [102] used gold decoration to show the deformation and transformation of a thin, spherulitic PE film, cast from decalin onto a hot mica sheet, into a microfibrillar morphology.

Scanning electron microscopy studies of PET films showed that a range of spherulitic morphologies could be induced by casting from trifluoroacetic acid. Differences in morphology resulting from different solvent evaporation rates were studied on cast and stretched films by polarized light microscopy (PLM), SEM, and small angle light scattering [103]. Polarized light microscopy of model films has often been conducted using the hot stage, which permits the determination of the growth habit and crystallization kinetics of polymer systems [104]. The disappearance of certain features as the temperature is raised can be correlated with melting peaks at the same temperature in DSC traces. Thin PE samples that would be affected by adhering to the glass microscope slide or coverslip were embedded in a matrix of low melting point, e.g. low molecular weight PE.

Petermann and coworkers [105–111] conducted a series of model studies on the morphology and crystallization mechanisms of thin films produced by crystallization under a high rate of extensional flow. Two preparation methods were developed to produce oriented thin films (see Sections 4.1.3.6 and 4.1.3.7). An early method [105] involved drawing oriented fibers from molten film surfaces and annealing. Lamellar thickening occurred during annealing of PE [106]. A method still used to form high modulus thermoplastic films was developed to produce highly oriented films by a longitudinal flow gradient, which compares with the industrial use of high speed spinning processes. The method has been applied to the study of HDPE and isotactic PP. The structure of the oriented films is that of a stack of parallel lamellae on edge, which penetrate through the entire film. This makes the interpretation of transmission images particularly simple, as there is no overlapping of adjacent structural regions. Thin spherulitic films were formed [112] from a drop of a dilute PE solution in xylene placed on the surface of glycerol at 140°C. After solvent evaporation, the films were transferred to a hot stage (Mettler-Toledo, Inc. Colombus, OH) placed on an optical microscope, crystallized, and quenched.

Block copolymers are also cast or spin coated for evaluation by a range of techniques, and these are discussed later in this chapter (see Section 5.3.4.6). Chen and Thomas [113] used force modulation microscopy (FMM), a SPM technique that measures relative elasticity across a surface, to study the block copolymer morphology of roll cast and spin coated films. Three model samples were investigated: unannealed poly(styrene-butadiene-styrene) triblock copolymer, fabricated from solution using roll casting, cut perpendicular to the oriented cylinders with a razor blade; annealed and unannealed poly(styrene-b-methyl methacrylate) diblock copolymer spin coated from solution; and an ultrathin film of a rod-coil diblock copolymer cast from a dilute solution onto a carbon coated mica sheet. Triblock copolymers were strained in tension using a copper grid as support and examined by OM, TEM, and AFM for morphological details of the deformed regions [114]. In this study, samples were stained in a RuO_4 solution after microtomy for study of deformation and fracture properties of the films.

Cheng et al. [115] studied a polystyrene (PS)-based block copolymer in which organometallic material, (PS)-*block*-polyferrocenyldimethylsilane (PFS), forms spherical domains within a PS matrix. The polymer was spin cast from solution to cover thermally oxidized silicon with profile grooves 50 nm deep and then annealed and etched with oxygen plasma to remove the exposed PS, revealing the PFS spherical domains. Figure 5.24A and Fig. 5.24B

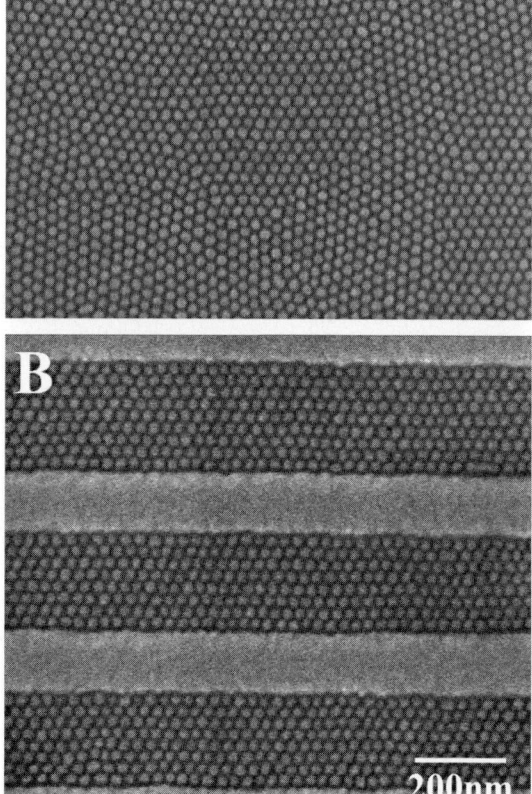

FIGURE 5.24. Scanning electron microscopy images of PS-PFS polymer spin coated onto a smooth silica surface (A) and confined in grooves on a silica substrate. (From Cheng et al. [115], © (2003) Wiley-VCH; used with permission.)

are SEM images of the polymer on a smooth silica surface and confined within grooves in a silica substrate, respectively. The difference in the images is to some degree a function of the substrate topography, which can be used for nanofabrication [115]. Recently, propylene/ethylene copolymers were melted to form thin films in a DSC and rapidly cooled to freeze in the morphology, then microtomed at −75°C to expose the bulk for ICAFM [116]. Phase images at different blend compositions clearly showed the morphology changes.

5.2.2.3 Semicrystalline Films

Thomas and coworkers investigated the structure of PE films by techniques that include TEM and STEM imaging. Yang and Thomas [117] studied the crystallization mechanisms and morphology of semicrystalline polymers by defocus imaging of oriented films [105, 110, 118] formed by the method described in the previous section using PE films formed from solutions in xylene at temperatures between 122°C and 130°C. Results showed that the melt-draw process does provide films containing highly oriented lamellar structure. Figure 5.25 shows bright field defocus phase contrast electron micrographs of as-drawn and annealed PE films [117]. The films consist of crystallites with the molecular axis aligned in the draw direction, as shown in the electron diffraction pattern inset. The bright field defocus phase contrast images consist of bright interlamellar regions and gray crystalline lamellae. The as-drawn films (Fig. 5.25A) have shorter, less well oriented structures, whereas annealing (Fig. 5.25B) causes an increase in crystallinity, orientation,

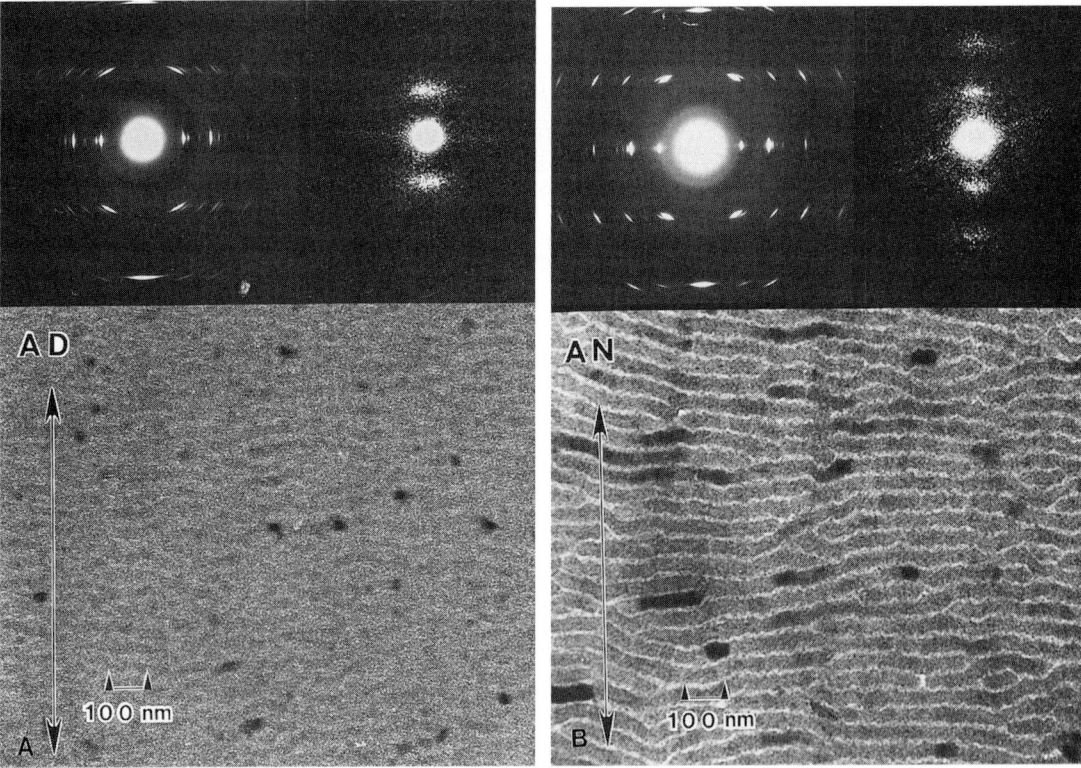

FIGURE 5.25. Electron diffraction (top left) and defocus phase contrast micrographs of melt-drawn polyethylene films. Optical diffraction patterns from the micrographs are at top right. As-drawn film (A) is well oriented; annealing (B) increases orientation and crystal size. Bright regions are interlamellar; the crystalline regions are gray or dark if they diffract. (From Yang and Thomas [117]; used with permission.)

and lateral lamellar size. The long period determined by x-ray diffraction corresponds with the lamellar spacing observed in the electron microscope.

In the model studies discussed thus far, thin films were formed from solutions or melts, thus limiting preparation artifacts. Important model studies have also been conducted by methods more commonly employed to study the morphology of commercial materials. Two types of studies have been conducted on samples prepared by (1) production of microtomed and stained sections (see Section 4.4) and (2) formation of replicas (see Section 4.6) of acid etched (see Section 4.5) materials. Both methods result in clear images of lamellar structures. Bassett and coworkers developed the second method and have reviewed its use for the study of crystallization and morphology in PE, other polyolefins, and isotactic polystyrene (iPS) [119–121]. For microtomy, polyolefins are stained first with chlorosulfonic acid [122], which makes them more rigid and easier to microtome [123, 124]. Quantitative results for the crystal size distribution can be obtained [125, 126] and compared with the results of SAXS or Raman spectroscopy [127]. These techniques show how the crystal morphology in melt cast films depends on the molecular weight, molecular weight distribution, and the crystallization temperature.

The microdomain model of crystallinity in poly(vinyl chloride) (PVC) by TEM has provided a controversy that is noted here simply to highlight the topic of image interpretation and specimen preparation issues that continue to be important, as the conclusions of Clark and Truss [128] were commented upon by Radzilowski and Thomas [129]. In this case, the HREM is the technique being used, and the issues relate to the microscope conditions, the filtering process, the radiation dose and damage, which with PVC would include loss of chlorine and specimen thickness. The issue here is simply that full disclosure of preparation methods and instrumental techniques are a requirement for research papers on polymers, as they clearly affect the images and interpretation.

5.2.2.4 Amorphous Films

The discussion thus far has dealt with the morphology of semicrystalline polymers formed into films. Although it is not clear where a discussion of the morphology of amorphous or glassy polymer films should be found, especially in a text primarily on the *applications* of polymer microscopy, it is clear that this controversial topic must at least be summarized. The controversy centers on the interpretation of images of thin amorphous or glassy polymer films in the TEM. Some microscopists have taken the fine structure in such images to represent true structure in amorphous polymers, whereas others take them to be artifacts of preparation or imaging. Most might today take the second position, but there are papers [130, 131] that show that the issue is still open. Grubb [132] summarized the major arguments involved, making an important point that the cause for the disagreement might well be due to not distinguishing the different cases. In fact, the amorphous material discussed may be a crystallizable polymer, such as PET or polycarbonate, quenched to a glass and annealed below the glass transition temperature, or it may be an amorphous material that never shows crystalline order, such as atactic polystyrene or poly(methyl methacrylate) (PMMA). Ordered regions in PET could possibly be explained by allowing that the TEM could detect incipient crystallization into nanometer-size ordered regions, whereas bulk measurements, such as x-ray diffraction linewidth, might show no change from the amorphous state. This, however, would not be a discovery of order in the amorphous state, and such an explanation could not hold for all atactic PS (aPS).

The earliest studies showing structures of some kind in amorphous films [133, 134] relied on observation of the film surface by replication, gold decoration, and shadowing. In addition, Yeh and Geil [134] and later Yeh [135, 136] used transmission bright field images of very thin films of aPS, among other materials, and claimed to see regions on the order of 3–10 nm across, which were called "nodules" and modeled as bundles of more or less parallel chains. Geil [137] summarized these studies, showing nodular structures by both surface and bulk preparation methods. The surface structures seen could not be misinterpretations of the image but might be artifacts of preparation or real structures that are associated with the surface. In either case, again, the observation does not prove there is order in amorphous bulk material.

The fine structure seen in TEM of unstained films is much more difficult to interpret, and it has been suggested that it is merely random phase noise in the films made visible by a small defocus [132, 138, 139]. It is certainly true that modern high resolution microscopy, and for polymers this is high resolution, requires detailed descriptions of the microscope parameters, when such an image is obtained. An image from a model structure of some electron density fluctuation must be calculated, and only if it agrees with the experimental image will the model structure be taken seriously. This level of analysis does not exist for the amorphous polymer images. Polymers have the further problem that amorphous materials, such as unshadowed films, exhibit radiation damage [140] as do crystalline materials, and thus it is possible that structures exist, but that they may not be seen by microscopy [141]. Grubb [142] studied annealed isotactic polystyrene that contained small crystals and determined that radiation damage would make crystals smaller than 4 nm across undetectable by their diffraction.

Uhlmann [139] conducted electron microscopy studies of thin amorphous films and observed what he termed a typical "pepper and salt" texture, characteristic of textures seen near the resolution limit in the electron microscope. For comparison, Uhlmann and coworkers [139, 143] obtained SAXS data that are not consistent with a nodular texture in glassy polymers. The SAXS intensity measurements of glassy polymers such as polycarbonate (PC), PMMA, PET, PVC, and PS do not support such a domain structure. Small angle x-ray scattering is a more suitable technique than TEM for detecting order, as a larger sample volume is statistically sampled.

Some authors do not interpret the textures observed for amorphous or glassy polymers as relating to any order, and the SAXS data do not support the idea of an ordered structure in these materials. The interpretation of the microstructures seen in amorphous glassy polymer films is clearly different in different laboratories. The issues are reviewed here, but no data supporting either view is fully described as this topic is beyond the scope of this book. This discussion is meant to draw attention to the issue of *interpretation*. Clearly, electron microscopy provides many useful observations; however, interpretation of the micrographs produced is nontrivial for structures near the resolution limit of the technique.

5.2.3 Industrial Films

Industrial films of such chemical composition as polyethylene, polypropylene, and polyester are manufactured for a wide range of applications as are high performance films made of such materials as polyimides and LCPs. Accordingly, the morphology of these materials is studied to determine structure-property relations, to understand how to improve properties, and also to control the quality of commercial products. Although model studies provide considerable detail relating to the structure, both before and after deformation of such films, model materials are generally thinner than commercial films, and thus the real product must also be evaluated. The types of preparation methods and instrumental techniques utilized closely parallel those described for polymer fibers. These techniques include: (1) measurement of birefringence and gel counting, (2) measurement of crystallinity and orientation, (3) TEM of unstained or stained ultrathin sections, (4) SEM of surfaces and bulk, and (5) SPM of surfaces.

Where the film is a coating, an added dimension to the study is the adhesion between the film and the substrate. Some industrial films have porous textures that are associated with the broad field of separation technology. These porous materials may be termed *films* or *membranes*, and they will be discussed separately below.

5.2.3.1 Optical Microscopy

The refractive index and birefringence of films can be measured in the optical microscope, which also allows details of texture in semicrystalline films to be resolved at the $0.2\,\mu\text{m}$ level. Birefringence is the more common technique as it permits measurement of the molecular orientation. Refractive index can be used to help identify an unknown material, and a technique has been described to determine the density from the refractive index. Density relates to crystallinity; however, x-ray diffraction, heat of fusion measurements, and direct density determination are all more common ways of obtaining measures of crystallinity. A biaxially oriented object has three refractive indices along its three axes (see Fig. 3.10). When the term "the birefringence of a film" is used loosely, it normally refers to the difference between the refractive index in the machine direction and that in the transverse direction, as observed by viewing the easy way, perpendicular to the film plane. There will normally be a larger difference between the refractive indices in the machine direction and in the perpendicular direction, but for this to be seen directly, a view in the film plane, along the transverse direction, would be required.

A major topic of interest relating to film structure is the effect of crystallinity on the deformation mechanism. The optical properties of biaxially oriented films were studied in 1957 by Stein [144], who determined the full set of birefringences by measuring the optical retardation as a function of the tilt of a PS film. Samuels [145] used complementary techniques of x-ray scattering, TEM of surface replicas, and birefringence measurement in a study of the microstructure and deformation of isotactic polypropylene films. The familiar theme of deformation of spherulites to a fibrillar structure was again observed in this study. The application of refractive index measurements for anisotropic films has been described for the evaluation of film properties and processing variables. Samuels [146] described methods for determining the percent crystallinity, birefringence, and refractive index distribution for commercial films. Bottle films have barrier characteristics related to the draw ratio used in processing. Paulos and Thomas [147] studied the effect of orientation on the structure and transport properties of a high density, blown PE film. Birefringence and crystallinity measurements revealed that the decrease in transport properties, and thus enhanced barrier properties, was related more to the high level of orientation than to crystallinity. An example of the spherulitic texture observed by polarized light microscopy of a thin cross section of a polyester film is shown in Fig. 5.26. This texture is related to the crystallinity of the polymer film, and a range and distribution of spherulite sizes can be related to both process variables and applications.

Blends of polymers (see Section 5.3) are also often used to form films, and such blown or extruded films can also benefit from examination by optical microscopy. One such study in which the blend is composed of LCP reinforced PE, a blown film used for balloon applications as different as weather balloons and angioplasty balloons, was examined by polarized optical

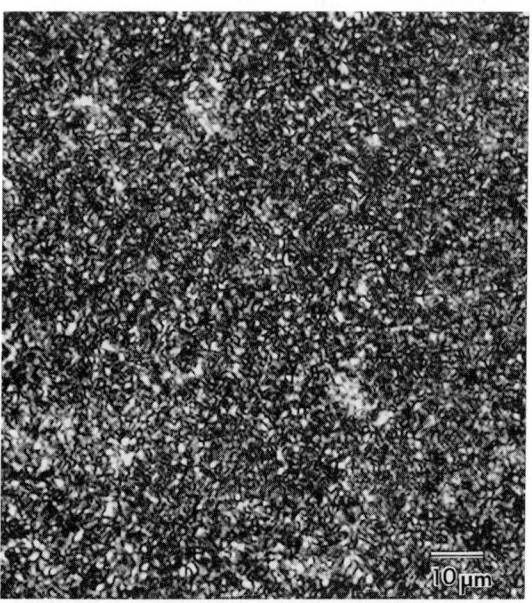

FIGURE 5.26. A fine spherulitic texture is observed in this optical micrograph of a polyester film cross section taken in polarized light.

microscopy to assess the size, shape, and distribution of the LCP in the PE (e.g., [148]). In this work, the morphology was observed in polarized light using a hot stage at a temperature of 160°C so that the PE phase was molten and the LCP fibrils were not. In many cases, the film is thin enough to press it between two glass slides for direct observation in bright field or polarized light.

Another topic of industrial interest is the evaluation of gels that relate to the cleanliness and processing of films. Generally, gels are discontinuities in the film and may be unmelted polymer, dirt of some sort surrounded by polymer, or some other material not fully blended with the film. Gel counting is commonly done using an optical microscope, such as in the work described by Huang and Wessel [149]. The technique involves cross polarization on stretched film and counting gels at various magnifications. The gel size distribution can be obtained for any film that is thin enough to transmit light, using a range of manual to automated image analysis methods. A similar method can be used to count "black specks" resulting from carbonaceous material during polymerization, or of carbon black size distribution.

5.2.3.2 Electron Microscopy

The SEM is often quite useful for the observation of the surface structure of films. In order to evaluate the initial and deformed morphologies, Sherman [150] studied plastic deformation and tearing in high density PE blown films with varied molecular weight and melt index. High resolution SEM studies [151] have directly shown the lamellae in a blown PE film (Fig. 5.27) prepared by drawing in either the machine direction or at right angles to the machine direction. Most commercial films are flat and smooth, and the surfaces have little structure or topography present. High tilt angles enhance the imaging of fine or shallow detail of film surfaces. In addition, the nature of the fillers or contaminants on the surface may be determined by x-ray microanalysis. In the blends study already mentioned [148], SEM of the film, frac-

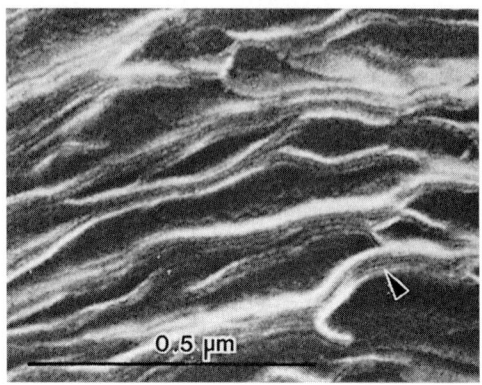

FIGURE 5.27. Direct imaging of lamellae is shown in this enlarged SEM micrograph of a blown polyethylene film surface. The arrowhead shows a single lamella. (From Tagawa and Ogura [151]; used with permission.)

tured in liquid nitrogen, broken perpendicular and parallel to the extrusion direction, and sputter coated with gold revealed the LCP fibrils quite clearly.

Complementary microscopic techniques are useful in the elucidation of polymer film microstructures. Optical techniques provide information relating to the orientation and crystallinity, and SEM can be used for surface detail relevant to end uses. Transmission electron microscopy techniques, similar to those used in model film studies and in fibers, are useful in describing the internal structures, especially of spherulites and their deformed counterparts, microfibrils. Transmission electron microscopy studies of films and fibers continue to provide fundamental observations relating the structure to properties and applications. Combination of SEM and TEM is often used, as for example a study of HDPE blown films [152] that also used SAXS data to obtain the lamellae orientation functions. Samples were treated using chlorosulfonic acid at 60°C for 6 h before being cryoultramicrotomed and stained for TEM; the samples were etched with heptane for SEM to make the lamellae more visible and then sputter coated with a thin layer of gold. The added use of SPM can provide very interesting details without as

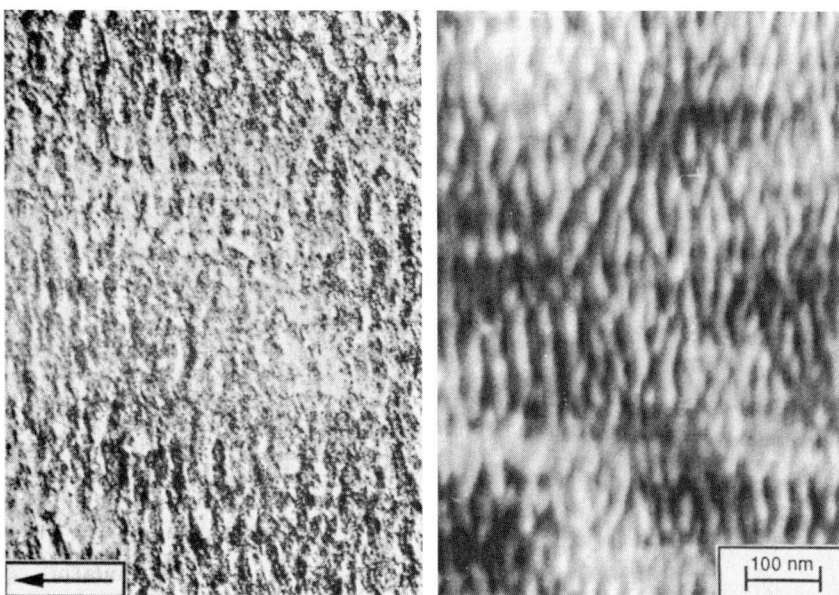

FIGURE 5.28. Comparison using TEM (left) and AFM (right) of a highly oriented PE film with lamellar structure. The arrow shows the molecular structure. Other than shadowing with Pt for TEM, the samples were identical. Both images can be analyzed to give the same values for lamellar thickness and their height above the surrounding film, but the AFM image is clearer and shows how the lamellae interlock and branch. (From Jandt [153]; used with permission.)

much specimen preparation. Direct comparison of AFM with TEM is shown in Fig. 5.28 of a highly oriented PE film with lamellar structure. The film was shadowed with Pt for TEM, so that the topography of one surface contributes most contrast to the image, and the height of the surface features can be measured by both techniques [153]; the AFM image is much easier to obtain and to interpret.

The direct visualization of the deformation processes in PE has been shown by HREM of thin films [154]. Adams et al. [154] used a STEM to study HDPE, formed by a melt drawing process and subsequently deformed at room temperature. That work shows the cavitation and formation of microfibers from the lamellae during deformation and the formation of fibrillar morphology under higher deformation.

The morphology of ionic aggregates in semicrystalline Zn and Na-neutralized poly-(ethylene-*ran*-methacrylic acid) (EMAA) ionomer blown films has been studied using STEM and SAXS by Benetatos and Winey [155] to assess the effect of film blowing on the nanoscale morphology. The advantage of STEM imaging is the ability to assess the ionic aggregates directly. Polyethylene based ionomers find application due to their extraordinary impact toughness and chemical and abrasion resistance. The ionic aggregates of Zn-EMAA are spherical, monodisperse, and uniformly distributed in the blown films prepared at low and high blow-up ratio as seen in images taken using high angle annular dark field (HAADF) (Fig. 5.29A) and bright field (BF) (Fig. 5.29B). Image contrast in STEM is based on average atomic number (Z) as the HAADF detector collects electrons that have been elastically scattered to high angles by high Z nuclei. Thus in HAADF STEM, the Zn-rich regions appear bright on a dark background corresponding to lower-Z elements (hydrocarbon). In BF STEM, the electrons scattered to high angles are not detected, and the high-Z features appear dark.

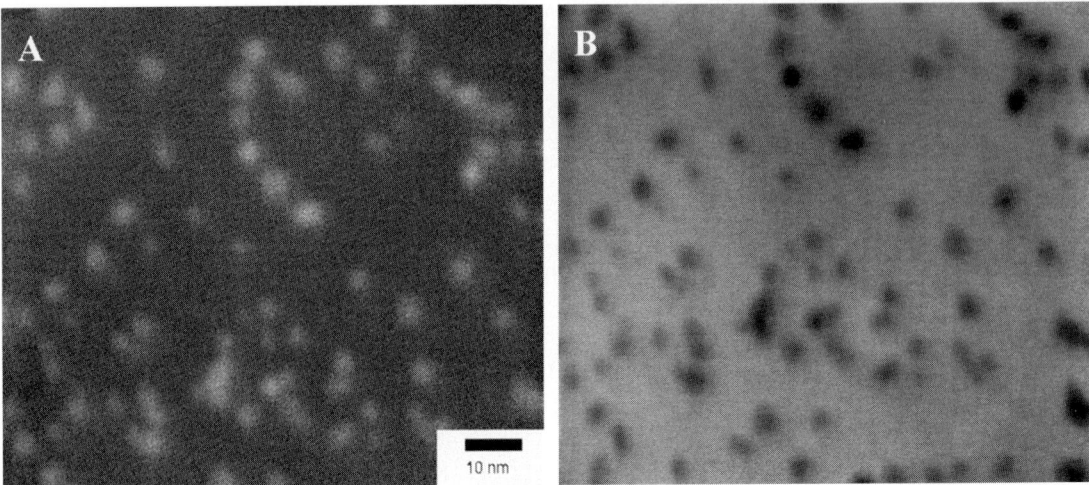

FIGURE 5.29. High angle annular dark field (A) and bright field (B) STEM images of nanoscale Zn-rich aggregates present in Zn-neutralized EMMA ionomer blown films. (From Benetatos and Winey [155], © (2005) Wiley-Interscience; used with permission.)

5.2.3.3 Scanning Probe Microscopies

The surface textures of films examined by SEM are often uninformative, because small changes in height of the surface do not give rise to significant image contrast. Scanning probe microscopy is quite different as the height of the film surface can be the primary output signal and height resolution is extremely good, so that an accurate exaggerated relief map of the surface can be produced. The magnification perpendicular to the plane of the film can be 10 or 100 times the magnification in the plane of the film, and this brings out surface structure not otherwise visible. Many studies have been made of very thin or monolayer films of organic materials deposited on a flat substrate, which has commonly been highly oriented pyrolytic graphite (HOPG). These films may be deposited directly or made using Langmuir Blodgett (LB) techniques in a trough. For example, Albrecht et al. [156] used LB methods to prepare extremely thin films of PMMA on graphite and characterized them by STM and AFM. These authors were interested in nanometer scale fabrication and information recording; they found that a voltage pulse applied to the STM tip caused a local modification of polymer fibrils.

Another example of the use of AFM on very thin films is the imaging of thin layers of poly (tetrafluoroethylene) (PTFE) that are deposited on glass simply by rubbing it with the solid polymer. Electron diffraction has shown that these films are highly ordered and very well oriented. Atomic force microscopy provides direct measurement of the film thickness and continuity [157]. Individual fibrils that were not attached to the substrate had different appearances when scanning in different directions. This is because the AFM tip pushes them around during scanning. Arrays of parallel rods with the intermolecular spacing of PTFE crystals are seen. The authors also claim to distinguish the helical structure of the individual molecules and compare it with models derived from electron diffraction.

Two reviews provide further details of SPM of organic surfaces [158] and thin films [159]. These techniques must continue to be compared with more conventional methods in order to be able to interpret the image and to ensure that the sizes measured on structural details have not been modified, such as by the tip in AFM. Topics such as imaging of individual chemisorbed molecules, supported physisorbed molecular assemblies, biopolymers, and bulk surfaces of polymers are shown imaged under

Films and Membranes

vacuum, fluids, and air, and the local surface modifications imposed by the tip were described [158]. Hues et al. [159] reviewed many instrumental issues in regard to the imaging of thin films by AFM, describing principles of the technique in some detail for the interested reader.

Tsukruk et al. [160, 161] conducted studies using SPM of organic and polymeric films, from self-assembled monolayers to composite molecular multilayers. Aspects of these films are described, including the surface morphology, surface defects, and molecular scale ordering. Surface modification to the materials during scanning with the AFM tip is also considered. Topics such as measurements of the forces between surfaces, surface stability, wear, adhesion, and elasticity are studied by AFM. Tsukruk [160] discussed molecular ordering, phase transformations in monolayer molecular films, fibrillar surface textures of polymers, such as PE, cellulose, aramids, polyimide fibers, latex dispersions, and polymer blends. The stability and modification of polyglutamate LB bilayer films in the AFM were also discussed [161]. The possibilities for surface modification by the AFM tip were explored as holes were fabricated or written into the film surface. Bilayers were deposited on polished substrates cut from silicon wafers, and AFM images were obtained. In this case, the data on average thickness and macroscopic roughness was also shown by x-ray reflectivity measurements for comparison with AFM. The ability to use the AFM tip for lithography and also the potentially adverse effects of AFM imaging have been discussed (e.g., [162]).

Atomic force microscopy in intermittent contact and contrast modes has been used to study the surface topography of films such as PET containing isophthalate, which shows the fibrillar surfaces, whereas the addition of isophthalate creates a granular texture [163]. Films and razor cut bottle sidewall samples were glued onto steel disks with epoxy for imaging. This study also shows deformation induced by the imaging force (see Section 3.3). Combined SEM and AFM were used to study segmented block copolymer films after gas plasma etching with carbon dioxide (CO_2) or argon [164] and compared with XPS and contact angle measurements. A granular nanostructure was formed after CO_2 plasma etching although the argon etching did not change the surface structure; both treatments increased the wettability of the films.

Atomic force microscopy and SEM, with thermal and x-ray techniques, were used to study the effect of different film production methods, such as extrusion and compression molding [165]. Contact mode AFM images of calcite filled composite films were compared with SEM, which indicates there is a correlation of crystallinity to surface roughness of the films. Thin polymer films are used in demanding applications, such as diffusion barriers, dielectric coatings, and electronic packaging, which require methods to test their mechanical properties. Stafford et al. [166] developed such a buckling-based metrology for measuring the elastic moduli of polymer thin films by using the spacing of periodic wrinkles in thin films coated with a soft thick substrate. Silicon wafers are used for spin casting from dilute polymer solutions and thickness measured by interferometry and surface detail by AFM images showing the periodic wrinkles. The mechanical properties of biocompatible protein polymer thin films, used as coatings for implantable devices for the central nervous system, have been studied by scratch testing, tensile testing, and nanoindentation. Scanning electron microscopy and SPM showed the properties were a function of the microstructure.

5.2.3.4 SEM and SPM Examples

The value of LVSEM is shown, for example in Fig. 5.30 of a PE biaxially blown film [167]. The film was cut, pulled until it necked, and imaged uncoated at 800 eV with an undeformed region showing lamellae (Fig. 5.30A) and an image of the transition toward deformation (Fig. 5.30B); Fig. 5.30C is a necked region with fibrils parallel to the applied load (P is direction of applied load).

Early application of SPM imaging of lamellar structures in melt extruded polyethylene films was shown by Chen et al. with FESEM, birefringence, and x-ray scattering measurements [168, 169]. Polyethylene and PP extruded films with row lamellar structures are film precursors

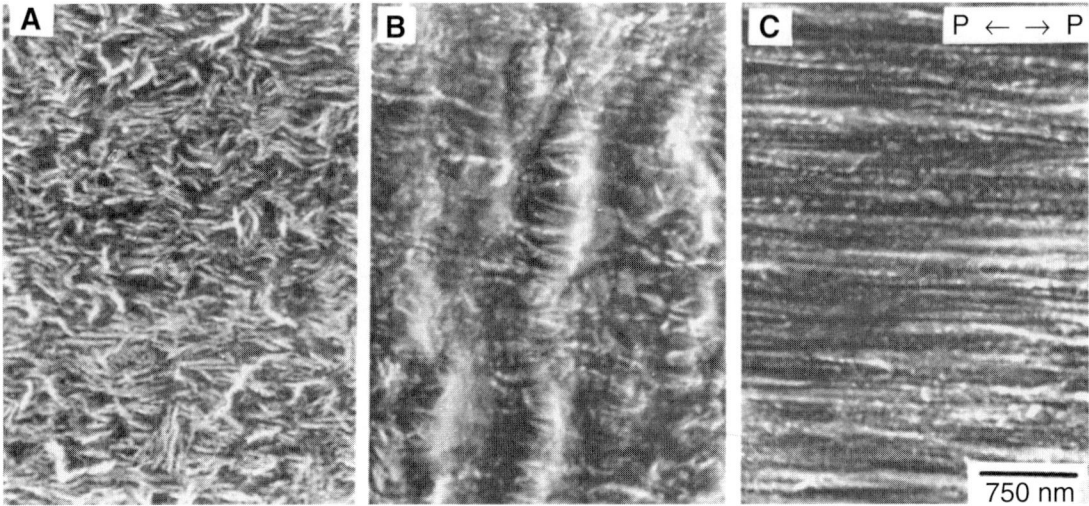

FIGURE 5.30. Low voltage SEM of an uncoated PE biaxially blown film, pulled until it necked and imaged in an undeformed region showing lamellar deformation (B) and in a region showing necking (C). (Used with permission, T. Reilly; unpublished [167].)

of flat sheet microporous membranes (see Section 5.2.4). Birefringence measurements, using optical microscopy of the melt extruded films, shows that improved film orientation can be achieved by one of several methods, annealing, extruding at high speed, or by the use of high molecular weight polymers. Imaging by the various scanning methods all clearly reveal the lamellar structures in the PE and PP films. X-ray scattering relates the increase of molecular alignment to changes in lamellar perfection and lamellar alignment during annealing. These techniques provide a means to establish structure-process-property relationships for the manufacture of microporous membranes. Examples shown in Fig. 5.31 are all annealed and processed at high extrusion rate with the molecular weight varied. Field emission SEM of samples sputtered with 2 nm Pt and imaged at ca. 5 kV (Fig. 5.31A) show fine lamellar structures in a lower molecular weight, annealed PE film precursor. The STM samples were mounted on silicon substrates and coated with 5 nm Pt using ion beam sputtering (see Section 4.7.3.2). The STM image (Fig. 5.31B) is of a low molecular weight, annealed PE film precursor, illustrating the lamellar structures and spacings, which are similar to those seen in the FESEM image. Finally, AFM imaging was performed by contact AFM [168, 169] (Fig. 5.31C) of a higher molecular weight sample, revealing similar detail to that shown by FESEM and STM. The study showed the effect of polymer molecular weight, film extrusion rate, and annealing on lamellar textures was found in precursor films. The resulting microporous structures were shown to be a direct reflection of these precursor effects.

Transmission electron microscopy is generally the preferred method for study of blends and block copolymer morphologies, but AFM provides a complement to such studies [170]. Thin films of triblock copolymers were spin cast onto silicon substrates, annealed, and examined by ICAFM, which clearly showed the end-on cylindrical structures formed in the height and phase images (Fig. 5.32). Characterization of pentablock material of compression molded parts was also conducted by TEM and AFM with the latter images shown in Fig. 5.33, again showing a similar cylindrical morphology better resolved in the phase image (right) than in the height image (left).

5.2.3.5 Electronic Films and Devices

Pentacene and other conjugated molecules are of interest for use as the active layer in organic

Films and Membranes

FIGURE 5.31. Film precursors of polyethylene flat sheet microporous membranes, all produced using a high extrusion rate, and annealed, but with different molecular weights. An FESEM image taken at ca. 5 kV (A) shows fine lamellar structures in a lower molecular weight, annealed PE film. The STM image (B), taken using a bias voltage of 100 mV and a tunneling current of 1 nA, of a low molecular weight, annealed PE film precursor reveals three dimensional lamellar structures similar overall to the textures in the FESEM image. An AFM contact image of a higher molecular weight PE film (C), taken using a long range scanner with a pyramidal Si_3N_4 tip, with the force monitored using a laser beam and a position sensitive photodetector, reveals similar detail as was shown by FESEM and STM. (From Chen et al. [168, 169]; used with permission.)

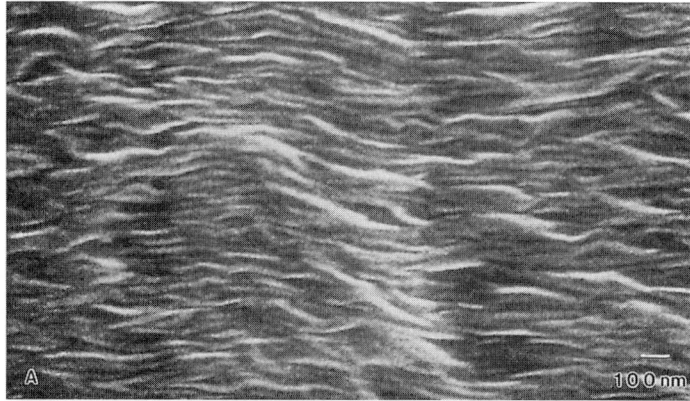
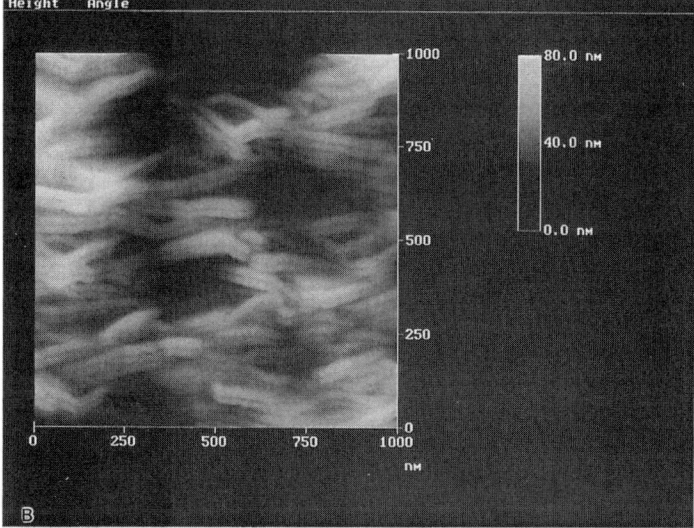
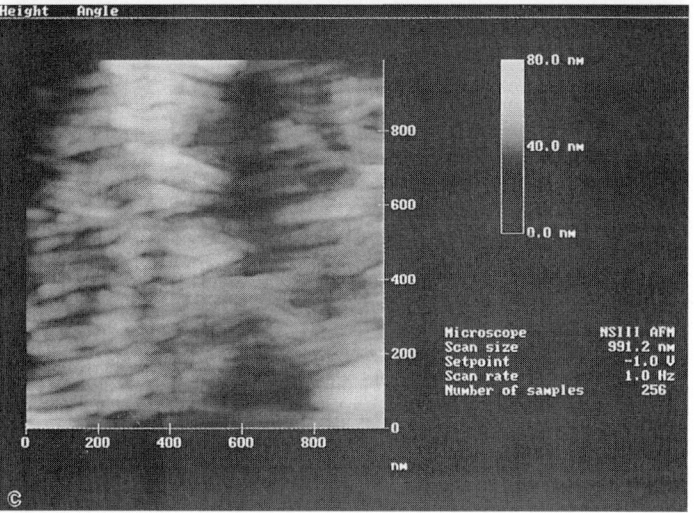

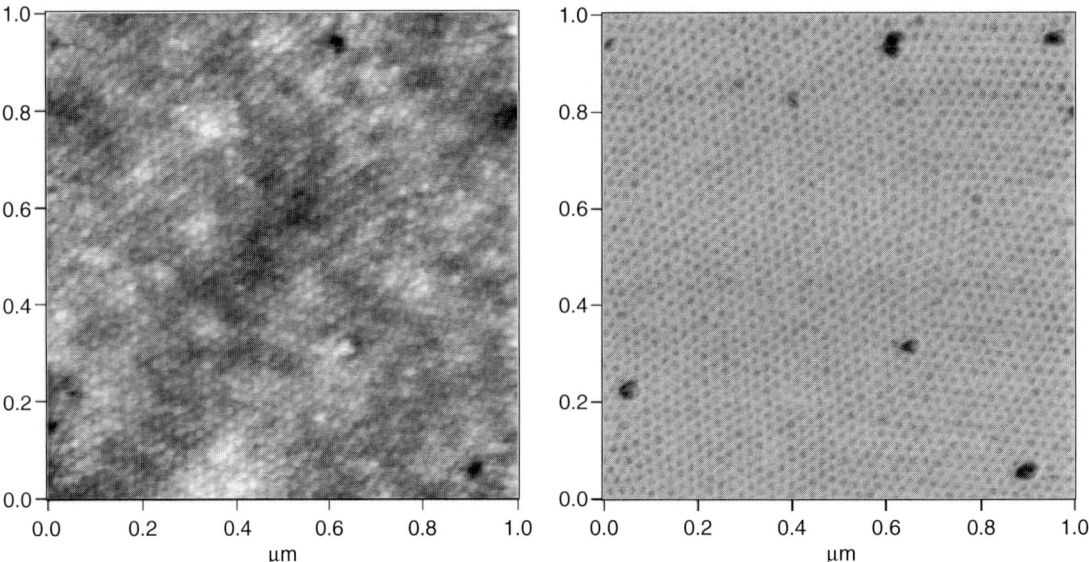

FIGURE 5.32. Height (left) and phase (right) ICAFM images of a hydrogenated SBS (styrene-butadiene-styrene) triblock thin film. (From Bar and Meyers [170]; used with permission of the *MRS Bulletin*.)

field effect transitors [171, 172]. Pentacene is a highly crystalline material and has been characterized as a thin film, made by thermal evaporation of the powder onto carbon coated and uncoated mica substrates, by OM, FESEM, TEM (under low dose conditions), and diffraction techniques.

Control of the orientation of organic molecules with electric fields plays an important role in commercial devices such as liquid crystal

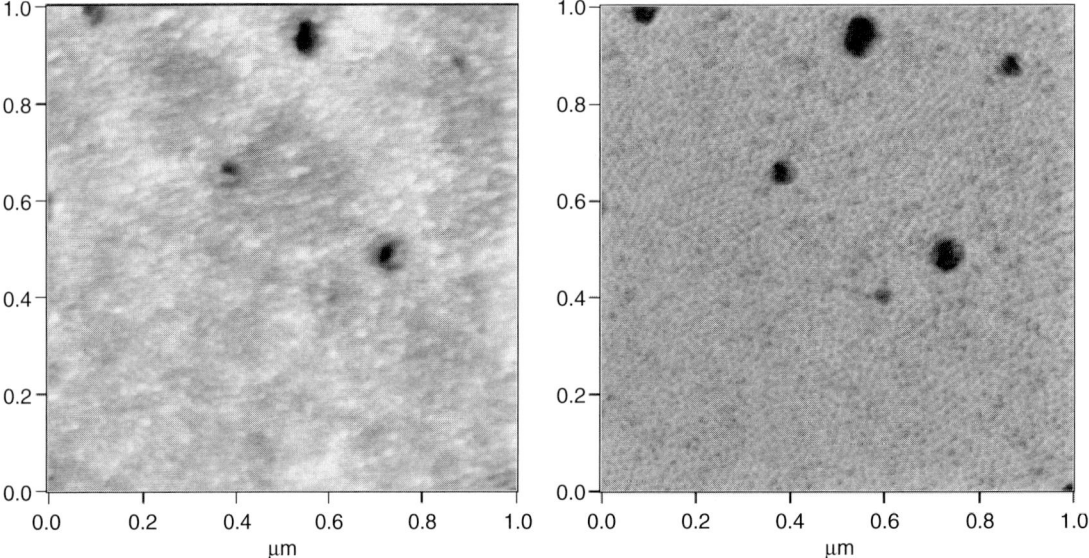

FIGURE 5.33. Height (left) and phase (right) ICAFM images of a hydrogenated SBSBS pentablock thin film. (From Bar and Meyers [170]; used with permission of the *MRS Bulletin*.)

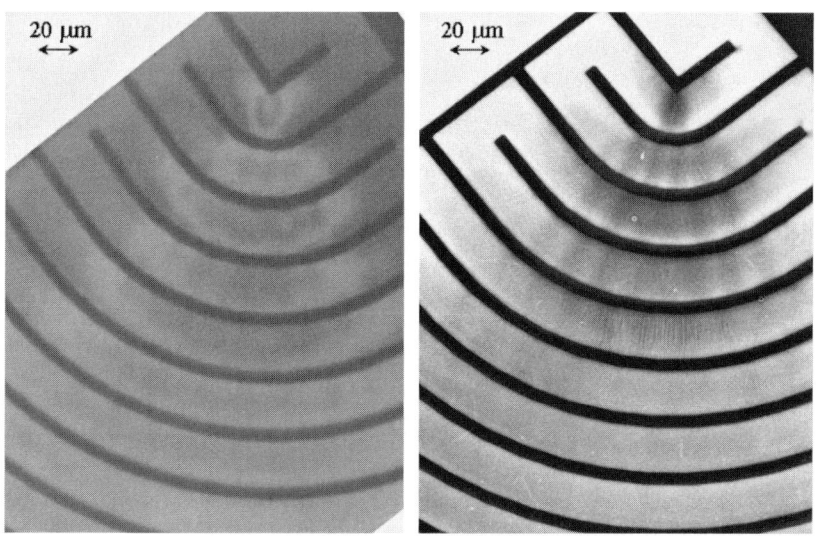

FIGURE 5.34. Optical micrograph (left) and bright field TEM image (right) of PHIC on a bend device, taken from the same region of the same specimen. (From Martin [173], © (2002) Elsevier; used with permission.)

displays and also has the potential to form other devices. Martin [173] studied poly(hexyl isocyanate) (PHIC) and poly(benzyl L-glutamate) after orientation by the electric field while in solution and solidified into a stable, oriented structure by solvent evaporation. Images in OM and TEM provided a means to directly image the disclinations that mediate molecular alignment. Films were cast from solution onto a glass slide and an electric field used for orientation while optical images were recorded on film and with a video camera. Transmission electron microscopy was performed at 120 kV using a LaB_6 filament. Since the PHIC is on a bend device that is 3 mm × 3 mm with a small 1 mm window, the same region of the sample is used for imaging by OM (left) and TEM (right), as seen in Fig. 5.34 [173]. These images permit determination of the relation between image contrast in OM and TEM for samples of known thickness and orientation.

Scanning probe microscopy techniques have been used to characterize surfaces related to the processing of benzocyclobutene (BCB) dielectric thin films. Thermally cured resins and photodefinable resins are used for electronic applications, such as multichip modules, printed circuit boards, and active matrix liquid crystal displays [174, 175]. Scanning thermal microscopy (SThM) is being used to provide property information about polymer surfaces, such as the study of the processing of photodefinable benzocyclobutene (Photo-BCB) dielectric films to make vias to conductive copper substrates [174]. After development and hard thermal cure of the photodefined material, an oxygen plasma is used to remove the residues in the via holes. Perspective view ICAFM images of 30 μm holes before and after plasma treatment are shown in Fig. 5.35. Topview SThM images before plasma treatment are shown in Fig. 5.36 with the DC images (left) giving an indication of thermal conductivity where the brighter areas are more conductive. The AC phase images (right) are the result of a modulation on the 40°C DC signal at 1 kHz (top) or 30 kHz (bottom) frequency. Intermittent contact mode AFM is also used to follow changes in adhesion promoter morphology to help explain adhesion performance between the polymer dielectric and silicon oxide wafer substrates. The surface morphology of

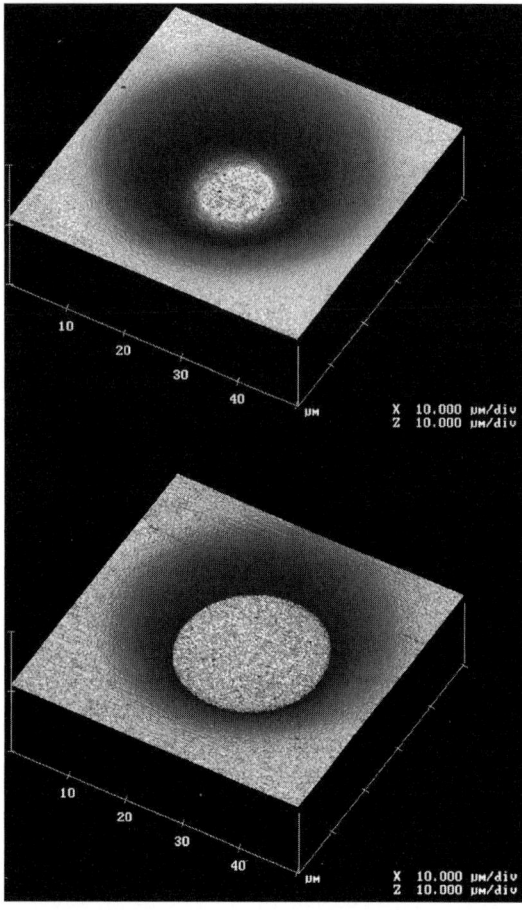

FIGURE 5.35. Intermittent contact AFM image of 35 μm via holes patterned in a photoBCB thin film over a copper substrate before plasma treatment to remove polymer residues (top) and after treatment (bottom). (From Meyers et al. [174], © (2000) American Chemical Society; used with permission.)

the promoter layer after deposition at room temperature and after baking is shown in Fig. 5.37 [175].

5.2.3.6 Multilayered Films

Multilayered films can be formed by various processes to create a structure with specific mechanical and physical properties. Microlayers of foam/film were formed by microlayer coextrusion technology [176], resulting in unique mechanical properties studied by optical microscopy. Cohen et al. [177] studied the deformation mechanism of a structure of alternating glassy rubbery layers, at different orientations of the deformation. Films of a polystyrene-polybutadiene-polystyrene (PS/PB/PS) triblock copolymer were prepared by casting from solution, using a roll casting technique in which the solution is processed between rotating cylinders while the solvent is evaporated. The morphology was studied using TEM of samples stretched to 300% and cross-linked by electron irradiation (energy 2.6 MeV, dosage 200 Mrad) at the MIT High Voltage Lab. Cryosectioning was done at −90°C, and sections were stained in the vapor above a 4% aqueous solution of osmium tetroxide for 2 h to stain the PB component prior to examination by TEM at 200 kV.

Hiltner, Baer, and coworkers have conducted extensive work on the production and observation of microlayers, assemblies of thousands of alternating layers of two polymers, with individual layer thickness on the nanometer size scale, that create an "interphase" region by layer multiplying coextrusion (see Section 1.3.2) (e.g., [178–180]). Multilayered tapes consisting of two ductile and incompatible polymers (PET and PC) were made by coextrusion of both components as uniform laminates with thousands of alternating layers [178]. The multilayered PET/PC samples were annealed at high temperature, and semithin sections (ca. 1 μm) were cut using glass knives parallel to the extrusion direction and perpendicular to the layers. Sections were strained in tension, stained in RuO$_4$ vapor, and imaged by high voltage electron microscopy (HVEM) at 1000 kV. Adhikari et al. [179] also studied PET/PC multilayer composites, in this case by TEM (120 kV) of ultrathin sections made using a diamond knife and treated in RuO$_4$ vapor for several hours. Intermittent contact mode AFM was conducted on the cryoultramicrotomed block face to show the strain-induced structural changes. An extruded film of an amorphous polyester, poly(ethylene terephthalate-co-1,4-cyclohexanedimethylene terephthalate) (PETG), and PS was embedded in 5 min epoxy, cured over night at room temperature, sectioned perpendicular to the plane of the film, and the flat block face was examined in air by AFM, as seen

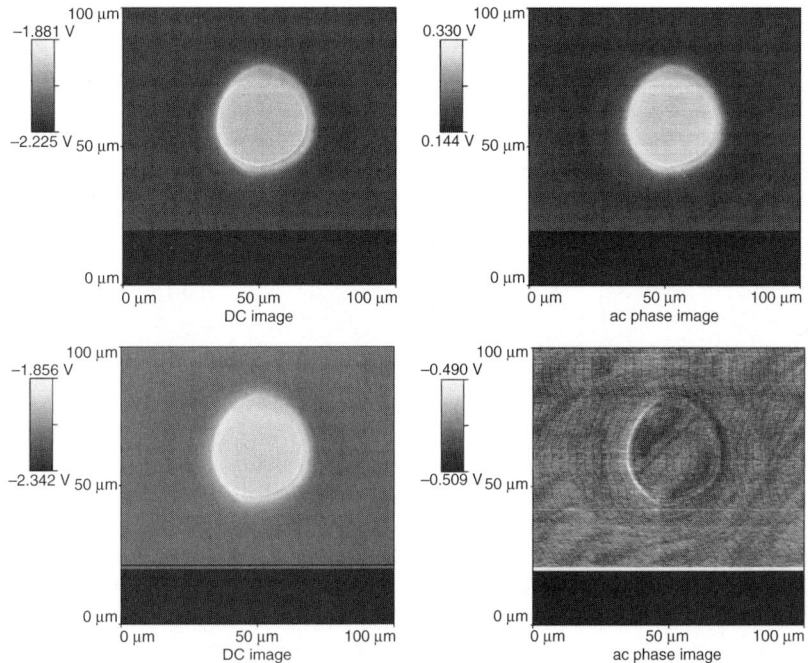

FIGURE 5.36. Scanning thermal microscopy images of 35 μm via holes patterned in a photoBCB thin film over a copper substrate before plasma treatment to remove polymer residues. DC thermal images (left) and AC phase images (right) are shown for frequencies of 1 kHz (top) and 30 kHz (bottom). (From Meyers et al. [174], © (2000) American Chemical Society; used with permission.) (See color insert.)

in Fig. 5.38, showing the layer thickness in the assemblies [180].

Chaotic advection, described and reviewed by Aref [181], has been used as a method to build *in situ* blend morphologies by the recursive stretching and folding of melt domains in response to shear flows by Zumbrunnen and coworkers at Clemson University to produce multilayer films and interpenetrating blends [182]. Chaotic mixing of immiscible binary components was used in both batch and continuous flow process to obtain extruded films with many internal layers, the former to enable study of the formation and the latter for industrial films [183]. Polystyrene and low density polyethylene (LDPE) were used as a model binary system and

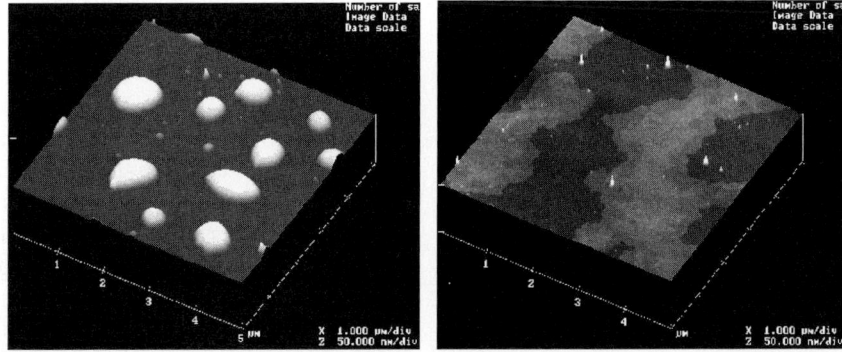

FIGURE 5.37. Intermittent contact AFM images of adhesion promoter after spin coating at room temperature (left) and after a hot plate baking step at 100°C (right). (From Meyers et al. [175], © (2001) Wiley-VCH; used with permission.)

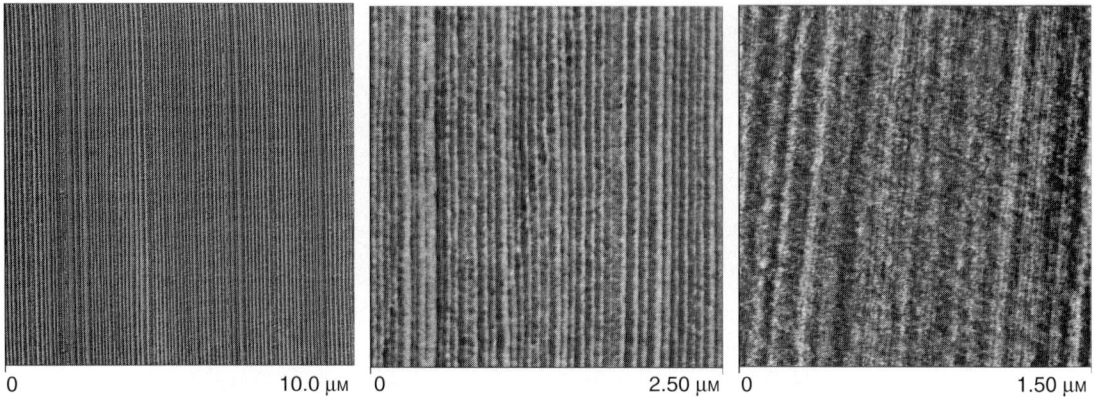

FIGURE 5.38. Atomic force microscopy phase images showing cross sections of PETG/PS assemblies with different layer thicknesses (left is 60 nm, center is 30 nm, right is 8 nm). (From Liu et al. [180], © (2004) American Chemical Society; used with permission.) (See color insert.)

samples cut to 300 μm thickness and then polished with a silk lap and 0.05 μm alumina powder in water for OM study. For blends produced for longer mixing times and containing thinner films, samples were fractured after immersion in liquid nitrogen for study by LVSEM in orientations parallel and perpendicular to the axes of the mixing cavity. Barrier films can also be produced by chaotic mixing as was done for ethylene vinyl alcohol copolymer, LDPE, and maleic anhydride modified PE, the latter used as a compatibilizing agent [184]. Morphology study was conducted after freeze fracturing and gold sputter coating for SEM; in early work, delamination of the extruded films was observed. Nanoscale structures have also been developed by chaotic advection [185]. A multilayer blend morphology that has a hierarchical structure and intrinsic mechanical interlocking was formed by chaotic advection of immiscible polymer melts. In this study [182], a continuous chaotic advection blender (CCAB), or "smart blender," was used to investigate influences of these morphologies on tensile and impact toughness properties of PP-LDPE blends. Extruded blends provided improved properties relative to properties associated with droplet morphologies typically obtained with conventional compounding equipment. Film samples produced using different specific rod rotation sequences (N) were examined by LVSEM. Films were immersed in liquid nitrogen for 5 min or longer, fractured in the transverse or machine direction, and evaluated as a function of N to show progressive morphology development. A comparison of morphologies in the machine direction (Fig. 5.39A–C) and transverse view (Fig. 5.39D) is shown for 500 μm films at N = 12 for various LDPE volumes [182].

Finally, an industrial example of end use appearance in a film with a modifier is shown in Fig. 5.40 [186]. For hazy or cloudy films, the resolution of TEM may be needed as in this example of a multilayer film that appeared clear as extruded but cloudy after deep drawing to 80% of its original thickness. A dispersed polymeric phase in the symmetrical nylon layers surrounding a central barrier layer deforms from mostly submicron spheres too small to scatter light (Fig. 5.40A) into mostly micrometer sized flat "pancakes" that scatter much more light (Fig. 5.40B) [186].

5.2.4 Flat Film Membranes

5.2.4.1 Introduction

The technology of membrane separations continues to be a growing field where the polymer membrane contributes unique separation properties based on its structure and on its chemical composition. Various manufacturing processes are used to create special structures in forms such as flat films and hollow fibers. Lonsdale [187] provides a review of the history and status of separation media and their applications, and a text [188] provides a discussion of the materials science of synthetic membranes.

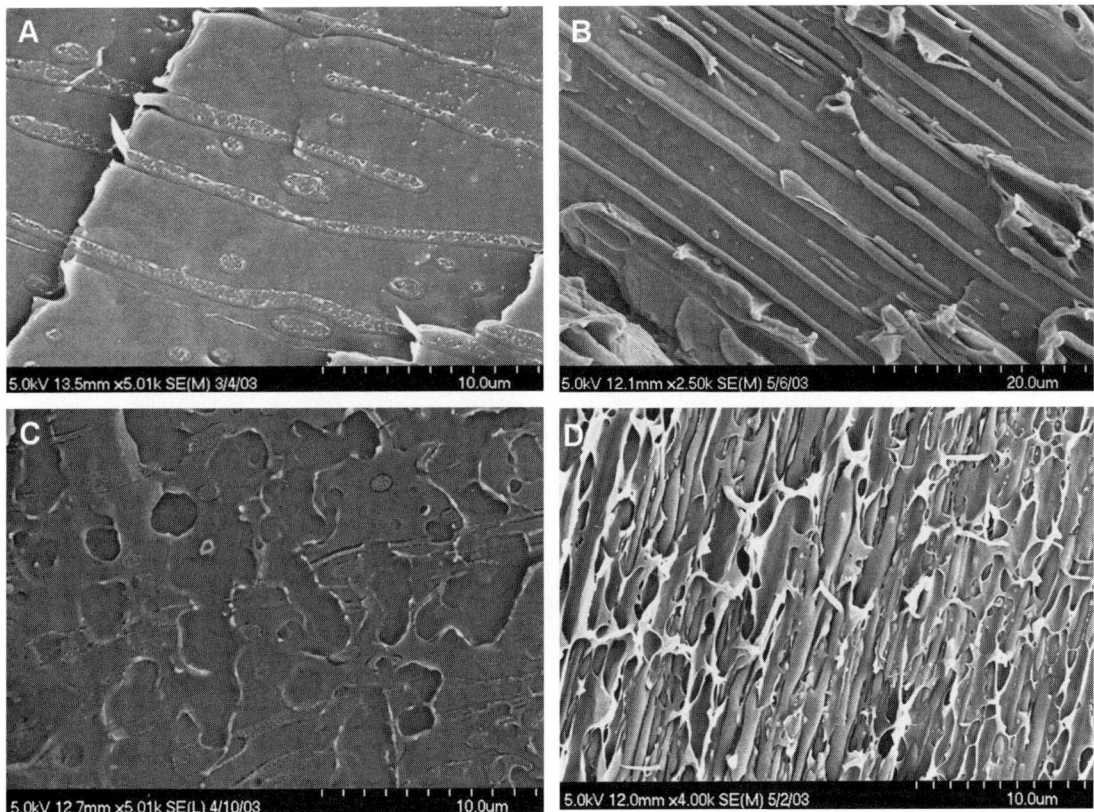

FIGURE 5.39. Comparison of morphologies in machine direction views for 500 μm films at N = 12 for (A) 10%, (B) 20%, and (C) 30% by volume LDPE shows the fibrous structures at lower volumes and a dual continuous phase morphology at higher volumes. A transverse view of the interpenetrating morphology in (C) is shown in (D). (From Dhoble et al. [182], © (2005) Elsevier; used with permission.)

Conventional filters provide separation of particles in the range of 10 to 1000 μm. Microfilters, ultrafilters, nanofilters, and reverse osmosis membranes provide separation on the scale from 1 μm down to less than a nanometer, as shown in Table 5.2. In order to provide such varied separation properties, the pore size, shape, and distribution are significantly different in these membranes. Microfiltration involves the passage of water and dissolved materials while retaining micrometer sized suspended materials using homogeneous membranes. *Microporous membranes* or microfiltration (MF) can meet many separation demands when the pore size range is 0.05–1 μm, providing a range of applications [189]. *Ultrafilters* (UFs) are surface permeable, passing water and salts while retaining macromolecular sized particles. The relation between the surface pores and flux for ultrafilters has been described [190]. *Nanofiltration* (NF) is the newest named membrane, enabling the removal of monovalent ions such as chlorides at the 50% to 90% level and often used for water softeners and raw water treatment. *Reverse osmosis* (RO) filters pass water and retain both dissolved and particulate materials in the ionic size range. Reverse osmosis membranes are generally asymmetric and anisotropic; that is, they have a gradient of pores ranging from a dense surface layer to a porous substructure, which provides mechanical strength, and they only work in one direction. *Composite membranes* are also asymmetric with a thin surface film sprayed or coated onto a porous substructure, supported by a synthetic fabric.

5.2.4.2 Literature Review

Electron microscopy has been applied to the determination of the structure of membranes for

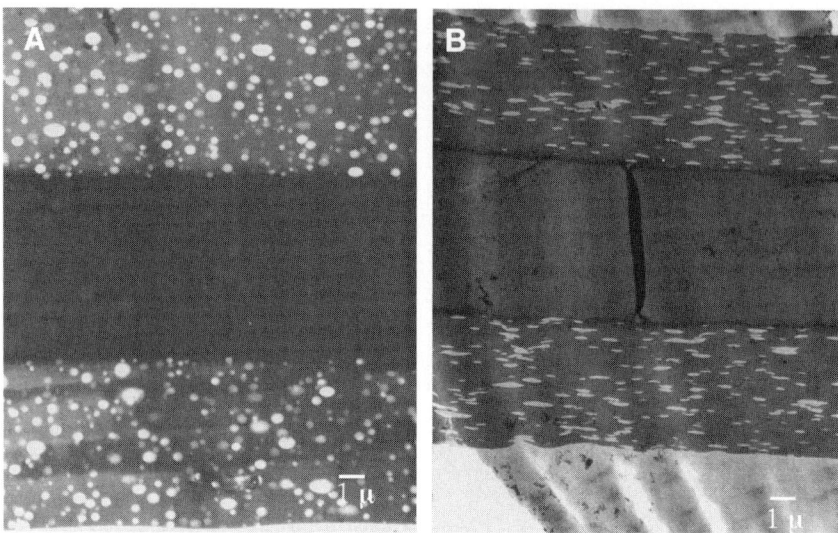

FIGURE 5.40. Transmission electron microscopy thin section of phosphotungstic acid stained cryosections of an as extruded (A) and deep drawn (B) coextruded film constructed of toughened nylon/barrier resin/toughened nylon. (From Wood [186]; used with permission of the American Chemical Society Rubber Division.)

correlation with transport properties. The SEM provides both the best overall view and detailed three dimensional images of these structures [191]. Optical microscopy can give a rapid overview without the possibility of a change in structure caused by an electron beam or vacuum. Essentially, three structural types of membranes have been described: homogeneous, asymmetric, and composite. In the homogeneous membrane, the pore structures are uniform throughout the cross section, whereas asymmetric membranes exhibit a pore gradient from a dense surface layer to macrovoids. Polycarbonate and polyacrylonitrile are examples of homogeneous membranes, and polybenzimidazole (PBI) and cellulose acetate (CA) are examples of asymmetric membranes. Composite membranes have a dense surface layer with a support structure composed of another polymer, often polysulfone.

Transmission electron microscopy was used to study high temperature polyimide-based nanofoams derived from triblock copolymers for applications in microelectronics packaging, storage cells, and high temperature polymer membranes [192]. Films before and after forming were prepared for microtomy by sub-

TABLE 5.2. Membrane separation processes

Process	Filtration	Materials retained	Pore size
Conventional filtration	Coarse filters	Large particles	>2 (10–1000) μm
Microfiltration (MF)	Microporous membranes	Suspended matter	0.1–20 μm
Ultrafiltration (UF)	Membrane	Macromolecules, colloids (passes salts)	0.01–0.5 μm
Nanofiltration (NF)	NF membrane Not as pure as RO	Dissolved and suspended materials (ionic) saline at 50–90% at 200 psi	1–80 nm
Reverse osmosis (RO)	Semipermeable membrane	Dissolved and suspended materials (ionic) saline at 98–99% at 200 psi	1–80 nm
Gas separation	Semipermeable membrane	Gases and vapors	0.2–1.5 nm

merging only part of them in the epoxy making it easier to locate them for trimming and leaving the protruding film free of embedded epoxy. Samples were microtomed at room temperature using a diamond knife and the sections picked up and stained in the vapor of a 1% RuO_4 solution. Transmission electron microscopy imaging at 200 kV showed irregularly shaped voids with nanometer size pores.

Field emission SEM and AFM have been used to image surface topography of membranes. Imaging with FESEM at low voltages does not require a metal coating, which might fill very fine pores if a thick layer is used. Low voltage is important to minimize beam damage of fine topography, again as the finer pores might be easily damaged or filled with carbon prior to the operator even being aware it has taken place. Comparison of AFM with SEM is most useful to ensure that interpretation of the AFM images is accurate. Atomic force microscopy does not require that the membrane be dry nor is the sample placed in a high vacuum system, although this is required for SEM. Variable pressure SEMs permit hydrated materials to be imaged, however, this technique is not very useful at very high resolution, so its value depends on the level of detail that is of interest. Fritsche et al. [193, 194] studied the structure of polyethersulfone ultrafiltration membranes using SEM and AFM, concluding that size differences in the topography was due to the metal coating and vacuum used in the SEM, although they prepared the samples by different methods. The SEM samples were dried, frozen in liquid nitrogen, and fractured, followed by Au coating and imaging at 80,000× using a 25 kV electron beam. The SEM images clearly showed cracks in the thick metal coating, not mentioned by the authors, but significantly different sized textures were observed in the AFM images of hydrated, uncoated samples. For such comparisons to be valid, the specimen preparation must be similar and the conditions used should not produce artifacts.

Ultrafiltration membranes made of different materials—polyethersulfone (PES), poly(vinylidene fluoride) (PVDF), and polyacrylonitrile (PAN)—have also been characterized by FESEM and image analysis (e.g., [195]). Samples were coated with a 5 nm layer of sputtered Pt for FESEM at 2 or 3 kV, and images were digitized for analysis using the public domain program NIH Image (developed by the National Institutes of Health, Division of Computer Research and Technology). Results, combined with porosity and water permeability values, gave material information. Nanofiltration type electronically conductive membranes, formed by deposition of a gold coating (150 nm thick), under a commercial NF membrane, have also been studied by FESEM (e.g., [196]).

Atomic force microscopy is finding more use in examination of membranes, but artifacts must be addressed, as was done by Bowen and Doneva [197], who noted changes in pore size and structure and used Fast Fourier Transform (FFT) filtering to show the true pore shape. Samples for AFM were prepared by attaching them to steel disks with double sided tape. These same authors used AFM to characterize ultrafiltration membranes [198, 199] and characterized the pore dimensions and quantified the interaction or adhesion of cellulose with two polymeric UF membranes. Atomic force microscopy was also used to characterize molecularly imprinted composite polyethersulfone membranes for quantification of the pore size and surface roughness [200].

Proton exchange membranes (PEMs) are being developed using rigid rod polymers as the supporting substrates as they have the potential to meet the high strength, dimensional stability, chemical resistance, good barrier properties, and high temperature properties required for fuel cell applications. Fuel cells have the potential to produce cleaner more efficient energy. Rigid rod polymer membranes, such as poly(p-phenylenebenzobisoxazole) (PBZO) and poly(p-phenylenebenzobisthiazole) (PBZT), have been extruded and studied by wide angle x-ray diffraction (WAXD), SAXS, AFM, SEM, and TEM [201]. Fracture surfaces were prepared after immersion in liquid nitrogen for low voltage high resolution SEM. Samples for TEM were microtomed from an epoxy mounted sample, followed by surface etching with oxygen plasma. Atomic force microscopy was conducted in the repulsive contact mode on small pieces of the membranes mounted on the sample holder with double sided tape. Layers were peeled from both surfaces with tape and the thickness measured with an AFM. The study showed the orientation of the rings with respect to the surface and suggested their potential as microporous PEMs.

Yang and Martin [202] have been working on developing biocompatible polymer coatings on the surface of neural implants to lengthen their life when they are implanted in the brain. Microporous films of the conducting polymer developed for this application, poly(3,4-ethylenedioxythiophene) (PEDOT) and polypyrrole (PPy), were electrochemically deposited on the microelectrodes of neural probes using different sized polystyrene latex spheres as templates. Scanning electron microscopy revealed a three dimensional microporous structure consisting of voids with interconnected channels; the surface morphology varied with the particle size of latex spheres and coating thickness.

5.2.4.3 SEM: Surface and Bulk

The SEM is used for the study of the surface and bulk structures of membranes. Membranes are prepared by attaching them to the specimen stub and applying a conductive surface coating. Bulk structures are observed for membranes fractured in air or liquid nitrogen, sectioned, or critical point dried. Figures 5.41 and 5.42 are

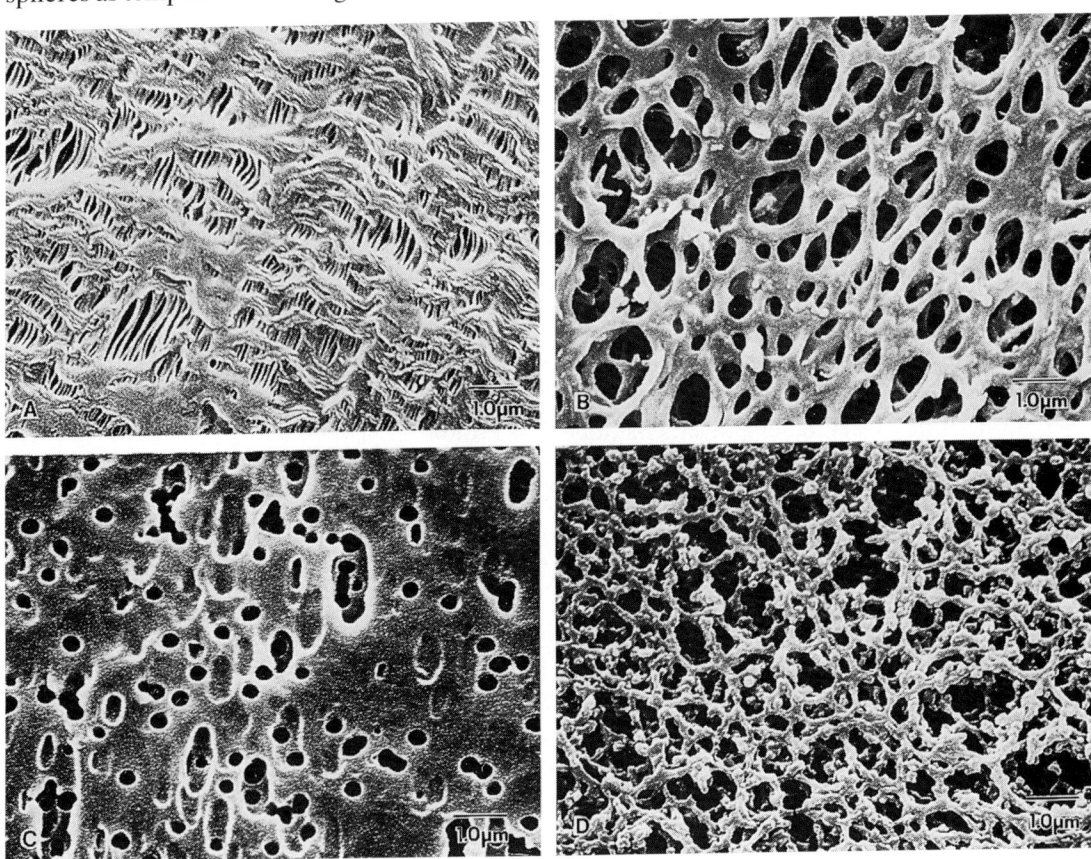

FIGURE 5.41. Scanning electron microscopy micrographs of several membrane surfaces reveal a range of pore structures that in turn result in a range of separation applications. An experimental, microporous, polyethylene membrane is shown (A) with elongated, stretched porous regions of various sizes, separated by fibrils, in the draw direction and unstretched lamellae normal to the draw direction. This surface structure is quite different from three commercial membranes (B–D). One membrane (B) consists of a low density network of rounded pores, many of which are larger than $1\,\mu m$ across. A nucleopore membrane (C) has more defined pore structure with rounded pores bored through from one side to the other. The morphology in (D) is an open network structure with the polymer in the form of strings of particles.

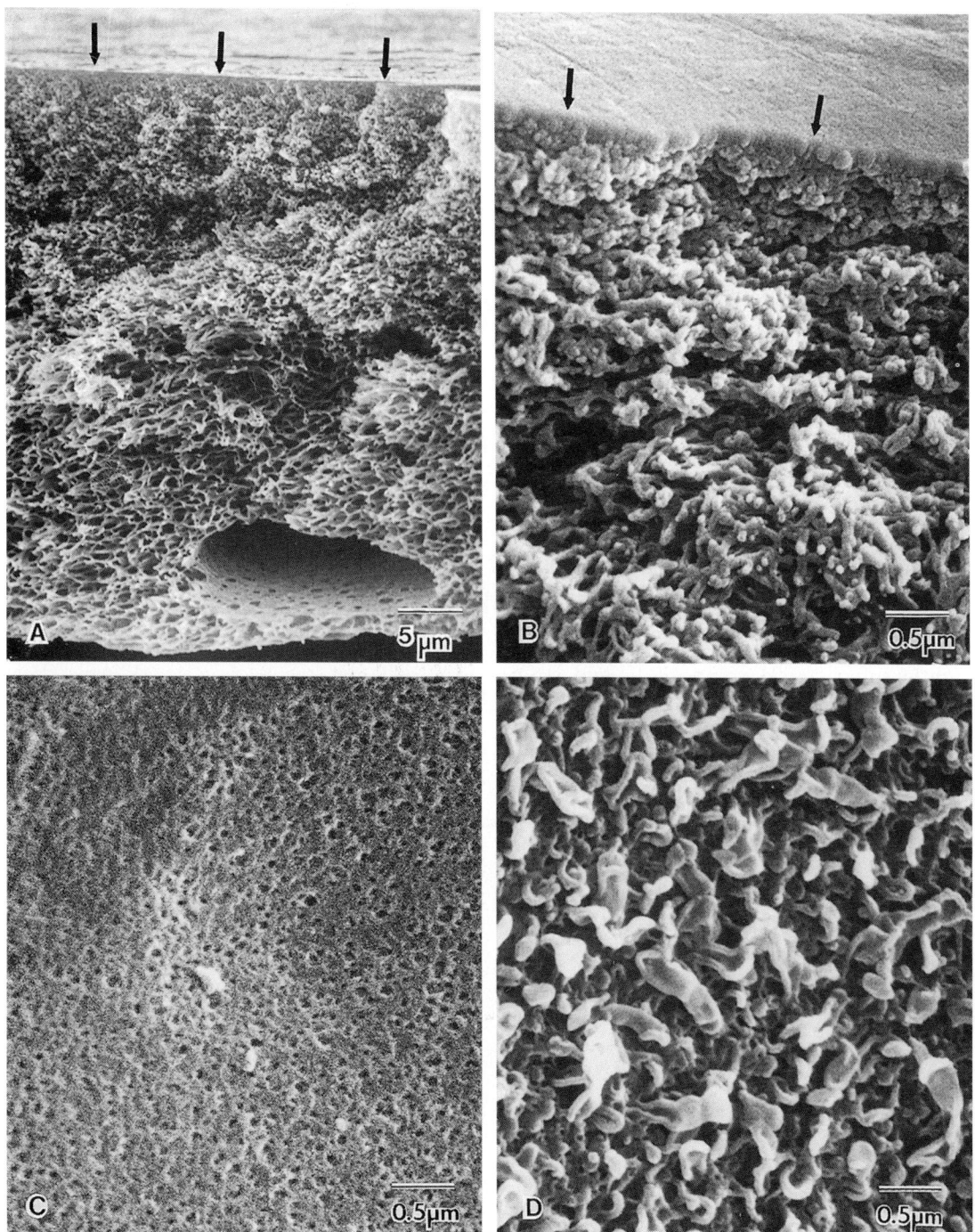

FIGURE 5.42. A polysulfone composite membrane is shown in both cross section (A and B) and surface views (C). Scanning electron microscopy images reveal large macrovoids on the bottom side of the membrane within a porous texture support layer. A dense surface layer (arrow) appears to be composed of granular particles of polymer with little pore volume. Some surface porosity is seen (C), but these pores are considerably smaller than those observed in the bulk of the membrane. Chemical etching of the top surface results in removal of the active surface (D), which gives another view of the bulk porous morphology.

examples of the varied structures of typical membranes that can be imaged by this technique. An experimental HDPE microporous membrane has pores elongated in the draw direction (Fig. 5.41A) where the pores are less than 1 μm wide and have a range of lengths to about 2 μm. The pore volume is formed by stretching lamellae; the remaining unstretched lamellae are seen as flatter regions, perpendicular to the draw direction. Another example (Fig. 5.41B) is a membrane with a large, rounded, and stretched porous network. A nucleopore membrane (Fig. 5.41C) has very discrete and rounded pores etched randomly into the film surface. This structure is similar to the polycarbonate nucleopore films [187], which have circular pores of constant cross section that run from the top to the bottom of the membrane. A cast membrane (Fig. 5.41D) has an open three dimensional network structure that appears formed by polymer in coated, particulate strings.

Polysulfone composite membranes provide a different chemical composition and structure compared with some of the examples shown. A polysulfone composite membrane is shown by SEM of cross sections (Fig. 5.42A, B) and of the top surface (Fig. 5.42C). A porous texture is seen (Fig. 5.42A) with larger macrovoids near the bottom surface. There is an open porous structure with a pore gradient, with smaller pores nearer the dense top surface (Fig. 5.42B). The asymmetric membrane has very fine surface pores, about 0.05–0.2 μm across (Fig. 5.42C) with an underlying open network composed of strings of polymer. The surfaces of composite membranes are generally dense, and SEM micrographs may not reveal any resolvable surface pores. Chemical etching of this dense surface layer is useful to observe the porous substructure (Fig. 5.42D).

5.2.4.4 Reverse Osmosis Membranes

Cellulose nitrate and cellulose acetate (CA) were among the first asymmetric, reverse osmosis membranes to be produced [203]. Plummer et al. [204] described 13 specimen preparation methods for the observation of CA membrane structures. They pointed out the lack of contrast in epoxy embedded sections and that one of the best stains, osmium tetroxide, reacts with the polymer. Freeze fractured membranes were found by these authors to be of questionable value. In our experience, if care is taken, SEM study of fractured membranes can provide an informative view of the structure even though some structures collapse and their sizes cannot be accurately determined. A method found acceptable was ultrathin sectioning of gelatin embedded wet membranes (TEM). The structure of CA membranes was shown by replication [205] and SEM [206].

Optical, scanning, and transmission electron micrographs of a commercial cellulose acetate asymmetric membrane are shown in Fig. 5.43. Each view provides a different perspective on the membrane structure while, together, they give the complete structural model. Specimen preparation for OM and TEM cross sections was by microtomy of embedded membrane strips using a method developed to limit structural collapse (see Section 4.3.4). An optical micrograph (Fig. 5.43A) shows the membrane cast on a woven support fabric with an active surface layer (top), which appears as a "skin" several micrometers thick, and a support structure of rounded macrovoids. Scanning electron microscopy images (Fig. 5.43B, C) provide higher magnification views of the membrane cross section formed by fracturing in liquid nitrogen; this bulk view shows the network of submicrometer sized pores (Fig. 5.43C). Transmission electron microscopy micrographs (Fig. 5.43D, E) show greater detail of the microstructure, although there is little contrast between the polymer and the epoxy embedding media. Smaller pores, not clearly resolved, are observed near the membrane surface (arrows), and larger pores are seen deeper within the asymmetric membrane (Fig. 5.43E). Resolution of the finest pores is limited due to the section thickness.

Asymmetric polybenzimidazole membranes have been developed for RO applications, in the form of hollow fibers [207] and flat film membranes [208] for water transport. By comparison with cellulose acetate, PBI has very attractive chemical, flammability, and thermal properties. There are two problems encountered in attempt-

Films and Membranes

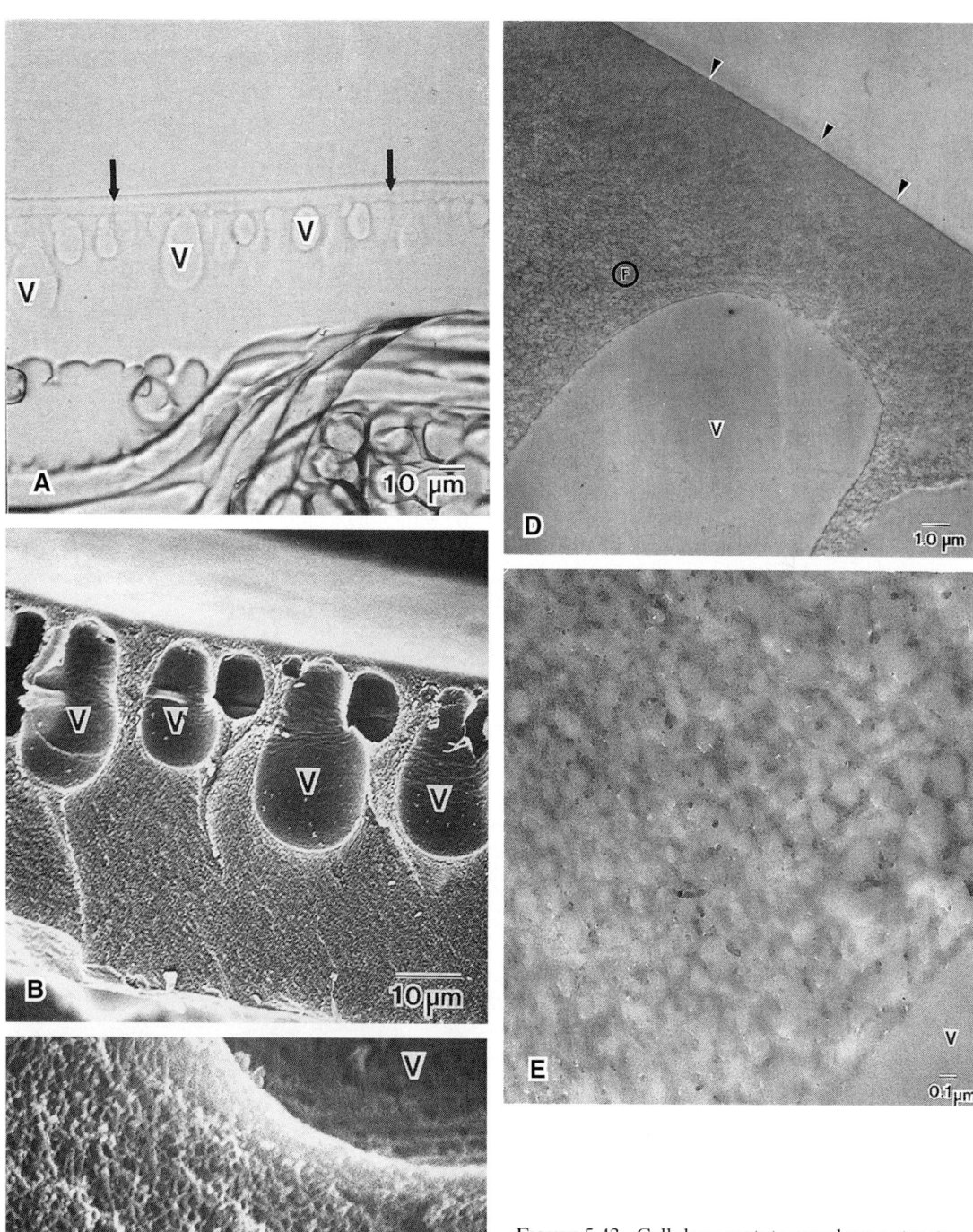

FIGURE 5.43. Cellulose acetate membrane structures are shown by complementary techniques. The optical micrograph (A) shows an overview of the membrane, cast on a woven fabric support (bottom). A surface layer (arrows) is observed above large, rounded macrovoids (V). Scanning electron microscopy cross sections reveal these macrovoids in more detail (B) and also show the nature of the fine pores (C). A TEM micrograph (D) of a section near the surface (arrowheads) reveals a dense layer, with a porous microstructure, shown more clearly at higher magnification (E).

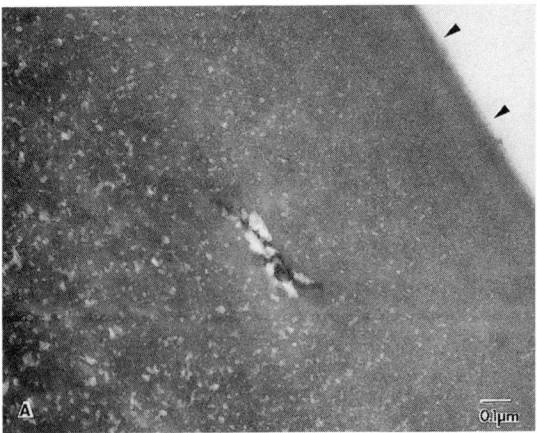 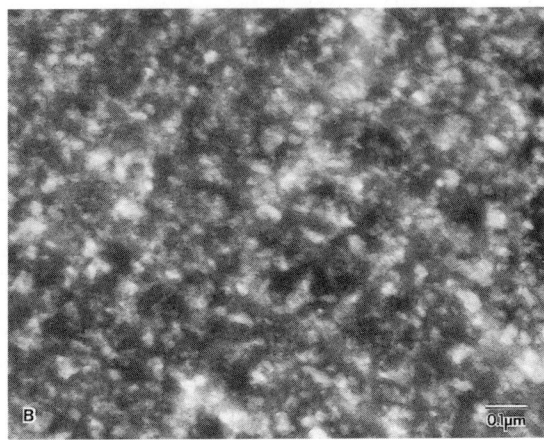

FIGURE 5.44. The fine structure of a PBI asymmetric membrane is shown in TEM micrographs of cross sections. A dense surface layer (arrowheads) is observed in a micrograph (A) taken with the high brightness lanthanum hexaboride gun that shows no pores are resolved in the top 50 nm of the dense surface layer. Pores on the order of about 0.05 μm are clearly shown (B) within the membrane support structure.

ing the preparation of such membranes for TEM: (1) deformation during drying and (2) lack of contrast. Often, specific methods must be developed for each membrane type, although method development is quite time consuming.

A general method was developed [209] to limit drying deformation by directly embedding the wet membrane and removing water during resin infiltration (see Section 4.3.4). Transmission electron microscopy micrographs taken with a lanthanum hexaboride high brightness gun, for enhanced resolution, show pores less than 5 nm, but no pores are resolved in the top 50 nm of the surface layer (Fig. 5.44A) of this PBI membrane. Within the bulk membrane (Fig. 5.44B), there are much larger pores (about 50 nm). Scanning electron microscopy images of a fractured, critical point dried membrane (Fig. 5.45) show robust, macrovoid structures with the top dense surface layer clearly composed of a monolayer of densely packed, deformed particles, about 80 nm in diameter, packed so closely as to limit surface porosity. Less well packed particles form the more open bulk membrane texture. The structure shown confirms those hypothesized from earlier TEM replica studies of wet poly(amide-hydrazine) and dry polyimide asymmetric membranes [210, 211].

Thin film composite polyamide membranes are used for reverse osmosis and nanofiltration

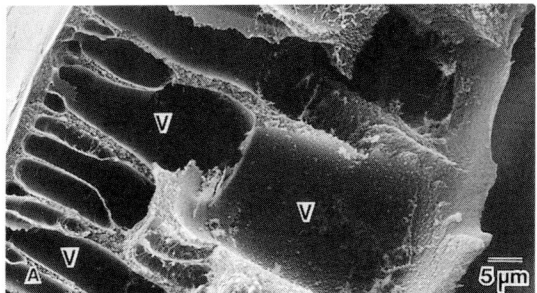
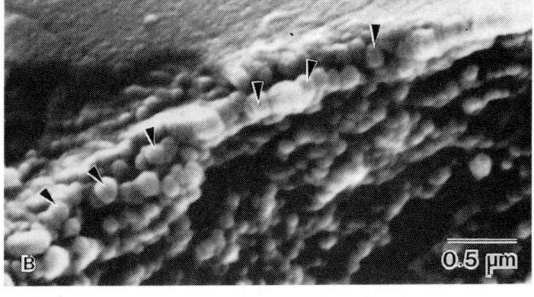

FIGURE 5.45. High resolution SEM images of a critical point dried and fractured PBI membrane reveal the fine structure quite clearly. The overview micrograph (A) shows the macrovoids (V) and the porous walls within the membrane. The robustness of the macrovoids suggests that the method is useful for observation of the *in situ* structure. The dense surface layer is composed of spherical particles (arrowheads) that are deformed into a dense monolayer, whereas the support structure below is formed by a more open network of these spherical particles (B).

due to their excellent performance and economics for water treatment. Freger et al. [212] modified the surface of a polyamide membrane by grafting of a hydrophilic polymer onto the surface to improve the fouling properties of thin film composite membranes. The structure of the NF membranes was modified using graft polymerization of acrylic (AA) monomers and was characterized by AFM, TEM, and Fourier transform infrared spectroscopy. The NF-200 membrane (Filmtec; trademark of Dow Chemical Co., Midland MI) is composed of a semiaromatic piperazine-based PA layer on top of a polysulfone support reinforced with a polyester nonwoven backing. Atomic force microscopy images were taken of dry membranes, although the RO membranes in this study were measured in a liquid cell under water on the AFM. Membranes were stained by treatment with dilute NaOH solution, followed by immersion in an excess of uranyl nitrate for about 15 min, washed, and dried in a vacuum at 40°C. For TEM, dry treated samples were mechanically separated from the backing and small pieces embedded in resin, cut to 60–100 nm thickness using an ultramicrotome, and placed on a carbon/collodion covered copper grid for imaging at 120 kV.

Figure 5.46 shows an AFM image (left) of modified NF-270 with an average surface roughness of 4.9 nm and a TEM cross sectional image (right) of a similarly modified membrane [212].

5.2.4.5 Microporous Membranes

A method using staining and ultramicrotomy (see Section 4.4.2) has been demonstrated that shows the three dimensional structure of microporous membranes such as Celgard 2400 and 2500 membranes (trademark of Celgard USA, Charlotte, NC) [213]. Celgard is formed by film extrusion, annealing, and stretching isotactic polypropylene. This produces an oriented crystalline structure with parallel arrays of pores [214]. The surface view of Celgard 2400 was shown by high resolution FESEM (see Fig. 4.31) where rows of pores are aligned parallel to the machine direction and drawn fibrils separate regions of undrawn crystalline lamellae and define the pore volume. The surface structure of Celgard 2400 has much smaller pores than Celgard 2500, but the overall three dimensional structures are similar. Transmission electron microscopy sections of Celgard 2400, prepared along the three axes, are shown in Fig. 5.47.

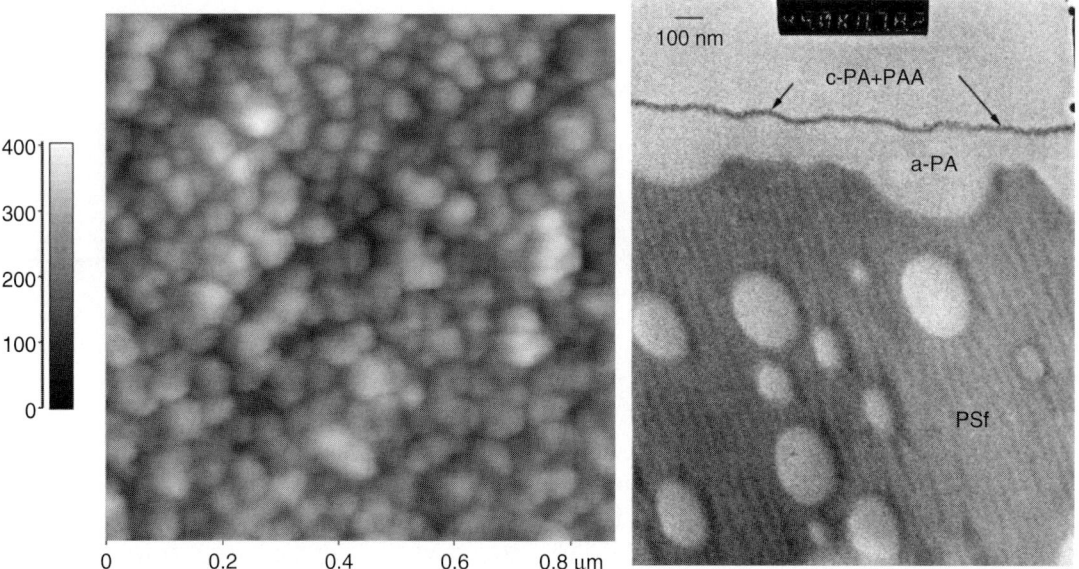

FIGURE 5.46. Atomic force microscopy image (left) of a modified NF-270 membrane (size of image is ca. 1 × 1 μm) and TEM cross sectional image (right) of the same membrane. (From Freger et al. [212], © (2002) Elsevier; used with permission.)

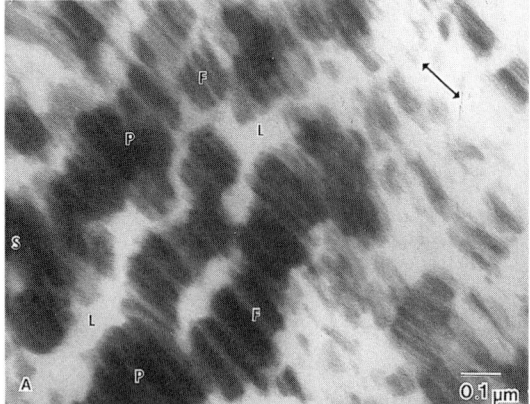

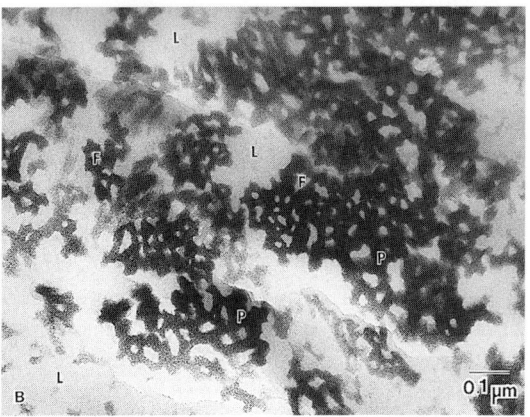

FIGURE 5.47. Transmission electron microscopy micrographs are shown of a surfactant treated and osmium tetroxide stained Celgard 2400 membrane cut along the three dimensions of the membrane: along the machine direction, across the machine direction, and in the plane of the membrane along the face. A section cut along the machine direction (A) reveals fibrils (*F*) separating electron dense pores (*P*), filled with stained surfactant, arranged in rows elongated in the machine direction (arrow). Unstained lamellae (*L*) are white regions between these pores. The cross axial section (B) shows that these pores are arranged in networks that do not run straight across the film but have a tortuous path.

Ultrathin sections cut along the axis, in the longitudinal direction (Fig. 5.47A), show the pores are oriented parallel to the machine direction. Dense regions are surfactant stained, nonporous regions are white, and gray regions result from the effect of the section thickness [213]. The fibrils separating the pores do stain but not as much as the surfactant filled pores. The cross section (Fig. 5.47B) is composed of a network of pores, with little order, in agreement with the axial view. The TEM micrographs clearly show short parallel rows of pore channels, separated by unstretched lamellae and defined by the drawn fibrils. There is a random, tortuous pore volume that provides unique microporous membrane applications. These micrographs could be combined to formulate a three dimensional model as was done earlier for Celgard 2500 (Fig. 5.48).

Field emission SEM, using a cold cathode field emission gun (FEG) SEM, of Celgard 2400 (samples supplied by B.A. Petrey, Celgard USA) lightly coated with carbon [215] were compared with ICAFM images of the same membrane, prepared by attachment to carbon tape on a steel flat [216]. The FESEM images are shown in Fig. 5.49 with fine, aligned pore structures. Atomic force microscopy images, in

FIGURE 5.48. A three dimensional model of Celgard 2500 (trademark of Celgard LLC) is shown composed of sections cut along, across, and in the plane of the machine direction viewed by TEM. The surface is shown by an SEM micrograph. (From Sarada et al. [213]; used with permission.)

Films and Membranes

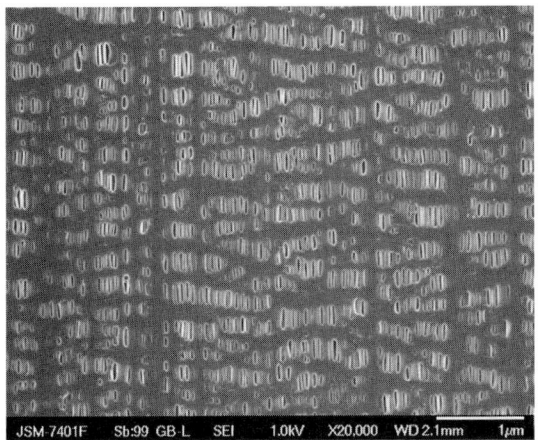
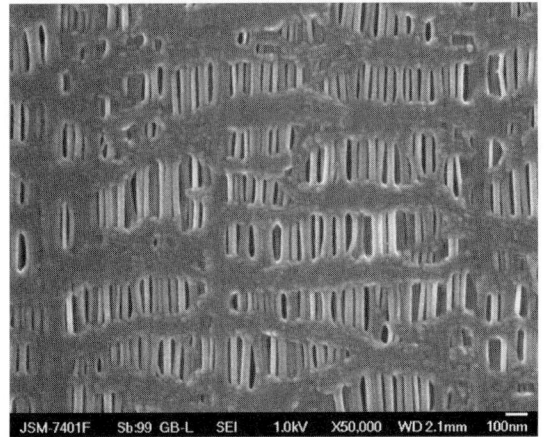
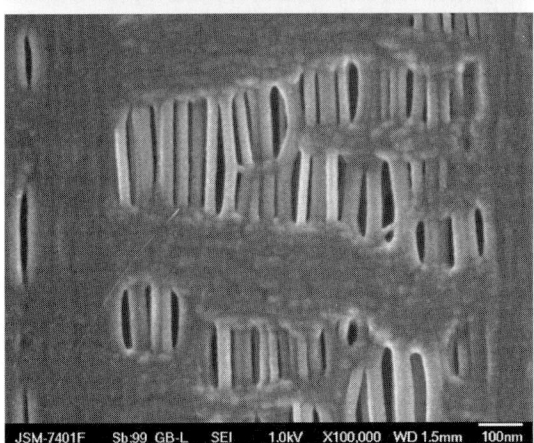

FIGURE 5.49. Field emission SEM images of Celgard 2400 clearly showing the porous network. (From Robertson, JEOL USA, Peabody MA [215]; unpublished.)

Fig. 5.50, in height mode (left) and phase mode (right), clearly show the same structure overall with detail resolved in the phase images of the undeformed lamellae, perpendicular to the draw direction (fibrils).

Field emission SEM has provided increased information in the case of several experimental membranes [217] with very different pore structures. In the first case, the surface of a PTFE membrane exhibits a nonuniform series of rounded pores (Fig. 5.51A), and an image taken at much higher magnification reveals the three dimensional nature of the pores as they extend into the bulk membrane (Fig. 5.51B). The surface of the membrane is seen to be wrinkled in texture, but various attempts to view the image at lower magnifications, after high magnification imaging, did not reveal picture frame contrast that would suggest that this texture is due to beam damage. In a final example [217], a two phase system was processed into a film that was fractured and imaged. Details of the texture shown in the FESEM images in Fig. 5.52 reveal very coarse, irregularly shaped pores and fine, dispersed second phase particles.

5.2.5 Hollow Fiber Membranes

The structures of hollow fiber membranes are somewhat analogous to those described for flat film or sheet membranes produced from similar polymers. Hollow fiber membranes [187, 218] have been produced from CA, PP, PBI, polysulfone, aromatic polyamides, and polyacrylonitrile. Polysulfone hollow fibers were described [219] in an SEM study that showed a dense skin formed on the surface with a porous or spongy subsurface support structure. The fibers were prepared for SEM by breaking at liquid nitrogen temperature. The porosity of the hollow fibers was shown to be complex and asymmetric. Cabasso and Tamvakis [220] described composite hollow fiber membranes in some detail, showing the surface structure was composed of a polysulfone porous substrate coated with cross linked polyethyleneimine or furan resin. A dense, semipermeable layer on top of the porous substructure is responsible for high salt rejection in this

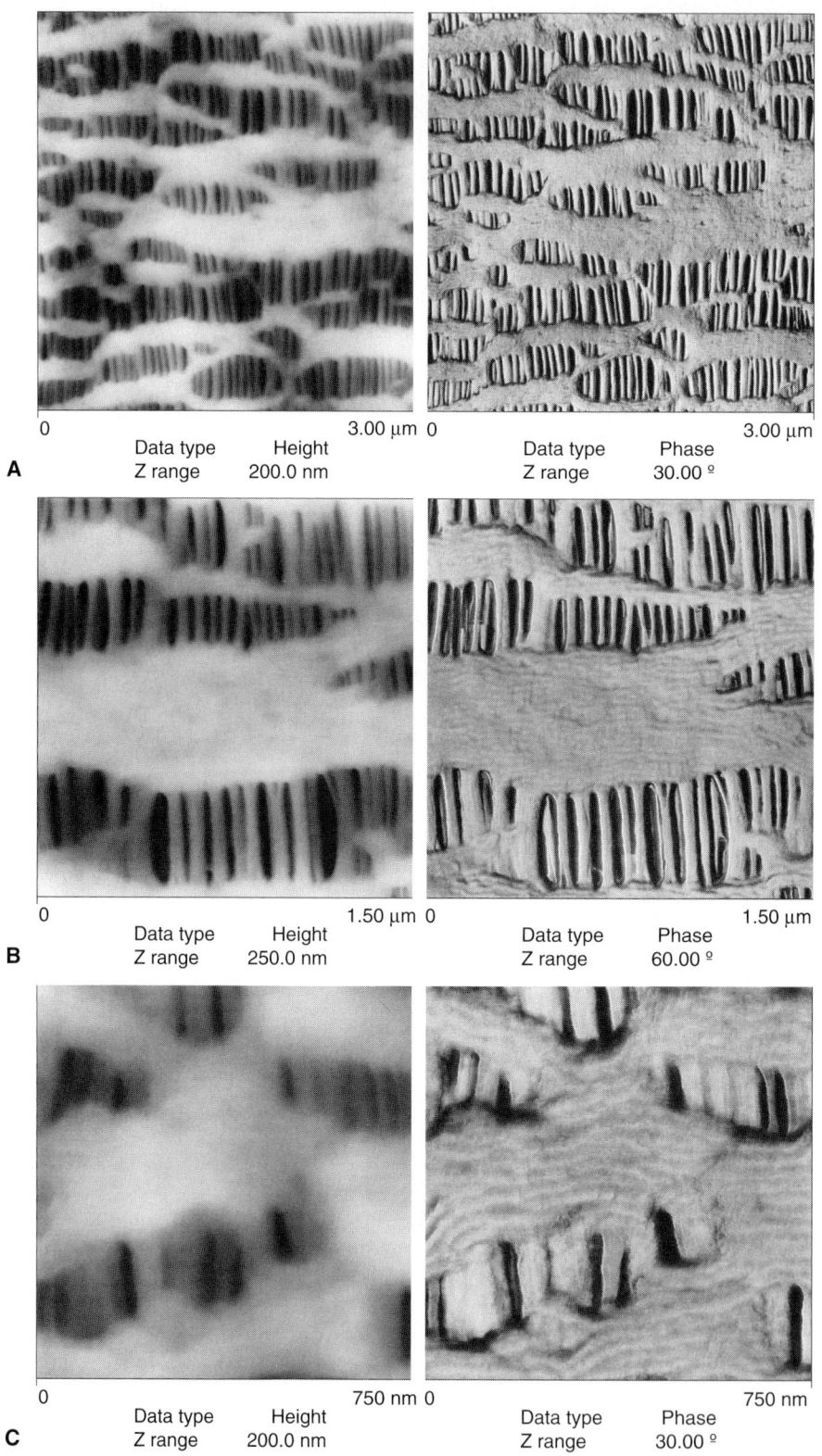

FIGURE 5.50. Atomic force microscopy images taken in intermittent contact of Celgard 2400 taken using moderate amplitude, light tapping, clearly show consistent details to the TEM and FESEM images with detail of the undeformed lamellae running perpendicular to the draw direction (fibrils).

Color Plate I

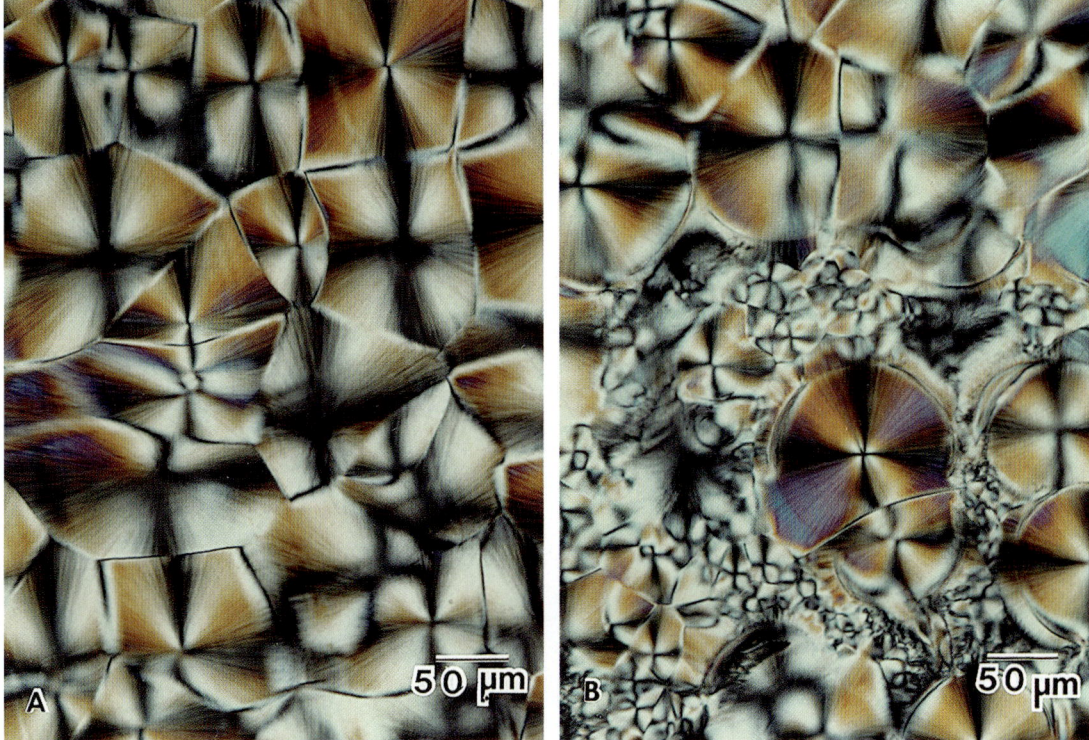

FIGURE 1.3. A thin section of bulk crystallized nylon, in polarized light, reveals a bright, birefringent and spherulitic texture. (A) At high magnification, a classic Maltese cross pattern is seen, with black crossed arms aligned in the position of the crossed polarizers; the sample was isothermally crystallized, and exhibits large spherulites. (B) The sample quenched during crystallization yielding large spherulites surrounded by smaller ones.

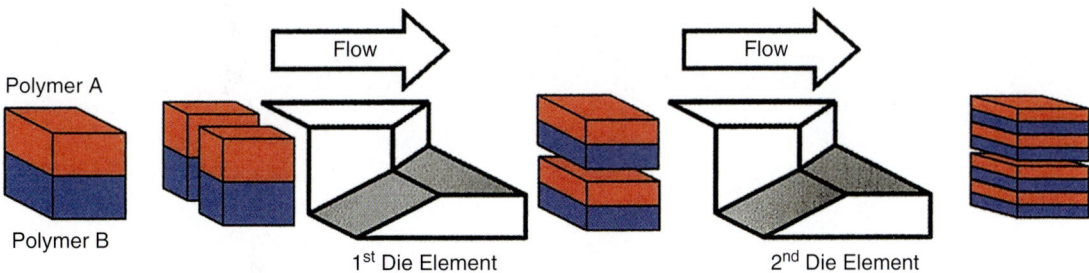

FIGURE 1.8. Schematic of layer multiplying coextrusion used for forced assembly of polymer nanolayers shows that two die elements multiply the number of layers from 2 to 8. (From Liu et al. [66], © (2004) American Chemical Society; used with permission.)

Color Plate II

FIGURE 3.15. Two images of a thin melt cast film of high density polyethylene: the region is $200 \times 250\,\mu m$. The left hand image was taken in crossed polars. The radial "Maltese cross" is due to the extinction position. The spherulites in this material have dark circumferential bands. The crystals twist as they grow, and their orientation in these bands has the optic axis perpendicular to the specimen plane. The right hand image is the same area when a first order red plate is also used. The blue and yellow colors show that the spherulites are negative.

FIGURE 3.16. Two images of a thin melt cast film of polycaprolactone. As with Fig. 3.15, the region is $200 \times 250\,\mu m$ and the left hand image was taken in crossed polars. The spherulites in this material are much less regular and some show colors under crossed polars, indicating a thicker film or a larger birefringence. The right hand image is the same area when a first order red plate is also used. The colors can still show that these spherulites are also negative.

Color Plate III

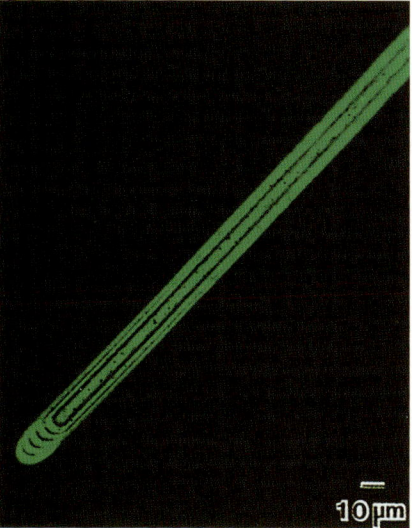

FIGURE 5.4. A polyester fiber in polarized light, aligned at 45° to the crossed polarizers. The dark bands are fringes that reflect the high-order birefringence. In the orthogonal position, the same fiber would exhibit extinction and thus appear black.

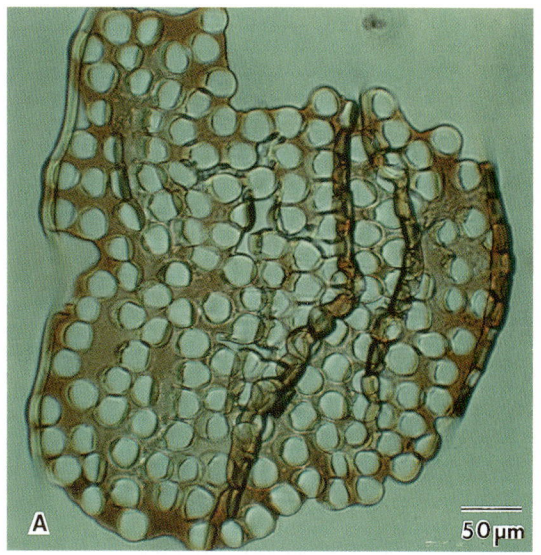

FIGURE 5.19. Tire cords are shown by several complementary microscopic methods. A cross section of an adhesive coated yarn is shown by optical microscopy (A).

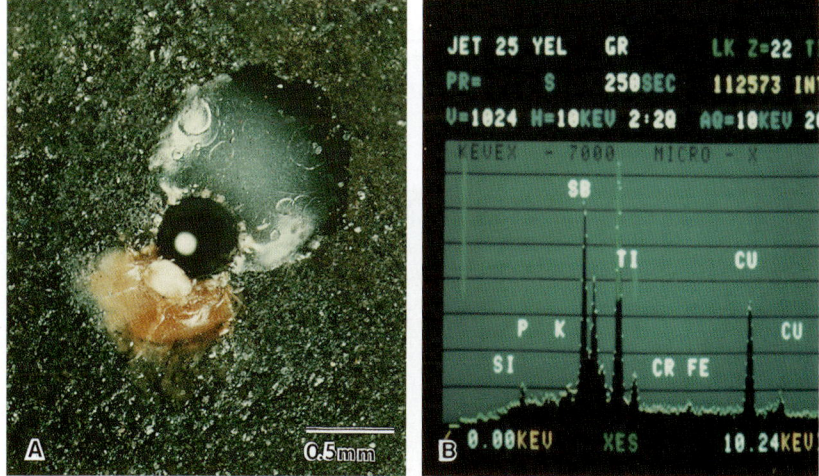

FIGURE 5.18. Combination of optical microscopy and EDS analysis permits the identification of contaminants plugging a spinneret. Rust colored material present on the spinneret (A) was scraped off to give the EDS spectra (B). The plugging material contains silicon, phosphorus, antimony, titanium, chromium, and iron. The copper is due to the specimen holder.

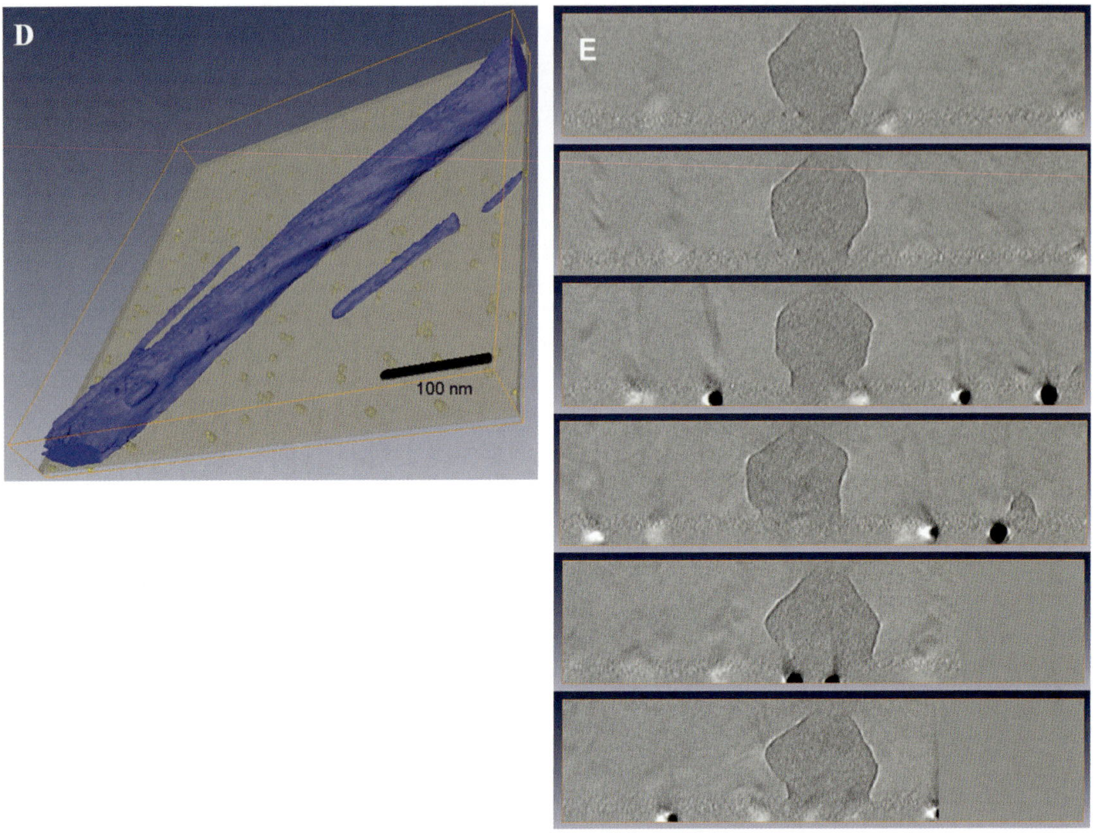

Color Plate IV

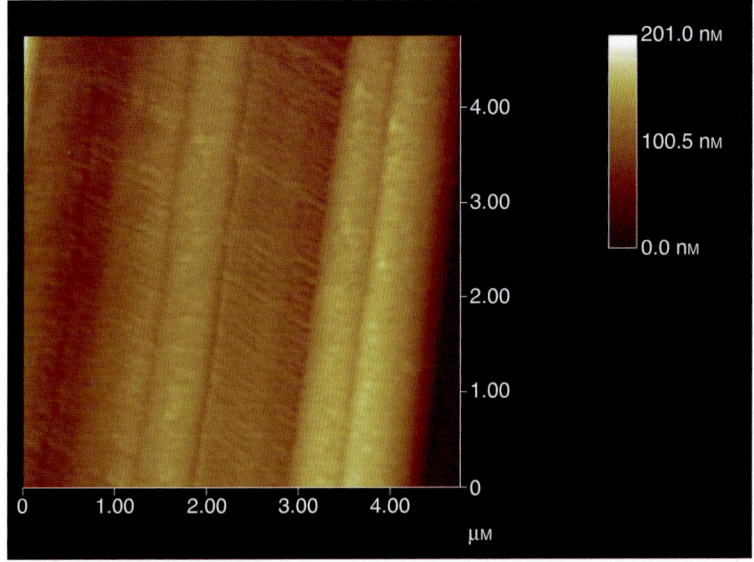

FIGURE 5.22. The three dimensional reconstruction of the fiber by electron tomography [87] shows the overall shape of the fiber (D) and the cross sections indicating the strands are in close contact throughout their length (E). (From Kübel et al. [86], © (2001) American Chemical Society, and Kübel [87]; used with permission.)

FIGURE 5.23. Large scale AFM image of the surface of unstretched threads shows fibers and fibrils. Image dimension is 4.8 × 4.8 μm. (From Gould et al. [94], © (1999) Elsevier; used with permission.)

Color Plate V

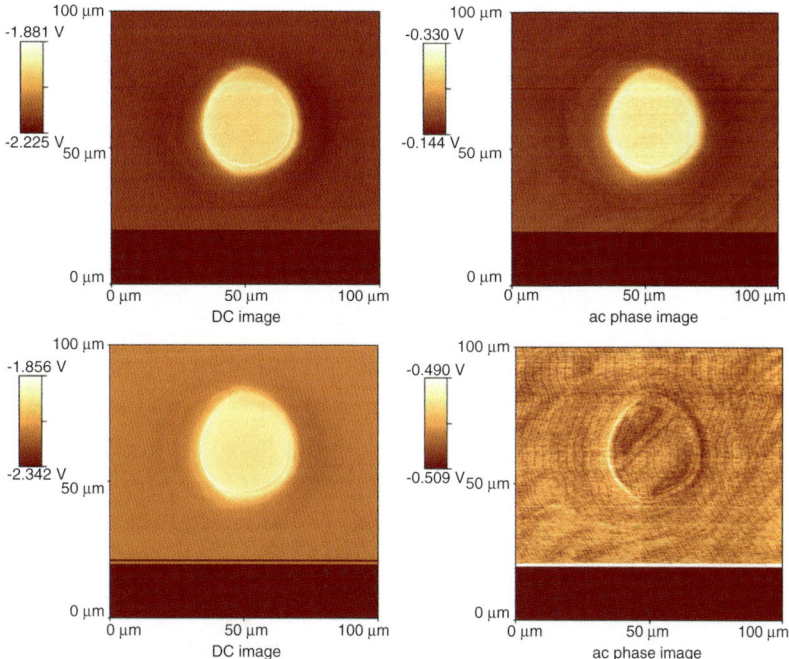

FIGURE 5.36. Scanning thermal microscopy images of 35 μm via holes patterned in a photoBCB thin film over a copper substrate before plasma treatment to remove polymer residues. DC thermal images (left) and AC phase images (right) are shown for frequencies of 1 kHz (top) and 30 kHz (bottom). (From Meyers et al. [174], © (2000) American Chemical Society; used with permission.)

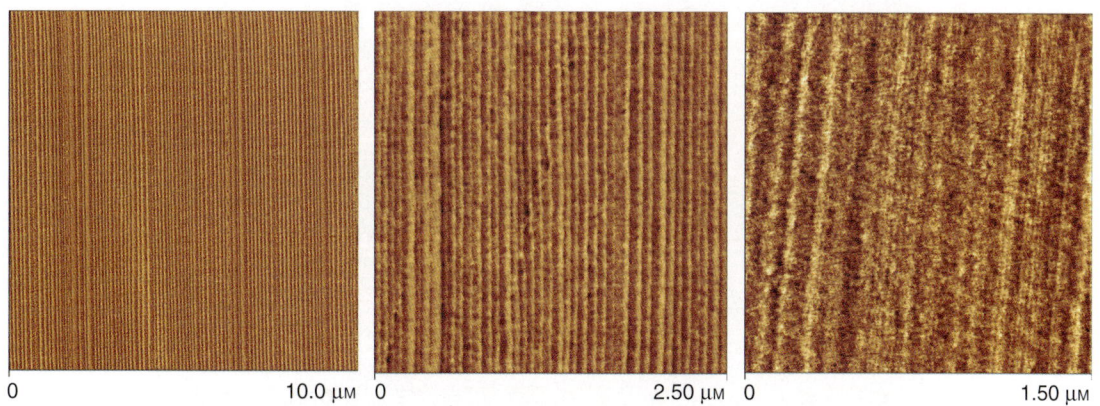

FIGURE 5.38. Atomic force microscopy phase images showing cross sections of PETG/PS assemblies with different layer thicknesses (left is 60 nm, center is 30 nm, right is 8 nm). (From Liu et al. [180], © (2004) American Chemical Society; used with permission.)

Color Plate VI

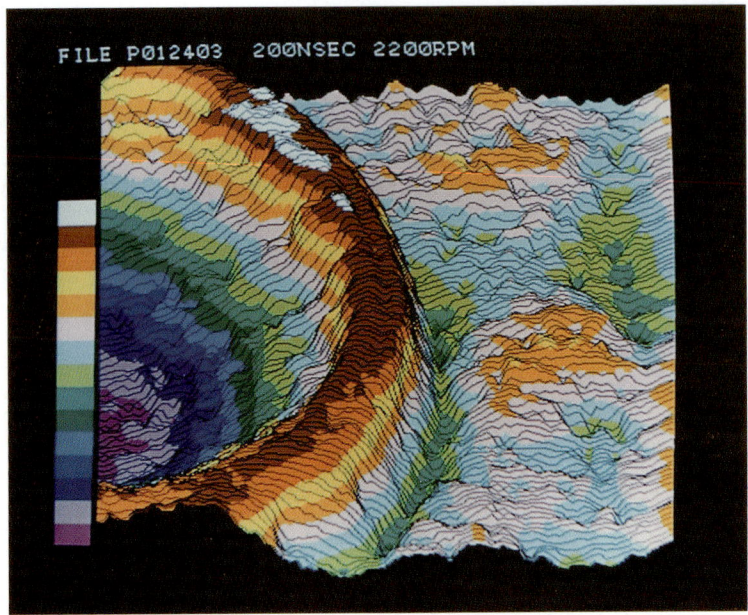

FIGURE 5.58. Scanning tunneling microscopy provides a direct measurement of the depth of a pit. This is the height difference between the media surface and the center of the pit, as seen in this image, especially where the depths of structures are very shallow compared with the three dimensional geometry. The total height variation in the image is 77 nm divided into 15 different colors; the depth of the pit is ca. 41 nm. (From Goldberg et al. [281]; used with permission.)

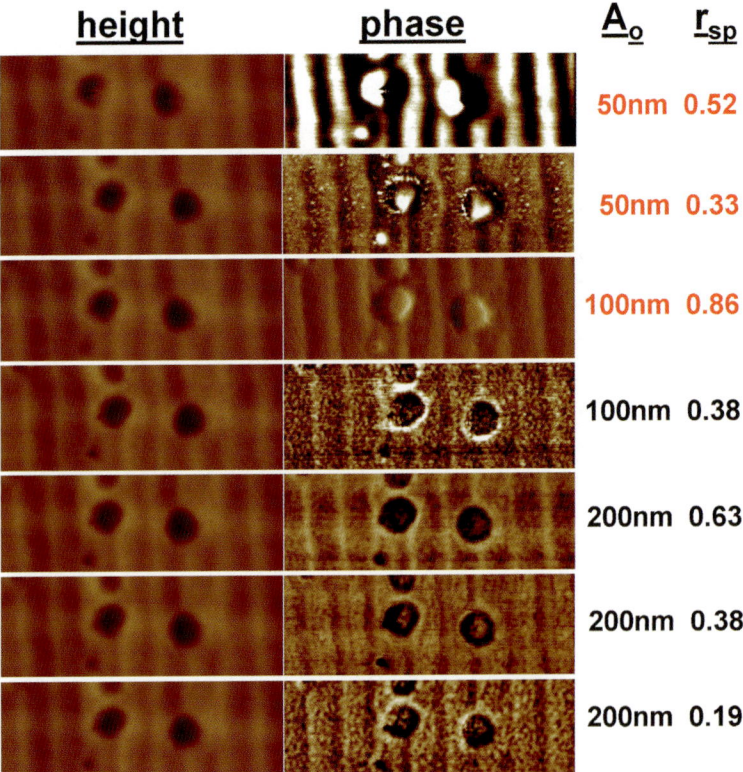

FIGURE 5.67. Intermittent contact mode AFM height and phase images of CET-GRC (5 wt.% rubber, 0.1 μm diameter) after cryo-polishing (images are $1 \times 0.3\,\mu m$). The table at the right indicates the parameters used to generate each pair of images. (From Meyers et al. [174], © (2000) American Chemical Society; used with permission.)

Color Plate VII

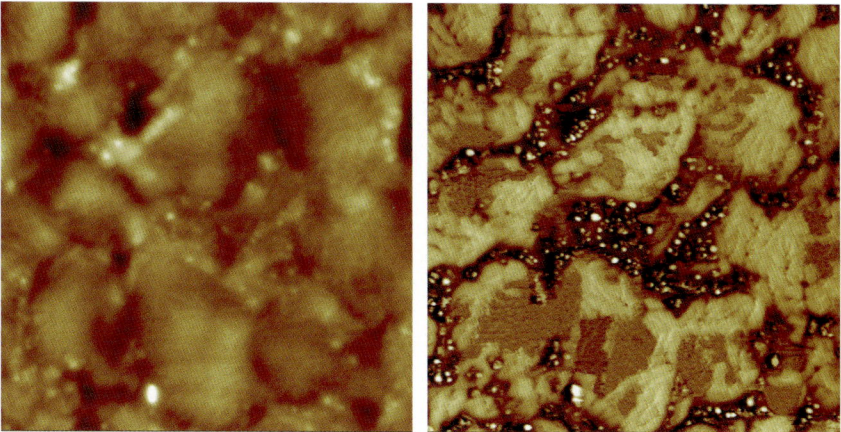

FIGURE 5.76. Height (left) and phase (right) images (5 μm on a side) of a thermoplastic vulcanizate made by mixing EPDM, iPP, and carbon black show the detailed morphology of these complex blends. (From Magonov and Yerina [253], © (2005) Springer; used with permission.)

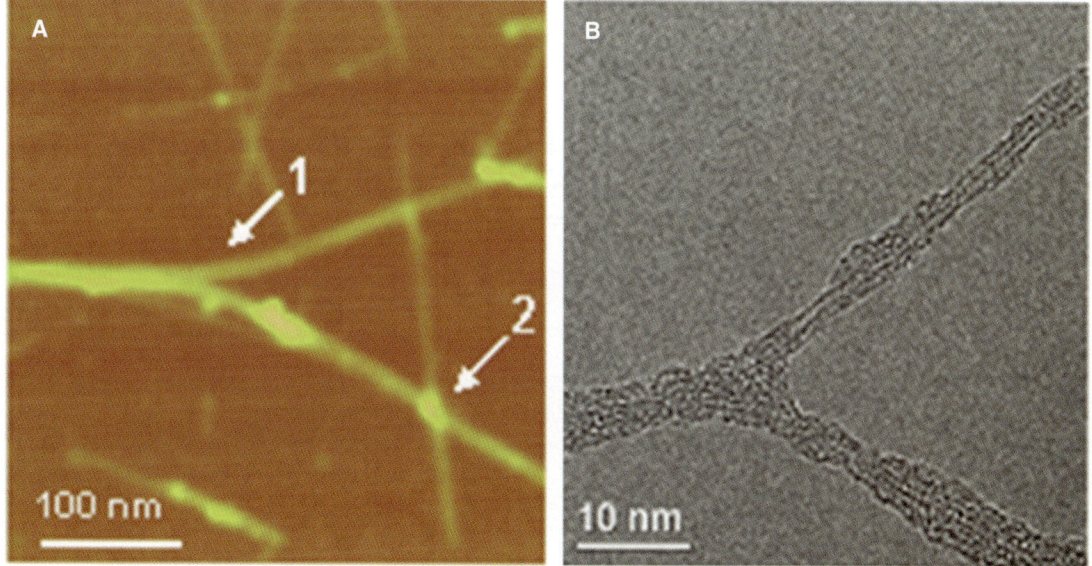

FIGURE 5.112. Atomic force microscopy image of nano-1/SWCNT sample exhibiting a Y-junction (1) and an X-junction (2)(A). Transmission electron microscopy image of peptide-coated SWCNTs exhibiting Y-junction apparently created through peptide–peptide interactions provided complementary evidence that individual peptide-wrapped SWCNTs could be isolated using an amphiphilic α-helical peptide (B). (From Musselman et al. [506], © (2004) American Chemical Society; used with permission.)

Color Plate VIII

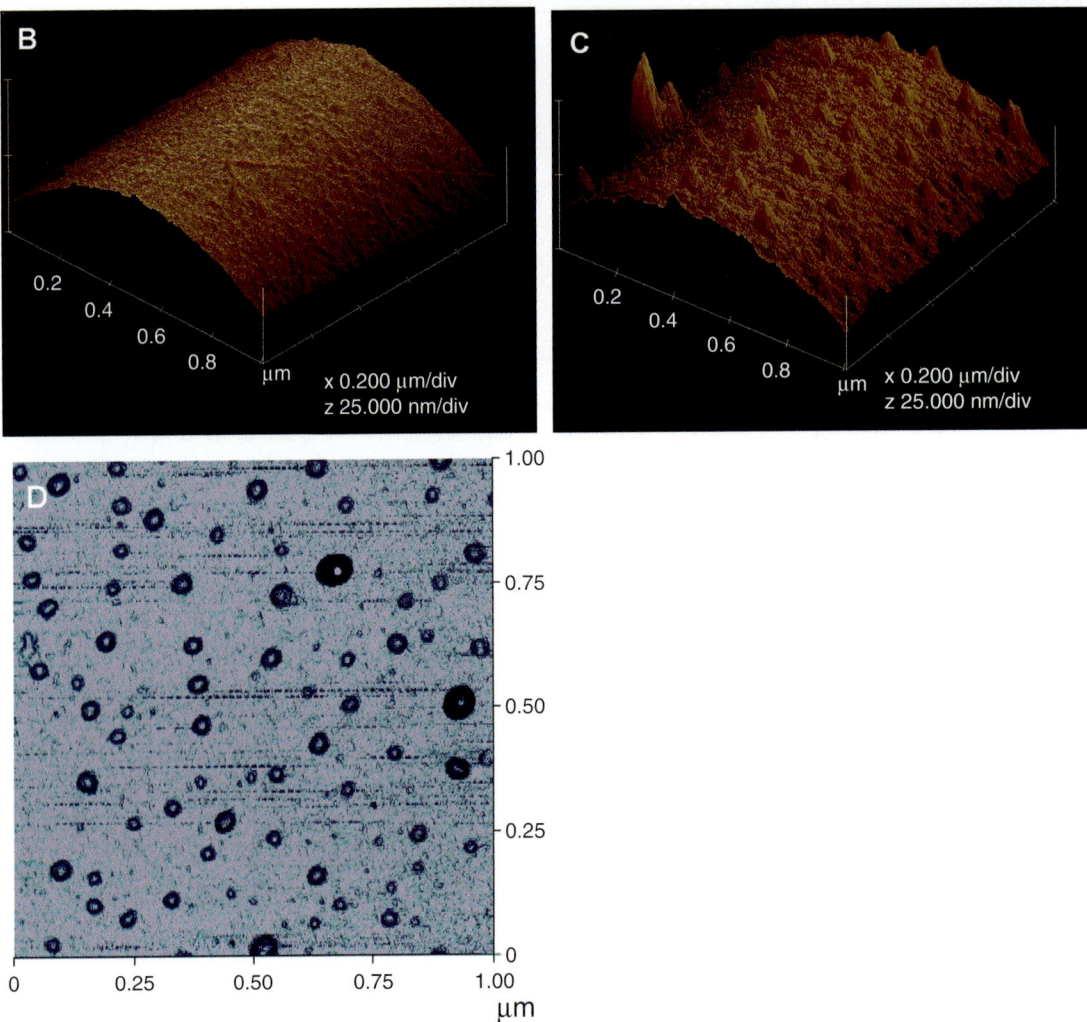

FIGURE 5.121. Atomic force microscopy images of glass fiber surfaces heat cleaned and acid treated (B) show little surface detail, whereas a silanized glass fiber surface (C) shows detail with height information. The top view (D) after treatment with acetonitrile solvent shows detail of the particle size. (From Turrion et al. [584], © (2005) Elsevier; used with permission.)

Color Plate IX

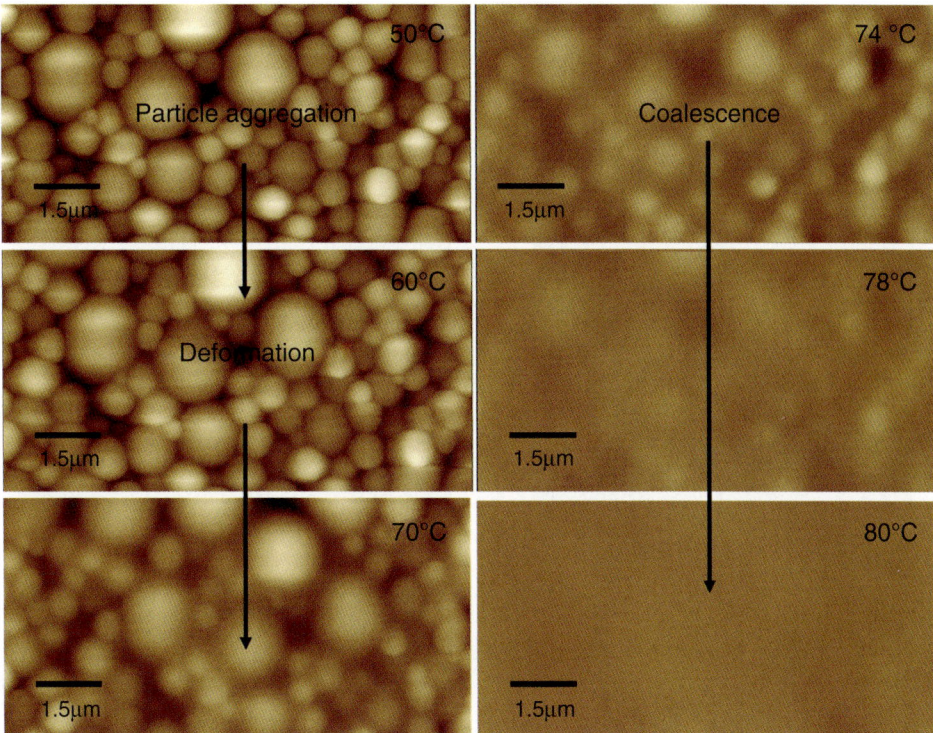

FIGURE 5.122. Height images of ethylene octane (EO) copolymer dispersions, dried on glass slides at room temperature, were heated and imaged to show the effect of particle size on film formation temperature using an AFM with a miniature hot stage. (From Li et al. [586]; unpublished.)

Color Plate X

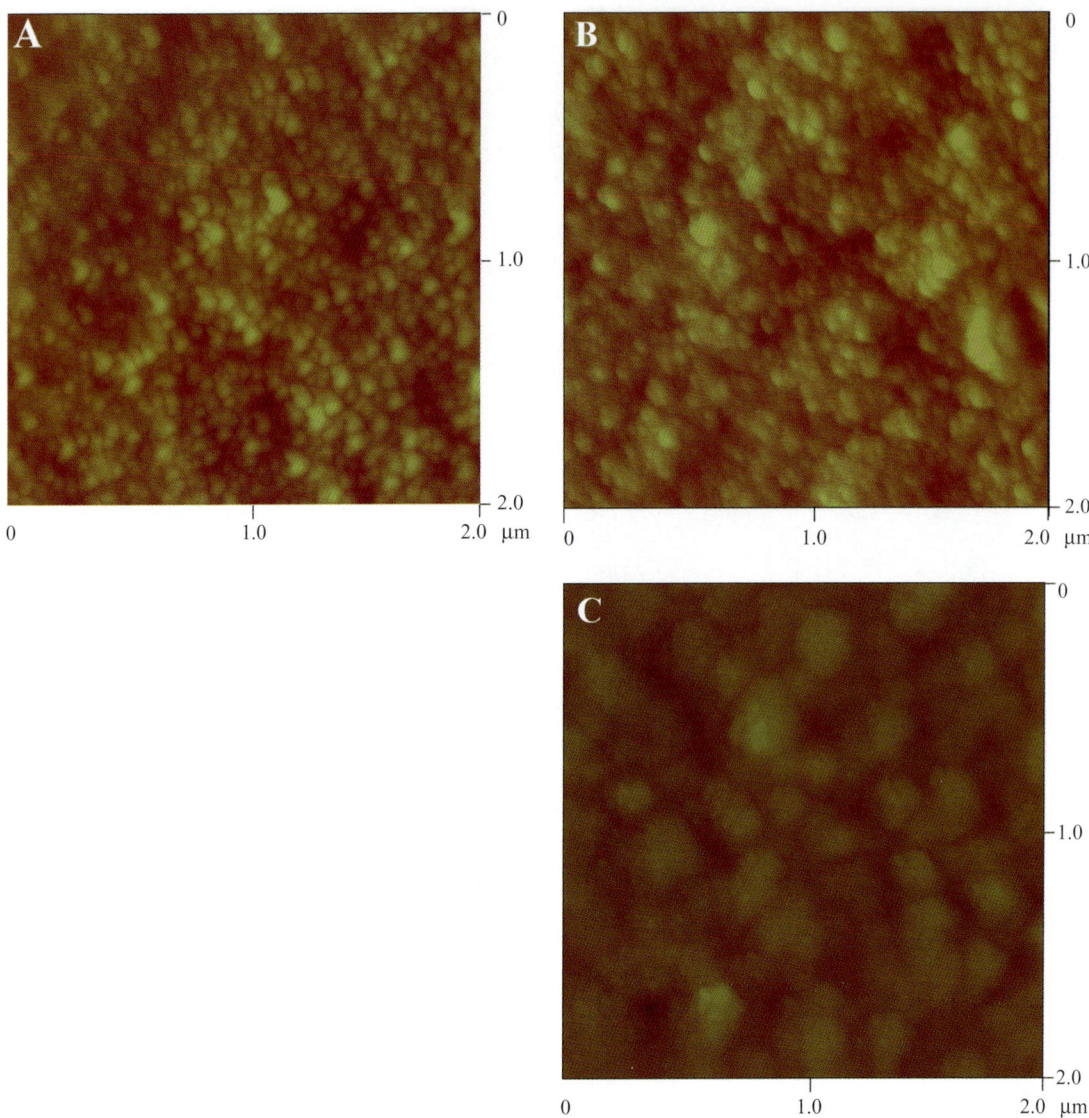

FIGURE 5.124. Atomic force microscopy images of PEDOT/LiClO$_4$ coatings deposited on gold-coated sites with deposition charge (A–C) of 1.8, 7.2, 43.2 mC. The scanning length was 2 μm. (From Yang and Martin [587]; used with permission, Materials Research Society.)

Color Plate XI

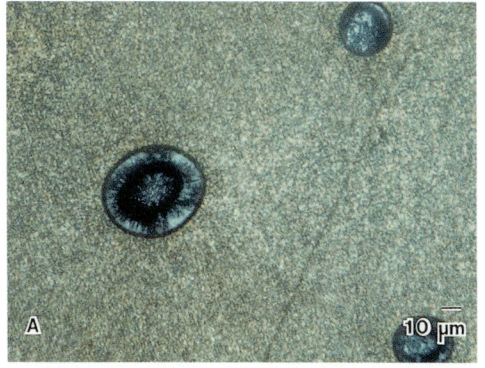

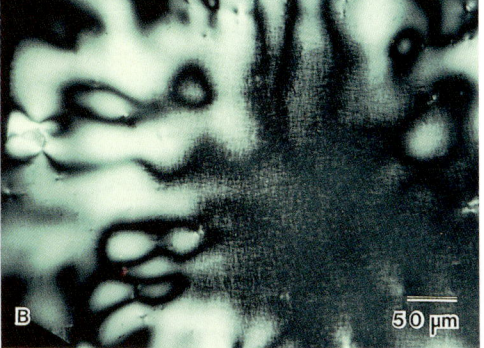

FIGURE 5.129. Thick and thin regions of a thermotropic melt structure in polarized light. In a thick region (A), the fine structure is not too clear but the onset of decomposition is shown by the round bubbles. A thinner region (B) shows thread-like detail and a nematic texture with four brushes.

FIGURE 5.131. Incomplete extinction in uniaxially oriented TLCP fiber, ribbons and films give a "salt and pepper" texture that is seen as individual domains less than 0.5 μm across. The similar polarization colors in polarized light suggest the domains are within the same order and thus have similar birefringence.

FIGURE 5.130. A polarized light micrograph of a section cut from a molded article reveals a complex, fine nematic texture with no obvious orientation. Color in the image enhances the detail.

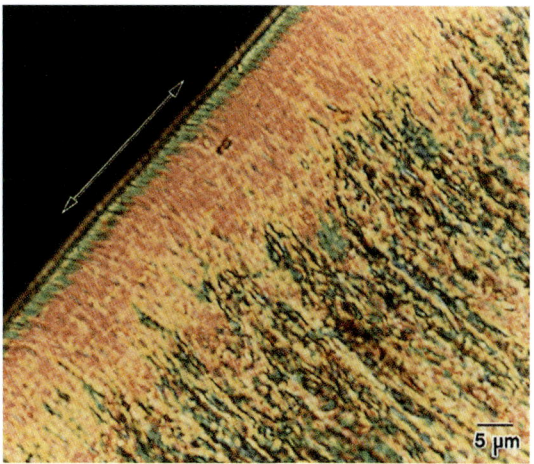

FIGURE 5.133. Polished thin sections of a low orientation extrudate observed in polarized light. An unoriented free-fall strand shows banding normal to the strand axis (arrow) and away from the slightly oriented skin.

Color Plate XII

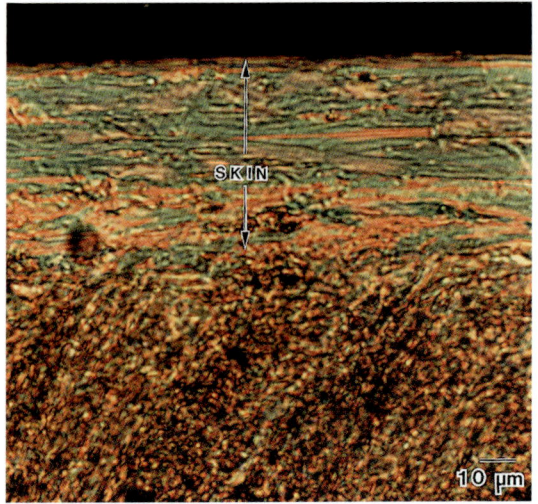

FIGURE 5.137. Skin-core structures are shown in more detail in a highly magnified polarized light micrograph with the specimen at 45° to the crossed polarizers. The skin is seen clearly to be more oriented than the core.

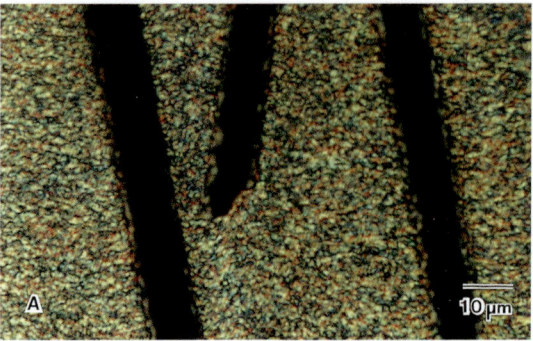

FIGURE 5.141. A glass fiber reinforced LCP composite is shown to have interesting morphology. A polished thin section is shown in polarized light (A) to exhibit a fine domain texture with some orientation of the polymer on the glass surfaces.

FIGURE 5.138. Free-fall TLCP strands are shown in polished sections in circularly polarized light (A)

FIGURE 5.145. Optical micrograph of an ultrathin longitudinal section of an extruded experimental PBZT film (on a TEM grid) in polarized light with the fiber at 45° to the crossed polarizers. A definite skin-core texture is observed as the skin appears bright yellow and the core appears blue and less oriented.

Color Plate XIII

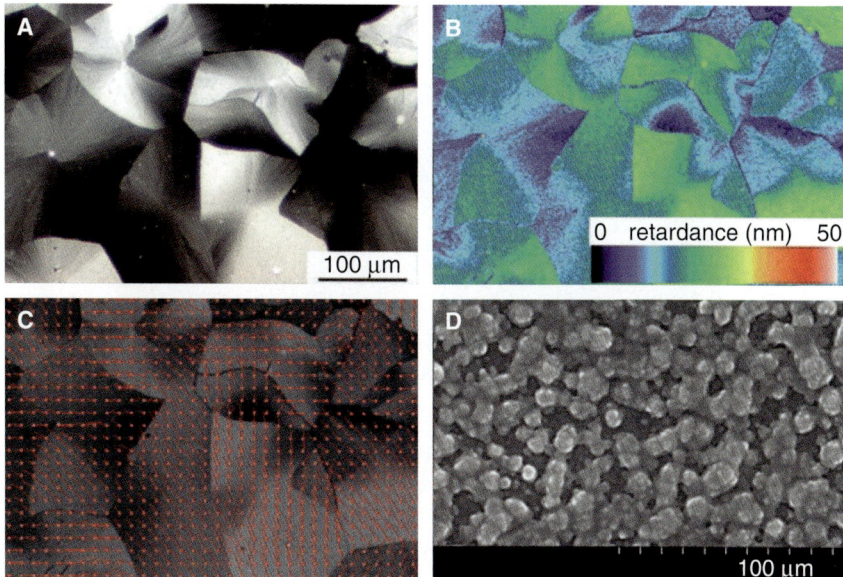

FIGURE 6.2. Thin calcite films obtained by heating a PMAA brush stabilized ACC film for 2 h at 250°C. (A) Polarized optical micrograph; (B) LC-PolScope image (retardance values are indicated as false color); (C) retardance gray-scale image (red vector overlay indicates the orientation of the slow birefringence axis); (D) high magnification SEM image. (From Tugulu et al. [38]; reproduced with permission.)

Color Plate XIV

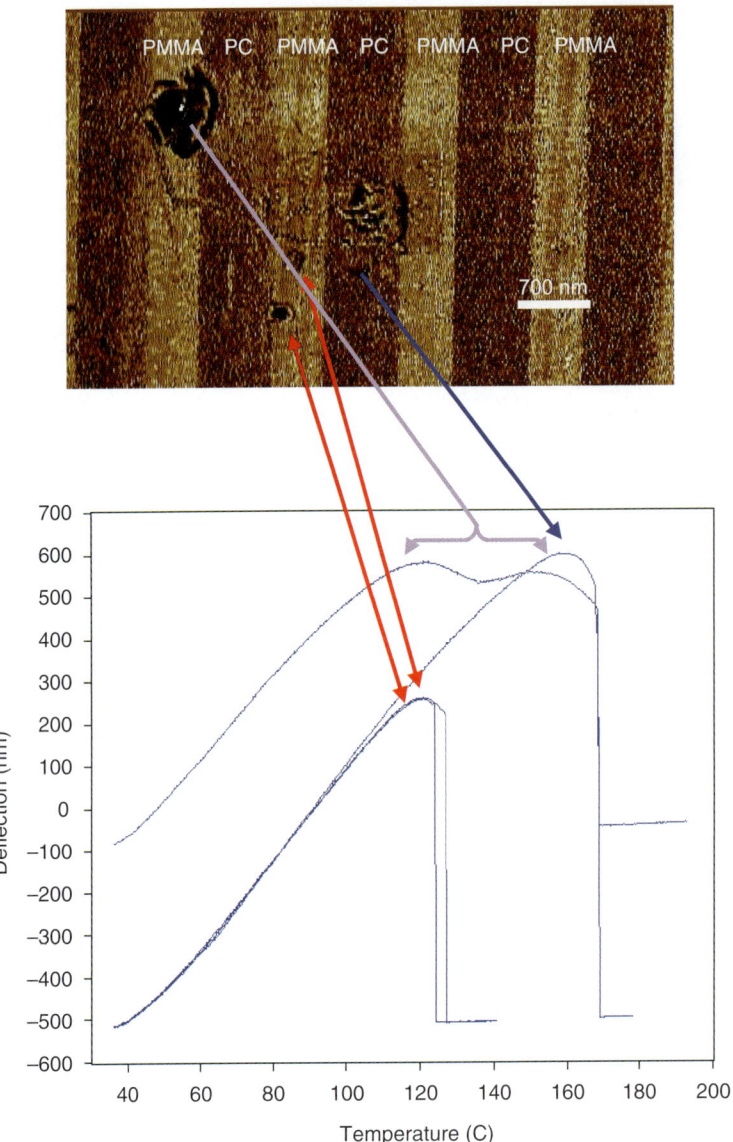

FIGURE 6.8. Thermal analysis of a multilayer polymer using a nanoTA heated tip. The sensor deflection was used for detection of the glass transition temperature of each component when the penetration was confined to each layer. For the PMMA, this was 120°C, and for the PC, it was 160°C.

Color Plate XV

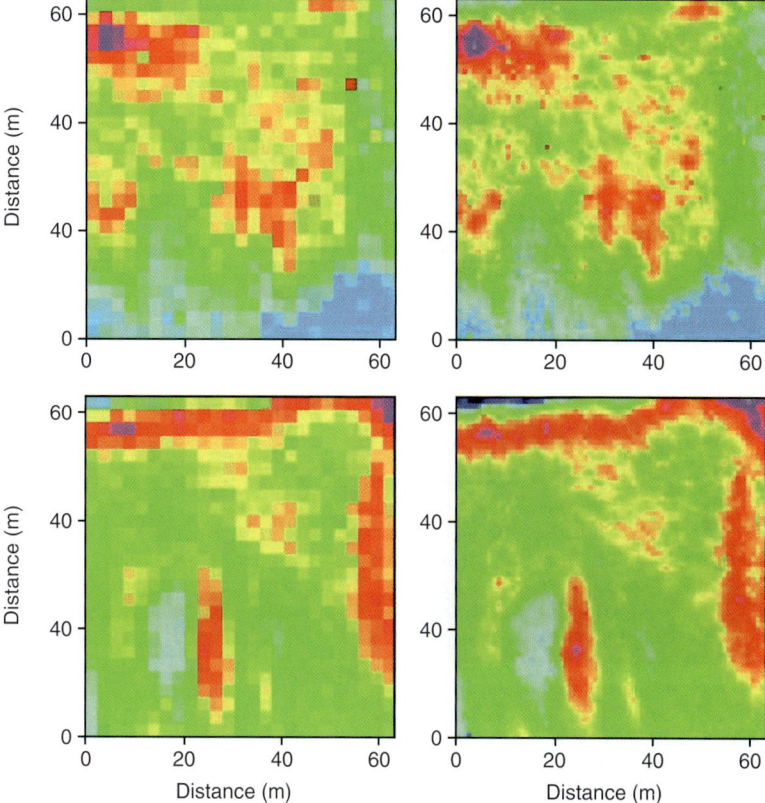

FIGURE 6.15. Comparison of IR mapping and IR imaging of a thin tissue section measured in transmission. Integrated intensities of a band typical of sugar polymers in tissues (at 1,155 cm^{-1}) are plotted at the top and those for methylene (at 2,850 cm^{-1}) at the bottom. Mapping with 12 μm aperture and 10 μm stepping is shown on the left (32 × 32); IR imaging (64 × 64) on the right. Images are 250 μm square. (From Schultz [307]; reproduced with permission.)

Color Plate XVI

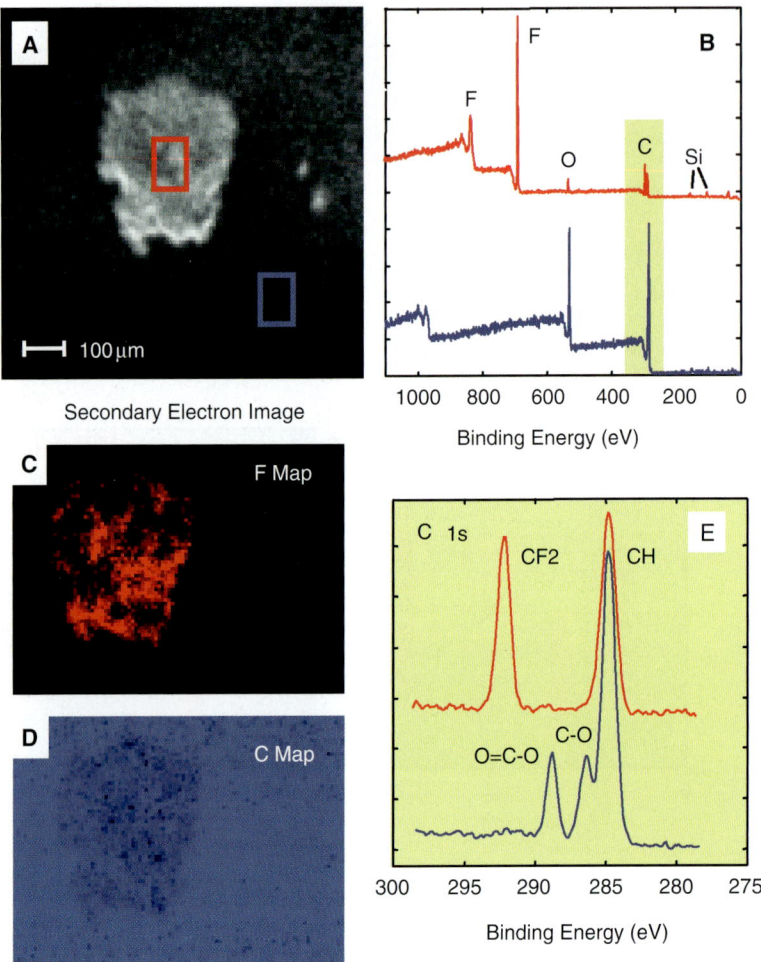

FIGURE 6.23. X-ray photoelectron spectroscopy microprobe images of contamination on a polyester sheet. (A) A secondary electron image; 20 μm x-ray beams on the indicated area gave (B) the survey electron spectrum and (E) the high energy resolution spectra. These show the presence of fluorine in the contaminant and by the presence of CF_2 that it is a fluorocarbon. The maps of (D) carbon and (C) fluorine confirm this and also show that the other smaller contaminants seen in the secondary electron image are not of the same material. (From PHI [367]; reproduced with permission.)

Films and Membranes

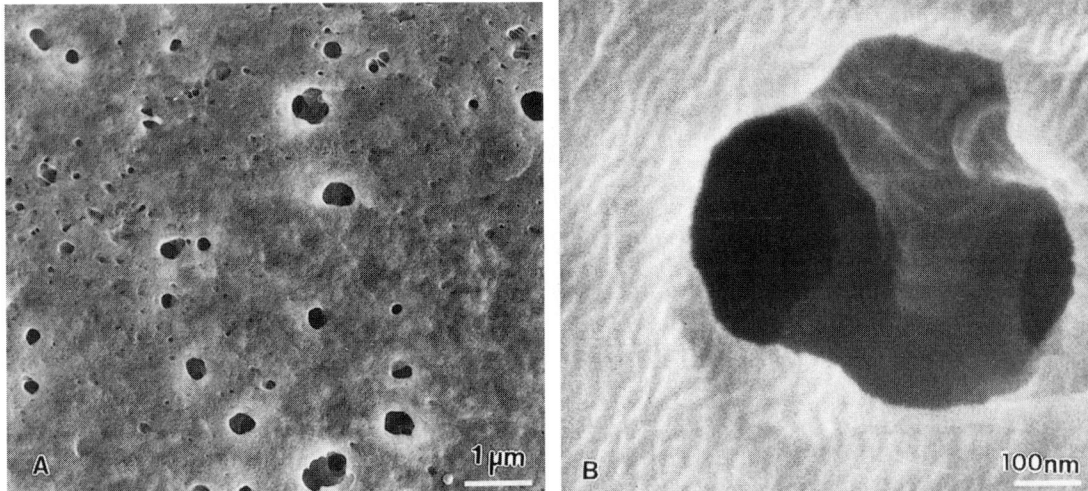

FIGURE 5.51. The surface of a PTFE membrane imaged at 5 kV in an FESEM exhibits a nonuniform series of rounded pores and a three dimensional nature as the pores extend into the bulk membrane. The surface of the membrane appears wrinkled in texture, but imaging at lower magnifications, after high magnification imaging, did not reveal picture frame contrast that would suggest this texture is due to beam damage. (From M. Jamieson, unpublished [217].)

asymmetric RO membrane. Scanning transmission electron microscopy was used to image ionomers used in novel polyimide membranes such as in the production of asymmetric hollow fiber gas separation membranes requiring high thermal stability [221]. Hollow fiber gas separation membrane cross sections were prepared for FESEM imaging by cryofracturing the fiber in liquid nitrogen and sputter coating with elemental chromium or Au/Pd; the dense surface layer was found to be key to properties [222]. Scanning electron microscopy images of freeze fractured hollow fiber membranes produced by dry-jet wet spinning from PEEKWC, a modified poly(ether ether ketone), were used to make pore size measurements as a function of different spinning conditions [223].

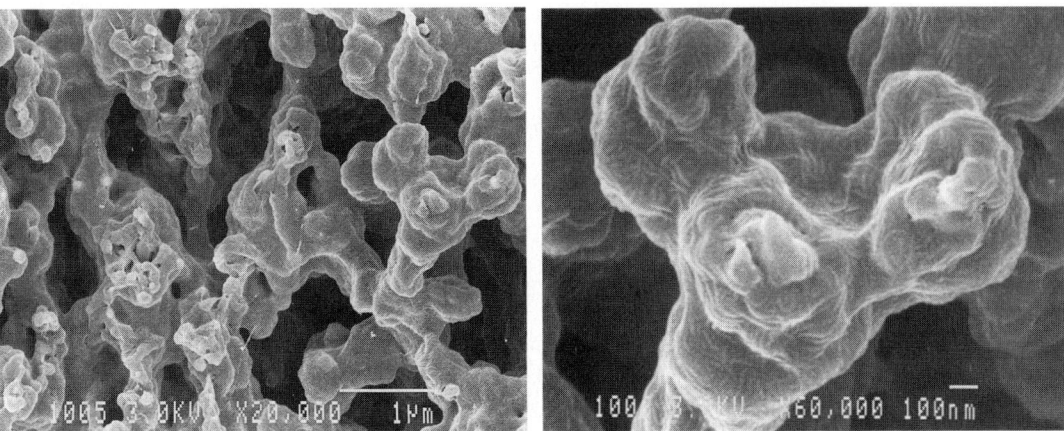

FIGURE 5.52. Field emission SEM images of a two phase polymer system, processed into a film that was fractured and imaged, reveals coarse, irregularly shaped pores and fine dispersed phase particles. (From M. Jamieson, unpublished [217].)

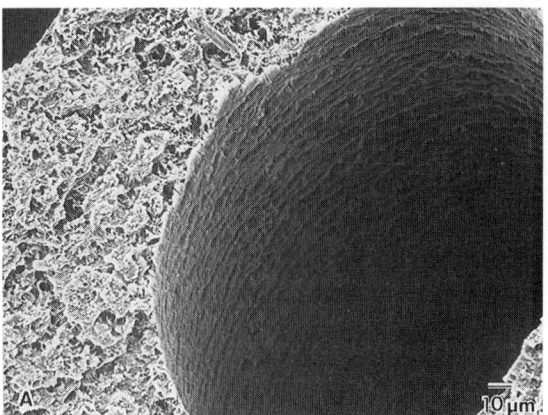 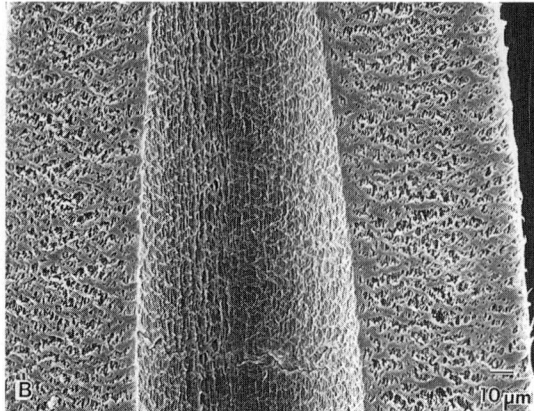

FIGURE 5.53. Hollow microporous polyethylene fibers are shown in SEM images of cross sectional (A) and longitudinal (B) views, which permit assessment of the structures.

An example of a PE hollow fiber membrane was prepared for SEM by fracturing in liquid nitrogen to show the bulk cross sectional structure (Fig. 5.53A), and a cold razor blade was used to fracture the fiber for the longitudinal view (Fig. 5.53B). Combination of both views shows the dimensions and the porous structure.

5.3 ENGINEERING RESINS AND PLASTICS

5.3.1 Introduction

It is well known that the microstructure and the mechanical properties of engineering resins and plastics are determined by the chemistry of the polymers and the manufacturing process. Thus, resins and plastics produced by such processes as injection and compression molding, extrusion, and thermoforming are evaluated by microscopy techniques in order to determine their structure and to provide an understanding of structure-property-process relations. The range of polymers that are considered engineering resins and plastics is very large, and their applications are even broader. A listing of common commercial plastics and resins is provided in Appendix IV with trademarks and some applications. Information such as processing data, mechanical properties, and other detailed information can be found on the company Web sites and in many texts; for example, Daijal [224], Margolis [225], Paul and Sperling [226], and Paul et al. [227]. Sperling [228] is an excellent teaching text that provides all the basic concepts of multicomponent polymers, modeling of their behavior, fracture behavior, and characterization by many techniques, including microscopy. The more recent edited volume by Michler and Baltá-Calleja [229] covers the broad range of work on mechanical properties resulting from specific polymer nanostructure and morphology with review chapters on polymer morphology, characterization techniques, processing, and mechanical property development.

5.3.1.1 Resins and Plastics

Polymers in the category of engineering resins and plastics may be classified in several ways. They may be thermoplastic or thermoset. They may be crystalline or amorphous, and they may be single phase or multiphase systems. This would allow for eight types of materials except that thermosets, because of their irregular cross-linked structure, are never crystalline. *Single phase* polymers do not have discernible second phase structures of different chemical composition. Thus, homopolymers and random copolymers are single phase polymers, even if they are semicrystalline and so contain amorphous and crystalline regions. Blends of the few pairs of polymers that mix well (miscible blends) can also produce single phase material. The single phase polymers have a wide range of mechanical properties and morphologies, which depend on their specific characteristics, such as flow behavior and melting point, if they are crystalline. Examples of single phase thermoplastics are PE, iPP, PA, and POM, which are all crystalline, and PS, PMMA, poly(vinyl acetate) (PVAC), and polystyrene-poly(phenylene oxide) (PS-PPO) blends, which are all amorphous. Epoxies, unsaturated polyesters, and phenol formaldehydes are examples of single phase materials that are amorphous thermosets. Much information can be obtained by microscopy of crystalline thermoplastics, whereas microstructural study of single phase amorphous materials is not usually of much practical interest. This is why most microscopy studies of single phase polymers relate to crystalline materials, and amorphous polymers are mostly described as part of multiphase systems.

Multiphase or multicomponent polymers can clearly be more complex structurally than single phase materials, for there is the distribution of the various phases to describe as well as their internal structure. Most polymer blends, block and graft copolymers, and interpenetrating networks are multiphase systems. A major commercial set of multiphase polymer systems is the toughened, high impact or impact modified polymers. These are combinations of polymers with dispersed *elastomer* (rubber) particles in a continuous matrix. Most commonly, the matrix is a glassy amorphous thermoplastic, but it can also be crystalline or a thermoset. The impact modified materials may be blends, block or graft copolymers, or even all of these at once.

As may be guessed from the names for these systems, the rubber particles are added to improve the mechanical properties of the matrix material, particularly to improve their impact strength or toughness. The size of the rubber particles, their distribution, composition, and compatibility with the matrix all influence the mechanical properties of the final engineering resin.

Typical multiphase polymers that include elastomers are:

(1) high impact polystyrene (HIPS);
(2) acrylonitrile-butadiene-styrene (ABS);
(3) poly (styrene-acrylonitrile) (SAN);
(4) acrylonitrile-chlorinated poly(ethylene-styrene) (ACS);
(5) poly(styrene-butadiene-styrene) (SBS);
(6) ethylene-propylene terpolymer (EPDM).

5.3.1.2 Characterization

A wide range of microscopy techniques are applied to the characterization of engineering resins and plastics. For example, crystalline polymers are viewed by polarized light microscopy (PLM) to reveal the size and distribution of spherulites and the nature of the local orientation. Surface details, such as wear and abrasion, are best viewed by SEM or SPM. Vaziri et al. [230] for example conducted a detailed investigation of the wear of polymer materials. Scanning electron microscopy of fractured and/or etched materials provides additional information on multicomponent structures. Combined OM and EM and more recently AFM are applied to measure the size and distribution of the dispersed phases for correlation with mechanical properties such as impact strength. Particle size distributions are now routinely obtained with commercial image analyzers. Critical parameters, such as calibration and statistical sampling, were considered in a study of latex particle size distribution (see Section 4.9.2) using automated image analysis [231]. Dispersed phase particle size measurements of acrylonitrile blends were made directly from PLM micrographs and from TEM negatives [232] with a similar system. A morphological parameter, "CoContinuity," has been developed for quantitative measurement of morphology in co-continuous polymer blends [233].

Wu [234] reviewed the basic principles of optical and electron microscopes and their applications to characterization and investigation of polymer toughening mechanisms. Multiphase polymers are prepared for SEM by methods such as fractography, etching, and extraction and for OM and TEM by thin sectioning methods. The dispersed phase in a multiphase polymer is often directly examined by SEM study of fractured surfaces. Polymers with large dispersed phases that adhere poorly to the matrix are the best candidates for direct analysis but are unfortunately the worst engineering materials. Well adhered and small dispersed phases are often not visible directly in fractured samples viewed in the SEM. Electron microscopy of polymer blends has been reviewed by Bassett [235]. Chemical staining, chemical and electron beam etching [236], differential interference contrast [237, 238], and phase contrast TEM are all methods to increase image contrast. X-ray microanalysis and backscattered electron imaging provide contrast based on the presence of high atomic number materials, such as chlorine in poly(vinyl chloride). Bucknall [239] reviewed applications of microscopy to the deformation and fracture of rubber toughened polymers, pointing out the need for complementary techniques to be used especially for study of deformation micromechanics. In this review, PLM is said to be important for study of crazing, TEM for study of cavitation behavior of complex rubber particles, and SEM to study all micromechanisms.

The method of differential radiation induced contrast depends on enhancement of contrast in multicomponent polymers where the components have different electron beam-polymer interactions [236]. Contrast has been observed in sections of styrene-acrylonitrile/poly(methyl methacrylate) (SAN/PMMA) polymers where the PMMA exhibits a high rate of mass loss compared with SAN, creating contrast between the phases. It is well known that electron irradiation results in chain scission and crosslinking, loss of mass, and crystallinity [95]. Polystyrene, polyacrylonitrile, and SAN crosslink and thus are stable in the electron beam, whereas polymers exhibiting chain scission, for example, PMMA, POM and poly(vinyl methyl ether), degrade in the beam. It is suggested that experiments be conducted on the homopolymers to determine the expected irradiation damage mechanism in the multicomponent system [236].

There are multiphase polymers where OM and SEM techniques cannot fully describe the microstructure due to a combination of small particle size (less than $0.5 \mu m$) and good adhesion between the dispersed phase and the matrix. Additionally, broad particle size distributions are often encountered, and in these cases a combination of techniques is required to describe the microstructure. Transmission electron microscopy requires ultrathin specimens, about 50–500 nm or less in thickness, which are prepared by film casting or ultrathin sectioning. Films formed by casting or dipping methods provide a much easier specimen preparation method than ultrathin sectioning of bulk plastics; however, a major question in such studies is always whether the microstructure is the same as in the bulk polymer. Specific stains are often required to provide contrast between the dispersed phase and the matrix polymer (see Section 4.4).

Newer techniques can aid imaging and interpretation of information for engineering resins, blends, and plastics. Field emission SEM at low voltages can replace conventional SEM, providing similar information but with much more detail. From early work in the late 1980s showing the utility of improved contrast and reduced beam damage, even with metal coated samples [240, 241], advances have been made that permit imaging of uncoated specimens with excellent resolution [242, 243]. Imaging of polymer blends and copolymers has benefited from LVSEM imaging. Schwark et al. [244] imaged the surface morphology of styrene-butadiene block copolymers by LVSEM. Himelfarb and Labat [245] characterized polymer blends and block copolymers using both conventional and LVSEM and TEM of stained polymer blends. In this work, they used preferential staining with ruthenium tetroxide and suggested that higher accelerating voltages (10–25 kV) are preferred for the measurement of particle size and shape. For high resolution images of surface topography, in this case 20 nm domains in hydrogenated styrene-butadiene-styrene block copolymers, the spatial resolution in LVSEM is comparable with conventional TEM. Imaging of polymer blend systems has also been done using variable pressure SEM and STEM (e.g., [246, 247]). The advantage of this technique is that samples do not require metal coatings or modification, and controllable charging effects can provide novel voltage contrast. Often, high resolution SEM imaging reveals smaller dispersed phase sizes, interfacial regions, and other details that preclude the need for the more laborious microtomy and staining required for TEM. In the cases of "molecular composites," of structures formed by spinodal decomposition, HREM can often image domains that were not seen previously, and such materials may have been thought to be composed of a single phase.

The scanning probe microscopies are also important to engineering resins because such imaging does not require a high vacuum system, and finer surface detail may be imaged than by SEM, FESEM, or TEM. Annis et al. [248] and Vezie et al. [249] provided the first detailed, complementary studies of diblock copolymers using conventional TEM, low voltage high resolution SEM, and AFM. Bar and Meyers [170] reviewed SPM applied to polymers. Magonov described AFM techniques in a number of review papers [250–253], including imaging of a range of polymer blends. Cross-linkable epoxy thermoplastics modified with 5 wt.% grafted rubber concentrate were "cryo-polished," that is, cryomicrotomed, at $-90°C$ using a diamond knife to provide a smooth surface (block face) for SPM [174]. intermittent contact mode AFM was applied to the study of both adhesion and mechanical deformation on this sample. Pfau et al. [254] used TEM and force modulation AFM to map elastic properties for a variety of blends, including rubber toughened PP and HIPS, preparing samples by microtomy. Thomann et al. [255] studied blends of iPP with random poly-(ethene-co-1-butene) (PEB) using phase imaging ICAFM, OM, and TEM. Intermittent contact mode AFM was used to study polydimethylsiloxane (PDMS) samples of different crosslink densities [256]. Michler et al. [257] studied the toughness enhancement of nanostructured amorphous and semicrystalline polymers using SEM, TEM, HVEM, and SFM of a variety of blends by *in situ* deformation,

One of the major problems with AFM characterization is the use of microtomy to prepare a flat surface (block face), which can result in artifacts and surface roughness, even debonding the interfacial regions. This limitation has been addressed by using focused ion beam (FIB) preparation (see Section 4.5.5) followed by ICAFM [258] to analyze model interface thicknesses in HDPE/PS/PMMA ternary blends. The SEM and AFM images were analyzed to quantify the volume average diameters; it was found that the polymers had different etching rates resulting in topologic contrast in ICAFM. Clearly, SPM imaging requires the same careful consideration and complementary techniques that are required for all imaging techniques. Examples of the various imaging techniques will be provided in the sections that follow.

5.3.2 Process-Structure Considerations

5.3.2.1 Extrusion and Molding

A brief description of the relation of the extrusion and molding process to engineering resin or plastic morphology is intended to provide a basis for structure-property studies and is not a complete description of polymer manufacturing processes. The focus is on examples of the structure of crystalline or crystallizable thermoplastics formed by these processes. There is a wide range of processes than can be considered for manufacturing polymer products including: extrusion, injection and compression molding, RIM (reaction injection molding), blow molding, thermoforming, and casting, among many others (e.g., [259, 260]). The structure formed by any process affects the mechanical and physical properties. The structural heterogeneity resulting from injection molding of plastics results in a higher degree of anisotropy for instance compared with compression molding and extrusion, but similar techniques are used to evaluate the microstructures formed by these processes. Specialized processes for blends will be noted in Section 5.3.4. In the following discussion, most of the background information has been taken from the earlier edition of this book and, for example, from Griskey [260], who should be considered if greater details are required. Further details regarding these processes are found in the reference books and on the polymer manufacturers' Web sites.

Extrusion can be defined as the shaping of a material by forcing it through a die by means of a rotating screw using a machine called an *extruder*. The goal is to deliver thermally homogeneous polymer melts at a uniformly high rate whether the machine is an extruder, blow molding machine, or in cases in which the melt is injected into a mold, as in injection molding. Polymer material in the form of pellets is generally fed into the hopper of the extruder and then to the screw channel where thermal energy is supplied by external heaters and the flowing polymer. The plastic is melted and conveyed through a series of screw components and ultimately to a die, which results in mixing of the polymer or polymers and other materials. This process causes deformation, resulting in molecular orientation in the extrusion direction. The amount of orientation depends on the polymer, temperature and flow rate of the melt, among other factors. Single screw, twin and multiple screw configurations all affect the melt flow and the final morphology of the extruded and/or molded product. Extruded products include fibers, films, tubes, rods, and pipes and pellets used for further processing (e.g., injection molding).

In the case of injection molding, generally the polymer pellets, of either neat or filled polymer (composites), are heated until they are melted or thermally softened and forced into a mold and then cooled to a specific shape. There are clearly many more engineering details to fully understand these processes, but overall the polymer morphology in extrudates and moldings is affected by the major process variables, such as melt and mold temperature, pressure, shear, and elongational flow. The injection process causes deformation of the polymer and orientation in the flow direction. The flow pattern during mold filling [261] has a semicircular shaped advancing front, curving toward the mold wall, where the macromolecules orient parallel to the wall. Orientation induced by elongation and shear flow is found in the flow direction [262], especially near the surface. As with extrusion, the injection molding process causes formation of anisotropic structures. Anisotropic morphologies formed due to such flow fields are termed skin-core, multilayered, or banded to describe the variation in orientation in the final specimen. Molded parts are used in all major industries today.

Bowman [263] conducted a systematic study of the structure-property-process relations of injection molded polyacetals (polyoxymethylene, or POM) and observed correlations between process conditions, structures, and mechanical properties. Barrel temperature effects were studied as they are known to influence both microstructure and mechanical properties [264]. Increased barrel temperature was shown to reduce the outer skin layer while increasing the extent of the equiaxed, unoriented core, resulting in decreased tensile yield strength parallel to the injection direction.

Compression molding is used with thermosetting resins, finding limited use today and it is not useful for intricate parts. Compression molding involves placing polymer powder or granules into a mold and softening by heating. This process consists of forcing a resin and a curing or crosslinking agent into a mold using pressure resulting in thermosetting the product. There is little orientation because polymer flow is limited. In injection molding, the polymer melt is injected into a cooled or heated mold. Compression molding is generally used for very high temperature parts and for thermosets. Blow molding is a process that is used to produce hollow objects, and it may be extrusion blow molding, injection blow molding, or stretch blow molding with the former being used in much commercial production. Blow molded parts include bottles and fuel tanks.

5.3.2.2 Spherulitic Structures

The units of organization in polymers are lamellae or crystals and spherulites. Bulk polymers are composed of lamellar crystals that are typically arranged as spherulites when cooled from the melt. Polymer melts solidified without deformation form structural units, termed *spherulites*, composed of a central nucleus with a radiating array of lamellae. Process variables can affect the physical properties of the material during processing, which in turn affects the resulting morphology. The morphology of the fabricated product in turn influences final mechanical properties and part performance. Pressure increases during processing, for instance, can increase both the melting temperature and the glass transition temperature of a polymer, with the result that the polymer solidifies more quickly. In a crystalline polymer, the nucleation density can increase, resulting in a decrease in spherulite size with increased pressure in injection molding. A schematic image of a spherulite was shown in Chapter 1 (see Fig. 1.1), and a PLM image of an acetal cooled from the melt (see Fig. 1.2) shows the recrystallization and formation of spherulites (see Section 1.2.2). The size and nature of spherulites is well known to be affected by the temperature of crystallization, as was shown for example for PET [265]. Standard production processes, such as extrusion and molding, however, often produce deformation of the spherulitic structure, and the local polymer orientation is frozen in the final product. The crystallization of polymers is also affected by their composition and whether they are homopolymers or multiphase polymers. More details on the morphology of crystalline polymers has been published by Bassett (e.g., [266–268]).

Keith et al. [269] reported that small concentrations of compatible polar polymers change the morphology of aliphatic polyesters. Crystalline polyesters such as poly(ε-caprolactone) have much larger spherulites when about 1% PVC or poly(vinyl butyral) is added. These amorphous polar polymers act as antinucleating agents; typically these are polar, low molecular weight compounds that preferentially adsorb on the nucleating impurities and have a low melting point. At the polymer crystallization temperature, they keep a liquid surface on the particles that usually act as nuclei, suppressing their effect. This study, using transmitted light interference contrast microscopy to observe the banded spherulites, shows that miscible polymers can act in the same way.

Polarized light microscope images show details of the spherulitic structure in molded nylon. The nonspherulitic skin, a transition zone, and a spherulitic core region are observed in Fig. 5.54A. This is as expected, as the quench rate declines away from the surface, for example, in

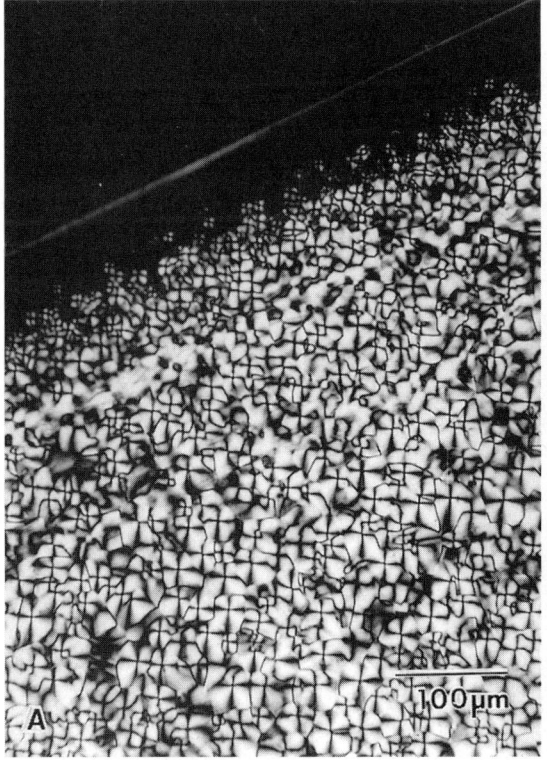 

FIGURE 5.54. Polarized light micrographs of a molded nylon cross section show a nonspherulitic skin (top in A) and rounded isolated spherulites in the transition zone. A classical Maltese cross extinction pattern is observed with black brushes showing the radial texture within the spherulites (B). (See also Fig. 1.3 color insert.)

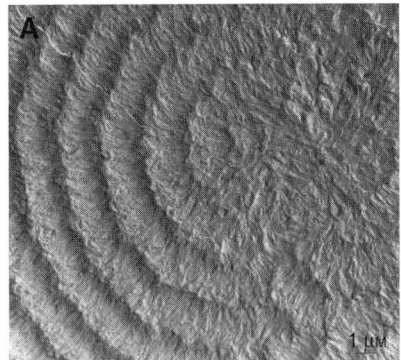
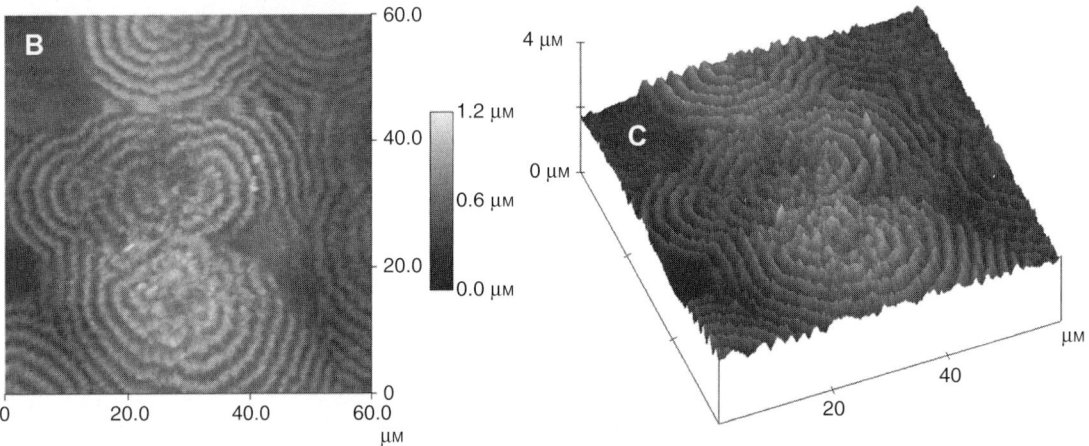

FIGURE 5.55. Banded spherulites are observed in TEM of permanganically etched, high density copolymer exhibiting a mature banded structure (A). Intermittent contact mode AFM images of the same polymer are shown in a normal height image (B) and in a three dimensional height image (C). (From Janimak et al. [271], © (2001) Elsevier; used with permission.)

the transition zone, and thus there is some nucleation of spherulites. A view of the transition zone (Fig. 5.54B) shows round spherulites in a fine textured matrix. The equiaxed spherulites, especially adjacent to the skin, indicate there is little or no orientation. Poly(butylene terephthalate) (PBT) skin thickness has been shown to increase with decreasing melt and mold temperatures, resulting in increased impact properties in bars with molded-in notches [270].

As with all microscopy techniques and preparation methods, artifacts can form and be imaged by AFM. Examples of AFM images are shown here, using permanganic etching of a thin film [271] and microtomy followed by permanganic etching [84]. Images of PE spherulites have been made using AFM to examine the three dimensional profile of permanganically etched, HDPE [271, 272]. The advantage of AFM imaging is the ability to obtain direct three dimensional information without the need for heavy element staining and replication for TEM. The first example of AFM imaging of spherulite structures is a comparison with the standard TEM method [271]. Specimens were melt crystallized under dry nitrogen at 170°C in a hot stage, cooled at 15°C/min to room temperature, and microtomed in a rotary microtome and slices examined with interference contrast OM followed by etching for 3h in a 1% w/v solution of potassium permanganate dissolved in 2:1 sulfuric and dry ortho-phosphoric acids (see Section 4.5). For TEM standard, two stage indirect carbon replicas were made of the etched specimen using cellulose acetate film moistened with acetone. The etched high density copolymer observed by TEM of a replica (Fig. 5.55A) shows the banded structure as is also seen directly using ICAFM

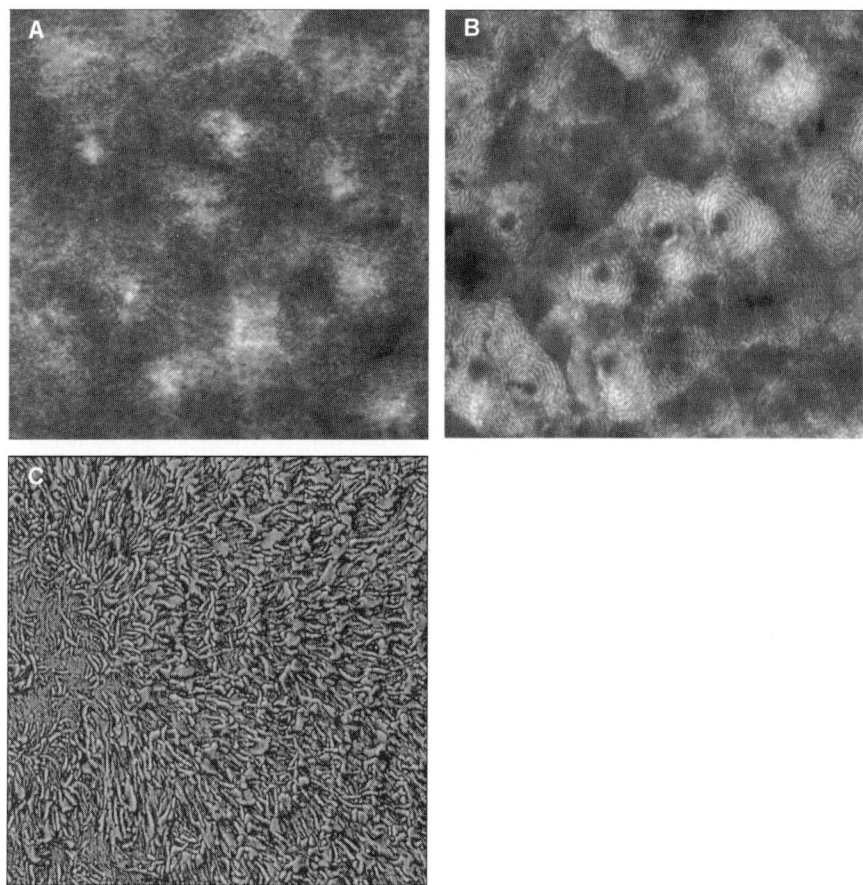

FIGURE 5.56. Intermittent contact mode AFM height images of a microtomed surface of a commercial LDPE pellet, before and after permanganate etching (A and B, respectively), and in a phase image (C) of the spherulite. Size of images are 20 μm on a side. (From Chernoff and Magonov [84], © (2003) American Chemical Society; used with permission.)

height images (Fig. 5.55B, C), showing several impinged spherulites in a normal height image and three dimensional height image, respectively. In another example, commercial LDPE was microtomed to form a flat block face and it is shown in height images, before and after permanganate etching (Fig. 5.56 A and B, respectively), used to clean up surface debris, and in a phase image of the spherulite (Fig. 5.56C) shown by the arrow [84]. Clearly, there was debris on the original surface, obscuring details that the etching process was able to remove.

5.3.2.3 Skin-Core Structures

Anisotropic structures are typically observed in molded parts and extrudates. The higher orientation in extrusion can result in highly oriented rods or strands, at high draw ratios and/or small diameters, or in structures with an oriented skin and a less oriented central core in thicker strands. This *skin-core* texture is due to a combination of temperature variations between the surface and the bulk and the flow field in both extrusion and molding processes. Extensional flow along the melt front causes orientation. Solidification of the polymer on the cold mold surface freezes in this orientation. Flow between the solid layers is affected by the temperature gradient in the mold, and the resulting flow effects result in a rapidly cooled and well oriented skin structure and a slowly cooled, randomly oriented core. Extensional flow along the melt front results in molecular orientation parallel to the knit lines when two melt fronts meet. The resulting knit, or *weld line*, is a region of weakness in the molded part, and weld line fractures are commonly encountered. Malguarnera and Manisali [273] reviewed the topic of weld line formation in injection molded thermoplastics. The local orientation is quite important as tensile strength and impact strength properties are known to be higher in the orientation direction. The generally expected correlation of Young's modulus values increasing with orientation along the tensile axis has been observed [274, 275]. Izod impact strength values have been shown to be higher for specimens with increased skin and shear layer thickness, for postnotched PP [275, 276], and for increased skin thickness for molded-in notched PBT [270].

Typically, in a semicrystalline polymer there are three zones within the molded part: an oriented, nonspherulitic skin; a subsurface region with high shear orientation, or a transcrystalline region; and a randomly oriented spherulitic core. The thickness of the skin and shear zone is known to be an inverse function of the melt and mold temperature with decreased temperatures resulting in increased layer thickness. In the skin, the lamellae are oriented parallel to the injection direction and perpendicular to the surface of the mold. Amorphous polymers also show a thin surface oriented skin on injection molding. When amorphous polymers are heated to the glass transition temperature and then relaxed, they exhibit shrinkage in the orientation direction and swelling in the other directions.

In contrast with polyacetals, PE, and PP, multilayered textures have not been observed for PBT [270] or nylon, and moldings exhibit a low crystallinity or amorphous surface layer, with little or no orientation, and a crystalline core that depends on mold conditions [263]. This amorphous skin is due to the rapid quenching of the polymer at the surface and the high glass transition temperature of these materials, not to the flow of the polymer. Thus, a simple skin-core texture rather than a multilayered texture has been observed for these polymers. An example was described in the previous section for nylon (see Fig. 5.54).

The structures of injection molded semicrystalline polymers are quite heterogeneous, resulting in substantial differences in mechanical and thermal properties at different points within a single molding. The interrelationships among processing, microstructure, and properties of thermoplastics have been reviewed by Katti and Schultz [262]. The thickness of the various layers, especially the oriented surface skin, is affected by process variables, such as temperature and pressure. These authors described three temperatures that are known to be important: the mold or wall temperature, the melt or barrel temperature, and the freezing point. A cooler wall temperature will result in thicker skin and shear zones while a mold temperature near the melt temperature will lead to higher degrees of orientation. Melt and mold temperature are the most important processing parameters for both crystalline polypropylene and for amorphous polystyrene and HIPS. In general, higher melt temperature improves mechanical properties. Apparently, increases in pressure result in thicker surface layers [277]. The effect of crystallinity [278] and the cooling rate from the melt [279], as expected, confirmed that differences in microstructure result in differences in tensile properties.

5.3.3 Single Phase Polymers

5.3.3.1 Amorphous Polymers

Polymers are considered to be either *amorphous* or *crystalline* although they may not be completely one or the other. There is no measurable order seen by x-ray scattering, an absence of crystallographic reflections, in noncrystalline or amorphous polymers. There is a range of amorphous polymers that are commercially significant, including PS, PC, and PMMA. Studies of these materials include observation of their orientation and texture, but as with amorphous films, these textures are not generally very interesting to image. Addition of additives, particles, or fibers and contamination and failure analysis can be imaged as for semicrystalline polymers, and this is shown in other sections of this chapter. A commercially important product will be used to describe the various microscopy techniques and preparation methods commonly used for amorphous polymers.

Compact Disks (CDs) with digitally encoded music have taken a major share of the music recording business from more traditional analog media. In addition, CD-ROM, CD-RW, and DVD disks are also being marketed that permit inexpensive information storage at high density. A technical advantage of optical recording is that reading and writing with a focused laser beam has spot sizes less than $1\,\mu$m in size, thus there is a high density of information. Grooves or other features on the disk permit tracking of the data. The groove structure in PC substrates commonly used in the fabrication of optical disks has been studied [280], as have the pit structures in organic write once (WORM) optical data storage media [281], by SEM, TEM, FESEM, STM, and AFM. In the WORM disks, organic thin films are spin coated onto the PC and then marked with diode lasers. Critical parameters in optical recording are the size, shape, and depth of the features in which the information is coded. The symmetry of the groove geometry is important to the tracking as is the flatness of the land or groove bottom. Average sizes of the grooves are obtained by diffraction patterns of incident light, but these are bulk average data and often there is a need to characterize the geometry in more detail, especially during development of new products.

A cross section of a PC disk would appear as a regular series of grooves and lands (raised, flat regions) with some periodicity. Depths can be on the order of 60 nm with periodicities around $1.5\,\mu$m and groove:land ratio of 1:3. The substrates [280] were injection molded commercial disks with grooves created by the mold insert. Samples were cut and coated with thin gold and platinum films using ion beam sputtering (IBS) (see Section 4.7.3) to form a conducting layer. Initially, STM images were obtained using a "pocket-sized" STM [280]. Complementary images were obtained from single stage carbon replicas of the sample surfaces in the TEM. The PC was shadowed with Au/Pd at a shallow angle (ca. 30°), and then a thin carbon layer was deposited in a vacuum evaporator. The disk was dissolved using methylene chloride, and the replicas were placed on copper grids for TEM evaluation. In addition, substrates were thin sectioned about 100 nm thick perpendicular to the grooves for TEM study. Scanning electron microscopy of the IBS coated disk showed general details, and these were compared with FESEM images of uncoated disks.

The morphology of the grooves in a PC substrate is shown in Fig. 5.57. A SEM micrograph in Fig. 5.57A [280] of a gold coated disk reveals the general groove and land morphology of the disk with ca. $0.5\,\mu$m width grooves and $1.5\,\mu$m

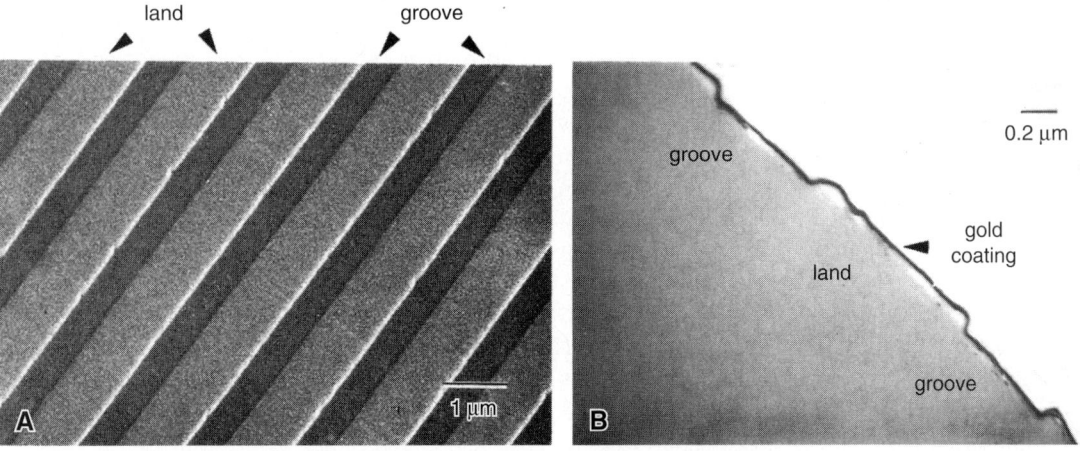

FIGURE 5.57. The morphology of the grooves in a polycarbonate substrate is shown in an SEM micrograph (A) of a gold coated disk, which reveals the general groove and land morphology with ca. $0.5\,\mu$m width grooves and $1.5\,\mu$m periodicity. Transmission electron microscopy of ultrathin cross sections of gold coated disks (B) show a heavy dark line of the continuously coated surface; the thickness is in good agreement with the thin film thickness monitor. (From Baro et al. [280]; used with permission.)

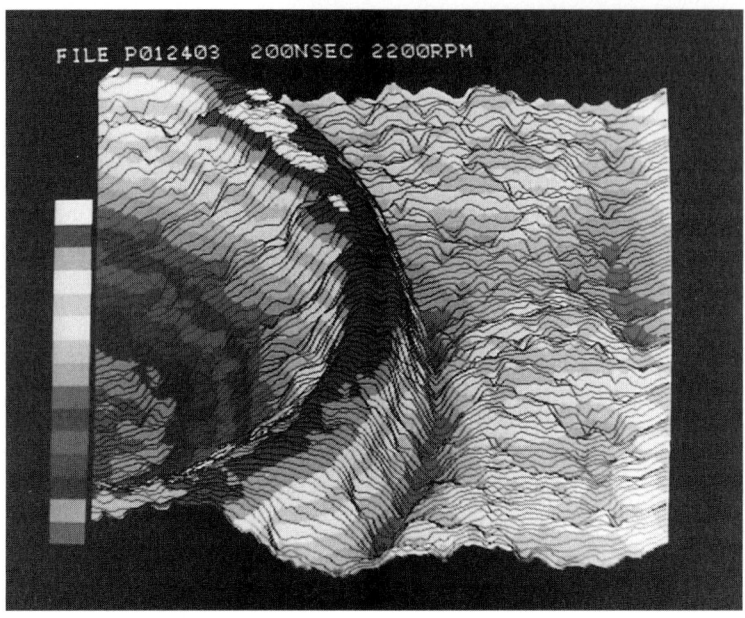

FIGURE 5.58. Scanning tunneling microscopy provides a direct measurement of the depth of a pit. This is the height difference between the media surface and the center of the pit, as seen in this image, especially where the depths of structures are very shallow compared with the three dimensional geometry. The total height variation in the image is 77 nm divided into 15 different colors; the depth of the pit is ca. 41 nm. (See color insert.) (From Goldberg et al. [281]; used with permission.)

periodicity. Complementary TEM of ultrathin cross sections of gold coated disks are shown in Fig. 5.57B [280]). The heavy dark line shows the continuously coated surface; the thickness is in good agreement with the thin film thickness monitor. Transmission electron microscopy of carbon replicas of the IBS gold coated disk showed the grain structure and the groove asymmetry (in Fig. 4.28 [280]). Large particles (see arrow) are defects in the substrate surface. Scanning tunneling microscope (STM) imaging was conducted as part of that work to show the groove depth [280]. An additional conclusion of this study was that platinum coated samples produced much finer textures than the gold coatings, in agreement with prior expectations (see Section 4.7.3). The pits made in organic WORM media were also examined using STM and evaluated versus media performance [281]. Disks were prepared by spin coating an organic medium from an organic solvent onto polycarbonate substrates at varying spin speeds. Samples were laser marked at various energies and pulse times and then IBS coated. The STM was used to determine the thickness of the organic layer and also the pit geometry. The thickness of a "soft" organic layer is very difficult to determine by profilometry or even TEM of microtomed sections, as there is no sharp step to aid measurement. As seen in Fig. 5.58, STM provides a direct measurement of the depth of a pit, that is, the height difference between the media surface and the center of the pit [281], although STM has been substantially replaced today by AFM imaging.

A comparison of pit sizes in PC disks in Fig. 5.59 shows ICAFM images of audio CD, DVD, and high density DVD surfaces with submicrometer sized pits [216]. A stained audio CD was also examined by ICAFM (Fig. 5.60), which showed various regions to exhibit surface damage that provided a direct correlation of playability with morphology [216].

5.3.3.2 Semicrystalline Polymers

Crystalline polymers are more correctly termed *semicrystalline* as their measured densities differ from those obtained for perfect materi-

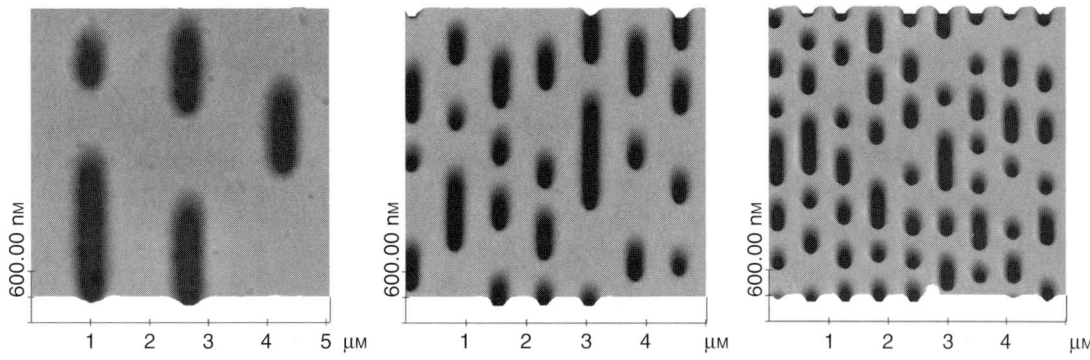

FIGURE 5.59. A comparison of pit sizes in PC disks is shown in ICAFM images of audio CD (680 MB) on far left, DVD (4.7 GB) in the center, and high density DVD (7.5 GB) on the right, with submicrometer sized pits.

als. The degree of crystallinity, measured by x-ray scattering, also shows these polymers are less than completely crystalline. The general morphology of crystalline polymers is now well known and understood and has been described by Geil [260], Keller [282], Wunderlich [283], Grubb [284, 285], Uhlmann and Kolbeck [286], Bassett [287, 288], and Seymour [289].

The morphology of molded articles depends on the chemical composition of the polymer, the process variables, and the mold geometry. Standard molded tensile bars are discussed here for simplicity, but the principles are the same for any molding, although the nature of the specific flow field must be taken into account. The relationships between process conditions, microstructure, and mechanical properties of an injection molded thermoplastic have been reviewed [236, 290, 291]. Semicrystalline moldings and extrudates are most often imaged by PLM, SEM, and TEM and more recently by SPM techniques as well. Preparation is done by sectioning, cutting and polishing, staining and etching. The morphologies of injection molded tensile test bars of PE [292], POM [274, 277, 293, 294], and PP [275, 276] are similar and can be described as complex, multilayered, skin-core structures. This structure is shown in the SEM micrograph of a molded POM test bar in Fig. 5.61. The molding shows the orientation of the polymer, emphasized here by the presence of voids that are highly oriented at the bar surface and less oriented within the core. Several intermediate layers are seen between the skin and the core.

An example of the multilayered structures common in polyacetals is shown in the polarized light micrographs (Fig. 5.62) that depict a uniformly nucleated crystalline structure formed due to mold filling and variations in the

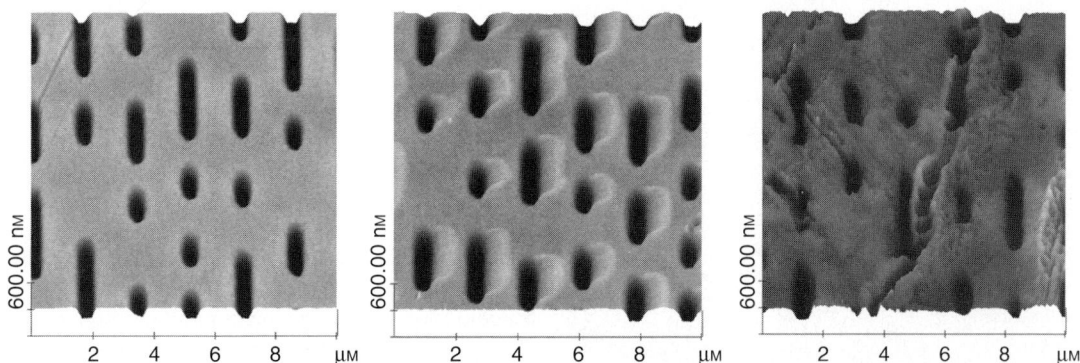

FIGURE 5.60. A stained audio CD was also examined by ICAFM, which showed various regions to exhibit surface damage (especially far right) that provided a direct correlation of playability with morphology.

320　　　　　　　　　　　　　　　　　　　　　　　　Applications of Microscopy to Polymers

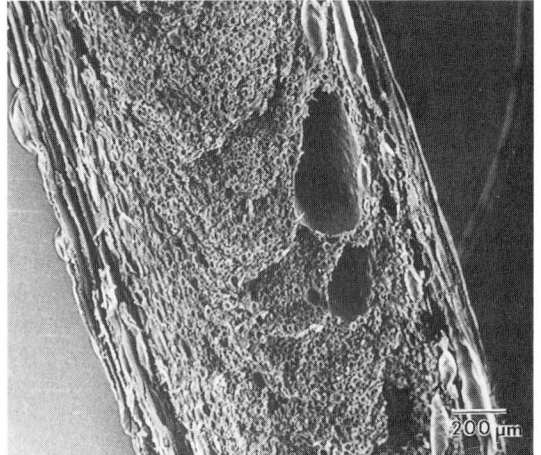

FIGURE 5.61. Scanning electron microscopy of a fractured, molded POM test bar, containing a high void level shows a skin-core morphology. Elongation of the voids at the skin surface is due to high orientation, whereas the more rounded voids and the semicircular flow front in the core results from less orientation in that region of the mold.

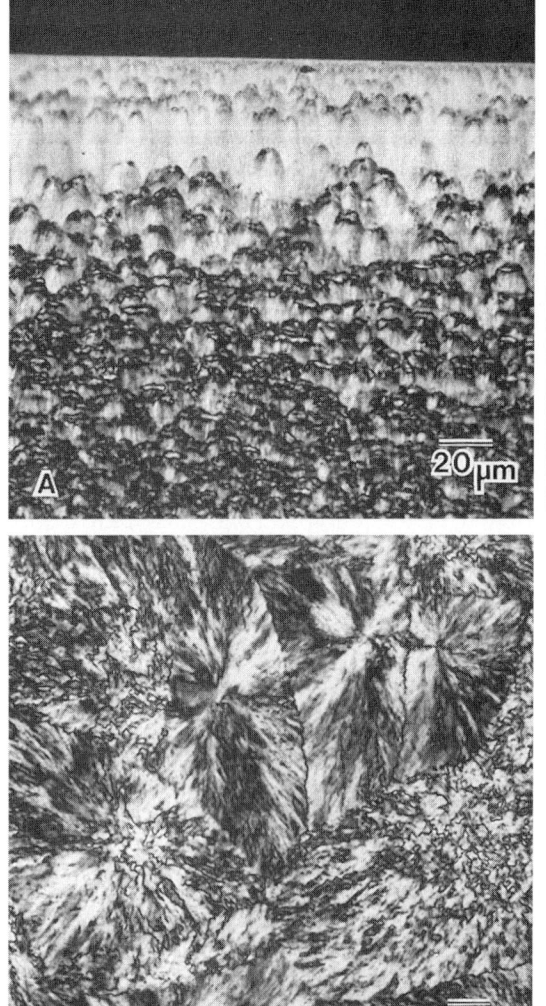

FIGURE 5.62. Optical micrographs of polyacetal section show a spherulitic texture in polarized light. An overview of the outer mold region shows a birefringent skin (top) and an unoriented spherulitic core (A). Between the skin and core is a transition zone composed of spherulites with parabolic boundaries (B). Spherulites that are polygonal in shape due to impinging one another are seen in the core (C).

rate of cooling of the melt. The skin surface in the microtomed section (Fig. 5.62A, top) is birefringent, nonspherulitic, and highly oriented; the molecular chains are oriented parallel to the injection direction. The central portion of the bar consists of a core (Fig. 5.62A, bottom) with randomly oriented spherulites (Fig. 5.62C). It has no preferred molecular or lamellae orientation. There are usually one or more layers between the skin and core that are transitional shear zones with intermediate structure. In polyacetals, this has been termed *transcrystallinity* [262, 295]. Transcrystalline growth is controlled by the heat flow to the mold wall and is initiated by the cold mold wall; the melt at the wall cools rapidly, and dense spherulite nucleation takes place adjacent to the wall. Spherulites nucleated close to the wall, in the thermal gradient, have parabolic boundaries (Fig. 5.62B). Their continued growth inward can lead to a layer of columnar structure. The number and extent of the layers depend upon the specific processing conditions. For example, Bowman [263] identified five layers in acetal copolymers, but where the mold is thin there may be no core, and in a hot mold there may be no skin. Multilayered structures are also observed in other molded plastics. Poly(butylene terephthalate) has been observed with four zones or layers: a nonspherulitic skin, regions with and without flow lines, and a central core with many flow lines [296]. Rapid solidification of the polymer while filling the mold could explain the presence of the flow lines in this banded texture. Injection molded polypropylene was shown with four layers [297–299].

A polarized light micrograph of an extruded polyacetal pellet section (Fig. 5.63A) has an oriented skin and a spherulitic core. Etching for 15 min in an oxygen plasma (162°C) results in about 5% weight loss, and a spherulite texture is seen by SEM (Fig. 5.63B) with a 4 h treatment resulting in a coarser texture (Fig. 5.63C), and only 10% of the original weight of the pellet remains. Smaller spherulites at the surface are revealed by the short etch, and larger spherulites are uncovered by further etching. Spherulites too small to measure by optical microscopy can be seen in the SEM, although untreated molded bar surfaces or fracture surfaces often do not reveal the spherulitic texture. Ion, plasma, or chemical etching can reveal spherulites by differential etching of crystalline and amorphous material.

Molded plastics generally have relatively smooth surfaces, although more detail may be seen by AFM than by SEM. Surface topography and molecular organization of a molded flexural test bar of a copolymer was shown by contact mode AFM to have surface indentations and fibrils (Fig. 5.64) [300]. Samples were prepared by gluing sections of the molded test bar onto steel disks with epoxy resin. Scanning electron microscopy of polyacetal shows a rather smooth texture as well (Fig. 5.65A). Etching is often quite useful in manufacturing processes to promote adhesion between the plastic and a surface coating, such as by electroplating [301]. Chemical etching for short times results in surface pit formation with pit shapes reflecting the microstructure (Fig. 5.65B, C). The effect of acid etching was to produce elongated pits in the direction of polymer flow into the mold. Longer etching times resulted in etching deeper into the unoriented core; thus, the bottoms of the pits are larger and more rounded than at the surfaces (Fig. 5.65C). Cross sectioned, etched, and electroplated materials provide more information on the etch depth. Low magnification SEM (Fig. 5.65D) did not show the etched structure, but the penetration of the electroplating metal is shown in an EDS map (Fig. 5.65E), and more detail is seen at higher magnification (Fig. 5.65F). Such undercut structures are quite important for well adhered, plated plastic parts.

5.3.4 Multiphase Polymers

5.3.4.1 Introduction

The topic of multiphase polymers is vast with books, reviews [224–227, 229, 230, 235, 302–311], and hundreds of research papers describing the processes, morphologies, properties, and applications of these important materials. The field of multiphase polymers has been driven by the realization that wholly new molecules are not always required for new applications and that blends can provide a rapid and economical means of development.

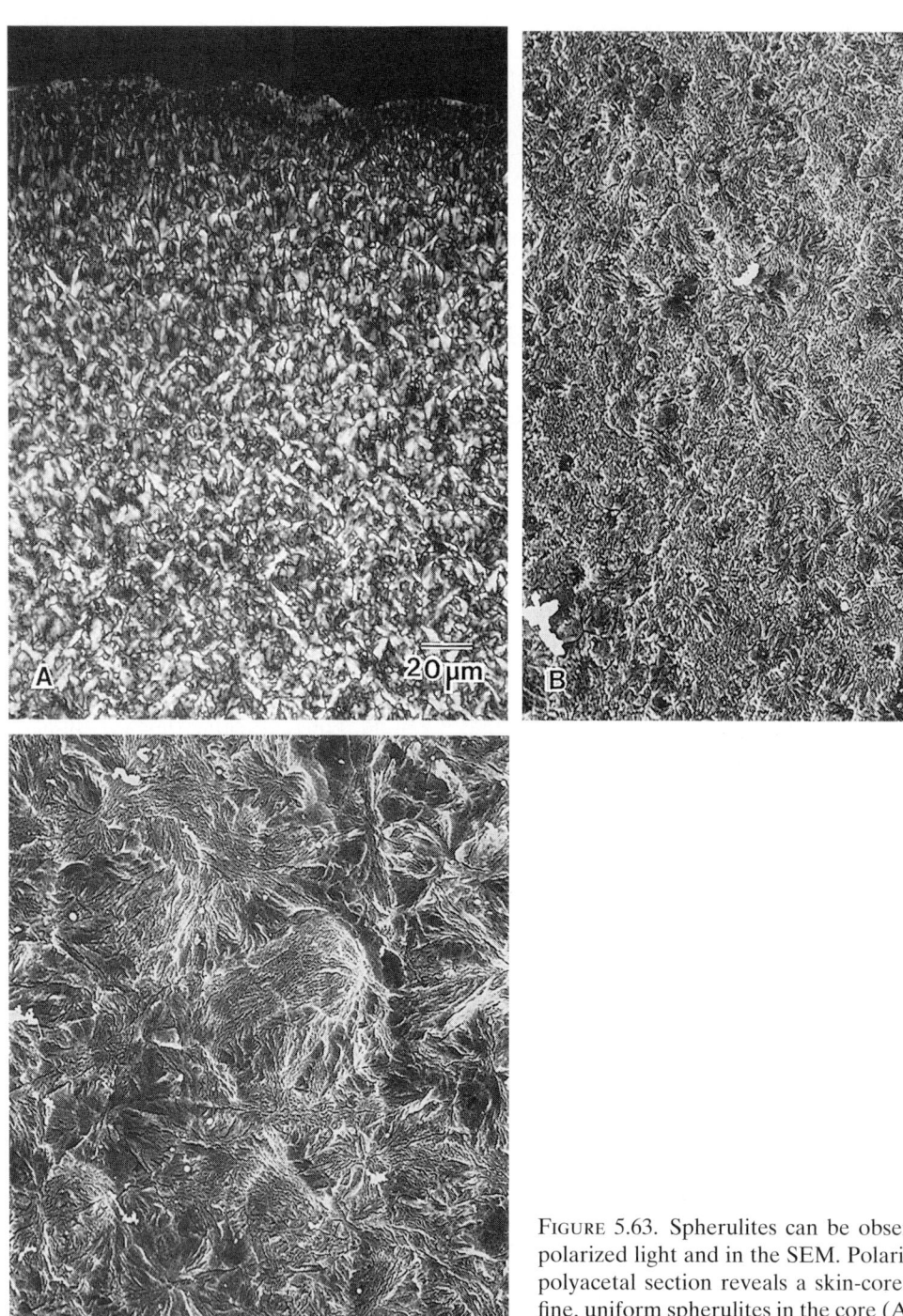

FIGURE 5.63. Spherulites can be observed by both polarized light and in the SEM. Polarized light of a polyacetal section reveals a skin-core texture with fine, uniform spherulites in the core (A). Treatment in an oxygen plasma at 162°C for 15 min uncovers the spherulite texture (B), and after a 4h treatment, larger spherulites are observed (C) within the molded specimen by SEM.

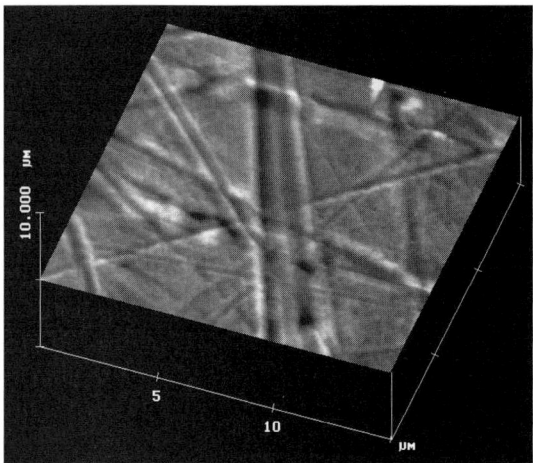

FIGURE 5.64. Surface topography and molecular organization of a molded flexural test bar of a copolymer was shown by contact mode AFM to have surface indentations (A) and surface fibrils (B). (From Schiraldi et al. [300], © (2001) Wiley-Interscience; used with permission.)

A recent review of block copolymer patterns provides an interesting view of the use of these special polymers for nanostructure fabrication [312]. As expected, processing of blends is critical to the microstructure and properties. Typical processing involves melt mixing using single and twin screw extruders, and then the blends are extruded, injection molded, or blow molded by similar methods used for engineering resins.

The morphology of these multiphase polymers depends on a range of parameters: the polymer molecules, interfacial tension, viscosity, shear mixing, and phase separation kinetics. In the resulting blends, the size of the dispersed phase, the drop breakup and coalescence are all governed by the deformation process. A wide spectrum of properties can be obtained by the appropriate blending of polymers. White and Min [313], for example, investigated the development of polymer blend morphology during processing and the role of interfacial tension and viscosity ratio as it affects phase morphology. Random copolymerization is used to modify the properties of a single phase, whereas graft and block copolymerization is used to modify the adhesion properties of the interface between the matrix and the dispersed phase.

Improvement in mechanical properties often accompanies addition of one polymer to another; for instance, the tensile strength and modulus of PE are increased by the addition of PP [314]. There has also been considerable work in the area of interpenetrating polymer networks [227]. The features of major interest have been multiphase morphologies, including the size, shape, and complexity of the "phase within a phase" structure. Paul et al. [227] has shown that domains are formed by nucleation and growth mechanisms, and then they are modified by diffusion to lower energy structures.

Multiphase polymers are commonly toughened plastics that contain a soft, elastomer or rubbery component in a hard glassy matrix or in a thermoplastic matrix. The modification depends on the composition and deformation mechanism of the material [302, 315], but normally it increases the fracture toughness and strength from that of the unmodified matrix resin. Impact strength, as measured for instance by an Izod impact or Charpy testing apparatus, is affected by the dispersed phase particle size, the rubber composition, and adhesion to the matrix. Dispersed rubber particles 0.1–10 μm in size [316] and typically in the range 0.1–2 μm [302] are generally best, and good adhesion between the particles and the matrix is required for enhanced impact strength. Dispersed phase morphology can include wholly compatible phases: phases that are too compatible may not result in impact enhancement. However, there is no single particle size, chemical composition, or elastomer content that provides a formula for a successful polymeric system. The microstructure of multiphase polymers formed by extrusion and molding is discussed below.

5.3.4.2 Toughened Resins

Rubber toughening was discovered nearly 60 years ago, and the original theories to explain the phenomenon were proposed 35 years ago and have been reviewed (e.g., [317]). Since that time, however, numerous deformation studies have been conducted in the TEM [43, 318–320] (see Section 4.8.3). Control of the toughening

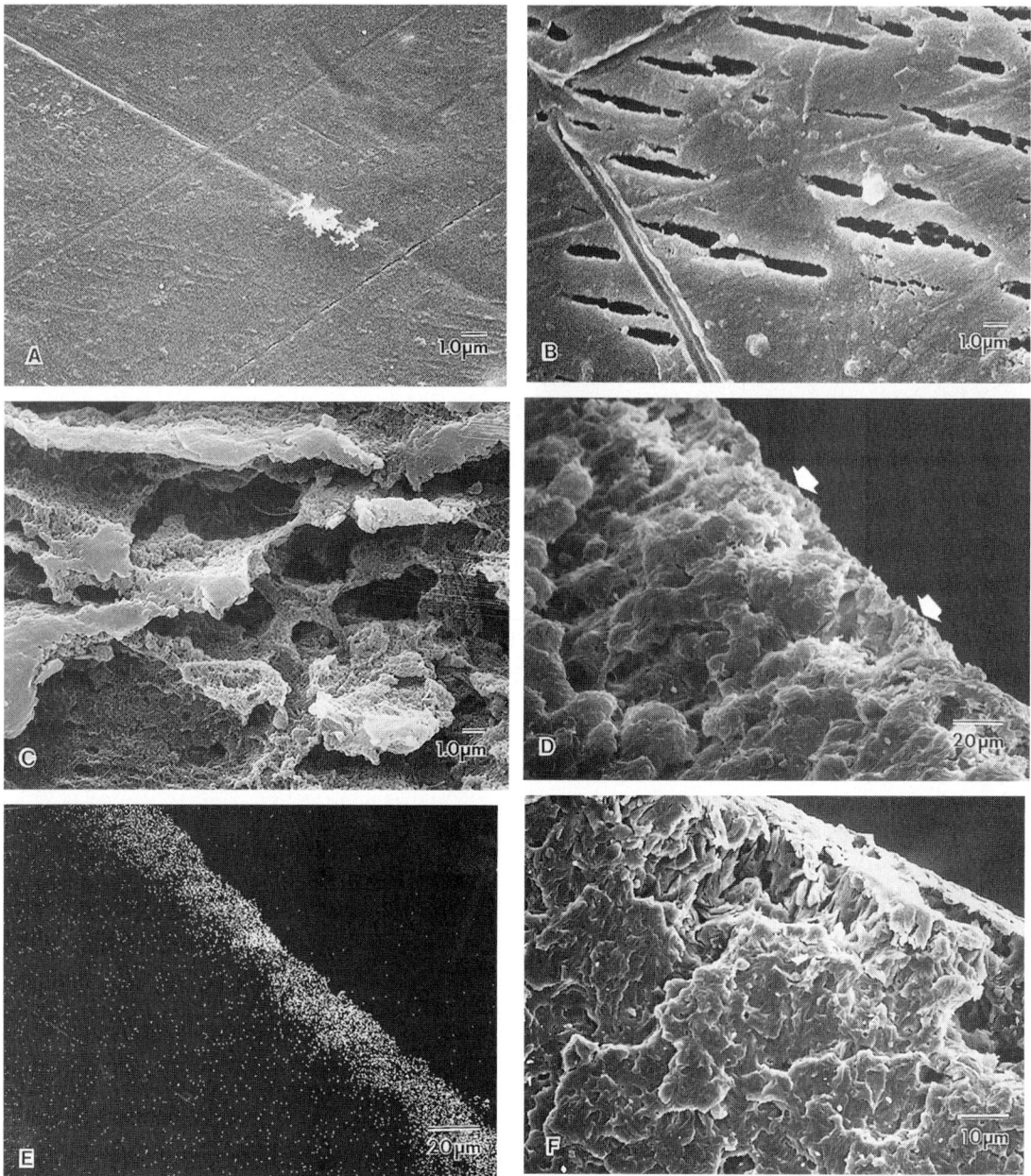

FIGURE 5.65. Scanning electron microscopy of a molded polyacetal surface shows a smooth texture (A) with little surface detail. Etching for short times results in elongated pits, oriented in the direction of polymer flow (B). Longer etching times result in surface pits deeper below the surface, due to etching larger spherulites in the core (C). Fractured cross sections of plated and etched surfaces do not show the structure near the surface (arrows) (D) except in EDS maps of the plating material (E) or at higher magnification (F).

process depends to some degree on the ability to characterize the structures resulting from the dispersion of the rubber particles, their grafting, crosslinking, and copolymerization. Microscopy techniques allow measurement of the size and shape of the dispersed phase particles and observation of the internal structures and the adhesion between the particles and the matrix. The particle size distribution is quite important as it is well known that there is an optimum "window" of particle sizes for each type of matrix; 0.1 μm particles provide good toughening for poly(vinyl chloride), whereas the rubber particle size for PS is more than 1 μm [321]. Particle size determination is simple if the polymer has discrete particles of the rubber dispersed in a matrix. However, the measurement is more difficult in cases where there is core-shell or graft copolymer morphology or where there are subinclusions of the matrix resin present in the rubber phase. Subinclusions are commonly observed in ABS, HIPS, and nylon blends and may be a result of the chemical process or the elastomer content.

Matrix polymers can be considered in two categories, brittle and ductile, each exhibiting specific requirements for reinforcing polymers. Brittle matrix polymers (PS, SAN, PMMA, and epoxy) have requirements for toughening as follows: interfacial adhesion, optimum particle sizes 0.1–0.3 μm depending on polymer matrix, chemical or physical crosslinking of the rubber [227]. The toughening mechanisms tend to be by crazing with some shear yielding and cavitation. These same authors [227] consider that ductile polymers (PC, polyamide, PP, PVC, PBT, and PBO) have the following requirements: adhesion is not always required, optimum particle sizes are less than 0.5 μm, crosslinking may be helpful, and at times trace levels of rubber have large effects (e.g., in PVC). Argon and Cohen [311] reviewed this topic claiming the most effective way of avoiding a ductile to brittle transition is to reduce the plastic resistance to delay reaching the brittle strength, which in unoriented polymers is governed by intrinsic cavitation. The most widely used techniques to meet this need involve incorporation of rubbery particles that can cavitate or rigid particles that can debond prior to plastic flow.

These authors studied six semicrystalline polymers to evaluate their potential for toughening under impact conditions and studied fractured blends by SEM.

Major factors that affect the size and distribution of dispersed phases in multiphase polymers are elastomer content, compatibility, processing, and viscosity. If the elastomer is present as the smaller volume fraction, it will most likely be the dispersed phase. As the volume fraction increases, the size of the dispersed phase can be larger, and there is more likelihood of subinclusions of the matrix polymer in the elastomer [318, 322]. These subinclusions appear to enhance impact properties. The shape of the dispersed phase often changes with differences in composition [323, 324], mixing conditions, relative viscosities, and with polymer orientation. For example, blends of 20% PE in PS form spherical domains, whereas at 50% PE, the domains are elongated and significantly larger in size [325]. Dispersed phase particles tend to be elongated parallel to the flow axis, and they are elongated in shape with the short axis normal to the surface of a molded or extruded part while remaining spherical near the core. The particles tend to be oriented at a 45° angle in the shear region. Generally, the finer the particle sizes the better, although particles that are too small (much less than 0.2 μm) do not generally affect properties.

Processing plays a major role in the nature of the dispersed phase in multiphase polymers. Changes in the shear forces and the temperature provide different structures. In the case of PS modified with polyisoprene, TEM studies showed that smaller particles, broken down in size by melt shearing, resulted in lowered impact strength and increased tensile strength [326]. Particle dimensions have also been shown to be affected by the viscosity of the molten polymer and the concentration of the modifier. Heikens et al. [327] investigated copolymer modified PS and LDPE and ethylene-propylene copolymer with block and graft copolymers and evaluated the mechanical properties and morphology, which showed that the graft and block copolymers increase the compatibility of the two phases.

After a brief review of processing, the three main types of multicomponent polymer systems will be discussed: (1) impact modified thermosets, (2) impact modified thermoplastics, and (3) combinations of two or more semicrystalline polymers.

5.3.4.3 Processing of Multiphase Polymers

The topic of processing was discussed earlier in this section (see Section 5.3.2), in the discussion of films, especially multilayered films (see Section 5.2.6), and in the section on nanocomposites (see Section 5.4.6), all of which might be blends. A few additional comments are provided here as the role of processing in multiphase polymers is so important. Multiphase polymers are processed by all the standard processes, such as extrusion and molding, into useful structures such as films, moldings, rods, pipes, blown bottles, tanks, among many others. The nature of the equipment and especially their size is often a cause for major changes in morphology and properties. The use of supercompounding extruders, for instance, has resulted in changes in the size and dispersion of fillers and tougheners as the volume, pressure, residence time, and temperatures across the machine are very different compared with smaller machines. Process aids used to reduce melt viscosity also result in morphology changes as the viscosity ratio between polymers and rubbers, for instance, will result in changes in particle size and dispersion and thus the final product properties. The nature of the various processes also induces varied orientation of dispersed phases, which can result in anisotropic mechanical and physical properties.

Hiltner, Baer, and coworkers have conducted extensive work on the production and observation of microlayers (e.g., [180]) in which immiscible polymers are brought into intimate contact and by highly localized mixing of polymer chains create an "interphase" region. Assemblies of thousands of alternating layers of polymers, with individual layer thickness on the nanometer size scale, are created by layer multiplying coextrusion. An extruded film of an amorphous polyester, poly(ethylene terephthalate-*co*-1,4-cyclohexanedimethylene terephthalate (PETG), blended with PS was examined in air by AFM and was shown in the section on films (see Fig. 5.38 [180]), but the same process can be used with thicker objects.

Chaotic mixing was also noted earlier (see Section 5.2.6), and it is being used in the development of immiscible polymer systems by Zumbrunnen and his group at Clemson (e.g., [182, 328–330]) and also was reported by Sau and Jana [331] as part of their work on mixer design and its effect on morphology development in two immiscible polymers, PP and PA6. Liu and Zumbrunnen [328] showed improvements in impact properties of a PS matrix by the addition of only 9 vol.% of LDPE when combined using a three dimensional chaotic mixing process due to the formation of extended and interconnected structures rather than dispersed phases, as shown by SEM of fracture surfaces. A multilayer blend morphology that has a hierarchical structure and intrinsic mechanical interlocking was formed by chaotic advection of immiscible polymer melts of PP-LDPE [182]. Extruded blends provided improved properties relative to properties associated with droplet morphologies typically obtained with conventional compounding equipment. Film samples of these blends were examined by LVSEM and shown earlier (see Fig. 5.39) [182].

5.3.4.4 Toughened Thermoset Resins

Toughening of polymers with rubber has seen greatest application in thermoplastic resins. However, the technology has also been extended to thermosetting resins, such as epoxies [332, 333], which has been reviewed [334]. Epoxy resins toughened with rubber particles have enhanced properties like toughened thermoplastics, although the theory of such toughening is not as well understood. Bucknall [302] and Kunz-Douglass et al. [335] and others have discussed toughening models, however these are beyond the scope of the current discussion. In any model of toughness, a major parameter is the size distribution of the dispersed rubber particles. Microscopy provides this information in the same way as described for thermoplastic polymers.

Thermoset epoxy resins were toughened by small elastomer inclusions of a carboxyl terminated butadiene-acrylonitrile (CTBN) random copolymer by Visconti and Marchessault [332], who showed the variation in size as a function of CTBN content by TEM and light scattering. A major study of rubber modified epoxy resins has been reported by Manzione et al. [336, 337], who showed a range of morphologies that result in a range of mechanical properties, even for a single polymer. An amine-cured, rubber modified epoxy was characterized by STEM imaging, and quantitative methods were developed to determine the volume fraction of dispersed phase particles [338] as this is known to be critical to enhanced toughness. Sayre et al. [338] stained the samples in tetrahydrofuran (THF) containing osmium tetroxide. Bucknall [239] applied microscopy to the study of the deformation and fracture of rubber toughened polymers, including discussion of SEM of the fracture surfaces of rubber toughened epoxy resins that show the cavitation in the rubber particles.

The fracture morphology of a rubber toughened epoxy resin is shown in the SEM micrograph in Fig. 5.66 that is typical of glassy or brittle fracture with failure occurring across the well adhered dispersed phase particles. Subinclusions of the resin are observed within the dispersed phase particles, likely due to the high concentration of the rubber phase. Voids, large and small smooth holes, are observed within the matrix and also within the dispersed phase particles.

Cross-linkable epoxy thermoplastics modified with 5 wt.% grafted rubber concentrate (CET-GRC) have rubber domains ca. 0.1 μm in diameter and consist of a core-shell structure with an 0.03 μm diameter PS seed. Blocks of the CET-GRC were "cryo-polished," that is, cryo-microtomed, at $-90°$C using a diamond knife to provide a smooth surface for SPM imaging [174]. Intermittent contact mode AFM has been applied to the study of both adhesion and mechanical deformation on this sample. Careful control of the tapping parameters were used to obtain phase images in which the rubber phase contrast can invert relative to the phase signal of the epoxy matrix (Fig. 5.67) [174]. Contact conditions were controlled and the set-point voltage at which the phase signal was first observed to shift was noted (A_o) and then decreased to improve tapping until the desired operating set-point voltage (A_{sp}) was achieved; the ratio (r_{sp}) of A_{sp}/A_o determines the degree of contact. The force is sufficient to compress the rubber phase so it appears as holes in the height images with deeper holes noted for hardest tapping conditions (left side of Fig. 5.67). The phase angle of the tapping tip is

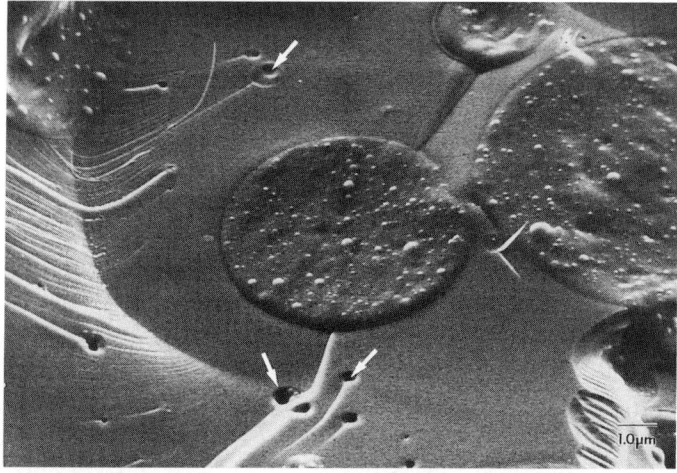

FIGURE 5.66. Scanning electron microscopy of a rubber toughened epoxy resin shows that brittle fracture occurs through both the matrix and the dispersed phases. Voids (arrows) are observed within the dispersed phase and also within the matrix. Small subinclusions are seen within the dispersed phases.

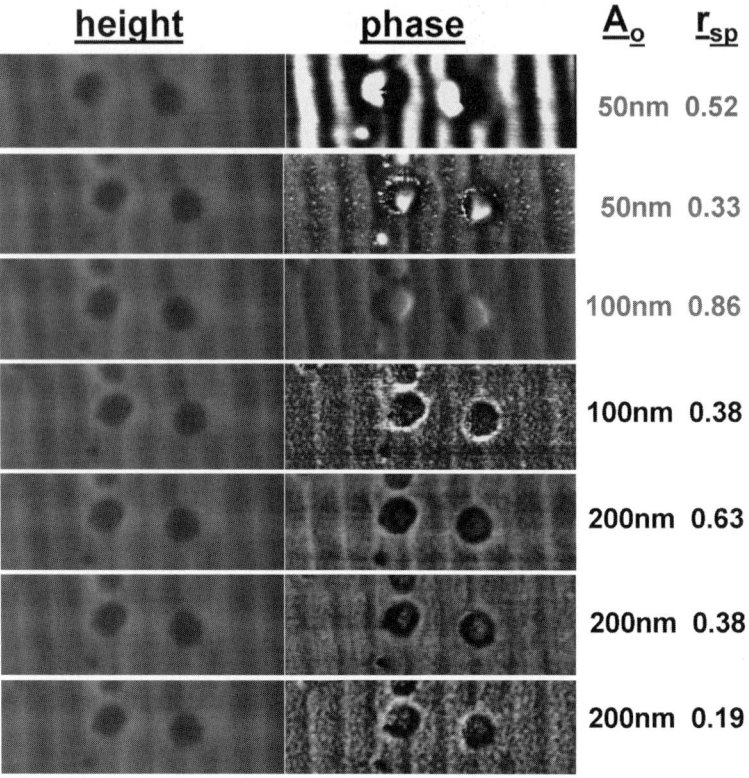

FIGURE 5.67. Intermittent contact AFM height and phase images of CET-GRC (5 wt.% rubber, 0.1 μm diameter) after cryo-polishing (images are 1 × 0.3 μm). The table at the right indicates the parameters used to generate each pair of images. (See color insert.) (From Meyers et al. [174], © (2000) American Chemical Society; used with permission.)

sensitive to the tapping conditions. The rubber exhibits positive contrast relative to the epoxy when tapping conditions are trapped in the adhesive regime (top three phase images) but negative contrast relative to the epoxy when conditions are in the repulsive regime (bottom five phase images). Clearly, SPM imaging requires careful consideration, and complementary techniques are warranted.

5.3.4.5 Impact Modified Thermoplastics

Much of the preceding discussion has included semicrystalline thermoplastics and blends that were used as examples for discussions of structure and process. A brief literature review of impact modified thermoplastics follows. Reich and Cohen [339] studied the phase separation of polymer blends in thin films and compared the behavior to that of the bulk material, as it is well known that phase transformations in thin, nonpolymeric, solid films differ from those in the bulk [340]. Handlin et al. [341] studied ionomer morphology and found that solvent casting produced artifacts but no information about ionic domains, whereas microtomed sections of sulfonated EPDM showed phase separated regions.

A TEM study of poly(vinyl chloride)/chlorinated polyethylene (PVC/CPE) assessed the dispersed phase morphology for correlation with impact properties [342]. Microtomed sections of the blend were stained by a two stage osmium tetroxide method to reveal the CPE phase [343]; as the concentration increased, the discrete two phase morphology changed to a continuous network resulting in a transition from brittle to ductile impact fracture (shown by SEM) and increased impact strength. The behavior of these blends was also studied [344] by deformation under an optical microscope and by TEM of stained sections. Complementary techniques of hot stage OM, TEM, and analysis of particle size were used for the determination of the nature of phase separation dynamics and morphology, especially the late stage mechanism of phase separation of polymer blends [345] showing that discrete particles formed early in the process but they may change by aggregation and coalescence.

Hobbs and coworkers [346, 347] studied blends of PBT and bisphenol A (BPA) polycarbonate (PC) toughened with a core-shell impact modifier by TEM and SEM. Selective staining with ruthenium and osmium tetroxide and etching with diethylene triamine were used to assess the distribution of the blend components and to investigate the effects of thermal history on morphology. The impact modifier was differentiated from the PBT and PC by reaction with OsO_4, which forms chemical complexes with double bonds; the PC was imaged preferentially by its greater ability to absorb RuO_4. Poly(butylene terephthalate) was observed as the continuous phase with the impact modifier observed isolated in islands of the PC with an interpenetrating network formed above 40% PC. Scanning electron microscopy studies of specimens etched with diethylene triamine showed information about melt miscibility and phase separation. The overall study provides evidence that supports the strong interfacial region and aids understanding of the properties of this blend system. The toughening mechanism of these blends was evaluated by combination of notched impact testing and morphological evaluation [346, 347]. Differences in behavior were discussed in terms of microscopic failure processes. Dekkers et al. [348] also studied the effect of morphology on the properties of blends of PBT, PC, and poly(phenylene ether) using electron microscopy. The dispersed particles of rubber modified PPE were observed encapsulated by thin envelopes of PC and embedded in a PBT matrix.

Many of the studies of multiphase polymers are conducted on unsaturated rubbers that are adequately stained by osmium tetroxide, which reveals the nature of the dispersed phase domains. Polymers with activated aromatic groups have been selectively stained by reaction with mercuric trifluoroacetate (see Section 4.4.8). Hobbs [349] successfully used this technique to provide contrast in blends of poly(2,6-dimethyl-1,4-phenylene oxide) and SBS block copolymer. Although this stain is effective in enhancing contrast, a drawback of the method is that the material is not hardened or fixed by the stain.

Changes in polymer morphology are expected upon addition of an elastomer; for instance, such addition is expected to cause a decrease in the spherulite size as the elastomer domains can act as nucleating sites [350]. This has been observed for many polymers including modified nylon [351]. Characterization of an EPDM impact modified nylon 6,6 has been reported [352] by the use of osmium tetroxide staining (1 week) followed by TEM imaging to show the core-shell microstructure. Transmission electron microscopy imaging of rubber tougheners for packaging materials was reviewed by Wood (e.g., [186]), who described the staining methods used (see Section 4.4). Environmental scanning electron microscope (ESEM) work on blends has been conducted by Donald and her group (e.g., [246, 247, 353]).

Microfibrillar reinforced composite blends of recycled PET, PP, and a compatibilizer (ethylene-glycidyl methacrylate; E-GMA) were prepared by melt extrusion, followed by continuous cold drawing. Test specimens were prepared by compression and injection molding. Samples characterized by SEM after fracture, WAXS, and mechanical testing showed that the extruded blends are isotropic but become highly oriented after drawing, and they are converted into microfibrillar composites during molding. Mechanisms of compatibilization of PO/LCP blends with graft copolymers have been studied by OM and SEM; OM micrographs of the melt formed films were analyzed to measure the size and size distribution of the dispersed phase particles as a function of time [354]. A novel image processing method was developed to extract interfacial area concentration measurements from two dimensional SEM or TEM micrographs of immiscible polymer blends [355]. The method operates by detecting edges within the images and using standard image processing to selectively eliminate false edges. Scanning electron microscopy images of polyethylene oxide/polystyrene (PEO/PS) blends were analyzed using this method to measure the amount of interfacial area in the samples. Such interfacial area measurements may be used in future investigations of blend dynamics, including coalescence, drop deformation, and to quantify the effects of compatibilizers on blend morphology.

More recently, AFM has been included as one of the complementary techniques for study of the morphology of blends. Pfau et al. [254] used TEM and force modulation AFM to map elastic properties for a variety of blends, including rubber toughened PP and HIPS, preparing samples by microtomy. Thomann et al. [255] studied blends of iPP with random poly-(ethene-co-1-butene) (PEB) using phase imaging ICAFM, OM, and TEM. The blends at some compositions were found to be miscible or partially miscible. Flat block faces were prepared for AFM using a microtome, and thin sections were cut with a diamond knife at −40°C for TEM after staining with RuO$_4$. Intermittent contact mode AFM was used to study polydimethylsiloxane (PDMS) samples of different crosslink densities [256].

One of the issues with AFM sample preparation is the use of a microtome to prepare a flat surface (block face), which can result in artifacts and surface roughness, even debonding the interfacial regions. This limitation has been addressed by using focused ion beam (FIB) preparation followed by ICAFM [258] to analyze model interface thicknesses in HDPE/PS/PMMA ternary blends. Focused ion beam is a physical process that uses an ion beam to etch away the surface of a polymer by control of the beam current and energy, although such a process is well known to cause changes in crystallinity, shrinkage, and other surface modifications (see Section 4.5.5). The preparation involves cutting a planar face using a microtome with samples and glass knives cooled to −160°C. Polystyrene was selectively dissolved with cyclohexane at room temperature for 1 day and then samples were plasma coated with Au/Pd for SEM study. Focused ion beam was conducted on the plasma coated samples using a 30 kV Ga+ beam; samples were fixed on a metallic support using silver glue or graphite tape for AFM imaging. Finally, the SEM and AFM images were analyzed to quantify the volume average diameters; it was found that the polymers had different etching rates resulting in topological contrast in ICAFM. Care must still be taken to use complementary microscopy methods to ensure measurements are correct.

Applications and examples of microscopy imaging and analysis to multicomponent polymers follow. The microstructure of semicrystalline multicomponent polymers can often be determined by polarized light microscopy of thin sections. A blend of two polyacetals, a homopolymer and a copolymer, is shown in the micrograph of a thin section (Fig. 5.68). The structure is rather interesting

FIGURE 5.68. A polarized light micrograph of a polymer blend cross section, composed of two different polyacetals, shows a nonuniform texture. A transcrystalline layer is seen adjacent to the skin, and larger spherulites are seen in a matrix of finer textures.

in that large spherulites of one phase are observed in finer spherulites of the other phase; a transcrystalline region is observed adjacent to the skin.

An example of the fracture surfaces of a polymer blend is shown imaged using a lanthanum hexaboride gun (Fig. 5.69A) and a FEG gun (Fig. 5.69B), showing significant differences in detail. The FEG gun results in improved detail, especially obvious at the blend interfaces. Addition of an elastomer modifies the brittle fracture behavior of the matrix, as shown in a fracture surface of a modified nylon. Phase contrast optical microscopy (Fig. 5.70A), TEM of a stained cryosection (Fig. 5.70B, C) and STEM imaging of unstained ultrathin sections (Fig. 5.70D) all show elliptically shaped, dispersed phase particles in a matrix. The contrast in STEM is a result of a difference in radiation damage between the two polymers; nylon likely crosslinks in the beam, whereas the elastomer phase exhibits mass loss due to chain scission [236]. Advantages of STEM

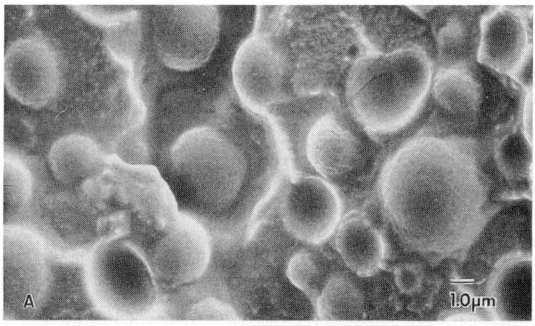
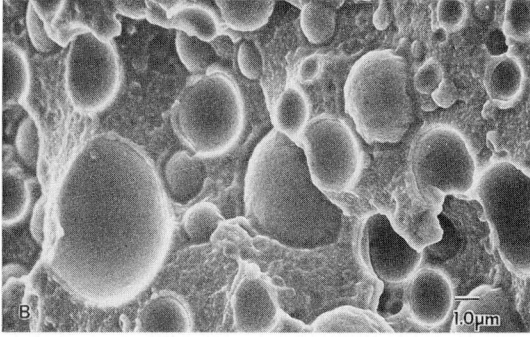

FIGURE 5.69. Similar fracture surfaces of a polymer blend were imaged at 5 kV accelerating voltage with a lanthanum hexaboride gun (A) and with a FEG gun (B). (From M. Jamieson [217]; unpublished.)

are that there is no need for any stain or etchant and the image can be rapidly obtained and processed. A disadvantage of STEM imaging is that the specimen is quickly changed and then destroyed by the electron beam unless great care is taken to limit the electron dose. The particle sizes and distribution are similar in the STEM images of room temperature sectioned nylon and the TEM sections of cryosectioned and stained polymer. The TEM image permits observation of the smaller sub-included nylon phase in the elastomer, due to the effect of staining [356].

The first example of AFM imaging is a comparison of TEM (Fig. 5.71) and AFM (Fig. 5.72) of ternary blends of nylon 6 with rubber: (1) nylon 6/maleated ethylene-propylene random copolymer/imidized acrylic polymer (IA) and (2) nylon 6/maleated styrene-(ethylene-co-butylene)-styrene/IA, both with 70/20/10 wt.% PA 6/EPR-g-MA/IA and PA 6/SEBS-g-MA/IA, respectively [357]. For the first ternary blend, preferential staining of the IA by RuO_4 vapor was followed by cryoultramicrotomy with a diamond knife at −45°C, and the PA phase was stained with phosphotongstic acid (PTA) (see Section 4.4.5). In the second blend, the sections were stained with RuO_4 and PTA to reveal the SEBS rubber and PA matrix, respectively. The surface of the microtomed blocks was used for phase contrast imaging by AFM. Digital image analysis was used to measure particle sizes from both TEM and AFM images. The values of particle average diameter and particle aspect ratio of rubber particles in the nylon 6 matrix were found to be similar in AFM and TEM. In the TEM images in Fig. 5.71, the rubber particles are white in the PA 6/EPR-g-MA/IA blend and black in the section due to the different staining responses. In the PA 6/SEBS-g-MA/IA, the larger SEBS-g-MA particles are elongated, whereas no elongation is observed in the EPR-g-MA, and the particles are not perfectly spherical. In Fig. 5.72, the topographic (A, B) and phase contrast (C, D) images from AFM of the two ternary blends, respectively, are shown. The hill marked in Fig. 5.72A was generated by the diamond knife, whereas this effect is not observed in the phase image in

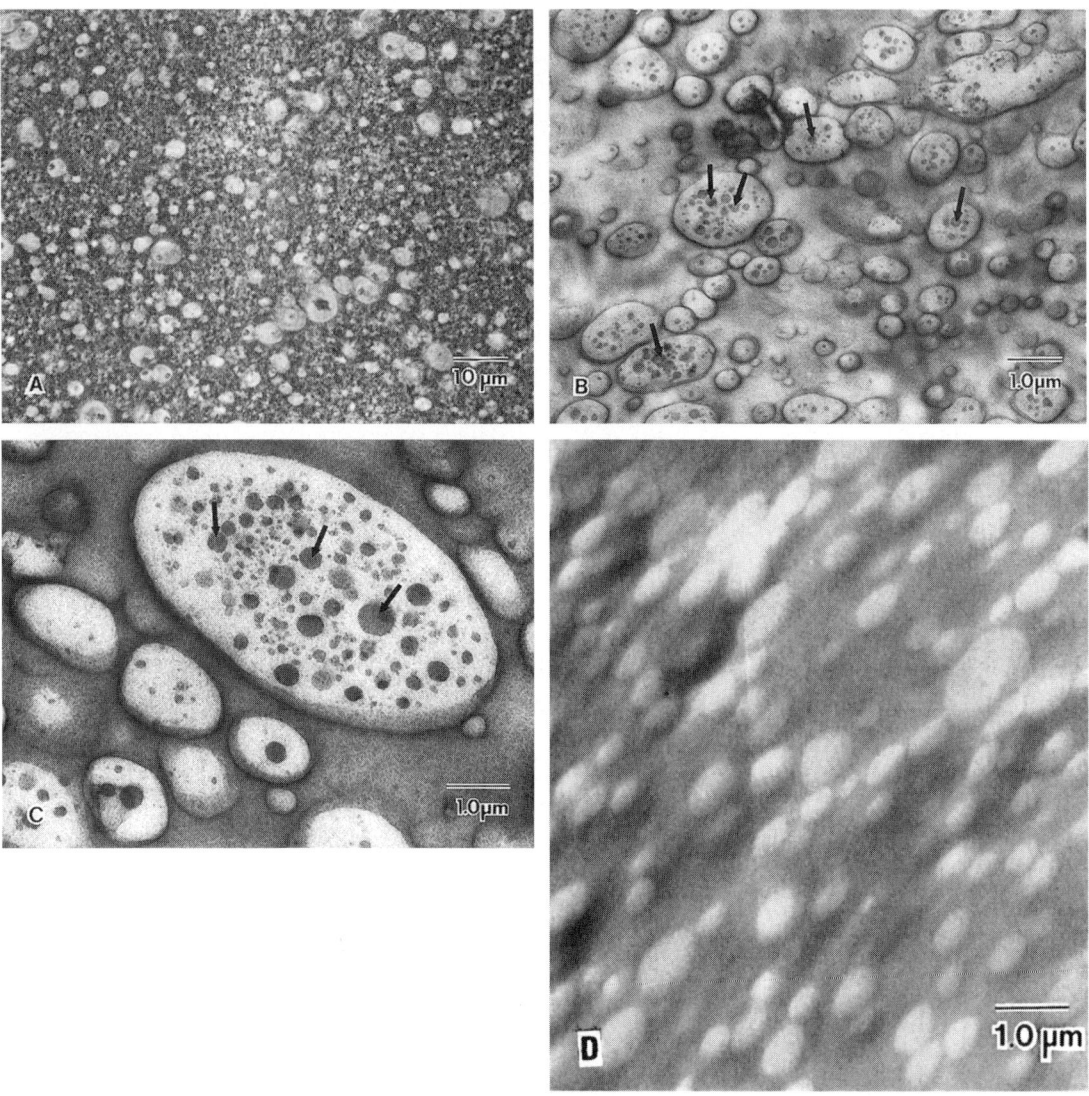

FIGURE 5.70. A phase contrast optical micrograph (A) of an impact modified nylon shows the fine dispersion of modifier in the matrix. Transmission electron microscopy micrographs of a cryosection, stained with ruthenium tetroxide (B and C), show more detail and finely dispersed subinclusions (arrows) within the elastomeric phase. Scanning transmission electron microscopy (D) of an unstained cryosection shows less dense regions in a darker background due to mass loss of the rubber phase during exposure to the electron beam, resulting in contrast enhancement.

Fig. 5.72C. The white regions in the phase images are the rubber domains, and the dark regions are the nylon 6 matrix. The IA phase was observed by TEM but could not be distinguished by AFM perhaps due to similar toughness as the matrix [357].

As seen here, and reviewed by Bar and Meyers [170], phase imaging by ICAFM is capable of mapping multiphase polymers with good resolution and contrast. Figure 5.73 shows ICAFM images of polypropylene/ethylene propylene rubber (PP/EPR) blends of different compositions: 85/15 wt.% (Fig. 5.73A), 70/30 wt.% (Fig. 5.73B), and 40/60 wt.% (Fig. 5.73C). Block faces of cryomicrotomed samples were used for imaging using a specially designed

Engineering Resins and Plastics 333

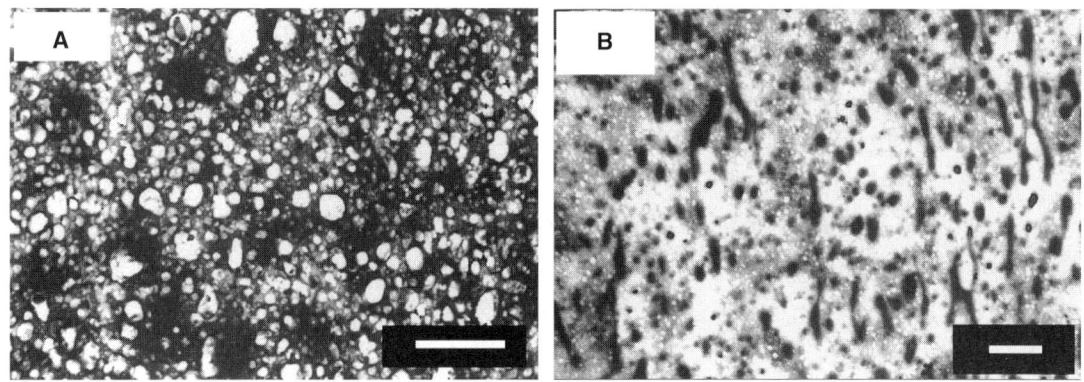

FIGURE 5.71. Transmission electron microscopy micrographs of ternary blends (A) PA 6/EPR-*g*-MA/IA and (B) PA 6/SEBS-*g*-MA/IA. Scale bars are 1 μm. (From Radovanovic et al. [357], © (2004) Elsevier; used with permission.)

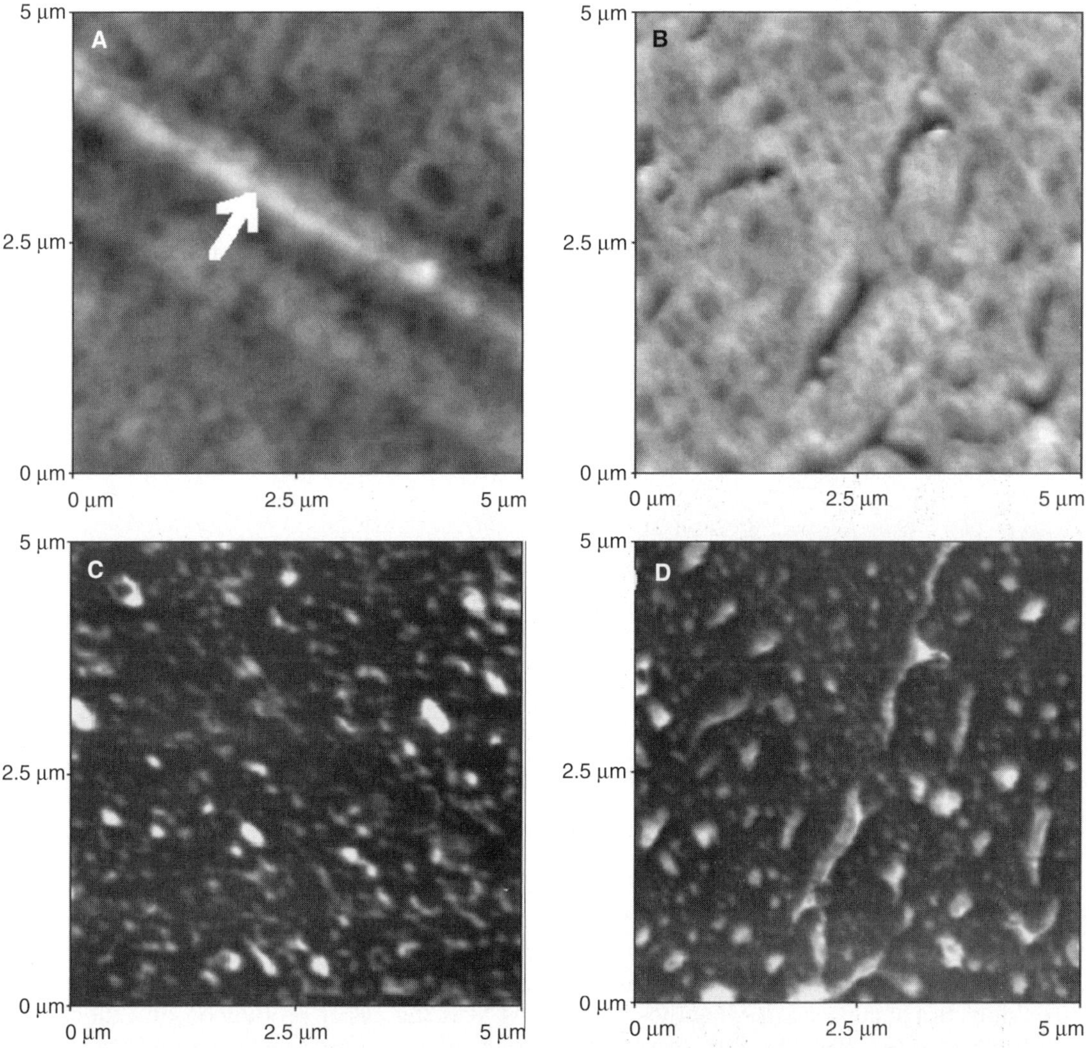

FIGURE 5.72. Atomic force microscopy images (5 × 5 μm) of ternary blends: (A) topographic and (C) phase contrast images of PA 6/EPR-*g*-MA/IA; (B) topographic and (D) phase contrast image of PA 6/SEBS-*g*-MA/IA. (From Radovanovic et al. [357], © (2004) Elsevier; used with permission.)

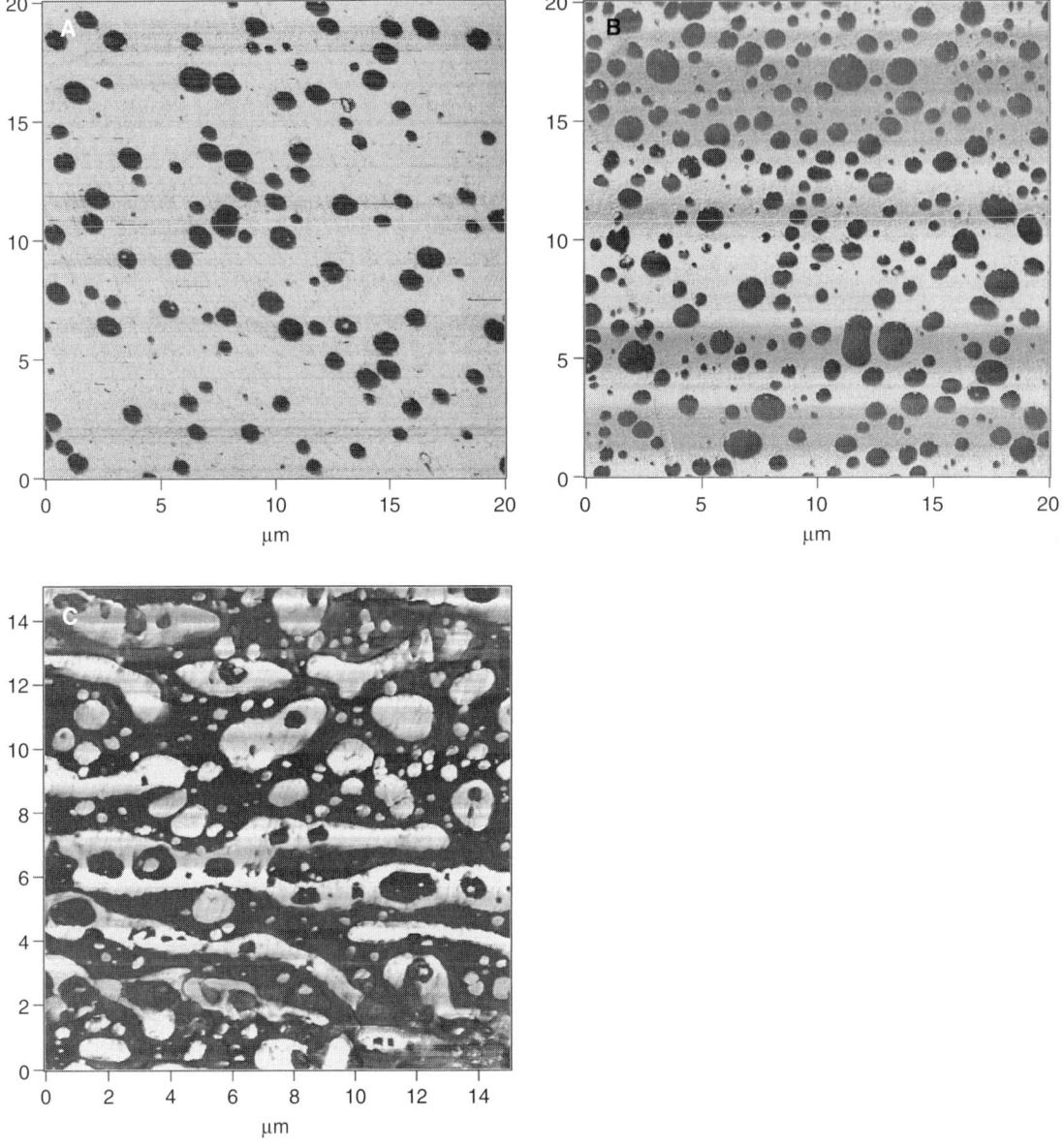

FIGURE 5.73. Phase images of polypropylene/ethylene propylene rubber (PP/EPR) blends of different compositions: 85/15 wt.% (A), 70/30 wt.% (B), and 40/60 wt.% (C). The regions of dark contrast in the images are the rubbery EPR phase, and the PP exhibits bright contrast showing the change in morphology as expected with changes in concentration. (From Bar and Meyers [170]; used with permission of the *MRS Bulletin*.)

sample holder [358] described earlier (see Section 4.3). The regions of dark contrast in the images are the rubbery EPR phase, and the PP exhibits bright contrast showing the change in morphology as expected with changes in concentration. Images are used for measurement of the diameter of the rubber domains and their distribution. The advantage of the AFM is the easier production of a block face than ultrathin sections required for TEM, although there is much less fine detail in the images; thus if that is required, TEM imaging is still the best technique to use [170]. A series of different polymer blends were also imaged by the group at Dow (e.g., [359]) for HIPS, impact PP, ethylene vinyl acetate (EVA)/PP, PP/ethylene-octene copolymer (EO), and PP/PMMA blends using these same sample preparation methods.

Magonov has described AFM techniques in a number of review papers [250–253], including imaging of a wide range of polymer blends. Multiphase polymer systems provide a different response in the AFM tip-sample interaction. The height images of PE blend samples, which consist of two polymers with different octene branch content and with different densities, are shown in Fig. 5.74 [250]. The polymer with higher density crystallizes with formation of a lamellar structure not seen in the lower density polymer. Light tapping (top images) shows slightly elevated domains over the surface compared with hard tapping (bottom images), which results in more pronounced morphology. Height profiles taken as part of this work show the different indentations on the soft and hard blend components.

Xenoy (trademark of GE Plastics, Schenectady NY) are blends of a polyester (typically PBT or PET) and PC. The Xenoy resin family offers good chemical resistance, great impact resistance even at low temperatures, heat resistance, and outstanding aesthetic and flow characteristics and is used in applications that are exposed to harsh conditions or that require a high degree of toughness. Images of a microtomed surface of a commercial three component Xenoy blend (PC, PBT, rubber) are seen in height and phase modes taken with light tapping (top images) and hard tapping (bottom images) in Fig. 5.75 [250]. The hard tapping phase image appears to be the most informative. The practical value of this work is to show that there is a vast difference in image contrast with specific tapping or contact conditions, and topographic studies should be performed in light tapping in which there is less change in the sample during imaging.

In another review [253], Magonov and Yerina described a wide range of polymers imaged by

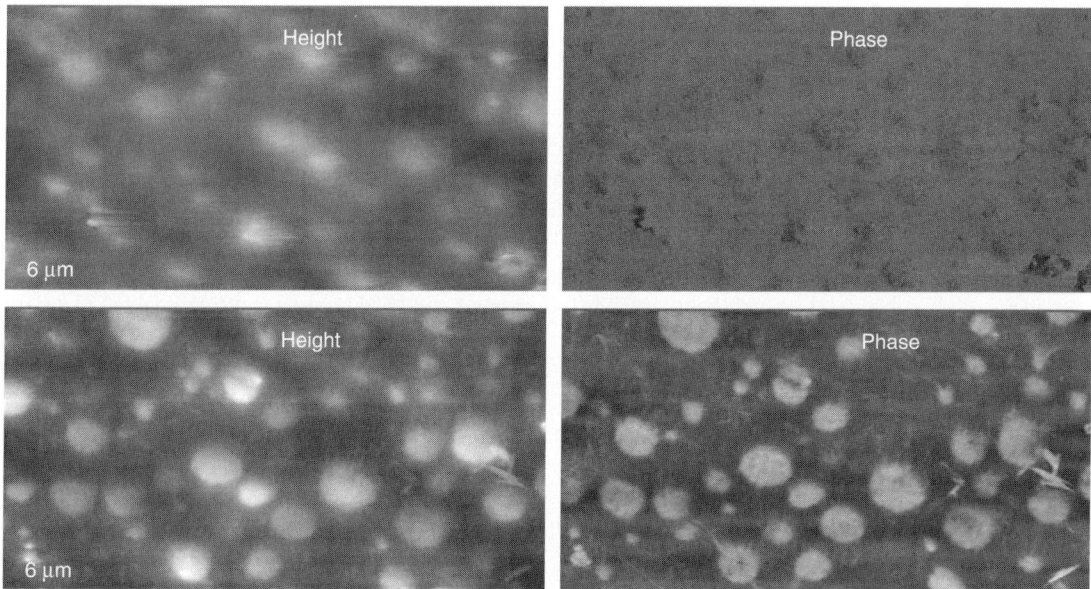

FIGURE 5.74. Height and phase images of a polyethylene blend sample with top images obtained with light tapping and bottom with hard tapping (6 μm on a side). (From Magonov [250], © (2001) Elsevier; used with permission.)

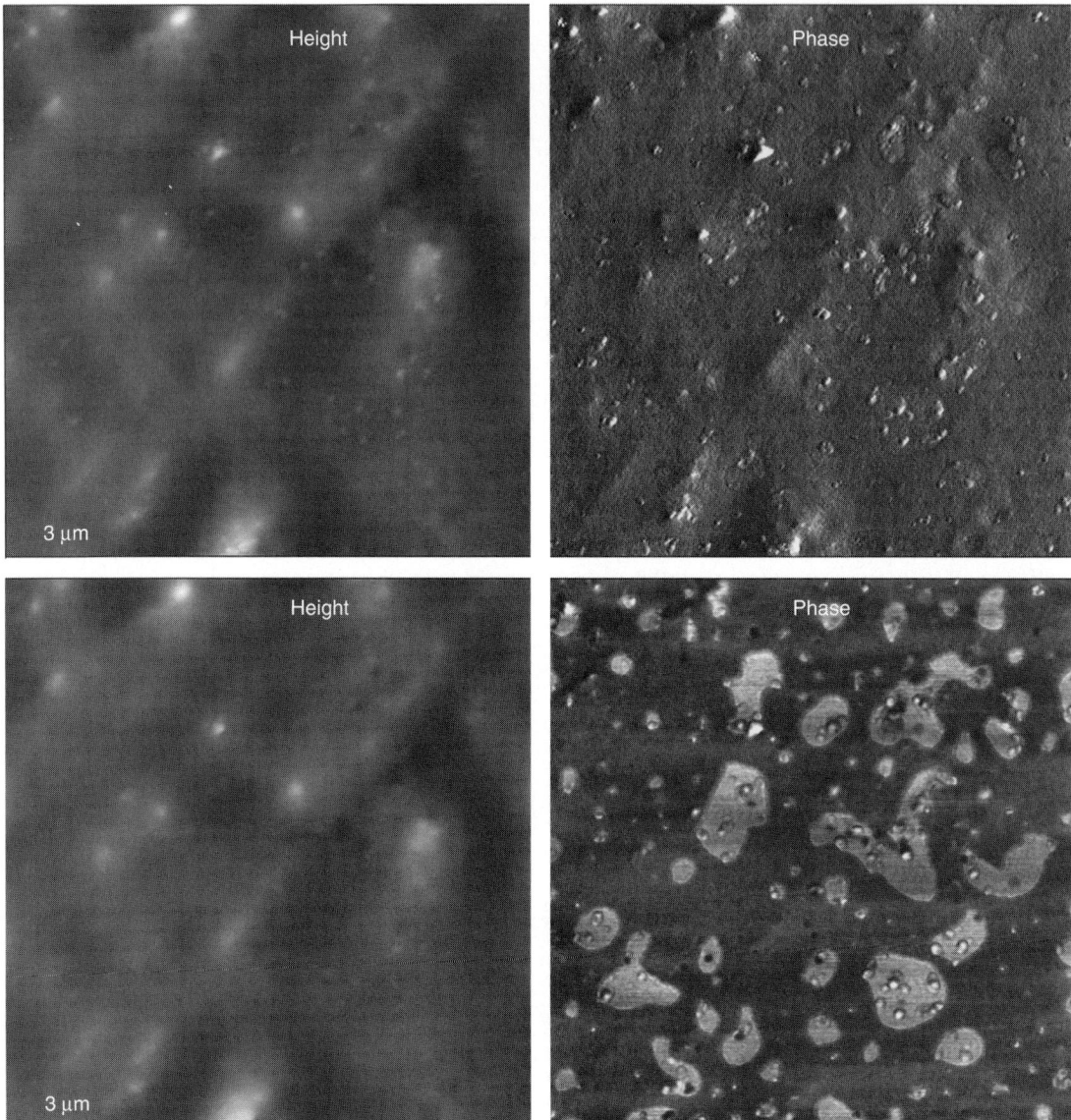

FIGURE 5.75. Height and phase images of a microtomed surface of a commercial three component Xenoy blend (PC, PBT, rubber) taken with light tapping (top images) and hard tapping (bottom images) (3 μm on a side). (From Magonov [250], © (2001) Elsevier; used with permission.)

AFM. A thermoplastic vulcanizate (TPV) made by mixing EPDM, iPP, and carbon black is shown in height and phase images in Fig. 5.76. Electrically conducting TPV filled with carbon black particles can be used as sensors, switches, and electromagnetic shields. Again, the phase image shows the best contrast with the distribution of the EPDM domains (bright regions), in which the contrast variations are said to reflect different crosslink densities, and the darker regions with 40 to 50 nm bright particles of iPP domains filled with carbon black. This complex morphology would benefit from examination by TEM to assess the carbon black dispersion.

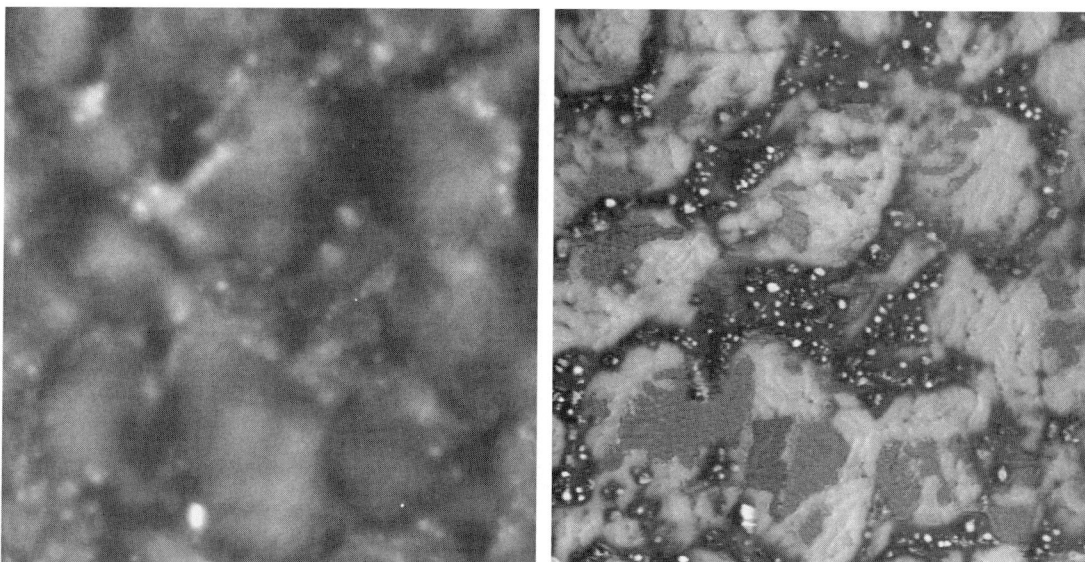

FIGURE 5.76. Height (left) and phase (right) images (5 μm on a side) of a thermoplastic vulcanizate made by mixing EPDM, iPP, and carbon black show the detailed morphology of these complex blends. (From Magonov and Yerina [253], © (2005) Springer; used with permission.) (See color insert.)

5.3.4.6 Block, Graft and Random Copolymers

Copolymers are a common form of heteropolymer that are formed from a sequence of two or more monomer units that can be alternating or include long sequences of one repeat unit, as found in block copolymers; copolymers were discussed in Chapter 1 in more detail (see Section 1.2.4). Each component of a block copolymer can be amorphous or crystalline [360]. Amorphous block copolymers generally form characteristic domain structures, such as those in SBS block copolymers. Crystalline block copolymers, such as polystyrene-poly(ethylene oxide), typically form structures more characteristic of the crystallizable component. Block copolymers that have the second component grafted onto the backbone chain are termed *graft copolymers*. Most commercial block copolymers have the glassy polystyrene (S) and rubbery polyisoprene (I) paired together, as reviewed by Bates [361]. Industrially significant graft copolymers are: high impact polystyrene (HIPS) and acrylonitrile-butadiene-styrene (ABS), both of which have rubber inclusions in a glassy matrix. Evaluation of blends and multilayer polymers in film form or by formation of films for model studies has been discussed (see Section 5.2) and will not be covered in any detail here.

Evaluation of bulk industrial material is best done by microtomy for OM, TEM, and SPM and by fracture of bulk molded or extruded samples for SEM and FESEM for determining microstructure. A brief literature review with examples of microscopy characterization of copolymers follows, but this review is not intended to reflect the thousands of studies and references on this important topic. Transmission electron microscopy is by far the most widely used characterization tool for the assessment of copolymers, and it has been used for several decades to uncover and provide understanding of copolymer microstructure.

The first example is using light microscopy to provide an overview of the crystalline structure and even the blend morphology. Semicrystalline multicomponent polymers can appear very confusing in polarized light as the spherulitic texture and the dispersed phase textures are superimposed and may not be distinguishable. Comparison of polarized light (Fig. 5.77A) and phase contrast (Fig.

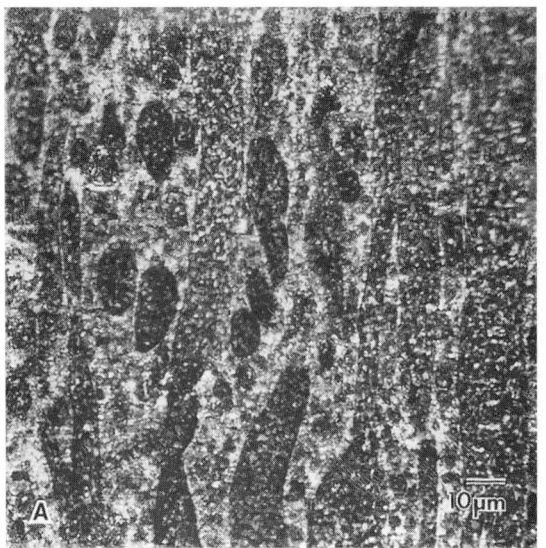
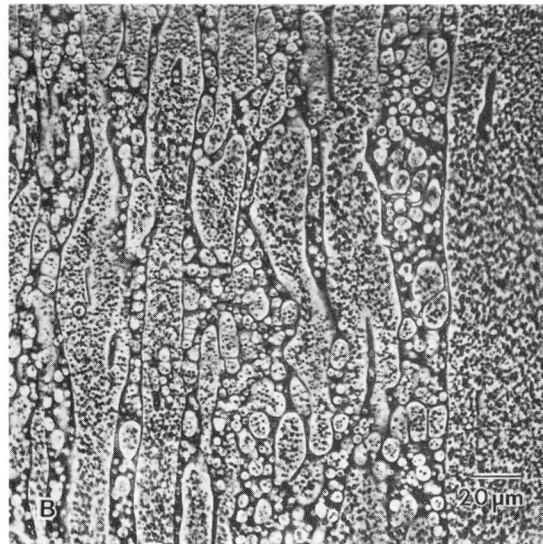

FIGURE 5.77. A polarized light micrograph (A) of a polyester-nylon copolymer shows bright and dark bands obscuring the spherulitic texture. Phase contrast optical microscopy (B) reveals the dispersed phase texture of the copolymer, which consists of multiple phases where the dispersed phase particles contain subinclusions of the matrix polymer.

5.77B) of a polyester-nylon copolymer shows that phase contrast more clearly reveals the nature of the dispersed domains in this complex microstructure.

The microstructure of the homopolymers should be examined for comparison with the multiphase polymer. Scanning electron microscopy of an Izod fracture surface of a POM/PP copolymer is shown in Fig. 5.78. The two phases are incompatible (i.e., they are present as two distinct phases). The dispersed phase particles range from less than 0.5 to 2 µm in diameter. The sample fracture path follows the particle matrix interface and holes remain where particles have pulled out of the matrix, showing there is little adhesion between the phases. The shape of dispersed phase particles is determined by the flow field and heat gradients that affect polymer orientation. For instance, the microstructure of copolymers of PE and PP is similar to the skin-core textures described for PE [362]. The orientation of the dispersed phase can affect the mechanical properties of the system. Spherical domains are more commonly formed in systems where phase separation occurs while the polymers are liquid. The SEM image appears to reveal spherical dispersed phase particles, although tilting can show they are actually elongated domains. As with typical fiber reinforcement (see Section 5.4.2), the length of the dispersed phase protruding from the matrix is an indicator

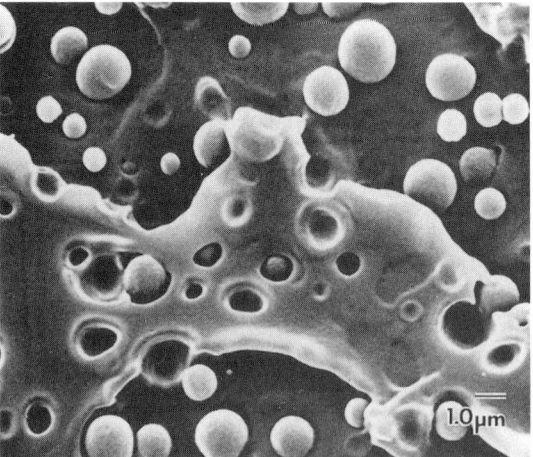

FIGURE 5.78. Dispersed phase particles are observed in a SEM image of a notched Izod fracture surface of a polyacetal/polypropylene copolymer. Dispersed phase particles about 0.5–2 µm across and pullouts, holes where particles were pulled out during fracture, are observed.

of the adhesion between that phase and the matrix.

Early morphological studies to determine the nature of multiphase polymers and blends were reviewed by Folkes and Keller [363]. Many studies were of extruded block copolymers of materials such as SBS where the dispersed phase, an unsaturated rubber stained with OsO_4 (see Section 4.4.2), was observed in the form of spheres, cylinders, or lamellae [364]. An excellent example is shown in a TEM micrograph of a thin section of a poly(styrene-butadiene) diblock copolymer, stained with OsO_4 [365], which depicts the (100) projection of a body centered cubic lattice (Fig. 5.79).

Toughening of a semicrystalline polymer by a phase segregated block copolymer introduces several levels of complexity, as shown in an example of a nylon 6,6 toughened with a Kraton G SEBS (Kraton Polymers US LLC) triblock copolymer in Fig. 5.80 [366]. Double staining was used to reveal the microstructure of both the nylon matrix with PTA and the styrene rich regions of the triblock polymer with RuO_4. There is selective absorption of PTA by the amorphous nylon phase so it becomes dense, and the individual crystalline lamellae are seen as meandering white regions in the matrix. Regions of Kraton up to 200 nm in size are seen to consist of styrene rich (black) cylinders in a matrix [366].

Adhikari et al. [367] also reviewed the morphology and micromechanical behavior of SBS block copolymer systems by using TEM and uniaxial tensile testing. Molded samples were ultramicrotomed to thin sections (ca. 70 nm thick) followed by staining the rubber phase with OsO_4. They found that the molecular

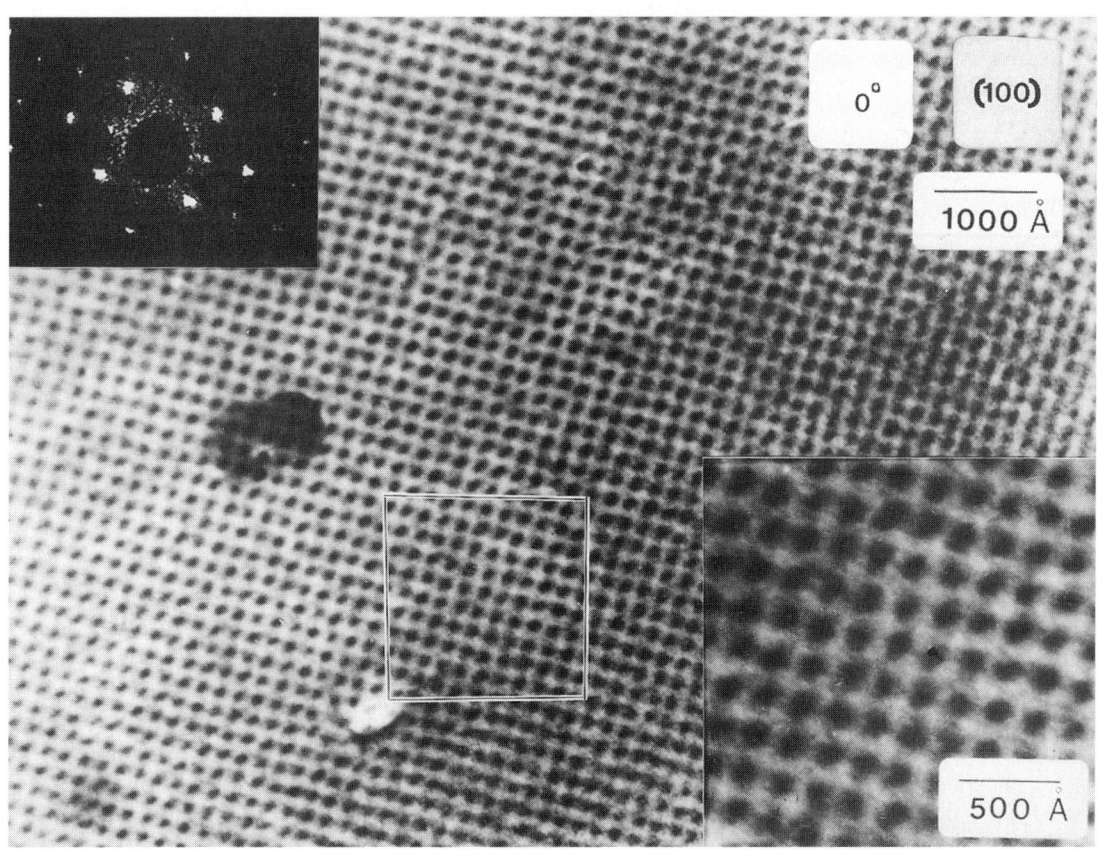

FIGURE 5.79. A transmission electron micrograph of an osmium tetroxide stained thin section of a poly(styrene butadiene) diblock copolymer (16.1 wt.% polybutadiene shows the (100) projection of a body centered cubic lattice. (From Kinning et al. [365]; used with permission.)

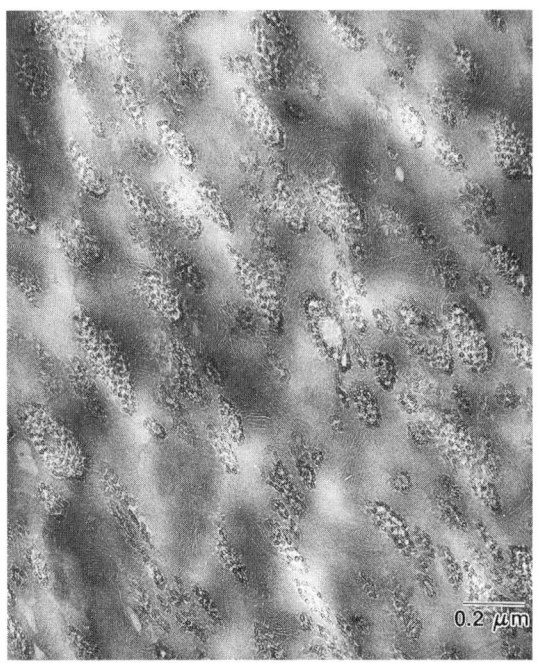

FIGURE 5.80. Transmission electron microscopy image of a thin section of nylon 6,6 with a thermoplastic elastomer, stained with RuO$_4$ and PTA. (From Wood [366]; used with permission.)

architecture influenced the morphology and properties. In another study, Adhikari et al. [368] showed a correlation between morphology and micromechanical deformation behavior of blends consisting of a lamellae forming linear SB block copolymer and PS homopolymer (hPS) by study of stained thin sections in TEM, fractured molded tensile bars by SEM, and FTIR. A change in morphology on addition of hPS suggested that the failure occurs at the interface between the added hPS and PS blocks of the block copolymer. Weng et al. [369, 370] studied the crystallization of a propylene-ethylene random copolymer by etching for 0.5 h in a permanganic reagent [371] (see Section 4.5.3) (1% w/v solution of potassium permanganate in an acid mixture of 2 vol.% of concentrated sulfuric acid and 1 vol.% of dried ortho-phosphoric acid) followed by making a two stage replica (see Section 4.6.2) and shadowing (Ta/W) carbon replicas for TEM.

The influence of the type and the content of cyclic monomers on the micromechanical deformation behavior of cyclic olefin copolymers (COC) of ethylene and different types of cyclic monomers was also studied by TEM [372] of ultrathin sections stained with RuO$_4$. HVEM at 1000 kV was used to study 1 µm semi-thin sections using a miniature tensile stage either under OM control or *in situ* inside the HVEM. Several deformation modes such as fibrillated crazes, homogeneous deformation zones, shear-bands, and combinations of some of these deformation structures were observed and correlated with the mechanical properties.

Atomic force microscopy is also used for the evaluation of multiphase polymer materials (e.g., [244, 248, 249, 373, 374]). Collin et al. [373] imaged specimen surfaces of copolymers prepared using a Mettler hot stage in a model study. The free surface images of samples prepared at varied annealing times were obtained using AFM in air in the constant force mode, which provided a better understanding of the formation of islands in the copolymer that was similar to previous optical microscopic observations but much more detailed and quantitative. Dikland et al. [374] used AFM to investigate the dispersion of low molecular weight compounds in ethylene-propylene copolymer rubbers. Light microscopy could not resolve the details, and the high vacuum in the electron microscope caused problems with volatility of the low molecular weight compounds.

Comparative studies of block copolymer morphology by TEM and/or SEM with AFM have been used to verify the structures observed. Schwark et al. [244], Annis et al. [248] and Vezie et al. [249] provided the first detailed, complementary studies of diblock copolymers using conventional TEM, low voltage, high resolution SEM, and AFM. Examples are shown in Fig. 5.81 (supplied by D. L. Vezie) of polystyrene-polybutadiene diblock copolymers, both of molecular weight 40,000 (designated SB 40/40). The bulk samples were cast from dilute solution in toluene over 7 days and then annealed at 115°C for 7 days. At this composition, the block copolymer phase separates into a lamellar structure with a 60 nm repeat with a corrugated surface structure. Samples were prepared by staining with OsO$_4$ (24 h), followed by cryoultramicrotomy perpendicular to the surface to

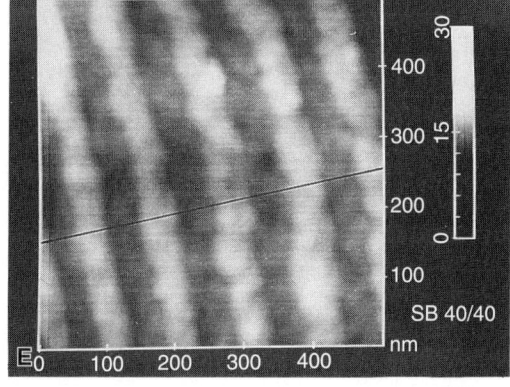

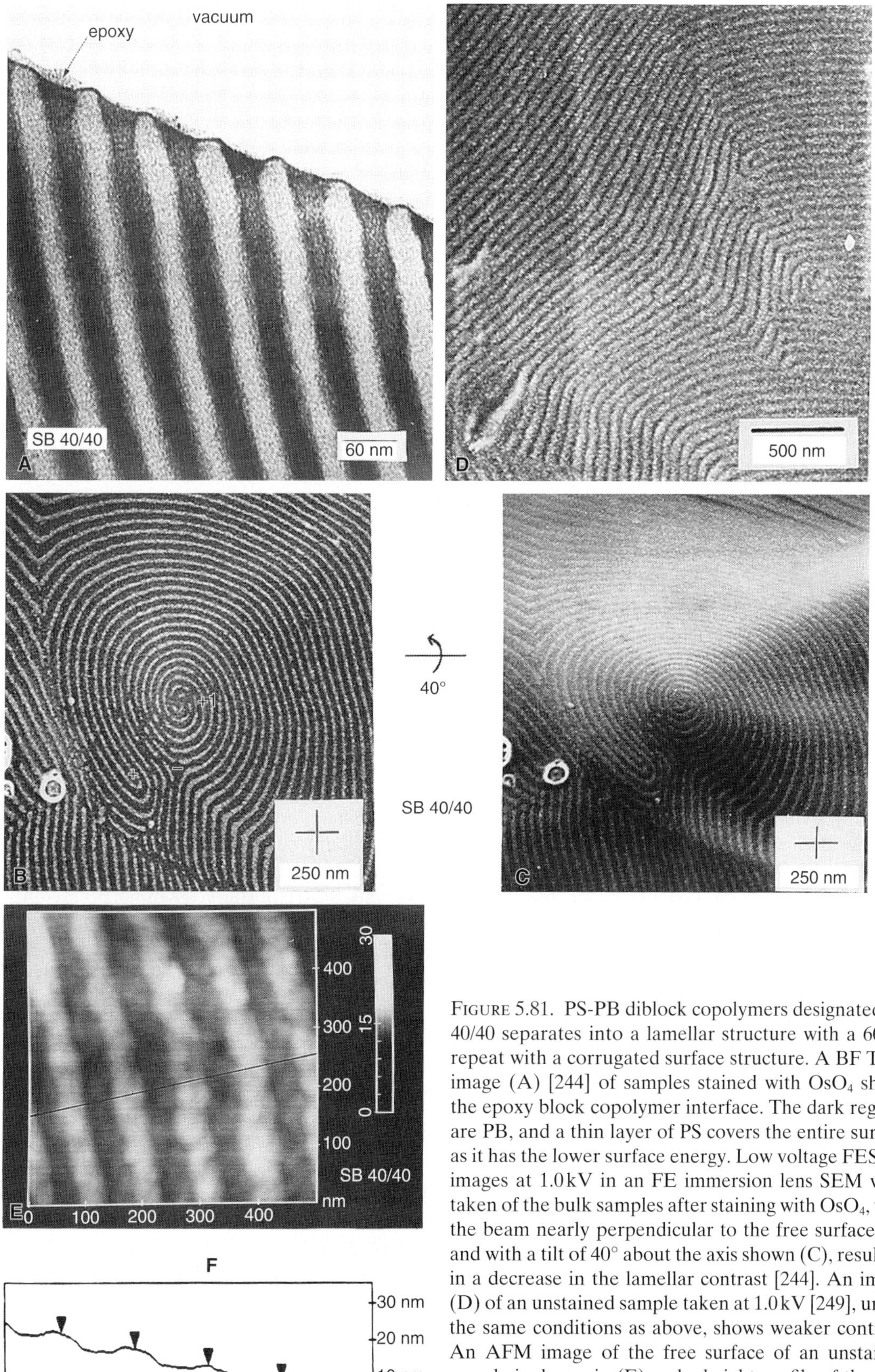

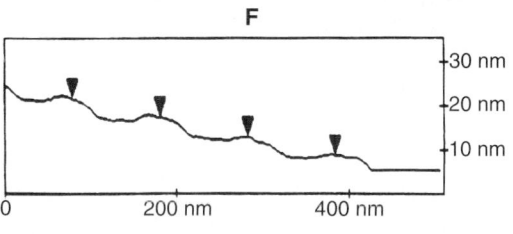

FIGURE 5.81. PS-PB diblock copolymers designated SB 40/40 separates into a lamellar structure with a 60 nm repeat with a corrugated surface structure. A BF TEM image (A) [244] of samples stained with OsO_4 shows the epoxy block copolymer interface. The dark regions are PB, and a thin layer of PS covers the entire surface as it has the lower surface energy. Low voltage FESEM images at 1.0 kV in an FE immersion lens SEM were taken of the bulk samples after staining with OsO_4, with the beam nearly perpendicular to the free surface (B) and with a tilt of 40° about the axis shown (C), resulting in a decrease in the lamellar contrast [244]. An image (D) of an unstained sample taken at 1.0 kV [249], under the same conditions as above, shows weaker contrast. An AFM image of the free surface of an unstained sample is shown in (E) and a height profile of the corrugated surface is shown in (F) [248]. (From Vezie et al. [244, 249]; used with permission.)

provide a cross section showing the epoxy block copolymer interface, as seen by bright field TEM (Fig. 5.81A) [244]. The image was taken at 200 kV at slight underfocus with the beam parallel to the lamellae. The dark regions are PB, and a thin layer of PB covers the entire surface as it has the lower surface energy. Figure 5.81B and Fig. 5.81C are low voltage FESEM images [244] taken of the bulk samples after staining and sputter coating with Au/Pd for 15 s. Images were taken at 1 kV in an FEG immersion lens SEM, in this context a high resolution SEM (HRSEM), with the beam nearly perpendicular to the free surface in Fig. 5.81B and with a tilt of 40° about the axis shown in Fig. 5.81C, resulting in a decrease in the lamellar contrast. Figure 5.81D is an image of an unstained sample taken at 1 kV [249], under the same conditions, showing contrast is present but is weaker without the stain. Figure 5.81E is an AFM image [248] of the free surface of an unstained sample, imaged using a silicon nitride pyramidal tip in repulsive mode with medium range forces. The height profile of the corrugated surface structure is shown in Fig. 5.81F [248]. The AFM is being used at comparatively low resolution here, so there are none of the difficulties of interpretation that can be true for molecular scale AFM images, and the surface profile can be accepted as accurate. The complementary nature of this outstanding study makes the results clearer and easier to understand.

Pfau et al. [254] noted that although the best characterization technique for the bulk morphology of multiphase polymer systems is TEM, AFM is also capable of mapping these systems with good resolution and contrast. This work reported on preparation techniques, the best of which was said to be cryomicrotomy to produce a flat block face for ICAFM study compared with TEM data. Studies included ABS, ASA, and PP blends. Ott et al. [375] also used TEM and SFM to verify structural features of a nanostructured bulk material such as the morphology in an ABC triblock copolymer with a "knitting pattern" morphology. Transmission electron microscopy preparation was by diamond knife ultramicrotomy followed by RuO_4 staining. Phase and height images by SFM were collected under ambient conditions on samples prepared by cryoultramicrotomy at −135°C with a diamond knife, as described earlier [358].

Characterization of the deformation processes in polymers *in situ*, by SEM, TEM, and SFM, used to understand the mechanisms of toughness in rubber modified amorphous polymers and butadiene-styrene block copolymers, has been reviewed [229, 376]. Scanning electron microscopy study can be done using special tensile devices and deformation tests (see Section 4.8) at temperatures from −150°C to +200°C. Samples of different thickness can be deformed in tensile devices or deformed samples examined directly by TEM or high voltage TEM, which enables study of thicker samples [377]. Radiation damage (see Section 3.4) is an issue with any EM technique, whereas SFM can be used for direct study of surfaces in air with no coating and no electron beam, and thus this technique is now being used especially using a deformation device for *in situ* study. Adhikari et al. [378, 379] studied cryomicrotomed (−120°C) thick films of different styrene-butadiene (SB) block copolymers with triblock architectures by ICAFM. The asymmetric SBS triblock copolymer (74 vol.%) forms, as expected, a cylindrical morphology with hexagonally packed PB cylinders in the PS matrix. Depending on the interfacial structure, block configuration, and the hard/soft phase ratio, other triblock copolymers (74 vol.% and 65 vol.%) show lamellae and randomly distributed PS cylinders in a random styrene-butadiene copolymer SB matrix, respectively. Michler et al. [377] used TEM and SFM to study asymmetric styrene-butadiene block copolymers and their blends with PS homopolymer and to explore the influence of phase morphology on the microindentation hardness and the nanomechanical deformation mechanisms using similar preparation methods. In contrast with polymer blends and random copolymers, in which microhardness generally is additive, the behavior of these block copolymer systems was not additive.

Thin films of homogeneous propylene-ethylene (P/E) copolymers of relatively narrow molecular weight distribution were studied as a function of constituent comonomer content by AFM and are mentioned here as an example of a model study for copolymers [116]. Samples were melted in an uncovered pan in a DSC, heated, and rapidly cooled to the desired melt temperature, held for 30 min, and then rapidly quenched in a dry ice/acetone bath. Specimens were then microtomed at −75°C and ICAFM performed to reveal the domain morphology.

Magonov and Yerina [253] reviewed AFM of a number of block copolymers. Phase ICAFM images on a film triblock copolymer of SBS immediately after spin casting (Fig. 5.82A) and after annealing (Fig. 5.82B) is compared with cross sections of a SBS rod cut perpendicular to the extrusion direction (Fig. 5.82C) and in a parallel cross section (Fig. 5.82D). The alternative lamellar structures of the PS glassy phase are bright compared with the darker rubbery PB blocks with the higher contrast observed after annealing. The extruded rod was annealed and the well ordered hexagonal pattern is seen in the image of the perpendicular cross section, whereas an extended liner structure is observed parallel to the flow.

High temperature *in situ* AFM is a powerful tool to study structure formation and melting behavior of propylene/ethylene copolymers at nanometer resolution as shown in the following

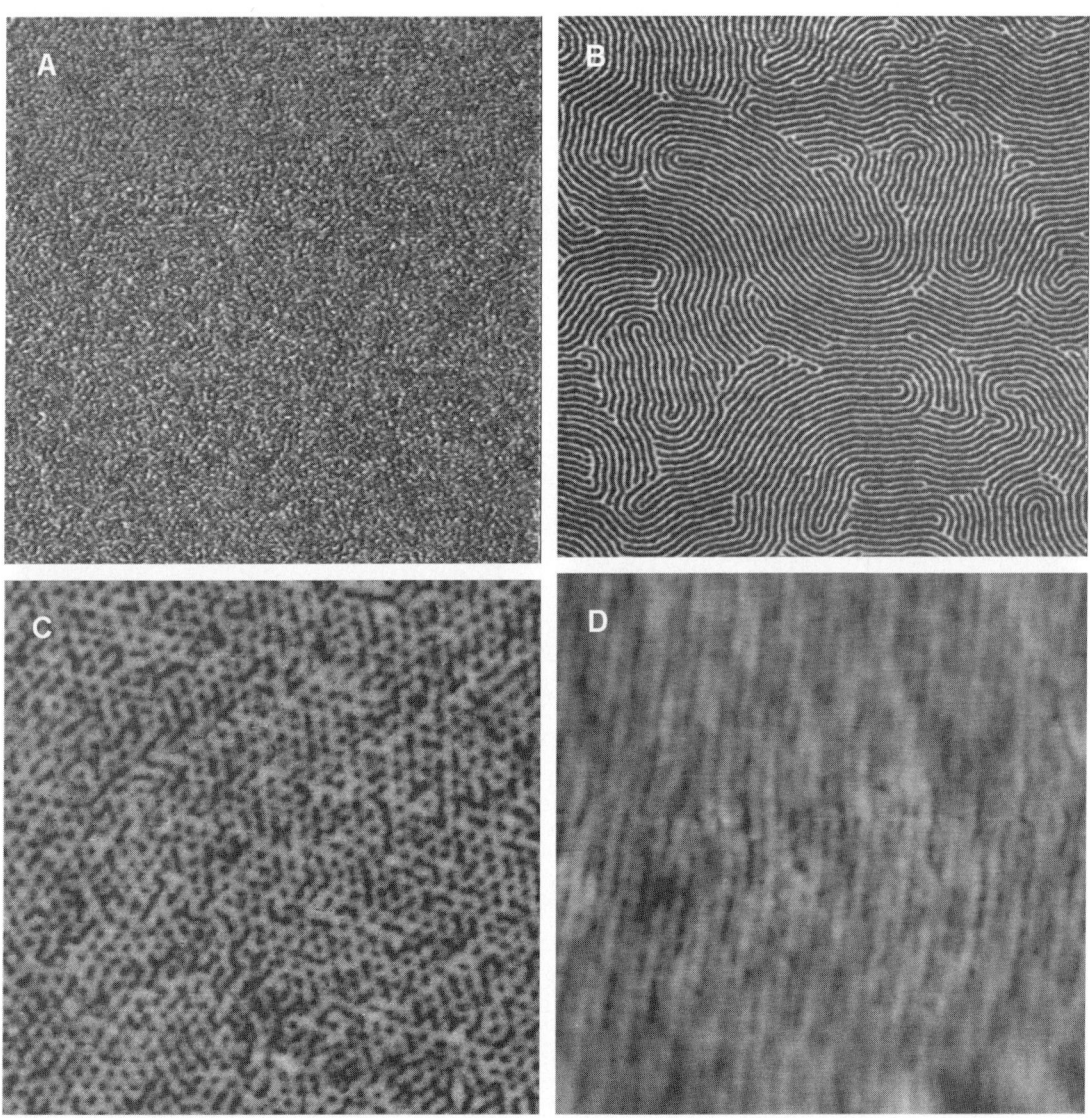

FIGURE 5.82. Phase images of SBS film just after spin casting (A) and after high temperature annealing (B) (2 μm on a side). Phase images of cross sections of an annealed SBS rod, made perpendicular (C) and parallel (D) to the flow direction (1 μm on a side). (From Magonov and Yerina [253], © (2005) Springer; used with permission.)

study by Bar and Meyers [170]. The phase image recorded at 90°C (Fig. 5.83A) reveals the existence of several different structures described as a cross-hatched web composed of long radial and transverse lamellae with other regions having smaller, less branched crystallites. Structural changes occur above 90°C (Fig. 5.83B) as the number of small crystallites decrease and dark "lakes" of molten polymer emerge. With time and increasing temperature, these molten regions grow and merge (Fig. 5.83C, D) with the cross-hatch web the last to melt upon further heating. Hot stage AFM allows monitoring of thermal transitions in much the way hot stage optical microscopy is done but at significantly higher resolution.

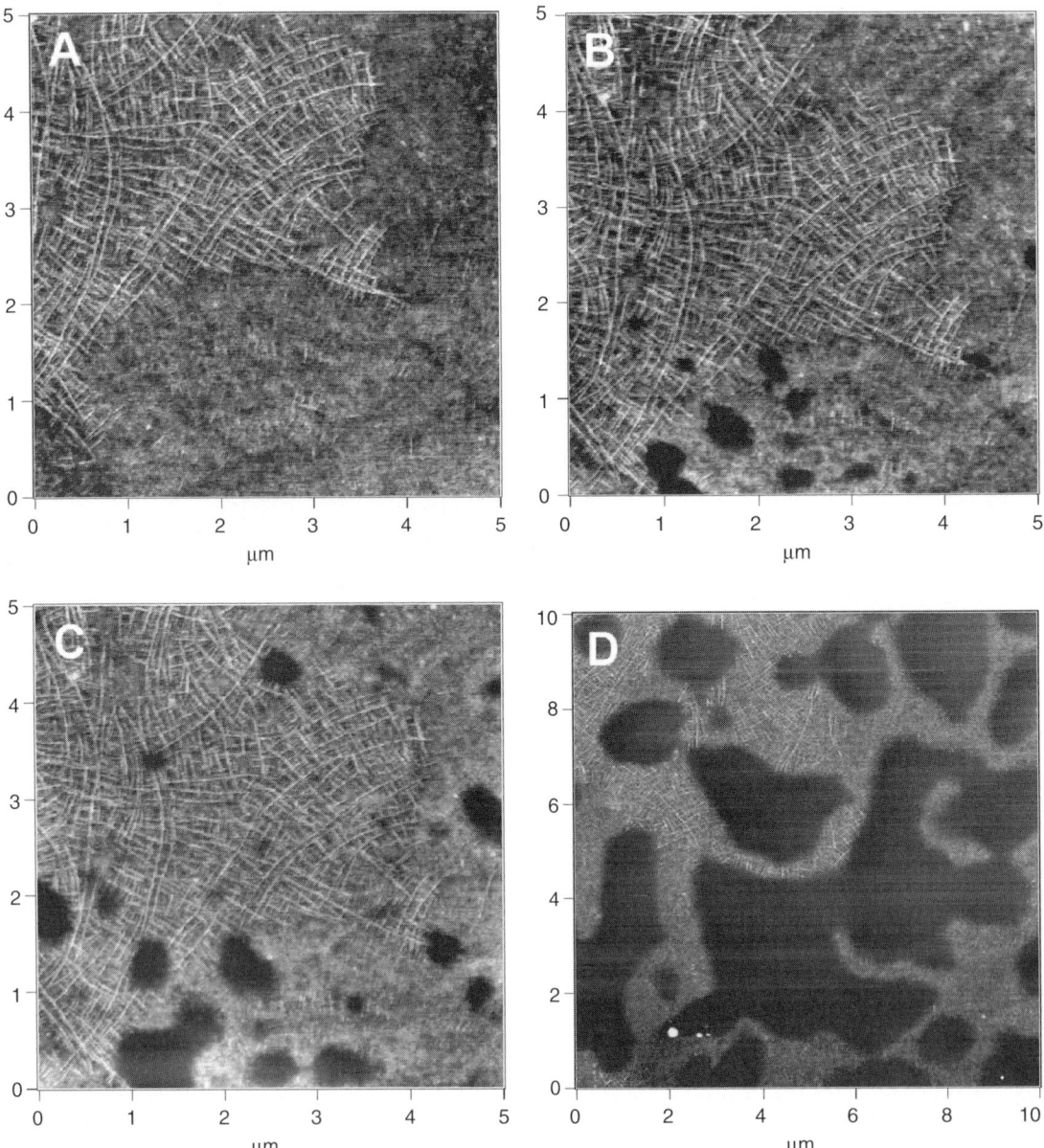

FIGURE 5.83. Atomic force microscopy phase images of a propylene/ethylene copolymer recorded at elevated temperatures of (A) 90°C, (B and C) 92.5°C, and (D) 100°C. In all images, the gray contrast covers phase signal variation of 20°. (From Bar and Meyers [170]; used with permission of the *MRS Bulletin*.)

Block copolymers have more recently been considered for nanotechnology applications stemming from the scale of the microdomains and the convenient tunability of size, shape, and periodicity; although most of the microscopy is done using thin films, the bulk state is also relevant for this topic as reviewed by Park et al. [380, 381]. The many applications of nanotechnology include membranes, photonic crystals, and high density information storage media, among many others. Transmission electron microscopy, SEM, and AFM techniques find applicability in studies of these copolymer structures using preparation methods already discussed.

High resolution high angle annular dark field (HAADF) STEM is a technique that is worth discussing to assess the nature of copolymer structure as has been done in work on ionomers [382]. Figure 5.84 shows a high magnification (600 kX) HAADF STEM image of CdSe nanoparticles present in a PS/PMMA block copolymer matrix while the lower image (60 kX) depicts the lamellar morphology of the block copolymer itself [383].

5.3.4.7 Polyurethanes

Multiphase polymers containing polyurethane are common toughened polymers in which it has long been known that phase separation is important in determining structure-property relationships [384]. In some cases, as with a polyurethane modified polyester, a semi-interpenetrating network is formed, as seen in TEM images of OsO_4 stained sections [385]. Cryosectioning has been used to study the morphology produced by a polymer process, known as reaction injection molding (RIM), in polyester-based polyurethanes [386]. The large temperature gradients in the mold were shown by Fridman et al. [386] to influence both molecular weight and morphology, and higher polymerization temperature resulted in better hard segment organization. A review of the structure of segmented polyurethanes has shown incompatibility to be a key factor in determining morphology. Chen et al. [387] used optical microscopy with a hot stage and video camera, and electron microscopy of cryosectioned specimens, to assess this morphology. Three crystal forms have been identified by complementary optical and TEM study of solution cast model films of polyurethane containing hard and soft segments [388].

Often, polymers such as polyurethanes do not fracture at room temperature, or the fracture is ductile, and both impact testing and specimen preparation must be conducted below room temperature. Additionally, etchants may be required in order to bring out the dispersed phase morphology. Demma et al. [389] studied the morphology and properties of blends of polyester-based thermoplastic polyurethanes with ABS, PS, SAN copolymer, and with an ASA (acrylonitrile-styrene-acrylic ester) terpolymer. Specimens did not break when Izod impact testing was performed at room temperature, and thus lower temperature was required. For SEM study, specimens were immersed in liquid nitrogen for 10 min and then fractured. Fractured ABS specimens did not reveal any particles. Chemical etching of the ABS blends was used in an attempt to show the size distribution of the phases. Etching with methyl ethyl ketone (4 h) at room temperature dissolved the SAN copolymers in ABS [390]. THF vapor treatment (1 h) was said to be a solvent for the thermoplastic polyurethane [391]. Micrographs of fractured and etched (methyl ethyl ketone) polyurethane blends revealed holes where the SAN copolymer was located. However, a major problem with etching preparation methods, as

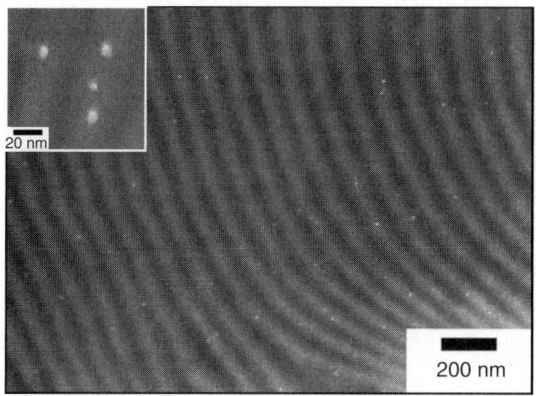

FIGURE 5.84. High magnification inset (600 kX) HAADF STEM image of CdSe nanoparticles present in a PS/PMMA block copolymer matrix; the lower image (60 kX) depicts the lamellar morphology of the block copolymer itself. (From Winey [383]; unpublished.)

noted by us and by these authors [389], is that often the etchant has adverse effects on the matrix as well as the dispersed phase.

Scanning electron microscopy was also used to examine the biocompatibility and biostability of modified polyurethanes used as catheters, gastric balloons, and prostheses [392]. However, oxidation needs to be reduced for these devices to be used *in vivo*. Modifications or substitutions of the soft segment have been studied to enhance biostability and evaluated by SEM and ATR-FTIR. Segmented copolymers contain hard blocks that form domains with high glass transitions or melting points and act as physical crosslinks with some of the most important properties associated with the rearrangement and deformation of the hard segment domains. AFM and SAXS were used to study segmented thermoplastic elastomers containing poly(tetramethylene oxide) soft segments and hard blocks of nylon 12; AFM was used to provide real-space resolution of the morphology during tensile elongation and after subsequent relaxation [393]. Samples were solvent cast onto glass slides from solution and films 10–50 µm thick were stretched for study by AFM under ambient conditions. The results were combined with DSC and birefringence data taken on films under strain to obtain insight into the microscopic basis for strain softening and plastic deformation in these segmented polymers.

A combination of TEM and *in situ* tensile testing in an environmental SEM was used to evaluate the static bulk and dynamic surface morphologies of medical polyurethanes [394]. Transmission electron microscopy results showed phase separated hard segment and soft segment structures. Surface morphology as a function of strain was studied using ESEM in conjunction with a tensometer. Samples were prepared by casting thin films from pellets using a hot press, quenching the films in a cold water bath, and then annealing them at elevated temperatures. Samples were stained with OsO_4 (stained the PS) and RuO_4 to distinguish soft segments from hard segments and then cryo-sectioned for TEM.

Scanning electron microscopy micrographs of a polyacetal/polyurethane multiphase polymer are shown in Fig. 5.85. The outer surface and a fractured internal surface of this extrudate were chemically etched in order to determine the nature of the dispersed phase. The surface (Fig. 5.85A) shows a complex structure due to etching. The fracture surface after solvent extraction (Fig. 5.85B) is complex, as it appears that both the matrix and the dispersed phase particles have been affected by the etchant. When using etching, other methods should also be used to confirm the nature of the microstructure. Combination of SEM and AFM has also been used to study various plastics applications, including a thermoplastic olefin (TPO) of 75 wt.% iPP and 25 wt.% EPR [395]. Etching with *n*-heptane at 60°C was done to remove the rubbery EPR domains from the crystalline PP matrix. The SEM image (Fig. 5.86A), taken

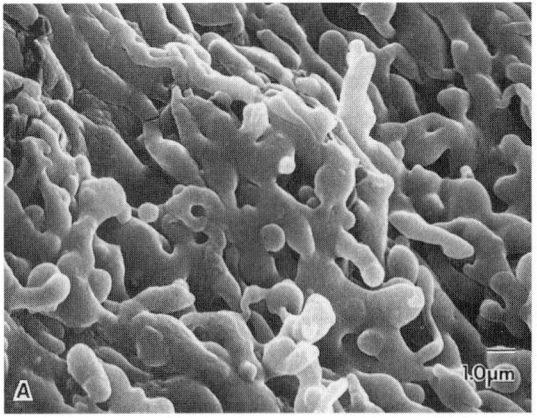 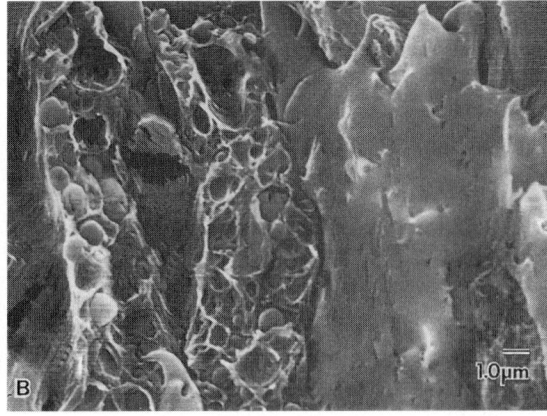

FIGURE 5.85. Scanning electron microscopy images of liquid nitrogen fractured polyacetal-polyurethane blend show a complex network morphology (A) made more complicated by chemical etching (B). Scanning electron microscopy of the etched fracture surface (B) suggests that the etchant has affected both the dispersed phase and the matrix.

Engineering Resins and Plastics

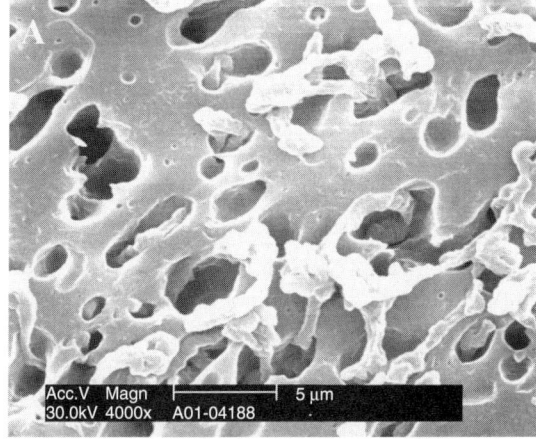

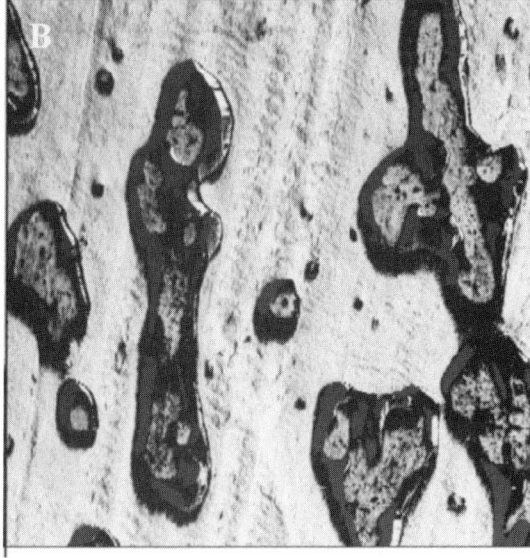

FIGURE 5.86. A compression molded TPO is imaged after etching with *n*-heptane by SEM (A) and after microtomy to produce a flat surface by AFM (B). Detailed internal structure is more clearly imaged by AFM than SEM due to the negative effect of the etching process. (From Mirabella and Weiskellel [395], © (2005) Taylor & Francis; used with permission.)

after etching and Au coating, shows the domain structure cannot be fully determined as there is incomplete etching and residue of the etched polymer left on the surface. White areas around the depressions are due to the etching leaving residues that charge in the electron beam, and because the rubber domains were etched away, no detailed structures within these domains can be seen. An AFM phase image (Fig. 5.86B) was produced from the same TPO compression molded sample after microtomy to produce a flat sample. More details of the soft domains are observed, including subinclusions within the rubber phase not observed by SEM.

5.3.4.8 Biodegradable Polymers

In recent years, there has been an increasing interest in the development of degradable polymers [396–400]. Much of this interest has been driven by increasing volumes and visibility of plastic waste from commodity and packaging products. Methods to reduce the volume of waste include reduction of use, increased recycling of polymers, and the development of biodegradable materials. Starch based blends can be used as a biodegradable alternative in disposable packaging, cups, composting bags, and so forth. Biodegradable materials have another use in biomedical products, such as degradable sutures that do not need to be removed, screws and pins to repair broken bones that degrade as the bone heals, and devices for the controlled release of drugs into tissue [397]. The definition of "biodegradable" is in dispute, but fully degradable materials and materials that quickly break down into more stable, small fragments can be distinguished. The latter would remove the visibility of plastic waste but would not be suitable for medical devices described here.

The fully degradable materials that are of interest include polyesters that are sensitive to hydrolysis. This is the reverse of the condensation polymerization reaction, and these materials will depolymerize in the presence of water back to oligomers that can be digested by microbes. Biopol (trademark of Metabolix, Inc., Cambridge MA) is a material of this type, a copolyester of poly 3-hydroxybutyrate (PHB) and poly 3-hydroxyvalerate (PHV), and is produced by bacterial fermentation and will slowly degrade and is compostible [398]. This company develops PHAs (polyhydroxyalkanoates; chemically related to polyesters). These polyesters, because of their biological preparation, are much purer than the average commercial polymer. The homopolymers are very highly crystalline and have very large spherulites, and even samples for optical microscopy may contain no effective nuclei [399]. The commercial product is a copolymer because it has reduced crystallinity and spherulite size and therefore better mechanical properties. Other polyesters used for biomedical applications are poly(lactic acid), poly(glycolic acid), and poly(p-dioxanone). These generally degrade faster; degradable sutures, for example, must be packed individually in vacuum sealed foil and used immediately on opening.

Starches, a class of natural polysaccharides, have been modified and used combined with thermoplastic materials to form blends that degrade into small, stable particles. These blends can often be processed by standard melt forming polymer techniques, by compounding and extrusion into films and fibers, and injection or blow molded. Accordingly, process and material variables that affect multiphase polymers affect these degradable blends. Naturally, blends can also be designed that are fully biodegradable, for example by blending poly(hydroxybutyrate)-based materials with natural poly(saccharides) and synthetic poly(caprolactone) polymers, as described by Yasin and Tighe [400].

Irradiation-modification of blends of various starches with a synthetic polymer, poly(ethylene-co-vinyl alcohol) (EVOH), was carried out using an electron beam [401]. The effect of irradiation on the starch blend filaments was evaluated by SEM after freeze fracturing under liquid nitrogen. The irradiation was said to increase the beam damage of the starch but not the EVOH under regular SEM conditions, and thus lower beam energy and higher chamber pressures were preferred. Enzymatic etching was used to prepare the cross sections for study and remove the starch. Transmission electron microscopy has also been used to examine the morphology of extruded thermoplastic starch/EVOH blends [402]. The blend composition is tailored to provide the properties needed for thermoplastic processing by injection molding, fiber spinning, or making blown film. Simmons and Thomas [402] reviewed this topic and details on a method for evaluation by TEM. Samples were cryoultramicrotomed at a sample temperature of −20°C to 15°C and a knife temperature from −20°C to 0°C, with temperatures decreasing as the EVOH level increased, using a 35° diamond knife to prepare sections 30–50 nm thick. Sections were collected with an eyelash tool and deposited onto a droplet of ethanol on copper grids. No water is used during this entire process to eliminate swelling of the starch. Sections were stained with elemental iodine vapor for 1 h to stain the amorphous starch and carbon coated. These authors discussed the imaging issues with these materials that warrant review prior to conducting such studies. Issues include beam sensitivity, contrast mechanisms for the starch versus EVOH, and contrast changes as a function of electron beam exposure. Low dose techniques were used with conventional TEM to reliably image the blends.

Blends of thermoplastic starch and natural rubber were prepared using natural latex and cornstarch [403], fractured in liquid nitrogen, vacuum coated with a thick layer of gold (20 nm), and studied by SEM to reveal the dispersed phase morphology. Recycled LDPE/cornstarch blends were extruded and characterized by SEM [404]. Atomic force microscopy has also been used to image the internal structure of starch granules [405]. The starch was embedded in melamine or Araldite resin after drying and then sectioned 1.5 μm thick onto water using glass knives. Sections were stained with very dilute iodine in potassium iodide for light microscopy and directly for AFM. The rapid setting Araldite, which is nonpenetrating, allowed the ultrastructure to be viewed better than melamine resin, which obscured the detail.

The change in the structure as a result of the hydration or degradation process is of major interest for biodegradable materials, therefore examination of the material both before and after hydration or degradation process is appropriate. Dynamic imaging techniques such as environmental or high pressure SEM (HPSEM) to image wet specimens during water induced swelling and degradation would seem to be the preferred technique for such studies. Video taping images of dynamic experiments has value for understanding the failure of biodegradable systems and understanding the mechanisms of degradation. An example of the effect

Engineering Resins and Plastics

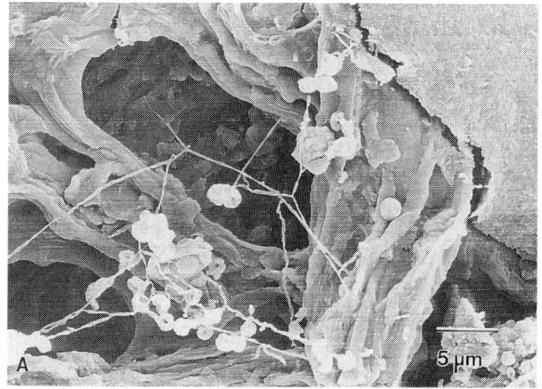

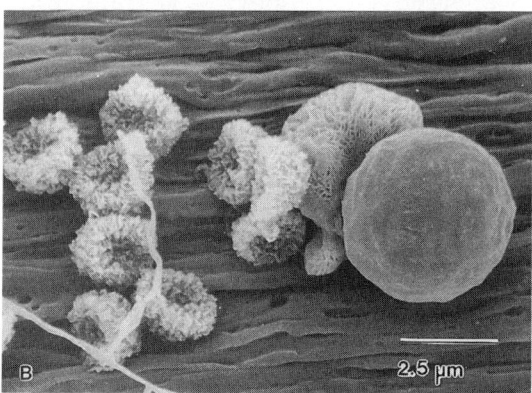

FIGURE 5.87. The HPSEM images of compost surfaces that were evaporated with chromium were taken at 10 kV. An HPSEM image (A) of the film has a porous surface due to the removal of one of the blend components during the composting stage. Microorganisms and white rot fungus remain on the moderately hydrophilic surface. A porous extruded film surface (B) in more detail has pores due to composting; robust microorganisms and yeast colonies adhere to the hydrophilic film surface. (From Loomis et al., unpublished [406].)

of degradation on a starch based polymer blend film imaged in an HPSEM is shown in Fig. 5.87 [406]. The biocomponent films were prepared and then slightly oxidized using a plasma etcher and exposed to a thermophilic compost for 60 days. Upon completion of the compost cycle, the samples were rinsed with isopropanol to remove loose debris and dried at room temperature. The HPSEM images were all taken at 10 kV of surfaces that were evaporated with chromium. Figure 5.87A is an HPSEM image of the film, which has a porous surface due to the removal of one of the blend components during the composting stage. Microorganisms and white rot fungus remain on the moderately hydrophilic surface. Figure 5.87B shows the porous extruded film surface in more detail, again with pores due to the removal of one of the blend components during composting; robust microorganisms and yeast colonies adhere to the hydrophilic film surface. Although the HPSEM is not a technique for high resolution imaging, it does allow imaging of wet materials and dynamic experiments.

5.3.5 Failure or Competitive Analysis

A major objective of microstructural analyses on multiphase and other polymers is failure analysis, that is, determination of the mode or cause of failure, especially of a product in use in the marketplace. Failure analysis generally involves characterization of a material that has failed, either in service or in a physical test. Controls are not always available, and timing is often critical. In some cases, the types of analysis required may well be similar to those described above: phase contrast light microscopy, PLM, SEM, TEM, and AFM. Other microstructural techniques that are valuable in solving such materials problems are chemical contrast imaging and elemental x-ray mapping.

Plastic pipes made from PVC and high density PE are used in many applications, including pipes for water and gas transport where brittle failure obviously limits their use. A method for inducing brittle failure for testing was developed as a plane strain fracture toughness test [407, 408]. The test involved notching the specimen and using a razor blade induced fracture to replace fatigue cycling tests. The fracture toughness measured was a function of the resistance to brittle failure in a sharp crack region. Microscopy of the fracture zone was used to characterize the nature of the fracture surface.

A toughened nylon with low impact strength is seen in the SEM images in Fig. 5.88 with a brittle surface exhibiting a classical flaw-mirror-mist-hackle fracture pattern. At higher magnification, the fracture is shown to have propagated from a defect site and has formed fracture ridges where the fracture accelerated (Fig. 5.88B). An enlarged view of the defect site (Fig. 5.88C)

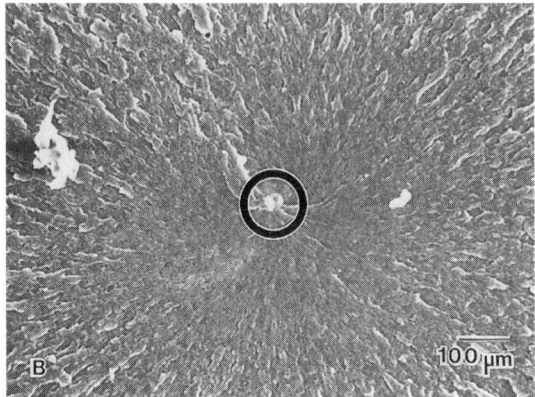

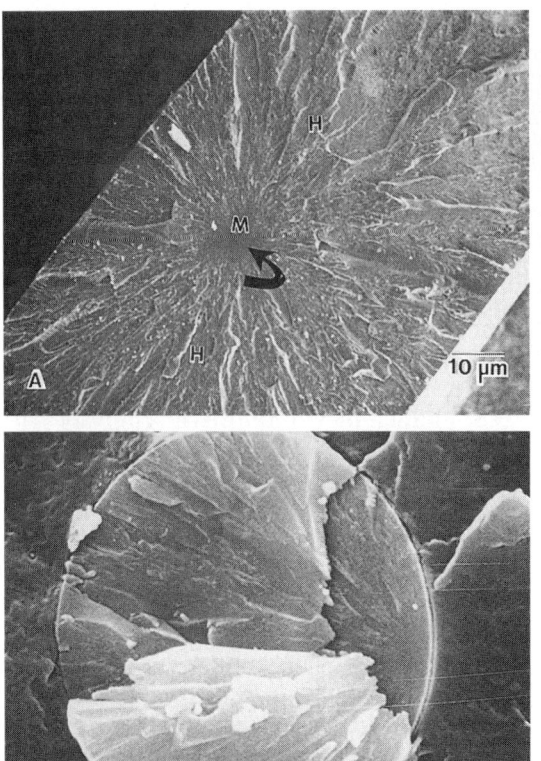

FIGURE 5.88. Scanning electron microscopy study of brittle failure in an impact modified nylon molded article reveals classic fracture morphology. The locus of failure (arrow) is seen (A) with surrounding mirror (*M*), mist and ridged hackle (*H*) regions propagating out into the bar. A higher magnification view of the flaw and mirror is shown (B). The flaw (C) appears to be a round fiber, likely a contaminant.

shows it to be a round contaminant fiber, which itself exhibits brittle failure.

X-ray microanalysis and BEI provide information regarding the elemental composition of dispersed phase particles and also of the contaminants responsible for failure. These techniques are useful for the study of multiphase polymers, if one of the phases contains an element not contained by the other. For example, Price et al. [409] used SEM/EDS to study the interface between two polymers, one of which was PVC, to measure the local composition during interdiffusion. Failure of elastomer compounds used in tank track pads, natural rubber, SBR, EPDM, and 50/50 SBR/BR was studied as a function of elastomer cure and degradation in service [410]. Andrade et al. [411] used BEI of osmium tetroxide stained PS/PB to provide images that showed the dispersed and stained PB as a result of atomic number contrast. Hobbs and Watkins [412] used the inherent differences in atomic number of multiphase polymers for imaging by SEM and BEI. They stained polymers of low inherent chemical contrast with either osmium tetroxide or 15% bromine in methanol (30 s).

The combination of SEM, BEI, and elemental mapping provides identification of the chemical composition and distribution of elements within polymers that is quite useful for failure analysis. For example, SEM of a multiphase polymer (Fig. 5.89A) shows a fractured polymer surface containing large dispersed phase particles, about $10\,\mu m$ across, and submicrometer sized particles as well as large holes. The BEI image (Fig. 5.89B) shows that both the large and the smaller particles are brighter than the background polymer, and thus they are composed of elements of higher atomic number than the polymer. Elemental mapping shows that the small particles, at and below the surface, contain antimony (Fig. 5.89C), whereas the larger particles contain chlorine (Fig. 5.89D). The subsurface particles are not observed in SEM but *are* sampled in the x-ray detection volume. Care must be taken in elemental mapping and x-ray analysis, in general, to allow for the fact that the detection volume for x-rays is always greater than that for scattered electrons, and the actual detection volume is a function of both the atomic number of the matrix and the accelerating voltage of the electrons.

Another use of microscopy is the study of competitive materials, which is similar to doing failure analysis as the full nature of the material is not known. In this example, the question was to learn the nature of the process used to make a dual ovenable food tray. Transmission electron microscopy cross sections, stained with RuO$_4$ vapor (Fig. 5.90), revealed a bilayer construction with neat PET (left side) at the shiny surface and submicrometer modifier particles with a sharp interface outlined by the stain formed by coextruding a multilayer construction [52]. A skin ailment in a molded auto resin part, made of Noryl GTX (trademark of GE Plastics), an inherently conductive blend of PA and PPE designed for electrostatic painting, is shown in Fig. 5.91. The sample was sectioned perpendicular to the melt flow direction using cryoultramicrotomy with a diamond knife, and sections were accumulated in cold ethanol, transferred to a water surface, and placed on copper grids for TEM at 200kV [52]. The domain structure is fairly uniform in the center of the part (Fig. 5.91A), but coarser domains have been pulled apart at the surface (Fig. 5.91B).

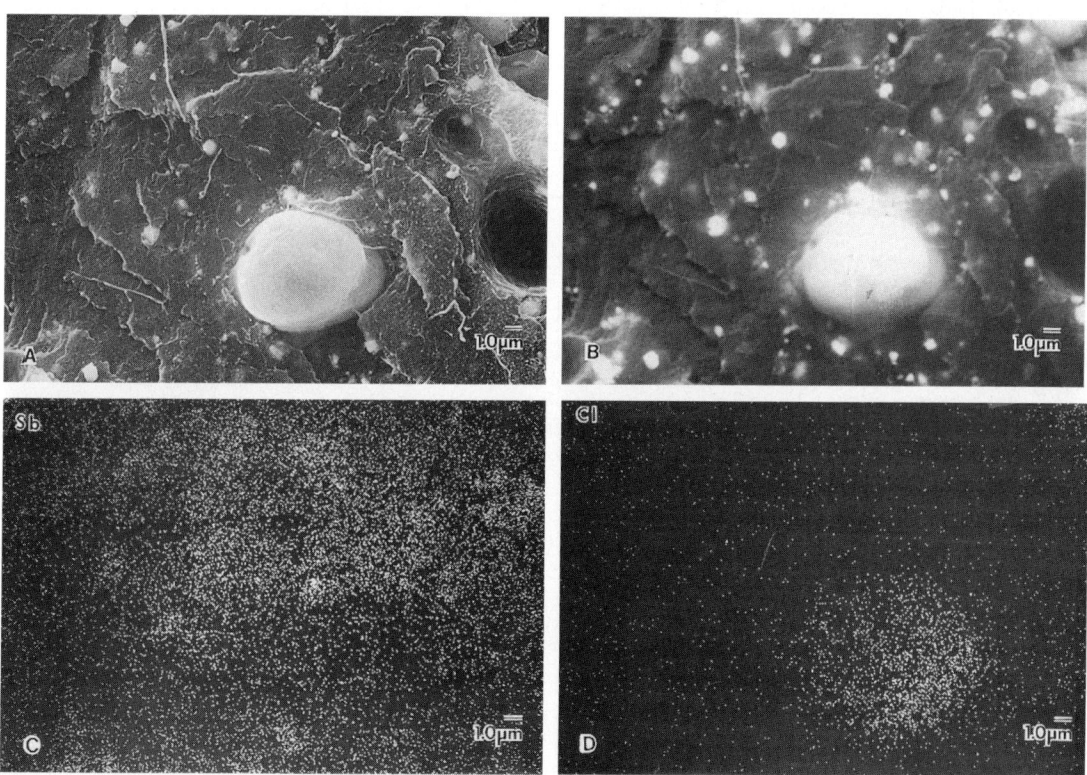

FIGURE 5.89. Scanning electron microscopy of a multiphase polymer (A) shows large dispersed phase particles and submicrometer sized particles and holes. BEI (B) shows that the dispersed phase particles have higher atomic number than the matrix. Elemental mapping shows that the small particles contain antimony (C), whereas the larger particles contain chlorine (D). Once the specific size/shape of the particles are identified by mapping, BEI imaging can be used to study the specimen surface.

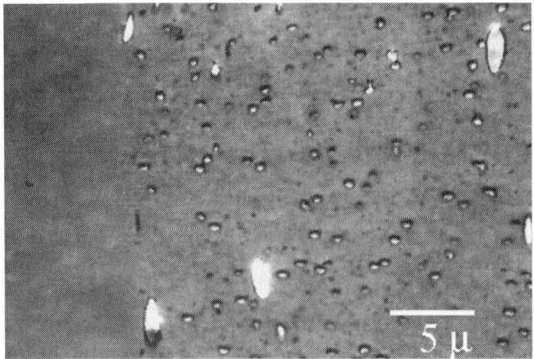

FIGURE 5.90. Transmission electron microscopy cross section of a dual ovenable food tray reveals a bilayer construction with neat PET (left side) at the shiny surface and submicrometer modifier particles with a sharp interface outlined by the stain formed by coextruding a multilayer construction. (From Wood [52]; unpublished.)

Atomic force microscopy techniques including scanning thermal microscopy have been used in research and failure analysis, as reviewed by Bar and Meyers [170]. The insulation material for high tension cables can be composed of a mixture of a cross-linkable long fiber PE (LFPE) with ethylene-butylacrylate (EBA). Extruded defects or lumps on the surface of the outside cladding were observed between 50 and 600 μm in size; these defects were apparently traced to the insulation layer below the outside cladding material. The defects were excised from the cable insulation and thick sections cryomicrotomed for identification, as shown in the reflected light images in Fig. 5.92A, B; the defect has a different texture than the matrix. Intermittent contact mode AFM phase images indicated that the EBA was well dispersed in the crosslinked PE matrix as 50 to 100 nm phases (Fig. 5.92C), but the defect region was rougher in texture and devoid of discrete EBA phases (Fig. 5.92D). Scanning thermal microscopy in the local thermal analysis (LTA) mode was used to determine the melting behavior of the defect and matrix (see regions marked in Fig. 5.92B). The final figure, Fig. 5.92E, shows the LTA on the gel defects to have a different melting temperature than the matrix, which was also confirmed by hot stage optical microscopy [170].

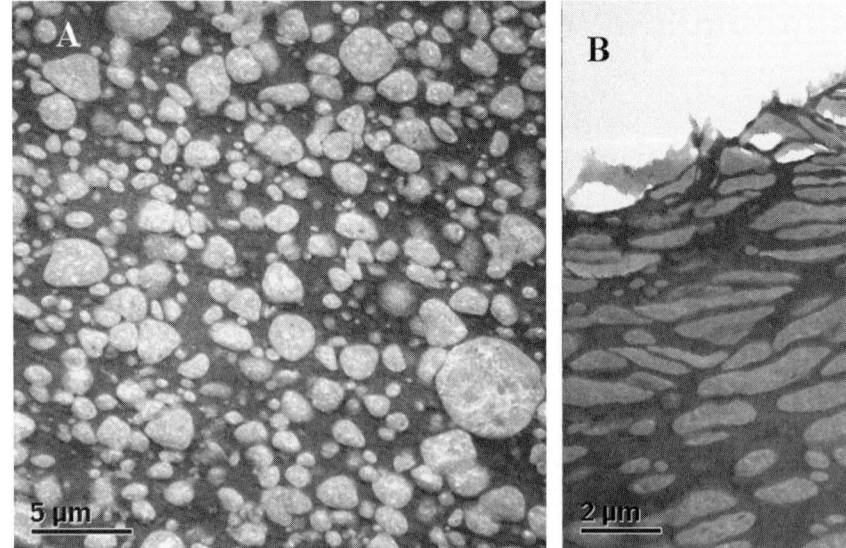

FIGURE 5.91. Transmission electron microscopy cryoultrathin cross sections, cut perpendicular to the melt flow direction and stained with RuO$_4$ vapor, revealed a skin ailment in a molded auto resin part, made with Noryl GTX (A) compared with the same blend near the skin surface (B). (From Wood [52]; unpublished.)

Engineering Resins and Plastics

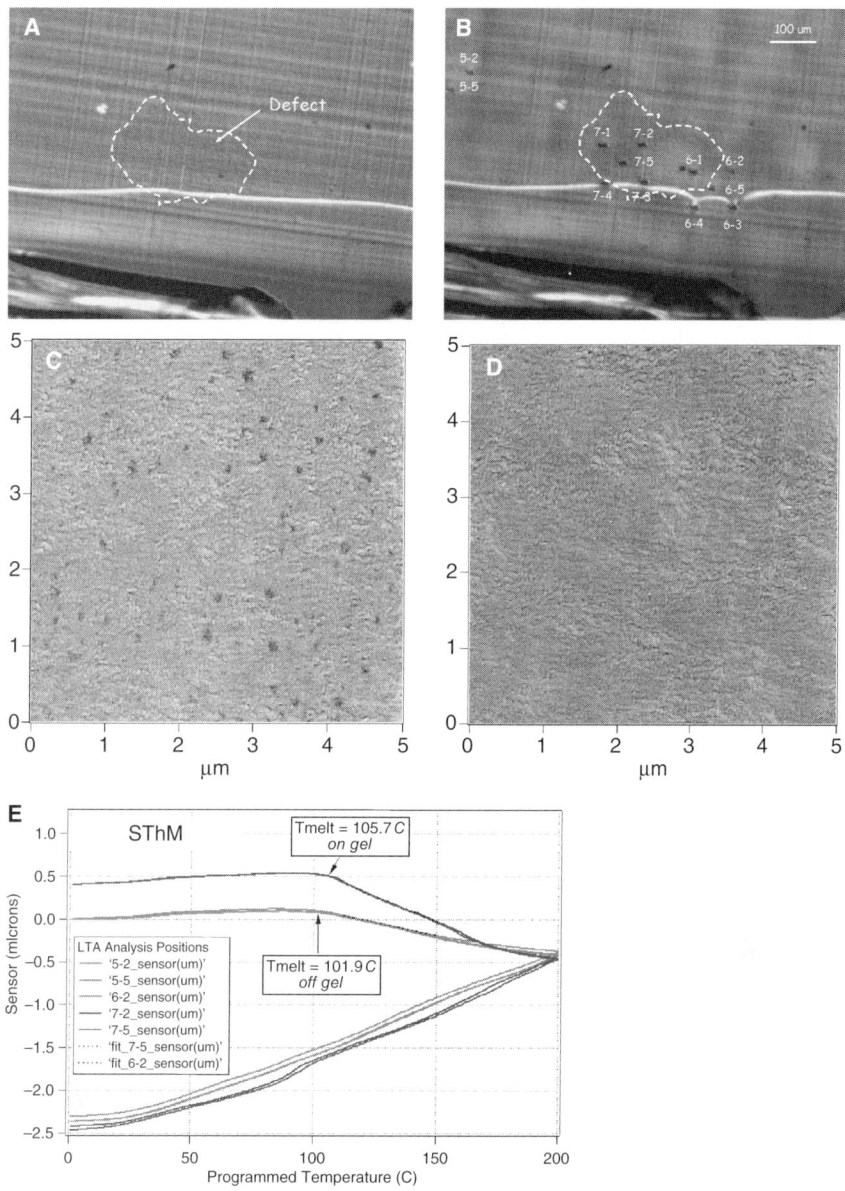

FIGURE 5.92. Reflected light image of a defect in cable insulation. The defect appears rougher in texture than the surrounding matrix (dotted outline) (A). Reflected light image of the thick section shown in (A) following three local thermal analysis cluster measurements using SThM (B). Intermittent contact mode AFM phase images of (C) the cross linked PE/EBA matrix and (D) the defect shown in (A). The defect is devoid of discrete EBA phases. The thermomechanical response of the heated probe in contact with the defect (solid curves) and matrix (dashed curves of [A]) are seen to melt at different temperatures (E). (From Bar and Meyers [170]; used with permission of the *MRS Bulletin*.)

5.4 COMPOSITES

5.4.1 Introduction

Polymers containing rigid fillers, both fibrous and particulate, are termed *composites*. Composites generally contain short or continuous fibers or nonfibrous particles, usually minerals. Fibers used in composites include both inorganic fibers, such as glass, carbon, graphite, ceramic and organic fibers, such as aramids, and naturally occurring fibers (e.g., flax and wood). Particles used in composites are generally minerals, such as mica, clay, talc, wollastonite, calcium carbonate, and carbon black. Polymer composites have seen a major upheaval in the past decade with the surge in the literature and the marketplace of new materials termed *nanocomposites* based on their use of nanometer-sized fillers (e.g., clays and carbon nanotubes). The development of these materials has paralleled the development of nanostructure characterization tools, such as SPM. These new materials provide a source of improvements in mechanical and barrier properties that offer major opportunities in many materials fields. Accordingly, a section of this chapter will be devoted to nanocomposites.

Thermoplastic homopolymers, multiphase polymers, and thermosets are all used as the matrix for composite materials. Rigid fillers are added to composites either to *fill* the polymer and perhaps modify some physical property or to *reinforce* the matrix, that is, to bear some of the load. Fiber composites are used in applications such as in small boat hulls and automotive bodies because of their light weight and high strength. Applications of carbon fiber reinforced composites are in tennis and racket ball rackets, fishing poles, and in the medical/orthopedic field for joint replacement [413], and more recently in aircraft [414] and ships designed for fast responses. Continuous graphite (carbon), glass, and Kevlar fibers in epoxy resins have been the major composite materials applied to such components.

Incorporation of short or long glass or carbon fibers into many engineering materials improves their strength, stiffness, and heat deflection. The strength and fatigue crack resistance depend on the volume fraction of fiber in the composite, the mechanical properties of the fiber and polymer, and particularly the adhesion between the two components. Adhesion is controlled by a compatible surface finish, or "size," put on the fibers to enhance bonding. In fact, in the case of glass fibers, the fibers require protection so they do not break during handling. The complex interface between the size and the fiber is termed the *interphase region*, in much the same way as the interface of polymer blends. Sizing developments are well guarded secrets by the various glass fiber manufacturers, who try to provide products that are easy to handle and form into composites. Nonfibrous fillers find application in low cost composites where they may enhance physical properties or simply replace the more expensive polymer with minimal loss of mechanical integrity. There is increased demand for additives for electronics with conductive and antistatic additives to limit charge buildup during fabrication, storage, and use (e.g., auto fuel systems and medical applications). Additives are also used as flame retardants, colorants, thermal and chemical stabilizers, to enhance conductivity, and to improve electrical properties of the thermoplastics or thermosetting plastics used as a matrix.

Cellulosic materials have been used as fillers in polymer composites (Bakelite) and as cellulose derivatives (collodion) since the very beginning of the plastics industry. More recent applications for cellulosic fillers include the use of wood flour [415] or fibers with isotactic PP for interior automotive panels, and wood polymer composites for construction material, especially decking. Improved modulus composites based on cellulosic fiber with thermosets and thermoplastics are being used for automotive, furniture, and road making, to name a few applications [416]. As with all fillers, a major reason for using these natural fibers is to reduce the cost of the product, although they are of high modulus and may increase the stiffness if their aspect ratio is kept high enough during processing. Reviews describe current trends in the use of cellulosics in general [416], prospects for the use of wood cellulose as reinforcements in polymer composites [417], composites that contain natural polymers, including cellulose derivatives [418], wood consumption compared with plastics [419], and natural fibers, plastics, and composites [420].

The processing of composites is quite complex as the nature of the polymer matrix, the filler or reinforcement, the sizing or surface treatment of the filler, and the process all contribute to the microstructure and properties. Key to the assembly is the dispersion of the filler, the wetting of the filler surface, and the chemical bonding, if any, at the interface. The lack of fiber to resin bonding can result in a decrease in the stress transfer from the resin to the fiber and limited reinforcement. Important topics in processing composites are found in the literature (e.g., [421–425]). All of the structures, from the macro to the micro and nanoscale, are relevant to control and therefore must be observed using a range of microscopes. The relevant microscopy techniques and examples of their application to morphological study are described here and elsewhere (e.g., [426–429]).

5.4.2 Literature Review

5.4.2.1 Composites in General

The strength of a short fiber reinforced composite depends upon the fiber length and orientation, the volume fraction of fibers (typically 15–40%), the matrix and its mechanical properties, and the interfacial bond between the fiber and the matrix resin. Long fiber composites are dependent upon the method of preparation, whether by extrusion of long glass fiber composites or the use of carbon or graphite fabrics or laminates. The properties are also dependent of course on the specific fiber and polymers and the type of composite. Fiber composites include multilayered composites, both laminates and hybrids, and single layer composites, using either continuous or discontinuous fibers [421]. Natural fibers are also established options for composites. The most common are short fiber composites with random orientation although long fiber composites are becoming more common. End uses for these composites include the full range of materials—electronics, medical, lighting, automotive, construction, and so forth. A text by Folkes [430] provides an excellent review. De and White [431] review short fiber composites, of all types, including design, process, and property considerations.

Optical microscopy and SEM are the techniques usually employed to evaluate the structure of composite materials. The hardness of composites generally (but not always) precludes microtomy, so polishing is used to provide thin sections for transmitted light microscopy or thick sections for reflected light study. Since about 1967 [432], fractured composites have been studied by SEM techniques, primarily due to the great depth of focus of this technique. Specimens are fractured by normal physical testing procedures or in use, and SEM is used to evaluate such features as the size, distribution, and adhesion of the fibers or particles. The length of protruding fibers and their adhesion to the matrix play a major role in revealing the strength and toughness enhancement. Quantitative microscopy techniques can be used to determine the length distribution of the fibers in a composite or the length of fibers protruding from a fracture surface. These are important parameters that relate to mechanical properties. Fiber length distribution information is important, as there is a *critical fiber length* required for reinforcement [433, 434]. After compounding and extrusion and/or injection molding, there is a distribution of lengths with many fibers that may be less than this critical length, resulting in poor composite properties. Earlier studies [435] described a method of removal of fibers from composites, without damage, followed by quantitative analysis of the fiber length distribution. Such methods are used to understand and optimize the glass fiber length during processing.

Oron [436] described dynamic evaluation of fracture mechanisms in composite materials by direct observation in the SEM. In this systematic study, tensile loading of notched carbon fiber epoxy composites was conducted in the SEM. The three stages of fracture described by Tetelman [437] are: (1) the tip of the notch is blunted and microcracks form there, (2) microcracks grow with fiber debonding, and (3) failure occurs with cracks widening followed by catastrophic failure. The effect of short glass fibers and particulate fillers on fatigue cracking in polyamides was evaluated in the SEM [438].

Improvement in fatigue crack propagation was found with increased glass fiber loading and reinforcement. The percolation network in polyolefins containing antistatic additives was imaged by low voltage SEM, revealing the conductive pathway in the volume beneath the sample surface [439]. Fiber matrix interactions were studied in aramid short fiber reinforced thermoplastic polyurethane composites using SEM to compare with mechanical properties [440]. Composites with small metal or metal oxide particles were studied using variable pressure SEM in which BEI images showed the morphology and size distribution, which was quantified using image analysis techniques [441].

Electronic devices are moving to ever higher frequencies and speeds resulting in a need for better understanding of the materials construction and microstructure as subtle effects in particle size, dispersion, and part and layer thickness affect dielectric and other electronic properties. Compounds made using conductive fillers, such as graphite fibers and ceramic particles, enable a material to transfer heat through conduction, a necessary feature in electronic components, lighting, and other devices in which heat can limit service. High temperature resistant thermoplastics, such as PPS, PEI, PPO, and liquid crystal polyesters, have been combined with conductive fibers, generally chopped graphite and nickel coated graphite. Fiber distribution, orientation, and fracture morphology are determined by optical and SEM techniques. Poor bonding in a nickel coated graphite composite, as observed in SEM tensile fractures, resulted in lowered tensile and impact properties [414].

5.4.2.2 Contact Microradiography

Contact microradiography is a method that permits assessment of the length and orientation of glass fibers in composites and thus is included here for completeness. Sections are cut 50–150 μm thick with a low speed diamond cutting saw and exposed to an x-ray beam providing an image on an underlying photographic plate. The resulting photograph has more contrast than an optical micrograph [430, 442]. The effect of molding conditions on fiber length and distribution has been studied by contact microradiography. Flow geometry and injection speed have been shown to affect fiber orientation and skin thickness of the molding. Bright et al. [443] used contact microradiography of glass filled PP to show the effect of the flow field on the fibers. They explained the flow field according to the "fountain" flow model [261], which shows that skin orientation is due to extensional flow, whereas shear flow is observed in the core. This extensional flow causes a high degree of molecular orientation along the flow axis in fiber filled composites in much the same way as was described for unfilled polymers (see Section 5.3). A high injection rate causes an increase in skin thickness, and fibers in the core are aligned transverse to the flow axis. Low injection rates result in a thicker core with fibers parallel to the flow axis. Mold temperature also affects the nature of the skin and core, and thus fiber orientations at colder mold temperatures give thicker skin layers. Fiber alignment resulting from three flow conditions was shown using contact microradiography [444]. Convergent flow causes high fiber alignment, and diverging flow causes fibers to align perpendicular to the flow direction. Shear flow leads to alignment nearly parallel to the flow direction. Overall, process conditions affect polymer orientation and filler fibers are generally oriented with the polymers. The interested reader is directed to the references for details relating to the method.

5.4.2.3 Carbon and Graphite Fiber Composites

Specific surface features identified by SEM examination of fractured composites have been more widely documented in the case of carbon fiber than for glass fiber composites. Fatigue damage in graphite/epoxy composites was investigated by Whitcomb [445] where delamination and cracking were found to result during fatigue tests. This is an important result as such testing relates to long term behavior of the composite. Richards-Frandsen and Naerheim [446] described graphite/epoxy composites where three point bending induced delamination fatigue failure. Fracture surfaces were examined in the SEM in order to characterize the structure. Several common features were noted: matrix cleavage, hackle formation (in the epoxy), and wear. Defects causing crack propagation were caused by resin-rich inclusions and voids. Tensile fracture in high modulus graphite fiber composites, subjected to off-axis tensile loading, showed several fracture features [447], such as matrix lacerations or hackle, fiber pullout, resin-free fibers, matrix cleavage, and matrix debris. Fracture modes were identified [447], and smooth surfaces with matrix lacerations in intralaminar shear and smooth surfaces with matrix cleavage were shown in transverse tensile fracture. Although fractography studies of brittle fibers can be conducted to determine the locus and cause of failure by classical fracture mechanics, such a simple determination cannot be made in the case of a composite [448]. Accordingly, single fiber testing of graphite fibers in thermosetting resins has been conducted in an optical microscope to determine the nature of the stress concentration at the fiber matrix interface [449].

Composites may also be examined by transmission EM methods, such as by bright field, dark field, and electron diffraction of ultrathin sections. Sections of carbon fiber composites are quite difficult to obtain, but the technique is possible [450, 451]. Important information that can be obtained relates to the fiber-resin interface, which is known to be critical to composite properties and is often adversely affected by environmental conditions.

5.4.2.4 Hybrid Composites

Hybrid composites are formed from a mixture of fibers and a bonding matrix [452], and these include carbon-glass and carbon-aramid fibers. Hybrid composites have economic, physical, and mechanical property advantages that are exploited for various applications. For instance, composites containing expensive carbon fibers, for added strength and stiffness, may include glass fibers for increased toughness. Kalnin [453] evaluated glass-graphite fiber/epoxy resin composites. Nylon 6 polymerized around unidirectionally aligned carbon and glass fibers showed reinforcement properties [454]. Kirk et al. [455] measured the fiber debonding length in a model hybrid carbon and glass fiber composite and evaluated the fracture energy as a function of the carbon fiber/glass fiber ratio. They determined that the rule of mixtures prediction overestimates the fracture energy in the hybrid composites. Hardaker and Richardson [452] described the theoretical and empirical developments in this area.

5.4.3 Composite Characterization

The major topics in this section are the microscopy techniques and preparation methods used for fiber composites (e.g., [429]). There are three imaging techniques and methods that are typically applied to assessment of composite microstructure: (1) optical microscopy, by reflected light of polished surfaces and transmitted light through thin sections, (2) SEM analysis of both polished and fractured specimens, and (3) TEM of ultrathin sections generally of particle filled composites. Scanning probe microscopy can also be used to study polished surfaces and SPM and TEM techniques are most often used to assess the microstructure and nanostructure of nanocomposites due to the small size of the dispersed particles, generally clays and carbon nanotubes. Specimen preparation for these techniques has been discussed in Chapter 4, and more detail on applications will be given later in this section. Applications of microscopy to the study of glass, carbon, or graphite fiber composites and mineral filled composites will be explored, especially relating to the effects of processing, fiber length, and interfacial bonding.

5.4.3.1 OM Characterization

Fiber and particulate fillers can be observed prior to compounding or they can be removed from the composite by high or low temperature ashing, plasma ashing, microwave ashing, or etching. High temperature ashing in a muffle furnace can change the phase of the filler or cause embrittlement and breakage. Low temperature ashing with radio frequency (RF) plasma is used to remove the organic matter without such phase changes or embrittlement [435]. It is important to know the morphology of the fillers before studying them in a composite as an aid in understanding the complex morphologies. Examples of fiber and particulate fillers, removed or free from resin, are shown in Figs 5.93 and 5.94. A transmitted light micrograph (Fig. 5.93A) shows the general nature of short glass fibers. Glass fibers removed from a filled resin by low temperature ashing are shown by SEM in Fig. 5.93B. Flat mica flakes (Fig. 5.94A) are quite different in morphology from irregularly shaped talc particles (Fig. 5.94B) and finely divided clay particles (Fig. 5.94C). Polarized light microscopy of such minerals also aids their identification if the filler in the composite is an unknown. In that case, the *Particle Atlas* by McCrone and Delly (the book [456] and online reference [457]) is an invaluable source of representative optical micrographs of common minerals, fibers, and particles.

Reflected light microscopy is a technique that provides an overview of composite microstructures. Specimens are cut, polished (see Section 4.2), and examined in reflected light, generally at magnifications about 30–500×. Glass fibers appear as light colored round or elliptical sections. The orientation and loading of glass fibers in various regions of a composite can be shown by such studies. It is worth noting that the ratio of the major and minor axes of the ellipse produced when the cylindrical fibers are intersected during cutting and polishing can be used to determine the angle of the fibers (their orientation) within the matrix. Image analyzing computer systems can be used to rapidly measure the orientation of many fibers within the matrix in this way, as long as the fibers are exactly cylindrical. For most fibers, a thin section viewed in transmission is required, and this is difficult with a normal optical microscope. A very thin section is difficult to prepare and includes long sections of few fibers; on the other hand a thicker section is not all in focus at one time and may contain overlapping fibers. However, only a clear image can give good results, as the computer system must be able to distinguish fiber boundaries and discard overlaps. The laser confocal scanning microscope (LCSM) can overcome these problems. The instrument is described in Section 6.2.1 (see also [458]). It has the ability to form "optical sections," that is, sharp images of a particular plane within a composite, as long as the matrix material is transparent. Observation of one such sectional image gives information on the orientation of fibers within that plane, but the associated computer systems can put together the information in many such images, formed as the sample is scanned along the optic axis of the instrument. The result is a full determination of fiber orientation, for a depth of up to 20 or 30 μm into the composite [459, 460].

Transmitted light microscopy permits assessment of the polymer matrix as well as the fiber orientation. Polished thin sections can be used for bright field, polarized light, and phase contrast microscopy, as shown for a glass fiber reinforced polyamide (Fig. 5.95). In bright field (Fig. 5.95A), in a section cut perpendicular to

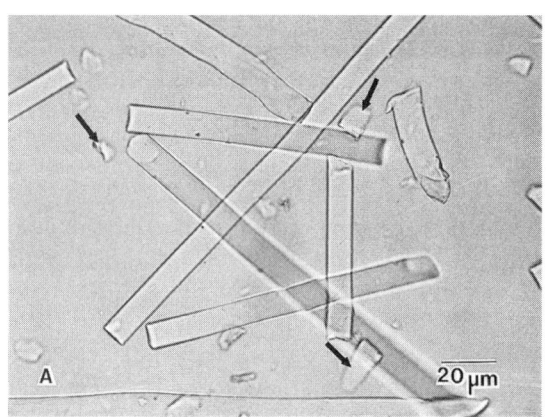
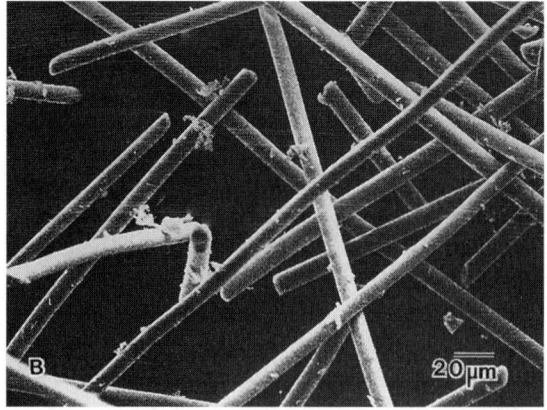

FIGURE 5.93. An optical micrograph (A) of glass fibers removed from a filled engineering resin by a low temperature ashing method shows that they are isotropic, clear cylinders that tend to break into bits or shards (arrows). A low magnification SEM (B) shows the surface texture and shape of the fibers.

Composites

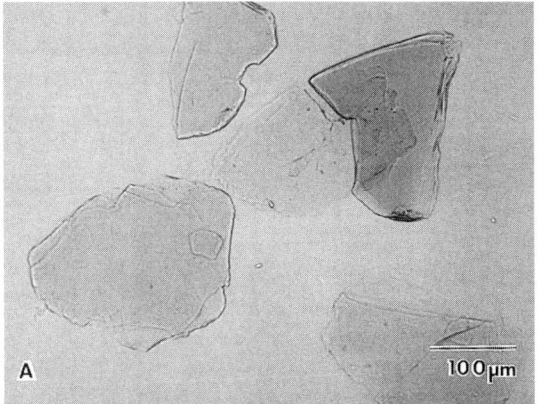
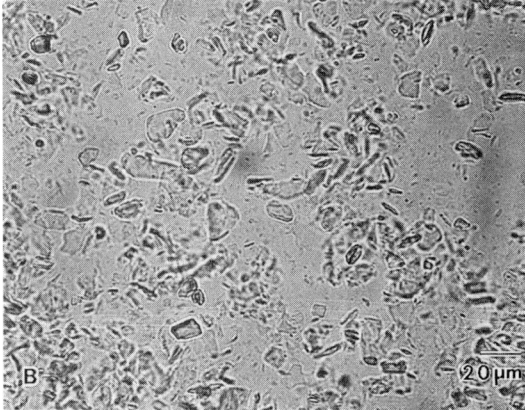
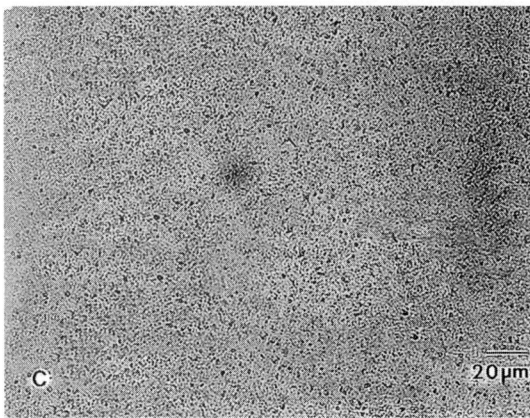

FIGURE 5.94. Optical micrographs show the size and shape of several minerals used as fillers for composites: (A) mica flakes appear platy in shape with irregular boundaries, (B) talc particles have a much finer, platy texture and the particles exhibit a range of shapes from nearly fibrous to platy; and (C) clay particles are very fine grained with no characteristic shape.

the flow axis of the test bar, the fibers are round, showing that they are oriented along the bar axis. The regions between the fibers contain the fine textured polymer. In polarized light (Fig. 5.95B), the polymer is birefringent and composed of fine spherulites whereas the fibers are isotropic, as expected. Phase contrast imaging (Fig. 5.95C) shows particles (white) are present that are different in refractive index compared with the polymer. Overall, combination of the three optical microscopies shows that the reinforced polyamide is composed of glass fibers, oriented with the flow axis of the mold, and that dispersed, second phase particles are also present.

5.4.3.2 SEM of Composites

Scanning electron microscopy is the most widely used imaging technique for the study of both short and continuous fiber composites. The nature of the adhesion between the matrix and the resin and information relating structure to mechanical properties can be obtained by SEM assessment of the composite fracture surface. Voloshin and Arcan [461] showed by SEM inspection that debonding of the fiber and matrix in unidirectional glass fiber/epoxy composites is the cause of shear failure under bidimensional stress. Wu et al. [462] studied the impact behavior of short fiber/liquid crystalline polymer composites by SEM evaluation of instrumented impact tested and modular falling weight tested materials. They observed a high degree of anisotropy in both the morphology and mechanical properties of the unfilled polymer. Adding short fibers reduced this anisotropy. The SEM observations also suggested failure mechanisms in these composites. The mechanisms of fatigue in short glass fiber reinforced polyamide 6 were studied by SEM of specimens after environmental conditioning and testing, providing a correlation between tensile and fatigue strength [463]. Even relatively low magnification SEM can be used to evaluate the distribution of fibers and orientation with the direction of flow, as was done for extrusion blow molded, long fiber reinforced polyolefins [464].

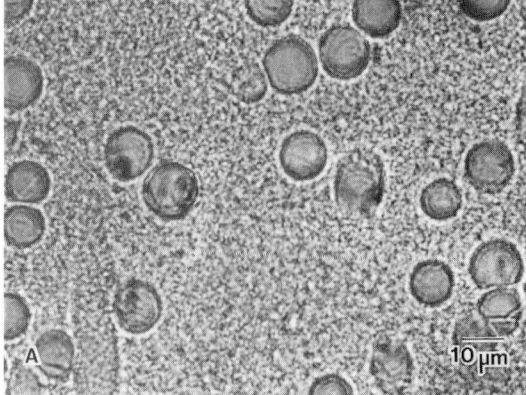
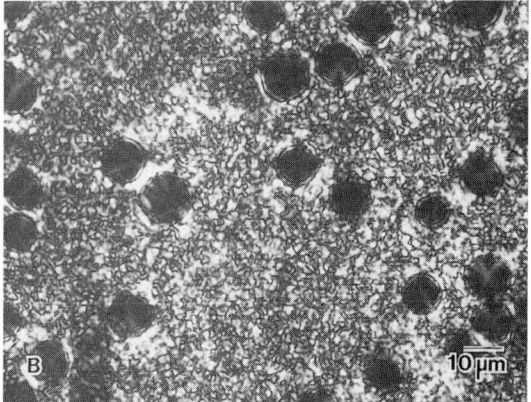
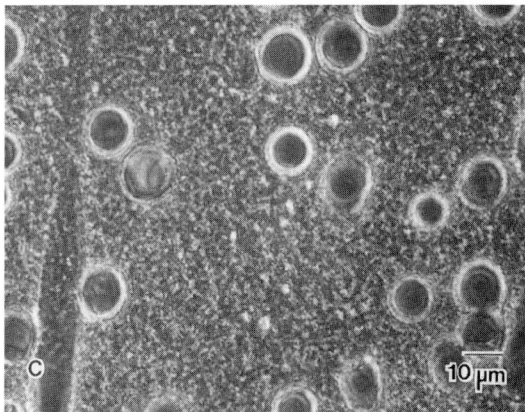

FIGURE 5.95. Transmitted light micrographs of a glass fiber reinforced polyamide polished thin section is shown by three optical techniques. Imaging in bright field (A) reveals clear, round fibers aligned perpendicular to the section plane and a mottled textured matrix. Polarized light (B) shows the glass fibers are isotropic (black), whereas the polymer is birefringent and composed of finely textured spherulites. Phase contrast (C) shows that there are small white regions of different refractive index than the matrix.

The fiber lengths of fillers or reinforcements are dramatically diminished during compounding and injection molding, and many studies have been conducted on both minimizing that breakage and determining the critical fiber length required for reinforcement [433]. Most of these studies have been conducted on glass fibers, although the principles are similar for all reinforcements. There is also a great range of variability of orientation and fiber length in composites. However, critical applications require precise predictions of mechanical behavior, which has been addressed [430, 465]. A model of microstructure-property relations has been developed to predict mechanical behavior. Quantitative microscopy was applied to determine the location and orientation of the fibers in a polished specimen and to measure fiber pullout lengths in SEM fracture surfaces. The critical fiber length was assumed to be about four times the mean observed pullout length. It is important to know the length of the fibers in the composite compared with the critical fiber length for that system in order to know if the fibers are providing reinforcement.

Scanning electron microscopy images of composite surfaces of two different specimens resulting from notched Izod impact testing of glass fiber filled thermoplastics are shown in Fig. 5.96. Fibers are aligned parallel to the surface in the skin, and they are not aligned in the core (Fig. 5.96A). Fibers are shown that exhibit poor fiber wetting, as seen by the clumps of fibers with no resin rather than single fibers distributed within the matrix (Fig. 5.96B). A typical "hackle" morphology is observed in the fractured resin (Fig. 5.96C), between fibers and larger hackle or ridged patterns within the matrix. Short fibers with resin coating (Fig. 5.96D) suggest there is some good bonding, whereas fibers are seen debonded from the fracture surface (Fig. 5.96E). The surface of an individual glass fiber shows a thin layer of resin bonded to the fiber surface (Fig. 5.96F). Adhesion is controlled by a compatible surface finish, or "size," put on the fibers to enhance bonding. Classical brittle fracture of the glass fibers is seen (Fig. 5.97A, B), and the bonding at the fiber surface is shown in

Composites

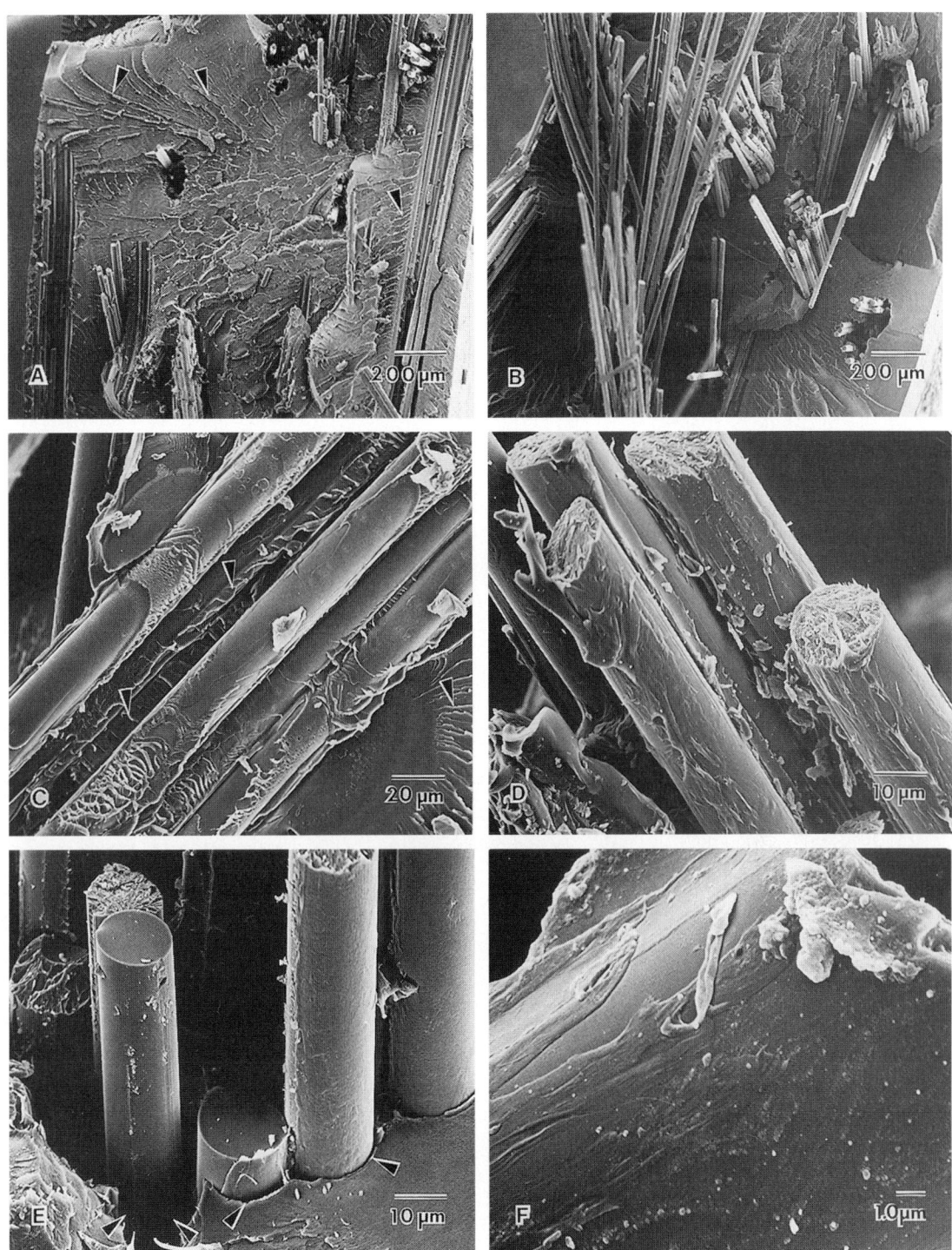

FIGURE 5.96. Scanning electron microscopy of Izod impact fractured, glass fiber filled thermoplastic text specimens show nonuniform distribution of fibers in the two different specimens (A, C, D and B, D, F). The fibers (A) appear aligned parallel to the skin, and the matrix exhibits brittle failure as hackle marks (arrowheads) are seen. The fibers (B) protruding appear long and poorly wetted with the resin. Hackle or ridged patterns (arrowheads) are observed (C). Resin is also seen on the fiber surfaces in some regions (D and F), whereas cleaner fiber surfaces and less well bonded regions are also observed (E).

detail (Fig. 5.97C) in another glass fiber reinforced resin.

A comparative study using different types of SEM techniques is worthwhile when imaging fiber filled composites. An uncoated, glass fiber filled nylon is shown in a BSE image taken at low vacuum and 5 kV in Fig. 5.98A and in SEI at high vacuum and low kV (1.2 kV) in Fig. 5.98B [215]. The atomic number contrast in the BSE image shows the fiber and the polymer coating more clearly than in the poorly resolved SE image taken at high vacuum. Finally, it should be remembered that failure analysis often requires several techniques be used for evaluation with the most common being OM and SEM, coupled with EDS (e.g., [466]). Figure 5.99 shows a fracture surface of a molded sample with a defect particle present, found by EDS spectra and elemental map to contain Ti and to likely be an undispersed TiO_2 particle.

5.4.3.3 Problem Solving Application

The application of TEM to imaging rubber modifiers in polymers was reviewed (see Section

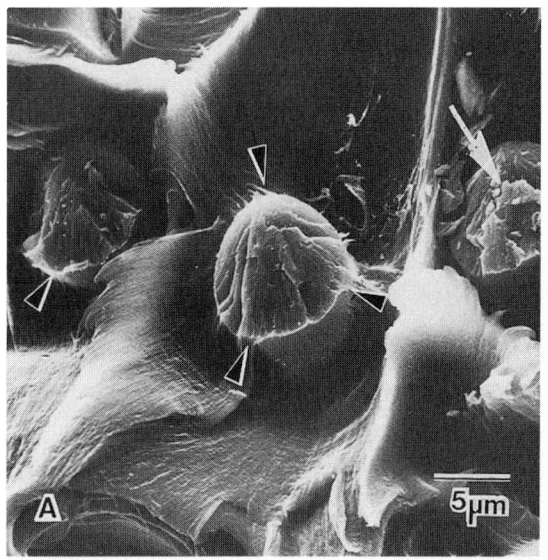
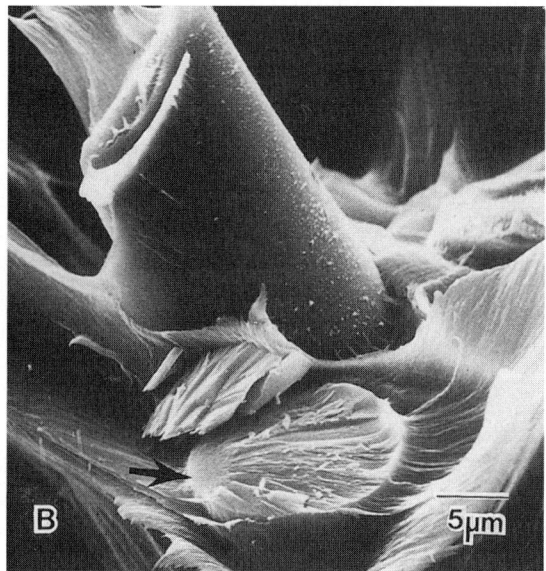
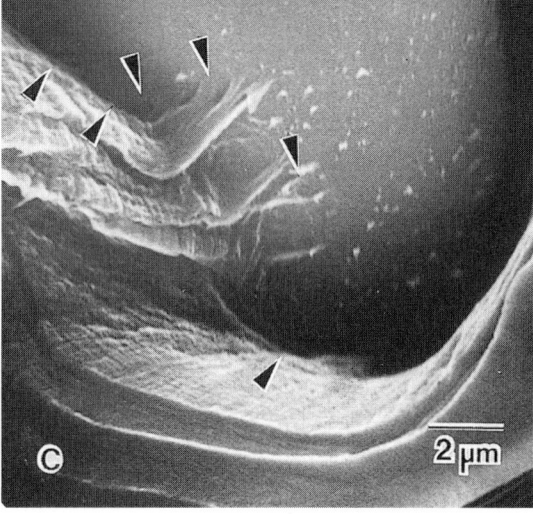

FIGURE 5.97. Scanning electron microscopy of a glass fiber reinforced nylon shows excellent compatibility between the fiber surfaces and the matrix. Short, nearly lateral failure across the matrix and fiber is observed (A). Some glass fibers exhibit classical brittle fracture (arrow) (B). Adhesion of the matrix resin is seen (arrowheads) at the fiber surfaces (C).

Composites

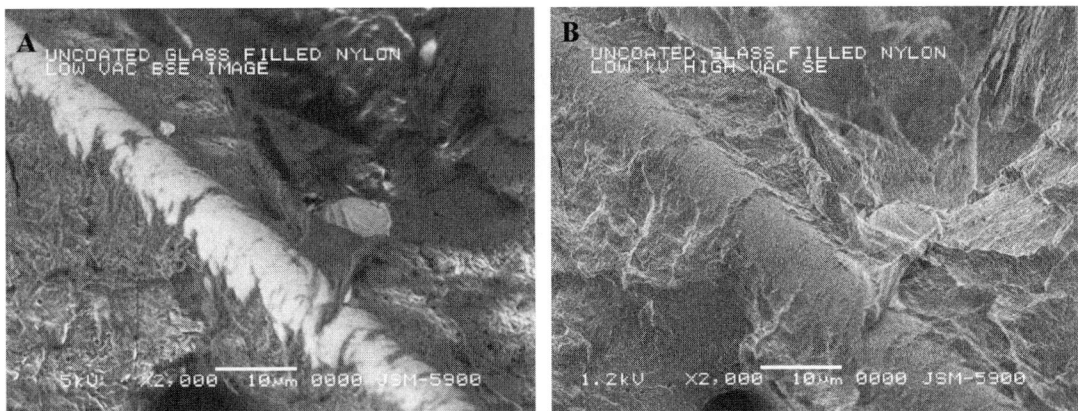

FIGURE 5.98. An uncoated glass fiber filled nylon is shown in a BSE image taken at low vacuum and 5 kV (A) and in SEI at high vacuum and low kV (1.2 kV) (B). (From V. E. Robertson, JEOL USA [215], unpublished.)

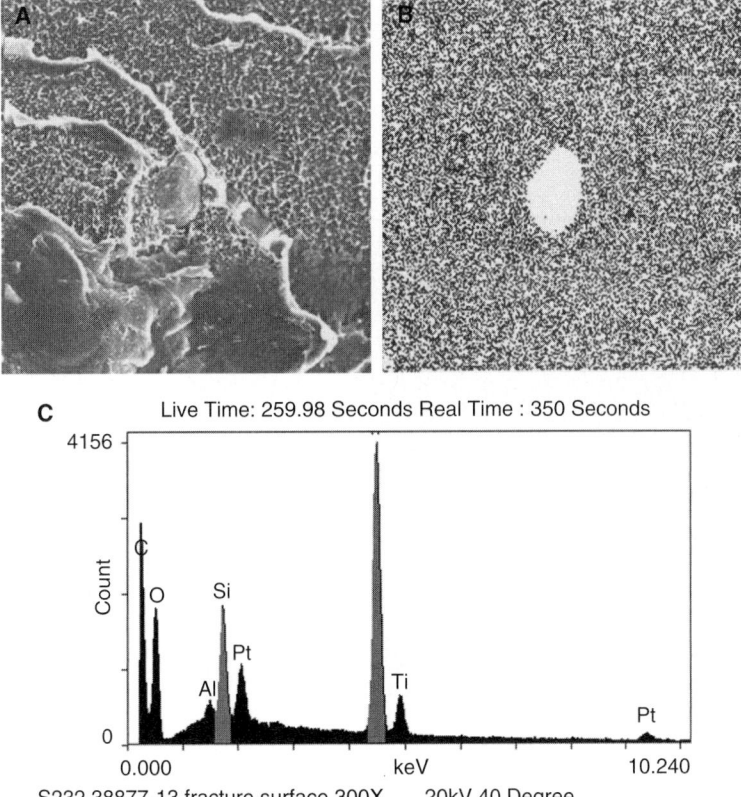

FIGURE 5.99. Scanning electron microscopy of a fracture surface of a molded sample with a defect particle present (A), found by EDS spectra (C) and map (B) to contain Ti and to likely be an undispersed TiO_2 particle.

5.3). However, in many cases TEM is required for identification of the details of an unknown product in the marketplace, and in some cases the composite is complex and includes glass fibers or other additives. In the example here, a glass fiber reinforced polyamide contained elastomer, but the type and amount were unknown [186]. The samples were cryoultrathin sectioned with an older diamond knife, because of the presence of the glass fibers, and images of the sections unstained (Fig. 5.100A) and stained (Fig. 5.100B) with RuO_4 vapor showed the dark dispersed phase of the elastomer in the latter and irregularly shaped mineral particles in the former. The broken glass fibers are the larger structures in the image, with holes in the sections, but they do not detract from the ability to image and measure the other components. At higher magnification (Fig. 5.100C), the dark dispersed phase is more clearly resolved as fairly round particles about 0.2 μm in diameter. Particle shapes are

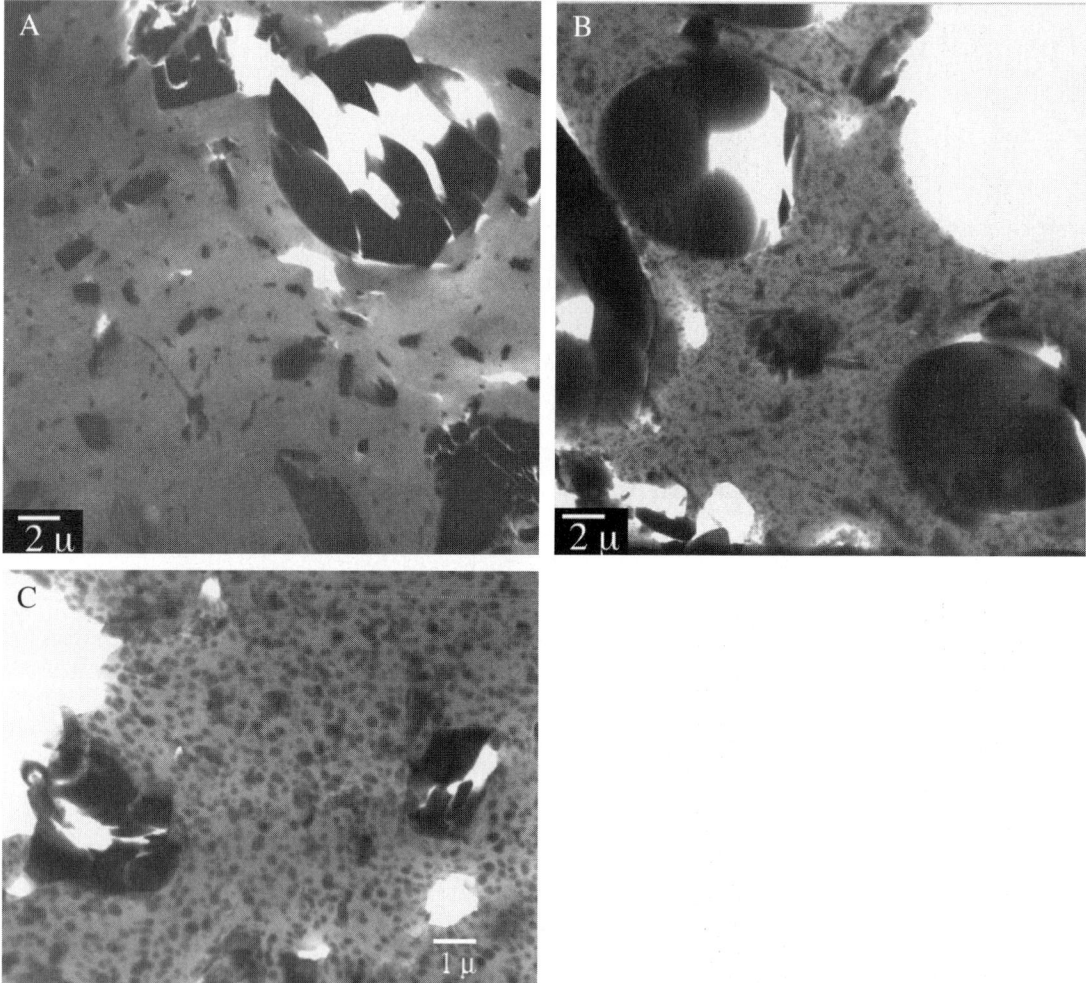

FIGURE 5.100. A glass fiber reinforced polyamide with elastomer was cryoultrathin sectioned with an older diamond knife, due to the presence of the glass fibers, and images of the sections unstained (A) and stained with RuO_4 vapor (B) showed the dark dispersed phase of the elastomer in the latter and the irregularly shaped mineral particles in the former. The broken glass fibers are the larger structures in the image. At higher magnification (C), the dark dispersed phase is more clearly resolved as fairly round particles about 0.2 μm in diameter. (From Wood [186]; used with permission of the American Chemical Society Rubber Division.)

Composites

generally determined by sectioning both parallel and perpendicular to the melt flow direction [186].

5.4.4 Carbon and Graphite Fiber Composites

The microscopy techniques described for the evaluation of glass fiber composites are widely used to determine the microstructure of carbon and graphite fiber composites. Microscopy of crack propagation in carbon fiber reinforced composites is also very important in understanding mechanical properties. Test specimens and actual composite products are often evaluated to determine the distribution of the fibers in the resin, typically epoxy, and the degree of resin wetting of the fibers. Voids in the composite can be the locus of failure, and their identification and cause are quite important to mechanical property evaluation. Studies of the effects of surface treatments have been done by SEM, but also HREM has been used to assess the interfacial phenomena (e.g., [467]).

An example of reflected light microscopy of a polished specimen is shown in Fig. 5.101, which elucidates the complex engineering involved in fabrication of a tennis racket. The micrographs show the nature of the continuous fabric, or *prepreg*, used for fabrication with layers of fiber bundles arranged at angles to one another (Fig. 5.101A). Black regions are

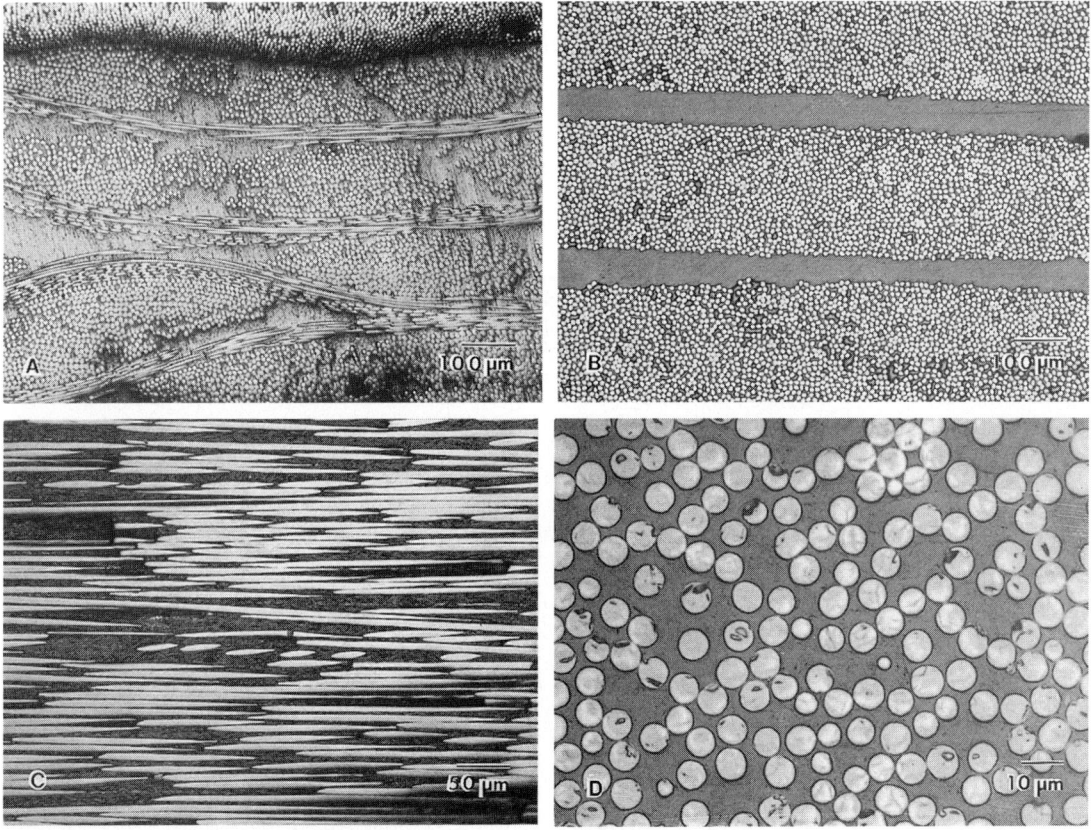

FIGURE 5.101. Reflected light micrograph of polished carbon fiber composites are shown: fibers are white, voids (holes) are black, and the resin is gray in color due to surface reflectivity. A section of a graphite fiber reinforced composite (A) shows a complex arrangement of fibers with uniform wetting and packing. Single ply carbon fiber panels (B) embedded in epoxy and polished are uniformly packed. Higher magnification micrographs show detail is possible by this technique for fibers oriented parallel (C) and perpendicular (D) to the polished surface.

voids where there is no reflection of light. The polished fiber cross sections in the micrographs appear white due to their high reflectivity. Gray regions are devoid of fibers and are termed *resin-rich*. Single panels of unidirectionally oriented yarn bundles, or *tows*, are uniform in both distribution and resin wetting (Fig. 5.101B). Higher magnification of fibers that are nearly parallel (Fig. 5.101C) and perpendicular (Fig. 5.101D) to the polished surface are also shown.

The polished surface of a specimen with carbon fibers and a resin contains several voids and some regions show poor fiber wetting, whereas adjacent regions exhibit good wetting and fiber distribution (Fig. 5.102A, B). Fibers that exhibit poor resin wetting are shown by SEM of carbon fiber epoxy fracture surfaces (Fig. 5.103). The carbon fiber fabric has long fibers protruding from the surface (Fig. 5.103A). Fracture along the bundles shows hackle formation in the resin matrix (Fig. 5.103B). Higher magnification micrographs (Fig. 5.103C, D) show fibers protruding from the surface with little resin on the striated surfaces, and resin is seen pulled away from the fiber surfaces. Overall, this particular composite exhibits poor wetting of the fibers with resin, and thus fracture occurs at the fiber-matrix interface. Although SPM imaging could be conducted on the polished sample, the three dimensional fracture surfaces would not be appropriate for this technique.

5.4.5 Particle Filled Composites

5.4.5.1 Introduction

Minerals and other particles used as fillers in plastic moldings either to enhance mechanical properties or to reduce shrinkage and flammability include mica, clay, talc, silica, wollastonite, glass beads, carbon black, and calcium carbonate. Theberge [468] summarized product advances in thermoplastic composites. Small mica flakes are known to boost mechanical properties in composites, but color changes are caused by their brown color. Mica was used to reinforce polypropylene with optimum performance obtained for particles in the range 80–280 μm diameter [469]. Factors found important in this study were concentration, aspect ratio, surface treatment, processing, mean size and size distribution of the mica flakes. Garton et al. [470] discussed modification of the interface in mica reinforced PP. Adhesion of the filler is quite important to improving properties, and therefore fillers are usually surface treated with coupling agents to enhance the interfacial bonding and wetting.

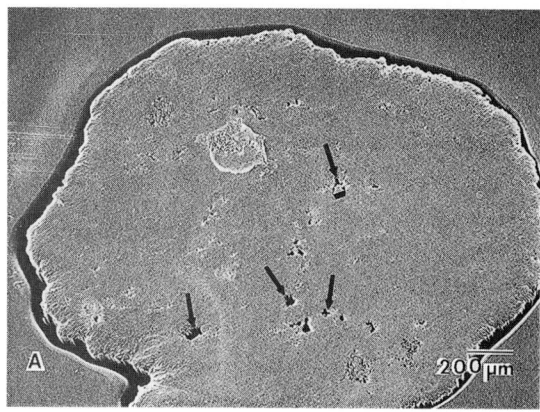 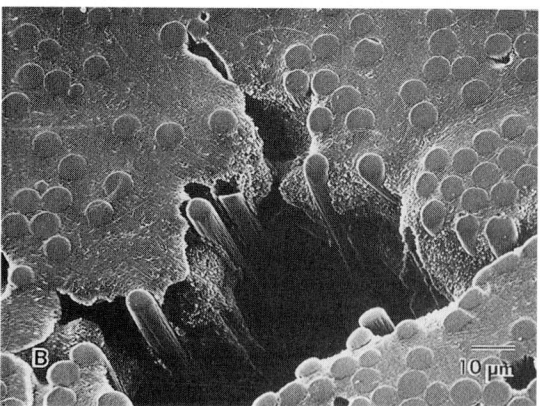

FIGURE 5.102. Scanning electron microscopy of a polished carbon fiber composite specimen reveals details relating to fiber uniformity and packing. Wetting of the fibers with resin is seen clearly. The overview (A) shows several voids (arrows) in the specimen, which are seen to be regions where the fibers are poorly wetted with resin (B).

Composites

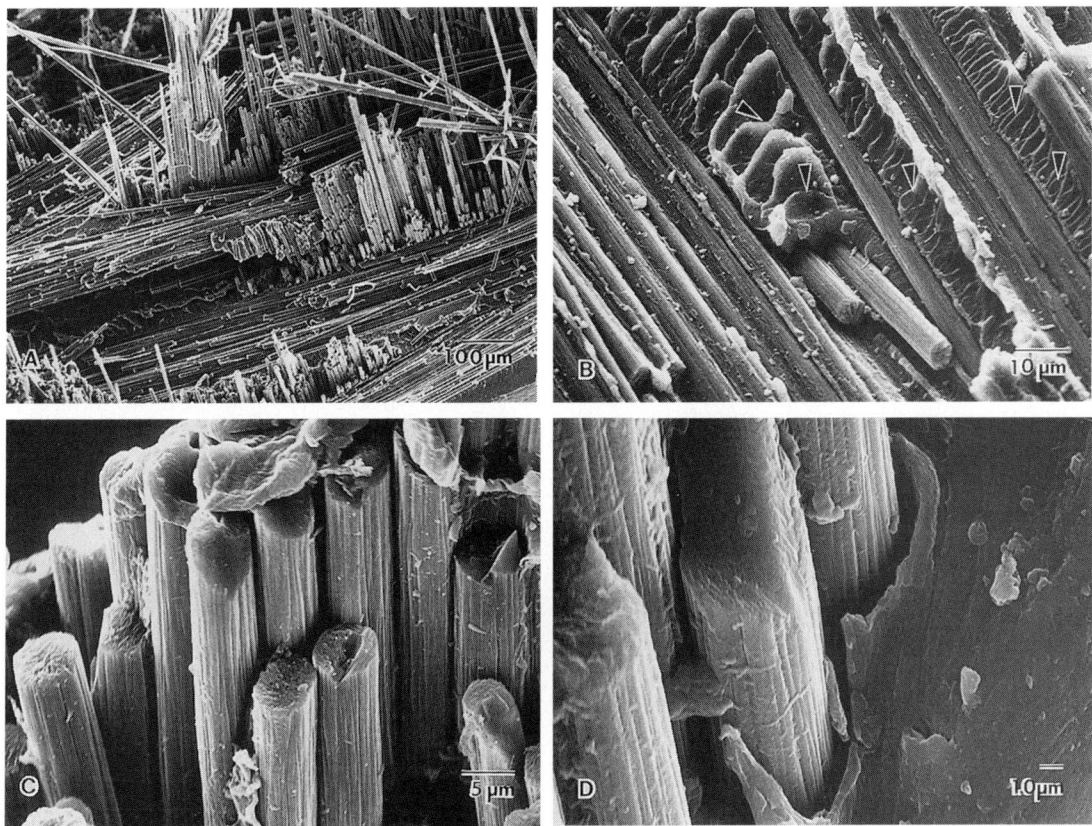

FIGURE 5.103. Fracture surfaces of a carbon fiber composite are shown in the SEM images. An overview (A) shows the fibers are in the form of a fabric, with yarns aligned at 90° to one another. Matrix hackle, or ridges (arrowheads), are seen adjacent to the fibers in one orientation (B). Poor fiber wetting is shown as clean, striated fiber surfaces (C) pulled away from the matrix (D).

Mineral fibers have also been used in the reinforcement of polypropylene and nylon 6 [471]. In some cases, the properties are similar to those of glass fiber composites. An important factor is the extent of fiber length loss during processing. Mineral fillers that are not fibers, with an aspect ratio (length/width) less than 10:1, reduce warpage of thermoplastics but they also result in a loss of tensile strength and other mechanical properties. Miller et al. [472] studied the role of the interface on properties in model composites using SEM to examine fracture surfaces of glass sphere filled polyethylene. The SEM provided a qualitative view of the bonding process and the nature of the region of modified matrix surrounding the glass spheres. High degrees of chemical modification to the glass surfaces resulted in the formation of an interfacial layer and a change in failure mechanism from adhesive to cohesive failure.

This section would not be complete without some mention of flame retardants, as the fire safety of plastics continues to increase in importance. Electrical and electronic end uses are increasing as thin wall connectors are used in miniaturized applications. Although halogenated additives continue to be used, some applications and regions of the world are demanding halogen-free phosphorus compounds and others with cost-performance being major issues. Dispersion of the fillers is a major requirement for the actual properties of the composite, and thus microscopy plays a role, generally OM and SEM.

5.4.5.2 Carbon Black Filled Rubber

Carbon black is a common particulate filler added to stiffen rubbers, for conductivity, and also simply to provide black coloration. The microscopy of rubbers generally involves the characterization of the particle size distribution of such additives. Microscopy of rubbers is found in many sections of this chapter and in many references (e.g., [186, 366, 429, 473]). Often, the rubber particles are composites or blends themselves containing additives and fillers required to modify properties. The behavior of the rubber is very dependent upon the size distribution of the carbon black and its adhesion to the rubber matrix. Optical microscopy and TEM are employed to study the structure of filled rubbers, as has been reviewed here and by Kruse [474] and Hobbs [475]. Preparation methods are similar to those discussed for rubber in tire cords, the ebonite method (see Section 4.4.6), ultrathin cryosectioning (see Section 4.3.5) and staining methods (see Section 4.4). Scanning electron microscopy analysis is more difficult than might be assumed, as often the filler is well adhered to the rubber and thus is not obvious. The fracture morphology of elastomers has been studied by SEM; for example, the flaw controlled fatigue fracture of styrene-butadiene rubber (SBR) and EPDM rubber [476].

Carbon black particles are generally aggregates that appear as fine, dense particles when observed in sections viewed in an optical microscope (Fig. 5.104). However, processing problems can occur that result in significantly larger particles (Fig. 5.104A), which can be the locus of failure in the molded part. Particle size distribution is especially critical for conducting polymers, although it can be difficult to determine the locus of failure, as larger particles might have poor adhesion to the matrix and a fracture surface might well contain only a hole where the particle was located before failure. If such failures occur, optical assessment can provide leads to the problem if larger than usual aggregates are observed. At times, black molded parts are seen that are streaked, that is, the black color appears nonuniform, and different gray levels may be observed (Fig. 5.104B). A more subtle surface variation, matte and gloss surfaces on the same part, may also be caused by poor or improper mixing of the carbon black during processing. Poor distribution can be a problem in any fiber filled or particle composite and is likely to cause variations in both physical and mechanical properties as a result. Optical microscopy shows the carbon black distribution and processing variants, and optimization can be monitored by optical techniques for both quality control and problem solving.

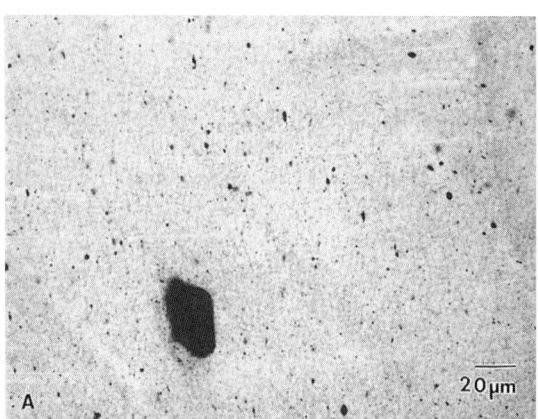 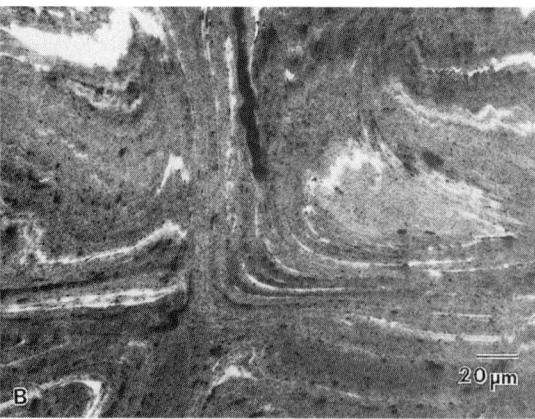

FIGURE 5.104. In bright field optical micrographs, carbon black particles appear as dense particles (A). Such particles can aggregate and provide a locus of failure. Streaks in molded parts, due to poor mixing of the carbon black, are shown in a section of a molded specimen (B).

Composites

It is well known that carbon black is found as aggregates best imaged by TEM of ultrathin sections. The dense particles in Fig. 5.105A are aggregates of individual carbon black particles. This micrograph is a good control for study of a multiphase polymer containing carbon black as shown in a TEM micrograph of a black, polyurethane-filled polyacetal in Fig. 5.105B. Interestingly, the carbon black particles have enhanced the image contrast as they are located within the dispersed polyurethane, and thus the dispersed phase is observed without staining.

5.4.5.3 Examples of Particle Filled Composites

Mica is often used as a filler or reinforcement as this mineral can provide enhanced properties. Scanning electron microscopy fracture surfaces of a mica composite and a glass fiber-mica composite are shown in Fig. 5.106. The mica composite has a fibrillar fracture surface texture, due to the thermotropic copolyester matrix, with the platy mica flakes aligned with the polymer and split along these plates during fracture (Fig. 5.106A). A glass fiber-mica composite fracture surface (Fig. 5.106B) shows the mica has frac-

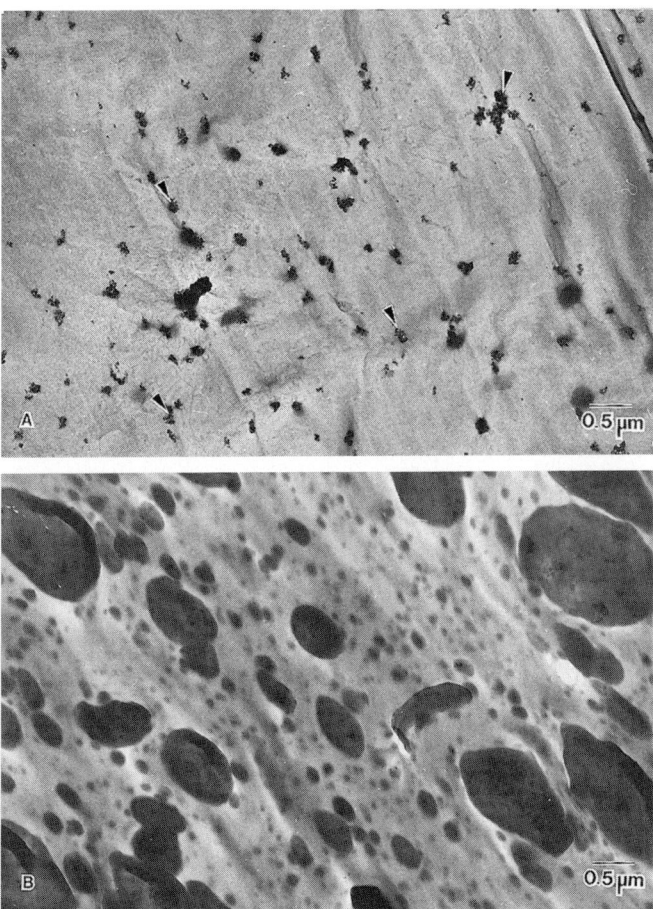

FIGURE 5.105. Carbon black particles (arrowheads) are shown in a sectioned carbon black filled polymer (A) to be aggregates of smaller particles less than $0.1\,\mu m$ in diameter. Interestingly, a black, multiphase polymer, shown in a TEM micrograph of an ultrathin section (B), has carbon black particles within the dispersed phase domains.

tured and the particles and the glass fibers appear well adhered to the matrix. These factors suggest good adhesion and reinforcement, leading to enhanced mechanical properties. Mica flakes can be dozens of micrometers in diameter, and thus they lend themselves to optical observation. Sections of a mica filled polyester are shown in an optical bright field micrograph in Fig. 5.107A. The size and shape of the mica flakes after processing may be determined from such images. Measurement of the thickness of the thin, platy flakes is much more difficult, as they tend to align in any matrix, and they are much thinner than they are wide. Transmission electron microscopy of an ultrathin section (Fig. 5.107B) shows the mica in cross sectional view to be dense and broken during sectioning.

Particulate fillers are known to cause changes to the polymer matrix, for example calcium carbonate affects the nucleation of some thermoplastics. The addition of any particulate filler has the potential of causing nucleation, although specific nucleating agents are also added to polymers to affect crystallization. Chacko et al. [477] observed no large scale order, i.e., there are no spherulites present upon addition of calcium carbonate to PE. An optical micrograph of POM appears spherulitic (Fig. 5.108A), whereas addition of calcium carbonate appears to result in a loss of spherulitic texture (Fig. 5.108B). Addition of fillers can also have deleterious effects, for example, the formation of a nonuniform or matte surface finish on a molded part, as shown in the SEM micrograph (Fig. 5.108C). Such particles in craters could be caused by a problem with the mold surface, but EDS analysis provided confirmation of the presence of calcium, as shown in the EDS map (Fig. 5.108D). Apparently, the calcium carbonate particles at the surface of the mold were not fully wetted with the polymer under the specific molding conditions used, and therefore a nonuniform matte surface finish resulted. Thus, processing polymers with particulate or fibrous materials does not preclude changes in the polymer morphology during the process, especially if the filler is added either prior to polymerization or before crystallization.

5.4.6 Nanocomposites

5.4.6.1 Introduction

The marketplace is always looking for new products that have better performance at a lower cost, and such demands are very clearly seen in the composites arena where materials development is closely followed, or led, by newly developing microscopes. The fields of nanocomposites and nanostructure have seen parallel development during the past two

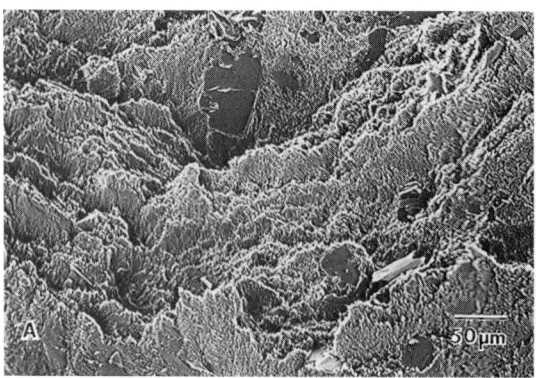
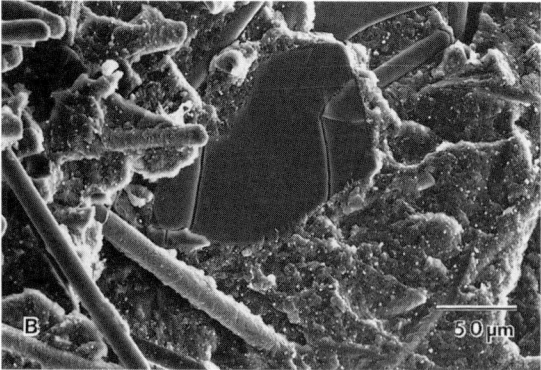

FIGURE 5.106. Scanning electron microscopy of a mica filled plastic bottle (A) and a mica-glass fiber composite (B) both show the platy shape of mica. Although the mica fracture surfaces do not appear resin coated, there is good adhesion of these particles with the matrix. The mica is aligned with the oriented polymer (A).

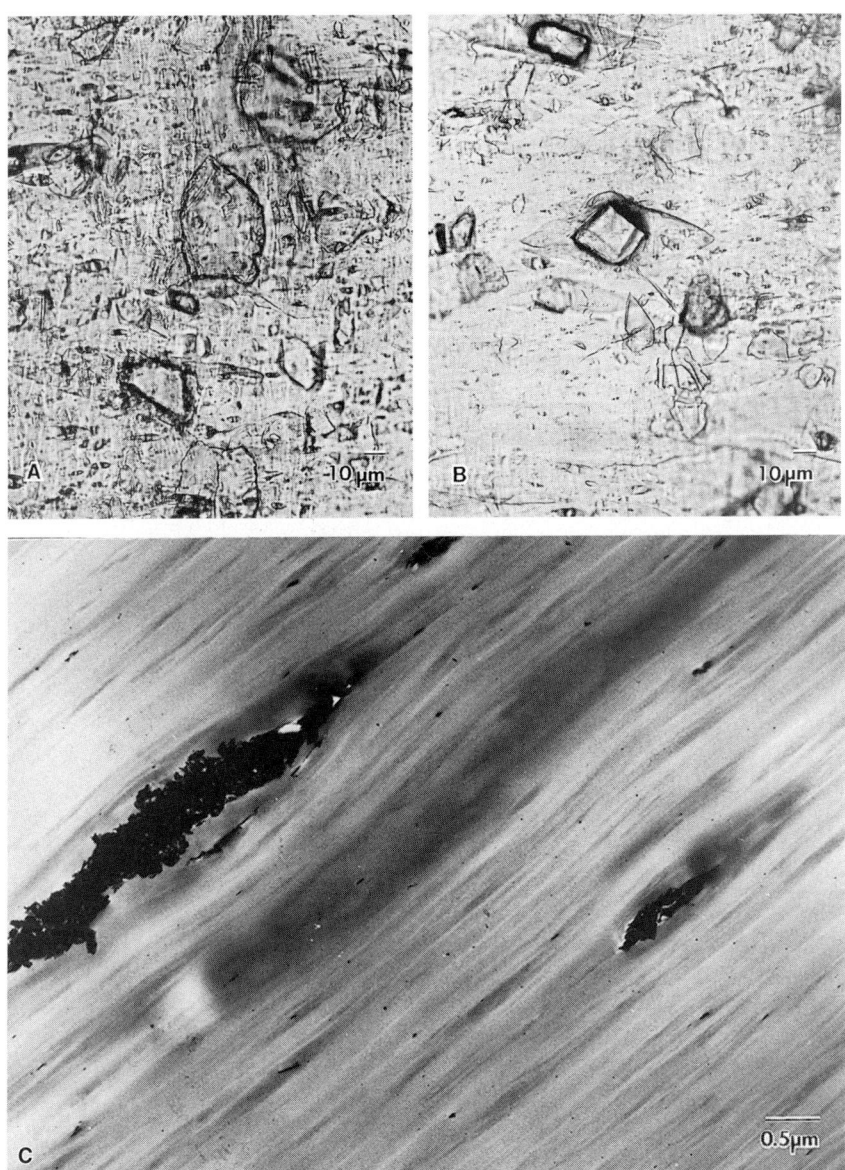

FIGURE 5.107. Sections of a mica filled thermoplastic are shown in the optical (A and B) and TEM (C) micrographs. The platy mineral filler particles are aligned with the polymer. The particulate texture of the mica (black) in the TEM cross section reflects the effects of diamond knife fracture of individual mica flakes.

decades, both with a focus on materials on the scale of nanometers (10^{-9} m). The electrical and electronics markets demand for miniaturization and for higher conductivity, antistatic additives and flame retardants is continuing. Whereas many argue that the cost for nanoadditives and their processing is too high, others point out the need and great promise of such new products. Every scientific journal and magazine has articles on this topic from every vantage point, and every major scientific conference has reported on the challenges and the opportunities sometime during the past few years (e.g., [478, 479]) making it impractical to review these references

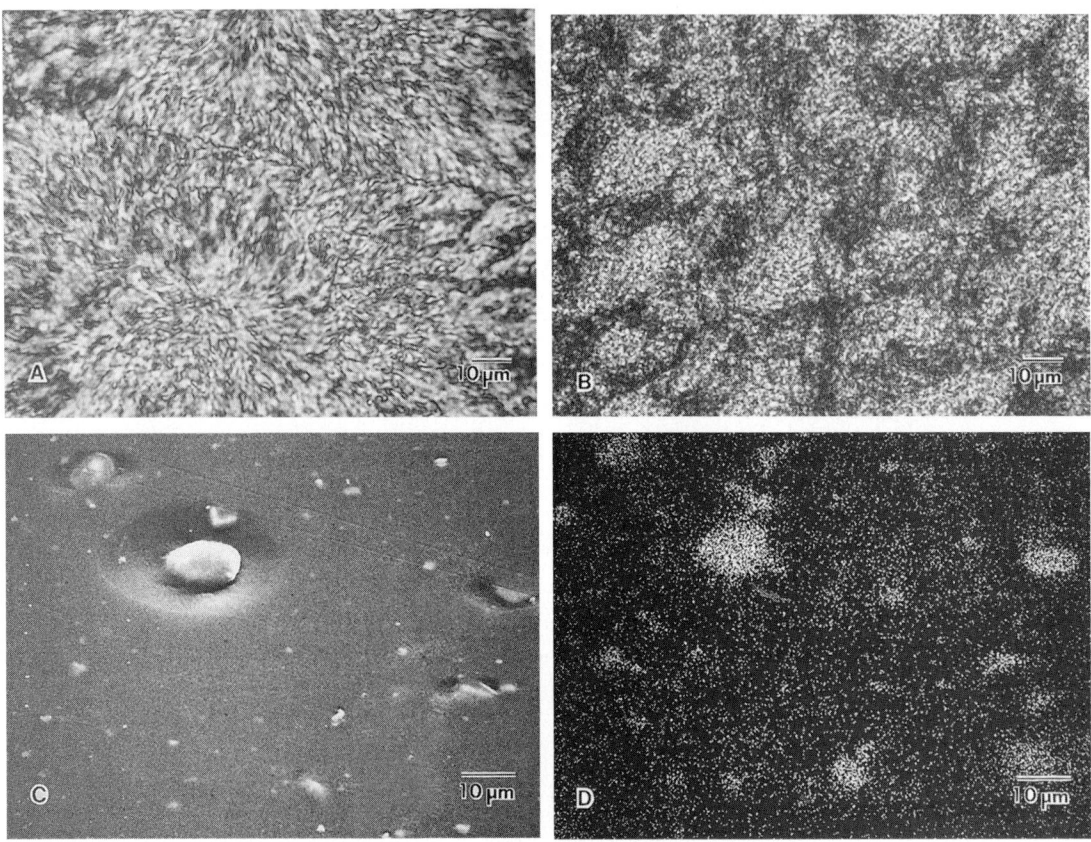

FIGURE 5.108. Semicrystalline thermoplastics, such as a polyacetal, shown in polarized light (A) do not appear spherulitic after the addition of calcium carbonate (B). The surface of such a filled, molded part can exhibit a matte finish due to poor wetting of the particles with the polymer. Scanning electron microscopy observation of the molded surface (C) shows pock marks, which are particles in cavities not filled with polymer. EDS mapping (D) shows the particles contain calcium.

in this rapidly changing field. Whether the advance in microscopes came first or the nanomaterials themselves is not the issue here but simply that they have developed and this multidisciplinary topic is a must for the polymer microscopist to understand. Several texts and articles that describe information relevant to microscopy are cited here by way of example [84, 185, 229, 253, 420, 480–486] as current work is best found by conducting an appropriate search on the specifics of interest.

Polymer nanocomposites are a class of materials that generally contain less than 6% of nanometer sized additives. The nanofillers are of two general forms, plates (layered silicate nanoclays; mainly montmorrillonite) or carbon nanotubes (e.g., single and multi-walled carbon nanotubes; SWCNTs and MWCNTs, respectively). The increasingly available nanoparticles create the opportunity for many different products in the future. The major differences between nanocomposites and traditional filled composites are the significantly smaller additive size and the lower concentration used in the product resulting in weight reduction, especially noteworthy for aerospace applications. In the case of nanometer sized clay particles, they are known to improve fire resistance, barrier properties, stiffness in films, chemical resistance, and mechanical properties (e.g., strength, modulus, and dimensional stability). However, processing to fully disperse these tightly packed clays (exfoliation) is a major issue. Carbon nanotubes are conductive additives that are thermally stable when compounded into engineering resins for electronic and medical applications. Polymers used for these applications include nylon 6 and nylon 12, polypropylene, polyesters, polycarbonate, TPO, and blends. Over time, the number and type of nanofillers and polymers that can be used has increased. Fabrication methods generally used for making nanocomposites are during polymerization, melt compounding, or solution blending. Forming techniques for nanocomposites are similar to those used for all polymers: injection molding, blow molding, extrusion, film or fiber making, and coatings.

Applications that benefit from these improved properties include gas barriers for bottles (beer bottles), food packaging (boil-in bags, stand-up pouches), fuel tanks, automotive applications, electronics and electrical applications (components, printed circuit boards), electrically conductive parts, wires, cables, and many others. Most major plastics companies and compounders are developing, manufacturing, and marketing such products.

The materials science focus is on the balance of the materials (polymer and filler), method of cost-effective processing for good dispersion, fabrication of the product, and the final performance and thus the evaluation of the interaction of the matrix and filler (e.g., [479]). Research has included the manufacture of nanomaterials such as clays by exfoliation and other methods to form finer, high aspect ratio particles. The range of polymers used has certainly increased with recent work done using natural or biopolymers [487]. The smaller size of the fillers results in significantly higher particle density and most importantly a larger interfacial area. This interface is responsible for the final properties. Thus, the nanofillers provide a means to "engineer and tailor morphology to achieve a desired property suite from the polymer nanocomposites" [479]. Characterization methods used include TEM, SEM, and SPM, rheology, thermal analysis, and mechanical analyses.

Although there can be major improvement in physical and mechanical properties, as with all materials the nature of the materials and the process used are key to limiting downsides, such as loss of toughness. Another potential downside is the fine sizes of the fillers themselves, and the possibility of their ingestion in the human lung and possible threat of disease such as is known for asbestiform materials. Finally, the patent literature on nanocomposites gives a view of the breadth and importance of this new technology.

5.4.6.2 Literature Review of Clay Nanocomposites

A common class of nanomaterials in use today is montmorrillonite, a layered silicate clay composed of stacked ultrathin platelets with higher surface areas than conventional particle reinforcements. When polymer nanocomposites are formed, the polymer fits into the spaces between the clay platelets resulting in swelling of the clay causing the plates to be exfoliated or dispersed throughout the polymer (e.g., [478]). Thus, the process for forming the nanocomposites is key to the dispersion and properties. Karger-Kocsis and Zhang [485] provide one of the many book chapters reviewing manufacturing methods, structure-property relationships, and characterization of these materials. The focus for this section is on the microscopy techniques and preparation methods used to study the complex structure of nanocomposites. Important elements of that structure include the dispersion of the particles, changes in the bulk matrix, and the formation and nature of the interphase between the particles and the matrix. The techniques commonly used for these studies are TEM, AFM, and SEM. Transmission electron microscopy is the preferred method to examine the dispersion of these particles, and as has been noted in this text many times, the higher the magnification used, the more care must be taken to ensure the images are representative (or not) of the bulk structure. Changes in the matrix polymer may be assessed by polarized light microscopy, but there is usually a need for TEM and/or AFM with combination of microtomy and staining or etching to image detail of the crystalline microstructure. Scanning electron microscopy fracture analysis continues to be the best method for assessing structure-property relations, especially for toughness. Complementary techniques include x-ray diffraction (XRD) and light scattering, and the suggestion is made to conduct XRD and TEM for silicate dispersion and to use various spectroscopy techniques, especially dynamic mechanical thermal analysis, to assess the formation of an interphase material [485].

Ethylene-vinyl alcohol copolymer (EVOH)/clay nanocomposites were prepared by dynamic melt blending [488], and the blend morphology was studied by SEM of microtomed surfaces after gold sputtering; freeze fracturing was not useful to distinguish between the two phases. Ultrathin sections were cut at room temperature using a glass knife for TEM imaging. X-ray diffraction showed advanced EVOH intercalation within the galleries, whereas TEM images indicated exfoliation, thereby complementing the XRD data. Addition of the clay was shown to change the thermal and viscosity properties of the composite. This study shows that complementary studies are important to understanding new materials. Poly(phenylene oxide) (PPO)/PA6 (50/50 w/w) blend nanocomposites were prepared by melt mixing of PPO, PA6, and organically modified clay and the morphology was studied by SEM, TEM, rheology, and WAXD [489]. The extruded nanocomposite was fractured in liquid nitrogen and then etched with chlorobenzene for 4h at room temperature to dissolve the PPO phase and kept in a vacuum for 8h before coating with gold for SEM imaging. Samples were ultramicrotomed at room temperature to about 60nm thickness and sections stained with OsO_4 for 20min for TEM. Careful comparison was done with SEM of the blend with various levels of clay, showing the change in the domain size of the dispersed PPO to decrease at 2% clay, but a co-continuous morphology formed at 5% clay. The TEM observation shows that all the organoclay is dispersed only in the PA6 phase with a high degree of exfoliation, and there is no clay detectable in the PPO phase for the nanocomposites regardless of the amount present. Therefore, the clay was shown to have significant effects on the blend morphology.

Apparently, relatively few attempts have been made to address the thermal stability of nanocomposites and thus research has been conducted using a pure, natural, and commercially available layered silicate, treated with 7-octenyltrichlorosilane in order to chemically graft the functional pendent organic group containing a C=C bond and using a pure, natural, and layered silicate, treated with synthesized imidazolium salts [490]. The nanodispersion of the treated clay in an epoxy matrix was evaluated qualitatively by XRD, TEM, confocal laser microscopy, and laser-induced fluorescence spectroscopy enabling complimentary characterization of the clay platelets over several length scales.

Polymer-clay nanocomposites as a class of flame retardant materials have a balance of mechanical, thermal, and flammability properties. Nanocomposites appear to offer advantages for flame retardants, especially in regions of the world where brominated compounds are being assessed negatively [491]. Although mechanical properties are improved with addition of nanoclays, they often also show reduction of heat release rates that can be important in a fire. Complex composites with organoclay and, for example, magnesium hydroxide, produces sufficient flame retardancy and allows for a reduction in the total filler content [491], giving hope for future research in this area. Natural and synthetic clays in polystyrene were studied by TEM of room temperature ultrathin sections (to show the size and dispersion of the clay), XRD, thermal analysis, and various heat release tests. Although the data suggest the synthetic clay does slightly better at reducing the heat release rate, the nanocomposites continue to burn once ignited.

The crystallization behavior of PA6 and its nanocomposites undergoing a microcellular injection molding process was studied using TEM, XRD, SEM, and PLM; the addition of polarized light microscopy was important to study crystal formation [492]. A synthesis approach to make PP nanocomposites by *in situ* polymerization was studied by OM, TEM, and

XRD [493]. The OM images showed the overall clay dispersion and TEM images supported exfoliation of the clay into very small stacks. Nanocomposites based on iPP and montmorillonite were studied by TEM and XRD to assess the polymorphism of the polymer and the interaction between the iPP and the clay [494]. Polypropylene is well known to be a commodity polymer with low cost and good mechanical performance and thus it lends itself to development of nanocomposites. Generally, the materials need to be modified to form compatible systems. Samples were prepared for TEM by staining with RuO_4 and cryomicrotomed to assess the clay morphology and distribution. Wide angle x-ray diffraction was used to study the influence of the additives on polymer crystallization, and SAXS was employed to assess the lamellar morphology. Transmission electron microscopy confirmed the XRD results, and image analysis was used to measure the particle sizes.

The phase structure and clay dispersion in PA6/PP/organoclay (70/30/4) systems with and without 5 parts of maleated polypropylene (MAH-g-PP) compatibilizer were studied by AFM [495]. Polished surfaces of specimens that were chemically and physically etched with formic acid and argon ion bombardment, respectively, were examined. Argon ion etching showed the PP was more resistant than PA6. In the absence of the compatibilizer, the organoclay was seen in the PA6 phase, as was shown also by TEM. The addition of MAH-g-PP resulted in a markedly finer PP dispersion and good interfacial bonding between PA6 and PP. In this blend, the organoclay was likely dispersed in the PA6-grafted PP phase. Nanocomposites of PA6 and clay were made at different clay loadings by melt processing, and the mechanical structure and microstructure were explored [496]. Scanning electron microscopy of fractured specimens showed a rounded "cabbage-like sheet" when the clay was exfoliated and absent in the poorly dispersed material. Transmission electron microscopy of cryomicrotomed samples also showed the clay dispersion.

Research is being conducted that shows that the thermomechanical properties of polymer nanocomposites are quantitatively equivalent to planar polymer films [497]. This work is focused on drawing a direct analogy between film thickness and an appropriate experimental interparticle spacing. Silica/polystyrene nanocomposites were made with varied loadings and the thermal and morphological properties assessed using SEM of fracture surfaces and TEM of microtomed samples to permit assessment of particle dispersion and interparticle spacing. The changes in glass transition temperature with decreasing interparticle spacing for two filler surface treatments were shown to be quantitatively equivalent to the corresponding thin film data with a nonwetting and a wetting polymer-particle interface. It appears from this work that the glass transition process requires that the interphase regions surrounding different particles interact. Clearly, more research is needed to understand these novel materials.

Natural polymer research has included use of these alternative materials with nanoparticles [487] because of three significant properties: multifunctionality, biodegradability, and biocompatibility. Breakthroughs in cost of production and property profiles for biomaterials will be needed before they become reasonable to market. Research has been conducted on melt formation of a starch-clay nanocomposite for bioplastic applications [487]; however, an issue is the high water uptake and thus loss in mechanical properties requiring modification of the clay and the composite process. As with other nanocomposites, microstructural characterization is typically by TEM and AFM.

5.4.6.3 Literature Review of Carbon Nanotube Composites

Single wall carbon nanotubes (SWCNTs) are an arrangement of a sheet of carbon atoms joined in a pattern of hexagons and rolled into a cylinder. The conductivity of the nanotubes depends on the way in which the ends wrap around and meet. The mass-strength ratio and exceptional mechanical properties are of importance for space applications as critical parts of both shuttle vehicles and satellites depend on strength and toughness of the materials, while there are strict limitations on the weight of the components. The significant potential of nanotubes as a material for space applications is being investigated as are new processes in an attempt to realize these potential properties by limiting phase segregation (e.g., [498]). Scanning electron microscopy and TEM evaluation is critical to development of new process-structure understanding.

Carbon nanotubes have been used to toughen high performance PE to produce improved electrical and mechanical properties. Many studies have been conducted to develop these materials with the best dispersion to exploit these new applications. In one study, the addition of 1 wt.% multiwalled carbon nanotubes (MWCNTs) to high modulus ultrahigh molecular weight polyethylene (UHMWPE) films drastically enhanced toughness [499]. The morphology of the nanotubes was observed by SEM and TEM and the surface of the nanocomposite films was shown by OM, SEM, and AFM. A combination of tensile and Raman spectroscopic measurements showed that the presence of MWCNTs in the composite can lead to a ~150% increase in strain energy density in comparison with the pure UHMWPE film at similar draw ratios. This is accompanied by an increase of ~140% in ductility and up to 25% in tensile strength. Plasma coating of carbon nanofibers with ultrathin films of polystyrene has been used to enhance the dispersion and interfacial bonding when used in a polystyrene nanocomposite. Scanning electron microscopy and HRTEM were used to assess the coating interface [500].

Single wall carbon nanotube/poly(methyl methacrylate) (PMMA) nanocomposites were prepared by a coagulation method, providing uniform dispersion of the nanotubes in the polymer matrix, as observed by OM, SEM, AFM, and Raman imaging [481]. Intermittent contact mode AFM amplitude images of SWCNTs deposited from a dimethylformamide (DMF) suspension onto an amine-terminated silicon surface showed the nanotubes overlapped very little and thus the length and height could be measured. Multiple methods were said to be critical for understanding the complex morphologies on a range of size scales.

Schulte and Nolte [501] reviewed the topic of carbon nanotubes and carbon nanofiber polymer composites. They reviewed the field of research in carbon materials since the 1985 discovery of C_{60} structures and described the nature of the two types of carbon nanotubes: SWCNT and MWCNT, their production, and imaging the structures by high resolution TEM. Carbon nanotubes are characterized by an extremely high specific surface area that can interact with the matrix material and form a potentially strong interface for good load transfer [501], but also important is the wide variation in properties that depend on the process for dispersion. As with nanoclays, the effect of these nanoparticles on the crystalline morphology of the matrix has been studied, generally using TEM. For the most part, commercial applications for polymer composites with nanotubes are based on their electrical properties; further research is ongoing on the cost-performance of nanotubes and the best methods for their dispersion into polymers.

5.4.6.4 Problem Solving Applications

Several published problem solving applications have been chosen as examples of microscopy of nanocomposites. Research is being conducted on the process to form nanoclay composites that will be effective for clay particle exfoliation and improved property enhancements. One of the newer process methods being developed at Clemson University is the use of chaotic blenders (e.g., [184, 185, 329, 502]). Examples of this work on PA6/nanoclay nanocomposites having platelets volumetrically oriented and localized within alternating platelet-rich and virgin PA layers, produced with a continuous chaotic blender (see Chapter 1), are shown in Fig. 5.109 [502, 503]. The nylon and clay were first compounded with a twin screw extruder to form a master batch. A nanocomposite film with overall volume composition of 2% platelets was obtained using the continuous chaotic blender. Thin transverse sections were cryomicrotomed for TEM investigation, seen in Fig. 5.109, which shows the layers of oriented platelets and matrix polymer. Sufficient chaotic blending results in minimization of the matrix

Composites

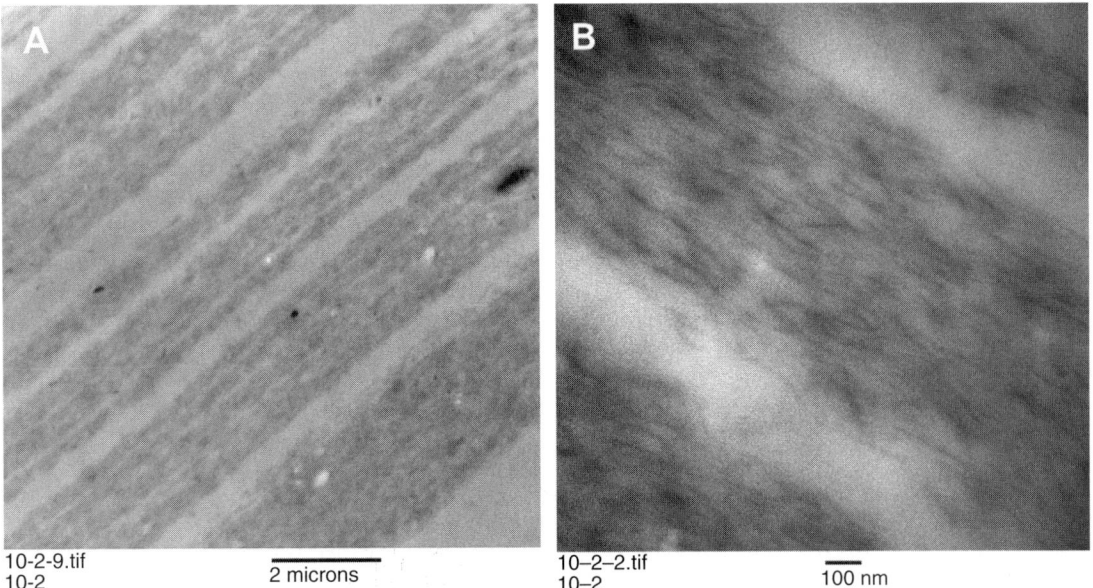

FIGURE 5.109. Transmission electron microscopy micrographs of cryomicrotomed sections of a nylon/clay nanocomposite produced by continuous chaotic blending forms layers of oriented platelets and matrix polymer. (From Zumbrunnen et al. [502, 503]; used with permission of the Society of Plastics Engineering.)

layers and better dispersed nanoclay with improved mechanical properties.

The morphology of thermoplastic olefin (TPO)/clay nanocomposites with clay loadings of 0.6–6.7 wt.% was investigated by AFM, TEM, and XRD and compared with their mechanical behavior [504]. A master batch of TPO (25%), maleic anhydride grafted PP (25%), and clay (50%) was compounded and used with TPO to vary the clay loading of the nanocomposite. Atomic force microscopy was conducted on cryomicrotomed surfaces, as a function of clay loading as shown in Fig. 5.110 [504]; the polymer particle diameter was seen to decrease as the clay loading increases. Cryomicrotomy with a diamond knife was conducted after end block staining with RuO_4 for 8 h to produce 40 nm thick sections for TEM, of the same samples, as shown in Fig. 5.111 [504]. Transmission electron microscopy micrographs showed that the clay platelets preferentially segregated to the rubber-particle interface. The ethylene-propylene rubber (EPR) particle morphology in the TPO underwent progressive particle breakup and decreased in particle size as the clay loading increased from 0.6 to 5.6 wt.% resulting in an increase in the flexural modulus without a loss in toughness.

The group at the Nano Tech Institute and Department of Chemistry at the University of Texas at Dallas, led by Musselman, has conducted research on the separation of SWCNTs to enable their use in biological applications such as artificial muscles (actuators) and biomedical sensors, among many other electrical and electronic applications [505–509]. Two challenges for effectively exploiting the remarkable properties of SWCNTs are the isolation of intact individual nanotubes from the raw material and their assembly into useful structures. In an early work, OM, SEM, and TEM were used to image carbon nanotubes coated to control their assembly into macromolecular structures [505]. Scanning electron microscopy sample preparation was by placing a drop of the nanotube solution on a precleaned Si chip or carbon substrate, wicking away the solvent and drying the sample in air, and imaging without metal coating. Transmission electron microscopy sample preparation was conducted on a sample

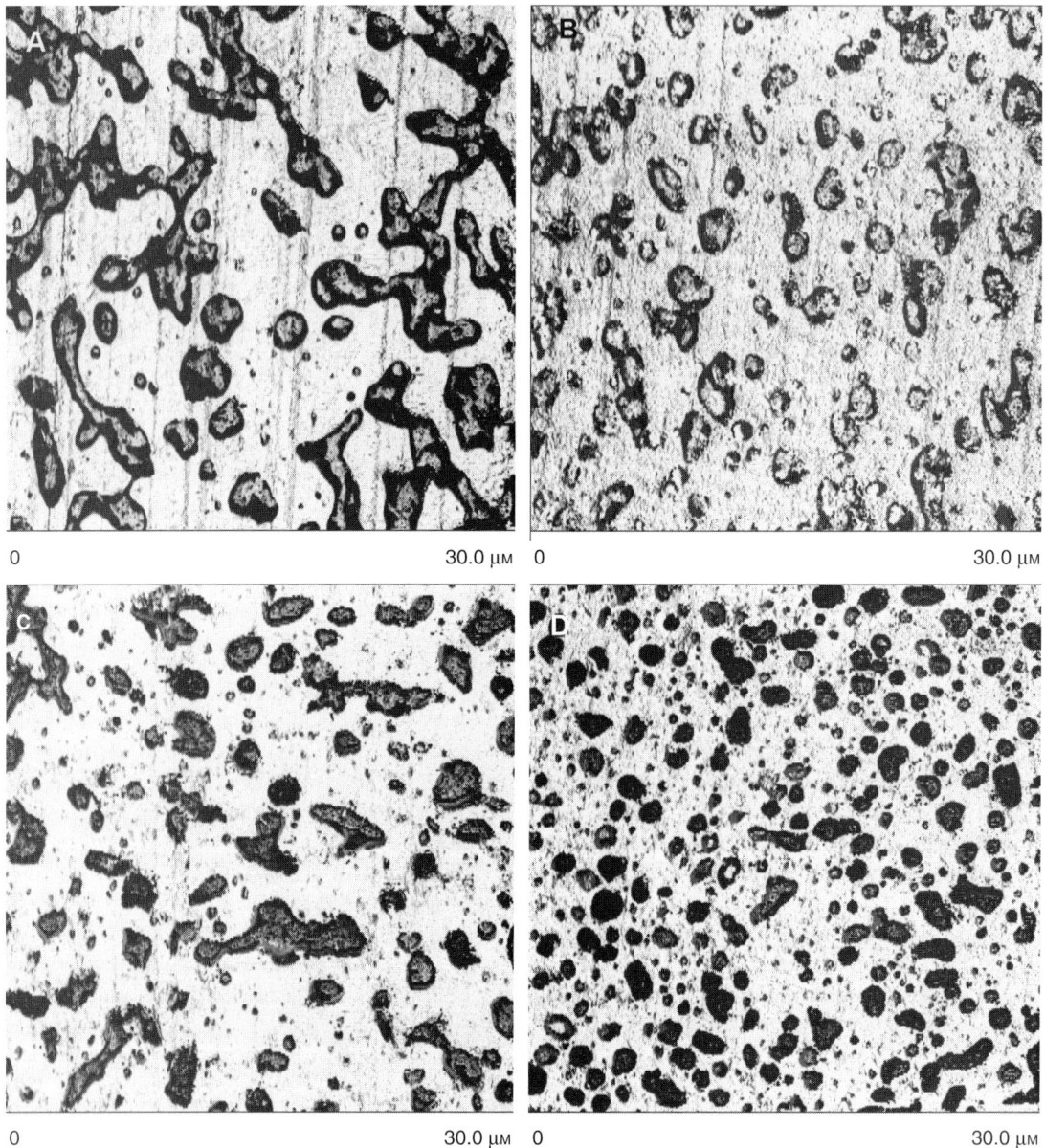

FIGURE 5.110. Atomic force microscopy phase images of (A) TPO-0 (no clay), (B) TPO-1 (0.6% clay), (C) TPO-3 (2.3% clay), (D) TPO-6 (5.6% clay). (From Mirabella et al. [504], © (2004) Wiley-Interscience; used with permission.)

made using one drop of solution on a Cu TEM grid with a holey carbon support film. In a more recent study [506], TEM sample preparation was as described, and AFM images were acquired in air of samples deposited on mica, rinsed, and dried. In this study, AFM (Fig. 5.112a) and TEM (Fig. 5.112b) images provided complementary evidence that individual peptide-wrapped SWCNTs could be isolated using an amphiphilic α-helical peptide [506].

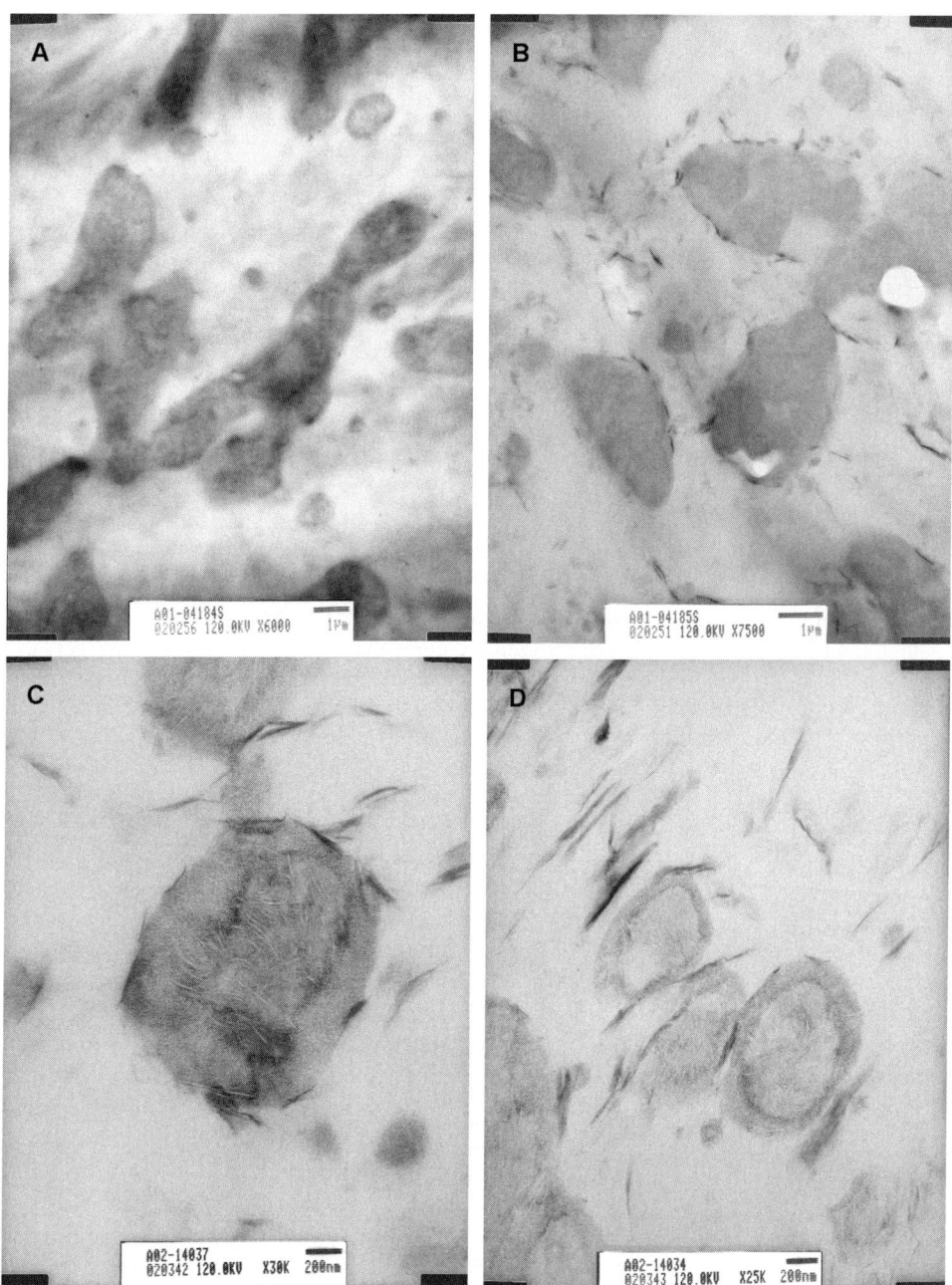

FIGURE 5.111. Transmission electron microscopy images of cryomicrotomed and RuO_4 stained sections of (A) TPO-0 (no clay), (B) TPO-1 (0.6% clay), (C) TPO-3 (2.3% clay), (D) TPO-6 (5.6% clay). Rubbery domains (elliptically shaped) are surrounded by clay platelets (dark rod-like structures). Note the images are taken at different magnifications. (From Mirabella et al. [504], © (2004) Wiley-Interscience; used with permission.)

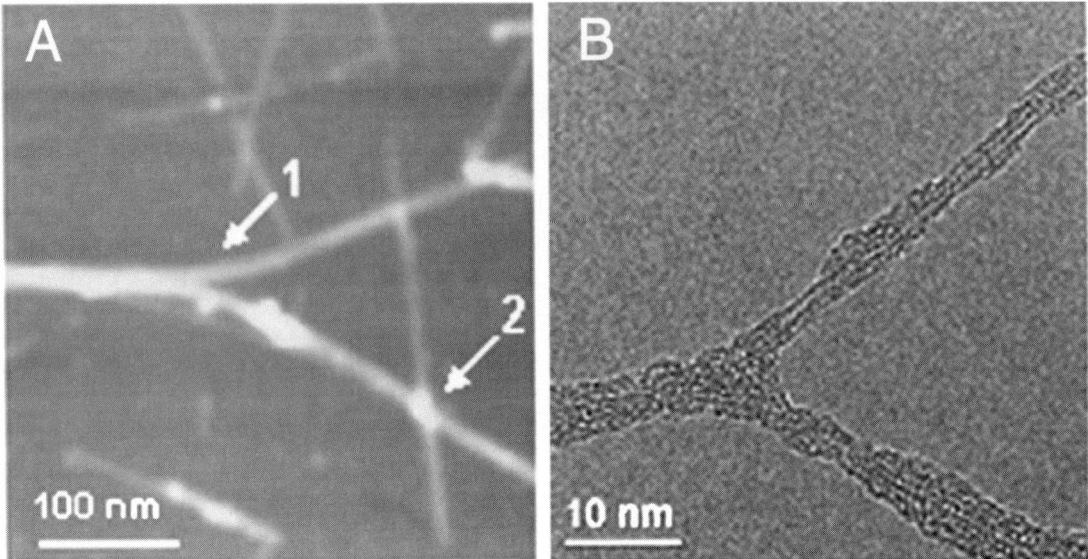

FIGURE 5.112. Atomic force microscopy image of nano-1/SWCNT sample exhibiting a Y-junction (*1*) and an X-junction (*2*) (A). Transmission electron microscopy image of peptide coated SWCNTs exhibiting Y-junction apparently created through peptide-peptide interactions provided complementary evidence that individual peptide-wrapped SWCNTs could be isolated using an amphiphilic α-helical peptide (B). (See color insert.) (From Musselman et al. [506], © (2004) American Chemical Society; used with permission.)

5.5 EMULSIONS, COATINGS AND ADHESIVES

5.5.1 Introduction

Emulsions, dispersions of one liquid in another liquid, find broad application in the fields of paints and coatings, paper, printing, food, medicine, pesticides, and cosmetics. Emulsions include a broad range of liquids that consist of a stable, continuous liquid phase in which a second discontinuous immiscible liquid phase is present [510, 511]. Broadly, these polymers can be classified as macroemulsions, latexes, colloids, and microemulsions. The major structural unit of interest is the "particle." The two types of emulsion, based on the size of the dispersed particles, are *macroemulsions*, where particles range from 0.2 to 50 µm, and *microemulsions*, with particles from 10 to 200 nm. Particle size determines the optical clarity of the emulsion. Macroemulsions tend to be milky white, with particles less than 1 µm ranging from blue-white to semitransparent; and microemulsions, with particles less than 50 nm, are generally transparent. A *latex* is a special case of a water emulsion of rubber or polymer particles that is applied extensively in paints and coatings. *Colloids* are defined as any particle, liquid or solid, which has some linear dimension between 1 nm and 1 µm and which when dissolved in a liquid will form a suspension. Thus, either "emulsion" or "colloid" can represent this entire group of liquids, and in fact, emulsions can be considered as colloidal suspensions.

Microemulsions [512, 513] are special types of emulsions that form spontaneously and have very small particles. Microemulsions are optically clear, thermodynamically stable dispersions of two immiscible liquids obtained by the use of carefully adjusted surface-active molecules (surfactants). Both liquids in a microemulsion will be present in regions of the same order of magnitude, with the "dispersed" phase on the order of 10–100 nm. Aggregates of surface-active molecules, or *micelles*, form into colloidal-sized clusters in such a way that hydrophilic groups are directed toward the water. These definitions [514] are general in nature, but they suffice for the current purpose; the interested reader is directed to texts on this complex topic for a more rigorous discussion [510, 513, 515–518].

Adhesives are polymers in this general class of materials used in many applications, from the back of postage stamps to demanding military applications cementing metal joints in military aircraft with polymers such as epoxy resins. The interfaces of such materials must be characterized to determine the strength of the adhesive bonds and the relation of properties such as peel strength with morphology. Adhesion science has been described in the literature (e.g., [517, 519–523]).

It is well known that the particle shape, size, and distribution of a latex or emulsion control the properties and end-use applications. Many types of latex are manufactured with a controlled and sometimes monodisperse distribution of particle sizes. These polymer liquids are wet and sticky, making specimen preparation for microscopy very difficult. Because particle size and shape are so important to properties, the preparation must focus on not changing the particles as found in the fluid state. Preparation includes simple methods (see Section 4.1) such as dropping a solution onto a specimen holder, staining/fixation (see Section 4.4), microtomy (see Section 4.3), and special cryo methods (see Section 4.9). All microscopy techniques can be used for these studies. This section is meant to provide a brief survey of the types of microscopy applications that have been found useful in the evaluation of emulsions, latexes, and their use as coatings and adhesives.

5.5.2 Emulsions and Latexes

The structure and morphology of multiphase polymers have been discussed (see Section 5.3), and the particle size and distribution have been shown to be quite important for mechanical properties and applications. In many polymers (e.g., ABS), the particle size distribution is determined during the rubber manufacturing process. The surfactant concentration during emulsion polymerization controls the size distribution of the rubber latex, and subsequent grafting increases the size further. Particle sizes can be controlled to yield a range of sizes or a monodisperse latex. Particles that are larger than $1\,\mu$m in diameter are difficult to produce because they tend to coagulate. Crosslinking of butadiene can occur during the process, but this can be controlled somewhat by the addition of inhibitors. Styrene and acrylonitrile monomers are added to the polybutadiene latex in the second stage of the emulsion polymerization process, and new particles can form or polymer can be deposited on the surface of the polybutadiene where grafting can take place. The end result of such processes is a range of particle compositions, sizes, and morphologies. Core-shell morphologies are commonly encountered in ABS latex where the polybutadiene is the core and the shell is styrene-acrylonitrile copolymer (SAN). Subinclusions of SAN can also form within the rubber phase. Subinclusions have also been shown to form in other polymers, such as in nylon containing polymer blends with rubber (see Section 5.3).

5.5.2.1 General Literature Review

An overview of polymer latex film formation and properties with extensive references on this important topic has been provided by Steward et al. [524], who state that the current market driver is the need to find alternatives for solvent-based systems with their adverse environmental impact. This field has been studied for more than 50 years, and yet the mechanisms of formation of water insoluble coatings from polymers in aqueous solution is still a subject of interest, aided in no small part by various microscopies, especially the advent of SPM.

Molau and Keskkula [525] were among the first to study the mechanism of particle formation in rubber containing polymers. They showed that phase separation occurs between the rubber and a vinyl polymer during the polymerization of solutions of rubber in vinyl monomers, which is followed by formation of an oil-in-oil emulsion. Structural investigations by phase contrast optical microscopy (see Section 5.3) reveal dispersed particle size and distribution. Ugelstad and Mork [526] reported on diffusion methods for the preparation of emulsions and polymer dispersions where the size and distribution of the latex particles were monitored by very simple optical, SEM, and

TEM methods. A microemulsion polymerization was reported [527] with spherical latex particles produced about 20–40 nm in diameter.

The formation of micelles, or colloidal particles, by block copolymers in organic solvents was reviewed by Price [528]. The molecular weight of polystyrene was estimated from specimens prepared by spraying and evaporation for TEM. Freeze etching a drop of solution rapidly frozen with liquid nitrogen [529] was described (see Section 4.9.4) where the solvent was allowed to evaporate and a replica produced of the fracture surface. Another method [530] was to allow a drop of an osmium tetroxide stained micellar solution to spread and evaporate onto a carbon film for TEM. Price et al. [531] investigated micelles from a polystyrene-poly(ethylene/propylene) block copolymer in a lubricating oil. The specimens were prepared for TEM either by casting a film on water and picking up the section on a carbon coated grid, followed by shadowing, or by painting the solution onto a freshly cleaved mica surface and coating the surface with carbon/platinum. Lee and Ishikawa [532] prepared ultrathin, osmium tetroxide stained cross sections for TEM examination of "inverted," core-shell latexes. Replicas of microemulsions have shown micelles to be spherical particles using TEM [533]. A lengthy discussion of microemulsions [534] provides the key types of experimental results from such studies: phase diagrams, structures, thermodynamic considerations, and a discussion of interfacial tension.

Shaffer et al. [535] modified staining techniques for TEM of latex particles by combining a few drops of the latex with a few drops of a 2% uranyl acetate solution, which serves as a negative stain. A drop of that mixture was deposited on a stainless steel formvar coated grid, dried, and stained with RuO_4 to differentiate the rubbery core, which is not stained, from the dark shell, which stains due to its ring structure. A method was more fully described and a figure shown (see Fig. 4.18) of latex particles prepared by staining with OsO_4, RuO_4, and phosphotungstic acid to reveal full details of the core-shell particles.

Cryomicrotomy and chemical fixation were reviewed as methods for imaging the morphology of rubbery latex particles [536]. In one method, a drop of latex was frozen in liquid nitrogen, sectioned with a diamond knife, and stained with OsO_4 vapor for TEM. When applied to latexes made by emulsion polymerization of methyl methacrylate in a natural rubber latex seed, inclusions are clearly visible. A chemical fixation method was described for imaging the morphology of rubbery latex particles by addition of glutaraldehyde, followed by OsO_4. The sample is then dehydrated in ethanol, epoxy resin added, and the sample cured, ultramicrotomed, and imaged by TEM.

The film formation of latexes to form coatings is well known and has been followed by SEM [537] and TEM [538] and more recently by AFM as it is important in understanding how latex polymers form films in coatings applications. Films of varied thickness, of 60/40 poly(styrene-butadiene), were aged for various times and the aging subsequently stopped by bromination [537]. They were then examined in the SEM [539] and the "further gradual coalescence" process [540, 541] was shown to result in a change of particle shape as a function of aging [539]. A detailed TEM study of the effect of surfactants on the film formation of latexes was reported where the rate of film coalescence was monitored by film replication [538]. Coating layers prepared on nonabsorbent substrates, composed of mineral pigments, latex binders, and polymeric thickeners and dispersants, typical of those used in paper coatings, were studied using conventional SEM (gold coated) and SEM at higher pressures and humidity [246]. Wang et al. [542] used TEM to study the morphology of films containing polystyrene after staining with uranyl acetate, and AFM was used on purely acrylic materials and their blends or hybrid latexes with polyurethane. The effect of the degree of blockiness and molecular weight of four different poly(vinyl alcohol)s used in the emulsion polymerization of vinyl acetate (VAc) on the latex film morphology development, imaged by AFM, showed that the surface morphology is strongly dependent on the degree of blockiness and molecular weight of the PVOH used [543].

5.5.2.2 Literature Review: SEM

Imaging in a low voltage (high vacuum) SEM, as has been shown throughout the various examples in this text, and by Vezie et al. [544], provides an alternative for high resolution imaging complementary to TEM and SPM. Practically, lateral resolution on the order of 5 nm at 1 kV accelerating voltage can be obtained in polymer samples. Low voltage SEM has been reviewed as a practical method for study of nanosized core-shell latexes embedded in a polymethylmethacrylate matrix and semicrystalline polypropylene/ethylene-propylene rubber [545]. Three methods were compared: (1) a diamond knife block face was stained with RuO_4 and observed directly; (2) ultrathin sections from the stained block were picked up on TEM grids, covered with a discontinuous carbon film; and (3) ultrathin stained sections were picked up onto an SEM stub and water was removed. The effects of a range of accelerating voltages versus magnification (about 1–3 kV is best for about 40,000× and a working distance and spot size of 3 mm and 3, respectively) and evaluation of the effect of the metallic support and comparison with TEM make this study a valuable resource. For thin specimens (about 100 μm thick), a copper support was best suited for charge dissipation in SEI images, and the lower atomic number elements, such as Al, were best for BEI imaging.

The direct study of wet polymer latex systems and other liquid mixtures has been conducted in low vacuum, variable pressure, and "environmental" SEMs, which permit the vacuum to be controlled so that the samples need not be dried or coated (e.g., [246, 546–551]. Meredith and Donald [546] were among the first to study wet, electrically insulating latexes in different environments, such as water vapor, nitrous oxide, and nitrogen, and documented various contrast features. These include bright particle edges, or "haloes," due to retention of water around the hydrophilic shells; "puddles" on the surfaces of film-formed samples; condensation of water into fine interstices of particles; and masking of some features when imaged in nitrogen. Stokes [550] reviewed *in situ* experiments in such SEMs that may be carried out including mechanical deformation and the observation of dynamic processes such as wetting and swelling behavior of materials, thermal responses, the effects of hydration, dehydration, and rehydration, and film formation. However, as with all new techniques, the presence of artifacts or imaging difficulties has to be studied, and these have shown there is significant electron beam damage in hydrated specimens. Kitching et al. [547] studied beam damage mechanisms by exposing polypropylene films to the electron beam at varying doses and exposure times under both hydrating and dehydrating conditions. Fourier transform infrared (FTIR) results showed crosslinking occurred during the direct interaction of the electron beam with the polymer. Royall et al. [549] also found acute damage in the environmental SEMs, especially due to the nature of the beam sensitive specimens that are imaged and more importantly due to the water, which acts as a source of highly mobile free radicals.

Theoretically, variable pressure SEMs permit the study of *in situ* drying of latexes [552], however the larger latex particles are typically around 500 nm in diameter, and this results in radiation damage of the specimen due to the electron beam, so that the evolution of particular features cannot be followed. The change from ambient temperature and pressure to the conditions of temperature and pressure in such an SEM can subject the specimen to a very high evaporation rate, which can disrupt film formation. The inclusion of a drop of water in the specimen chamber apparently largely alleviated this, enabling successful imaging of film formation. Craven et al. [551, 553] discussed the effects caused by an excess quantity of ionized gas molecules within variable pressure SEMs on specimen charging, recombination, and development of space charge, and they described a new device for removal of excess charge. Thiel et al. [553] presented a framework for understanding charging processes in low vacuum SEMs. Overall, it is clear that the technique of low vacuum SEM as with all other techniques must be compared with standard high vacuum SEM and other complementary techniques to ensure a true understanding of the polymer morphology. Advantages and disadvantages must be considered, and further research is likely on this topic.

5.5.2.3 OM, SEM and TEM Characterization

Optical and SEM techniques have been used to image micelles and latex particles by some straightforward preparation methods. Particles in suspension or air dried may be imaged directly by optical microscopy when the particles are well over 1 μm in diameter, as seen in images of rather large poly(vinyl acetate) particles (Fig. 5.113A) and poly(vinylidene chloride) particles (Fig. 5.113B). Film forming latexes cannot generally be examined by simple air drying as they tend to form a film, and thus freeze drying may be used (see Section 4.9.2). Katoh [554] freeze dried poly(ethyl acrylate), which has a softening temperature below room temperature, and coated the sample with carbon and gold while at low temperature. Examination in the SEM showed the particles to be spherical. Direct observation of monodisperse latexes has been shown by a special optical technique [555] developed for understanding the stability behavior of monodisperse systems. A metallurgical microscope with a pinhole plate of aluminum foil positioned at the field aperture iris of the illumination tube was used to increase the resolving power. The concentrated sample was placed into a Pyrex glass tube (15 mm diameter) with a thin glass window. After equilibration, the cell was examined using an oil immersion objective. The image was displayed on a TV monitor and video recorded for measurement of interparticle distances. Kachar et al. [517] developed a technique of video enhanced differential interference contrast to study aggregates and interactions of colloids by real time experiments with a high resolution TV camera connected to an optical microscope equipped for Nomarski differential contrast.

Transmission electron microscopy has proved to be the most effective technique for the characterization of the particle size distribution in emulsions. A dilute solution cast on a carbon film and metal shadowed shows an agglomerate of latex particles in a commonly encountered drying pattern where the shadowing method shows the particles are flat and obviously not separate or discrete (Fig. 5.114A). The area in Fig. 5.114B would also be difficult to use for particle size measurement, whereas in Fig.

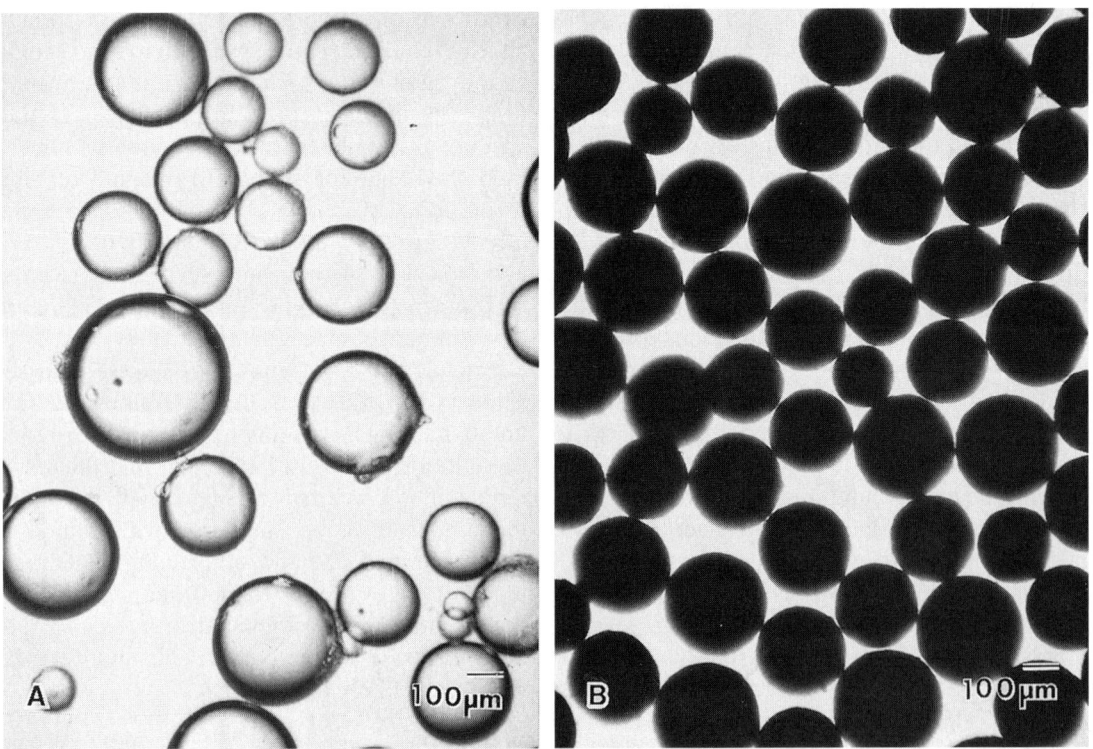

FIGURE 5.113. Optical micrographs show the size and shape of poly(vinyl acetate) (A) and poly(vinylidene chloride) (B) beads.

Emulsions, Coatings and Adhesives

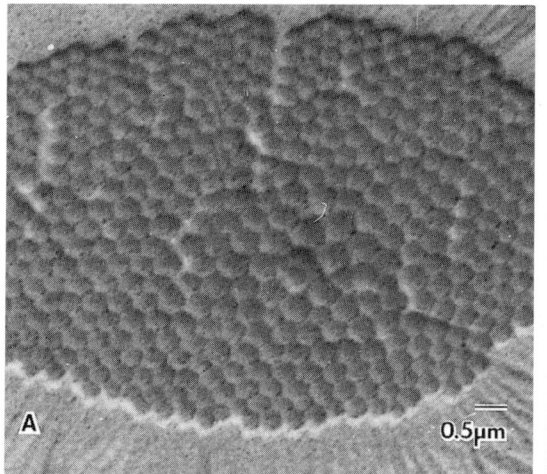
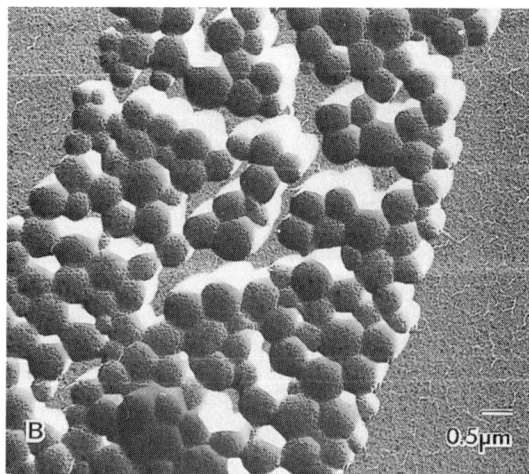
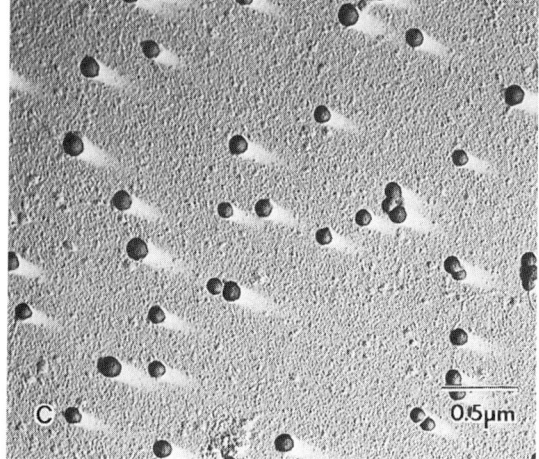

FIGURE 5.114. Transmission electron microscopy micrographs of several emulsion particle samples show a range of aggregation. An air dried droplet (A) resulted in agglomerated flat particles. More three dimensional particles would still be difficult to measure as they are touching (B). The emulsion particles (C) are well dispersed, and shadowing with chromium clearly shows that they are discrete spheres after freeze drying.

5.114C the emulsion particles are well separated and the shadows show they are three dimensional in shape due to freeze drying. A combination of staining, to enhance contrast, and a cold stage in the microscope has been applied to the study of latexes by TEM in order to limit such flattening and aggregation. Shaffer et al. [556] developed a phosphotungstic acid (PTA) staining method (see Section 4.4.5) where the latex was added to a 2% PTA stain and then dropped onto a TEM grid for imaging. A cold stage was used in the microscope to limit any change in the particles during examination. The effect of the cold stage and PTA staining was shown previously (see Fig. 4.21A, B).

Transmission electron microscopy micrographs are shown as an example of both the replication method and the effect of aging on film formation. The surface of a film of PVAC/BA latex is shown after 8 h (Fig. 5.115A, B) and after 15 days (Fig. 5.115C); both were prepared with the same surfactant. Clearly, the particulate nature of the film is still obvious after 8 h, but the film texture has changed with time, and after 15 days there were no obvious surface details remaining.

5.5.3 Particle Size Measurements

Latex particle size measurements are generally conducted by either light scattering or by electron microscopy. Rowell et al. [557] reported rigorous measurements of polystyrene latex particles by both techniques, and the average values agreed within 1%. The preparation for TEM involved the drying of a drop of the

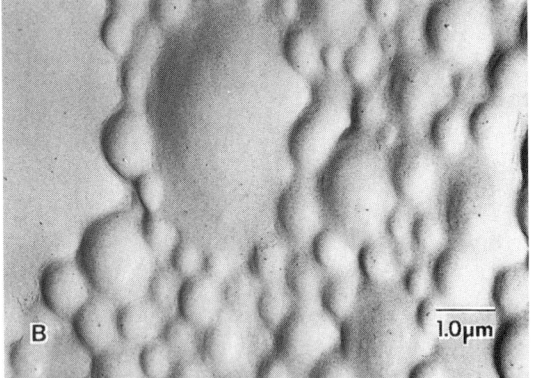
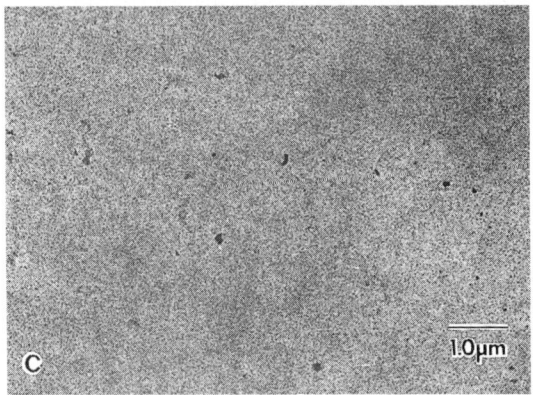

FIGURE 5.115. Vinyl acetate latex film coalescence is shown by TEM of platinum–palladium–carbon replicas from aged latex films cast on glass. Transmission electron microscopy micrographs show the particulate nature of a film aged for 8 h (A and B) compared with the flat film observed after 15 days aging (C).

diluted latex onto a replica of a diffraction grating, shadowing it with carbon, and taking direct measurement from enlarged micrographs. Micelles of block copolymers of polystyrene and poly(dimethylsiloxane) in n-alkanes were shown by a similar preparation when the diluted solution was dropped onto a carbon coated grid for TEM [558].

Image analyzing computers are now routinely applied to the measurement of structures observed in microscope images (see Section 2.8). Quantitative microscopy involves several steps that all must be considered as part of the analysis. Key factors are to ensure that the specimen does not change dimension during preparation or during microscopy, that a representative sample is prepared and analyzed, and that the calibration of the microscope is accurate. Gratings are used to calibrate the microscope, and standard polystyrene latex is used to have a check on the change in particle size as a function of the method, the vacuum, and other microscope conditions. Analysis directly from TEM negatives avoids an additional step of printing micrographs. The actual measurement of particle dimensions with an automatic image analyzer is trivial compared with these earlier steps, although due consideration must be given to statistical experimentation, sampling, and presentation of the distribution data. An example of the determination of particle size distribution that is generally applicable is the study of a latex prepared by freeze drying (see Section 4.9.2), shown in Fig. 4.49, in which it is clear that the film forming nature of air dried latex precludes direct measurement of particle size and distribution. Freeze drying results in three dimensional particles and shadows with well dispersed particles, if low enough concentration is used in the preparation.

5.5.4 Adhesives and Adhesion

5.5.4.1 Literature Review

Adhesives are polymers that are used in many ways, for example in composites, automotive tire cords, plywood, tapes, and labels. A particularly demanding application is the cementing of metal joints in military aircraft with polymers such as epoxy resins. The interfaces must be characterized to determine the strength of the adhesive bond and the relation of such properties as peel strength with morphology. The morphology of the adhesive fracture surfaces is generally investigated in the SEM. The topic of adhesion science and adhesion and absorption of polymers has been described (e.g., [519, 520, 559]).

Smith and Kaelble [560] conducted a study to determine the adhesive failure of a metal-polymer system. A multidisciplinary study was used to describe the aluminum alloy and the epoxy adhesive. Ellipsometry was used to estimate oxide film thickness and optical properties. Optical microscopy, SEM, and TEM established the topography, and wettability parameters were calculated from contact angle measurements and bond strengths. Morphological assessments of the polymer adhesive-metal joint were made by SEM of the fracture surface or by production of transverse sections. Hahn and Kotting [561] prepared transverse sections by machining followed by ion etching with argon to remove the smeared structure of the phenolic and epoxy resins. The SEM showed a variation in adhesive morphology as a function of location with respect to the metal part that further depended upon the adhesive type, cure, condition, and nature of the metal surface.

Brewis and Briggs [562] showed pretreatments to be important in adhesion. Surface pretreatments for polyolefins include chlorine, ultraviolet radiation, dichromate/sulfuric acid, hot chlorinated solvents, and corona discharge. These authors reviewed the nature of the changes and mechanisms associated with pretreatments, including references to SEM and TEM studies of the resulting structures. Surface modification of PE by radiation induced grafting resulted in improved wet peel strength, and optical and SEM techniques showed a change in bond failure in a wet environment [563]. A surface graft with good adhesive bonding to epoxy adhesives was produced by vapor phase grafting of methyl acrylate onto PE. The SEM was also used to observe the surfaces of molded, roughened, and etched polyolefins, PVC and ABS both before and after metallization [564]. The study showed that the metal bonds to the plastic mainly by mechanical anchoring.

Two papers, by separate groups working simultaneously, described the structure of the resorcinol-formaldehyde-latex (RFL) adhesives used for bonding rubber to tire cords. Rahrig [565] suggested that the RFL adhesives are two phase systems with an interpenetrating network morphology based on his dynamic mechanical and thermal analysis studies. Meantime, Sawyer et al. [566] reported on a method and its application to actual automotive tires, whereby the fine structure of the RFL and, more importantly, the interfacial morphology were clearly shown. Micrographs were described previously (see Fig. 4.22 and Fig. 5.20); these reveal the complex RFL morphology by an ebonite staining (see Section 4.4.6) microtomy method.

Much study of adhesion has focused on developing improved adhesives and composites for aerospace applications [567, 568]. Surface analysis techniques such as SEM and XPS are commonly employed to analyze fracture surfaces, and contact angle measurements have also supported such studies. Adhesives and adhesion are quite important in two other major areas of application: as adhesives bonding rubber in automotive tires and as sticky tapes and labels. Pressure-sensitive adhesives or self-adhesive materials have been described by Creton [569]. Peel tests are an industry standard used to assess the properties of pressure-sensitive adhesives. A recent design for a high throughput peel test with optical probe imaging of the result during or after the test is worth noting here for the interested reader to review [570]. An array of microlenses is used to measure adhesion at multiple points in another high throughput adhesion measurement test [571].

5.5.4.2 Problem Solving Example

Sticky tapes and labels are obvious uses of adhesives that cover a wide range of everyday applications. An adhesive used in the manufacture of Post-it (trademark of 3M, St. Paul, MN) products—small slips of paper with a reusable adhesive strip at one edge—is seen in the SEM images (Fig. 5.116) [572]. The adhesive partly coats the paper fibers and rounded domains about 5–50 μm form a contact on the applied surface (Fig. 5.116A). Regions between these rounded domains have a fine particulate structure that does not adhere, permitting the label to be removed easily. In order to image the adhesive nature of these materials, two adhesive strips were attached and then peeled back and examined in the SEM. (The adhesive strip and a piece of paper peel away from each other too easily to permit imaging of the adhesive mode.) Figure 5.116B–D shows this adhesive action, with an overview of the specimen inset in Fig. 5.116B. Thus, the adhesive has a low contact area compared with the total surface area, permitting easy removal. The SEM clearly shows the morphology of such adhesives.

5.5.5 Wettability and Coatings

5.5.5.1 Introduction

The spreading of a liquid onto a substrate relates to such applications as the coating of liquids on paper and the spreading of binders and finishes on fibers. Such wettability studies are affected by the roughness of the solid surface and the manner in which the liquid spreads on the surface. Polymers may provide the solid substrate, the spreading liquid, or both. Mason and coworkers [573–576] described interesting results relating the effect of surface roughness to wetting. These authors explored the concept of wettability theoretically [574] and described the equilibrium contact angle that a liquid surface makes with a solid it contacts as a measure of that wettability [575]. The SEM was used to demonstrate and confirm these theoretical projections. In the early study [573], molten drops of PE and PMMA were allowed to spread and solidify on a paper substrate prior to standard preparation for SEM. Later, poly(phenyl ether) (PPE) vacuum pump oil (Santovac-5; Monsanto Chemicals, St. Louis,

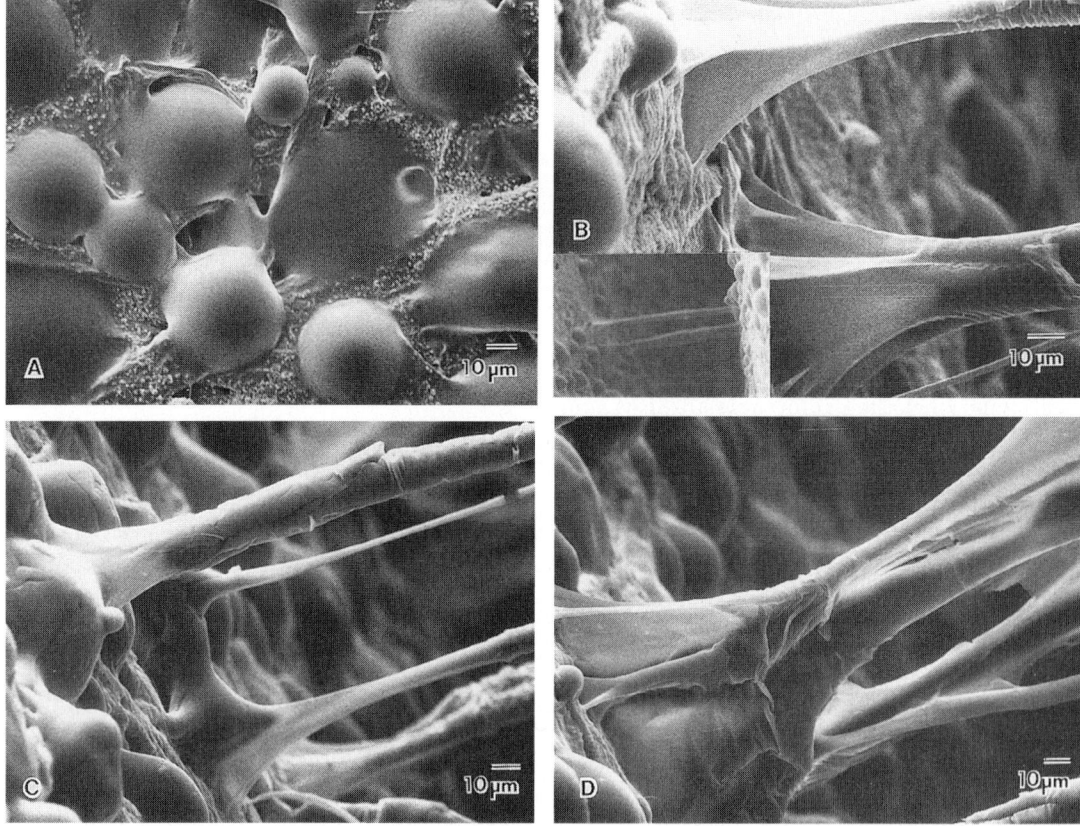

FIGURE 5.116. The morphology of an adhesive layer on a Post-it product is shown in these SEM micrographs. The surface of the adhesive coating appears to be composed of oblate spheroidal shaped particles (A). A thinner, particulate filled coating appears to cover the paper fibers between these adhesive particles. Two strips of adhesive were attached to one another and partially pulled apart, as shown in the insert (B). Strings of adhesive are seen to connect the two strips (B–D). (From D.R. Sawyer, unpublished [572].)

MO), which is not volatile, was used for dynamic studies. The PPE was fed through a hole in the sample stub, from outside the specimen chamber, and the wetting experiment was recorded on video tape [576]. Mori et al. [577] used freshly cleaved mica surfaces as steps 60 nm in height to inhibit wetting. Surface roughness has a major effect on the local contact angle between the liquid and the substrate of interest. The surface of coatings is more commonly assessed by various SPM techniques as will be described below.

5.5.5.2 AFM Characterization

An overview of polymer latex film formation and properties by Steward et al. [524] includes many techniques, especially AFM, and suggests the mechanisms involved in deforming spherical particles into void-free films are still the subject of controversy and debate. With that in mind, however, the specimen preparation required for AFM is minimal, and the ability to examine a wet specimen eliminates artifact formation due to drying, an electron beam, and the effect of vacuum. In the case of latex samples, the measurement of the particle diameter and distribution is also enhanced by the ability to resolve fine details and to easily make digital measurements. Issues with AFM imaging relate to instrumental parameters, especially whether the AFM is in the contact or noncontact mode (see Section 3.3). Karbach and Drechsler [578] described the use of AFM as a tool for high resolution and high contrast of materials such as coatings under controllable ambient conditions, providing examples using thermoplastic polyurethane, which they cryomicrotomed and stained with OsO_4 and used the block face for AFM and SEM (BEI) and the sections for TEM. This paper is an excellent example of the use of complementary techniques. Characterization of polymer coatings on metals was studied using ICAFM and FTIR of a multiphase polymer blend system exposed to a hydrolytic acidic environment to study the degradation process [579]. Lee [580] studied the structure of model coatings by AFM and SEM of freeze fractured surfaces and TEM of microtomed cross sections of the dispersions of plastic pigment and latex particles. These complementary studies aided the calculations.

Results of an early study of polystyrene and poly(ethyl methacrylate) (PS/PEMA) latex, imaged in the noncontact mode by AFM, are shown in Fig. 5.117. Figure 5.117A shows uncleaned latex to have some surfactant remaining on the surface of the particles and interfering with packing, whereas Fig. 5.117B shows a cleaned surface with little extraneous detail and excellent packing.

Another example of AFM imaging compared with FESEM imaging [581] is shown in Figs. 5.118 and 5.119 of latex particles that are used for modifying epoxy to improve toughness [582]. Figure 5.118 is an FESEM image [581] of a whitened region from a three-point bend fracture surface of epoxy modified with carboxyl-terminated butyl nitrile rubber (CTBN) particles. Atomic force microscopy of the same specimen shows the particle size correlates well with the FESEM image. The AFM image in Fig. 5.119A [581] provides much more detail at the particle-epoxy interface, which can be compared with fracture toughness measurements. Figure 5.119B, C is a line scan analysis of an interesting feature that shows the epoxy forming a bridge between the rubber particles. The height of the feature is about 530 nm. A more recent ICAFM image showing the top view of a multilayer PS latex (822 nm) is seen in Fig. 5.120. [216].

5.5.5.3 Problem Solving Examples

Glass fibers are usually pretreated with a polymer coating, or "finish," which protects the fibers during handling and which may be designed to maximize adhesion in composites. The SEM was used extensively during the early 1970s, first to evaluate and then to control such finish application [583] on glass fibers. An example is shown by SEM of glass fiber surfaces (Fig. 5.121A) of coatings ranging from thin and uniform to "lumpy"; these coatings hold fibers together by ductile "strings." More recently, complementary characterization (Fig. 5.121B–D) has been conducted on polyaminosiloxane glass fiber coatings showing the effect of different solvents and the nanomorphology by ICAFM [584].

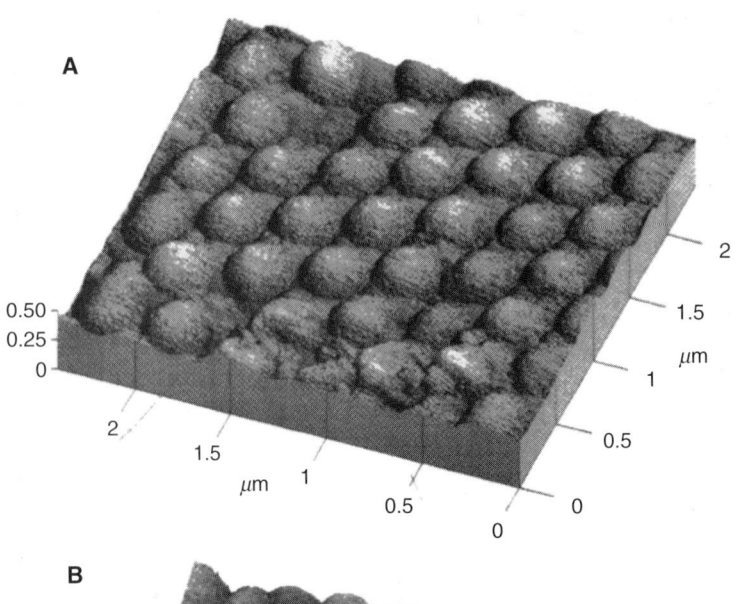

FIGURE 5.117. Atomic force microscopy in the noncontact mode of an uncleaned PS/PEMA latex (A) and a cleaned PS/PEMA latex (B). (From O. L. Shaffer [581]; unpublished.)

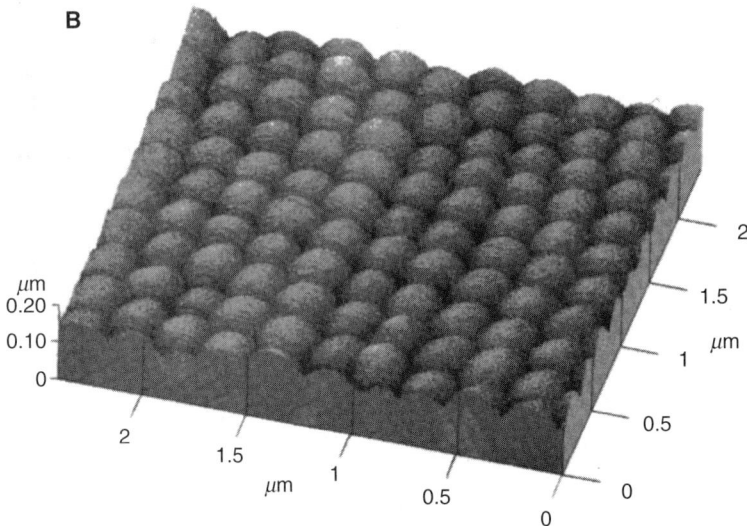

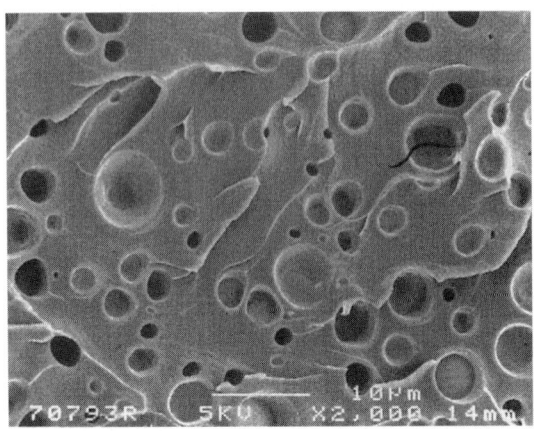

FIGURE 5.118. Field emission SEM (5 kV) image of a whitened region from a three-point bend fracture surface of epoxy modified with carboxyl-terminated butyl nitrile rubber (CTBN) particles. (From O.L. Shaffer [581]; unpublished.)

Emulsions, Coatings and Adhesives 391

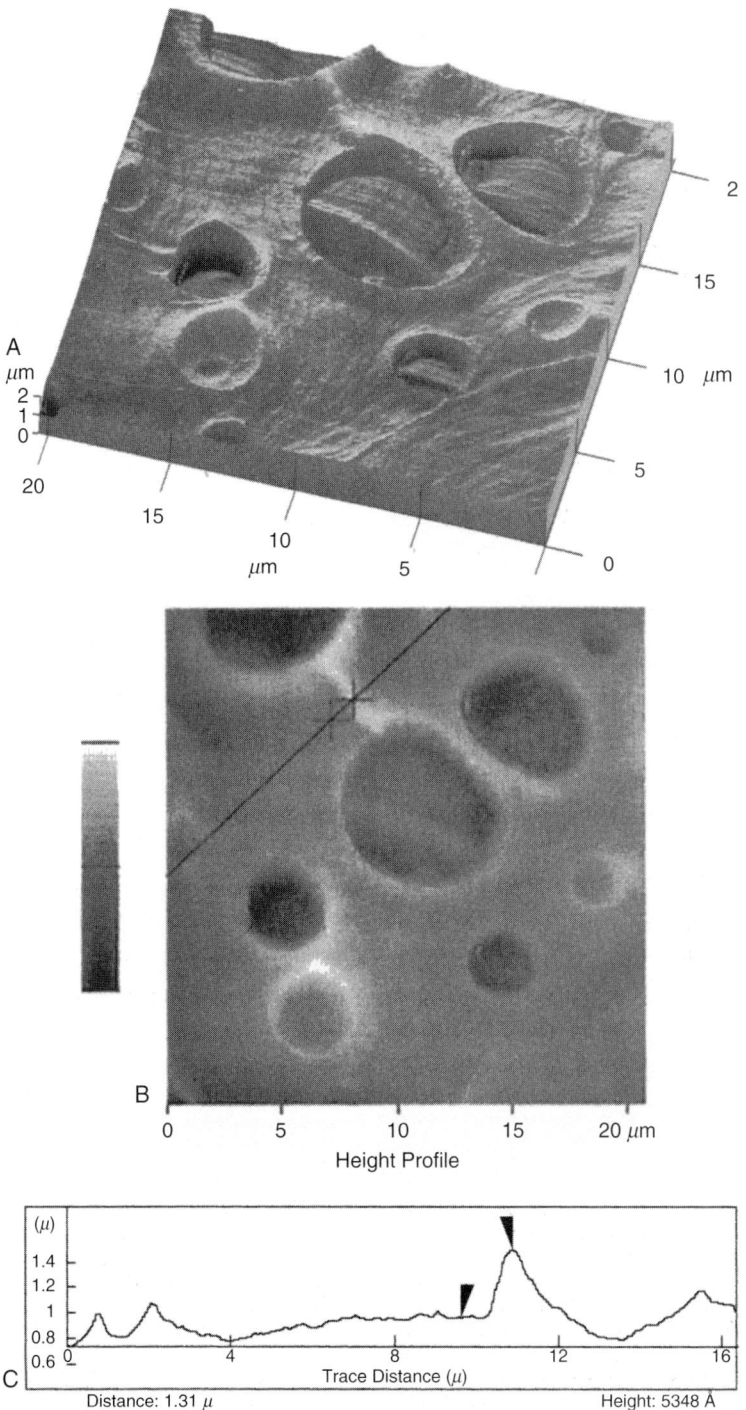

FIGURE 5.119. Atomic force microscopy of the same specimen as in Fig. 5.118 shows the particle size correlates well with the FESEM image. The AFM image (A) [581] provides much more detail at the particle-epoxy interface, which can be compared with the fracture toughness measurements. A line scan analysis (B and C) of an interesting feature shows the epoxy forming a bridge between the rubber particles. The height of the feature is about 530 nm. (From O.L. Shaffer, unpublished [581].)

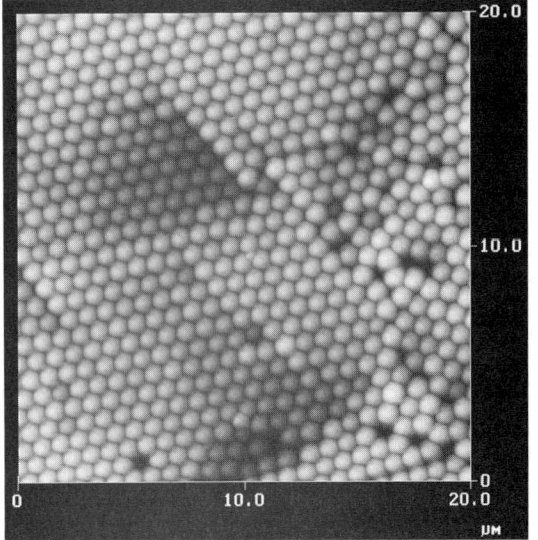

FIGURE 5.120. Tapping mode AFM image of 822 nm PS latex multilayers taken with a silicon tip (k = 48 N/m).

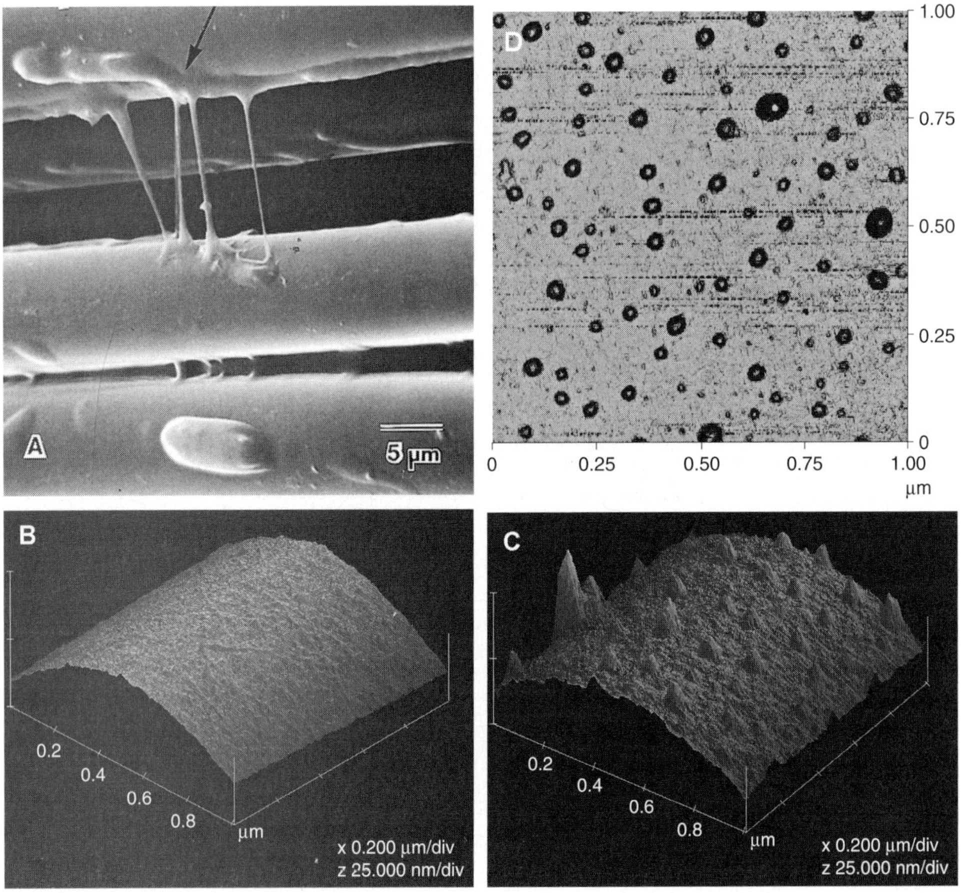

FIGURE 5.121. An SEM image (A) shows the morphology of surface finish coated glass fibers. The finish is seen in several morphologies; as a thin film coating on the surfaces, as lumps of material on the surfaces and connecting fibers, and as etchings or fibrils spanning across the fibers. Atomic force microscopy images of glass fiber surfaces heat cleaned and acid treated (B) show little surface detail, whereas a silanized glass fiber surface (C) shows detail with height information. The top view (D) after treatment with acetonitrile solvent shows detail of the particle size. (See color insert.) (From Turrion et al. [584], © (2005) Elsevier; used with permission.)

Film formation of polymer latex continues to be the subject of interest as the formation of void free films and the mechanisms of their deformation are continuing areas of research [524, 585, 586]. A recent example using AFM with a miniature hot stage is of ethylene vinyl acetate (EVA) and ethylene octane (EO) copolymer dispersions, dried on glass slides at room temperature, heated, and imaged to show the effect of particle size on film formation temperature. Figure 5.122 shows height images of EO as a function of temperature providing quantitative detail of the particle aggregation, deformation, and coalescence. Images at room temperature show clear particle boundaries (not shown), whereas the 0.1 to 3 μm particles clearly deform at elevated temperature, aiding understanding of the formation mechanism.

Correlations between SEM, AFM, nanoindentation, and impedance spectroscopy of some conducting polymer films of interest for biomedical devices has been conducted by Yang and Martin [587]. The conducting polymer used for biomedical applications, poly(3,4-ethylenethiophene) (PEDOT), is used as a coating on microfabricated neural prosthetic devices to improve long term performance of implants. Images taken by SEM (Fig. 5.123) and AFM (Fig. 5.124) of films electrochemically deposited on the electrode sites showed that the thickness of the PEDOT coatings increased systematically with deposition charge. Thickness measurements were made, and the work

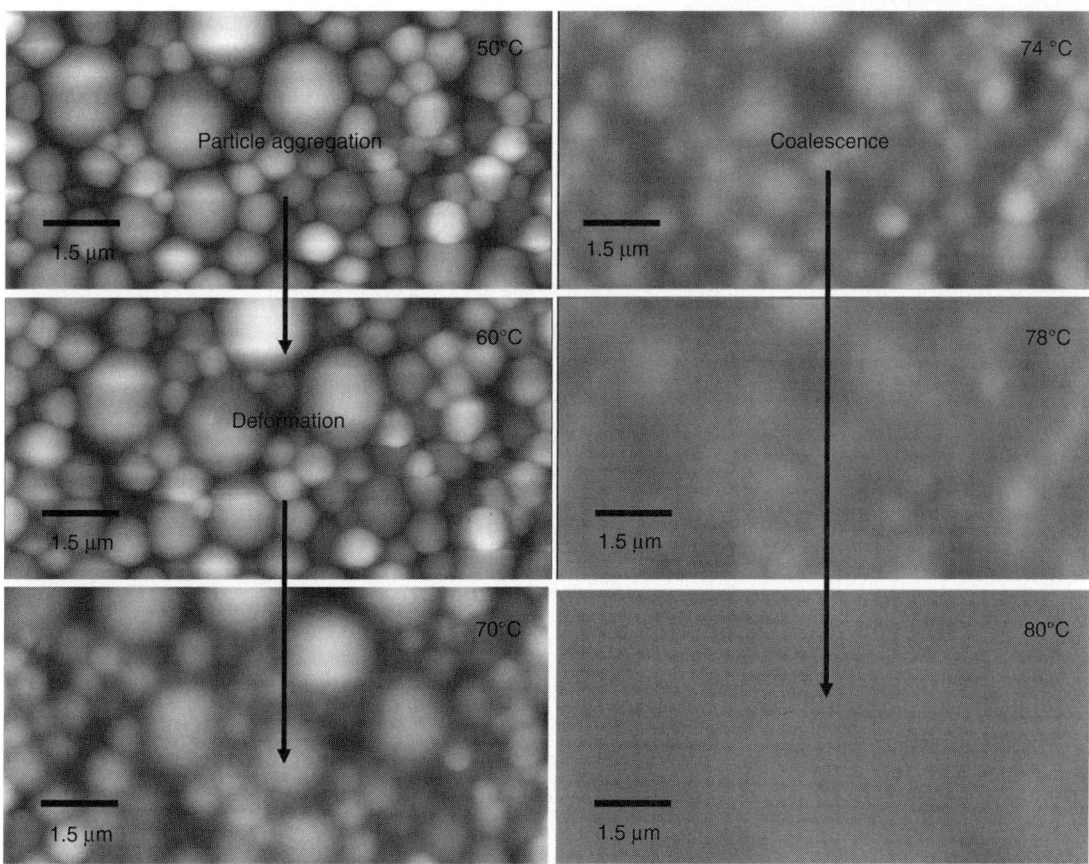

FIGURE 5.122. Height images of ethylene octane (EO) copolymer dispersions, dried on glass slides at room temperature, were heated and imaged to show the effect of particle size on film formation temperature using an AFM with a miniature hot stage. (See color insert.) (From Li et al. [586]; unpublished.)

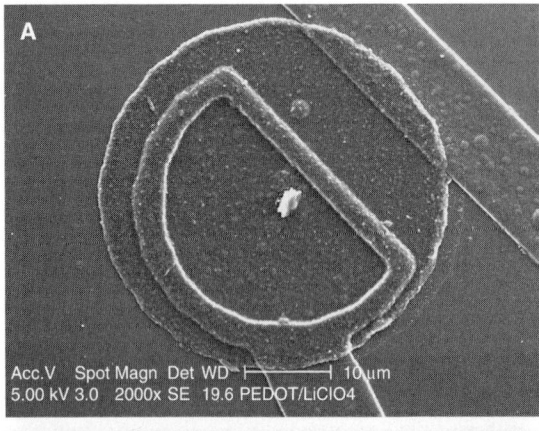
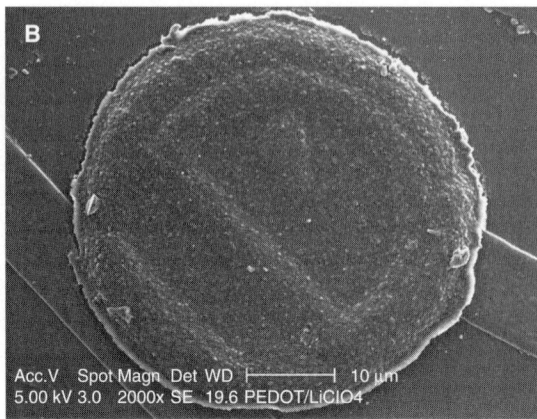
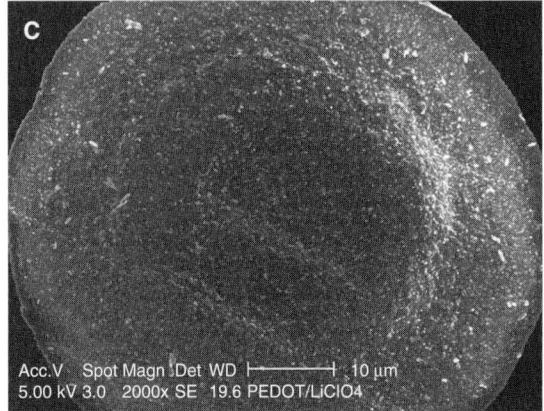

FIGURE 5.123. Scanning electron microscopy micrographs of PEDOT/LiClO$_4$ coatings deposited on gold coated sites with deposition charge (A–C) of 1.8, 7.2, 43.2 μC. The concentration of EDOT and LiClO$_4$ was 0.01 M, the growth current density was 0.5 mA/cm^2. (From Yang and Martin [587]; used with permission, Materials Research Society.)

showed that the lowest impedance films are those that are the softest, with an increase in the effective surface area of the coatings and improvement in mechanical properties.

Tang and Martin [588] studied the interfacial structure and deformation between a chlorinated polyolefin (CPO) adhesion promoter and automotive thermoplastic polyolefin (TPO) substrates, which have inherently poor adhesion with paints, by optical, SEM, and TEM techniques. Interfacial adhesion between paints, other coatings, and substrates is important to their performance. Optical microscopy imaging was performed using a video camera and also using microtomed sections, 2–4 μm thick. The surface of the block face was used for SEM imaging, and RuO$_4$ staining was followed by ultrathin sectioning with a diamond knife for HRTEM at 400 kV. The samples generally have a three layer paint structure: a top coat, a base coat, and an adhesion promoter layer (Fig. 5.125A). The diffusion of the rubber phase into the CPO layer was observed and an interphase thickness around 200 nm was seen by TEM imaging (Fig. 5.125B). After the tensile cracking test, paints tend to delaminate from the substrate at the edges of cracks, and the titanium in the pigment and the chlorine in the adhesion promoter can be used to trace the crack propagation. Delamination was shown to occur near the adhesion promoter and substrate interface by SEI (Fig. 5.125C) and elemental mapping of titanium (Fig. 5.125D) and chlorine (Fig. 5.125E). Defective adhesion was observed by SEM (Fig. 5.125F) due to material tearing during delamination. This summary provides an example of how complementary microscopy studies are used in research in the important area of coatings [588].

5.5.5.4 Cryo-TEM Characterization

The interaction between polymers and surfactants and colloidal systems in general has gained interest in many fields in recent years due to

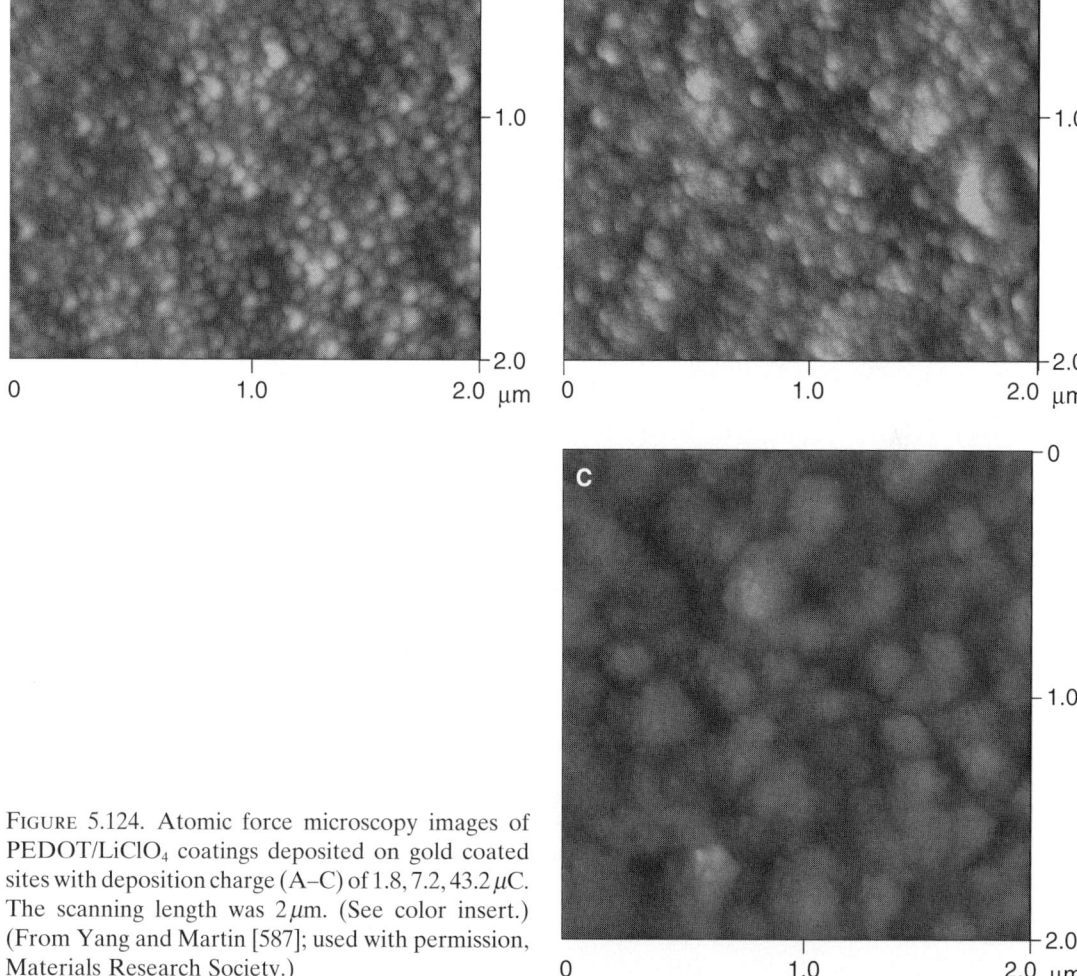

FIGURE 5.124. Atomic force microscopy images of PEDOT/LiClO$_4$ coatings deposited on gold coated sites with deposition charge (A–C) of 1.8, 7.2, 43.2 μC. The scanning length was 2 μm. (See color insert.) (From Yang and Martin [587]; used with permission, Materials Research Society.)

their use as detergents, hair care products, foams, and emulsions. Cryo-TEM is a relatively new technique (see Section 4.9.5) that has been described for imaging of these materials at high resolution (e.g., [589–592]). The first example is of a study of the nano and microparticles formed by complexation of poly(diallyldimethylammoniumchloride) (PDAC) and sodium dodecyl sulfate (SDS) [589]. The complexation was characterized by several techniques, including direct imaging by cryo-TEM at –180°C, which shows details of the complexes formed. The images also reveal the evolution of the nanostructure as a function of excess surfactant into lace-like aggregates and finally into spheroidal micelles (Fig. 5.126). According to the authors, the nanostructure of the complexes strongly suggests they are made of a hexagonal liquid crystalline phase, further supported by SAXS.

An example of the conformation of long polyelectrolyte chains attached to colloidal latex particles is also shown here by cryo-TEM [590]. The dense grafting of the polyelectrolyte chains ("spherical polyelectrolyte brush," or SPB) leads to a confinement of the counterions

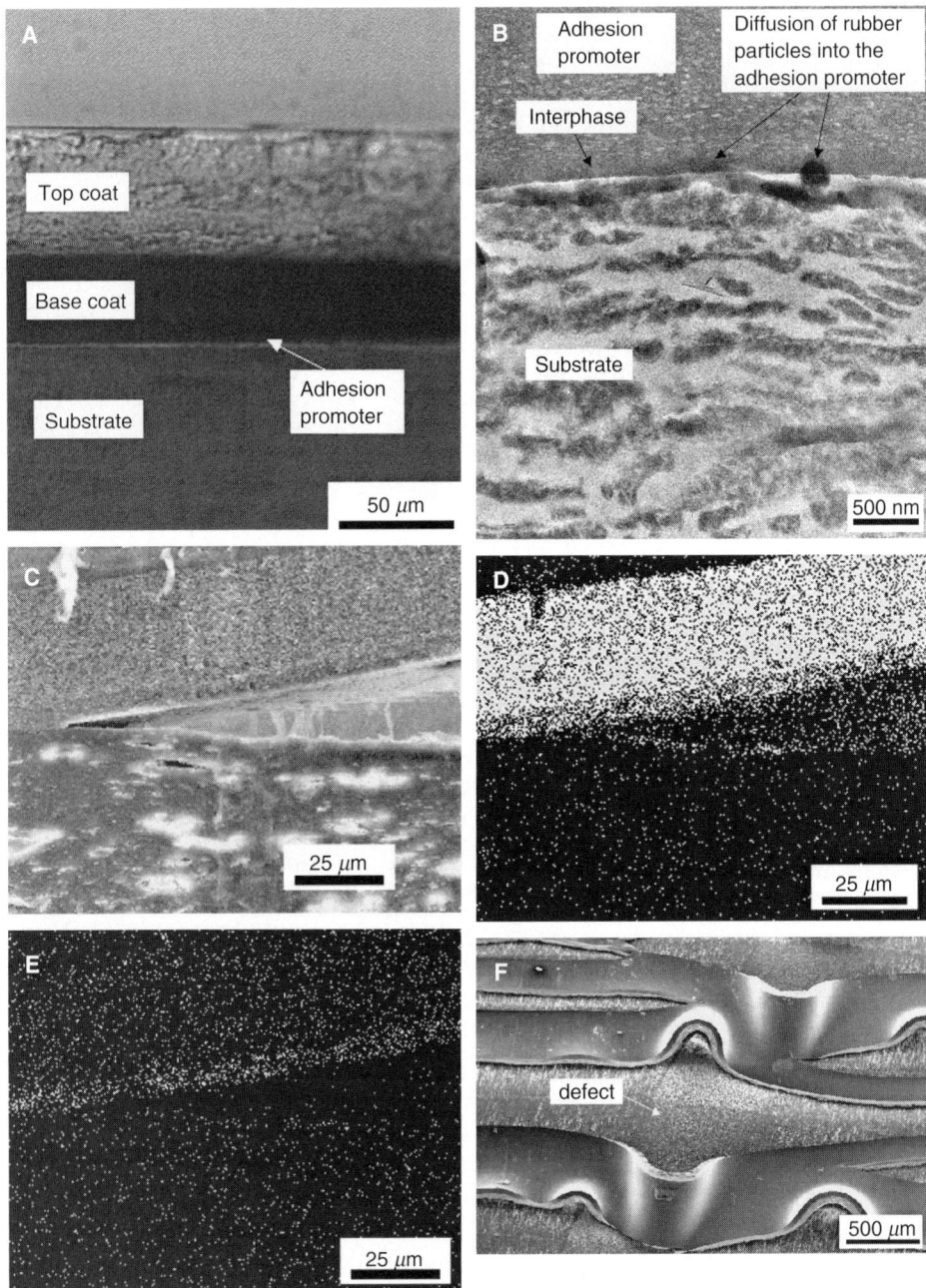

FIGURE 5.125. The three layer paint structure is shown by OM: a top coat, a base coat, and an adhesion promoter layer (A). The diffusion of the rubber phase into the CPO layer was observed, and an interphase thickness around 200 nm was seen by TEM imaging of a RuO_4 stained section (B). Delamination was shown to occur near the adhesion promoter and substrate interface by SEI (C) and elemental mapping of titanium (D) and chlorine (E). The SEM image (F) shows the rippling arising from tearing during the tensile cracking test that appears to arise from an area of poor adhesion. (From Tang and Martin [588], © (2003) Springer; used with permission.)

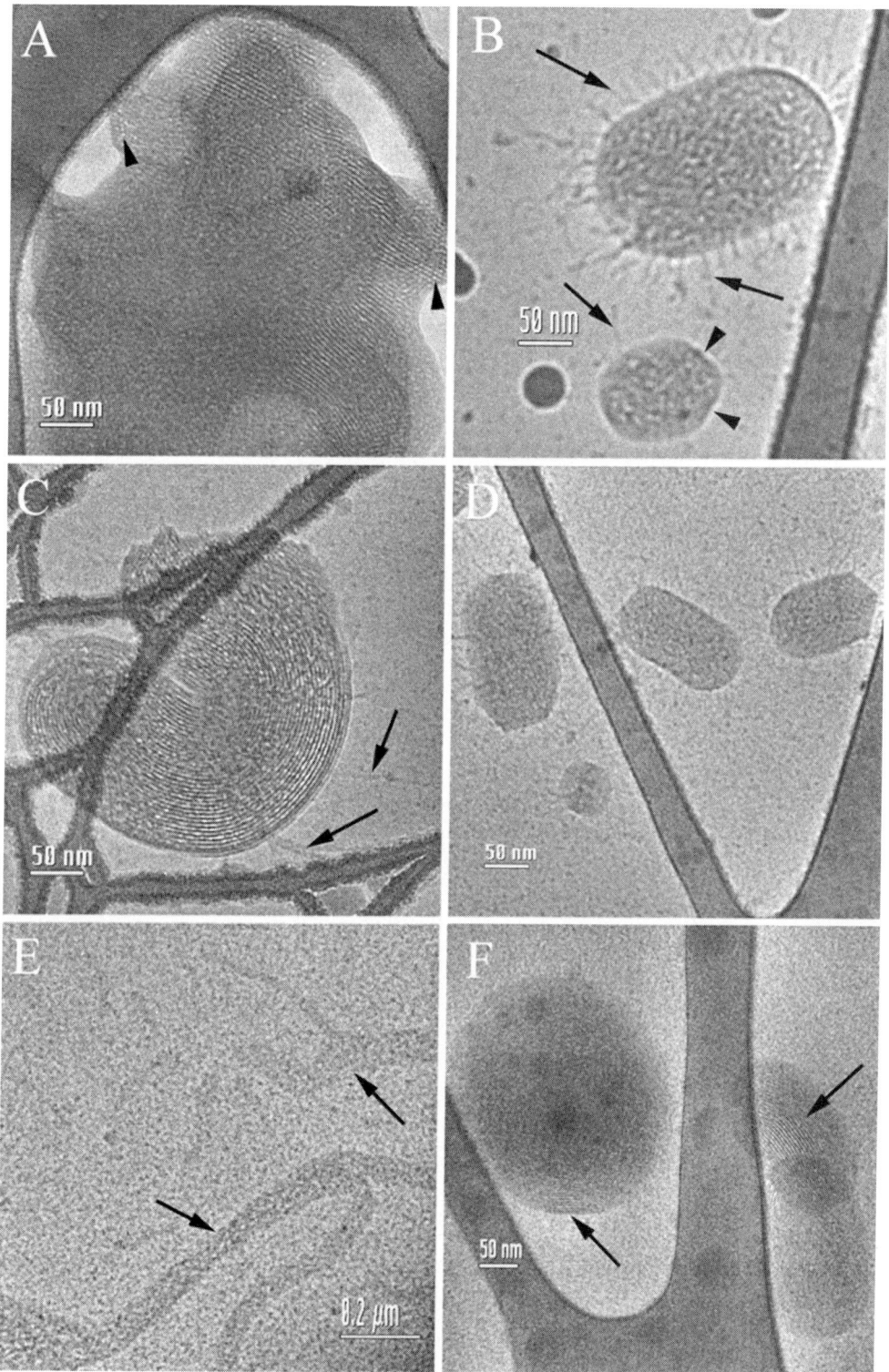

FIGURE 5.126. Cryo-TEM images show the evolution of the nanostructure and microstructure with changes in the SDS-to-PDAC molar ratio, r, at fixed PDAC concentration of 6.8×10^{-3} M (0.1% w/w). (A) r = 0.738; arrowheads point to round objects, possibly cross section views of the hexagonal phase; (B and C) r = 4.43, arrows show thread-like micelles, arrowheads point to a liquid crystalline aggregate; (D) r = 8.49; (E) r = 13.3, arrows indicate partially solubilized complexes; (F) r = 29.5, with excess surfactant, some liquid crystalline aggregate showing fringes is still seen. (From Nizri et al. [589], © (2004) American Chemical Society; used with permission.)

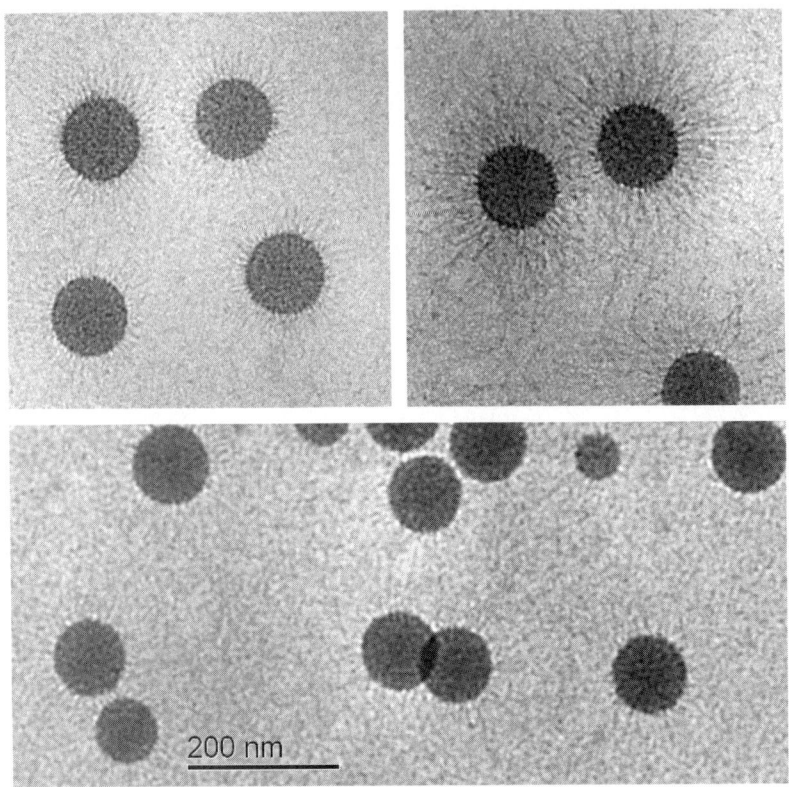

FIGURE 5.127. Cryo-TEM images of vitrified 1 wt.% SPB suspensions. The contrast is enhanced compared with the original particles (bottom) by replacing the sodium counterions of the polyelectrolyte chains by cesium ions (top left) and, additionally, by BSA molecules that are adsorbed in close correlation with the polyelectrolyte chains (top right). (From Wittemann et al. [590], © (2005) American Chemical Society; used with permission.)

within the polyelectrolyte layer attached to the core particles. Figure 5.127 is a high resolution TEM image showing the interaction of these brushes in suspension. This study showed that cryo-TEM can be used for *in situ* analysis of the spatial structure of colloidal particles.

5.6 HIGH PERFORMANCE POLYMERS

5.6.1 Introduction

5.6.1.1 Introduction to Liquid Crystalline Polymers

The development of high performance polymers, such as high performance plastics, high modulus fibers, and super-tough polymer blends, has accelerated in recent years as a direct result of increased knowledge of process-structure-property relationships. Highly oriented materials have been produced by modification of conventional polymers [70, 593] and by the design of rod-like, liquid crystalline polymers. A liquid crystalline polymer (LCP) is one that forms a partially ordered state on heating (*thermotropic* LC) or in solution (*lyotropic* LC). The term *liquid crystal* when applied to polymers defines a state that has one or more of the following characteristics, in common with low molecular weight crystals (reviewed in [594]: (a) anisotropy of properties (e.g., optical anisotropy in the absence of three dimensional order); (b) anomalously low solution or melt viscosity [595]; (c) molecular orientation by magnetic or electric fields [596]; (d) an endotherm, detectable by differential scanning calorimetry (DSC) at the temperature where a thermotropic mesophase first flows freely.

Liquid crystalline states in polymers are generally classified in the same way as LC states in small molecules [597, 598]. The degree of molecular order in liquid crystals is intermediate between the three dimensional order in solid crystals and the disorder of an isotropic liquid. A liquid crystal can be nematic, cholesteric, or smectic due to their degree of molecular order. Nematic crystals are ordered in one dimension; the long axes of the molecules are parallel and the local direction of alignment is called the "director." Cholesteric crystals have the director ordered in a spiral fashion, and colors appear if the twist of the period of the spiral is the wavelength of light. Smectic crystals have their molecules parallel and arranged in layers.

Heating a thermotropic liquid crystal results in decreasing molecular order. The general pattern is shown in Scheme 5.1, but not all possible phases may appear, and there are many types of smectic crystals. In addition, the LC phase may appear upon cooling rather than upon heating.

In many cases, unique optical textures are observed for the various orientations and structures of the three classes of liquid crystals. Thin films of nematic crystals, for example, can be identified by the pattern of dark threads (isogyres) that can appear in the optical microscope in transmission with crossed polarizers. Hot stage polarized light microscopy (PLM) is often used to identify the phases and the transition temperatures. In some cases, the optical texture is not uniquely identifiable, and x-ray diffraction and thermal analysis by DSC are used to complement microscopy.

Liquid crystalline polymers have been discussed and reviewed [70, 594, 599–613] during the past several decades, in which the synthesis, processing, morphology, orientation, and structure-property relations are described. The major applications of these materials have been as high performance engineering resins, melt processable extrudates, and molded parts, generally reinforced with fibers and/or additives. High modulus fibers with unique properties due to the formation of ordered lyotropic solutions or thermotropic melts that transform easily into highly oriented, extended chain structures in the solid state are also found in many applications.

5.6.1.2 Chemistry of LCPs

There are thousands of LCPs that can be considered in three commercially important classes:

(1) aromatic polyamides;
(2) "rigid rod" polymers;
(3) aromatic copolyesters.

High modulus fibers from lyotropic aromatic polyamides, poly(*p*-phenylene terephthalamide) (PPTA), first commercialized under the Kevlar trademark by DuPont [614], find major applications as fibers in tire cords and heat and chemical resistant fabrics. Other fibers in this class of materials are used in important applications such as firefighter and military uniforms, gloves, and in many other hazardous applications. The aromatic polyamides, or *aramids*, are produced by a dry jet-wet spinning process where the nematic structure in solution is responsible for the high modulus fiber performance [615–619]. Another class of lyotropic fibers, also produced by dry jet-wet spinning, are the rigid-rod

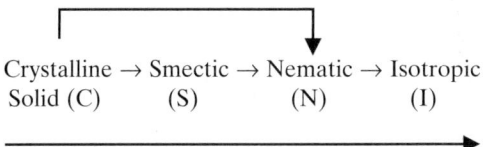

SCHEME 5.1.

polymers developed as part of the U.S. Air Force Ordered Polymers Program [620–624]. The most common of these ordered polymers, poly(p-phenylene benzobisthiazole) (PBZT), is difficult to process, but it exhibits among the highest tensile properties of all the LCP fibers produced to date. There are many other very high performance fibers developed during the past two decades.

Thermotropic aromatic copolyester fibers are produced by melt spinning as the combination of an aromatic backbone and flexible segments results in an LCP that can be melt processed. Earliest work was conducted on copolyesters, such as p-hydroxybenzoic acid (PHBA)-modified poly(ethylene terephthalate) (PET), 60/40 PHBA/PET (X7G) polymers, produced but not commercialized by the former Tennessee Eastman Company [595, 625, 626]. Other melt processable nematic thermotropic LCPs (TLCPs), based on combinations of 2,6-naphthalene dicarboxylic acid (NDA), 2,6-dihydroxynaphthalene (DHN), and 6-hydroxy-2-naphthoic acid (HNA), were developed by Celanese Corporation [605]. Liquid crystalline polymers have been commercially available for injection molding since the mid-1980s. Nearly every major chemical company in the world has produced LCPs, but the principal suppliers are Celanese, through Ticona, Florence, KY (Vectra®) and their affiliate, Polyplastics, Tokyo, JP (Vectra®); Xydar Solvay Advanced Polymers, Alpharetta GA (Xydar®); DuPont, Wilmington, DE (Zenite®); and Sumitomo, Tokyo, JP (Sumikasuper®). The majority of LCP products marketed are wholly aromatic polyesters and polyamides, although Toray and Idemitsu produce semiaromatic LCPs, based on the early Eastman technology. Thousands of patents have been granted for polymer compositions, reinforced and filled grades, polymerization methods, processing techniques, fibers, polymer blends, and end-use applications.

5.6.1.3 Microscopy of LCPs

Optical microscopy, SEM, and TEM studies of TLCPs have been reported (e.g., [612, 613]).

Mobile LCPs may be mounted between a glass slide and a glass coverslip for PLM. It is important to realize that the glass surfaces can have a strong effect on the orientation and structure of the LCP. Polishing the glass can align the director along the polishing grooves, and rigorous cleaning encourages the director to be normal to the glass surfaces. This orientation is called *homeotropic* because the specimen, viewed along the director, may appear dark, as if it were isotropic (e.g., [627]). On cooling TLCPs, the structure of the liquid crystalline state can be "frozen in" and studied in solid samples. Specimens for OM may then be prepared by microtomy and polishing. Scanning electron microscopy of LCP fibers follows standard preparation, which includes analysis of the surfaces, peelbacks (see Section 4.3.1), and fractures (see Section 4.8). Transmission electron microscopy requires thin specimens such as those produced during sonication, dispersion, and disintegration (see Section 4.1.3) or ultrathin sectioning (see Section 4.3.4).

Imaging techniques with higher spatial resolution have been applied to determine the size, shape, and organization of microfibrillar structures [628–633]. Field emission SEM at low voltage and SPM are capable of imaging regions from 1 nm to many micrometers on the same specimen. In the case of early work by STM, a fine grain metal coating, by ion beam microsputtering platinum at <5 nm thickness, worked well for samples placed on HOPG surfaces. Sections, ultrasonicated materials, peeled fibers or films can be imaged by FESEM, STM, or AFM techniques.

5.6.2 Microstructure of LCPs

5.6.2.1 Optical Textures

Characterization of the optical textures of LCPs is used in the identification of the specific phases present and in understanding the structure and its relation to the solid state properties. Dynamic hot stage microscopy experiments with videotape recording provide images of the texture associated with phase changes, as a function of temperature and time. The majority of the

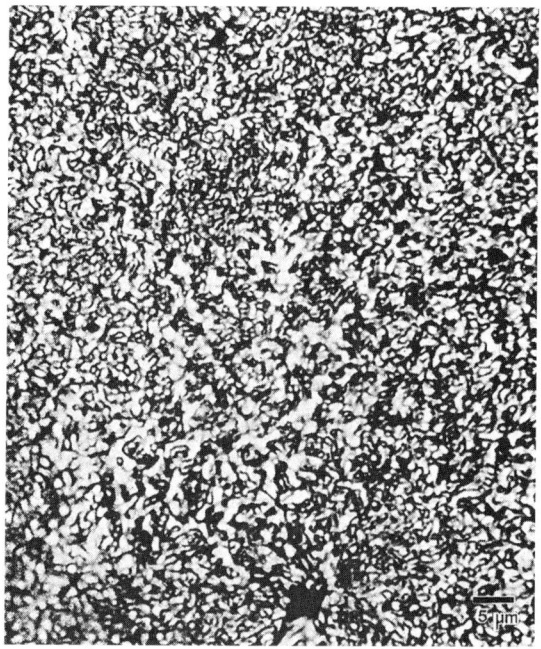

FIGURE 5.128. A polarized light micrograph of a sectioned nematic TLCP reveals a schlieren texture.

optical textures reported in the literature are of either melt or melt quenched structures, although the optical textures present in solid fibers, moldings, and extrudates have been described (e.g., [613]). A high pressure hot stage has been developed for optical microscopy and applied to liquid crystals and polymers [634]. Rheo-optical studies [635] of a slow cooled TLCP and a high density non-LCP polyethylene both exhibit similar textures in polarized light; as shown in a nematic LCP (Fig. 5.128) and when compared with polyethylene (see Fig. 5.3).

Mackley et al. [636] observed nematic threads, or isogyres, in X7G, and Viney et al. [637] observed changes as a function of temperature when heating thick sections between glass slides. Thick and thin regions form within the melt, with the thin regions frozen to one surface of the glass upon cooling. The differences in appearance are due to the superposition of structures. In crossed polarizers, a thick region (Fig. 5.129A) has a thread-like texture with little detail compared with a thin region (Fig. 5.129B), which reveals a definite nematic texture with line singularities, seen as points where generally two or four dark brushes meet; in this specimen, four brushes are observed. It is important to recognize the variation in textures with specimen thickness as sections of LCP molded articles are thick and quite complex (Fig. 5.130).

Optical studies of uniaxially aligned TLCP fibers, films, and ribbons observed in the orthogonal position in polarized light exhibit a "salt and pepper" texture and incomplete extinction. Close examination shows a fine *domain* texture with individual domains, about 0.5 µm in diameter, regions of local order (Fig. 5.131). There is a slight color variation between domains,

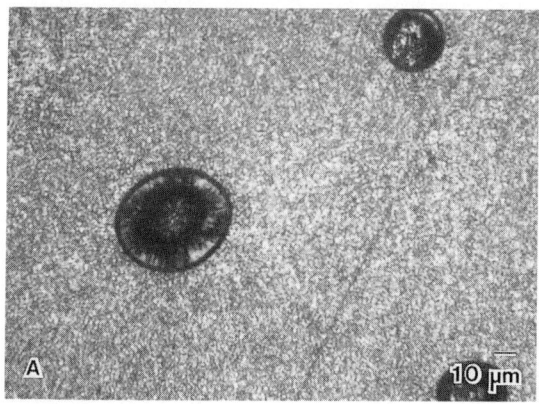 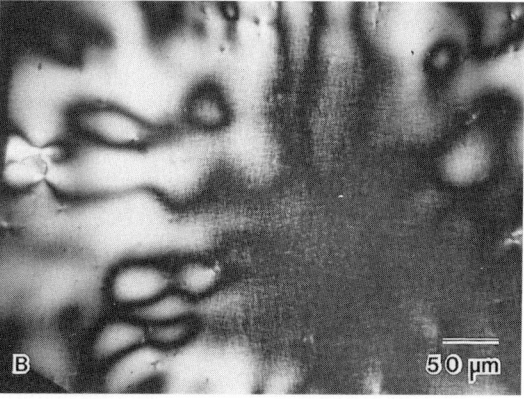

FIGURE 5.129. Thick and thin regions of a thermotropic melt structure in polarized light. In a thick region (A), the fine structure is not too clear but the onset of decomposition is shown by the round bubbles. A thinner region (B) shows thread-like detail and a nematic texture with four brushes. (See color insert.)

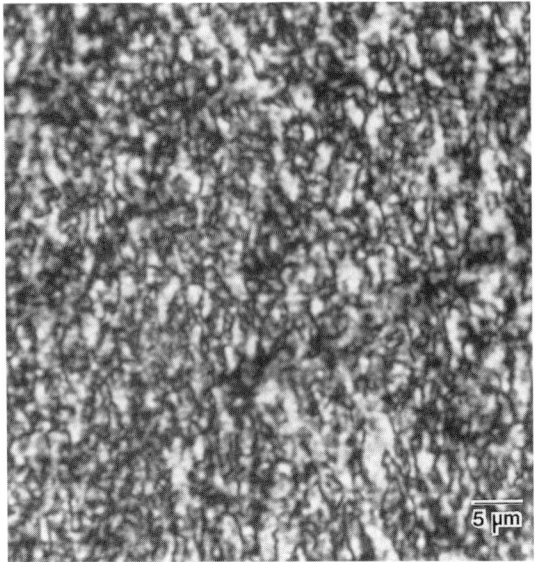

FIGURE 5.130. A polarized light micrograph of a section cut from a molded article reveals a complex, fine nematic texture with no obvious orientation. In the color insert, color in the image enhances the detail. (See color insert.)

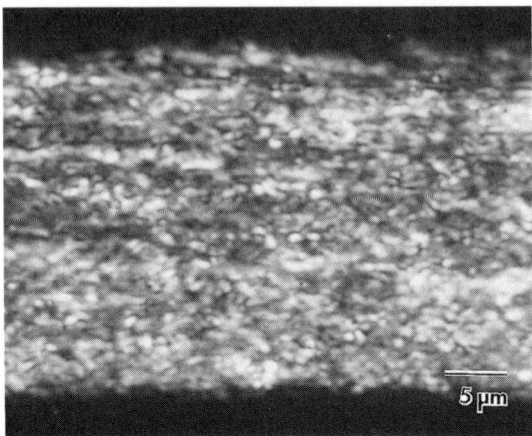

FIGURE 5.131. Incomplete extinction in uniaxially oriented TLCP fibers, ribbons and films gives a "salt and pepper" texture that is seen as individual domains less than 0.5 μm across. The similar polarization colors in polarized light suggest the domains are within the same order and thus have similar birefringence. (See color insert.)

which suggests they are distinct units with similar birefringence, and they may be the result of a serpentine molecular trajectory [613]. The fiber viewed at 45° to the polarization direction appears highly oriented, as expected.

An *in situ* rheo-optical and dynamic x-ray scattering study has given insight on formation of microstructures [646] showing the banded texture develops after cessation of shear. An example of the banded structures is shown in

5.6.2.2 Banded Structures in OM and TEM

Banding has been observed in both lyotropic and thermotropic polymers examined by optical and electron imaging techniques [613, 628, 637–644]; banding is a result of extensional or shear flow. Incomplete extinction has been observed for some of the aramids where "bands," normal to the fiber axis, are observed in polarized light [638] (Fig. 5.132). It is known that the aramids exhibit axial banding having periodicities of about 500 nm, observed by dark field TEM [645]. Simmens and Hearle [638] proposed that the optical observations and the pleated sheet model of Dobb et al. [645] are compatible and that the optical bands are the bends or folds between the pleats, which might well exhibit the local density differences observed by DF TEM.

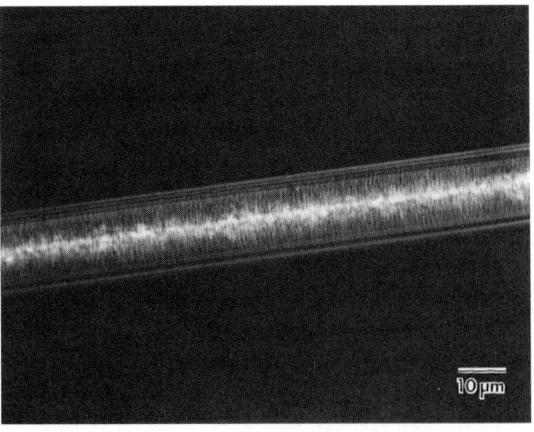

FIGURE 5.132. Incomplete extinction is also observed for PPTA fibers where bands arranged laterally across the fiber are thought to be a function of the pleated sheet structure.

a polished thin section of an extruded rod with low average orientation (Fig. 5.133). Banding is exhibited in regions away from the outer surface or skin with band widths of about 500 nm. Banding has also been seen in aramid fibrils in polarized light [645], and microbanding has been observed in TEM bright field images of spun aromatic copolyester fibrils [613], as shown in Fig. 5.134. This microbanding, on a scale less than 10 nm, has not been observed for heat treated Vectran LCP fibers, which had increased orientation. The interpretation of Donald and Windle [639, 640, 642] and Sawyer et al. [613, 632] is that the bands are associated with a serpentine path of the molecules along the shear direction, consistent with the fine domain structure in the highly oriented TLCP fibers. The analogy to the meander reported for PBZT [640] and the sharp path caused by the pleated sheet structures of some aramid fibers [645] are consistent. Finally, it is clear from TEM images of ultrathin sections of fibers and the STM images of peeled fibers, both shown in the next section, that the fibrils and microfibrils are not perfectly aligned within the sections.

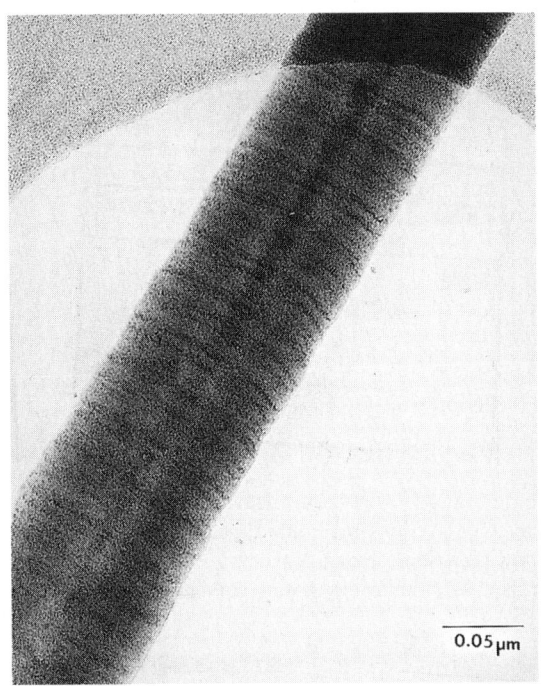

FIGURE 5.134. Microbanding is seen in a sonicated fibril of a spun TLCP fiber in a bright field TEM micrograph. Fibrils sonicated in ethanol are dropped onto holey carbon coated grids and carbon coated. Heat treated, high modulus fibers do not exhibit such banding.

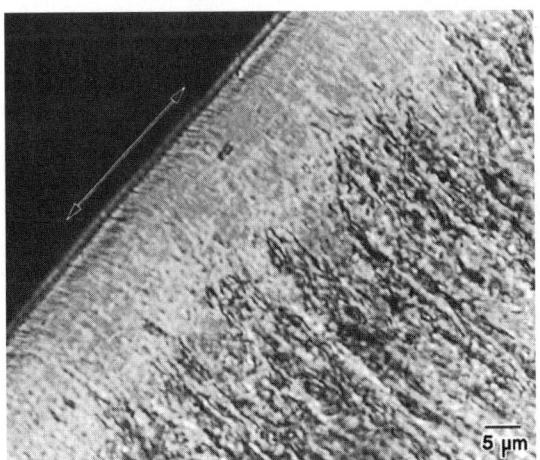

FIGURE 5.133. Polished thin sections of a low orientation extrudate observed in polarized light. An unoriented free-fall strand shows banding normal to the strand axis (arrow) and away from the slightly oriented skin. (See color insert.)

5.6.3 Molded Parts and Extrudates

5.6.3.1 Structure of Unfilled Moldings and Extrudates

Thermotropic LCPs are melt processed by injection molding and extrusion to form highly oriented rods, strands, and molded articles. It is well known [595, 647, 648] that extrudates have high molecular orientation that develops as a result of the effect of the flow field on the easily oriented extended chain molecules. Molded articles composed of thermotropic LCPs have properties that are better than short fiber reinforced composites and thus have been termed *self-reinforcing* [649, 650]. Highly anisotropic physical properties are explained by the highly anisotropic structures: layers, normal to the flow direction; bands, parallel to the flow; and a skin-core structure. The skin-core

FIGURE 5.135. A reflected light micrograph shows the layered structures in a molded bar aligned parallel to the flow direction (arrow). Variation in density and color reflect variation in orientation from layer to layer.

structure observed for LCPs is similar to the structure of typical thermoplastics (see Section 5.3) with orientation a maximum at the surface (skin), due to elongational flow, and at a minimum in the core, due to shear flow [261, 648, 650–652]. Microscopy permits assessment of structure-property correlations as the highly anisotropic structures are process dependent and relate to mechanical properties.

Thermotropic LCP molded bars exhibit a layered structure as shown by reflected light (Fig. 5.135) of a cut and polished bar. Skin-core morphologies are obvious in molded bars and extrudates with domains aligned in the flow direction. Complementary SEM of fractured injection molded bars provides a view of the layered structure (Fig. 5.136A), the surface skin (Fig. 5.136B), and the internal fibrillar struc-

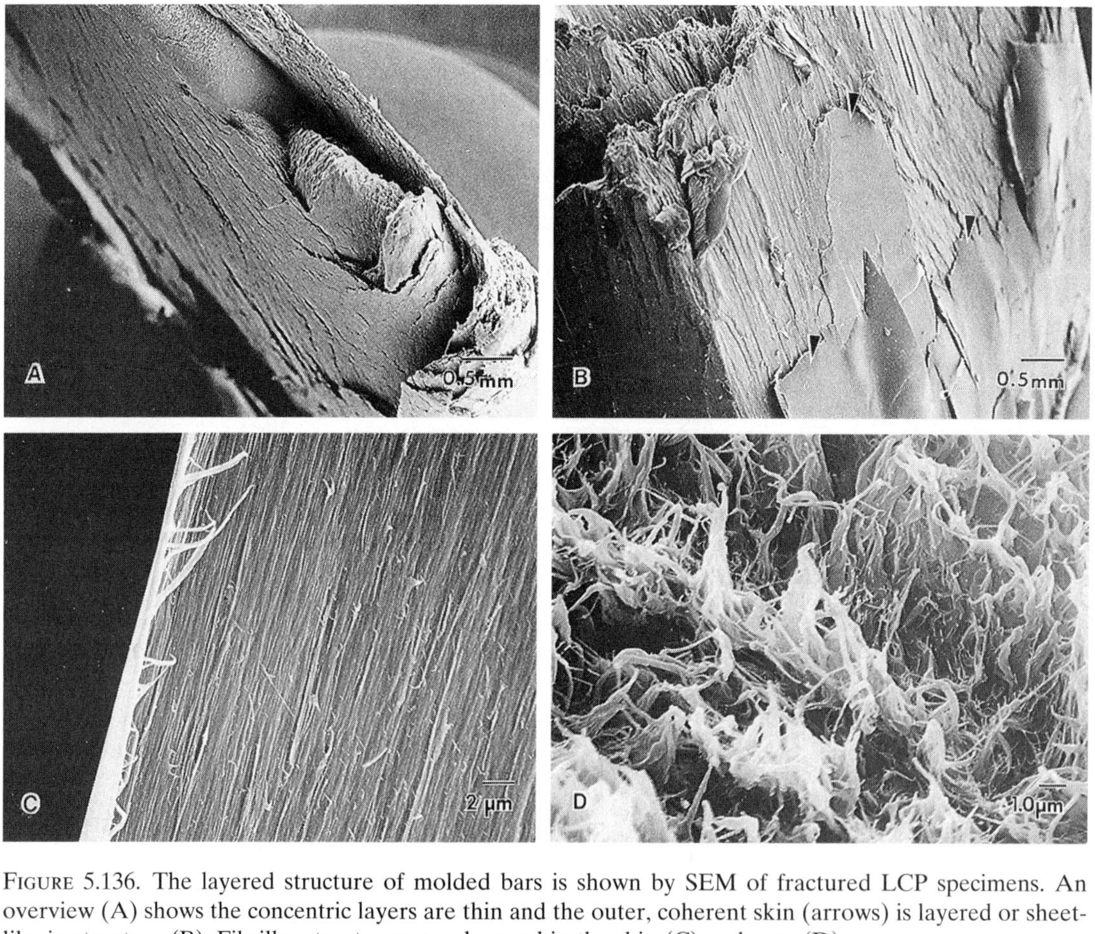

FIGURE 5.136. The layered structure of molded bars is shown by SEM of fractured LCP specimens. An overview (A) shows the concentric layers are thin and the outer, coherent skin (arrows) is layered or sheet-like in structure (B). Fibrillar structures are observed in the skin (C) and core (D).

tures of the inner skin (Fig. 5.136C) and core (Fig. 5.136D). Complex skin-core and banded textures are observed in extrudates, where the orientation is a function of the draw ratio and the final diameter, with higher orientation in finer strands. The orientation and incomplete extinction in the skin are shown in a section taken with the flow axis at 45° from this position (Fig. 5.137). Nematic domains are seen in the core, and to some extent in the skin, although the extinction (black) regions in the skin reflect higher orientation.

The common woody or fibrillar fracture of extrudates is clearly seen in the SEM micrographs in Fig. 5.138. This fracture morphology is controlled, to some extent, by the orientation of the strand. A micrograph of a polished thin section of a slightly oriented TLCP rod photographed in circularly polarized light reveals a nematic structure, banding, and evidence of skin orientation (Fig. 5.138A). Scanning electron microscopy fractures reveal a fibrillar structure (Fig. 5.138B, C). Etching experiments were also conducted in order to elucidate the fine structure of the extrudates.

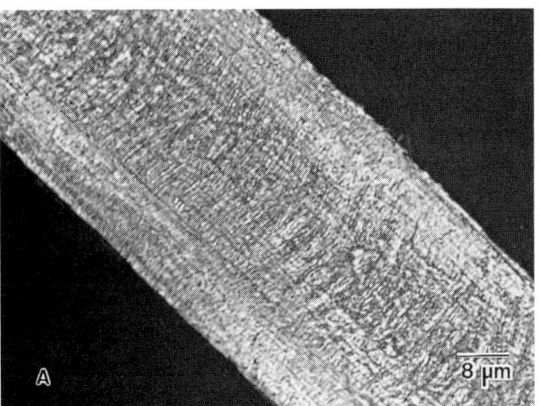

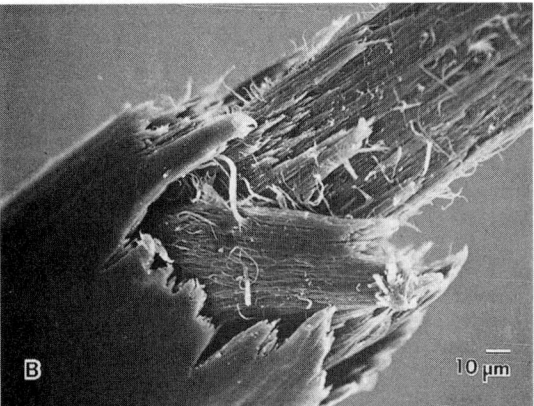

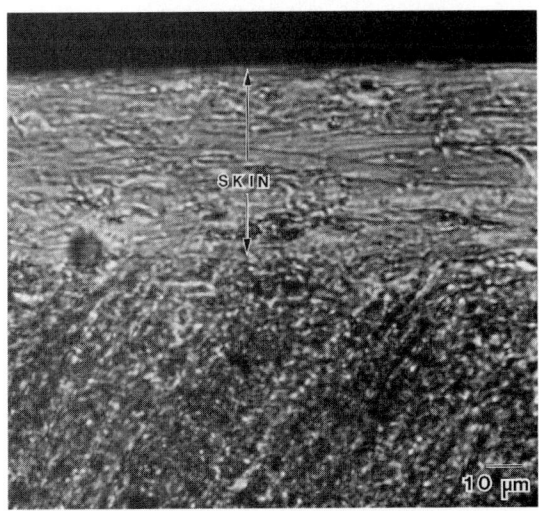

FIGURE 5.137. Skin-core are structures shown in more detail in a highly magnified polarized light micrograph with the specimen at 45° to the crossed polarizers. The skin is seen clearly to be more oriented than the core. (See color insert.)

FIGURE 5.138. Free-fall TLCP strands are shown in polished sections in circularly polarized light (A) (see color insert) and also by SEM of fractures (B and C). The less oriented strand appears more uniform in domain texture and also exhibits a coarser woody fracture (B). Some orientation is observed in the more highly oriented strand (A), and the fracture morphology is more uniform (C).

Imaging magnetically aligned thermotropic polyesters, using low voltage FESEM and AFM, has added information about the fine structure of LCPs [249, 653–655]. The sample shown in Fig. 5.139 is a random semiflexible terpolyester, heated to the nematic phase and aligned in a 13.2 Tesla magnetic field [653]. Samples were quenched to freeze in the order and then prepared according to the lamellar decoration technique developed by Thomas and Wood [654], in which the sample is annealed above the LCP glass transition but below the melting point. It then crystallizes in the form of lamellae, which grow perpendicular to the local chain axis and thus "decorate" the molecular director field. This method was used in the TEM study of the molecular director pattern in flow oriented thin films. When nematic LCPs are aligned in a magnetic field, they form defects known as *inversion walls*, in which the molecules on each side of the wall are rotated 180° with respect to one another. The study of such walls using low voltage HRSEM (Fig. 5.139A) has shown them to be three dimensional, confirmed by AFM (Fig. 5.139B) taken at the same magnification for comparison [249]; the upper micrograph is a higher magnification HRSEM image [249, 655], and the lower micrograph is an AFM image taken in the repulsive mode.

5.6.3.2 Structure Models of Unfilled Moldings and Extrudates

Structure models depicting LCP moldings and extrudates [613, 650, 652, 656–658] have been

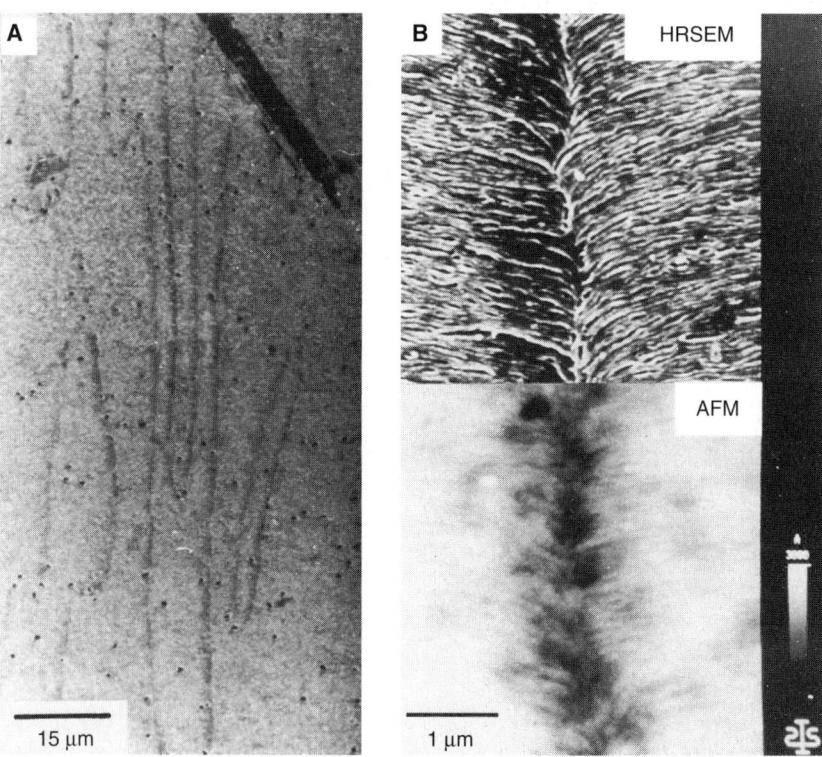

FIGURE 5.139. A random, semiflexible terpolyester, aligned in a magnetic field at the Francis Bitter National Magnet Lab at MIT [653] and decorated [654], is shown to form inversion walls, in which the molecules on each side of the wall are rotated 180° with respect to one another. A low magnification, low voltage HRSEM image (A) is shown of several inversion walls, which appear as dark lines as they are actually valleys [249]. The upper micrograph in (B) is a higher magnification HRSEM image [249, 655], and the lower micrograph in (B) is an AFM image taken at the same magnification [249]. Atomic force microscopy showed the wall is a valley ca. 200nm deep. (From Vezie et al. [249, 655]; used with permission.)

High Performance Polymers

derived from microscopy techniques. Thapar and Bevis [650, 652] showed a schematic of the skin-core and layered structures of injection molded thermotropic LCPs by polishing and etching methods for SEM. The samples were cut, polished, and etched with concentrated sulfuric acid for about 20 min, washed and cleaned ultrasonically in acetone. This work clearly showed the process dependence of the multiple layers, which exhibited continuous changes in topography from the sample edge to its center. The structure could be related to the observed increase in modulus with decreasing thickness. Baer and coworkers [657, 658] developed a similar model for an LCP extrudate by SEM fracture studies. Sawyer et al. [613, 628, 631, 632, 656] developed a general structure model after applying a wide range of microscopy techniques to the study of LCP extrudates and moldings as well as to highly oriented fibrous materials. This model (Fig. 5.140) shows skin-core, layered, and banded macrostructures and fine, hierarchical fibrillar microstructures.

5.6.3.3 Structure of Filled Moldings

Thermotropic LCPs in the form of filled engineering resins are generally marketed for molding into parts, such as interconnects, medical devices, automotive parts, and so forth. These end uses require high flow into the mold and high thermal stability to withstand post-molding processes and end uses at elevated temperatures. The resins are generally sold as composites in extruded pellet form, compounded with fillers, such as glass fibers, carbon fibers, minerals, and pigments. The product literature contains extensive information regarding formulations and properties—mechanical, physical, thermal, and electrical. Papers (e.g., [659–661]) and several texts discuss the orientation developed upon processing [594, 608–611]. Flow and process variables are known to affect the structure and properties of the products and have been studied by a range of techniques [661].

Samples molded under various injection conditions were examined by SEM of polished sur-

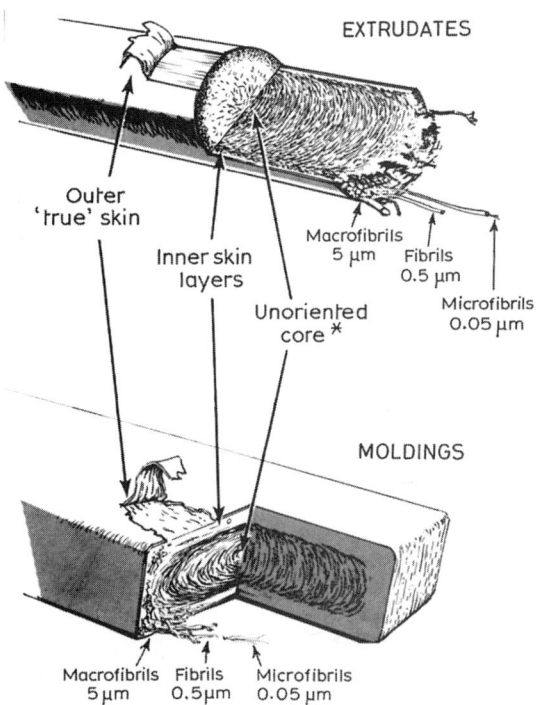

FIGURE 5.140. A structural model is shown that provides a schematic of the macrostructures in moldings and extrudates—structures such as layers, bands, and skin-core textures. Process changes appear to affect the macrostructures while the nature of the fine structures appears similar overall. (From Sawyer and Jaffe [613]; used with permission.)

faces etched in concentrated sulfuric acid for 10 min and compared with liquid nitrogen fractured samples. Removal of successive layers from moldings was followed by measurement of the modulus profile, and the orientation distribution was also measured by WAXS. Moldings with fillers, such as wollastonite, were also evaluated [661], and fillers were seen to affect the various layers and bands. Care must be taken in the interpretation of micrographs as the fine structures vary with the polymer, molding conditions and with the distance from the gate. Topics of interest for composites include adhesion, cohesion, interphase formation at the interface, the length or shape of the filler, and the arrangement of the fillers as determined by the flow pattern (see Section 5.4).

In LCP composites, the fibers can reduce local orientation of the polymer matrix. A polarized light micrograph of a polished thin section of a glass fiber reinforced Vectra molding shows fibers (black isotropic) in a nematic matrix (Fig. 5.141A). There is an indication of orientation of the polymer on the glass fiber surfaces, confirmed by SEM examination of a fracture surface (Fig. 5.141B–D). Additionally, there are composites that are reinforced either with LCP fibers in thermoplastic or thermoset matrices. For instance, aramid and Vectran fibers are used in composites with epoxies for aerospace applications that require high tensile strength and modulus.

5.6.3.4 Blends with LCPs

There have been numerous studies of blends of LCPs with thermoplastics and other LCPs during the past several decades, generally to reduce cost and to modify the physical properties. Scanning electron microscopy of fracture surfaces has been used to image the blend morphology and to reveal the nature of the dispersed phase, orientation, and anisotropy in much the same way as was described earlier (see Section 5.3.4). Processing affects the morphology of the blends, and a wide range of morphologies are observed that depend on the ratio of LCP to thermoplastic, process conditions, temperature, shear rate, and so forth [662–668]. The blend morphologies change dramatically in the core and skin region of the moldings with a tendency to elongated fibers in the skin layers and more globular domains in the core

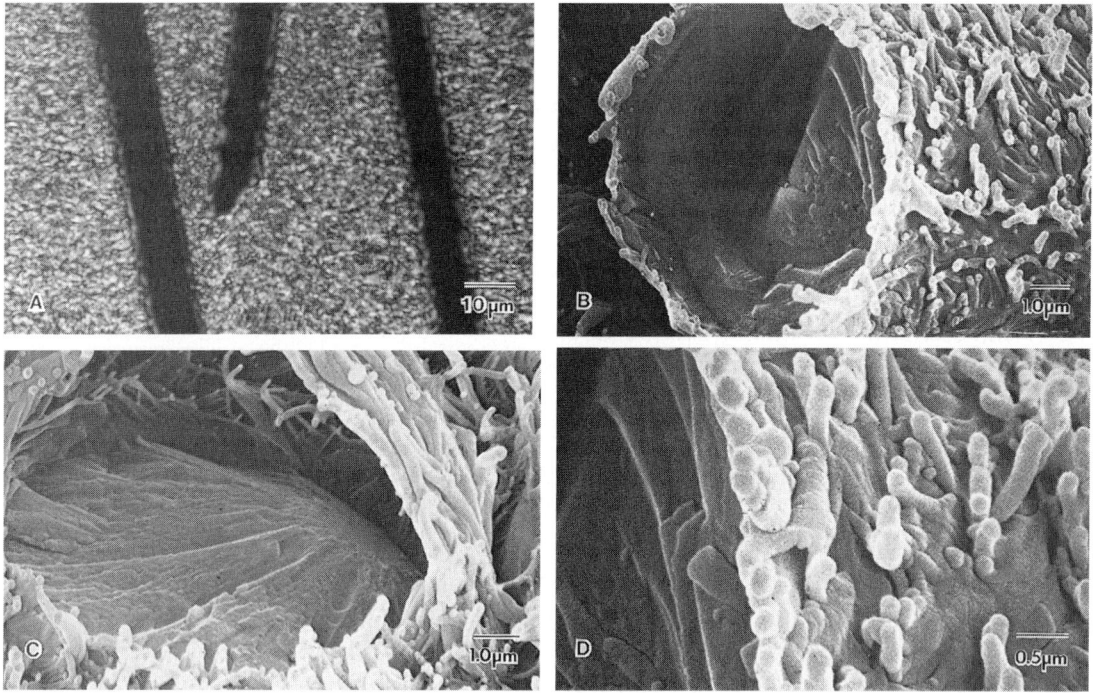

FIGURE 5.141. A glass fiber reinforced LCP composite is shown to have interesting morphology. A polished thin section is shown in polarized light (A) to exhibit a fine domain texture with some orientation of the polymer on the glass surfaces (see color insert). Scanning electron microscopy fracture views (B–D) show the tenacious adhesion of the LCP to the fibers. Fibrillar structures are oriented parallel to the fiber surface, and submicrometer sized domains are observed (D).

of the moldings. Blends in which the LCP is the major phase tend to exhibit globular domains of the thermoplastic within the more fibrillar LCP.

Akhtar et al. [667] studied blends of two TLCPs by thermal, rheological, mechanical, and morphology studies by SEM. Pracella et al. [668] studied PPS with a commercial LCP by blending in a Brabender mixer. The biphasic morphology was studied by SEM, which showed good contact between matrix and dispersed particles in blends with 5–20 wt.% LCP. An example of *in situ* composites of LCPs with PC produced by melt blending in a twin screw extruder was studied by TEM and SEM [669]. Spinnability of *in situ* composites fibers based on PEN and an LCP was conducted by rapid quenching of extrudates in ice water followed by freeze fracture in liquid nitrogen for SEM study and by collection of fibers from tensile testing [670]. The effects of a nematic LCP on crystallization and structure of PET/LCP blends was investigated by *in situ* studies of isothermal and non-isothermal melt crystallization by a range of thermal, x-ray scattering, and quantitative polarized light microscopy techniques [671].

A TEM thin section of an unstained LCP with 5% polycarbonate is shown as an example in Fig. 5.142, as prepared by Wood [366].

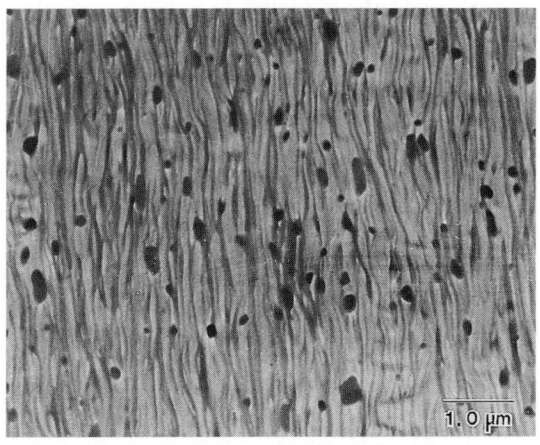

FIGURE 5.142. A TEM thin section of an unstained LCP with 5% PC is shown prepared by ultramicrotomy. (From Wood [366]; used with permission.)

Ultramicrotomy of the blend shows the typical banded texture of the LCP with dark submicrometer isotropic polycarbonate domains uniformly dispersed in the ordered matrix.

5.6.4 High Modulus Fibers

In the past two decades, major technological developments have occurred in the production of polymer fibers with high mechanical strength and stiffness. Materials science studies have been directed toward a better understanding of the relationship among chemical composition, physical structure, and mechanical properties.

5.6.4.1 Aromatic Polyamides

Aromatic polyamide fibers are produced by spinning solutions of PPTA-sulfuric acid dopes into a water coagulation bath [614], resulting in the formation of a crystalline fiber with a surface skin. Annealing at elevated temperature is known to increase the fiber modulus due to a more perfect alignment of the molecules [672]. The structure of the aramid fibers has been studied by Dobb and Johnson [645, 673–677] and summarized by many others [615, 617–619, 678]. Dobb et al. [673] first showed a lattice image for fibrillar fragments produced by sonication. Sonicated fibrils suspended over a holey carbon coated grid are shown with an electron diffraction pattern (Fig. 5.143A). A lattice image of an aramid formed from a fibrillar fragment [675] is shown in Fig. 5.143B. Bright field (Fig. 5.144A) and dark field (Fig. 5.144B) TEM micrographs of an ultrathin section show the fibrillar and banded textures in an aromatic polyamide fiber. Dobb et al. [675] described the structure of aramids by TEM and correlated increased modulus and preferred orientation, observing the general "hierarchies" of structure that are known to relate to mechanical properties. Measurement of crystallite sizes from bright and dark field images agreed with crystalline sizes calculated from x-ray diffraction data.

Aromatic polyamide fibers fail in tension by axial splitting, resulting in fine fibrils that can be

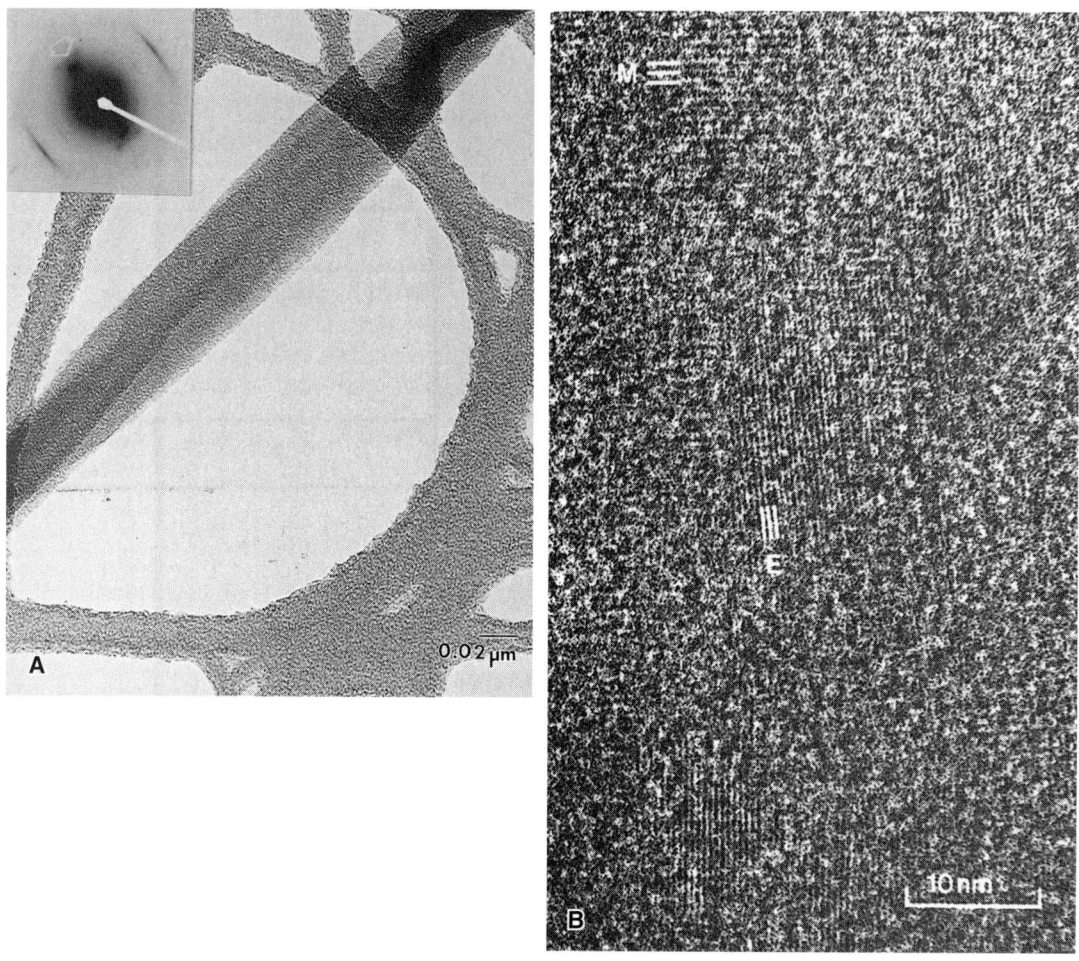

FIGURE 5.143. A TEM micrograph of sonicated fibrils from an aromatic polyamide distributed on a holey carbon grid shows the thin, tape-like fragments and corresponding electron diffraction pattern (A). A lattice image taken of sonicated fibrils shows the crystalline nature of the aramid structure (B). (Figure B from Dobb et al. [673]; used with permission.)

very long or may be woody or sheet-like in morphology depending on the process history. Cracks propagate parallel to the fiber axis due to rupture of hydrogen bonds. Pruneda et al. [679] suggested that tensile failure involves shear induced microvoid growth along the fiber axis leading to crack propagation, splitting, and failure. The fracture mechanism in fatigue reflects poor compressive properties [680–682] as failure is transverse to the fiber axis. Dobb et al. [677] studied the compression behavior of aramid fibers by SEM and TEM and proposed a mechanism for the formation of kink bands consistent with a loss of tensile strength after compression.

New types of aramid fibers have been studied to compare the tensile behavior and the structure by electron microscopy and x-ray diffraction studies [683]. Scanning electron microscopy images of fractured fiber surfaces, recovered after breaking in glycerol and cleaning in hot water, showed a range of internal defects in the fibers, and further information was gained from ultrathin sections of fibers impregnated with silver sulfide (see Section 4.4.7) [684]. Defects are well known to adversely affect tensile strength, and internal helical cracks in a high modulus variant were shown to result in decreased tensile strength [683–685]. Finally, the internal structure of aramid fibers has been studied by AFM methods (e.g., [631, 632, 686, 687]), which provides insights into the nanostructure.

5.6.4.2 Rigid Rod Polymers

Poly(p-phenylene benzobisthiazole) is one of a group of polymers with rod-like molecules, spun into high strength fibers. Allen et al. [688]

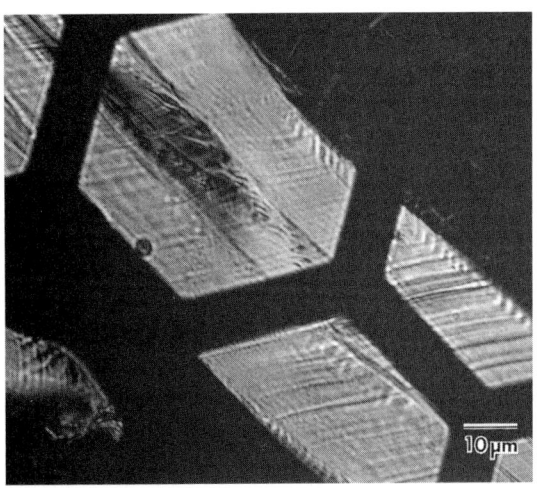

FIGURE 5.145. Optical micrograph of an ultrathin longitudinal section of an extruded experimental PBZT film (on a TEM grid) in polarized light with the fiber at 45° to the crossed polarizers. A definite skin-core texture is observed as the skin appears bright yellow and the core appears blue and less oriented. (See color insert.)

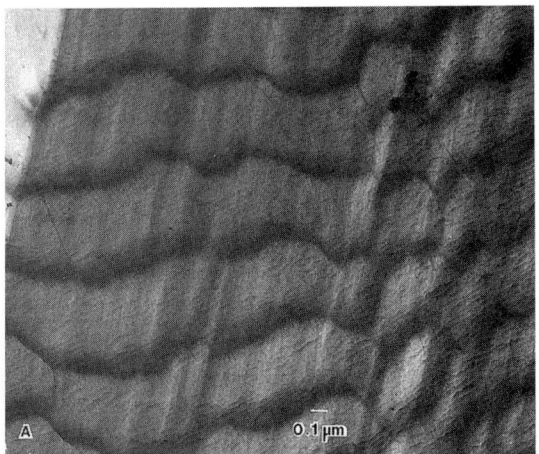

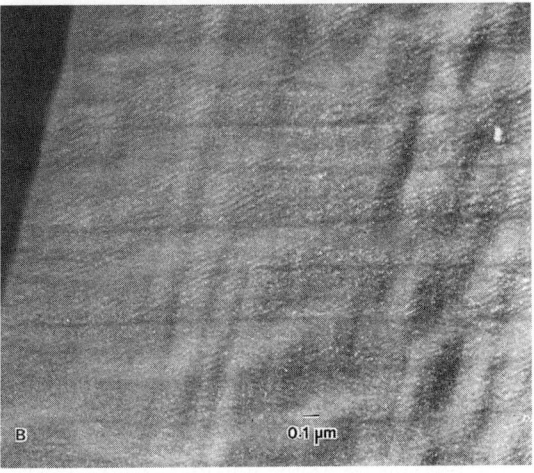

FIGURE 5.144. Transmission electron microscopy micrographs of ultrathin longitudinal sections of an aromatic polyamide fiber show the banded structures in bright field (A) and the crystallites in dark field (B).

described the development of high mechanical properties from anisotropic solutions of the polymer. The rigid rod molecules group to form fibrils, observed in peeled fibers in the SEM; optical microscopy has been important in defining the size and distribution of the macrovoids in the fibers [689]. Odell et al. [624] studied the role of heat treatment, which is known to enhance mechanical properties. Dark field EM studies show that the increase in modulus with heat treatment is likely due to the increase in crystal size and perfection. Ultrathin sections of PBZT extruded film on TEM grids were studied in polarized light (Fig. 5.145). Selected area electron diffraction patterns were taken of the sections in the core (Fig. 5.146A) and the skin (Fig. 5.146B). Diffraction results show a variation in orientation; the core is much less oriented than the skin, consistent with the results of Minter et al. [690], who used dark field imaging of spun and heat treated PBZT films.

Structural investigations have been carried out by many researchers, including Allen et al. [691], who used wide angle x-ray diffraction, mechanical testing, and SEM imaging of fractured as spun and heat treated PBZT fibers.

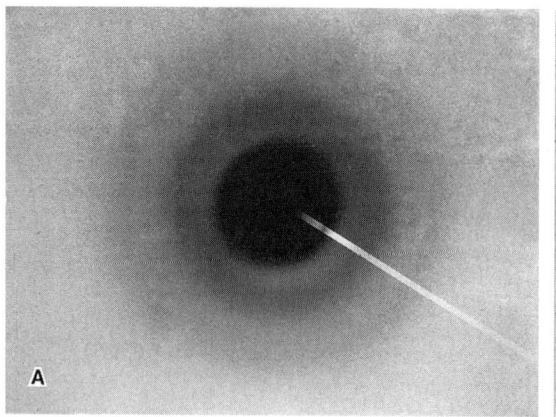 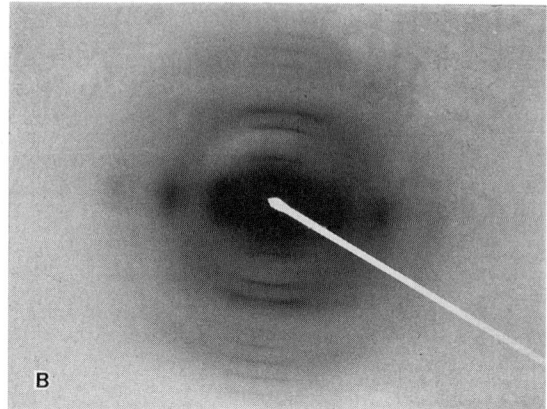

FIGURE 5.146. Selected area diffraction patterns of the sample in Fig. 5.145 show the core (A) is not oriented compared with the skin (B) in a PBZT thin section.

The aramids have a lower modulus because the chains are not straight compared with these stiffer polymers [692]. Details of the morphology of PBZT and PBO have also been shown using scattering techniques, electron microscopy, and HREM [693]. Molecules and crystallites are shown to be highly oriented, and a skin-core effect is observed. High performance fibers have also been studied using a microscale compression apparatus in an optical microscope [80]; with increasing compressive strain, kink band formation was observed and their density increased. The fundamental dimension of the buckling element for compression of PBO and PBZT fibers was calculated to be about $0.5\,\mu m$, the size of the fibril and not the smaller microfibril.

5.6.4.3 Aromatic Copolyesters

Thermotropic aromatic copolyesters have a major advantage over the lyotropes, as the former can be melt processed. Temperature affects the orientation and the mechanical properties, and the copolyesters have been shown to be biphasic by SEM [694–696], optical, and TEM [697–699] techniques. The biphasic structure of X7G has been reported [699] for extruded fibers by optical and EM imaging and microdiffraction. Transmission electron microscopy micrographs of ultrathin longitudinal sections reveal a dense dispersed phase elongated along the fiber axis (Fig. 5.147). Microdiffraction from regions 20–100 nm across show the dispersed phase domains (inset 1) and the oriented fiber matrix (inset 2).

A wide range of microscopy techniques has been applied to the characterization of oriented TLCP extrudates, including etching, sonication, ultramicrotomy, fracture, and peeling. Examples of the types of structures observed are shown in Figs. 5.147–5.149. Shear banding, typically at about 45° to the fiber axis, is common in all types of LCP fibers, as seem in polarized light in Fig. 5.148. A fibrillar texture is known to exist in all nematic LCPs and has been observed by several techniques. Sonicated fibrils examined in the TEM appear sheet-like and are wider than they are thick. Ultrathin sections examined in the TEM show fibrillar textures, as seen in a micrograph of a longitudinal section in Fig. 5.149.

5.6.5 Structure-Property Relations in LCPs

Microscopy techniques are used for the development of predictive structure-property-process models to develop marketable technologies, such as for TLCPs [613]. Models are used to fully describe the macro and microstructure of materials such as fibers, moldings, and extrudates. As with most polymers, process history and temperature affect these structures and the resulting properties. The discussion that follows includes examples of the types of microscopy techniques that can be helpful for

High Performance Polymers

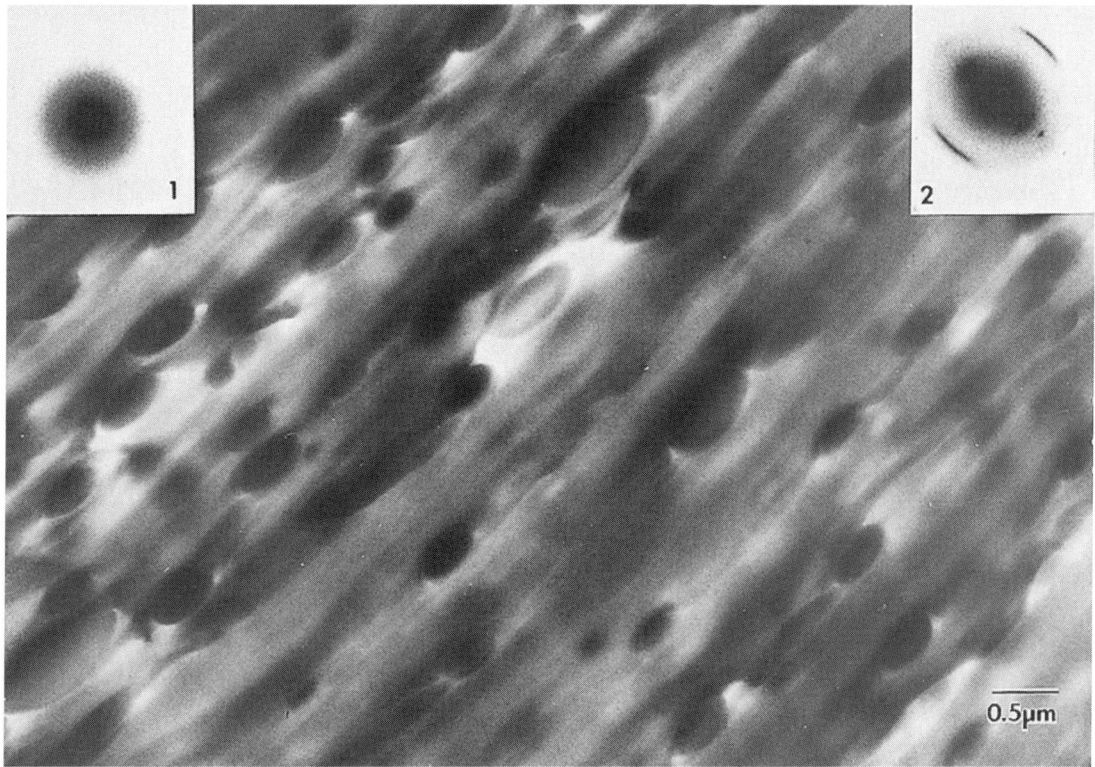

FIGURE 5.147. An example of the biphasic structure of X7G formed by specific thermal history is shown by TEM. An ultrathin, longitudinal section of the fiber is shown to have dense, elongated, dispersed phase particles present in a fibrillar matrix. Microdiffraction of the dispersed phase shows it is amorphous (inset 1) and likely PET, and a pattern for the matrix (inset 2) appears oriented and is likely the PHBA.

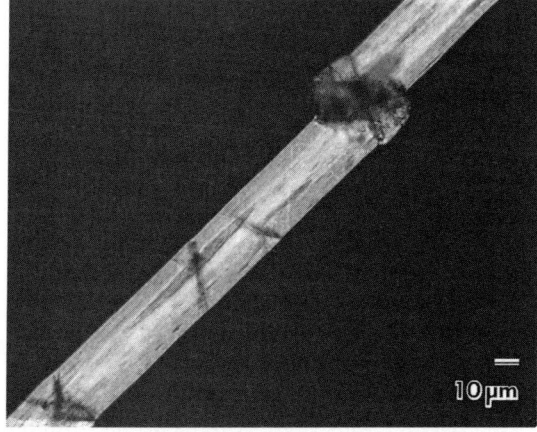

FIGURE 5.148. All LCP fibers have a tendency to form shear bands at an angle to the fiber surface. Shear banding is shown for a thermotropic copolyester in polarized light. Stressed fibers tend to fracture at such shear bands.

development of models in general. Polymer structure cannot be addressed simply by microscopy, and the interested reader should study the extensive literature on this topic for a more complete exposition.

5.6.5.1 Microstructure of LCPs

One general concept that has received much attention is the notion that the microfibril is the fundamental building block in polymers made from flexible linear molecules. This concept was proposed [700] for both natural and synthetic materials and has been known to exist in natural materials [701, 702] when they were imaged in the earliest TEMs. This minimum stable-sized structure appears to be the building block of polymers, and potentially this is the unit that can "aggregate" and account for

FIGURE 5.149. The *in situ* structure of a spun Vectran fiber is shown in a TEM micrograph of an ultrathin, longitudinal section. The orientation is shown by the microdiffraction pattern inset.

mechanical properties [703, 704]. One factor common to materials as dissimilar as cellulose, polyesters, aramids, and liquid crystal polymers is that the molecular chain is rather long compared with its width and thickness. Thus, it is possible that the microfibril is simply a replication of the molecular chain. Further discussion is outside the scope of this book.

A common feature of highly oriented fibers is poor compressive properties, demonstrated by kink bands that can be seen in the OM and the SEM. Kink bands have been studied [677, 679–682] and a mechanism for their formation, consistent with tensile loss, was proposed. Martin [693, 705, 706] addressed this issue by use of HREM, imaging the deformation and disorder in extended chain polymer fibers [705], showing that deformation occurs by strain localization into kink bands. High resolution EM images revealed the crystallite size, shape, orientation, and internal perfection. The nature of the disorder within a kink band was imaged, modeled, and compared with diffraction data. Kinks were observed [706] by TEM investigation of PBZT and PBO fiber fragments: Fig. 5.150A shows an equatorial DF image of a kink in PBO heat treated at 600°C. Bright areas in the image are crystallites that are scattering electrons into the objective aperture. Rotation of the sample showed that the material within the kink bands was misoriented but still crystalline. Figure 5.150B is a low dose HREM lattice image of a kink in a PBO heat treated at 665°C with the (200) lattice planes of spacing 0.55 nm.

Sawyer et al. [630–632] used high resolution FESEM and STM with other complementary techniques to describe the microfibrillar and kink structures (although work today would be by AFM to avoid the need for metal coating). Field emission SEM images of a highly oriented Vectran fiber were peeled to reveal kinked regions in the bulk (Fig. 5.151 left). Peeled back Vectran fibers imaged in the STM (Fig. 5.151 right) reveal a kink band with clear discontinuities across individual microfibrils suggesting bond breaking has occurred. The Y-shaped arrangement of the microfibrils is also confirmed, consistent with the structure observed by TEM

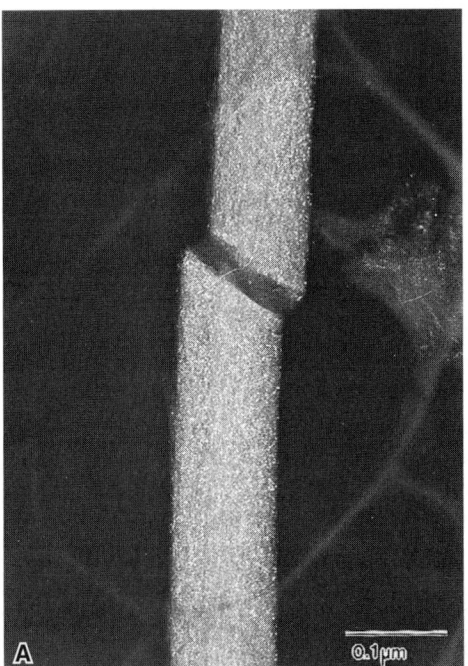

FIGURE 5.150. An equatorial dark field image of PBO fiber heat treated at 600°C (A) shows well defined and localized boundaries between the kink and the deformed fiber. A low dose HREM 0.55 nm (200) micrograph (B) of a kink in PBO heat treated at 665°C shows (200) lattice fringes of spacing 0.55 nm. The kink band clearly contains very localized high angle bending or buckling of the stiff polymer chains. (From Martin and Thomas [706]; used with permission.)

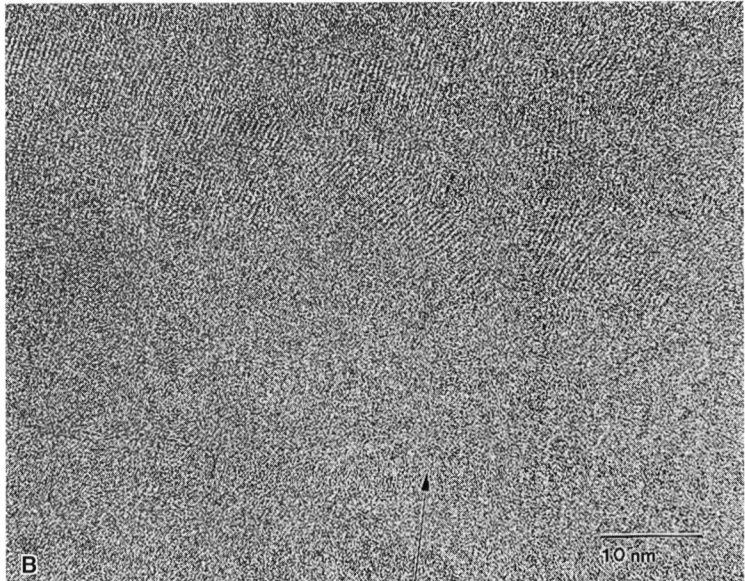

(Fig. 5.149). Sonicated Vectran (Fig. 5.152A) and aramid (Fig. 5.152B) fibers form into finer microfibrils, yet no clear interfibrillar tie fibrils were observed. Although the origin of the fibrils remains a question, these images appear to show microfibrils within larger fibrils. Scanning tunneling microscope images of sonicated Vectran LCP fiber (Fig. 5.153A) and Kevlar (Fig. 5.153B) have a periodic texture of about 50 nm observed across a group of microfibrils, arranged normal to the microfibril axis; this texture appears similar in size and spacing to the "nonperiodic layer (NPL) crystals" observed by Donald and Windle [707].

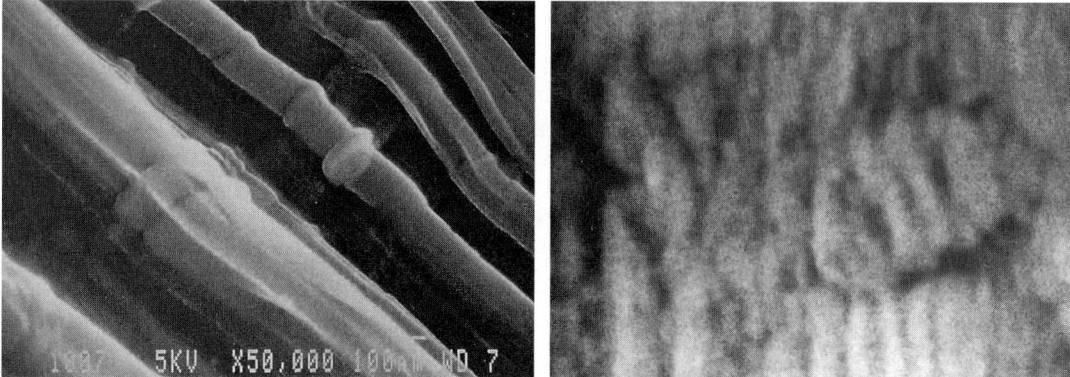

FIGURE 5.151. Field emission SEM image shows the internal texture in peeled back highly oriented Vectran fiber (left). Peeled back Vectran fibers are shown in a top-down view in the STM (right) of a region 500 nm in the x and y axes and a maximum z-axis range of 50 nm. The fibrillar texture and local disorder is observed. A kink band is observed as are clear discontinuities across individual microfibrils and damaged microfibrils within the kink band. (From Sawyer et al. [632]; used with permission.)

Microfibrils are also shown in Fig. 5.153, and measurements of the thickness of a few of the smallest microfibrils were ca. 3–5 nm wide and about 1 nm thick with a tape-like shape. Thus, the organization of the LCPs appears to be flat or tape-like microfibrils arranged within fibrils.

The failure mode in compression is via kink band formation, which is an intrinsic feature of highly oriented fibers. Adams and Eby [708] reported very high tensile moduli and strengths for PBO fibers, and much lower compressive strength, with failure occurring by buckling in local kink bands. The work of Martin [705, 706] and Lee [709] confirms the critical nature of these kink band defects and their relation to mechanical properties. Efforts to visualize microfibrils have been successful [632, 705,

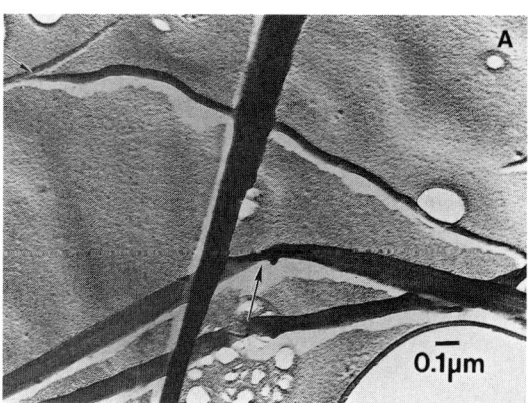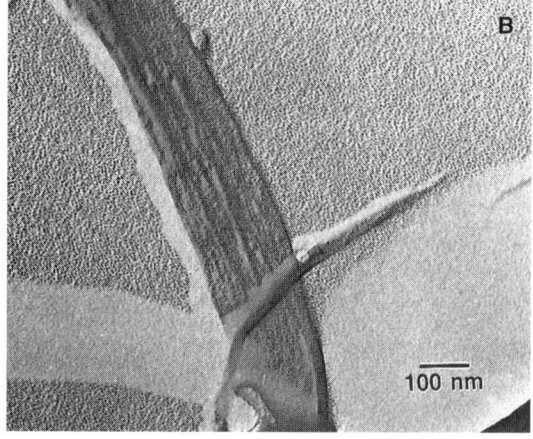

FIGURE 5.152. Transmission electron microscopy micrographs of sonicated fibrils, shadowed with metal in order to provide information about the three dimensional structure of (A) Vectran and (B) Kevlar fibers, illustrate the tape-like structure. A range of fibrils are observed that are long and are seen to fibrillate into finer fibrils. Twisting and a cotton-like flat, twisted, or tape-like fibril structure is observed. The aramid fiber (B) is shown to fibrillate into units less than 10 nm wide. (From Sawyer et al. [632]; used with permission.)

High Performance Polymers

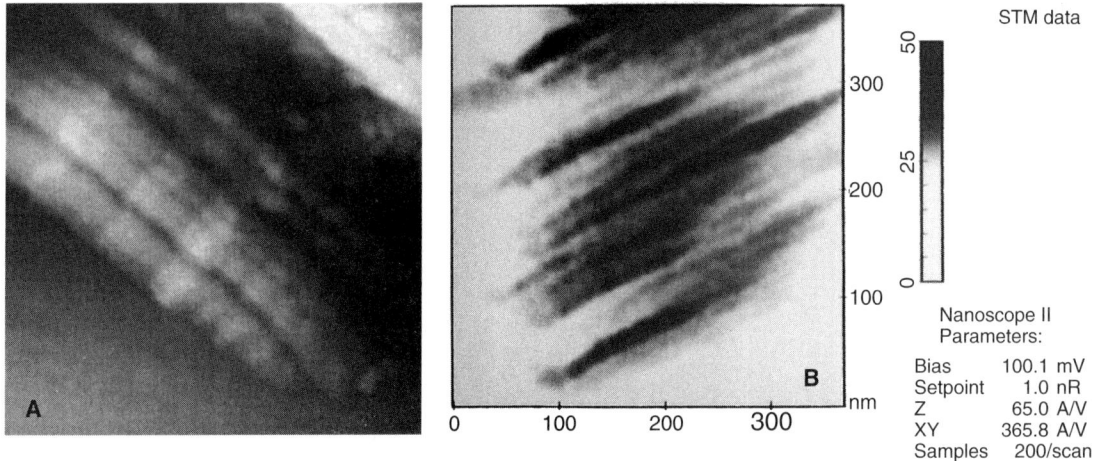

FIGURE 5.153. A STM image of a Vectran heat treated fiber (A) reveals microfibrils within a larger unit; area scanned is 500 nm across. A 50 nm periodicity is observed normal to the microfibril axis. A STM image of Kevlar (B) most clearly reveals the nature of the fibrillar hierarchy at the macromolecular level. A bundle of uniform microfibrils about 10 nm wide was imaged showing fine microfibrils. (From Sawyer et al. [632]; used with permission.)

710–712]. Martin [710] provided evidence of intermolecular twist defects in extended chain fibers, such as PBO, by use of molecular modeling. Spatially resolved electron diffraction also revealed the hierarchical structure in as spun LCP fibers at a resolution of 100–200 nm with bimodal orientation, whereas the heat treated fibers were more uniform in orientation [713]. These authors also used low dose HREM diffraction to calculate local orientation order parameters from the same TLCP fibers.

The macromolecules in as spun and annealed aramid fibers were imaged by AFM [712], showing the periodicities to be consistent with x-ray diffraction and computer simulation results. Gould et al. [714] used AFM to study LCP surfaces and suggested that microfibrils 2 nm in size formed that represent the basic structural unit in a model film composed of Vectra LCP. High resolution ICAFM was used to explore the morphology of Vectra LCP, before and after annealing with complementary thermal assessment [715]. These new, higher resolution imaging devices have been used to describe the microstructural textures in more detail and have permitted understanding of the role of the structure to the mechanical properties.

5.6.5.2 LCP Structure Model

The theme of this section was to provide an example of structure studies conducted by a range of microscopy techniques to develop structure-property-process models. A modified structure model incorporating the information gained from these techniques is shown in Fig. 5.154. The model suggests there are structures on the scale from 500 nm to <50 nm, specific to the liquid crystalline polymers. The key element is the microfibril, the same microstructural unit basic to melt spun and drawn flexible polymers. The orientation of the microfibrils is along the fiber or elongational axis, and this results in extremely high tensile modulus values for these materials. However, on a local scale, it is clear that the microfibrils meander along the path of the director and are not literally rigid rods. Structure models provide a means for understanding and developing materials using microscopy techniques to enable the microstructures to be observed.

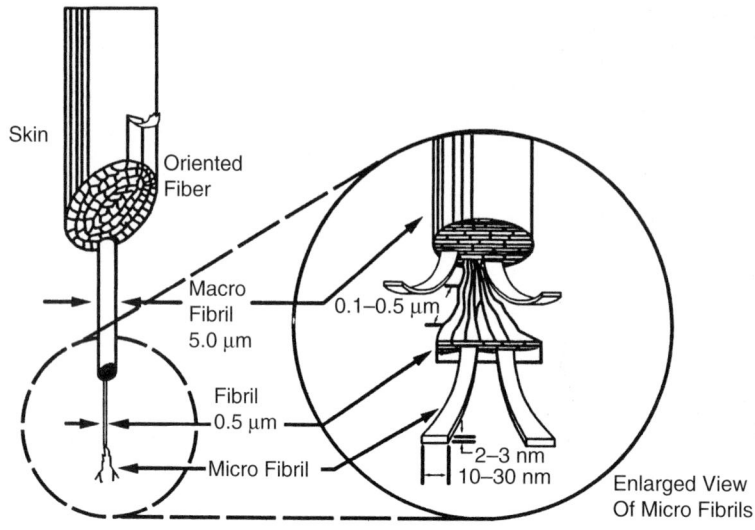

FIGURE 5.154. A new structure model is shown that has more detail than in Fig. 5.111 published earlier [613]. The model suggests there is a hierarchy, at least on the scale from 500nm to the 1nm size scale, specific to the liquid crystalline polymers. The key element shown is the microfibril, the same microstructural unit basic to melt spun and drawn flexible polymers. (From Sawyer et al. [632]; used with permission.)

References

1. N.S. Murthy and D.T. Grubb, *J. Polym. Sci. B Polym. Phys.* **41** (2003) 1538.
2. J.C. Guthrie, *J. Text. Inst.* **47** (1956) 248.
3. J.L. Stoves, *Fibre Microscopy* (Van Nostrand, Princeton, New Jersey, 1958).
4. R.D.V. Veld, G. Morris and H.R. Billica, *J. Appl. Polym. Sci.* **12** (1968) 2709.
5. J.W.S. Hearle and R.H. Peters, *Fiber Structure* (The Textile Institute, Manchester, 1963), p. 209.
6. W.O. Statton, *J. Polym. Sci.* **C20** (1967) 117.
7. A. Peterlin, *Polym. Sci. Symp.* **32** (1971) 297.
8. A. Peterlin, *J. Appl. Phys.* **48** (1977) 4099.
9. P. Barham and A. Keller, *J. Polym. Sci. Polym. Left. Edn.* **13** (1975) 197.
10. D.C. Prevorsek, R.H. Butler, Y.D. Kwon, G.E.R. Lamb and R.K. Sharma, *Text. Res. J.* **47** (1977) 107.
11. D.C. Prevorsek, Y.D. Kwon and R.K. Sharma, *J. Mater. Sci.* **12** (1977) 2310.
12. D.C. Prevorsek, P.J. Harget, R.K. Sharma and A.C. Reimschuessel, *J. Macromol. Sci. Phys.* **88** (1973) 127.
13. C. Reimschuessel and D.C. Prevorsek, *J. Polym. Sci. Polym. Phys. Edn.* **14** (1976) 485.
14. D.C. Prevorsek, G.A. Tirpak, P.J. Harget and A.C. Reimschuessel, *J. Macromol. Sci. Phys.* **89** (1974) 733.
15. J.W.S. Hearle, J.T. Sparrow and P.M. Cross, *The Use of the Scanning Electron Microscope* (Pergamon Press, Oxford, 1972).
16. J.W. Hearle and S.C. Simmens, *Polymer* **14** (1973) 273.
17. H.R. Billica and R.D.V. Veld, in *Surface Characteristics of Fibers and Textiles*, edited by M.J. Schick (Marcel Dekker, New York, 1975).
18. P.A. Annis and T.W. Quigley, Jr., in *Microstructural Characterisation of Fibre-Reinforced Composites*, edited by J. Summerscales (CRC Press, Boca Raton, 1998), p. 17.
19. *Textile World Manmade Fiber Chart*, in Textile World, Vol. 5. Billian Publishing. Available at www.textileworld.com.
20. R.G. Scott, *ASTM Special Publ.* **257** (1959) 121.
21. R.C. Faust, *Proc. Phys. Soc.* **68** (1955) 1081.
22. J.M. Preston and K.I. Narasimhan, *J. Text. Inst.* **40** (1949) T327.
23. E.F. Gurnee, *J. Appl. Phys.* **25** (1954) 1232.
24. M.A. Sieminski, *The Microscope* **23** (1975) 35.
25. R.D. Andrews, *J. Appl. Phys.* **25** (1954) 1223.
26. G.M. Bhatt and J.P. Bell, *J. Polym. Sci. Polym. Phys. Edn.* **14** (1976) 575.
27. V.B. Gupta and S. Kumar, *J. Appl. Polym. Sci.* **26** (1981) 1865.

28. M.K. Tomioka and S.H. Zeronian, *Text. Res. J.* **44** (1974) 1.
29. W.R. Goynes, J.H. Carra and J.D. Timpa, *Proc. 35th Annu. Mtg EMSA*, edited by W. Bailey (San Francisco Press, Boston, 1977), p. 168.
30. G. Gillberg and D. Kemp, *J. Appl. Polym. Sci.* **26** (1981) 2023.
31. J.W.S. Hearle, B. Lomas and A.R. Bunsell, *Appl. Polym. Symp.* **23** (1974) 147.
32. J.W.S. Hearle. *Proc. Textile Inst., Inst. Textile de France Conf.*, Paris, 1975, p. 60.
33. R.C. Laible, F. Figucia and B.H. Kirkwood, *Appl. Polym. Symp.* **23** (1974) 181.
34. V.B. Gupta, *J. Appl. Polym. Sci.* **20** (1976) 2005.
35. J.W.S. Hearle and B.S. Wong, *J. Text. Inst.* **68** (1977) 89.
36. C. Oudet and A.R. Bunsell, *J. Mater. Sci. Left.* **3** (1984) 295.
37. J. Smook, W. Hamersma and A.J. Pennings, *J. Mater. Sci.* **19** (1984) 1359.
38. J.W.S. Hearle and P.M. Cross, *J. Mater. Sci.* **5** (1970) 507.
39. J.W.S. Hearle and J.T. Sparrow, *Text. Res. J.* **41** (1971) 736.
40. A.R. Bunsell, J.W.S. Hearle and R.D. Hunter, *J. Sci. Instrum.* **4** (1971) 860.
41. R. Bunsell and J.W.S. Hearle, *J. Mater. Sci.* **6** (1971) 1303.
42. H.H. Kausch, *Polymer Fracture* (Springer Verlag, Berlin, 1978).
43. H.H. Kausch, Ed. *Crazing in Polymers, Adv. Polym. Sci. Ser. 52/3*, (Springer-Verlag, Berlin, 1983).
44. A.R. Bunsell, unpublished data.
45. P. Gupta and G.L. Wilkes, *Polymer* **44** (2003) 6353.
46. L. Fourt and N.R.S. Hollies, *Clothing: Comfort and Function* (Marcel Dekker, New York, 1970).
47. H.N. Yoon, L.C. Sawyer and A. Buckley, *Text. Res. J.* **54** (1984) 357.
48. H. Kirkwood, *Text. Res. J.* **44** (1974) 545.
49. J.H. Warren and N.S. Eiss, *Jr, Trans. ASME* **100** (1978) 92.
50. M. Mlynar, *International Nonwovens Journal* [online computer file] **9** (2000). www.inda.org/subscrip/
51. S. Thandavamoorthy, N. Gopinath and S.S. Ramkumar, *J. Appl. Polym. Sci.* **101** (2006) 3121.
52. B.A. Wood. *Eastern Analytical Symposium; figures not published*, 2005.
53. L.C. Sawyer, *Proc. 39th Annu. Mtg. EMSA*, (Claitors, Baton Rouge LA, 1981) p. 334.
54. A. Ziabicki and H. Kawai, Eds. *High-Speed Fiber Spinning: Science and Engineering Aspects* (Wiley-Interscience, New York, 1985).
55. W.G. Perkins and R.S. Porter, *J. Mater. Sci.* **16** (1981) 1458.
56. W.L. Wu, V.F. Holland and W.B. Black, *J. Mater. Sci. Left.* **14** (1979) 252.
57. G.M. Sze, J. Spruiell and J.L. White, *J. Appl. Polym. Sci.* **20** (1976) 1823.
58. B. Catorie, R. Hagege and D. Meimoun, *Bull. Inst. Text. Fr.* **23** (1969) 521.
59. R. Hagege, M. Jarrin and M. Sotton, *J. Microsc.* **115** (1979) 65.
60. R.E. Wilfong and J. Zimmerman, *J. Appl. Polym. Sci.* **17** (1973) 2039.
61. G. Gillberg, L.C. Sawyer and A. Promislow, *J. Appl. Polym. Sci.* **28** (1983) 3723.
62. M. Jaffe, in *Encyclopedia of Polymer Science and Engineering*, 2nd ed., edited by H.F. Mark, N.M. Bikales, C.G. Overberger and G. Menges (John Wiley, New York, 1986), p. 699.
63. E. Baer and A. Moet, Eds. *High-Performance Polymers* (Hanser, New York, 1991).
64. K. Hood, *Trends Polym. Sci.* **1** (1993) 129.
65. F. Lacroix, J. Loos and K. Schulte, *Polymer* **40** (1998) 843.
66. M.I. Abo El-Maaty, R.H. Olley and D.C. Bassett, *J. Mater. Sci.* **34** (1999) 1975.
67. P.J. Hine, I.M. Ward, N.D. Jordan, R.H. Olley and D.C. Basett, *J. Macromol. Sci. Phys.* **B40** (2001) 959.
68. I.M. Ward and P.J. Hine, *Polymer* **45** (2004) 1423.
69. I.M. Ward, *Structure and Properties of Oriented Polymers* (John Wiley, New York, 1975).
70. A. Zachariades and R.S. Porter, Eds. *The Strength and Stiffness of Polymers*, (Marcel Dekker, New York, 1983).
71. R.H. Olley, D.C. Bassett, P.J. Hine and I.M. Ward, *J. Mater. Sci.* **28** (1993) 1107.
72. M.A. Kabeel, D.C. Bassett, R.H. Olley, P.J. Hine and I.M. Ward, *J. Mater. Sci.* **30** (1995) 601.
73. A. Schaper, D. Zenke, E. Schulz, R. Hirte and M. Taege, *Phys. Status Solidi* **116** (1989) 179.
74. D.T. Grubb and M.J. Hill, *J. Cryst. Growth* **48** (1980) 407.
75. D.T. Grubb and K. Prasad, *Macromolecules* **25** (1992) 4575.
76. S.N. Magonov, K. Qvarnstrom, V. Elings and H.J. Cantow, *Polym. Bull.* **25** (1991) 689.

77. S.N. Magonov, K. Qvarnstrom, D.M. Huong, V. Elings and H.J. Cantow, *Polym. Mater. Sci. Eng.* (1991) 163.
78. L.M. Eng, H. Fuchs, K.D. Jandt and J. Petermann, *Helv. Phys. Acta* **65** (1992) 870.
79. B.K. Annis, J.R. Reffner and B. Wunderlich, *J. Polym. Sci. B Polym. Phys.* **31** (1993) 93.
80. W. Huh, S. Kumar, W. Wade Adams, *Polym. Eng. Sci.* **43** (2003) 684.
81. P.J. Hine and I.M. Ward, *J. Appl. Polym. Sci.* **91** (2004) 2223.
82. O.R. Hughes, D.K. Kurschus, C.K. Saw, J. Flint, T.P. Bruno and R.T. Chen, *J. Appl. Polym. Sci.* **74** (1999) 2335.
83. M.-C.G. Jones and D.C. Martin, *J. Mater. Sci.* **32** (1997) 2291.
84. D.A. Chernoff and S. Magonov, in *Comprehensive Desk Reference of Polymer Characterization and Analysis*, edited by R.F. Brady Jr. (ACS, Oxford Press, 2003), p. 490.
85. J.C. Bastidas, R. Venditti, J. Pawlak, R. Gilbert, S. Zauscher and J.F. Kadla, *Carbohydr. Polym.* **62** (2005) 369.
86. C. Kuebel, D.P. Lawrence and D.C. Martin, *Macromolecules* **34** (2001) 9053.
87. C. Kuebel, 2006, unpublished micrographs.
88. D.C. Martin, J. Chen, J. Yang, L.F. Drummy and C. Kübel, *J. Polym. Sci. B Polym. Phys.* **43** (2005) 1749.
89. S.B. Warner, M.B. Polk and K. Jacob, *J.M.S. Rev. Macromol. Chem. Phys.* **C39** (1999) 643.
90. S. Arcidiacono, C.M. Mello, M. Butler, E. Welsh, J.W. Soares, A. Allen, D. Ziegler, T. Laue and S. Chase, *Macromolecules* **35** (2002) 1262.
91. J.A. Coddington, H.D. Chanzy, C.L. Jackson, G. Raty and K.H. Gardner, *Biomacromolecules* **3** (2002) 5.
92. H.-J. Jin and D.L. Kaplan, *Nature* **424** (2003) 1057.
93. S. Zarkoob, R.K. Eby, D.H. Rencker, S.D. Hudson, D. Ertley and W.W. Adams, *Polymer* **45** (2004) 3973.
94. S.A.C. Gould, K.T. Tran, J.C. Spagna, A.M.F. Moore and J.B. Shulman, *Int. J. Biol. Macromol.* **24** (1999) 151.
95. D.T. Grubb, *J. Mater. Sci.* **9** (1974) 1715.
96. A. Keller and M.J. Machin, *J. Macromol. Sci. Phys.* **B1** (1967) 41.
97. C. Bonnebat, G. Roullet and A.J.D. Vries, *Polym. Eng. Sci.* **21** (1981) 189.
98. J. McDermott, *Industrial Membranes* (Noyes Data Corp., Park Ridge, NY, 1972).
99. E.H. Andrews, *Proc. R. Soc.* **A270** (1962) 232.
100. E.H. Andrews, *Proc. R. Soc.* **A277** (1964) 562.
101. K. Sakaoku and A. Peterlin, *J. Polym. Sci.* **9** (1971) 895.
102. P.M. Tarin and E.L. Thomas, *Polym. Eng. Sci.* **18** (1978) 472.
103. C.M. Chu and G.L. Wilkes, *J. Macromol. Sci. Phys.* **B10** (1974) 231.
104. J.D. Hoffman, L.J. Frolen, G.S. Ross and J.I. Lauritzen, *J. Res. Nat. Bur. Stand.* **75A** (1975) 671.
105. J. Petermann and H. Gleiter, *Philos. Mag.* **28** (1973) 1279.
106. J. Petermann, M.J. Miles and H. Gleiter, *J. Macromol. Sci. Phys.* **B12** (1976) 393.
107. J. Miles, J. Petermann and H. Gleiter, *J. Macromol. Sci. Phys.* **B12** (1976) 523.
108. J. Miles, J. Petermann and H. Gleiter, *Prog. Colloid Polym. Sci.* **62** (1977) 6.
109. R.M. Gohil and J. Petermann, *J. Polym. Sci Polym. Phys. Edn.* **17** (1979) 525.
110. M.J. Miles and J. Petermann, *J. Macromol. Sci. Phys.* **B16** (1979) 1.
111. J. Petermann and R.M. Gohil, *J. Mater. Sci.* **14** (1979) 2260.
112. V.P. Chacko, W.W. Adams and E.L. Thomas, *J. Mater. Sci.* **18** (1983) 1999.
113. J.T. Chen and E.L. Thomas, *J. Mater. Sci.* **31** (1996) 2531.
114. J. Ruokolainen, G.H. Fredrickson, E.J. Kramer, C.Y. Ryu, S.F. Hahn and S.N. Magonov, *Macromolecules* **35** (2002) 9391.
115. J.Y. Cheng, C.A. Ross, E.L. Thomas, H.I. Smith, G.J. Vancso, *Adv. Mater.* **15** (2003) 1599.
116. A.R. Kamdar, Y.S. Hu, P. Ansems, S.P. Chum, A. Hiltner and E. Baer, *Macromolecules* **39** (2006) 1496.
117. D.C. Yang and E.L. Thomas, *J. Mater. Sci.* **19** (1984) 2098.
118. E.L. Thomas, in *Structure of Crystalline Polymers*, edited by I.H. Hall (Elsevier-Applied Science, London, 1984), p. 79.
119. D.C. Bassett, *Principles of Polymer Morphology* (Cambridge University Press, Cambridge, 1981), p. 102.
120. D.C. Bassett, *CRC Crit. Rev. Solid State Mater. Sci.* **12** (1984) 97.
121. V. Vittoria and D.C. Bassett, *Coll. Polym. Sci.* **267** (1989) 661.
122. G. Kanig, *J. Crystal Growth* **18** (1980) 303.
123. D.T. Grubb and A. Keller, *J. Polym. Sci., Polym. Phys. Edn.* **18** (1980) 207.
124. I.G. Voigt-Martin, *J. Polym. Sci. Polym. Phys. Edn.* **18** (1980) 1513.

125. I.G. Voigt-Martin, *Adv. Polymer Sci.* **67** (1985) 195.
126. I.G. Voigt-Martin and L. Mandelkem, *J. Polym. Sci. B Polym. Phys.* **27** (1989) 967.
127. I.G. Voigt-Martin, G.M. Stack, A.J. Peacock and L. Mandelkem, *J. Polym. Sci. B Polym. Phys.* **27** (1989) 957.
128. D.J.M. Clark and R.W. Truss, *J. Polym. Sci. B Polym. Phys.* **34** (1996) 103.
129. L.H. Radzilowski and E.L. Thomas, *J. Polym. Sci. B Polym. Phys.* **35** (1997) 1009.
130. T. Takahama and P.H. Geil, *J. Polym. Sci. Polym. Phys. Edn.* **21** (1983) 1247.
131. S. Lee, H. Miyaji and P.H. Geil, *J. Macromol. Sci., Phys.* **B22** (1983) 489.
132. D.T. Grubb, in *Developments in Crystalline Polymers*, edited by D.C. Bassett (Applied Science, London, 1982), p. 26.
133. P.H. Geil, *J. Polym. Sci.* **20** (1967) 109.
134. G.S. Yeh and P.H. Geil, *J. Macromol. Sci. Phys* **Bl** (1967) 235.
135. G.S. Yeh, *Crit. Rev. Macromol. Chem.* **1** (1972) 173.
136. G.S. Yeh, *J. Macromol. Sci. Phys.* **B6** (1972) 451.
137. P.H. Geil, *J. Macromol. Sci. Phys.* **B12** (1976) 173.
138. E.L. Thomas and E.J. Roche, *Polymer* **20** (1979) 1413.
139. D.R. Uhlmann, *Disc. Faraday Soc.* **68** (1979) 87.
140. D. Vesely, A. Low and M. Bevis, in *Developments Electron Microscopy and Analysis*, edited by J. Venables (Academic Press, New York, 1976), p. 333.
141. M. Mayer, J.B.V. Sande and D.R. Uhlmann, *J. Polym. Sci. Polym. Phys. Edn.* **16** (1978) 2005.
142. D.T. Grubb, *Disc. Faraday Soc.* **68** (1979) 125.
143. D.R. Uhlmann, A.L. Renninger, G. Kritchevsky and J.B.V. Sande, *J. Macromol. Sci. Phys.* **B12** (1976) 153.
144. R.S. Stein, *J. Polym. Sci.* **24** (1957) 383.
145. R.S. Samuels, *J. Polym. Sci.* **3** (1965) 1741.
146. R.J. Samuels, *J. Appl. Polym. Sci.* **26** (1981) 1383.
147. J.P. Paulos and E.L. Thomas, *J. Appl. Polym. Sci.* **25** (1980) 15.
148. W. Chinsirikul, T.C. Hsu and I.R. Harrison, *Polym. Eng. Sci.* **36** (1996) 2708.
149. Y.-B. Huang and T.E. Wessel, *J. Appl. Polym. Sci.* **57** (1995) 1227.
150. E.S. Sherman, *Polym. Eng. Sci.* **24** (1984) 895.
151. T. Tagawa and K. Ogura, *J. Polym. Sci. Polym. Phys. Edn.* **18** (1980) 971.
152. A. Prasad, R. Shroff, S. Rane and G. Beaucage, *Polymer* **42** (2001) 3103.
153. K.D. Jandt, *Untersuchungen zur Polymer-Metall-Epitaxie in Computersimulation und Experiment* **5** (1993) 302.
154. W.W. Adams, D. Yang and E.L. Thomas, *J. Mater. Sci.* **21** (1986) 2239.
155. N.M. Benetatos and K.I. Winey, *J. Polym. Sci. B Polym. Phys.* **43** (2005) 3549.
156. T.R. Albrecht, M.M. Dovek, C.A. Lang, P. Grutter, C.F. Quate, S.W.J. Kuan, C.W. Frank and R.F.W. Pease, *J. Appl. Phys.* **64** (1988) 1178.
157. P. Dietz, P.K. Hansma, K.J. Ihn, F. Motamedi and P. Smith, *J. Mater. Sci.* **28** (1993) 1372.
158. H. Fuchs, *J. Mol. Struct.* **292** (1993) 29.
159. S.M. Hues, R.J. Colton, E. Meyer and H.J. Guntherodt, *Mater. Res. Soc. Bull.* **Jan** (1993) 41.
160. V.V. Tsukruk and D.H. Reneker, *Polymer* **36** (1995) 1791.
161. V.V. Tsukruk, M.D. Foster, D.H. Reneker, A. Schmidt, H. Wu and W. Knoll, *Macromolecules* **27** (1994) 1274.
162. R.J. Wilson, K.E. Johnson, D.D. Chambliss and B. Melior, *Langmuir* **9** (1993) 3478.
163. D.A. Schiraldi, S.A.C. Gould and M.L. Occelli, *J. Appl. Polym. Sci.* **80** (2001) 750.
164. M.B. Olde Riekerink, M.B. Claase, G.H.M. Engbers, D.W. Grijpma and J. Feijen, *J. Biomed. Mater. Res. A* **65A** (2003) 417.
165. P.P. Kundu, J. Biswas, H. Kim, C.-W. Chung and S. Choe, *J. Appl. Polym. Sci.* **91** (2004) 1427.
166. C.M. Stafford, C. Harrison, K.L. Beers, A. Karim, E.J. Amis, M.R. Vanlandingham, H.C. Kim, W. Volksen, R.D. Miller and E.E. Simon, *Nat. Mater.* **3** (2004) 545–550.
167. T. Reilly, unpublished data.
168. R.T. Chen, M.G. Jamieson and R. Callahan. *Proc. 50th Ann. Mtg. EMSA*, edited by W. Bailey (San Francisco Press, 1992), p. 1142.
169. R.T. Chen, C.K. Saw, M.G. Jamieson and T.R. Aversa, *J. Appl. Poly. Sci.* **53** (1994) 471.
170. G.K. Bar and G.F. Meyers, *MRS Bull.* **29** (2004) 464.
171. L.F. Drummy, J. Yang and D.C. Martin, *Ultramicroscopy* **99** (2004) 247.
172. L.F. Drummy, P.K. Miska, D. Alberts, N. Lee and D.C. Martin, *J. Phys. Chem. B* **110** (2006) 6066.
173. D.C. Martin, *Polymer* **43** (2002) 4421.

174. G.F. Meyers, B.M. DeKoven, M.T. Dineen, A. Strandjord, P.J. O'Connor, T. Hu, Y.-H. Chiao, H.M. Pollock and A. Hammiche, *ACS Symposium Series* **741** (2000) 190.
175. G.F. Meyers, M.T. Dineen, E.O. Shaffer, II, T. Stokich, Jr. and J.-H. Im, *Macromol. Symp.* **167** (2001) 213.
176. A. Ranade, A. Hiltner, E. Baer and D. Bland. *ANTEC 2004—Annual Technical Conference Proceedings, May 16–20 2004*, Chicago, IL, (SPE Brookfield CT) p. 3130.
177. Y. Cohen, R.J. Albalak, B.J. Dair, M.S. Capel and E.L. Thomas, *Macromolecules* **33** (2000) 6502.
178. E.M. Ivan'kova, G.H. Michler, A. Hiltner and E. Baer, *Macromol. Mater. Eng.* **289** (2004) 787.
179. R. Adhikari, W. Lebek, R. Godehardt, S. Henning, G.H. Michler, E. Baer and A. Hiltner, *Polym. Adv. Technol.* **16** (2005) 95.
180. R.Y.F. Liu, T.E. Bernal-Lara, A. Hiltner and E. Baer, *Macromolecules* **37** (2004) 6972.
181. H. Aref, *Phys. Fluids* **14** (2002) 1315.
182. A. Dhoble, B. Kulshreshtha, S. Ramaswami and D.A. Zumbrunnen, *Polymer* **46** (2005) 2244.
183. O. Kwon and D.A. Zumbrunnen, *J. Appl. Polym. Sci.* **82** (2001) 1569.
184. O. Kwon and D.A. Zumbrunnen, *Polym. Eng. Sci.* **43** (2003) 1443.
185. D.A. Zumbrunnen, S. Inamdar, O. Kwon and P. Verma, *Nano Lett.* **2** (2002) 1143.
186. B.A. Wood, *Technical Papers—American Chemical Society, Rubber Division, Spring Technical Program, 161st, Savannah, GA, Apr. 29-May 1, 2002*, (ACS Washington DC, 2002), p. 669.
187. H.K. Lonsdale, *J. Membrane Sci.* **10** (1982) 81.
188. D.R. Lloyd, Ed. *Membrane Materials for Liquid Separations*, ACS Symp. Ser. 269 (American Chemical Society, Washington, DC, 1985).
189. P. Meares, Ed. *Membrane Separation Processes*, (Elsevier Scientific, Amsterdam, 1976).
190. A.G. Fane, C.J.D. Fell and A.G. Waters, *J. Membrane Sci.* **9** (1981) 245.
191. W. Pusch, *J. Membrane Sci.* **10** (1982) 325.
192. J.S. Fodor, R.M. Briber, T.P. Russell, K.R. Carter, J.L. Hedrick and R.D. Miller, *J. Polym. Sci. B Polym. Phys.* **35** (1997) 1067.
193. A.K. Fritsche, A.R. Arevalo, A.F. Connolly, M.D. Moore, V.B. Elings and C.M. Wu, *J. Appl. Poly. Sci.* **45** (1992) 1945.
194. A.K. Fritsche, A.R. Arevalo, M.D. Moore, C.J. Weber, V.B. Elings, K. Kjoller and C.M. Wu, *J. Appl. Poly. Sci.* **46** (1992) 167.
195. I. Masselin, L. Durand-Bourlier, J.-M. Laine, P.-Y. Sizaret, X. Chasseray and D. Lemordant, *J. Membr. Sci.* **186** (2001) 85.
196. H. Essis-Tome, C.K. Diawara, L. Robbiola, G. Cote, A. Kossir, K. El Kacemi, Z. Qafas and M. Pontie, *Electrochem. Commun.* **6** (2004) 1061.
197. W.R. Bowen and T.A. Doneva, *J. Membr. Sci.* **171** (2000) 141.
198. W.R. Bowen and T.A. Doneva, *Surf. Interface Anal.* **29** (2000) 544.
199. W.R. Bowen, J.A.G. Stoton and T.A. Doneva, *Surf. Interface Anal.* **33** (2002) 7.
200. N. Hilal, V. Kochkodan, L. Al-Khatib and G. Busca, *Surf. Interface Anal.* **33** (2002) 672.
201. S.-Y. Park, H. Koerner, R. Ozisik, S. Juhl, S. Putthanarat, B.L. Farmer and R.K. Eby, *Polymer* **45** (2004) 49.
202. J. Yang and D.C. Martin, *Sensors Actuators B* **101** (2004) 133.
203. S. Loeb and S. Sourirajan, in ACS *Adv. Chem. Ser. 38* (American Chemical Society, Washington, DC, 1962), p. 117.
204. H.K. Plummer, C.D. Melvin and E. Eichen, *Desalination* **15** (1974) 93.
205. M. Katoh and S. Suzuli, in *Synthetic Membranes*, Vol. 1, *Desalination*, ACS Symp Ser. 153 (American Chemical Society, Washington, DC, 1981), p. 247.
206. K.N. Kapadia and V.P. Pandya, *Desalination* **34** (1980) 199.
207. F.S. Model and L.A. Lee, *Am. Chem. Soc. Div. Org. Coatings Plast. Chem., Prepr.* **32** (1972) 384.
208. F.S. Model, H.J. Davis and J.E. Poist, in *Reverse Osmosis and Synthetic Membranes*, edited by S. Sourirajan (National Research Council Canada, 1977).
209. L.C. Sawyer and R.S. Jones, *J. Membr. Sci.* **12** (1984) 147.
210. M. Panar, H.H. Hoehn and R.R. Hebert, *Macromolecules* **6** (1973) 777.
211. C.W. Alegrane, D.G. Pye, H.H. Hoehn and M. Panar, *J. Appl. Polym. Sci.* **19** (1975) 1475.
212. V. Freger, J. Gilron and S. Belfer, *J. Membr. Sci.* **209** (2002) 283.
213. T. Sarada, L.C. Sawyer and M. Ostler, *J. Membr. Sci.* **15** (1983) 97.
214. H.S. Bierenbaum, R.B. Isaacson, M.L. Druin and S.G. Plovan, *Ind. Eng. Chem., Prod. Res. Dev.* **13** (1974) 2.

215. V.E. Robertson, unpublished, JEOL USA Inc., Peabody MA.
216. G.F. Meyers, unpublished data.
217. M. Jamieson, unpublished data.
218. H.K. Lonsdale and H.E. Podall, *Reverse Osmosis Membrane Research* (Plenum Press, New York, 1972).
219. I. Cabasso, K.Q. Robert, E. Klein and J.K. Smith, *J. Appl. Polym. Sci.* **21** (1977) 1883.
220. I. Cabasso and A.P. Tamvakis, *J. Appl. Polym. Sci.* **23** (1979) 1509.
221. A. Taubert, J.D. Wind, D.R. Paul, W.J. Koros and K.I. Winey, *Polymer* **44** (2003) 1881.
222. S.B. Carruthers, G.L. Ramos and W.J. Koros, *J. Appl. Polym. Sci.* **90** (2003) 399.
223. F. Tasselli, J.C. Jansen and E. Drioli, *J. Appl. Polym. Sci.* **91** (2004) 841.
224. M.D. Daijal, Ed. *Plastics Polymer Science and Technology* (Wiley-Interscience, New York, 1982).
225. J.M. Margolis, Ed. *Engineering Thermoplastics* (Marcel Dekker, New York, 1985).
226. D.R. Paul and L.H. Sperling, Eds. *Multicomponent Polymer Materials*, Adv. Chem. Ser. No. 211 (American Chemical Society, Washington, DC, 1986).
227. D.R. Paul, J.W. Barlow and H. Keskkula, in *Encyclopedia of Polymer Science*, 2nd ed., edited by H.F. Mark, N.M. Bikales, C.G. Overberger and G. Menges (Wiley, New York, 1988), p. 399.
228. L.H. Sperling, *Polymeric Multicomponent Materials* (John Wiley, New York, 1997).
229. G.H. Michler and F.J. Balta-Calleja, Eds. *Mechanical Properties of Polymers Based on Nanostructure and Morphology* (CRC Press Taylor & Francis, Boca Raton and New York, 2005).
230. M. Vaziri, R.T. Spurr and F.H. Stott, *Wear* **122** (1988) 329.
231. L.C. Sawyer, B. Strassle and D.J. Palatini. *Proc. 37th Annu. Mtg EMSA*, San Antonio, edited by W. Bailey (Claitors Baton Rouge LA, 1979), p. 620.
232. E.L. Lawton, T. Murayama, V.F. Holland and D.C. Felty, *J. Appl. Polym. Sci.* **25** (1980) 187.
233. W.A. Heeschen, *Polymer* **36** (1995) 1835.
234. J. Wu, C.-M. Chan and Y.-W. Mai, *Plastics Eng. (New York)* **52** (1999) 505.
235. D.C. Bassett and A.S. Vaughan, in *Polymer Characterization Techniques and Their Application to Blends*, edited by G.P. Simon (American Chemical Society, Washington, DC, 2003), p. 436.
236. E.L. Thomas and Y. Talmon, *Polymer* **19** (1978) 225.
237. P. Prentice and J.G. Williams, *Plast. Rubber Process. App.* **2** (1982) 27.
238. P. Prentice, *Polymer* **23** (1982) 1189.
239. C.B. Bucknall, *Journal of Microscopy, Fifth International Conference on Microscopy of Composite Materials (April 2000)* **201** (2001) 221.
240. S.J. Krause, W.W. Adams, S. Kumar, T. Reilly and T. Suzuki, *EMSA Proc.* **45** (1987) 466.
241. L.C. Sawyer and M. Jamieson, *EMSA Proc.* edited by W. Bailey (San Francisco Press) **47** (1989) 334.
242. J. Pawley and D.C. Joy, *EMSA Proc.* **50** (1992) 1278.
243. S.J. Krause and W.W. Adams, *MSA Proc.* **51** (1993) 866.
244. D.W. Schwark, D.L. Vezie, J.R. Reffner, E.L. Thomas and B.K. Annis, *J. Mater. Sci. Lett.* **11** (1992) 352.
245. P.B. Himelfarb and K.B. Labat, *Scanning* **12** (1990) 148.
246. A.M. Donald and B.L. Thiel, *Structure and Dynamics of Materials in the Mesoscopic Domain, Proceedings of the Royal Society-Unilever Indo-UK Forum in Materials Science and Engineering, 4th, Pune, India, Dec. 8–12, 1997* (1999) 1.
247. S.J. Williams, D.E. Morrison, B.L. Thiel and A.M. Donald, *Scanning* **27** (2005) 190.
248. B.K. Annis, D.W. Schwark, J.R. Reffner, E.L. Thomas and B. Wunderlich, *Makromol. Chem.* **193** (1992) 2589.
249. D.L. Vezie, W.W. Adams and E.L. Thomas, *Polymer* **36** (1995) 1761.
250. S. Magonov, in *Handbook of Surfaces and Interfaces of Materials*, edited by H.R. Nalwa (Academic Press, New York, 2001), p. 393.
251. D.A. Ivanov and S.I. Magonov, in *Polymer Crystallization. Observations, Concepts and Interpretations (Lecture Notes in Physics Vol.606)*, edited by J.-U. Sommer and G. Reiter (Springer-Verlag, New York, 2003), p. 98.
252. S. Magonov, in *Applied Scanning Probe Methods*, edited by P. Avouris, D. Klitzing, H. Sakaki and R. Wiesndanger (Springer, Berlin, Heidelberg, 2004), p. 207.
253. S. Magonov and N.A. Yerina, in *Microscopy for Nanotechnology*, edited by N. Yao and Z.L. Wang (Kluwer Academic Press, New York, 2005), p. 113.
254. A. Pfau, A. Janke and W. Heckmann, *Surf. Interface Anal.* **27** (1999) 410.

255. Y. Thomann, J. Suhm, R. Thomann, G. Bar, R.-D. Maier and R. Muelhaupt, *Macromolecules* **31** (1998) 5441.
256. G. Bar, R. Brandsch, M. Bruch, L. Delineau and M.-H. Whangbo, *Surf. Sci.* **444** (2000) 11.
257. G.H. Michler, Adhikari, Rameshwar, Henning, Sven, *Macromol. Symp.* **214** (2004) 47.
258. N. Virgilio, B.D. Favis, M.-F. Pepin, P. Desjardins and G. L'Esperance, *Macromolecules* **38** (2005) 2368.
259. J. Kresta, Ed. *Reaction Injection Molding Polymer Chemistry and Engineering*, ACS Symp. Ser. 270, (American Chemical Society, Washington, DC, 1985).
260. R.G. Griskey, *Polymer Process Engineering* (Chapman & Hall, New York, 1995).
261. Z. Tadmor, *J. Appl. Polym. Sci.* **18** (1974) 1753.
262. S.S. Katti and J.M. Schultz, *Polym. Eng. Sci.* **22** (1982) 1001.
263. J. Bowman, *J. Mater. Sci.* **16** (1981) 1151.
264. D.E. Scherpereel, *Plast. Eng.* **Dec** (1976) 46.
265. N.C. Watkins and D. Hansen, *Text. Res. J.* **68** (1968) 388.
266. D.C. Bassett, *J. Macromol. Sci. Phys.* **B42** (2003) 227.
267. D.C. Bassett, *Macromol. Symp.* **214** (2004) 5.
268. D.C. Bassett, in *Mechanical Properties of Polymers Based on Nanostructure and Morphology*, edited by G.H. Michler and F.J. Balta-Calleja (CRC Press Taylor & Francis, Boca Raton and New York, 2005), p. 3.
269. H.D. Keith, F.J. Padden, Jr. and T.P. Russell, *Macromolecules* **22** (1989) 666.
270. S.Y. Hobbs and C.F. Pratt, *J. Appl. Polym. Sci.* **19** (1975) 1701.
271. J.J. Janimak, L. Markey and G.C. Stevens, *Polymer* **42** (2001) 4675.
272. L. Markey, J.J. Janimak and G.C. Stevens, *Polymer* **42** (2001) 6221.
273. S.C. Malguarnera and A. Manisali, *Polym. Eng. Sci.* **21** (1981) 586.
274. E.S. Clark, *Appl. Polym. Symp.* **24** (1974) 45.
275. M.R. Kantz, H.D. Newman and F.H. Stigale, *J. Appl. Polym. Sci.* **16** (1972) 1249.
276. M.R. Kantz, *Int. J. Polym. Mater.* **3** (1974) 245.
277. E.S. Clark, *SPEJ* **23** (1967) 46.
278. D.G.M. Wright, R. Dunk, D. Bouvart and M. Autran, *Polymer* **29** (1988) 793.
279. W.Y. Chiang and M.S. Lo, *J. Appl. Poly. Sci.* **34** (1987) 1997.
280. A.M. Baro, L. Vazquez, A. Bartolame, J. Gomez, N. Garcia, H.A. Goldberg, L.C. Sawyer, R.T. Chen, R.S. Kohn and R. Reifenberger, *J. Mater. Sci.* **24** (1989) 1739.
281. H.A. Goldberg, F.J. Onorato, K. Chiang, M. Jamieson, Y.Z. Li, R. Piner and R. Reifenberger. 1989, p. 170.
282. A. Keller, *Rep. Prog. Phys.* **31** (1968) 623.
283. B. Wunderlich, *Macromolecular Physics, Vol. 1, Crystal Structure, Morphology, Defects, and Vol. 2, Crystal Nucleation, Growth, Annealing* (Academic Press, New York, 1973).
284. D.T. Grubb, *J. Microsc. Spectrosc. Electron.* **1** (1976) 671.
285. D.T. Grubb, in *Developments in Crystalline Polymers—1*, edited by D.C. Bassett (Applied Science, London, 1982).
286. D.R. Uhlmann and A.G. Kolbeck, *Sci. Am.* **233** (1975) 96.
287. D.C. Bassett, *Principles of Polymer Morphology* (Cambridge University Press, Cambridge, 1981).
288. D.C. Bassett, *CRC Crit. Rev. Solid State Mater. Sci.* **12** (1984) 97.
289. R.B. Seymour, Ed. *The History of Polymer Science and Technology* (Marcel Dekker, New York, 1982).
290. J. Bowman, N. Harris and M. Bevis, *J. Mater. Sci.* **10** (1975) 63.
291. J. Bowman and M. Bevis, *Plast. Rubber Mater. Appl.* (1976) 177.
292. V. Tan and M.R. Kamal, *J. Appl. Polym. Sci.* **22** (1978) 2341.
293. E.S. Clark and C.A. Garber, *Int. J. Polym. Mater.* **1** (1971) 31.
294. E.S. Clark, *Appl. Polym. Symp.* **20** (1973) 325.
295. D.R. Fitchmun and S. Newman, *J. Polym. Sci. Polym. Left. Edn.* **7** (1969) 301.
296. J.F. Callcar and J.B. Shortall, *J. Mater. Sci.* **12** (1977) 141.
297. D.R. Fitchmun and Z. Mencik, *J. Polym. Sci., Polym. Phys. Edn.* **11** (1974) 951.
298. Z. Mencik and D.R. Fitchmun, *J. Polym. Sci. Polym. Phys. Edn.* **11** (1974) 973.
299. G. Menges, A. Wubken and B. Horn, *Colloid Polym. Sci.* **254** (1976) 267.
300. D.A. Schiraldi, M.L. Occelli and S.A.C. Gould, *J. Appl. Polym. Sci.* **82** (2001) 2616.
301. C.B. Bucknall, *Toughened Plastics* (Applied Science, London, 1977), p. 333.
302. C.B. Bucknall, *Toughened Plastics* (Applied Science, London, 1977).

303. D.R. Paul and S. Newman, Eds. *Polymers Blends, Vols. I and II* (Academic Press, New York, 1978).
304. J.A. Manson and L.H. Sperling, *Polymer Blends and Composites* (Plenum, New York, 1976).
305. L.H. Sperling, Ed. *Recent Advances in Polymer Blends, Grafts and Blocks* (Plenum, New York, 1974).
306. N.A.J. Platzer, Ed. *Multicomponent Polymel Systems*, ACS Adv. Chern. Ser. 99 (American Chemical Society, Washington, DC, 1971).
307. N.A.J. Platzer, Ed. *Copolymers, Polyblends, and Composites*, ACS Adv. Chern. Ser. 142 (American Chemical Society, Washington, DC, 1975).
308. D. Klempner and K.C. Frisch, Eds. *Polymer Alloys: Blends, Blocks, Grafts and Interpenetrating Networks* (Plenum, New York, 1977).
309. S.L. Cooper and G.M. Estes, Eds. *Multiphase Polymers*, ACS Adv. Chern. Ser. 176 (American Chemical Society, Washington, DC, 1979).
310. E.V. Thompson, in *Polymer Alloys 2: Blends, Blocks, Grafts, Interpenetrating Networks*, edited by D. Klempner and K.C. Frisch (Plenum, New York, 1977, 1980).
311. A.S. Argon and R.E. Cohen, *Polymer* **44** (2003) 6013.
312. M. Li and C.K. Ober, *Materialstoday* **9** (2006) 30.
313. J.L. White and K. Min, *Makromol. Chem., Macromol. Symp.* **16** (1988) 19.
314. A.P. Plochocki, *Polimeri* **10** (1965) 23.
315. W.J. Coumans, D. Heikens and S.D. Sjoerdsma, *Polymer* **21** (1980) 103.
316. H. Keskkula and W.J. Frazer, *J. Polym. Sci.* **7** (1968) 1.
317. C.B. Bucknall, *Toughened Plastics* (Applied Science, London, 1977). p. 9
318. A.M. Donald and E.J. Kramer, *J. Mater. Sci.* **17** (1982) 2351.
319. B.D. Lauterwasser and E.J. Kramer, *Philos. Mag.* **39A** (1979) 469.
320. A.M. Donald, T. Chan and E.J. Kramer, *J. Mater. Sci.* **16** (1981) 669.
321. C.B. Bucknall, *Toughened Plastics* (Applied Science, London, 1977), chap. 7.
322. C.B. Bucknall, *Toughened Plastics* (Applied Science, London, 1977), p. 9, p. 185.
323. F.J. McGarry, *Proc. R. Soc. Lond.* **A319** (1970) 59. March 29–April, 1976.
324. C.B. Bucknall and T. Yoshii. *Third Int. Conf. on Deformation Yield and Fracture of Polymers*, March 29–April 1, 1976, p. 13.
325. A. Aref-azar, J.N. Hay, B.J. Marsden and N. Walker, *J. Polym. Sci. Polym. Phys. Edn.* **18** (1980) 637.
326. H. Keskkula and S.G. Turley, *Polymer* **19** (1978) 797.
327. D. Heikens, N. Hoen, W. Barentsen, P. Piet and H. Laden, *J. Polym. Sci. Polym. Symp.* **62** (1978) 309.
328. Y.H. Liu and D.A. Zumbrunnen, *J. Mater. Sci.* **34** (1999) 1921.
329. S. Inamdar and D.A. Zumbrunnen, *ANTEC* (conf. Proc., 2001), pp. 5.
330. A. LaCoste, K.M. Schaich, D. Zumbrunnen and K.L. Yam, *Packaging Technol. Sci.* **18** (2005) 77.
331. M. Sau and S.C. Jana, *Polym. Eng. Sci.* **44** (2004) 407.
332. S. Visconti and R.H. Marchessault, *Macromolecules* **7** (1974) 913.
333. C.B. Bucknall, *Toughened Plastics* (Applied Science, London, 1977), p. 31, p. 82.
334. C.K. Riew and J.K. Gillham, Eds. *Rubber-Modified Thermoset Resins*, ACS Adv. Chem. Ser. 208 (American Chemical Society, Washington, DC, 1984).
335. S. Kunz-Douglass, P.W.R. Beaumont and M.F. Ashby, *J. Mater. Sci.* **15** (1980) 1109.
336. L.T. Manzione, J.K. Gillham and C.A. McPherson, *J. Appl. Polym. Sci.* **26** (1981) 889.
337. L.T. Manzione, J.K. Gillham and C.A. McPherson, *J. Appl. Polym. Sci.* **26** (1981) 907.
338. J.A. Sayre, R.A. Assink and R.R. Lagasse, *Polymer* **22** (1981) 87.
339. S. Reich and Y. Cohen, *J. Polym. Sci. Polym. Phys. Edn.* **19** (1981) 1255.
340. S. Mader, *Thin Solid Films* **35** (1976) 195.
341. D.L. Handlin, W.J. MacKnight and E.L. Thomas, *Macromolecules* **14** (1981) 795.
342. A. Siegmann and A. Hiltner, *Polym. Eng. Sci.* **24** (1984) 869.
343. D. Fleischer, E. Fischer and J. Brandrup, *J. Macromol. Sci. Phys.* **B14** (1977) 17.
344. A. Siegmann, L.K. English, E. Baer and A. Hiltner, *Polym. Eng. Sci.* **24** (1984) 877.
345. B.Z. Jang, D.R. Uhlmann and J.B.V. Sande, *Rubber Chem. Technol.* **57** (1983) 291.
346. S.Y. Hobbs, M.E.J. Dekkers and V.H. Watkins, *J. Mater. Sci.* **23** (1988) 1219.
347. M.E.J. Dekkers, S.Y. Hobbs and V.H. Watkins, *J. Mater. Sci.* **23** (1988) 1225.
348. M.E.J. Dekkers, S.Y. Hobbs and V.H. Watkins, *Polymer* **32** (1991) 2150.

349. S.Y. Hobbs, *J. Macromol. Sci. Rev. Macromol. Chem.* **C19** (1980) 221.
350. E. Martuscelli, C. Silvestre and G. Abate, *Polymer* **23** (1982) 229.
351. R.W. Hertzberg and J.A. Manson, *Fatigue of Engineering Plastics* (Academic Press, New York, 1980).
352. M.T. Hahn, R.W. Hertzberg and J.A. Manson, *J. Mater. Sci.* **18** (1983) 3551.
353. I.C. Bache, C.M. Ramsdale, D.S. Thomas, A.-C. Arias, J.D. MacKenzie, R.H. Friend, N.C. Greenham and A.M. Donald. *Polymer Interfaces and Thin Films, Nov 26–30 2001*, Boston, MA, United States, Materials Res. Soc Boston MA, 2002, p. 185.
354. F.P. La Mantia, R. Scaffaro, P.L. Magagnini and M. Paci, *J. Appl. Polym. Sci.* **77** (2000) 3027.
355. J.A. Galloway, M.D. Montminy and C.W. Macosko, *Polymer* **43** (2002) 4715.
356. C.B. Bucknall, *Toughened Plastics* (Applied Science, London, 1977), p. 117.
357. E. Radovanovic, E. Carone, Jr. and M.C. Goncalves, *Polym. Testing* **23** (2004) 231.
358. Y. Thomann, R. Thomann, G. Bar, M. Ganter, B. Machutta and R. Mulhaupt, *J. Microsc.* **195** (1999) 161.
359. J. Li, W. Liang, G.F. Meyers and W.A. Heeschen, *Polym. News* **29** (2004) 335.
360. I.W. Hamley, *The Physics of Block Copolymers* (Oxford Science Pubs., Oxford, 1998).
361. F.S. Bates, *MRS Bull.* **30** (2005) 525.
362. M.J. Henke, C.E. Smith and R.F. Abbott, *Polym. Eng. Sci.* **15** (1975) 79.
363. M.J. Folkes and A. Keller, in *Physics of Glassy Polymers*, edited by R.N. Haward (Applied Science, London, 1973).
364. J. Dlugosz, M.J. Folkes and A. Keller, *J. Polym. Sci. Polym. Phys. Edn.* **11** (1973) 929.
365. E.L. Thomas, D.J. Kinning, D.B. Alward and C.S. Henkee, *Macromolecules* **20** (1987) 2934.
366. B.A. Wood, in *Advances Polymer Blends and Alloys Technology*, Vol. 3 (Technomic Publishing, Lancaster, 1992), p. 24.
367. R. Adhikari, R. Godehardt, W. Lebek, S. Goerlitz, G.H. Michler and K. Knoll, *6th Annual UNESCO School & IUPAC Conference on Macromolecules & Materials Science, 2003*, Berg-en-Dal, South Africa, (Wiley-VCH Verlag 2004), p. 173.
368. R. Adhikari, T.A. Huy, S. Henning, G.H. Michler and K. Knoll, *Colloid Polym. Sci.* **282** (2004) 1381.
369. J. Weng, R.H. Olley, D.C. Bassett and P. Jaaskelainen, *J. Polym. Sci. B Polym. Phys.* **41** (2003) 2342.
370. J. Weng, R.H. Olley, D.C. Bassett and P. Jaaskelainen, *J. Polym. Sci. B Polym. Phys.* **42** (2004) 3318.
371. R.H. Olley and D.C. Bassett, *Polymer* **23** (1982) 1707.
372. V. Seydewitz, M. Krumova, G.H. Michler, J.Y. Park and S.C. Kim, *Polymer* **46** (2005) 5608.
373. B. Collin, D. Chatenay, G. Coulon, D. Ausserre and Y. Gallot, *Macromolecules* **25** (1992) 1621.
374. H.G. Dikland, S.S. Sheiko, L.v.d. Does, M. Moller and A. Bantjes, *Polymer* **34** (1993) 1773.
375. H. Ott, V. Abetz, V. Altstadt, Y. Thomann and A. Pfau, *J. Microsc.* **205** (2002) 106.
376. G.H. Michler, *J. Macromol. Sci. Phys.* **B40** (2001) 277.
377. G.H. Michler, F.J. Balta-Calleja, R. Adhikari and K. Knoll, *J. Mater. Sci.* **38** (2003) 4713.
378. R. Adhikari, R. Godehardt, W. Lebek, R. Weidisch, G.H. Michler and K. Knoll, *J. Macromol. Sci. Phys.* **B40** (2001) 833.
379. R. Adhikari and T.A.H. Goerg H. Michler, Elena Ivan'kova, Reinhold Godehardt, Werner Lebek, Konrad Knoll, *Macromol. Chem. Phys.* **204** (2003) 488.
380. C. Park, J. Yoon and E.L. Thomas, *Polymer* **44** (2003) 6725.
381. J. Yoon, W. Lee and E.L. Thomas, *MRS Bull.* **30** (2005) 721.
382. N.M. Benetatos, B.W. Smith, P.A. Heiney and K.I. Winey, *Macromolecules* **38** (2005) 9251.
383. K.I. Winey, unpublished data.
384. S.L. Cooper and A.V. Tobolsky, *J. Appl. Polym. Sci.* **10** (1966) 1837.
385. D.J. Hourston and Y. Zia, *Polymer* **20** (1979) 1497.
386. I.D. Fridman, E.L. Thomas, L.J. Lee and C.W. Macosko, *Polymer* **21** (1980) 393.
387. C.H.Y. Chen, R.M. Briber, E.L. Thomas, M. Xu and W.J. MacKnight, *Polymer* **24** (1983) 1333.
388. R.M. Briber and E.L. Thomas, *J. Macromol. Sci. Phys.* **B22** (1983) 509.
389. G. Demma, E. Martuscelli, A. Zanetti and M. Zorzetto, *J. Mater. Sci.* **18** (1983) 89.
390. B. Chavvel and J.C. Daniel, in *Copolymers, Polyblends, Composites*, edited by A.P.Z. Platzer (American Chemical Society, Washington, DC, 1975), p. 159.

391. J.A. Koutsky, N.V. Hien and S.L. Cooper, *Polym. Lett.* **8** (1970) 353.
392. A.B. Mathur, T.O. Collier, W.J. Kao, M. Wiggins, M.A. Schubert, A. Hiltner and J.M. Anderson, *J. Biomed. Mater. Res.* **36** (1997) 246.
393. B.B. Sauer, R.S. McLean, D.J. Brill and D.J. Londono, *J. Polym. Sci. B Polym. Phys.* **40** (2002) 1727.
394. J.E. Taylor, P.R. Laity, S.S. Wong, K. Norris, P. Khunkamchoo, M. Cable, G. Andrews, A.F. Johnson, R.E. Cameron *Microsc. Microanal.* **12** (2006) 151.
395. F.M. Mirabella Jr. and A. Weiskellel, *Polym. News* **30** (2005) 1.
396. R.D. Gilbert, V. Stannett, C.G. Pitt and A. Schindler, in *Developments in Polymer Degradation*, edited by N. Grassie (Applied Science, London, 1982), p. 259.
397. M. Chasin and R. Langer, *Biodegradable Polymers as Drug Delivery Systems* (Marcel Dekker, New York, 1990).
398. E.R. Howells, *Chem and Ind.* (1982) 502.
399. P.J. Barham, *J. Mater. Sci.* **19** (1984) 3826.
400. M. Yasin and B.J. Tighe, *Plastics, Rubber and Composites Processing and Applications* **19** (1993) 15.
401. A.D. Sagar, M.A. Villar, E.L. Thomas, R.C. Armstrong, E.W. Merrill, *J. Appl. Polym. Sci.* **61** (1996) 139.
402. S. Simmons and E.L. Thomas, *Polymer* **39** (1998) 5587.
403. A.J.F. Carvalho, A.E. Job, N. Alves, A.A.S. Curvelo and A. Gandini, *Carbohydr. Polym.* **53** (2003) 95.
404. A.G. Pedroso and D.S. Rosa, *Carbohydr. Polym.* **59** (2005) 1.
405. M.J. Ridout, A.P. Gunning, M.L. Parker, R.H. Wilson and V.J. Morris, *Carbohydr. Polym.* **50** (2002) 123.
406. G. Loomis, M. Izbicki, C. Kliewer and D.R. Sawyer, unpublished data.
407. J.F. Mandell, A.Y. Darwish and F.J. McGarry, *Polym. Eng. Sci.* **22** (1982) 826.
408. J.F. Mandell, D.R. Roberts and F.J. McGarry, *Polym. Eng. Sci.* **23** (1983) 404.
409. F.P. Price, P.T. Gilmore, E.L. Thomas and R. L. Laurence, *J. Polym. Sci. Polym. Symp.* **63** (1978) 33.
410. D.W. Dwight and J.E. McGrath, *Final Report for Dept. Army Contract DAAK30–78-C-0098, US Army Tank, Automotive Research and Development Command, Warren, MI Dec. 1979*, in Vol. 5 (1979).
411. J.D. Andrade, D.L. Coleman and D.E. Gregonis, *Makromol. Chem. Rapid Commun.* **1** (1980) 101.
412. S.Y. Hobbs and V.H. Watkins, *J. Polym. Sci. Polym. Phys. Edn.* **20** (1982) 651.
413. L.N. Gilbertson. *Proc. of SEM Conf*, Vol. 1, 1977, p. 109.
414. L.J. Buckley, I. Shaffer and R.E. Trabocco, *SAMPE Q.* **16** (1984) 1.
415. T.G. Vladkova, P.D. Dineff and D.N. Gospodinova, *J. Appl. Polym. Sci.* **91** (2004) 883.
416. D. Maldas and B.V. Kotka, *TRIP* **1** (1993) 174.
417. P. Zadorecki and A.J. Michell, *Polym. Compos.* **10** (1989) 69.
418. B. Westerlind, M. Rigdahl and A. Larson, in *Composite Systems from Natural and Synthetic Polymers*, edited by L. Salmen, A.d. Ruvo, J.C. Seferis and E.B. Stark (Elsevier Science Publishers, Amsterdam, 1986), p. 83.
419. D.D. Stokke, *Mater. Res. Soc. Symp. Proc.* **266** (1992) 47.
420. F.T. Wallenberger and N.E. Weston, Eds. *Natural Fibers, Plastics and Composites*, (Kluwer Academic, Boston, 2004).
421. L.J. Broutman and R.H. Krock, Eds. *Modern Composite Materials* (Addison Wesley, Reading, MA, 1967).
422. D. Hull, *An Introduction to Composite Materials* (Cambridge University Press, New York,1981).
423. J.C. Seferis and L. Nicolais, Eds. *The Role of the Polymeric Matrix in the Processing and Structural Properties of Composite Materials* (Plenum, New York, 1983).
424. A.R. Bunsell, C. Bathias, A. Martrenchar, D. Menkes and G. Verchery, Eds. *Advances in Composite Materials, Proc. Third Int. Conf. on Composite Materials, Paris, 26–29 August 1980*, (Pergamon, Oxford, 1980).
425. J. Summerscales, D. Green and F.J. Guild, *J. Microsc.* **169** (1993) 173.
426. K. Friedrich, *Colloid. Polym. Sci.* **259** (1981) 808.
427. J.F. Mandell, D.D. Huang and F.J. McGarry, *Polym. Compos.* **2** (1981) 137.
428. J.F. Mandell, F.J. McGarry, D.D. Huang and C.G. Li, *Polym. Compos.* **4** (1983) 32.
429. J. Summerscales, Editor, *Microstructural Characterisation of Fibre-Reinforced Composites* (CRC Press, Boca Raton, 1998).
430. M.J. Folkes, *Short Fibre Reinforced Thermoplastics* (Research Studies Press, New York, 1982).

431. S. De and J.R. White, *Short Fibre-Polymer Composites* (Woodhead Publishing, Cambridge England, 1996).
432. J.D. Fairing, *J. Compos. Mater.* **1** (1967) 208.
433. H.L. Cox, *Br. J. Appl. Phys.* **3** (1952) 72.
434. M.J. Folkes, Short Fibre Reinforced Thermoplastics (Research Studies Press, New York, 1982), p. 19.
435. L.C. Sawyer, *Polym. Eng. Sci.* **19** (1979) 377.
436. M. Oron, *Proc. of the SEM, IITRI* **6** (1973) 94.
437. A.S. Tetelman, Composite Materials: Testing and Design (ASTM STP, 1969), p. 473.
438. R.W. Lang, J.A. Manson and R.W. Hertzberg, *Polym. Eng. Sci.* **22** (1982) 982.
439. V. Dudler, M.C. Grob and D. Merian, *Polymer Degradation and Stability* **68** (2000) 373.
440. C. Vajrasthira, T. Amornsakchai and S. Bualek-Limcharoen, *J. Appl. Polym. Sci.* **87** (2003) 1059.
441. L.C. De Santa Maria, M.C.A.M. Leite, M.A.S. Costa, J.M.S. Ribeiro, L.F. Senna and M.R. Silva, *J. Microsc.* **213** (2004) 94.
442. M.W. Darlington, P.L. McGinley and G.R. Smith, *J. Mater. Sci. Left.* **11** (1976) 877.
443. P.F. Bright, R.J. Crowson and M.J. Folkes, *J. Mater. Sci.* **13** (1973) 2497.
444. R.J. Crowson, M.J. Folkes and P.F. Bright, *Polym. Eng. Sci.* **20** (1980) 925.
445. J.D. Whitcomb. *ASTM Symp. on the Fatigue of Fibrous Composite Materials*, San Francisco, CA, ASTM Washington DC, May 22–23, 1979.
446. R. Richards-Frandsen and Y.N. Naerheim, *J. Compos. Mater.* **17** (1983) 105.
447. J.H. Sinclair and C.C. Chamis. *Proc. 34th Annu. SPI22-A*, SPI Washington DC, 1979, p. 1.
448. R.A. Kline and E.H. Chang, *J. Compos. Mater.* **14** (1980) 315.
449. K. Mizutani and T. Iwatsu, *J. Appl. Polym. Sci.* **25** (1980) 2649.
450. A. Oberlin, *Carbon* **17** (1979) 7.
451. M. Guigon, J. Ayache, A. Oberlin and M. Oberlin, in *Advances in Composite Materials*, edited by A.R. Bunsell, C. Bathias, A. Martrenchar, D. Mankes and G. Verchery (Pergamon Press, Oxford, 1980), p. 223.
452. K.M. Hardaker and M.O.W. Richardson, *Polym. Plast. Technol. Eng.* **15** (1980) 169.
453. I.L. Kalnin, *Composite Materials Testing Design, ASTM STP 497*, in Vol. 5 (Washington DC, 1972).
454. T.J. Bessell and J.B. Shortall, *J. Mater. Sci.* **12** (1977) 365.
455. J.N. Kirk, M. Munro and P.W.R. Beaumont, *J. Mater. Sci.* **13** (1978) 2197.
456. W.C. McCrone and J.G. Delly, *The Particle Atlas, 2nd ed.*, 6 volumes; 1–4, 1973; volume 5, 1979; volume 6, 1980; The Particle Atlas, Electronic Edition, 1992 (Ann Arbor Science, Ann Arbor, MI, 1973).
457. W.C. McCrone and J.G. Delly, *McCrone Atlas of Microscopic Particles, and an online database* (McCrone Associates, Inc., 2005). www.mccroneatlas.com
458. T. Wilson, *Confocal Microscopy* (Academic Press, London, 1992).
459. A. Clarke, N. Davidson and G. Archenhold, *J. Microsc.* **171** (1993) 69.
460. J.L. Thomason and A. Knoester, *J. Mater. Sci. Left.* **9** (1990) 258.
461. A. Voloshin and L. Arcan, *J. Compos. Mater.* **13** (1979) 240.
462. J.S. Wu, K. Friedrich and M. Grosso, *Composites* **20** (1989) 223.
463. J.J. Horst and J.L. Spoormaker, *Polym. Eng. Sci.* **36** (1996) 2718.
464. A. Garcia-Rejon, A. Meddad, E. Turcott and M. Carmel, *Polym. Eng. Sci.* **42** (2002) 346.
465. F.J. Guild and B. Ralph, *J. Mater. Sci.* **14** (1979) 2555.
466. K.P. Battjes. *ANTEC 2004—Annual Technical Conference Proceedings, May 16–20 2004*, Chicago, IL (SPE, Brookfield, CN), p. 3034.
467. M. Guigon, in *Microstructural Characterisation of Fibre-Reinforced Composites*, edited by J. Summerscales (Woodhead Pub, Cambridge, 1998) 204.
468. J.E. Theberge, *Polym. Plast. Technol. Eng.* **16** (1981).
469. J.P. Trotignon, B. Sanschagrin, M. Piperaud and J. Verdu, *Polym. Compos.* **3** (1982) 230.
470. A. Garton, S.W. Kim and D.M. Wiles, *J. Polym. Sci. Polym. Left. Edn.* **20** (1982) 273.
471. J. Kubat and H.E. Stromvall, *Plast. Rubber: Process.* **June** (1980) 45.
472. J.D. Miller, H. Ishida and F.H.J. Maurer, *J. Mater. Sci.* **24** (1989) 2555.
473. B.A. Wood, in *Polymer Alloys and Blends*, edited by G. Shonaike and G.P. Simon (Marcel Dekker, New York, 1999), p. 496.
474. J. Kruse, *Rubber Chern. Technol.* **46** (1973) 1.
475. S.Y. Hobbs, in *Plastics Polymer Science and Technology*, edited by M.D. Bayal (Wiley-Interscience, New York, 1982), p. 239.

References

476. R.J. Eldred, *J. Polym. Sci. Polym. Lett. Edn.* **10** (1972) 391.
477. V.P. Chacko, F.E. Karasz, R.J. Farris and E.L. Thomas, *J. Polym. Sci. Polym. Phys. Edn.* **20** (1982) 2177.
478. R. Stewart, in *Plastics Engineering*, SPE Brookfield CN, April 2006, p. 12.
479. R.A. Vaia and H.D. Wagner, *Mater. Today* Nov. (2004) 32.
480. M.R. Bockstaller, R.A. Mickiewicz and E.L. Thomas, *Adv. Mater.* **17** (2005) 1331.
481. F. Du, R.C. Scogna, W. Zhou, S. Brand, J.E. Fischer and K.I. Winey, *Macromolecules* **37** (2004) 9048.
482. J. Njuguna and K. Pielichowski, *Adv. Eng. Mater.* **6** (2004) 204.
483. T.J. Pinnavaia and G.W. Beall, Eds. *Polymer-Clay Nanocomposites* (Wiley, New York, 2001).
484. H.-W. Wang, K.-C. Chang and H.-C. Chu, *Polym. Int.* **54** (2005) 114.
485. J. Karger-Kocsis and Z. Zhang, in *Mechanical Properties of Polymers Based on Nanostructure and Morphology*, edited by G.H. Michler and F.J. Balta-Calleja (CRC Press Taylor & Francis, Boca Raton and New York, 2005), p. 553.
486. K. Friedrich, S. Fakirov and A.E. Zhang, Eds. *Polymer Composites* (Springer, New York, 2005).
487. S. Fischer, in *Natural Fibers, Plastics and Composites*, edited by F.T. Wallenberger and N.E. Weston (Kluwer Academic, Boston, 2004), p. 345.
488. N. Artzi, Y. Nir, M. Narkis and A. Siegmann, *J. Polym. Sci. B Polym. Phys.* **40** (2002) 1741.
489. Y. Li and H. Shimizu, *Polymer* **45** (2004) 7381.
490. D. Raghavan and C. Chen, *NTIS Report: To Investigate the Impact of Tailorable Interface on the Morphology and Performance Characteristics of High Temperature Nanocomposites*, in NTIS report, U. S. Dept. of Commerce. (2004).
491. G. Beyer, in *Plastics Additives & Compounding* **Sept/Oct** (2005) 32.
492. M. Yuan, L.-S. Turng, S. Gong, A. Winardi and D. Caulfield, *J Cellular Plastics* **40** (2004) 397.
493. T. Sun and J.M. Garces, *Advanced Materials (Weinheim, Germany)* **14** (2002) 128.
494. E.M. Benetti, V. Causin, C. Marega, A. Marigo, G. Ferrara, A. Ferraro, M. Consalvi and F. Fantinel, *Polymer* **46** (2005) 8275.
495. W.S. Chow, Z.A.M. Ishak and J. Karger-Kocsis, *J. Polym. Sci. B Polym. Phys.* **43** (2005) 1198.
496. S.Z. Xie, Shimin; Zhao,Bin; Qin, Huaili; Wang, Fosong; Yang, Mingshu, *Polym. Int.* **54** (2005) 1673.
497. A.Y. Bansal, H.; LI, C.; Cho, K.; Benicewicz, B.; Kumar, S.K.; Schadler, L.S., *Nat. Mater.* **4** (2005) 693–698.
498. N.A. Kotov, A.A. Mamedov, M. Prato, D.M. Guldi, J.P. Wicksted and A. Hirsch, *Proceedings of the SPIE — The International Society for Optical Engineering; UV/Optical/IR Space Telescopes: Innovative Technologies and Concepts, 3–5 Aug. 2003* **5166** (2004) 228.
499. S.L. Ruan, P. Gao, X.G. Yang and T.X. Yu, *Polymer* **44** (2003) 5643.
500. D. Shi, J. Lian, P. He, L.M. Wang, F. Xiao, L. Yang, M.J. Schulz and D.B. Mast, *Appl. Phys. Lett.* **83** (2003) 5301.
501. K. Schulte and M.C.M. Nolte, in *Mechanical Properties of Polymers Based on Nanostructure and Morphology*, edited by G.H. Michler and F.J. Balta-Calleja (CRC Press Taylor & Francis, Boca Raton and New York, 2005).
502. C. Mahesha, D.A. Zumbrunnen and Y. Parulekar. *63rd Ann. Tech. Conf.*, Boston, 2005 (SPE, Brookfield, CN), p. 1920.
503. C. Mahesha and D.A. Zumbrunnen. *64th Ann. Tech. Conf.*, (SPE, Brookfield, CN), p. 491.
504. S. Mehta, F.M. Mirabella, K. Rufener and A. Bafna, *J. Appl. Polym. Sci.* **92** (2004) 928.
505. G.R. Dieckmann, A.B. Dalton, P.A. Johnson, J. Razal, J. Chen, G.M. Giordano, E. Munoz, I.H. Musselman, R.H. Baughman and R.K. Draper, *J. Am. Chem. Soc.* **125** (2003) 1770.
506. V. Zorbas, A. Ortiz-Acevedo, A.B. Dalton, M.M. Yoshida, G.R. Dieckmann, R.K. Draper, R.H. Baughman, M. Jose-Yacaman and I.H. Musselman, *J. Am. Chem. Soc.* **126** (2004) 7222.
507. A. Ortiz-Acevedo, H. Xie, V. Zorbas, W.M. Sampson, A.B. Dalton, R.H. Baughman, R.K. Draper, I.H. Musselman and G.R. Dieckmann, *J. Am. Chem. Soc.* **127** (2005) 9512.
508. V. Zorbas, A.L. Smith, H. Xie, A. Ortiz-Acevedo, A.B. Dalton, G.R. Dieckmann, R.K. Draper, R.H. Baughman and I.H. Musselman, *J. Am. Chem. Soc.* **127** (2005) 12323.
509. V.Z. Poenitzsch and I.H. Musselman, *Microsc. Microanal.* **12** (2006) 221.

510. K.J. Lissant, Ed. *Emulsions and Emulsion Technology* (Marcel Dekker, New York, 1974).
511. M.J. Rosen, *Surfactants and Interfacial Phenomena* (Wiley-Interscience, New York, 1978).
512. L.M. Prince, Ed. *Microernulsions: Theory and Practice* (Academic Press, New York, 1977).
513. I.D. Robb, Ed. *Microemulsions* (Plenum, New York, 1982, 1982).
514. G. Gillberg, personal communication.
515. S. Friberg, Ed. *Food Emulsion* (Marcel Dekker, New York, 1976).
516. M. El-Aasser, *Advances in Emulsion Polymerization and Latex Technology* (Lehigh University, Bethlehem, PA, 1979).
517. B. Kachar, D.F. Evans and B.W. Ninham, *J. Colloid Interface Sci.* **100** (1984) 287.
518. B.W. Ninham, *J. Phys. Chern.* **84** (1980) 1423.
519. L.H. Lee, Ed. *Adhesion Science and Technology* (Plenum, New York, 1975).
520. L.H. Lee, Ed. *Adhesion and Adsorption of Polymers* (Plenum, New York, 1980).
521. S. Horiuchi, T. Hamanaka, T. Aoki, T. Miyakawa, R. Narita and H. Wakabayashi, *J. Electron Microsc.* **52** (2003) 255.
522. J. Im, T. Stokich, Jr., J. Hetzner, G. Buske, J. Curphy, E.O. Shaffer and G. Meyers, *Proceedings—International Symposium on Advanced Packaging Materials: Processes, Properties and Interfaces, Braselton, GA, Mar. 14–17, 1999* p. 53.
523. G.F. Meyers, B.M. DeKoven, M.T. Dineen, A. Strandjord, P.J. O'Connor, T. Hu, Y.-H. Chiao, H. Pollock and A. Hammiche, *Polymer Preprints (American Chemical Society, Division of Polymer Chemistry)* **39** (1998) 1222.
524. P.A. Steward, J. Hearn and M.C. Wilkinson, *Adv. Colloid Interface Sci.* **86** (2000) 195.
525. G.E. Molau and H. Keskkula, *Appl. Polym. Symp.* **7** (1968) 35.
526. J. Ugelstad and P.C. Mork, *Adv. Colloid Interface Sci.* **13** (1980) 101.
527. S.S. Atik and J.K. Thomas, *J. Am. Chem. Soc.* **104** (1982) 5868.
528. C. Price, in *Developments in Block Copolymers*, edited by I. Goodman (Applied Science, London, 1982), p. 39.
529. C. Price and D. Woods, *Eur. Polym. J.* **9** (1973) 827.
530. C. Booth, V.T.D. Naylor, C. Price, N.S. Rajab and R.B. Stubbersfield, *J. Chem. Soc. Faraday Trans.* **74** (1978) 2352.
531. C. Price, A.L. Hudd, R.B. Stubbersfield and B. Wright, *Polymer* **21** (1980) 9.
532. D.I. Lee and T. Ishikawa, *J. Polym. Sci. Polym. Chem. Edn.* **21** (1983) 147.
533. J. Biais, M. Mercier, P. Lalarme, B. Clin, A.M. Bellocq and B. Lemanceau, *C. R. Acad. Sci. Paris* **C285** (1977) 213.
534. A.M. Bellocq, J. Biais, P. Botherol, B. Coo, G. Fourche, P. Lalanne, B. Lemaire, B. Lemanceau and D. Roux, *Adv. Colloid Interface Sci.* **20** (1984) 167.
535. I. Segall, O.L. Shaffer, V.L. Dimonie and M.S. El-Aasser. *Proc. 51st MSA*, edited by W. Bailey (San Francisco Press, 1993), p. 882.
536. N. Subramaniam, J. Simpson, M.J. Monteiro, O. Shaffer, C.M. Fellows and R.G. Gilbert, *Microsc. Res. Techn.* **63** (2004) 111.
537. M.S. El-Aasser and A.A. Robertson, *J. Paint Technol.* **17** (1975) 50.
538. B.R. Vijayendran, T. Bone and L.C. Sawyer, *J. Dispersion Sci. Technol.* **3** (1982) 81.
539. M.S. El-Aasser and A.A. Robertson, *Kolloid Z. Z. Polym.* **252** (1973) 241.
540. E.B. Bradford and J.W. Vanderhoff, *J. Macromol. Chem.* **1** (1966) 335.
541. E.B. Bradford and J.W. Vanderhoff, *J. Macromol. Sci., Phys.* **B6** (1972) 671.
542. C. Wang, F. Chu, C. Graillat, A. Guyot, C. Gauthier and J.P. Chapel, *Polymer* **46** (2005) 1113.
543. B.M. Budhlall, O.L. Shaffer, E.D. Sudol, V.L. Dimonie and M.S. El-Aasser, *Langmuir* **19** (2003) 9968.
544. D.L. Vezie, E.L. Thomas and W.W. Adams, *Polymer* **36** (1995) 1761.
545. C. Gaillard, P.A. Stadelmann, C.J.G. Plummer and G. Fuchs, *Scanning* **26** (2004) 122.
546. P. Meredith and A.M. Donald, *J. Microsc.* **181** (1996) 23.
547. S. Kitching and A.M. Donald, *J. Microsc.* **190** (1998) 357.
548. C.P. Royall, D.J. Stokes, I. Hopkinson and A.M. Donald, *Polym. News* **26** (2001) 226.
549. C.P. Royall, B.L. Thiel and A.M. Donald, *J. Microsc.* **204** (2001) 185.
550. D.J. Stokes, *Adv. Eng. Mater.* **3** (2001) 126.
551. J.P. Craven, F.S. Baker, B.L. Thiel and A.M. Donald, *J. Microsc.* **205** (2002) 96.
552. C.P. Royall and A.M. Donald, *Scanning* **24** (2002) 305.
553. B.L. Thiel, M. Toth and J.P. Craven, *Microsc. Microanal.* **10** (2004) 711.
554. M. Katoh, *J. Electron Microsc.* **28** (1979) 197.
555. K. Furusawa and N. Tomotsu, *J. Colloid Interface Sci.* **93** (1983) 504.

References

556. O.L. Shaffer, M.S. El-Aasser and J.W. Vanderhoff. *Proc. 41st Annu. Mtg. EMSA*, edited by W. Bailey (San Francisco Press, 1983), p. 30.
557. R.L. Rowell, R.S. Farinato, J.W. Parsons, J.R. Ford, K.H. Langley, J.R. Stone, T.R. Marshall, C.S. Parmenter, M. Seaver and E.B. Bradford, *J. Colloid Interface Sci.* **69** (1979) 590.
558. J.V. Dawkins and G. Taylor, *Macromol. Chem.* **180** (1979) 1737.
559. L.H. Lee, Ed. *Characterization of Metal and Polymer Surfaces: Polymer Surfaces* (Academic Press, New York, 1977).
560. T. Smith and D.H. Kaelble, in *Treatise on Adhesion and Adhesives*, edited by R.L. Patrick (Marcel Dekker, New York, 1981).
561. O. Hahn and G. Kotting, *Kunstatstoffe* **74** (1984) 238.
562. D.M. Brewis and D. Briggs, *Polymer* **22** (1981) 7.
563. S. Yamakawa and F. Yamamoto, *J. Appl. Polym. Sci.* **25** (1980) 25.
564. M. Kadreva, in *Physicochemical Aspects of Polymer Surfaces*, edited by K.L. Mittal (Plenum, New York, 1983), p. 125.
565. D.B. Rahrig, *J. Adhesion* **16** (1984) 179.
566. G. Gillberg, L.C. Sawyer and A.L. Promislow, *J. Appl. Polym. Sci* **28** (1983) 3723.
567. J.P. Wightman, *SAMPE Q.* **13** (1981) 1.
568. B. Beck, *An Investigation of Adhesive/Adherend and Fiber/Matrix Interactions, Part B, SEM/ESCA Analysis of Fracture Surfaces, NASA Report NAG1–127, January 1983*, (1983).
569. C. Creton, E.J. Kramer and G. Hadziioannou, *Colloid Polym. Sci.* **270** (1992) 399.
570. A. Chiche, W. Zhang, C.M. Stafford and A. Karim, *Measurement Sci. Technol.* **16** (2005) 183.
571. A.M. Forster, W. Zhang, A.J. Crosby and C.M. Stafford, *Measurement Sci. Technol.* (2005) 81.
572. D.R. Sawyer, unpublished data.
573. J.F. Oliver and S.G. Mason. *The Fundamental Properties of Paper Related to its Uses; Trans. Symp. Mtg.*, Conf., Vol. 2, 1973 (Ernest Berm, London, 1976) p. 428.
574. C. Huh, M. Inoue and S.G. Mason, *Can. J. Chem. Eng.* **53** (1975) 367.
575. c. Huh and S.G. Mason, *J. Colloid Interface Sci.* **60** (1977) 11.
576. J.F. Oliver and S.G. Mason, *J. Colloid Interface Sci.* **60** (1977) 480.
577. Y.H. Mori, T.G.M.V.d. Ven and S.G. Mason, *Colloids Surf.* **4** (1982) 1.
578. A. Karbach and D. Drechsler, *Surf. Interface Anal.* **27** (1999) 401.
579. D. Raghavan, X. Gu, T. Nguyen and M. Vanlandingham, *J. Polym. Sci. B Polym. Phys.* **39** (2001) 1460.
580. D.I. Lee, *Prog. Organic Coatings* **45** (2002) 341.
581. O.L. Shaffer, unpublished micrograph.
582. J. Qian, R. Pearson, M.S. El-Aasser and V. Dimonie, *SAMPE* **25** (1993) 40.
583. L.C. Sawyer, unpublished data.
584. S.G. Turrion, D. Olmos and J. Gonzalez-Benito, *Polym. Testing* **24** (2005) 301.
585. D. Urban and K. Takamura, Eds. *Polymer Dispersions and Their Industrial Applications* (Wiley-VCH Verlag, Weinheim Germany, 2002).
586. J. Li, W. Liang and S. Chum. *Fall MRS*, 2004, unpublished micrographs.
587. J. Yang and D.C. Martin, *J. Mater. Res.* **21** (2006) 1124.
588. H. Tang and D.C. Martin, *J. Mater. Sci.* **38** (2003) 803.
589. G. Nizri, S. Magdassi, J. Schmidt, Y. Cohen and Y. Talmon, *Langmuir* **20** (2004) 4380.
590. A. Wittemann, M. Drechsler, Y. Talmon and M. Ballauff, *J. Am. Chem. Soc.* **127** (2005) 9688.
591. D. Danino, A. Bernheim-Groswassen and Y. Talmon, *Colloids Surfaces A* **183–185** (2001) 113.
592. S. Nilsson, M. Goldraich, B. Lindman and Y. Talmon, *Langmuir* **16** (2000) 6825.
593. A. Ciferri and I.M. Ward, Eds. *Ultra-high Modulus Polymers* (Applied Science, London, 1979).
594. A.M. Donald and A.H. Windle, *Liquid Crystalline Polymers* (Cambridge University Press, Cambridge, 1992).
595. W.J. Jackson, Jr and H.F. Kuhfuss, *J. Polym. Sci. Polym. Chem. Edn.* **14** (1976) 2043.
596. W.R. Krigbaum, A. Ciferri, J. Asrar, H. Toriumi and J. Preston, *Molecular Crystals and Liquid Crystals* **76** (1981) 79.
597. P.G. deGennes, *The Physics of Liquid Crystals* (Oxford University Press, Oxford, 1979).
598. D. Demus and L. Richter, *Texture of Liquid Crystals* (Verlag Chemie, New York, 1978).
599. P.J. Flory, in *Polymer Liquid Crystals*, edited by A. Ciferri, W.R. Krigbaum and R.B. Meyer (Academic Press, New York, 1982), Chap. 4.
600. A. Ciferri, W.R. Krigbaum and R.B. Meyer, Eds. *Polymer Liquid Crystals* (Academic Press, New York, 1982).

601. J.F. Johnson and R.S. Porter, Eds. *Liquid Crystals and Ordered Fluids* (Plenum, New York, 1970).
602. G.W. Gray and P.A. Winsor, *Liquid Crystals and Plastics, Vols 1 and 2* (Horwood, Chichester, 1974).
603. J.H. Wendorff, in *Liquid Crystalline Order in Polymers*, edited by A. Blumstein (Academic Press, New York, 1978), p. 41.
604. J.L. White and J.F. Fellers, in Fiber Structure and Properties, edited by J.L. White, *Appl. Polym. Symp.* 33 (1978) 137.
605. G. Calundann and M. Jaffe. *Proc. Robert A. Welch Conf. on Chemical Research, XXVI, Synthetic Polymers*, Robert A Welch Found. Houston TX, 1982, p. 247.
606. M.G. Dobb and J.E. McIntyre, *Properties and Applications of Liquid-Crystalline Main-Chain Polymers*, Adv. in Polym. Sci. Ser. 60/61 (Springer-Verlag, Berlin, 1984).
607. G. Calundann, M. Jaffe, R.S. Jones and H.N. Yoon, in *Fibre Reinforcements for Composite Materials*, edited by A.R. Bunsell (Elsevier, Amsterdam, 1988), Chap. 5.
608. A.E. Zachariades and R.S. Porter, Eds. *High Modulus Polymers* (Marcel Dekker, New York, 1988).
609. R.A. Weiss and C.K. Ober, Eds. *Liquid Crystalline Polymers*, ACS Symposium Series 435 (American Chemical Society, Washington, DC, 1990).
610. C.L. Jackson and M.T. Shaw, *Int. Mater. Rev.* **36** (1991) 165.
611. W.A. MacDonald, in *Liquid Crystalline Polymers*, edited by A.A. Collyer (Elsevier, London, 1992), p. 407.
612. L.C. Sawyer and A. Kaslusky, in *Encyclopedia of Materials: Science and Technology*, edited by A.H. Windle (Elsevier, London, 2001).
613. L.C. Sawyer and M. Jaffe, *J. Mater. Sci.* **21** (1986) 1897.
614. S.L. Kwolek, U. S. Patent 3,600,350, 1971.
615. M. Panar, P. Avakian, R.C. Blume, K.H. Gardner, T.D. Gierke and H.H. Yang, *J. Polym. Sci. Polym. Phys. Edn.* **21** (1983) 1955.
616. M.G. Northolt, *Polymer* **21** (1980) 1199.
617. J.R. Schaefgen, T.I. Bair, J.W. Ballou, S.L. Kwolek, P.W. Morgan, M. Panar and J. Zimmerman, in *Ultra-high Modulus Polymers*, edited by A. Cifferri and I.M. Ward (Applied Science, London, 1979), p. 173.
618. J.R. Schaefgen, in *The Strength and Stiffness of Polymers*, edited by A. Zachariades and R.S. Porter (Marcel Dekker, New York, 1984), p. 327.
619. M. Jaffe and R.S. Jones, in *Handbook of Fiber Science and Technology*, edited by M. Lewin and J. Preston (Marcel Dekker, New York, 1985), p. 349.
620. T.E. Helminiak, *Am. Chem. Soc., Div. Org. Coatings Plast. Chem. Prepr.* **40** (1979) 475.
621. E.J. Roche, T. Takahashi and E.L. Thomas, in *Fiber Diffraction Methods*, edited by A.D. French and K.H. Gardner (American Chemical Society, Washington, DC, 1980), p. 303.
622. E.J. Roche, R.S. Stein and E.L. Thomas, *J. Polym. Sci. Polym. Phys. Edn.* **18** (1980) 1145.
623. S.R. Allen, A.G. Fillippov, R.J. Farris and E.L. Thomas, in *The Strength and Stiffness of Polymers*, edited by A. Zachariades and R.S. Porter (Marcel Dekker, New York, 1983), p. 357.
624. J.A. Odell, A. Keller, E.D.T. Atkins and M.J. Miles, *J. Mater. Sci.* **16** (1981) 3309.
625. J. Wooten, W.C., F.E. McFarlane, J.T.F. Gray and J.W.J. Jackson, in *Ultra-high Modulus Polymers*, edited by A. Cifferri and I.M. Ward (Applied Science, London, 1979), p. 227.
626. J. Economy and W. Volksen, in *The Strength and Stiffness of Polymers*, edited by A. Zachariades and R.S. Porter (Marcel Dekker, New York, 1983), p. 293.
627. C. Noel, C. Fridrich, F. Laupretre, J. Billard, L. Bosio and C. Strazielle, *Polymer* **25** (1984) 263.
628. L.C. Sawyer and M. Jaffe, in *High Performance Polymers*, edited by E. Baer and A. Moet (Carl Hanser Verlag, Germany, 1991), p. 56.
629. I.H. Musselman and P.E. Russell, in *Microbeam Analysis-1989*, edited by P.E. Russell (San Francisco Press, San Francisco, CA, 1989), p. 535.
630. I.H. Musselman, P.E. Russell, R.T. Chen, M.G. Jamieson and L.C. Sawyer. *Proc. XIIth International Congress for Electron Microscopy*, edited by W. Bailey (San Francisco Press, 1990), p. 866.
631. L.C. Sawyer, R.T. Chen, M. Jamieson, I.H. Musselman and P.E. Russell, *J. Mater. Sci. Lett.* (1992) 69.
632. L.C. Sawyer, R.T. Chen, M. Jamieson, I.H. Musselman and P.E. Russell, *J. Mater. Sci.* **28** (1993) 225.
633. S.D. Hudson, *Curr. Opin. Colloid Interface Sci.* **3** (1998) 125.
634. Y. Maeda and M. Koizumi, *Rev. Sci. Instrum.* **67** (1996) 2030.

635. T. Asada, in *Polymer Liquid Crystals*, edited by A. Ciferri, W.R. Krigbaum and R.B. Meyer (Academic Press, New York, 1982), p. 247.
636. M.R. Mackley, F. Pinaud and G. Siekmann, *Polymer* **22** (1981) 437.
637. C. Viney, A.M. Donald and A.H. Windle, *J. Mater. Sci.* **18** (1983) 1136.
638. S.C. Simmens and J.W.S. Hearle, *J. Polym. Sci. Polym. Phys. Edn.* **18** (1980) 871.
639. A.M. Donald and A.H. Windle, *Colloid Polym. Sci.* **261** (1983) 793.
640. A.M. Donald and A.H. Windle, *J. Mater. Sci.* **18** (1983) 1143.
641. A. Zachariades, P. Navard and J.A. Logan, *Mol. Cryst. Liq. Cryst.* **110** (1984) 93.
642. A.M. Donald and A.H. Windle, *J. Mater. Sci.* **19** (1984) 2085.
643. C. Viney and W.S. Putnam. *Proc. 52nd. MSA*, edited by W. Bailey (San Francisco Press, 1993) p. 864.
644. W.S. Putnam and C. Viney, *Molecular Crystals and Liquid Crystals* **199** (1991) 189.
645. M.G. Dobb, D.J. Johnson and B.P. Saville, *J. Polym. Sci. Polym. Phys. Edn.* **15** (1977) 2201.
646. A. Romo-Uribe and A.H. Windle, *Proc. Roy. Soc. London A* **455** (1999) 1175.
647. Y. Ide and Z. Ophir, *Polym. Eng. Sci.* **23** (1983) 261.
648. S. Garg and S. Kenig, *Proc. ACS Div. Polym. Mater. Sci. Eng.* **52** (1985) 90.
649. Z. Ophir and Y. Ide, *Polym. Eng. Sci.* **23** (1983) 792.
650. H. Thapar and M. Bevis., *J. Mater. Sci. Lett.* **2** (1983) 733.
651. Z. Tadmor and C.G. Gogos, *Principles of Polymer Processing* (Wiley-Interscience, New York, 1979).
652. H. Thapar and M.J. Bevis, *Plastics and Rubber Processing and Applications* **12** (1989) 39.
653. S.D. Hudson, D.L. Vezie and E.L. Thomas, *Makromol. Chern. Rapid Commun.* **11** (1990) 657.
654. E.L. Thomas and B.A. Wood, *Faraday Discuss. Chern. Soc.* **79** (1985) 229.
655. W.W. Adams, D.L. Vezie and E.L. Thomas. *Proc. of 50th Annu. EMSA*, edited by W. Bailey (San Francisco Press, 1992) p. 266.
656. L.C. Sawyer and M. Jaffe, *Proc. ACS Div. Polym. Mater. Sci. Eng.* **53** (1985) 485.
657. E. Baer, A. Hiltner, T. Weng, L.C. Sawyer and M. Jaffe, *Proc. ACS Div. Polym. Mater. Sci. Eng.* **52** (1985) 88.
658. T. Weng, A. Hiltner and E. Baer, *J. Mater. Sci.* **21** (1986) 744.
659. P.D. Frayer, *Polym. Compos.* **8** (1987) 379.
660. S. Kenig, B. Trattner, H. and erman, *Polym. Compos.* **9** (1988) 20.
661. C.J.G. Plummer, B. Zulle, A. Demarmels and H.H. Kausch, *J. Appl. Poly. Sci.* **48** (1993) 751.
662. S. Garg and S. Kenig, in *The Strength and Stiffness of Polymers*, edited by A. Zachariades and R.S. Porter (Marcel Dekker, New York, 1983).
663. R.A. Weiss, W. Hue and L. Nicholais, *Polym. Eng. Sci.* **27** (1987) 684.
664. D. Beery, S. Kenig and A. Siegmann, *Poly. Eng. Sci.* **31** (1991) 459.
665. D. Dutta, R.A. Weiss and K. Kristal, *Polym. Compos.* **13** (1992) 394.
666. H.H. Chuah, T. Kyu and T.E. Helminiak, *Polymer* **28** (1987) 2130.
667. S. Akhtar and A.I. Isayev, *Polym. Eng. Sci.* **33** (1993) 32.
668. M. Pracella, P. Magagnini and L.L. Minkova, *Polym. Networks Blends* **2** (1992) 225.
669. X. Hu, Q. Lin, A.F. Yee and D. Lu, *J. Microsc.* **185** (1997) 109.
670. X. He, M.S. Ellison and R.P. Paradkar, *J. Appl. Polym. Sci.* **86** (2002) 795.
671. G. Georgiev, P. Cebe and M. Capel, *J. Mater. Sci.* **40** (2005) 1141.
672. H. Blades, U. S. Patent 3,747,756, 1973.
673. M.G. Dobb, A.M. Hendeleh, D.J. Johnson and B.P. Saville, *Nature* **253** (1975) 189.
674. M.G. Dobb, D.J. Johnson and B.P. Saville, *Phil. Trans. R. Soc.* **A294** (1980) 483.
675. M.G. Dobb, D.J. Johnson and B.P. Saville, *J. Polym. Sci. Polym. Symp.* **58** (1977) 237.
676. M.G. Dobb, D.J. Johnson, A. Majeed and B.P. Saville, *Polymer* **20** (1979) 1284.
677. M.G. Dobb, D.J. Johnson and B.P. Saville, *Polymer* **22** (1981) 960.
678. P. Aviakian, R.C. Blume, T.D. Gierke, H.H. Yang and M. Panar, *Am. Chem. Soc., Div. Polym. Chern., Polym. Prepr.* **21** (1980) 8.
679. C.O. Pruneda, R.J. Morgan and F.M. Kong. *29th Nat. SAMPE Symp.*, April 3–5, 1984, p. 1213.
680. A.R. Bunsell, *J. Mater. Sci.* **10** (1975) 1300.
681. M.M. Lafite and A.R. Bunsell, *J. Mater. Sci.* **17** (1982) 2391.
682. J.H. Greenwood and P.G. Rose, *J. Mater. Sci.* **9** (1974) 1804.

683. M.G. Dobb and R.M. Robson, *J. Mater. Sci.* **25** (1990) 459.
684. M. Sotton and A.M. Vialard, *Textile Res. J.* (1981) 842.
685. M.G. Dobb, C.R. Park and R.M. Robson, *J. Mater. Sci.* **27** (1992) 3876.
686. D. Snetivy, G.J. Vancso and G.C. Rutledge, *Macromolecules* **25** (1992) 7037.
687. S.F.Y. Li, A.J. McGhie and S.L. Tang, *Polymer* **34** (1993) 4573.
688. S.R. Allen, A.G. Fillippov, R.J. Farris, E.L. Thomas, C.P. Wong, G.C. Berry and E.C. Chenevey, *Macromolecules* **14** (1981) 1135.
689. S.R. Allen, A.G. Fillippov, R.J. Farris and E.L. Thomas, *J. Appl. Polym. Sci.* **26** (1981) 291.
690. J.R. Minter, K. Shimamura and E.L. Thomas, *J. Mater. Sci.* **16** (1981) 3303.
691. S.R. Allen, R.J. Farris and E.L. Thomas, *J. Mater. Sci.* **20** (1985) 2727.
692. D.T. Grubb, in *Materials Science and Technology*, edited by E.L. Thomas (VCH Publishers, Weinheim, 1993).
693. D.C. Martin and E.L. Thomas, in *Mater. Sci. and Eng. of Rigid Rod Polymers*, edited by W.W. Adams, R. Eby and D. McLemore (Mat. Res. Soc. Symp. Proc., Boston, MA, 1989).
694. B.P. Griffin and M.K. Cox, *Br. Polym. J.* **December** (1980) 147.
695. F.N. Cogswell, *Br. Polym. J.* **December** (1980) 170.
696. A.E. Zachariades, J. Economy and J.A. Logan, *J. Appl. Polym. Sci.* **27** (1982) 2009.
697. D.G. Baird and G.L. Wilkes, *Polym. Eng. Sci.* **23** (1983) 632.
698. A. Zachariades and J.A. Logan, *Polym. Eng. Sci.* **23** (1983) 797.
699. L.C. Sawyer, *J. Polym. Sci., Polym. Phys. Edn.* **22** (1984) 347.
700. L.H. Sawyer and W. George, in *Cellulose and Other Natural Polymer Systems: Biogenesis, Structure, and Degradation*, edited by J.R. Malcolm Brown (Plenum, New York, 1982), p. 429.
701. E. Baer and A. Moet, Eds. *High Performance Polymers* (Hanser, New York, 1991).
702. A. Frey-Wyssling, *Science* **119** (1954) 80.
703. M.J. Troughten, A.P. Unwin, G.R. Davies and I.M. Ward, *Polymer* **29** (1988) 1389.
704. D.I. Green, A.P. Unwin, G.R. Davies and I.M. Ward, *Polymer* **31** (1990) 579.
705. D.C. Martin, *Direct imaging of deformation and disorder in extended chain polymer fibers*, U. Mass, Amherst, PhD. dissertation (1991).
706. D.C. Martin and E.L. Thomas, *J. Mater. Sci.* **26** (1991) 5171.
707. A.H. Donald and A.H. Windle, *J. Mater. Sci. Lett.* **4** (1985) 58.
708. W.W. Adams and R.K. Eby, *Mat. Res. Soc. Bull.* **XII** (1987) 22.
709. C.Y.C. Lee, *Poly. Eng. Sci.* **33** (1993) 907.
710. D.C. Martin, *Macromolecules* **25** (1992) 5171.
711. Y. Cohen and E.L. Thomas, *Macromolecules* **21** (1988) 433.
712. D. Snetivy, G.J. Vancso and G.C. Rutledge, *Macromolecules* **25** (1992) 7037.
713. J.E. Taylor, Romo-Uribe, A., Libera, M.R., *Macromolecules* **35** (2002) 1751.
714. S.A.C. Gould, J.B. Shulman, D.A. Schiraldi and M.L. Occelli, *J. Appl. Polym. Sci.* **74** (1999) 2243.
715. B.B. Sauer, W.G. Kampert and R.S. McLean, *Polymer* **44** (2003) 2721.

Chapter 6
Emerging Techniques in Polymer Microscopy

6.1 INTRODUCTION 435
6.2 OPTICAL AND ELECTRON
    MICROSCOPY 436
    6.2.1 Confocal Scanning
          Microscopy 436
    6.2.2 Optical Profilometry 437
    6.2.3 Birefringence Imaging 438
    6.2.4 Aberration Corrected
          Electron Microscopy 438
    6.2.5 Ion Microscopy 440
6.3 SCANNING PROBE
    MICROSCOPY 441
    6.3.1 Chemical Force Microscopy ... 441
    6.3.2 Harmonic Imaging 443
    6.3.3 Fast Scanning SPM 444
    6.3.4 Scanning Thermal
          Microscopy 445
    6.3.5 Near Field Scanning Optical
          Microscopy 449
    6.3.6 Automated SPM 449
6.4 THREE DIMENSIONAL
    IMAGING 451
    6.4.1 Introduction 451
    6.4.2 Physical Sectioning 452
    6.4.3 Optical Sectioning 454
    6.4.4 Tomography 455
          6.4.4.1 Electron
                  Tomography 456
          6.4.4.2 X-ray
                  Microtomography ... 457
6.5 ANALYTICAL IMAGING 459
    6.5.1 FTIR Microscopy 459
    6.5.2 Raman Microscopy 460
    6.5.3 Electron Energy Loss
          Microscopy 461

6.5.4 X-ray Microscopy 462
6.5.5 Imaging Surface Analysis 464
      6.5.5.1 Imaging SIMS 464
      6.5.5.2 X-ray Photoelectron
              Microscopy 466
References 468

6.1 INTRODUCTION

In the previous edition of this book, this chapter was called "New Techniques in Polymer Microscopy." Now, nearly 10 years later, many of these once new techniques, like atomic force microscopy (AFM), have become standard for many polymer microscopists. Others, such as confocal scanning microscopy, or scanning near-field optical microscopy, have developed and become much more common, but they are rarely applied to polymeric systems. It is still worthwhile to mention these techniques as they might solve the problems arising in some new areas of polymer science and engineering. Therefore the chapter is now called "Emerging Techniques in Polymer Microscopy" and covers both the truly novel techniques that are emerging rapidly, and those that are not as new but have continued to emerge slowly for decades. As a wide range of techniques is covered in a relatively small space, there is no intention to be comprehensive; readers who come to use these new techniques will need more information than can be provided here.

One feature that continues to drive much of the innovation described in this chapter is the increasing amount of computing power embedded in microscopes. Sometimes this is fairly obvious; for example, any three dimensional (3D) microscopy must involve large data sets. Calculating a high resolution 3D image from 800 projected phase images will tax a network of servers. In other cases, for example the aberration correction of electron microscopes, it is natural to think of advances in the mechanical and electromagnetic design of the devices. But the instruments would not be feasible without computer control and fully automated alignment. As this trend continues, it may, for example, become reasonable to replace all phase contrast and other imaging hardware in the optical microscope with a technique that determines the phase of the image wavefront and then calculates what any imaging mode would do and what the image would look like [1]. The phase could be determined by some interferometric technique or by calculation from a through-focus series of images [2, 3].

6.2 OPTICAL AND ELECTRON MICROSCOPY

6.2.1 Confocal Scanning Microscopy

The *confocal scanning laser microscope* (CSLM; sometimes LCSM, for *laser confocal scanning microscope*) has many forms, but all are optical microscopes with some form of scanning added to the regular optics. A normal optical microscope produces poor images when the sample surface is rough or signal comes from a range of depths in a transparent sample. A CSLM does not have this limitation and has slightly better lateral resolution than the regular optical microscope. Because confocal microscopy is a standard biological technique, there are many sources of information, including manufacturers' Web sites [4, 5], book chapters [6, 7], an excellent practical guide [8], and a comprehensive handbook [9]. In a confocal microscope a small aperture, the *confocal pinhole*, is placed in the *confocal plane* where rays coming from a particular plane in the object are brought to focus, as shown in Fig. 6.1. When the light detector is placed behind the aperture, one point (x, y) in the plane is selected. The aper-

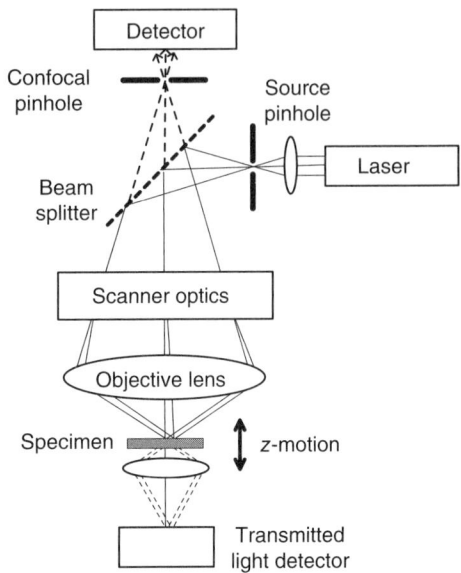

FIGURE 6.1. Schematic diagram of a laser scanning confocal microscope. The critical confocal feature is the pinhole aperture in front of the detector that rejects light coming from outside the focal plane. The transmitted light mode shown is not confocal.

ture also cuts out most of the light coming from other planes in the specimen. If the illumination is focused onto the selected point in the object, then information comes from the point (x, y, z) only. Scanning the illumination and the confocal aperture together over x and y builds up a scanned image of the selected plane.

Mechanically registering two scanning systems is difficult, so commercial CSLMs are reflection or fluorescence microscopes, where the beam passes through the same scanner twice as shown in Fig. 6.1. The beam is scanned on the specimen and then de-scanned onto a fixed confocal pinhole. Confocal microscopes may have a transmission mode, but generally this will not be confocal. There may be a separate regular illumination and imaging system for transmission; Fig. 6.1 shows the other case, where the scanned illumination is used with a transmitted light detector that has a large area and no aperture in front of it.

Another confocal scanning method uses a rapidly rotating disk containing many holes in the confocal plane (a *Nipkow disk*). These holes act as confocal apertures and mechanically scan over the object; the output is integrated by the eye or by CCD camera. Most conventional "point scanning" CSLMs use mechanical scanning of mirrors, limiting the acquisition speed to about 1 frame per second (fps). The disk scanning can easily give real-time video images of 30 fps or higher, up to 1000 fps. With pinholes in the disk, there is a trade-off between confocal selectivity (requiring small holes) and efficiency, as small holes block most of the light. The Yokagawa spinning disk system uses an array of microlenses to direct light into the array of pinholes [10], overcoming this limitation and allowing efficient high-speed imaging [11].

So far, the description of the instrument implies that the image formed is a slice or section of the specimen at a particular height. (The CSLM has been described as an "optical microtome"). If the specimen is scanned in the z direction—not continuously, but in steps between image scans—then storing all the images gives a three dimensional image of the specimen that can be processed to show and measure three dimensional features (see Section 6.4). If the specimen has a rough reflective surface, and the microscope is operated in reflection rather than fluorescence mode, each image will contain a bright contour line where the surface intersects the selected plane, with the rest of the field dark. Then it is not necessary to store all the data, the intensity values can be either added or set to the maximum obtained during the z-scan. In either case, at each point the bright contour line that appears when the surface is at the focus level dominates the image so the result is a single image that is in focus on the rough surface at every point (if the z-scan has a sufficient range).

This mode can be used for surface profiling [12, 13] with the value of each pixel set to the z value at which the image was brightest. A topographic map of the most reflective surface will be produced, whether external or internal. Line profiles can be selected in any direction, or surface roughness can be calculated from the whole profile. Since the CSLM can detect internal surfaces, it can be used to measure the thickness of thin polymer coatings [14, 15].

Although the major use of the CSLM is as a fluorescence microscope in biology [8], it is also used in a wide range of polymer studies [16]. These include internal interfaces in composites [17], biomaterials [18], emulsions [19], phase separation in reactive blends [20], hair [21], and diffusion in gels [22], films [23], and fibers [24]. Other applications are mentioned in Section 6.4.3, which concentrates on 3D imaging. This nondestructive, noncontact technique can give internal, 3D information and only requires that the sample is sufficiently transparent.

6.2.2 Optical Profilometry

As was mentioned in the previous section, the CSLM can be operated to give a topographical map of a reflective surface which is usually external but could be internal. It does so by scanning the sample in the z direction, on the optic axis, and finding the position of highest intensity of reflected light. There has been a report that the CSLM system has difficulty in profiling if there are steps or steep slopes on the surface [25]; more important is that vertical resolution depends on the optical sectioning power of the instrument. At low magnifications, the numerical aperture of the lens is low, the convergence angle of the rays is low, and so the vertical resolution is less. The vertical point spread function can have a width of several micrometers. The profiling software picks out the peak position (or the centroid) of this smooth point spread function, so the profiling resolution is much better than the width of the function. Even so, it is a limit which changes with the field of view selected. If the highest vertical resolution is to be maintained over a large area, smaller images must be collected and stitched together.

The alternative is to use an *interferometric optical profiler*, which modernizes the interference microscope (just as birefringence imaging, Section 6.2.3, modernizes the polarizing microscope). These devices use white light and one of a variety of methods to determine the position on the image of the central interference fringe. Some component is scanned along the optic axis—it could be the sample, the objective lens, or the reference mirror—and the software creates a topographic map. The vertical extent of the interference fringe does not depend on the image magnification, so neither does the profiling resolution, which can be as high as 100 pm. Compared to AFM surface profiling, the optical devices have a much lower lateral resolution but a much larger field of view and can deal with a wider range of heights [26]. If thin film thickness is measured by optical profilometry and by other techniques, comparison of the results can give the refractive index of the film [27].

6.2.3 Birefringence Imaging

Birefringence imaging is defined as forming an image where the orientation and/or the magnitude of the birefringence is determined and displayed at every point on the image. (Actually it is the retardation that is measured; turning this into birefringence requires measurement of the object thickness.) In the normal polarizing microscope the birefringence is measured manually, one point at a time. Indeed, apart from the polarizing dichroic filter "Polaroid"™ invented by Dr. Land in 1932, polarizing microscopy (see Section 3.1.7.2) has remained largely a 19th century technique. The polarization state of light is modified by manually inserting, rotating and tilting various cut and polished sections of natural crystals. Results, even quantitative results such as retardation, are obtained by visual inspection of the image. Devices that perform birefringence imaging and related analyses are bringing this up to date.

One such device is the Metripol (trademark of Oxford Cryosystems, Oxford, UK). This device has a rotating polarizer driven by a computer-controlled stepping motor and a fixed circular analyzer [28]. The polarizer steps around in 5, 10 or 50 steps, and the software calculates the slow axis orientation and retardation at every point from the set of images [29]. The instrument has been used to observe phase changes in materials [30] and structures in biological materials such as collagen. The 3D pattern of birefringence can be determined for a single point by applying the rotating polarizer method to a conoscopic view [31]. In a similar device, a rotating analyzer attached to a CSLM allows confocal birefringent imaging [32].

Another system, now called Abrio (trademark of CRi, Cambridge, MA), is more often referred to in the literature as the PolScope. This system has no moving parts but uses a fixed circular polarizer and analyzer and a compensator made from two liquid crystal components. The retardance of these varies with applied voltage, so that with two of them orientation and magnitude of retardation can be controlled [33]. The major use of this system today is in cell biology [34, 35], but it can be applied to a range of problems [36] including biomineralization [37, 38]. Figure 6.2 shows how these imaging modes help define the structure of a thin calcite layer [38].

Using more liquid crystal elements as shutters to change the direction of the cone of rays passing through the sample allows 3D information on birefringence to be collected. This has been applied to a biopolymer system [39, 40], but it assumes that the materials are uniaxial, and requires that there are a few birefringent objects in an isotropic matrix. Because optical anisotropy is a tensor property, the effects do not simply add up as they do in absorption, and projection tomography is not possible using polarized light. However, if polarized light is used in *optical coherence tomography* (see Section 6.4.3) it becomes "polarization sensitive" optical coherence tomography (PS-OCT) and can be used to make a birefringence image from a plane within the sample. Mostly used to investigate tissues such as skin or cartilage, the device has been used on polymer matrix composites [41] to see internal flow patterns in injection molded parts, such as seen in Fig. 6.3 [42], and to nondestructively measure internal stresses [43].

6.2.4 Aberration Corrected Electron Microscopy

After decades of work and anticipation, systems that correct for aberrations of the electromagnetic lenses used in electron microscopy have now been developed [44–46]. The resolution of the standard transmission electron microscope (TEM) was described in Section 3.1.4.3, and the spherical aberration coefficient (C_s) appears in

Optical and Electron Microscopy

FIGURE 6.2. Thin calcite films obtained by heating (for 2h at 250°C) an amorphous calcium carbonate film stabilized with poly(methacrylic acid) brushes. (A) Polarized optical micrograph; (B) LC-PolScope image (retardance values are indicated as false color); (C) retardance gray-scale image (red vector overlay indicates the orientation of the slow birefringence axis); (D) high-magnification SEM image. (See color insert.) (From Tugulu et al. [38]; reproduced with permission.)

all the formulas limiting resolution (Eqs. 3.7–3.12). If spherical aberration is corrected, and a field emission gun (FEG) or energy filtering system limits chromatic aberration, then these limits go away; previously ignored higher-order aberrations become important. Lattice images with resolution of 61 pm have already been obtained [47] and microscopes are planned with higher levels of correction [48]. Atoms are resolved with high contrast, and the structure of glassy materials may be fully determined by atomic-level tomography [49]. In the corrected scanning transmission electron microscopy (STEM), as the resolution improves, so does the optimum divergence angle. This means that the depth of focus is reduced, which can be a problem—or a feature if it is used to make optical sections and thus a 3D image [50].

There is still a noise limit to resolution, and this requires that a number of electrons pass through each resolved feature. Following the calculations in Section 3.4.4, if a 50 pm feature has 50% contrast and needs a 5:1 signal to noise ratio, it needs 50 electrons to pass through it, and this corresponds to $0.3\,C\,cm^{-2}$. From Fig. 3.38 this is just possible for some organic crystals, but not for most polymers. Even so, it would not be wise to rule out some applications of this new technique. Negative

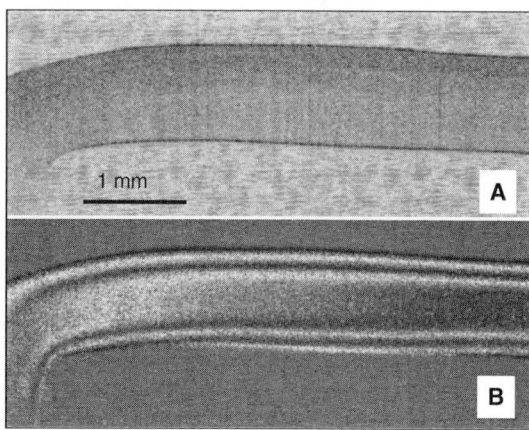

FIGURE 6.3. PS-OCT measurements of injection-molded polystyrene part: (A) OCT cross-section (intensity image); (B) polarization-sensitive retardation image exhibiting a homogenous core and anisotropic surface regions. (From Stifter et al. [42]; reproduced with permission.)

predictions were made in the early days of high resolution electron microscopy, but there have been a wide range of useful high resolution studies on polymers and organic thin films [51]. One possibility is that time-resolved electron microscopy, with exposure times of picoseconds to nanoseconds [52], can be used to form an image before the processes of radiation damage are complete [51].

In the scanning electron microscope (SEM), particularly at low voltages, the chromatic aberration is as important as the spherical aberration and an *aberration corrected SEM* (ACSEM) has both C_s and C_c reduced to very small values [53]. This instrument has a resolution of 0.6 nm at 5 keV. Correction decreases the minimum spot size at 1 keV from 18 nm to 2 nm, at the same time increasing the beam current by 30× [54]; Fig. 6.4 shows the improvement in image quality that results. The corrected lens can operate at a greater working distance, and this together with a greater beam current may make a corrected SEM useful in analytical work. As with the STEM, the depth of field is much reduced in the aberration-corrected microscope, to the point where the instrument forms thin optical sections that are in focus. Multiple exposures can be combined (in *through focus reconstruction*) to give an image with quantitative depth information [54].

6.2.5 Ion Microscopy

Ion microscopes include the field-ion microscope, not relevant to polymer studies, and the imaging secondary ion mass spectrometer or ion microprobe (e.g. the NanoSIMS, by CAMECA; see Appendix V). Since this is primarily an analytical instrument it is described in Section 6.5.5. The remaining type of ion microscope uses a focused ion beam to form an image by scanning it over a surface and is exactly analogous to the SEM. Currently available versions use gallium ions and may be called either a *scanning ion microscope* (SIM) or a *focused ion beam* (FIB) system. These tools are primarily for micromachining, to remove or deposit material [55]. The microscopy is vital to observe the machining operation but is otherwise secondary. The gallium ions are deposited in the specimen and cause it to sputter away. As the requirements for semiconductor device micromachining get more exacting, the resolution of FIB devices have increased, to 4 nm. *FIB-SEM* combined devices can be used to create higher resolution 3D images (see Section 6.4.2).

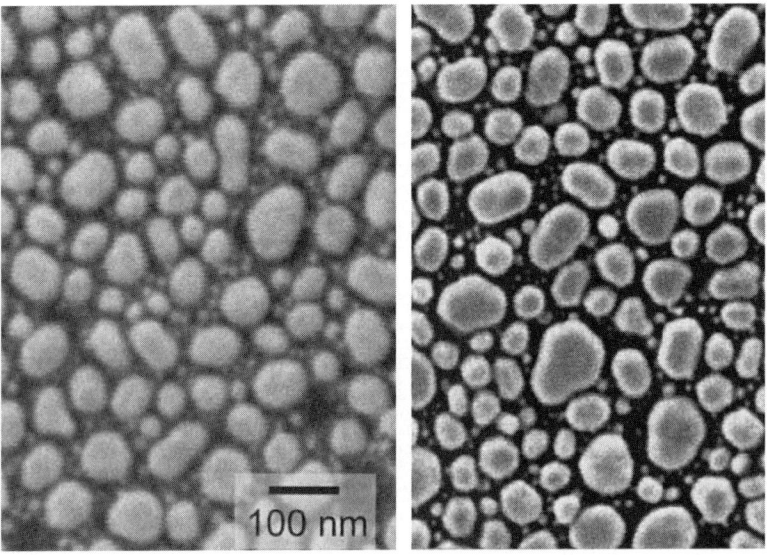

FIGURE 6.4. Uncorrected (left) and corrected (right) images from JEOL 7555S LSI inspection SEM at 1 keV and a working distance of 4 mm. The effective C_s and C_c values after correction are of the order of micrometers and the aperture is approximately 40 millirads. (From Joy [54]; reproduced with permission.)

A new microscope uses helium ions [56, 57]; until recently there has been no suitable high brightness source of such ions. A probe size of 1 nm is claimed, and although there are no such microscopes in use so far, there are some predictable advantages. One is that the interaction volume of ions is very small, so that the ion microscope at high beam voltages (20–50 kV) will behave like a low voltage SEM, with resolution controlled by probe size, and the signal depending only on the very top surface layer. One published image shows a fossil with no specimen preparation, so apparently charging will not be a problem [58]. The interaction of ions with materials is quite different from that for electrons, so the contrast will be different. There will be scattered ions as well as secondary electrons, so it may be possible to perform the equivalent of forward recoil spectroscopy (FRES) experiments [59] (distinguishing between deuterium and hydrogen in labeled polymers) but at high spatial resolution.

6.3 SCANNING PROBE MICROSCOPY

6.3.1 Chemical Force Microscopy

Chemical force microscopy (CFM) refers to the chemical modification of the AFM tip with specific functional groups to measure the forces involved in chemically specific interactions with a surface. It is implemented in lateral force microscopy, force spectroscopy (force curve), and force-volume imaging and has been in use since about 1994. For polymer surface characterization, CFM offers the potential to map functional groups on surfaces with high spatial resolution. However, for meaningful results it requires careful experimentation and controls and also significant correlated data from other surface analysis techniques such as x-ray photoelectron spectroscopy (XPS), SIMS, and contact-angle measurements. This is one reason why broad adoption of the technique has been slow. There are excellent reviews published by the pioneers of this technique, notably the Lieber group at Harvard [60, 61], the Vancso group in The Netherlands [62, 63], and the Tsukruk group at Iowa State [64].

The chemical interaction of interest will not be the only interaction with the surface, so there are always other forces involved (e.g. capillary, electrostatic and van der Waals forces). Not surprisingly, the first successful demonstrations of functional group imaging used specially prepared surfaces that maximized one specific chemical interaction. These were patterned mixtures of self-assembled monolayers (SAMs). These SAMs are alkane thiols with sulfhydryl (–SH) at one end and a functional group at the other end. Typical functional groups chosen are methyl (–CH$_3$), carboxylic acid (–COOH), hydroxyl (–OH), and amino (–NH$_2$). The lone pair of electrons from the sulfur of the thiol group binds strongly with gold, so gold-coated AFM tips and gold-coated substrates are used and coated with SAMs, such as shown in Fig. 6.5 [65]. These experiments were carried out in liquid to eliminate capillary forces. It was

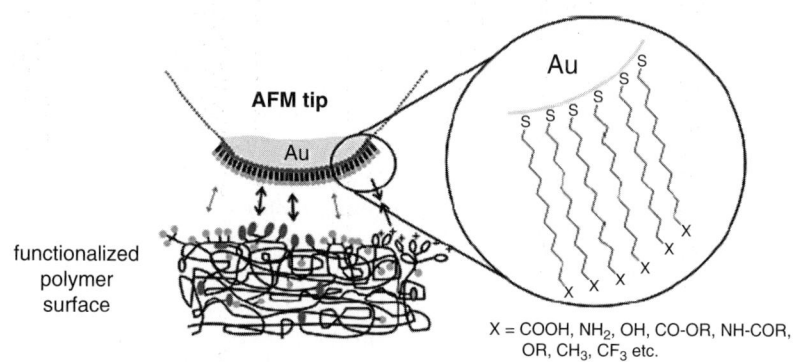

FIGURE 6.5. Schematic demonstrating the functionalization of gold-coated AFM tips with self-assembled monolayers. (From Schonherr et al. [65]; reproduced with permission.)

possible to rank the interactions between CH$_3$/CH$_3$, COOH/COOH, and CH$_3$/COOH tip/sample pairs based on statistical analysis of pull-off force measurements [66]. This variation in interaction force led to image contrast in frictional force imaging where micrometer scale patterning allowed clear distinction of the interacting regions. This approach was later extended to tapping mode in liquid [64]. With ionizable functional groups, pH affected the pull-off force so that the ionization pK$_a$ could be detected in a "force titration" experiment. It was recognized that the pull-off force was by its nature statistical as only few bonds were involved in bond-making and bond-breaking. Careful experimentation therefore requires sufficient repetition to obtain reliable mean values and standard deviations.

On real surfaces many other factors come into play (in addition to the chemistry) for accurate pull-off force measurements. These include mechanical deformation, surface roughness, tip radius effects, and contamination [67]. Some of the best and earliest CFM work on real polymer surfaces measured pull-off forces for a variety of polymer surfaces immersed in perfluorodecalin and ethanol [68]. Perfluorodecalin has such low polarizability it has been referred to as a "vacuum in a bottle." This work demonstrated the importance of the medium in which the measurements are made. The interaction was interpreted in terms of polar and nonpolar contributions to adhesion, analogous to macroscopic surface energy measurements on solid surfaces. Beake et al. [69] described the first application of CFM to plasma-treated poly(ethylene terephthalate) (PET) polymer in order to understand the reason for increased friction following argon plasma treatment. Later, force titration experiments on plasma-treated perfluorinated isotactic polypropylene (iPP) used hydroxyl (–OH) terminated tips in liquid. Titration of acid groups using pull-off force was extended to force-volume imaging where a map of pull-off force could be obtained with better than 50 nm spatial resolution. This showed lateral inhomogeneities in functional groups on the surface [70].

Chemical force microscopy showed the surface reconstruction of low density polyethylene (LDPE) oxidized with chromic acid (in conjunction with contact-angle measurements) [71] and the aging of melt-pressed films of PP using adhesion mapping to provide evidence of UV stabilizers at the surface [72]. Hydrophobic recovery of elastomeric poly(dimethylsiloxane) (PDMS) after UV/ozone oxidation could be followed by CFM in water using hydroxyl terminated tips. High spatial resolution (50 nm) adhesion mapping was correlated with modulus mapping of the surface, as highly adhesive regions were more hydrophobic and softer [73]. More recently, high resolution force-volume mapping of functional groups in the block copolymer of polystyrene and poly-t-butylacrylate (PS-b-PtBA) has been obtained using CFM after chemical treatments designed to modify the ester chemistry of the PtBA block [65].

Controlling the tip chemistry is useful for other purposes besides CFM. It has led to improvements in contact mode imaging by reducing the interfacial free energy; the resolution then became comparable to that in tapping mode [74, 75]. Similarly, modification of tip chemistry allowed high resolution scanning of highly drawn polyethylene (PE) and poly(tetrafluoroethylene) (PTFE) surfaces in air. The results were comparable to those obtained in liquid where the scanning forces were much lower [76].

Conventional silicon probes, not functionalized, can be used to detect hydrophilic and hydrophobic interactions by control of the humidity in the vicinity of the tip and sample during intermittent contact imaging. This work was pioneered by researchers at NIST [77, 78]. A specially designed cell was constructed to regulate the humidity in the vicinity of the tip and sample [79]. This method has been validated on specially prepared surface energy gradient specimens prepared by gradient oxidation of micropatterned SAM layers on silicon [80]. Phase imaging reveals improved contrast between hydrophilic and hydrophobic regions when the relative humidity (RH) is higher, about 30%, and then contrast falls off again above 90% RH [81]. The work has been extended to the characterization of block copolymers and blends [82]. The phase images are very sensitive to the wettability of the surfaces with phase contrast varying as a function of surface energy at high humidity. These experiments further explain why CFM requires careful ambient control. With so many possible competing forces, there is a large margin for error—measuring the wrong interaction.

6.3.2 Harmonic Imaging

In intermittent contact AFM (IC-AFM), also called tapping mode AFM (TMAFM), the cantilever is oscillating near resonance, but the tip experiences long range and short range attractive and repulsive interactions during each cycle as has been described in Section 3.3.4. The situation is very complicated and the total interaction is highly nonlinear making it difficult to model. With sufficient understanding and comprehensive modeling, it should be possible to extract the mechanical properties of the sample from the IC-AFM signal. These properties would include adhesion, modulus, and damping at high spatial resolution. Currently the phase shift of the oscillation is the standard imaging signal most related to the sample properties. However, the elastic properties are generally independent of the phase shift unless there is a strong dissipative component due to adhesion hysteresis and/or viscous damping [83].

Even without detailed modeling, it can be said that the attractive and repulsive interactions change the motion of the cantilever away from the simple sinusoidal motion of free resonance. By Fourier's theorem, any such altered motion can be broken down into the basic fundamental motion plus components at higher harmonic modes. Such harmonics will be more sensitive to the disturbance than the fundamental resonance frequency. Further, abruptly changing interactions such as the elastic resistance to tip penetration, or a sharp transition from attractive to repulsive interaction, will have the strongest effect at high frequencies. A tap that lasts for 5% of the cycle will show up strongly in high harmonics—those with frequencies associated with this shorter timescale.

Detailed analysis supports these general statements. When the quality factor, Q, of the vibrating cantilever is low, such as in liquid environments, the fundamental is relatively suppressed and there is a significant contribution of higher harmonics [84]. The full reconstruction of the tip-sample interaction must include the anharmonic contributions of the cantilever motion due to excitation of higher modes [85]. Fourier analysis of a single degree of freedom model predicts that the elastic properties of the surface can be distinguished more readily at greater than the 10th harmonic [86]. Spectroscopic analysis shows that the transition from net attractive to net repulsive interaction in IC-AFM is coincident with enhancement of these modes (Fig. 6.6) [87–90]. The transfer

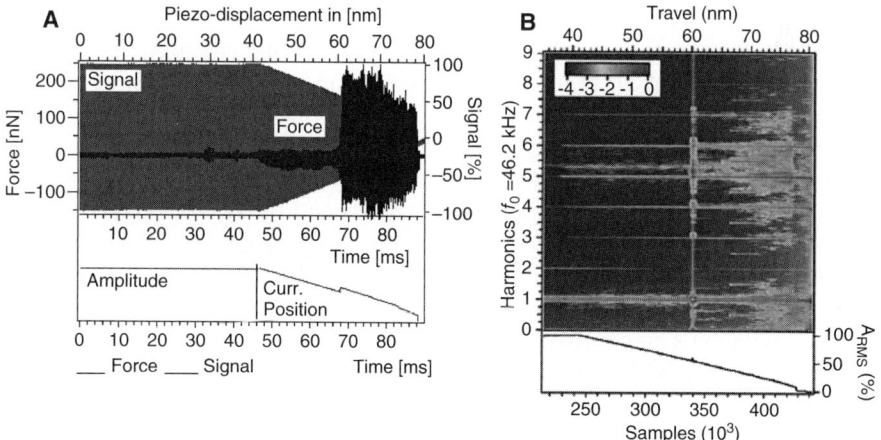

FIGURE 6.6. Development of harmonics in dynamic AFM. (A) Force and amplitude evolution with time during approach to a sample surface. Note the force is adhesive (noncontact) until about 50% of the amplitude has been attenuated where the interaction becomes repulsive (intermittent contact). (Adapted from Stark et al. [88]; used with permission.) (B) Logarithmic plot of harmonic intensities along the approach curve shown in (A). (Adapted from Stark et al. [89, 90]; used with permission.) Although there are harmonics in the attractive portion of the approach, the intensities of higher harmonics and subharmonics increase when the interaction is repulsive.

function of the cantilever vibration shows that the higher harmonics are 20–30 dB lower in amplitude than the fundamental motion and so are very weak.

The analysis described so far would seem to indicate the cantilever is a single oscillator (thus the "single degree of freedom model"), but it is a flexible beam, and when there is a force impulse on the tip, it flexes and bending modes of the beam are excited. Of the first three flexural (not resonant) modes of the cantilever vibration, the third is predicted to be the most sensitive to material properties. This is primarily because the contact time of the cantilever at the fundamental is approximately equal to the period of one cycle time of this flexural mode [91]. Improved image contrast was verified experimentally for patterned polyimide on glass for the third flexural mode [92] and for the third, fifth, and eighth harmonics for other composite materials using a dual lock-in system [93]. Harmonic analysis has been used to differentiate poly(methyl methacrylate) (PMMA) from PS, two polymers with similar bulk moduli [89]. Because the higher harmonic signals are generally very weak, special cantilever designs have been developed to produce resonant enhancement of harmonics by exciting higher flexural modes that are coincident with the harmonic of interest [94, 95].

Harmonic imaging, or *harmonic resonance imaging*, is currently a very new development. The theoretical work suggests that certain harmonics will favor specific interactions (e.g. higher harmonics for elastic material contrast). These approaches will require high speed electronics and digital lock-ins to monitor multiple signal channels that can detect the amplitude and/or phase of higher harmonics. Currently, two scanning probe microscope (SPM) vendors offer high speed digital lock-in control amplifiers for this purpose. Each of the harmonics will have its own amplitude, associated phase, and Q so there are many possible signals to explore, which will enable more rapid qualitative and quantitative property mapping at the spatial resolutions and timescales currently attained by conventional IC-AFM imaging.

Some of the harmonic resonance approaches may be alternatives to chemical force microscopy in the measurement of adhesive interactions. The higher harmonics are investigated in the adhesive regime before the transition to the repulsive regime. Here small amplitude, higher harmonic resonance interactions can be sensitive to van der Waals forces and less complicated by elastic or dissipative interactions. The amplitude response of the high harmonics, in particular the seventh harmonic, provided image contrast dominated only by adhesion when scanning a multilayer purple membrane on mica [96].

6.3.3 Fast Scanning SPM

Current SPM imaging speeds are limited by control electronics (feedback bandwidth), cantilever response, and scanning element resonances. The combined effect of these contributions is to limit high resolution images to line rates of about 1 Hz for current commercial systems. For a 512 × 512 pixel image, that means an image will take about 10 min to acquire. Force spectroscopy (force curves) can be obtained at 100 Hz (10 ms per point) before running into problems with bandwidth. Faster scanning rates, ideally at frame rates of at least 1 fps or faster, are desired [97]. This would make AFM more competitive with image capture in a high resolution SEM; real-time monitoring of dynamic processes requires AFM instruments that can image or measure at much higher rates. In fact, much of the driving force for high speed imaging has come from the biology community, which requires imaging of dynamic processes in fluid.

The need for fast imaging was recognized in the early development of both STM and contact mode AFM. As early as 1991, Barret and Quate [98] described high speed, constant height contact mode AFM imaging of semiconductor test structures at 3 fps. Even then it was clear that several issues limited high-speed scanning—scanning system resonances, tip contamination, and scanner nonlinearities. Most important, however, were the high contact forces, estimated to exceed yield stress, which damaged polymer resist layers like PMMA. The high forces could be used for thin film lithography [99] and thin film patterning [100] at high speeds. In order to obtain high speed constant force imaging, active cantilevers were developed where a second piezo, applied as a coating to the cantilever, was used in addition to the piezo controlling the probe motion. With this control of the cantilever the tip could respond more quickly to vertical height changes than with the piezo tube scanner alone. This approach requires control over two feedback loops, one for the slow scanning tube and the other for the high-speed lever. It has been applied to contact mode AFM [101, 102] and IC-AFM [103, 104].

For isothermal crystallization studies using a hot stage, moderate increases in scan speed were needed to follow lamellar growth kinetics. These were 10 Hz scan rates or 30 s frame rates, using phase imaging in IC-AFM [105, 106]. Design rules for cantilevers operating at these conditions require that they resonate at high frequency and have low quality factors for faster response. Low Q can easily be achieved when operating in a liquid environment; much of the high speed development has involved imaging in liquids. These developments have been led in large measure by the Hansma group at UCSB [97, 107, 108] and the Ando group at Kanazawa University in Japan [109–111]. For high speed tapping mode imaging in liquids, cantilevers were designed to be short, thin and of low mass [107, 108, 112]. A thousand fold decrease in mass translates to a factor of 30 times higher resonance frequency. Smaller cantilevers also require new head designs [113–115] for faster x, y scanning and smaller laser focusing for better signal sensitivity using the optical lever deflection detectors [116]. An alternative route to high-speed imaging is to prepare a large two dimensional (2D) array of active cantilevers that can be actuated simultaneously and operate in parallel. Such a device, the *Millipede* (Zurich), has been built by IBM for data storage applications [117].

For IC-AFM operation in air, active Q control may be employed to achieve faster rate scanning [103]. Alternatively, the cantilever can be tuned well below resonance to effectively reduce the cantilever Q without additional control electronics [118, 119]. These approaches can allow the line rate to increase by a factor of 5–10 times over conventional scanning speeds. The trade-off is a generally higher instantaneous peak force in tapping and reduced phase imaging contrast.

High speed tapping mode scanning places extra demands on the feedback control [120]. Approaches to maintain the set-point that is used to control the relative tapping amplitude have been developed [121]. Controlled experiments to investigate tip wear suggest that low set-point, harder tapping actually minimizes wear contrary to common assumptions [122]. Improvements in the speed of data acquisition [123] and signal conditioning [109] as well as implementation of feed-forward control algorithms [110] have been developed to increase bandwidth. The developments and improvements in these components were reviewed in 2005 [124] and 2006 [125]. Fast rates will be necessary for intermittent contact mode and other dynamic modes [111, 125] in order to more quickly assess morphological features of interest for further probing, such as nanoindentation.

One manufacturer is selling a video rate AFM, called the VideoAFM (trademark of Infinitesima, Ltd.; see Appendix VII). This AFM operates in contact mode using a tuning fork scanning element [126] where the sample or tip is affixed to one arm of a vibrating tuning fork. Scan rates of 15 fps are possible albeit at limited scan sizes. This technique has been used to study many aspects of polymer crystal growth [126]. In order to overcome the small image size at high rates, images may be tiled to produce a composite image of a much larger field [127].

6.3.4 Scanning Thermal Microscopy

Another mode of operation derived from contact AFM that is relevant to polymer studies is the *scanning thermal microscope* (SThM) [128–130]. Here the AFM probe is fabricated to make the tip temperature sensitive. If it is a resistive element, it can be either passive (temperature sensing) or active (heat applying). In SThM, heat injected by the probe affects a three dimensional volume around the contact point. It can therefore detect buried or subsurface features [131, 132]. There is some effort toward developing tomography using thermal imaging [133].

In the first SThM product, the micro-TA (now supplied by Anasys Instruments, Inc.; see Appendix VII), the probe consists of a bent loop of Wollaston wire. This is a 5 μm diameter Rh wire with a Pt/Ir sheath. The sheath is removed at the bend and this forms the tip, as shown in Fig. 6.7A [134]. The Wollaston probes have been used successfully for thermal analysis and thermal imaging of polymer systems. Thermal imaging can be conducted in a constant temperature DC mode, which provides an image related to thermal conductivity, or using AC (modulated) techniques to provide images related to diffusivity. Because of their low mass, the probes can be heated quickly and their temperature can be ramped at rates of ten to hundreds of degrees per second. Using a bridge balanced circuit, the differential current between a thermal probe in contact with the surface and a reference probe can be obtained during temperature ramping. This depends on the heat flow to the surface, providing a route to spatially localized differential thermal analysis.

The use of Wollaston probes for both imaging and local thermal analysis were reviewed extensively by Majumbdar in 1999 [129] and Pollock and Hammiche in 2001 [130]. For polymer characterization there has been notable work on imaging blend components in polymers [135–141], polymer composites [142, 143], and polymers for electronic applications, such as low-k dielectrics [144, 145] and polymer LEDs [146, 147]. An example of SThM imaging was shown in Fig. 5.36 [67] of 35 μm via holes patterned in a divinyl siloxane benzocyclobutene polymer(PhotoBCB) over a copper substrate before removing polymer residues. DC thermal images and AC phase images are shown in the figure at two frequencies. Fundamental studies of polymer thermal conductivity have been conducted using simultaneous force and temperature versus distance spectroscopy [148, 149] and by comparison of local thermal measurements with the results of bulk thermal

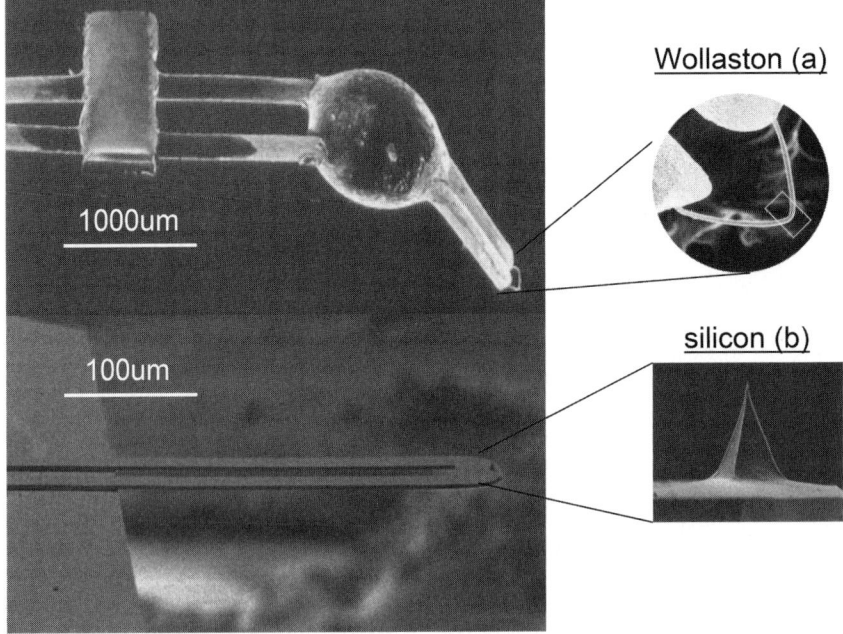

FIGURE 6.7. Resistively heated AFM probes used for thermal imaging and thermal analysis. (A) Wollaston wire probes used in the commercial microTA system. (B) Microfabricated silicon probes with integral heaters at the base of the tip used in the commercial nanoTA. (Figure 6.7A from Anasys Instruments [134]; reproduced with permission. Figure 6.7B from Hammiche and Pollack [157]; unpublished.)

analysis [150]. The latter study indicates the significant role of probe surface contact area changes through a T_g or melt temperature where the tip penetrates into the surface under the applied load for thermomechanical testing. The size of the Wollaston probes limits their application to the analysis of features that are several micrometers across. One notable use has been the measurement of thin film or surface glass transition temperatures for comparison with the bulk [151–153].

Recently there has been interest in developing high resolution thermal probes based on silicon microfabrication technology. This is work driven by two needs. The first is a need to keep the thermal interaction to the nanometer scale. This is required for thermomechanical writing and high-density data storage, for example, in the Millipede cantilever arrays [117, 154, 155] and also for materials characterization where heterogeneous components are often submicron in size. The second is a need to take advantage of dynamic AFM modes, such as phase contrast in IC-AFM, and not just contact mode. The design characteristics for these probes have been thoroughly evaluated [156]. Probes like these are being optimized for other applications, such as the nano-TA (trademark of Anasys Instruments, Inc.) thermal accessory for commercial AFM systems. The nano-TA takes advantage of silicon probe development to improve the spatial resolution of heated tips. Like conventional silicon probes used for IC-AFM, the nano-TA silicon probes are sharp (end radii <10 nm) and are of sufficient spring constant to work in either contact mode for thermal imaging or thermal analysis and tapping mode for dynamic imaging. The probes consist of a double-armed cantilever that is joined at the end where the tip sits, as seen in Fig. 6.7B [157]. Just under the tip is a region of silicon that has lower doping density to concentrate heat when current is passed through the arms [154, 156].

To demonstrate the improvement in lateral resolution, these probes have been used to measure the glass transition of coextruded polycarbonate (PC) and PMMA, which have alternating multilayers with submicron spacing. Figure 6.8 is an image of the material that was microtomed perpendicular to the layers. The tip was first used in IC-AFM mode to obtain a phase image that could distinguish the layers but not identify them. The same tip was then placed in contact with the material, and local thermal analysis was conducted using the tip deflection signal to detect the transition as the tip was heated at a rate of 10°C/s. If the tip is allowed to penetrate to significant depths, the T_g of both materials are sensed. If the tip is retracted very quickly after the transition, the size of the penetration mark is below 250 nm and the analysis is confined to each polymer layer.

A particularly exciting development for polymer characterization is the use of resistive probes as passive temperature sensors for *photo thermal infrared detection*. These experiments require a setup that combines a SThM with an infrared (IR) reflection instrument where the IR source can be focused on the tip-sample contact. When infrared radiation of appropriate frequency is absorbed by the material in the tip-sample contact region, there is a temperature rise that can be detected. An interferogram of the thermal fluctuations is obtained that can be Fourier analyzed to produce a spectrum as a function of wavelength. Early studies using Wollaston-based probes (micro-TA) demonstrated that it was possible to obtain localized Fourier transform IR (FTIR) spectra of several polymers including PtBA, polycarbonate, polyethylene, and polyimide with resolution comparable to far-field data using conventional IR sources [158, 159]. The configuration for this technique is shown in Fig. 6.9 [160]. The intensity of the IR source is low, and reasonable quality spectra with a signal to noise ratio (SNR) of greater than 5 required about 10 min of collection [159]. Exploratory studies using tunable and brighter sources, such as an optical parametric resonator [159] or synchrotron [161], lead to significant improvements, but these are not economical or practical for general use.

Other studies have been conducted to try to understand the lateral resolution and the depth sensitivity of the probe for IR detection. A polystyrene film was coated with layers of polyisobutylene ranging from $2\mu m$ to

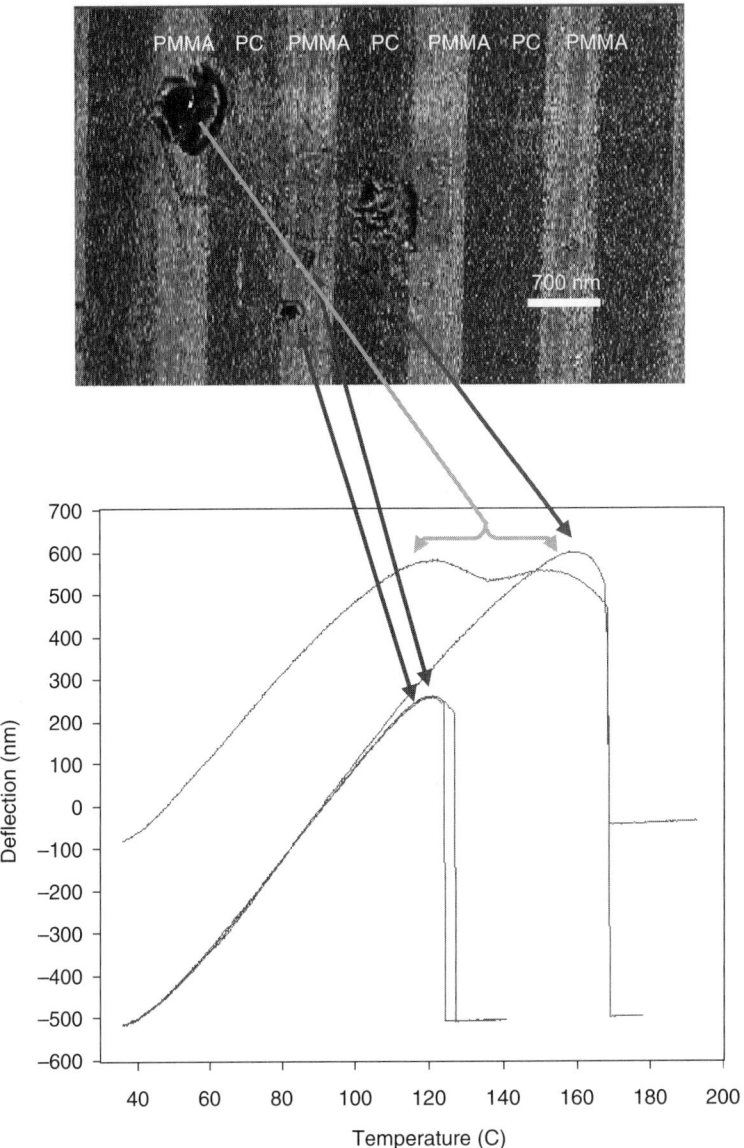

FIGURE 6.8. Thermal analysis of a multilayer polymer using a nanoTA heated tip. The sensor deflection was used for detection of the glass transition temperature of each component when the penetration was confined to each layer. For the PMMA, this was 120°C, and for the PC, it was 160°C. (See color insert.)

50 μm, and the attenuation of signal from the PS was measured and modeled [159]. The sensitivity of the technique was further explored by comparing the spectral intensity of PET films ranging in thickness from 0.9 μm to 250 μm. The thinner film gave more intense spectra (higher surface temperature) as the film thickness approached the thermal diffusion length; for thicker films the heat diffuses into the polymer and thus reduces the surface temperature [162]. This technique has been used to study the life cycle of cells, following the turnover of cellular components that are IR active [163].

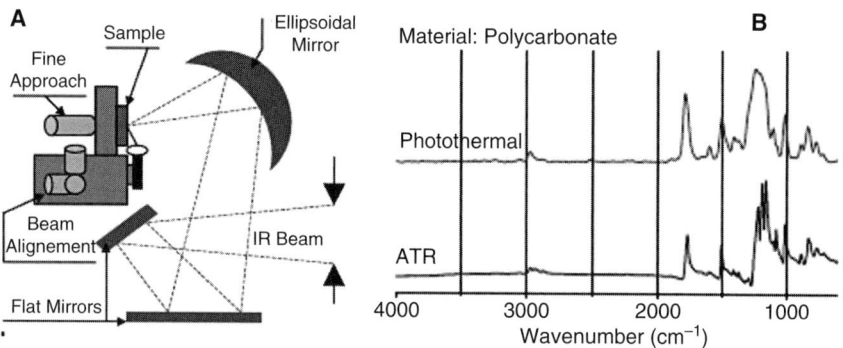

FIGURE 6.9. Configuration for photothermal infrared spectroscopy using a SThM inside an FTIR instrument. (A) Schematic of the setup showing how the IR source is focused on the tip sample contact. (B) Comparison of photothermal IR with conventional far-field FTIR for a polycarbonate surface. (From Bozec et al. [160]; reproduced with permission.)

6.3.5 Near Field Scanning Optical Microscopy

Near-field optical microscopy uses a flexible optical fiber to focus the illumination on the object and collect the scattered light. If the fiber or the specimen is scanned, it is a scanning optical microscope, and if there is some depth selectivity, it is a kind of confocal scanning microscope [164]. The near-field scanning optical microscope is similar to this scanning optical microscope but with the extra requirement that the fiber is extremely small and very close to the surface of the sample [165, 166]. 'Very close' here means at a much smaller distance than the wavelength of light. The advantage gained is that the resolution can be much better than in other optical microscopes, less than one tenth of the wavelength, about 30 nm for visible light. The equipment used to provide this close surface scanning is extensively borrowed from scanning probe microscopy.

As the fine optical fiber can be inserted close to an object in water, the first applications of this instrument have been directed toward high resolution observation of tissue sections and living cells [167]. The associated *photon tunneling microscope* has been applied to polymer surfaces [168]. In principle all the contrast modes of normal optical microscopy can be used in the near-field and there may be other unique contrast mechanisms [169]. The light signal can be analyzed spectroscopically to give detectability limits down to a single molecule, or nano-aggregates in doped polymers, going beyond the realm of the micro-Raman analyzer to the pico- or atto-Raman [170] (see Section 6.5.2).

Successful applications of the near-field scanning optical microscope (NSOM) to polymer materials require that there are components or additives that absorb or fluoresce in the visible range of wavelengths. The polarized NSOM can also be used to study molecular orientation with single-crystal polymers. Applications that have been reviewed include Langmuir-Blodgett films [171], photosensitizers in liquid crystal polymers [172], liquid crystal polymer dynamics [173], orientation in polyethylene single crystals [174], phase separation in polymeric light emitting diodes [175, 176], dendrite materials [177], and nonlinear optical polymers [178]. The field was reviewed by Ito and Aoki in 2005 [179].

6.3.6 Automated SPM

The workflow for AFM is becoming more automated, and this is expected to continue with new developments already underway. Borrowing from the success of AFM technology in the semiconductor industry [180], there will be a move to integrate automated sample preparation, sample loading, image acquisition, data/image reporting, and archiving for polymeric materials R&D and manufacturing.

The most challenging area for automation is sample preparation. Applying AFM to the internal morphology of polymer blends and composites requires microtomy or cryomicrotomy. Due to the wide range of glass transition temperatures and hardness of components, cryomicrotomy requires a highly skilled operator to prepare a good quality surface for AFM imaging. Automating this process will be a challenge. Integrated instruments are appearing; an AFM vendor and microtome vendor have recently allied to develop an AFM integrated with a room-temperature microtome [181]. A section is cut and the surface of the block face is imaged; the process is repeated with little need to align images for tomographic reconstruction to give a 3D image (see Section 6.4.2). Similarly, a manufacturer of accessories for electron microscopy has constructed an *in situ* microtome for a SEM (3View; trademark of Gatan; see Appendix V). For a complete solution it will be advantageous to incorporate improvements in microtomy. These include oscillating knife technology [182, 183] to better control compression (see Section 4.3.4) in the section and careful control over the cutting forces using force feedback [184]. Other approaches that could be readily automated include focused ion beam (FIB) milling [185] (see Section 4.5.5). This technique has produced high quality polymer blend surfaces for AFM imaging but only as an adjunct to microtomy [186] Sample preparation required hours of slow milling, compromising any time advantage of automation. Possible surface heating [187] and damage due to ion implantation must be considered when using FIB on polymer materials.

Once the polished surfaces are prepared, they will need to be mounted in automated stages for analysis. This is facilitated by cryomicrotomy holders that are already designed to fit into AFM stages [188] (Leica Microsystems; see Appendix VI). Automated image acquisition at prescribed locations is possible today with full and independent control over important imaging parameters used in IC-AFM, such as free amplitude, set-point ratio, scan size, scan speed, and signal gains. This allows automated acquisition of AFM data from combinatorial libraries. These have included inorganic/organic block copolymers [189], cured PDMS composites [190], photoembossed polymer lacquers [191, 192], and polymer libraries made by sector spinning [193]. Automated image acquisition can also be applied to arrays of samples, such as impact polypropylene and blown film surfaces [194] and to imaging of high-density optical disk masters [195]. Introduction of tip quality checking will be a necessary component of full automation with periodic review of the tip quality by scanning characterizers. This may even have to include the ability to exchange tips if quality criteria are not met. In polymer imaging it is not always tip wear that leads to degradation of image quality. Usually the tip picks up contaminants or its surface becomes too adhesive to obtain quality images. Characterizers for tip shape already exist, but new approaches to rapidly assess changes in the tip surface chemistry or to reproducibly control it could be needed.

Improvements in automated image processing and analysis will need to be developed to address the specific image analysis challenges of nanoscale polymer characterization. Glitch removal from AFM images has been automated [196]. Some of the image analysis routines for polymer characterization will be leveraged from applications in other fields such as the biological and life sciences. For example, robust methods are needed to statistically sample and measure lamellar thickness and curvature in semicrystalline polymers from phase images in IC-AFM. In the biological area, automated sizing of DNA fragments [197] for length [198, 199] and curvature [200] has been reported. In a similar manner, detailed analyses of carbon nanotubes has been achieved [201]. Other automated analysis routines for volume measurement of particles such as aerosols [202] and viruses [203] have been described as well as techniques to measure critical dimensions of pits and bumps in optical media like DVDs [204, 205]. Automated analysis of hair fiber images has been used to classify hair types by AFM [206]. Sheiko et al. reported on the automated analysis of contour length of large arrays of cylindrical brush macromolecules prepared by Langmuir–Blodgett techniques.

An AFM-based molecular weight distribution was calculated and compared to results from standard gel permeation chromatography techniques [207].

Many groups and software vendors (Image Metrology, Inc.; see Appendix VII) have developed automated routines to analyze force-distance curves resulting from force-volume experiments to map surface elasticity and adhesion of polymer coatings [208] and blocks [65] and rupture forces from force pulling experiments [209]. Automated image analysis is not restricted to direct analysis in the spatial domain and can include reciprocal space analysis in the frequency domain. Analysis routines have been written to convert AFM image data into a form that is strictly analogous to small-angle x-ray scattering (SAXS) data for direct comparison between both techniques [210].

6.4 THREE DIMENSIONAL IMAGING

6.4.1 Introduction

Three dimensional imaging in microscopy is most directly the imaging of the full 3D microstructure of an object, rather than the 2D surface, section or projection that "regular" optical, electron or scanning probe microscopy provides. The image is stored as a 3D array of numbers that correspond to the image signal from each point in x, y, z. Each element of this array is described as a *voxel* (volume element) instead of the pixel (picture element) of a 2D image. Three dimensional imaging and analysis of copolymer structures has been reviewed by Jinnai et al. [211]. If the 2D image is of a surface, from AFM for example, then removing material to expose a new surface, forming a new image, and repeating these steps will give a 3D image. If the 2D image is a projection, from TEM for example, then the specimen can be rotated and a new projection formed. The 3D image is calculated from a set of projections, a process usually called *tomography*. The resolution of the 3D image may be limited to reduce the data acquisition time or to keep the data set small; for example, a $256 \times 256 \times 256$ image has the same number of elements as a 4096×4096 2D image. Limits on the data set are less necessary as computing power becomes less costly.

Three dimensional imaging has been extended to define four dimensional (4D) microscopy, where a time sequence of x, y, z images is treated as a single object in x, y, z, and t. Although this has been mostly used in the optical microscopy of biological specimens [212, 213], the concept is beginning to be used in other microscopies and in materials science [214]. The data storage and analysis requirements increase sharply with the extra dimension. Even a single relatively low resolution 256^4 image with 256 gray levels will take more than 4 GB of data.

As images are now stored as numerical arrays, a 3D image may be defined more generally as one where the array is 3D. For example, if an analytical technique is applied to each point of a 2D image, giving an electron energy loss spectrum or a Raman spectrum at each point, then storing this image gives a 3D array in x, y and energy, or x, y and wavenumber. Another possibility is to take a time-series of 2D images; if the whole data set is treated as one object, this would be another 3D array and another kind of 3D image. This concept is easily extended to higher dimensionality, and it is usually described as *multidimensional microscopy* [215]. Individual techniques are likely to have been given their own names. *Volume imaging* (see Section 3.3.3.4) is an AFM technique where a force-displacement curve is stored at every image point [216]. *Hyperspectral imaging* refers to storing a spectrum at every image point, most commonly an optical spectrum. The term was originally applied to macroscopic imaging, but now there are hyperspectral imaging microscopes [217, 218]. *Tomographic spectral imaging* has been used to describe a 3D spatial image with an x-ray spectrum at each point [219].

In AFM where the signal at each image point is the height of the surface, the image can be displayed as in 2D where intensity corresponds to height, or as a projected representation of the 3D surface (for examples, see Fig. 4.11 and Fig. 5.67). However, in the terms described here, this is still a 2D image. Similarly, one of the primary attractions of the SEM is that the image can appear to show the 3D surface of the sample (for examples, see Fig. 4.40, Fig. 5.43B and Fig. 5.78), although it is strictly 2D. This appearance of 3D information can be turned into a true measure of surface height by taking a *stereo pair*, that is, a pair of images taken with the sample tilted by a small angle between them. Stereo pairs have also been common in the TEM. However obtained, the stereo pairs can be set with the tilt axis vertical and viewed to give an image with depth. Measurement of the parallax between features on the two images gives the height of the feature [220]. The height of a few individual features can be measured manually. Software that matches parts of the two images can give the surface roughness from a stereo pair [221].

The rapidly moving trend to 3D imaging in many fields is driven by the reduced cost of computing power and the integration of computers with microscopes. Display of the resultant images is not advancing as fast. Most screens and print remains 2D. Real-time color 3D imaging systems that do not require special glasses do exist [222], but they are not common. They may use *lenticular imaging* (a sheet of cylindrical lenses that direct vertical strips of an image to the left and right eyes [223]), time slicing [224], or other methods. To view a 3D image with such a system, the software displays images from different viewpoints either in vertical strips across the display or in sequence. These views may be calculated from a 3D image or taken directly from the instrument; one system gives real-time 3D TEM by synchronizing the display with beam tilt [225]. Moving the viewpoint makes the 3D image appear to rotate, and this can give the viewer a clear grasp of the structure. However, it may be difficult to pass this on to others. In print, a sequence of stereo pairs may give some idea, and if color is not needed the left and right images can be colored red and green and superimposed to give an anaglyph, which can be viewed through spectacles with one red and one green lens. The eye and brain can integrate a movie of the rotating object (each frame a single view without any stereo effect) to form an impression of the 3D structure, and this is often used in electronic publication.

6.4.2 Physical Sectioning

Forming a sequential set of slices using a microtome, then taking an image of each slice and combining the information to give 3D structural information is not at all a new technique. It has been used in biological microscopy as *serial sectioning* for the optical microscope for more than a century and in the TEM for at least 50 years. One report describes serial sectioning for the x-ray microscope [226]. Even when a ribbon of sections is mounted as a unit, the individual section images can be difficult to align correctly in x, y and rotation. Compressed or wrinkled sections require further correction and adjustment. Software, either part of an image analysis system or freely available [227], can deal with these problems and create a 3D image.

If the cut surface is imaged instead of the section, problems of alignment and correction are much less, allowing the images to be simply stacked into a 3D image. A modern variable pressure SEM (VPSEM) (see Section 3.2.5) can give high resolution images of polymer surfaces without any coating, so when an ultramicrotome is mounted in the specimen chamber, the SEM can be used to form a stack of images [228]. The commercial system, 3View (Gatan Inc.), is optimized for stained biological samples, giving a backscattered electron signal with strong atomic number contrast. It is claimed that sections 30 nm thick can be obtained.

Alternatively, an AFM may be used to form an image of the cut block surface. It is common to prepare a surface in a microtome and transfer it to an AFM for study while still mounted in the microtome sample holder. In principle it can be transferred back to the microtome, recut, and the operation repeated. Three dimensional image formation is obviously much easier if the AFM is mounted on the microtome and there is no sample transfer. Such a system is the Ntegra Tomo (trademark of Nanotech America/ NT-MDT; see Appendix VII), a Ntegra AFM mounted on a Leica UC6 ultramicrotome [181].

Three Dimensional Imaging

In this system the AFM does not prevent collection of the microtomed sections, so TEM of the cut sections can be combined with AFM 3D imaging of the block. The image signal can come from surface height variations that arise from mechanical property fluctuations in the block, or the AFM mode can be set to be sensitive to the specimen stiffness directly (see Section 3.3.2). This system is thus applicable to polymer samples including blends and filled systems, as shown in Fig. 6.10 [181].

When the surface of the remaining sample is imaged, there is no need for a section, and a microtome is not the only possible method for material removal. Techniques include micromilling with a diamond cutter (suitable for hard plastics, soft metals, or composites [229]), plasma etching [230–232], chemical etching [233], and ion beam etching. This last is the most developed technique; using a focused ion beam microscope to remove material and a FESEM to image the surfaces. Commercial microscopes that combine these in a single vacuum system may be called FIB-SEMS, *dual-beam* or *crossed-beam* instruments. The focused ion beam microscope has been used for about 20 years, starting in the semiconductor device industry [187], and since then extending to all sorts of materials [234] including frozen biological samples [235]. It is analogous in operation to the SEM, but the focused and scanned beam is of gallium ions instead of electrons. High currents of Ga^+ can be obtained; tens of nA that rapidly sputter away the sample. Alternatively, at much lower beam currents, images can be obtained using either secondary electrons or secondary ions, with a resolution of about 5 nm. A common application is the preparation of a thin TEM sample [236] (see Section 4.5.5) This technique permits a sample to be made at a specific location, such as the site of a device failure, and to make samples from composites such as polymer films on hard substrates [237, 238]. Surface damage and contamination by the gallium ions is an issue [239] but the dual-beam instrument has been used to

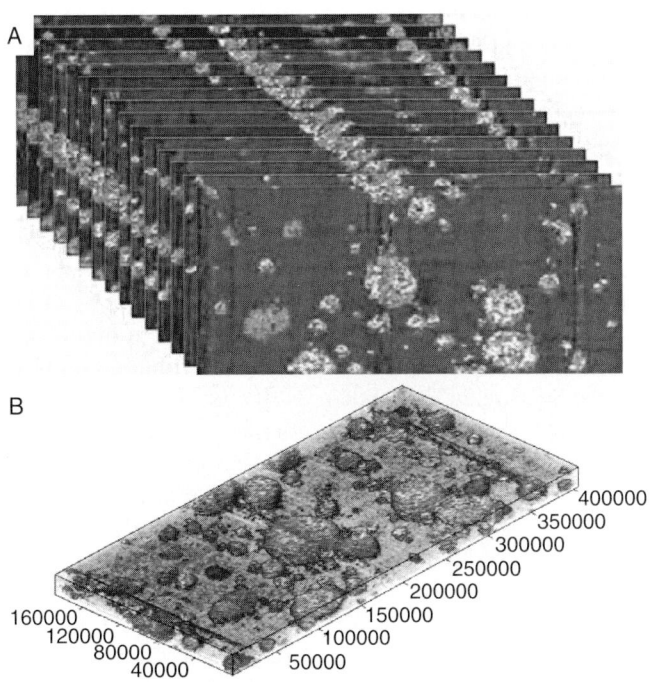

FIGURE 6.10. (A) Fifteen sequential images of polystyrene/high-impact polystyrene (HIPS) blended with silica (hard inclusions). Image size: $40 \times 20\,\mu m$ with 200 nm between sections. (B) Three dimensional reconstruction made from those 15 sections: $40 \times 20 \times 3\,\mu m$. (From Foster [181]; reproduced with permission.)

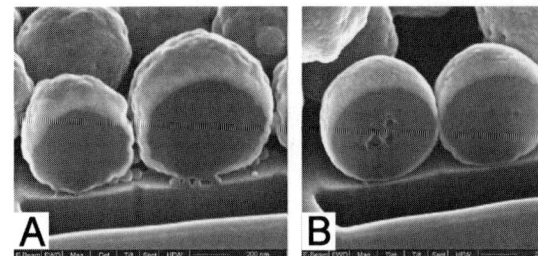 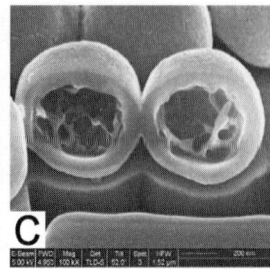

FIGURE 6.11. FIB-SEM images of cross-sections of latex particles, ~500 nm in diameter. (A) Solid core with shell latex particles. (B) Hollow latex particles with a small central void. (C) Set of particles with large central voids and thin shell material. Scanning electron microscopy images at 5 kV; time to cut through particles, 5–15 s. (From Beach et al. [240]; reproduced with permission.)

image the interior of delicate hollow latex particles (Fig. 6.11 [240]), and to form a 3D image of a 100 nm block copolymer structure after staining with OsO_4 [241]. A low voltage (1–2 keV) on the SEM side permits an insulating material to be imaged and keeps a good depth resolution because the penetration is limited.

6.4.3 Optical Sectioning

If the specimen is transparent and the microscope can be set to focus on one plane within the sample and reject or diminish signals from all others, this is called *optical sectioning*. Stacking these sectioned images will give a 3D image in-focus at all points. A very wide range of techniques can be used to form optical sections. Since out-of-focus signals are blurred, any system that acts as a high-pass filter, emphasizing edges and boundaries, will also form optical sections in a thicker sample. One such system in the optical microscope is differential interference contrast (DIC, see Section 2.2.3.3), but DIC images need further processing before 3D reconstruction [242]. An alternative approach is to consider the defocus as a 3D point spread function. In principle this can be removed by 3D deconvolution to give a fully focused 3D image [243]. This is most often applied to fluorescence images in biology, as rejection of out-of-focus signal is easier if the image contains relatively few bright regions on a dark background.

Another common method of optical sectioning is to use an aperture that is projected by the optical system onto the plane of interest—it is confocal with it. This limits illumination to one point on the plane. If another aperture in the imaging system allows only the signal coming from this point to enter the detector, the device is a confocal microscope, for example the CSLM (see Section 6.2.1). A 2D image is formed by scanning the laser spot over the plane. In the case of an optical reflection or fluorescence microscope, the returning light is de-scanned into the collection aperture by the same scanning system making alignment easier. To produce 3D images, the specimen is stepped a small distance along the optic axis by a motorized stage, and a new 2D image is collected at each step [244]. This is a well established technique in biology. There are relatively few reports of applications of 3D imaging to synthetic polymer systems, but they cover a very wide range. Some examples are a study of the deformation of polyurethane foam in compression [245], fibers, composites, and blends [16, 246], and 3D diffusion in hydrogels [247]. As a noninvasive method, it is well suited to extend into the time domain, to observe the dynamics of phase separation [248, 249].

The same confocal principle has been applied to the STEM [250]. This instrument gives improved performance when imaging thick samples, but the relatively small convergence angle of the electron probe means optical sectioning is not very effective. One possible advantage of the new aberration corrected electron microscopes (see Section 6.2.4) is that they allow higher convergence angles at high spatial resolution in the STEM, which should allow 3D imaging at the 1 nm level or better [251].

Three other very different techniques that use light are becoming available. The simplest in concept forces the optical signal to come from a single plane in the optical microscope by using illumination at 90° that has been focused (by a cylindrical lens) into a sheet. This is called *selective plane illumination microscopy* (SPIM) [252]. In the example given [252], the illumination was by laser, the imaging by fluorescence, and the vertical and lateral resolutions were both about 6 μm. So far this technique has only been applied to biological systems. A second method using localized illumination has been called *structured light imaging* [253] or *wide field optical sectioning* [254]. In this case, the sample is illuminated in strips, and the section is constructed from three overlapping strip images [255]. A newer design has the image processing built into the detector, so the direct output is an optical section [256]. Structured light imaging systems are now commercially available and give optical sections similar to CSLM.

Finally, *optical coherence tomography* (OCT) is an interferometric method of selecting slices in an optical image. In a 'regular' interference microscope that uses monochromatic light, the sample and reference beam can go in and out of phase, producing fringes, as the path difference changes. In polychromatic white light, the coherence length is very small, so that sample and reference beam can interfere constructively only if there is zero path difference. The OCT microscope operates in reflection, shining white light into the sample and collecting all that returns as the 'sample beam.' The path length of the reference beam is set so that only the portion returning from a given depth will interfere constructively [257]. In early systems the depth resolution was only 30 μm, but now it is submicron. A wide variety of optical systems can use this principle; a spot can be illuminated and scanned to form a depth slice, or the whole field can be illuminated at once. The depth scanning can be done by moving a mirror, or by transferring the modulation to the time domain and using rapid Fourier transforms to get the real-space information. Applications of this type of device to polymer systems have included polymer composites [258, 259] and tissue engineering scaffolds [260]. A comparison of OCT with x-ray tomography (see Section 6.4.4.2), both set up to have a resolution of 10–15 μm, found that OCT was as effective at a lower cost [261]. However, OCT was not suitable for carbon fiber composites, as too much light was absorbed. It is important to realize that OCT measures the optical path, that is, thickness times refractive index, and this can lead to distortions in the image, as shown in Fig. 6.12 [42].

6.4.4 Tomography

Tomography is technically any nondestructive means of obtaining images of slices of a 3D

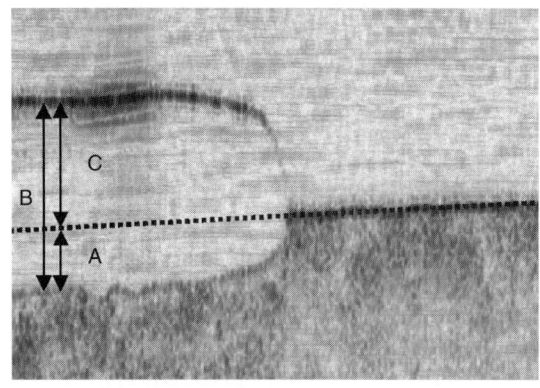

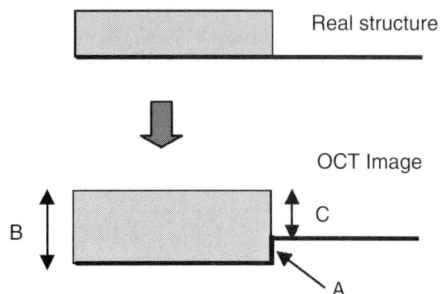

FIGURE 6.12. Optical coherence tomography image of a thin film of polyurethane elastomer partially covering a plane substrate surface (a fiber composite). The optical path difference in the film creates a virtual step in the substrate surface as schematically depicted above. (From Stifter et al. [42]; reproduced with permission.)

object. It should not apply if a full 3D image is obtained and displayed. However, it is usually taken to mean the reconstruction of a structure from a number of projections taken in different directions (and the presentation of the structural data as a set of slices). All such tomography relies on computation, but *computed tomography* has been taken as the name of the most common technique, x-ray tomography, used for medical purposes.

A point source and a line (1D) detector define a plane. A translucent sample placed between them gives a 1D projection, and if the sample is rotated about an axis perpendicular to the defined plane, a set of projections is obtained that can give the 2D structure of that slice of the sample. Of course, in medical devices it is the source and detector that move while the patient remains stationary. If a 2D detector is used, then 2D projections from a cone of rays will be obtained and calculations give the 3D structure of the illuminated volume. If the effective source is distant, the rays are parallel, and in this simpler case the direct result is a set of slices perpendicular to the axis of rotation. The mathematics of reconstruction care nothing for scale or wavelength, so any radiation that penetrates a sample can be used. If the detector size and geometry is suitable, microscopic 3D images can be produced. The following sections describe the more common techniques using electrons and x-rays, but tomography with light is also possible [262] and a commercial device that operates in white light, IR, and fluorescence mode is now available [263]. Reconstructive tomography can also be used to make 3D magnetic resonance imaging (MRI) images with micron scale resolution [264, 265] and to build a 3D model structure from SAXS data [266, 267].

6.4.4.1 Electron Tomography

The transmission electron microscope naturally illuminates samples with near-parallel illumination, it uses a 2D detector, and the samples are translucent. Therefore rotating the sample and collecting the projected images can give the basis for a 3D reconstruction at very high resolution. Difficulties with aligning and focusing the tilted specimen have been largely overcome, but there is normally a limited range of rotation. This will reduce vertical resolution and may also cause some oriented structures not to appear correctly. This problem has been overcome by tilting the sample about two orthogonal axes and combining the two 3D reconstructions [268]. More important for polymer and biological specimens is that radiation damage can make them change while the many images are acquired, and the reconstruction algorithms require that the object is stable. Reducing the electron dose to limit damage gives noisy images and low quality reconstructions. Most work has been done on biological systems [269] where cryo-techniques stabilize the specimen. In polymers, 3D reconstruction has been applied to block copolymer structures [211, 270–272]. The copolymer samples are often stained for these studies, increasing both contrast and stability. The 3D structure of a nanofiber of poly(m-phenylene isophthalamide) (MPDI), a highly stable aromatic polymer (Nomex; trademark of DuPont), has been obtained by TEM tomography [51, 273] and shown to be a twisted hexagonal rod, about 50 nm across (see Fig. 5.22D, E [273, 274]). Other systems studied include polymer blends [275] and nanocomposites [276, 277].

Electron tomography using regular brightfield images does not work well for crystalline materials, for which the image changes drastically as the sample is tilted through diffraction conditions. A signal that does not depend on crystalline orientation is required. Electrons scattered at high angles are suitable; they do not enter the imaging system of the TEM, but in the STEM they give the *high angle annular dark field* (HAADF) image. The HAADF signal depends strongly on the atomic number of the elements in the sample (Z-contrast) and such images can be combined to give a 3D tomographic reconstruction of a wide range of materials [278, 279]. Midgley [280] used STEM tomography to show that 1 nm catalyst particles distributed through a mesoporous silica support are mostly within the internal 3 nm pores. This involved 70 images with a total acquisition time of ~3 h. Another method for getting element-specific 3D structure is to use the energy filtering microscope (EFTEM, see Section 6.5) to obtain images at a core loss specific to one element and use these as the basis for tomographic reconstruction [281]. This technique requires very great stability and has so far been applied to inorganic materials only. However, zero-loss images in the EFTEM have been used to form tomographic images of complex ABC block copolymers in the unstained and stained state, as shown in Fig. 6.13 [282].

6.4.4.2 X-ray Microtomography

X-ray tomography is well developed as a medical technique with resolution ~1 mm, and it can be applied to study composite materials and internal damage in their structure [283]. Once the resolution is improved by arranging the source, geometry and detectors, it has many applications in materials science [284]. Micro-tomography can be done with a synchrotron source of x-rays

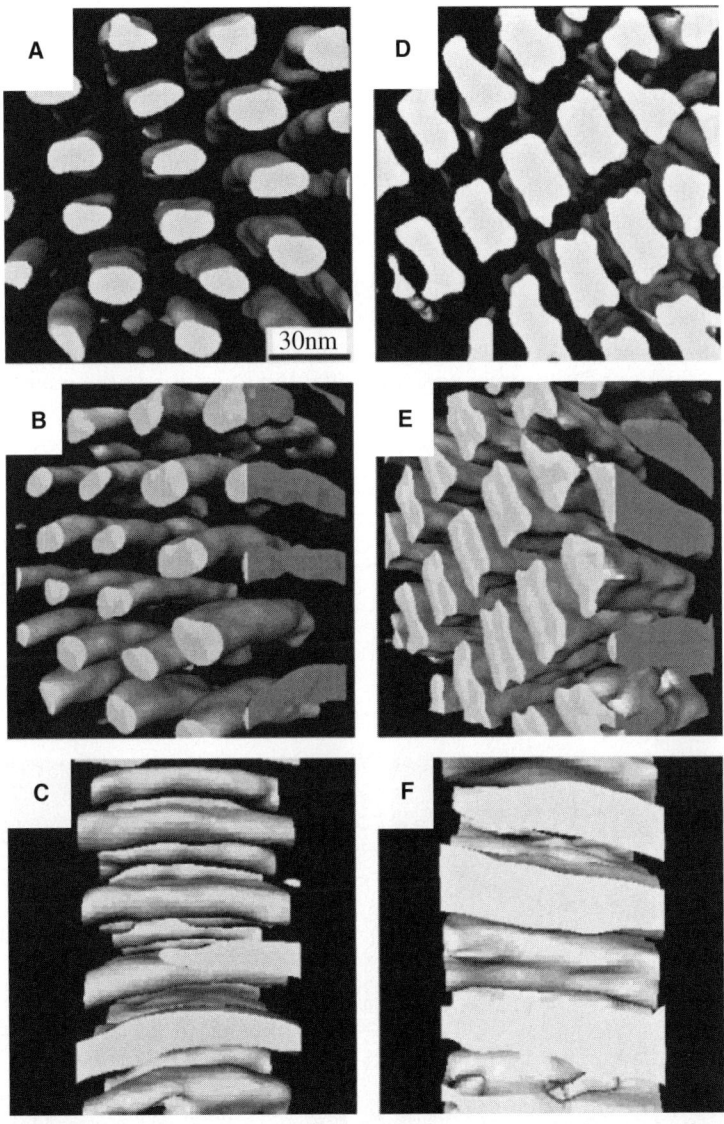

FIGURE 6.13. Three dimensional images of the (PI)(PS)(PDMS) 3-miktoarm star terpolymer obtained by electron tomography: (A–C) unstained specimen; (D–F) OsO_4-stained specimen. (From Yamauchi et al. [282]; reproduced with permission.)

or now with laboratory and even benchtop instruments that are available with resolution down to 0.4 μm. Weiss et al. [285] compared results from a synchrotron source with 1.4 μm resolution to results from a desktop system with 5 μm resolution and to SEM images of a surface. The synchrotron source was the ID22 beamline at ESRF (European Synchrotron Radiation Facility) with a 2048 × 2048 CCD detector; the desktop system was a SkyScan 1072 (Skyscan, Kontich, Belgium) with a 1024 × 1024 CCD detector. The sample was bone growing into porous calcium phosphate. They found that all methods gave good information and, not surprisingly, that the lower resolution instrument gave less clear images on the finest scale. The desktop instrument was easier to use than the synchrotron (assuming that one is available!). Exposure times were ~5 min for the synchrotron and ~40 min for the desktop instrument.

Polymers generally have a low absorption coefficient for x-rays. Thus for a pure polymer sample, contrast is more likely to be a limiting factor than the instrument resolution in normal absorption mode x-ray tomography. Most published x-ray tomography on polymer samples has involved higher contrast systems. These include determination of the distribution of catalyst residue in as-polymerized particles [286], fiber composites [287] and damage in these composites [288, 289], bone regrowth into biomaterial [290], and many porous structures such as foams [291] and biopolymer scaffolds [292]. The laboratory and tabletop instruments have incoherent sources, but the synchrotron sources are coherent. This means that they have a well-defined phase, and changes in phase as well as changes in transmitted amplitude can be used to form the image. Changes in the real part of the refractive index of the materials, that is, changes in density, cause a change in phase that can be a thousand times greater than the change in amplitude due to absorption—a large increase in contrast for polymers.

It is the intensity of the wave that is detected, so phase changes are not directly detectable (as with phase contrast in optical microscopy; see Section 3.1.5). In the simplest x-ray phase contrast mode, the detector is placed some distance behind the sample, and small angle scattering associated with local changes in density causes Fresnel fringes to appear around the interfaces [293, 294]. This is equivalent to increasing the contrast in the TEM by defocus. Increase the distance, and the image has higher contrast but gets further from a simple projection of the sample. The tomographic software can be adapted to interpret the edge-enhanced images and combine them [295], or if three images are taken at different distances from the sample, the exact waveform at the sample can be calculated. These can give a quantitative 3D map of sample density, and the technique is called *holographic tomography* [296]. Alternatively, using an x-ray interferometer with the sample in one beam gives an interference microscopy image, used to determine the structure of a PS-PMMA blend, shown in Fig. 6.14 [297].

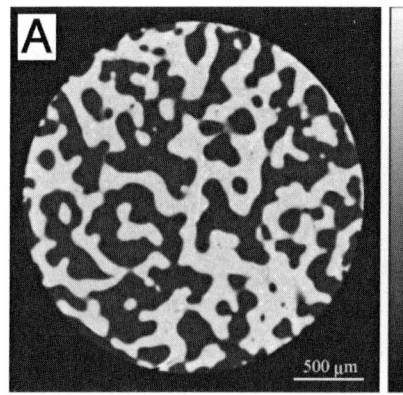 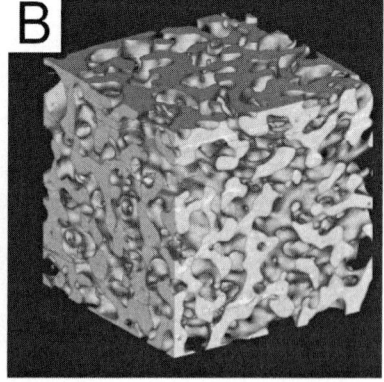

FIGURE 6.14. Reconstructed image of PS/PMMA blend by x-ray phase tomography: (A) phase tomogram and (B) volume-rendering view of reconstructed three dimensional data, where PS region has been made transparent. (From Momose et al. [297]; reproduced with permission.)

The high intensity x-ray beams at synchrotron sources allow real-time tomography and other advances [298].

Another way to increase contrast for organic materials is to use soft x-rays (x-rays of low energy, typically 500 eV). Using soft x-rays in transmission requires very thin specimens, which may be a problem. Tomography with soft x-rays has been performed on test objects [299] and on hydrated biological specimens [300] at 100 nm resolution. Both studies used synchrotron-based scanning transmission x-ray microscopes (STXMs), which have a limited depth of focus, making tomography challenging. Such devices have reached 15 nm resolution [301] but are more often used for microanalysis (see Section 6.5.4).

6.5 ANALYTICAL IMAGING

6.5.1 FTIR Microscopy

The principles of *infrared spectroscopy* are briefly described in Section 7.4.3 and more fully described in many texts. An introduction to the fundamentals as applied to polymers [302] and a reference text designed to help the reader solve polymer problems [303] may be useful to those without background in the field. Modern instruments are FTIR spectrometers that obtain the IR spectrum by Fourier transformation of the signal from an interferometer with a moving mirror. Attaching such a spectrometer to an IR microscope allows chemical analysis and other IR measurements to be made on a small spot, giving FTIR microscopy, which is also called *infrared microspectroscopy* (IMS). The size of the spot is controlled by the diffraction limit of the IR objective lens, which because of the long wavelength of IR may be as much as $20\,\mu m$. The position on the sample can be selected from a visible light image formed by another objective on the microscope or by the same objective—IR objectives are reflection devices, not refraction, so they work at all wavelengths. The microscope may operate in transmission or reflection, but a particularly valuable mode is *attenuated total reflection* (ATR). The ATR objective has a polished face of diamond, germanium or zinc selenide (ZnSe) that is pressed into contact with the sample. Infrared reflection is attenuated by absorption within a surface layer a few micrometers deep. Good contact is required, but that is easy for most polymers, and the advantage is that no sample preparation is required. The contact plate must have higher refractive index than the sample, which is a reason for using the relatively soft germanium (or ZnSe), but diamond is good for polymers and has the advantage that it is not easily damaged by hard inclusions. An image with an IR spectrum at every point can be built up by scanning the sample in x and y, but image acquisition times can be long (hours).

As the FTIR spectrum appears over time at a single point, parallel collection from every point on the image is possible with a 2D detector. The use of a 2D detector may be called *FTIR imaging*, or "chemical imaging," but the latter phrase has also been used to describe many different instruments; reviews of *chemical microscopy* cover dozens of techniques [304, 305]. Two dimensional IR detectors are now available; they are called *focal plane arrays,* and the detecting element is normally made from mercury cadmium telluride (MCT). Commercial devices may have 256×256 pixels, but much of the published work comes from 64×64 arrays. These were made in large quantities as the guidance system sensor for an anti-tank missile and were therefore available at reasonable cost [306]. Read-out time from the arrays is a limiting factor, and the moving mirror in the FTIR interferometer has to be moved in steps so that a data set is collected at each mirror position. Even so, collection of a 3D hyperspectral data set (see Section 6.4) is very much faster than with a single point detector. Figure 6.15 compares data obtained with a focal plane array with that from mapping for a biological sample. The information is basically the same, but the imaging is of higher resolution and took 50× less time [307]. Infrared imaging has been used to study defects in multilayer polymer films [308].

Synchrotron radiation can be used as a source of IR for FTIR imaging; it is hundreds or thousands of times brighter than the standard benchtop source; it and is also pulsed and polarized [309]. The extra brightness allows fast data collection at the diffraction-limited resolution

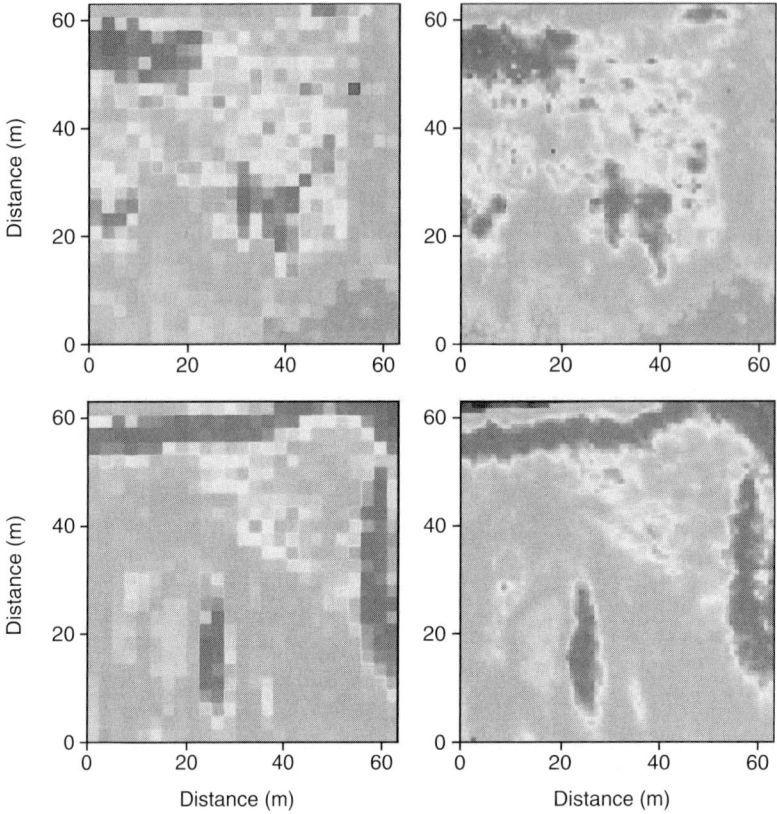

FIGURE 6.15. Comparison of IR mapping and IR imaging of a thin tissue section measured in transmission. Integrated intensities of a band typical of sugar polymers in tissues (at 1155 cm^{-1}) are plotted at the top and those for methylene (at 2850 cm^{-1}) at the bottom. Mapping with 12 μm aperture and 10 μm stepping is shown on the left (32 × 32); IR imaging (64 × 64) on the right. Images are 250 μm square. (See color insert.) (From Schultz [307]; reproduced with permission.)

with high quality spectra [310]. A range of applications in polymer science has included the study of polymorphism in iPP [311, 312].

6.5.2 Raman Microscopy

Raman is another vibrational spectroscopy, complementary to IR in that it is sensitive to the excitation of bonds that are non-polar but polarizable, while IR requires a dipole moment. The specimen is illuminated by a laser and most light is scattered elastically. Some small part is shifted down in frequency (Stokes) by a bond becoming vibrationally excited, and an even smaller part is shifted up (anti-Stokes) by vibrational de-excitation. The whole spectrum allows compounds and polymers to be identified and band details can be used to determine the crystallinity, orientation and stress. The lasers mostly operate in the visible, so a focused spot can be ~0.5 μm, much smaller than in IR; on the other hand, the Raman signal is usually very small.

Most Raman spectrometers are dispersive, not Fourier transform, so the Raman spectrum is spread out in space. A line detector gives the Raman spectrum of a point and a 2D detector can only give the spectra from a line. Thus most *Raman imaging* is obtained by scanning the sample. This can be relatively slow because of the small signal, with 1 s data acquisition time at each point. Even so there are many uses of Raman imaging, in gel structure for example [313, 314]. It may be more effective to use filtering rather than dispersion to pick out the signal of a specific Raman band. Instead of a hyperspectral image, a 2D image is formed with intensity related to the presence of some particular chemical group and this can be collected all at once with a regular CCD camera. For example, the spatial distribution of two or three known components in a blend or multilayer film can be determined from filtered images [315].

As the CSLM (see Section 6.2.1) uses an optical laser as its illumination source, it is relatively straightforward to attach a Raman spectrometer to its output to give a confocal Raman microscope. This further limits the region being analyzed and can produce optical sections and 3D images of chemical groups [316, 317]. Dichroism in a confocal Raman microscope has been used to measure polymer orientation in films [318, 319] and fibers [320].

A number of devices extend the use of Raman spectroscopy. A microscope that combines Raman and IR imaging allows better chemical identification, as the techniques are complementary. FT Raman uses a near-IR laser for illumination and is preferred if the sample fluorescence swamps the Raman signal. Both FT and regular Raman can be combined on one microscope, allowing the best illumination to be selected. Raman microspectroscopy can be combined with AFM; even a relatively low spatial resolution (200 nm) chemical analysis is useful in combination with the high resolution topographic imaging of the AFM [321]. Raman resolution can be increased in two ways using a scanning probe. In one, the probe is replaced with an optical fiber delivering the light, so that the microscope is a combination of NSOM and Raman [322, 323]. In the second, which is still in development, the tip is made from silver or gold. If a thin sample is placed on a silver film, the plasmons in the conductor interact with the vibrations in the thin film, increasing the Raman signal. This effect is called *surface enhanced Raman scattering* (SERS). If the SPM tip is the conductor, the signal is tip-enhanced (TERS) and effectively comes only from the tip region [324, 325].

A technique that is more complex optically and gives a large Raman signal is called *coherent anti-Stokes Raman scattering* (CARS). In a CARS microscope two laser pulses pump the sample bonds into vibrationally excited states. A third laser pulse triggers de-excitation (thus anti-Stokes), and light up-shifted from the probe frequency is detected [326]. In narrowband CARS, one vibrational state is excited, and the image formed is from this one excitation. Very short pulses give broadband CARS, and with clever pre-processing of these pulses, a single laser pulse can give the required three separate laser inputs [327]. The multiple photon interaction gives optical sectioning [328], so the ideal result of the CARS microscope is high speed, high resolution 3D chemical information with optical input, so that the workings of a living cell can be imaged without disrupting it [329]. However, the intensity of the pulses can be high, so that nonlinear effects are very significant and there may be damage.

6.5.3 Electron Energy Loss Microscopy

High energy electrons lose energy as they pass through a material, and this may happen in small (10–30 eV) amounts by interaction with valence electrons in the sample, or in larger amounts by interaction with inner shell electrons. The small losses are much more probable, so that the spectrum of energy loss peaks at this value (and at zero loss for thin specimens), with a long tail extending to higher energy loss. In *electron energy loss spectroscopy* (EELS), the loss spectrum of high energy electrons that have passed through a sample is determined. Electron energy loss spectroscopy is thus an accessory for TEM and STEM and is described in TEM texts [330–332]. There are good detailed specialist texts [333, 334] and a chapter that is an excellent review of the application of EELS to polymers [335].

An electron spectrometer is placed below the standard column in a TEM or STEM for EELS. In TEM, a defining aperture or slit selects the region contributing to the spectrum, while in STEM the probe naturally controls the area being analyzed. A 1D electron detector can collect the EELS data from a point, and a 2D detector can collect data from a line. These techniques are sometimes called PEELS, for parallel

EELS, as early systems had a single point detector, requiring the spectrum to be scanned. In any case, scanning the spot or line over the sample can form a 3D image in x, y and energy, which is usually called a *spectrum image*.

Alternatively, a slit placed after the dispersive element of the spectrometer will select only those electrons that have suffered a particular energy loss. The spectrometer introduces significant distortion into the paths of the electrons, but if this can be corrected, a 2D image can be formed from these selected electrons. Correction is by a set of multipole lenses (in the case of a spectrometer below the column) or by more magnetic sectors and a hexapole making up an omega filter that is built into the microscope column. This is energy filtering [336] giving an energy filtered image (EFI) and the whole system is called *energy filtering TEM*, or EFTEM. In principle, the energy selected can be scanned over the entire range, building the same 3D data set as in the spectrum image. If the polymer sample may damage or drift, limiting the total exposure time, this would not be a good idea. STEM PEELS has no extra apertures; all the electrons contribute to the spectrum and thus to the spectrum image. In EFTEM most electrons are discarded at the energy selecting slit so the efficiency is much less. Nevertheless, EFTEM is generally the preferred technique because images can be taken at energies interesting for specific chemical information, and there is no need to collect most of the possible spectrum [330, 336].

The energy spectrum is usually analyzed by looking at the core-loss regions; these are at the energy losses where a core electron from each element can just be excited. Figure 6.16 shows spectra for carbon and oxygen from two polymers [335]. The peaks are asymmetric, and the background is different on each side, so measuring their intensity requires some processing (see Chapter 39 in [330]). The potential of EFTEM for polymer systems has been reviewed [337, 338], and there are many reports of individual studies. In most, the analysis uses distinct heteroatoms, for example the interface between PMMA and SAN in a blend was followed by looking at the oxygen signal (PMMA) and the nitrogen signal (SAN), as shown in Fig. 6.17 [335].

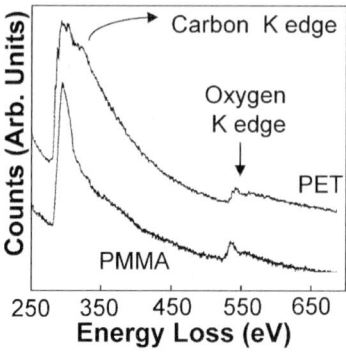

FIGURE 6.16. Core-loss spectra showing carbon and oxygen edges in poly(ethylene terephthalate) and poly(methyl methacrylate). (From Libera and Disko [335]; reproduced with permission.)

Note that the shapes of the two carbon peaks are quite different in Fig. 6.16. This relates to the different chemical bonding in the two polymers; in principle this information can be used in analysis, but it has been shown in PMMA that chemical changes due to electron beam damage rapidly alter the peak structure [339]. For many polymers, the low energy loss region may not contain any distinct features for use in analysis or imaging, but aromatic compounds do have a peak at about 7 eV, which has been used to identify polystyrene in a polystyrene/polybutadiene / poly(methyl methacrylate) sample [340].

Finally, EFTEM is placed in this section because of its use in chemical analysis, but filtering, especially to zero loss, can be an important method of enhancing "regular" TEM. Figure 3.2 showed how filtering at zero loss could improve a diffraction pattern; it will also generally improve contrast in images with a loss of brightness. Analysis might combine zero loss filtering with x-ray mapping to confirm the components [341]. Figure 6.13 was another example of EFTEM imaging; electron tomography using zero-loss images [282].

6.5.4 X-ray Microscopy

X-ray microscopes may magnify by projection from a small point source or use essentially parallel beams in the synchrotron relying on the small pixel size of 2D detectors for high

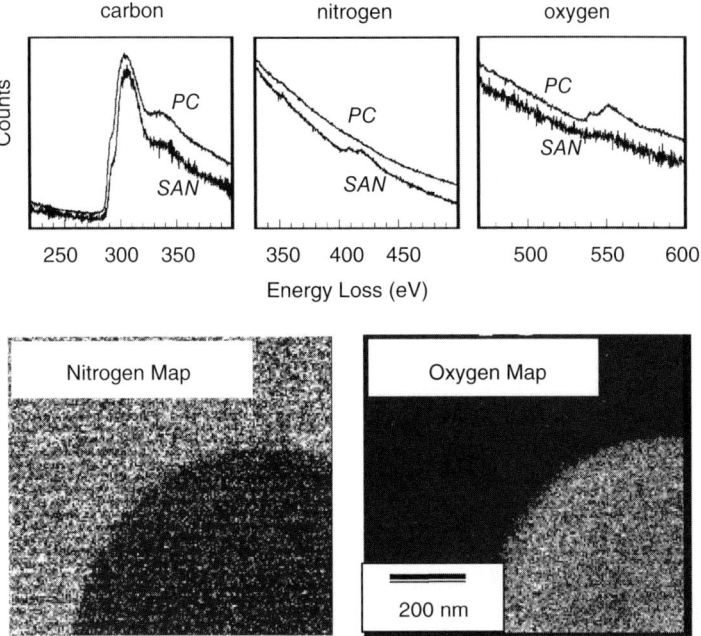

FIGURE 6.17. Energy filtered nitrogen and oxygen maps of a poly(styrene-co-acrylonitrile)-poly(carbonate) (SAN-PC) blend showing part of a PC particle in the SAN matrix. The TEM was a Zeiss 912 with an Omega filter (From Libera and Disko [335]; reproduced with permission.)

resolution. Microscopes that operate in this way are useful for tomography and were treated in Section 6.4.4.2. A third was briefly mentioned, the x-ray microscope using a synchrotron source and x-ray optics to focus the beam to a small spot. A transmission x-ray microscope (TXM) requires zone plates or other optics to take the place of both condenser and objective [342, 343], which is tricky. Most are STXMs where a very small spot is formed on the sample and the sample is scanned in x, y to form an image [344].

These instruments use soft x-rays and require a bright synchrotron source, but they can do chemical analysis at high spatial resolution (20–50 nm). The analysis important to polymers comes from the spectral details near to x-ray absorption edges (*near edge x-ray absorption fine structure*; NEXAFS). Figure 6.18 [345] shows that the same chemical information is present in the EELS spectrum, but the x-ray microscopes have higher energy resolution (note that the range of Fig. 6.18 is 10 eV while it is 400 eV in Fig. 6.16, showing the EELS

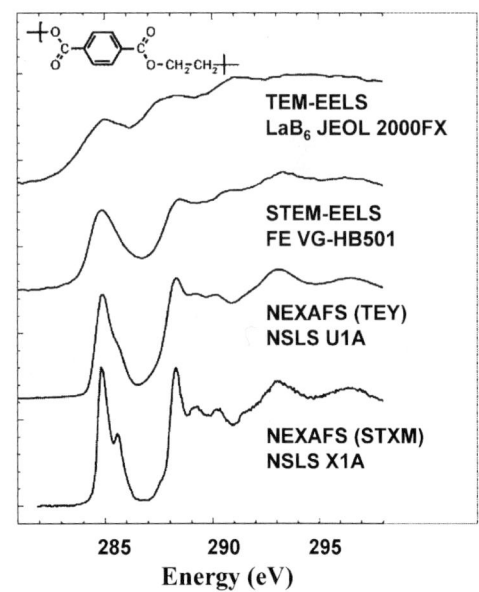

FIGURE 6.18. Comparison of carbon core-loss spectra from PET collected by electron energy loss spectroscopy (top two curves) and x-ray absorption spectroscopy (lower two curves). (From Ade and Urquhart [345]; reproduced with permission.)

spectrum). The synchrotron sources can tune to the any required energy, and images taken at key values can be used to differentiate components in a blend or copolymer. The application of STXM to polymers has been repeatedly reviewed [345–347]; in the last case, it was described as NEXAFS microscopy.

Figure 6.19 is a comparison of STXM and TEM (after staining) for particles with segregated regions of PS and PMMA [348]. The contrast in the STXM in Fig. 6.19C between the PS and PMMA is as high as in the TEM after staining (Fig. 6.19A). Transmission electron microscopy is the natural comparison, because STXM also requires a thin sample to be prepared (the sample in Fig. 6.19 is about 100nm thick [348]). For both polymers and biological materials, the advantages of this technique over the TEM are (i) no staining, (ii) the environment does not have to be high vacuum, and (iii) less radiation damage with x-rays than with electrons. However, at the highest resolution, the concentrated x-ray beam can cause damage, and there are not many such STXM devices in the world.

6.5.5 Imaging Surface Analysis

Any method of chemical analysis with high spatial resolution can form an image with chemical information, and for many purposes, chemical information from the top few micrometers of a sample would be considered surface analysis. For example, x-ray microanalysis in the SEM or microprobe can be used to give an image of the major elemental components (*x-ray mapping*, or if it is quantitative, *compositional mapping* [349]). Similarly, FTIR imaging with ATR (see Section 6.5.1) will give information on chemical groups. In both cases, the surface layer detected is microns thick. This section focuses on techniques that tell about the topmost atomic or molecular layers, the "true" surface techniques. Two techniques involving scanning probe microscopy have already been mentioned: chemical force microscopy (see Section 6.3.1), where an AFM tip is modified to be chemically selective; and TERS (see Section 6.5.2), where a Raman signal comes only from within 20nm of the metal tip.

6.5.5.1 Imaging SIMS

Secondary ion mass spectroscopy (SIMS) is a well-established surface analysis technique where an ion beam causes ions to sputter off of the specimen surface. Two texts are noteworthy: a very complete reference work [350] and a more practical guide [351]. Measuring the charge:mass ratio of these secondary ions identifies them, down to the isotopes involved, and because they come from within 1nm of the surface, this is true surface analysis. There are two standard forms to the instrument, dynamic SIMS and static SIMS; the basic difference is that the dynamic version has a high sputtering rate, so that it erodes away the sample. The signal evolves dynamically, giving a depth profile directly. The static device has a much lower incident beam current and gives a static analysis of the surface. The detectors on the two systems are different because of the different signal levels, with the "time of flight" (TOF) detector so strongly associated with the static instrument that it is often called TOF-SIMS. The damage cascade due to ion impact proceeds into the specimen, and ions ejected from the sample come from the surface and retain the original molecular structure (Fig. 6.20) [352]. The high flux of primary ions in dynamic SIMS means that organic molecules may be damaged by previous ion impacts, so typically this kind of system is used for elemental analysis and static SIMS for molecular analysis of polymers [353, 354].

More recent instrumental designs have brighter sources and better optics, so that the lateral resolution has gone from millimeters to micrometers to 50nm, and high resolution SIMS imaging is possible [355]. These new designs have also blurred the distinction between dynamic and static types. For example, in one design two ion beams may be used, one to image and one to sputter. This allows depth profiling with molecular-level chemical information (TOF dynamic SIMS) and has been used to characterize polymer solar cells [356] and multilayer polymer films [357]. Three dimensional structural information using elemental analysis has been obtained from conducting polymer films (Fe and Ru) [358], from polymer blends containing PVDF (F) and PMMA (O) [359], and from cancer cells (Ca) [360].

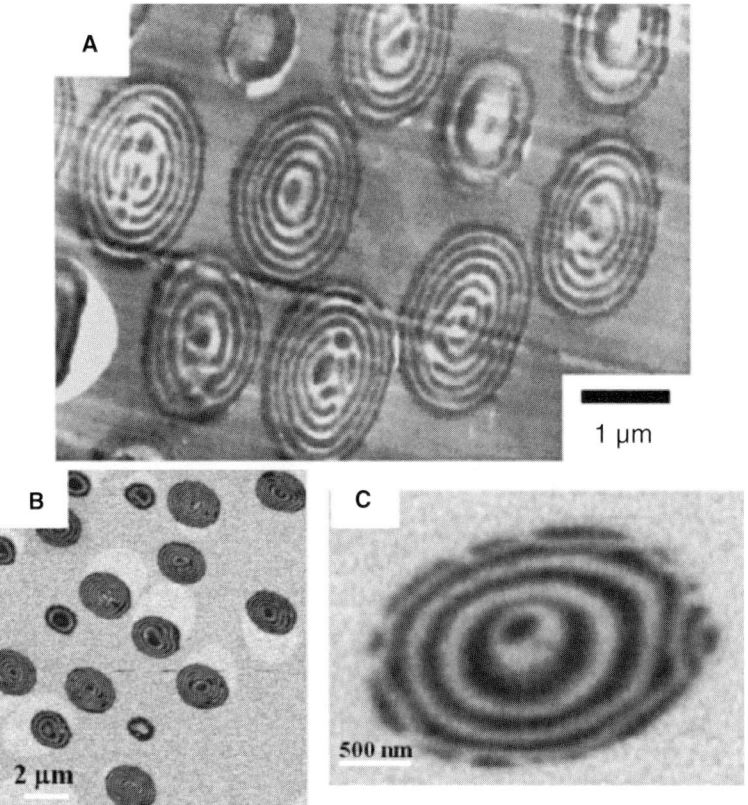

FIGURE 6.19. (A) Transmission electron microscopy image of "onion" PS/PMMA particles stained with RuO_4; (B) STXM image at 285.1 eV (only PS is strongly absorbing) of a group of particles embedded in epoxy resin; (C) STXM image of a cryomicrotomed sample at 285.1 eV. Note that the two STXM images are as-recorded transmission data in which the PS-rich regions appear dark. (From Takekoh et al. [348]; reproduced with permission.)

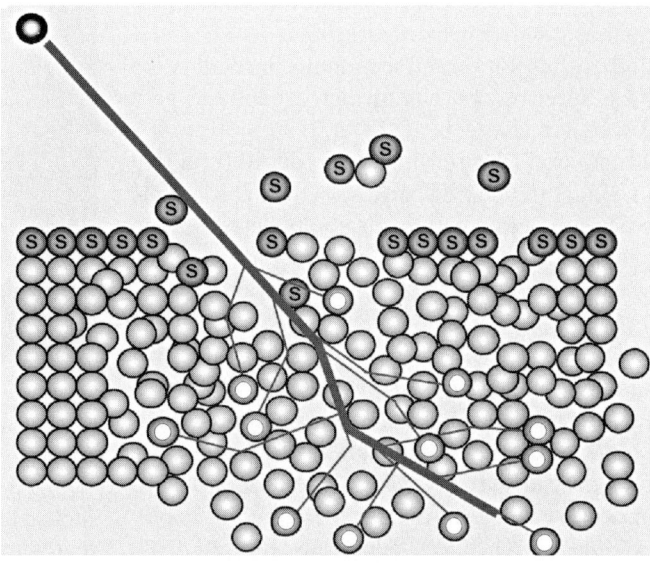

FIGURE 6.20. Schematic of the sputtering process in SIMS. The incident ion buries itself and causes a displacement cascade in the material, while the ions leaving the sample come from the surface and are not much disturbed. (Adapted from Cameca [352]; used with permission.)

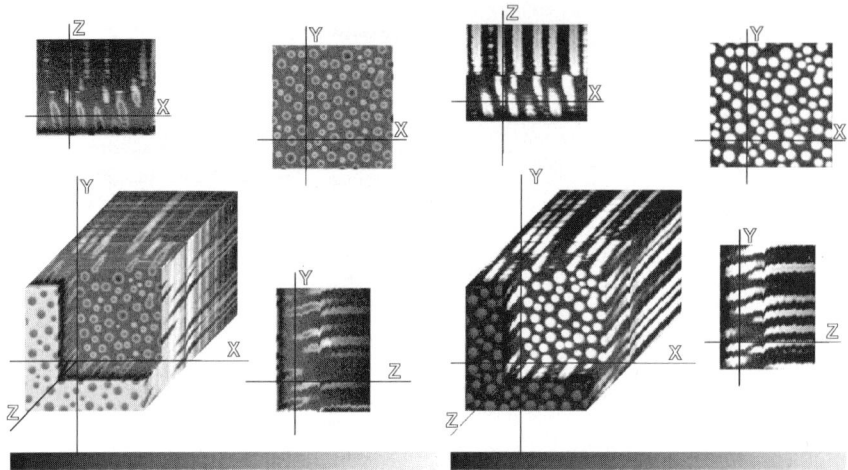

FIGURE 6.21. Sample of PS/PMMA blend. At left, ^{16}O map, and at right, ^{12}C. The X, Y area is $20 \times 20 \mu m$; the source does not state the depth, but it can be assumed that the phase-separated particles are roughly spherical, not the long streaks that they appear to be, so the Z axis is only ~$2 \mu m$. No correction was applied for specimen motion, which explains the shaky appearance of the Y, Z sections (lower right image of each map). (From CAMECA [364]; reproduced with permission.)

Imaging instruments range from the miniSIMS (trademark of Millbrook Instruments, Ltd., Blackburn, UK), a benchtop instrument with an efficient TOF detector that can quickly form maps for process control of surface contamination at ~$50 \mu m$ resolution [361–363], to the NanoSIMS (by Cameca Instruments Inc.; see Appendix VII), a complex instrument with 50 nm resolution. One result (Fig. 6.21) shows that the components of a phase-separated blend can be identified in 3D using a NanoSIMS [364]. Several other techniques could provide the same information as shown in this figure, but it demonstrates both the resolution and sectioning. Secondary ion mass spectroscopy has further capabilities as it can detect elements at 0.1 at.% in a $50 \times 50 \times 10$ nm voxel (this sensitivity is reduced from that of regular static SIMS because of the small volume analyzed), and it can for example locate isotopically labeled components. There are a wide range of experiments unique to this instrument.

6.5.5.2 X-ray Photoelectron Microscopy

In *x-ray photoelectron spectroscopy* (XPS), x-rays, typically Al Kα (1.48 keV) or Mg Kα (1.25 keV), strike the specimen, and by the photoelectric effect, electrons are emitted. This is also known as *electron spectroscopy for chemical analysis* (ESCA). Electrons emitted from the topmost surface layer have the x-ray energy less their binding energy, but those emitted from deeper lose more energy interacting with the sample. An electron spectrometer finds peaks at the binding energy of compounds at

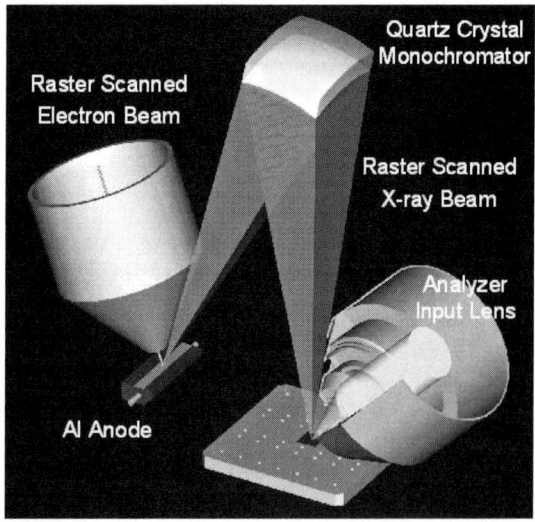

FIGURE 6.22. The source of the scanning x-ray spot in the PHI XPS microprobe is an electron beam scanning over an aluminum anode. Emitted x-rays are monochromated and focused onto the sample surface by a curved quartz crystal. (From PHI [366]; reproduced with permission.)

the surface. Thus the binding energy of the core electrons of oxygen, O1s, is about 532 eV, and details of peak position will distinguish between oxygen in an oxide, a hydroxide, and a phosphate in studies of corrosion. Determining the exact electron energy is easier if the source is grounded, which requires a conducting sample. Modern instruments have charge neutralizing systems that allow insulating specimens to be used and the technique is often applied to polymer surface analysis [353]. Since any contaminating layer would ruin the surface, XPS requires ultrahigh vacuum. Sputtering with Ar has been used to remove material so that depth profiles can be determined. For delicate samples like polymers, larger ions such as C_{60} give less damage to the remaining material [365].

Originally, XPS instruments had no lateral resolution; now images can formed by scanning a small (~10 μm) x-ray probe over the specimen. If this is combined with sputtering, 3D composition maps can be created. Figure 6.22 shows how the small x-ray probe can be created [366]. The primary source is a focused electron

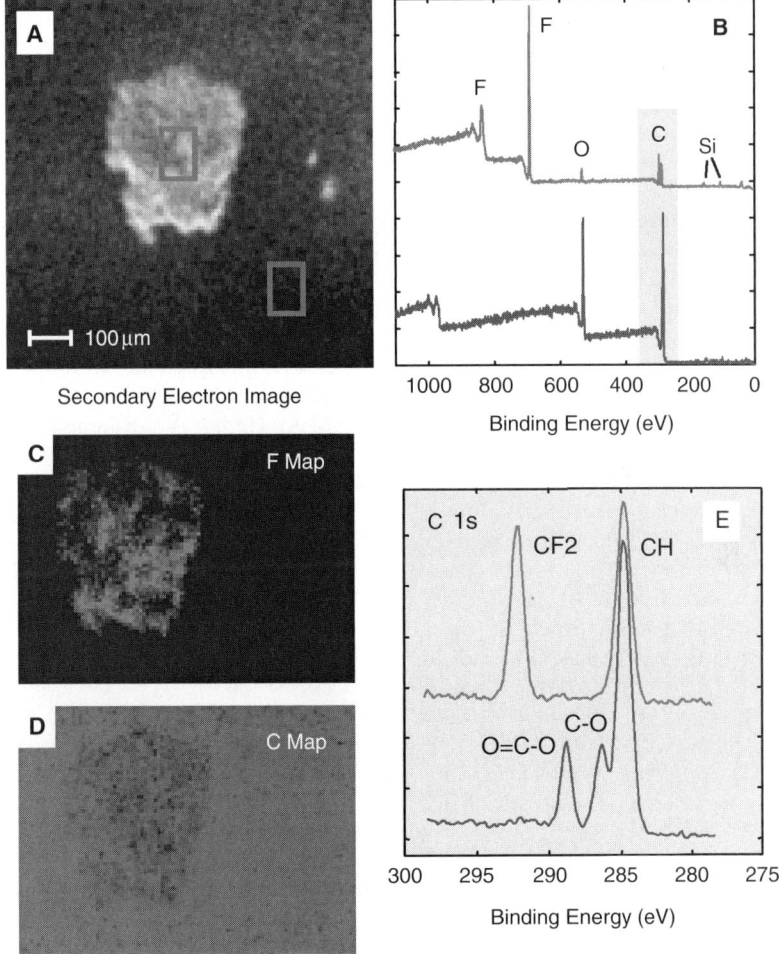

FIGURE 6.23. X-ray photoelectron spectroscopy microprobe images of contamination on a polyester sheet. (A) A secondary electron image; 20 μm x-ray beams on the indicated area gave (B) the survey electron spectrum and (E) the high resolution carbon spectra. These show the presence of fluorine in the contaminant and by the presence of CF_2 that it is a fluorocarbon. The maps of (D) carbon and (C) fluorine confirm this and also show that the other smaller contaminants seen in the secondary electron image are not of the same material. (See color insert.) (From PHI [367]; reproduced with permission.)

beam; a curved crystal acts both to make the beam monochromatic (Al Kα in this case) and to refocus the spot on the specimen. The area scanned is small enough for all parts of it to fall within the focus region of the crystal. An example is shown in Fig. 6.23, where an image of the fluorine content of a contaminant ~250 μm across is created [367]. The result is not only elemental analysis; it also shows which chemical groups are present. Analysis of samples with a thick contaminant layer can be done by other techniques, but XPS is the only technique that can provide this type of analysis on layers a few nanometers thick.

References

1. E. Cuche, P. Marquet and C. Depeursinge, *Appl. Optics* **38** (1999) 6994.
2. E.D. Barone-Nugent, A. Barty and K.A. Nugent, *J. Microsc.* **206** (2002) 194.
3. C.J. Bellair, C.L. Curl, B.E. Allman, P.J. Harris, A. Roberts, L.M.D. Delbridge and K.A. Nugent, *J. Microsc.* **214** (2004) 62.
4. *Microscopy Resource Center*. Available at http://www.olympusmicro.com/primer/index.html. Accessed June 2006.
5. *MicroscopyU: The Source for Microscopy Education*. Available at http://www.microscopyu.com/. Accessed June 2006.
6. D.B. Murphy, *Fundamentals of Light Microscopy and Electronic Imaging* (Wiley-Liss, New York, 2001).
7. D. Semwogerere and E.R. Weeks, in *Encyclopedia of Biomaterials and Biomedical Engineering*, edited by G.E. Wnek and G.L. Bowlin (Marcel Dekker, New York, 2004).
8. A.R. Hibbs, *Confocal Microscopy for Biologists* (Springer, New York, 2004).
9. J.B. Pawley, Ed. *Handbook of Biological Confocal Microscopy* (Plenum Press, New York, 2006).
10. S. Inoue and T. Inoue, in *Cell Biological Applications of Confocal Microscopy* (Methods in Cell Biology, 70), edited by B. Matsumoto (Academic Press, New York, 2002).
11. E. Wang, C.M. Babbey and K.W. Dunn, *J. Microsc.* **218** (2005) 148.
12. D. Hamilton and T. Wilson, *Appl. Phys. B* **27** (1982) 211.
13. H. Jordan, M. Wegner and H. Tiziani, *Measurement Sci. Technol* **9** (1998) 1142.
14. G. Cox and C.J.R. Sheppard, *Micron* **32** (2001) 701.
15. L.-P. Sung, J. Jasmin, X. Cu, T. Nguyen and J.W. Martins, *JCT Res.* **1** (2004) 267.
16. B.V.R. Tata and B. Raj, *Bull. Mater. Sci.* **21** (1998) 263.
17. A. Knoester and G.J. Brakenhoff, *J. Microsc.* **157** (1990) 105.
18. C.C. Chu, *Microscope* **46** (1998) 61.
19. A. Imberg and P. Hansson, *J. Phys. Chem. B* **109** (2005) 10830.
20. J.C. Cabanelas, B. Serrano, M.G. Gonzalez and J. Baselga, *Polymer* **46** (2005) 6633.
21. C. Hadjur, G. Daty, G. Madry and P. Corcuff, *Scanning* **24** (2002) 59.
22. M.D. Burke, J.O. Park, M. Srinivasarao and S.A. Khan, *Macromolecules* **33** (2000) 7500.
23. E.G. McFarland, S. Michielsen and W.W. Carr, *Appl. Spectrosc.* **55** (2001) 481.
24. K.D. Clerck, P.V. Oostveldt, H. Rahier, B.V. Mele, P. Westbroek and P. Kiekens, *Polymer* **45** (2004) 4105.
25. J.F. Aguilar and E.R. Méndez, *J. Modern Optics.* **42** (1995) 1785.
26. L. Wefers and E. Schollmeyer, *J. Polym. Sci. B Polym. Phys.* **31** (1993) 23.
27. P. Lemoine, J.P. Quinn, P.D. Maguire and J.A.D. McLaughlin, *Carbon* **44** (2006) 2617.
28. K. Pope, *Mater. World* **11** (2003) 19.
29. M.A. Geday, W. Kaminsky, J.G. Lewis and A.M. Glazer, *J. Microsc.* **198** (2000) 1.
30. D. Zekria, V.A. Shuvaeva and A.M. Glazer, *J. Phys. Condens. Matter* **17** (2005) 1593.
31. M.A. Geday and A.M. Glazer, *J. Appl. Crystallogr.* **35** (2002) 185.
32. F. Massoumian, R. Juskaitis, M.A.A. Neil and T. Wilson, *J. Microsc.* **209** (2003) 13.
33. R. Oldenbourg and G. Mei, *J. Microsc.* **180** (1995) 140.
34. R. Oldenbourg, in *Live Cell Imaging: A Laboratory Manual*, edited by R.D. Goldman and D.L. Spector (Cold Spring Harbor Laboratory Press, New York, 2004).
35. W.-H. Wang and D.L. Keefe, *Cloning and Stem Cells* **4** (2002) 269.
36. C. Hoyt and R. Oldenbourg, *Am. Lab.* **31** (1999) 34.
37. D. Volkmer, M. Harms, L. Gower and A. Ziegler, *Angew. Chemie Int. Ed.* **44** (2005) 639.
38. S. Tugulu, M. Harms, M. Fricke, D. Volkmer and H.-A. Klok, *Angew. Chemie Int. Ed.* **45** (2006) 7458.
39. M. Shribak and R. Oldenbourg, *Biol. Bull.* **205** (2003) 194.

References

40. M. Shribak and R. Oldenbourg, *Biophotonics Micro- and Nano-Imaging* **5** (2004) 57.
41. D. Stifter, P. Burgholzer, O. Hoeglinger, E. Goetzinger and C.K. Hitzenberger, *Appl. Phys. A* **76** (2003) 947.
42. D. Stifter, A.D. Sanchis Dufau, E. Breuer, K. Wiesauer, P. Burgholzer, O. Hoglinger, E. Gotzinger, M. Pircher and C.K. Hitzenberger, *Insight-Non-Destructive Testing and Condition Monitoring* **47** (2005) 209.
43. K. Wiesauer, A.D.S. Dufau, E. Gotzinger, M. Pircher, C.K. Hitzenberger and D. Stifter, *Acta Mater.* **53** (2005) 2785.
44. M. Haider, H. Rose, S. Uhlemann, B. Kabius and K. Urban, *J. Electron Microsc.* **47** (1998) 395.
45. N. Dellby, L. Krivaneka, D. Nellist, E. Batson and R. Lupini, *J. Electron Microsc.* **50** (2001) 177.
46. P.E.D. Batson, N; Krivanek, O L, *Nature* **418** (2002) 617.
47. P.D. Nellist, M.F. Chisholm, N. Dellby, O.L. Krivanek, M.F. Murfitt, Z.S. Szilagyi, A.R. Lupini, A. Borisevich, J.W.H. Sides and S.J. Pennycook, *Science* **305** (2004) 1741.
48. U. Dahmen, *Microsc. Microanal.* **11** (2005) 2142.
49. D. Van Dyck, S. Van Aert, A. den Dekker and A. van den Bos, *Ultramicroscopy* **98** (2003) 27.
50. A.Y. Borisevich, A.R. Lupini, S. Travaglini and S.J. Pennycook, *J. Electron Microsc.* **55** (2006) 7.
51. D.C. Martin, J. Chen, J. Yang, L.F. Drummy and C. Kübel, *J. Polym. Sci. B Polym. Phys.* **43** (2005) 1749.
52. H. Dömer and O. Bostanjoglo, *Rev. Sci. Instrum.* **74** (2003) 4369.
53. H. Kazumori, K. Honda, M. Matsuya, M. Date and C. Nielsen, *Microsc. Microanal.* **10** (2004) 1370.
54. D.C. Joy, in *Characterization and Metrology for ULSI Technology 2005; AIP Conference Proceedings Vol 788*, edited by D.G. Seiler, et al. (American Institute of Physics, 2005), p 535.
55. N. Yao, in *Handbook of Microscopy for Nanotechnology*, edited by N. Yao and Z.L. Wang (Kluwer Academic, New York, 2005), p. 247.
56. J. Notte, R. Hill, S. McVey, L. Farkas, R. Percival and B. Ward, *Microsc. Microanal.* **12** (2006) 126.
57. B.W. Ward, J.A. Notte and N.P. Economou, *J. Vac. Sci. Technol. B* **24** (2006), 287.
58. *Ultra High Resolution Ion Microscope Life Sciences Image Gallery*. Available at http://www.aliscorporation.com/solutions/life-sciences-gallery.asp?cat=7. Accessed November 2006.
59. R.J. Composto, R.M. Walters and J. Genzer, *Mater. Sci. Eng. R* **38** (2002) 174.
60. A. Noy, D.V. Vezenov and C.M. Lieber, *Annu. Rev. Mater. Sci.* **27** (1997) 381.
61. D.V. Vezenov, A. Noy and P. Ashby, *J. Adhes. Sci. Technol.* **19** (2005) 313.
62. G.J. Vancso. *223rd ACS National Meeting*, Orlando, FL, 2002, p. ANYL193.
63. G.J. Vancso, H. Hillborg and H. Schoenherr, *Adv. Polym. Sci.* **182** (2005) 55.
64. V.V. Tsukruk, V.N. Bliznyuk and J. Wu, *ACS Symp. Ser.* **694** (1998) 321.
65. H. Schonherr, C.L. Feng, N. Tomczak and G.J. Vancso, *Macromol. Symp.* **230** (2005) 149.
66. C.D. Frisbie, L.F. Rozsnyai, A. Noy, M.S. Wrighton and C.M. Lieber, *Science* **265** (1994) 2071.
67. G.F. Meyers, B.M. DeKoven, M.T. Dineen, A. Strandjord, P.J. O'Connor, T. Hu, Y.-H. Chiao, H.M. Pollock and A. Hammiche, *ACS Symp. Ser.* **741** (2000) 190.
68. K. Feldman, T. Tervoort, P. Smith and N.D. Spencer, *Langmuir* **14** (1998) 372.
69. B.D. Beake, J.S.G. Ling and G.J. Leggett, *J. Mater. Chem.* **8** (1998) 2845.
70. H. Schonherr, Z. Hruska and G.J. Vancso, *Macromolecules* **33** (2000) 4532.
71. H. Schonherr and J. Vansco, *J. Polym. Sci. B Polym. Phys.* **36** (1998) 2483.
72. A.S. Duwez and B. Nysten, *Langmuir* **17** (2001) 8287.
73. H. Hillborg, N. Tomczak, A. Olah, H. Schoenherr and G.J. Vancso, *Langmuir* **20** (2004) 785.
74. B.D. Beake and G.J. Leggett, *Polymer* **40** (1999) 5973.
75. B.D. Beake, N.J. Brewer and G.J. Leggett, *Macromol. Symp.* **167** (2001) 101.
76. H. Schonherr and J. Vancso, *Macromolecules* **30** (1997) 6391.
77. T. Nguyen, X. Gu, M. Fasolka, K. Briggman, J. Hwang, A. Karim and J. Martin, *PMSE Preprints* **90** (2004) 141.
78. X. Gu, T. Nguyen, M.J. Fasolka, D. Julthongpiput, L.J. Chen, M.R. VanLandingham, Y.C. Jean and J.W. Martin. *Proc. 27th Annual Meeting of the Adhesion Society*, (Adhesion Society Inc., Blacksburg, VA, 2004), p361.
79. X. Gu, M.R. VanLandingham, M. Fasolka, J.W. Martin, J.Y. Jean and T. Nguyen. *Proc. 26th Annual Meeting of the Adhesion Society*, (Adesion Society Inc., Blacksburg, VA, 2003), p185.
80. D. Julthongpiput, M.J. Fasolka, W. Zhang, T. Nguyen and E.J. Amis, *Nano Lett.* **5** (2005) 1535.
81. T. Nguyen, X. Gu, L. Chen, D. Julthongpiput, M. Fasolka, K. Briggman, J. Hwang and J.

82. X. Gu, T. Nguyen, L.-P. Sung, M.R. VanLandingham, M. Fasolka, J. Martin, Y. Jean, L. Nguyen, N.K. Chang and T.Y. Wu, *J. Coatings Technol. Res.* **1** (2004) 191.
83. J. Tamayo and R. Garcia, *Appl. Phys. Lett.* **71** (1997) 2394.
84. J. Tamayo, *Appl. Phys. Lett.* **75** (1999) 3569.
85. R. Hillenbrand, M. Stark and R. Guckenberger, *Appl. Phys. Lett.* **76** (2000) 3478.
86. R. Stark and W. Heckl, *Surf. Sci.* **457** (2000) 219.
87. M. Stark, R.W. Stark, W.M. Heckl and R. Guckenberger, *Appl. Phys. Lett.* **77** (2000) 3293.
88. M. Stark, R. Stark, W. Heckl and R. Guckenberger, *Proc. Natl. Acad. Sci. U.S.A.* **99** (2002) 8473.
89. R.W. Stark and M. Stark, in *Applied Scanning Probe Methods II Scanning Probe Microscopy Techniques*, edited by B. Bushan and H. Fuchs (Springer, Berlin, 2006), p. 1.
90. F. Jamitzky, M. Stark, W. Bunk, W. Heckl and R. Stark, *Nanotechnology* **17** (2006) S213.
91. O. Sahin and A. Atalar, *Appl. Phys. Lett.* **79** (2001) 4455.
92. A. Ulcinas and V. Snitka, *Ultramicroscopy* **86** (2001) 217.
93. R.W. Stark and W.M. Heckl, *Rev. Sci. Instrum.* **74** (2003) 5111.
94. O. Sahin, C.F. Quate, O. Solgaard and A. Atalar, *Phys. Rev. B* **69** (2004) 165416.
95. O. Sahin, C.F. Quate and O. Solgaard. *NSTI Nanotech*, (NSTI Cambridge, MA, Anaheim, CA, United States, 2005), p663.
96. S. Crittenden, A. Raman and R. Reifenberger, *Phys. Rev. B* **72** (2005) 235422/1.
97. P. Hansma, G. Schitter, G. Fantner and C. Prater, *Science* **314** (2006) 601.
98. R.C. Barrett and C.F. Quate, *J. Vac. Sci. Technol. B* **9** (1991) 302.
99. S.W. Park, H.T. Soh, C.F. Quate and S.I. Park, *Appl. Phys. Lett.* **67** (1995) 2415.
100. J.P. Bourgin, R.V. Sudiwala and S. Palacin, *J. Vac. Sci. Technol. B* **14** (1996) 3381.
101. S.R. Manalis, S.C. Minne and C.F. Quate, *Appl. Phys. Lett.* **68** (1996) 871.
102. S. Watanabe, T. Fujiu and T. Fujii, *ISAF '96, Proceedings of the IEEE International Symposium on Applications of Ferroelectrics, 10th, East Brunswick, NJ, Aug. 18–21, 1996*, **1** (IEEE NY, NY, 1996), 199.
103. T. Sulchek, R. Hsieh, J.D. Adams, G.G. Yaralioglu, S.C. Minne, C.F. Quate, J.P. Cleveland, A. Atalar and D.M. Adderton, *Appl. Phys. Lett.* **76** (2000) 1473.
104. N. Ookubo and S. Yumoto, *Appl. Phys. Lett.* **74** (1999) 2149.
105. T.J. McMaster, J.K. Hobbs, P.J. Barham and M.J. Miles, *Probe Microsc.* **1** (1997) 43.
106. J. Hobbs, T. McMaster, M. Miles and P. Barham, *Polymer* **39** (1998) 2437.
107. D.A. Walters, J.P. Cleveland, N.H. Thomson, P.K. Hansma, M.A. Wendman, G. Gurley and V. Elings, *Rev. Sci. Instrum.* **67** (1996) 3583.
108. M.B. Viani, T.E. Schaffer, A. Chand, M. Rief, H.E. Gaub and P.K. Hansma, *J. Appl. Phys.* **86** (1999) 2258.
109. N. Kodera, M. Sakashita and T. Ando, *Rev. Sci. Instrum.* **77** (2006) 083704/1.
110. T. Uchihashi, N. Kodera, H. Itoh, H. Yamashita and T. Ando, *Jpn J. Appl. Phys. 1* **45** (2006) 1904.
111. T. Ando, N. Kodera, E. Takai, D. Maruyama, K. Saito and A. Toda, *Proc. Natl. Acad. Sci. U.S.A.* **98** (2001) 12468.
112. A. Chand, M.B. Viani, T.E. Schaffer and P.K. Hansma, *J. Microelectromech. Systems* **9** (2000) 112.
113. T.E. Schaffer, M. Viani, D.A. Walters, B. Drake, E.K. Runge, J.P. Cleveland, M.A. Wendman and P.K. Hansma, *Proc. SPIE Int. Soc. Opt. Eng.* **3009** (1997) 48.
114. M.B. Viani, T.E. Schaffer, G.T. Paloczi, L.I. Pietrasanta, B.L. Smith, J.B. Thompson, M. Richter, M. Rief, H.E. Gaub, K.W. Plaxco, A.N. Cleland, H.G. Hansma and P.K. Hansma, *Rev. Sci. Instrum.* **70** (1999) 4300.
115. J.H. Kindt, G.E. Fantner, J.A. Cutroni and P.K. Hansma, *Ultramicroscopy* **100** (2004) 259.
116. T.E. Schaffer and P.K. Hansma, *J. Appl. Phys.* **84** (1998) 4661.
117. P. Vettiger, M. Despont, U. Drechsler, U. Durig, W. Haberle, M.I. Lutwyche, H.E. Rothuizen, R. Stutz, R. Widmer and G.K. Binnig, *IBM J. Res. Dev.* **44** (2000) 323.
118. J. Legleiter and T. Kowalewski, *Abstracts, 35th Central Regional Meeting of the American Chemical Society, Pittsburgh, PA, United States, October 19–22* (2003) 84.
119. T. Kowalewski and J. Legleiter, *J. Appl. Phys.* **99** (2006) 64903.
120. T. Sulchek, G.G. Yaralioglu, C.F. Quate and S.C. Minne, *Rev. Sci. Instrum.* **73** (2002) 2928.

121. J.H. Kindt, J.B. Thompson, M.B. Viani and P.K. Hansma, *Rev. Sci. Instrum.* **73** (2002) 2305.
122. C. Su, L. Huang, K. Kjoller and K. Babcock, *Ultramicroscopy* **97** (2003) 135.
123. G.E. Fantner, P. Hegarty, J.H. Kindt, G. Schitter, G.A.G. Cidade and P.K. Hansma, *Rev. Sci. Instrum.* **76** (2005) 026118/1.
124. G. Schitter, G. Fantner, J. Kindt, P. Thurner and P. Hansma. *2005 IEEE/ASME, International Conference on Advanced Intelligent Mechatronics*, Monterey, CA, July 24–28, 2005.
125. G.E. Fantner, G. Schitter, J.H. Kindt, T. Ivanov, K. Ivanova, R. Patel, N. Holten-Andersen, J. Adams, P.J. Thurner, I.W. Rangelow and P.K. Hansma, *Ultramicroscopy* **106** (2006) 881.
126. J.K. Hobbs, C. Vasilev and A.D.L. Humphris, *Polymer* **46** (2005) 10226.
127. J.K. Hobbs, C. Vasilev and A.D.L. Humphris, *Analyst (Cambridge)* **131** (2006) 251.
128. R. Pylkki, P. Moyer and P. West, *Jpn. J. Appl. Phys.* **33** (1994) 3785.
129. A. Majumdar, *Annu. Rev. Mater. Sci.* **29** (1999) 505.
130. H.M. Pollock and A. Hammiche, *J. Phys. D* **34** (2001) R23.
131. A. Hammiche, M. Song, H.M. Pollock, M. Reading and D.J. Hourston. *Polymer Preprints* **37** (2) (1996) 585.
132. A. Hammiche, H.M. Pollock, M. Song and D.J. Hourston, *Measurement Sci. Technol.* **7** (1996) 142.
133. R. Smallwood, P. Metherall, D. Hose, M. Delves, H. Pollock, A. Hammiche, C. Hodges, V. Mathot and P. Willcocks, *Thermochim. Acta* **385** (2002) 19.
134. *Anasys Instruments: nano-TA™ Probes*. Available at http://www.anasysinstruments.com/probes.php?fi=products. Accessed December 2006.
135. A. Murray and J. Leckenby, *Mater. World* **3** (1995) 227.
136. D.M. Price, M. Reading, A. Hammiche, H.M. Pollock and M.G. Branch, *Thermochim. Acta* **332** (1999) 143.
137. A. Hammiche, M. Song, H.M. Pollock, D.J. Hourston and M. Reading, *Polym. Mater. Sci. Eng.* **75** (1996) 275.
138. A. Hammiche, D.J. Hourston, H.M. Pollock, M. Reading and M. Song, *J. Vac. Sci. Technol. B* **14** (1996) 1486.
139. H.M. Pollock, A. Hammiche, M. Song, D.J. Hourston and M. Reading, *Polymer Preprints* **37** (2) (1996) 69.
140. H.M. Pollock, A. Hammiche, M. Song, D.J. Hourston and M. Reading, *J. Adhes.* **67** (1998) 217.
141. L. Barral, J. Cano, F.J. Diez, J. Lopez, C. Ramirez, M.J. Abad and A. Ares, *J. Polym. Sci. B Polym. Phys.* **40** (2001) 284.
142. A. Hammiche, L. Bozec, M. Conroy, H.M. Pollock, G. Mills, J.M.R. Weaver, D.M. Price, M. Reading, D.J. Hourston and M. Song, *J. Vac. Sci. Technol. B* **18** (2000) 1322.
143. M. Tillman, B. Hayes and J. Seferis, *J. Appl. Poly. Sci.* **80** (2001) 1643.
144. G.F. Meyers, M.T. Dineen, E.O. Shaffer, II, T. Stokich, Jr. and J.-H. Im, *Polymer Preprints* **41** (2) (2000) 1419.
145. G.F. Meyers, B.M. DeKoven, M.T. Dineen, A. Strandjord, P. O'Connor, T. Hu, Y.-H. Chiao, H.M. Pollock and A. Hammiche, in *Microstructure and Microtribology of Polymer Surfaces*, edited by V.V. Tsukruk and K.W. Wahl (American Chemical Society, Washington, DC, 2000), p. 190.
146. F.A. Boroumand, A. Hammiche, G. Hill and D.G. Lidzey, *Adv. Mater. (Weinheim)* **16** (2004) 252.
147. F.A. Boroumand, M. Voigt, D.G. Lidzey, A. Hammiche and G. Hill, *Appl. Phys. Lett.* **84** (2004) 4890.
148. V.V. Gorbunov, N. Fuchigami, J.L. Hazel and V.V. Tsukruk, *Langmuir* **15** (1999) 8340.
149. V.V. Gorbunov, N. Fuchigami and V.V. Tsukruk, *Polymer Preprints* **41** (2) (2000) 1495.
150. V.V. Tsukruk, V.V. Gorbunov and N. Fuchigami, *Thermochim. Acta* **395** (2003) 151.
151. D. Fryer and P.N.de. Pablo, *Macromolecules* **33** (2000) 6439.
152. V. Gorbunov, N. Fuchigami, I. Luzinov and V.V. Tsukruk, *Polymer Preprints* **41** (2) (2000) 1493.
153. H. Fischer, *Macromolecules* **38** (2005) 844.
154. B. Chui, T. Stowe, Y. Ju, K. Goodson, T. Kenny, H. Mamin, B. Terris, R. Ried and D. Rugar, *J. Microelectromech. Systems* **7** (1998) 69.
155. W. King, T. Kenny, K. Goodson, G. Cross, M. Despont, U. Dürig, H. Rothuizen, G. Binnig and P. Vettiger, *J. Microelectromech. Systems* **11** (2002) 765.
156. W. King, *J. Micromech. Microeng.* **15** (2005) 2441.
157. A. Hammiche and H. Pollock, unpublished data.
158. A. Hammiche, H.M. Pollock, M. Reading, M. Claybourn, P.H. Turner and K. Jewkes, *Appl. Spectrosc.* **53** (1999) 810.

159. L. Bozec, A. Hammiche, H.M. Pollock, M. Conroy, J.M. Chalmers, N.J. Everall and L. Turin, *J. Appl. Phys.* **90** (2001) 5159.
160. L. Bozec, A. Hammiche, H. Pollock and M. Conroy, *Anal. Sci.* **17** (2001) s494.
161. L. Bozec, A. Hammiche, M.J. Tobin, J.M. Chalmers, N.J. Everall and H.M. Pollock, *Measurement Sci. Technol.* **13** (2002) 1217.
162. A. Hammiche, L. Bozec, H.M. Pollock, M. German and M. Reading, *J. Microsc.* **213** (2004) 129.
163. A. Hammiche, M. German, R. Hewitt, H. Pollock and F. Martiny, *Biophys. J.* **88** (2005) 3699.
164. K.P. Ghiggino, M.R. Harris and P.G. Spizzitri, *Rev. Sci. Instrum.* **63** (1992) 2999.
165. M. Isaacson, A. Cline and H. Barshatzky, *J. Vac. Sci. Technol. B* **9** (1991) 103.
166. D.W. Pohl, in *Scanning Tunneling Microscopy II*, edited by M.S. Guntherodt and R. Wiesendanger (Springer-Verlag, Berlin, 1992).
167. E. Betzig and J.K. Trautman, *Science* **257** (1992) 189.
168. J.M. Guerra, M. Srinivasarao and R. Stein, *Science* **262** (1993) 1395.
169. J.K. Trautman, E. Betzig, F.S. Weinet, D.J. DiGiovanni, T.D. Harris, F. Hellman and E.M. Gyotgy, *J. Appl. Phys.* **71** (1992) 4659.
170. D. Birnbaum, S.-K. Kook and P.C.H.R. Kopelman, *Phys. Chem.* **97** (1993) 91.
171. J.A. DeAro, K.D. Weston and S.K. Buratto, *Phase Transitions* **68** (1999) 27.
172. P. Camorani, M. Labardi and M. Allegrini, *Mol. Cryst. Liquid Cryst. Sci. Technol. A* **372** (2002) 365.
173. J.E. Hall and D.A. Higgins, *ACS Symp. Ser.* **897** (2005) 25.
174. M. Fujihira, S. Itoh, A. Takahara, O. Karthaus, S. Okazaki and K. Kajikawa, *Springer Series in Optical Sciences* **84** (2002) 151.
175. D. Richards and F. Cacialli, *Philos. Trans. R. Soc. A* **362** (2004) 771.
176. D. Richards, *Philos. Trans. R. Soc. A* **361** (2003) 2843.
177. R.D. Schaller, L.F. Lee, T.-Q. Nguyen, P.T. Snee and R.J. Saykally, *Jpn. J. Appl. Phys. 1* **42** (2003) 4799.
178. H. Liang, W.L. Cao, I. Smolyaninov, W.N. Herman and C.H. Lee, *Polym. News* **29** (2004) 6.
179. S. Ito and H. Aoki, *Adv. Polym. Sci.* **182** (2005) 131.
180. S. Knight, R. Dixson, R.L. Jones, E.K. Lin, N.G. Orji, R. Silver, J. Villarrubia, A. Vladar and W.-L. Wu, *Comptes Rendus Physique* **7** (2006) 931–941.
181. B. Foster, *Am. Lab.* **37** (2005) 42.
182. D. Studer and H. Gnaegi, *J. Microsc.* **197** (2000) 94.
183. J.D. Harris and J.S. Vastenhout, *Microsc. Today* **14** (2006) 20.
184. M. Ericson and H. Lindberg, *Polymer* **38** (1997) 4485.
185. R. Langford, *Microsc. Res. Tech.* **69** (2006) 538.
186. N. Virgilio, B.D. Favis, M.-F. Pepin, P. Desjardins and G. L'Esperance, *Macromolecules* **38** (2005) 2368.
187. J. Melngailis, *J. Vac. Sci. Technol. B* **5** (1987) 469.
188. Y. Thomann, R. Thomann, G. Bar, M. Ganter, B. Machutta and R. Mulhaupt, *J. Microsc.* **195** (1999) 161.
189. D. Wouters, B.G.G. Lohmeijer, A.A. Alexeev, S. Saunin and U.S. Schubert, *PMSE Preprints* **90** (2004) 803.
190. R. Neffati, A. Alexeev, S. Saunin, J. Brokken-Zijp, D. Wouters, S. Schmatloch, U. Schubert and J. Loos, *Macromol. Rapid Commun.* **24** (2003) 113.
191. B.-J. De Gans, C. Sanchez, D. Kozodaev, D. Wouters, A. Alexeev, M.J. Escuti, C.W.M. Bastiaansen, D.J. Broer and U.S. Schubert, *J. Combinatorial Chem.* **8** (2006) 228.
192. N. Adams, B.-J.D. Gans, D. Kozodaev, C. Sanchez, C.W.M. Bastiaansen, D.J. Broer and U.S. Schubert, *J. Combinatorial Chem.* **8** (2006) 184.
193. B.-J. De Gans, S. Wijnans, D. Woutes and U.S. Schubert, *J. Combinatorial Chem.* **7** (2005) 952.
194. J. Li, G.F. Meyers, S. Werner and C. Reinhardt, *Abstracts, 38th Central Regional Meeting of the American Chemical Society, Frankenmuth, MI, May 16–20* (2006) CRM-226.
195. M. Kuwahara, P. Fons, J. Tominaga, K. Honma, A. Egawa, T. Miyatani, K. Nakajima, H. Abe and H. Tokumoto, *Rev. Sci. Instrum.* **76** (2005) 083706/1.
196. C. Hyon, S. Oh, H. Kim, S. Sull, S. Hwang, D. Ahn, Y. Park and E. Kim, *Rev. Sci. Instrum.* **73** (2002) 3245.
197. T.S. Spisz, Y. Fang, I.N. Bankman, R.H. Reeves and J.H. Hoh, *Johns Hopkins APL Technical Digest* **20** (1999) 135.
198. E. Ficarra, L. Benini, E. Macii and G. Zuccheri, *IEEE Trans. Information Technol. Biomed.* **9** (2005) 508.

References

199. T.S. Spisz, Y. Fang, R.H. Reeves, C.K. Seymour, I.N. Bankman and J.H. Hoh, *Med. Biol. Eng. Comput.* **36** (1998) 667.
200. E. Ficarra, D. Masotti, E. Macii, L. Benini, G. Zuccheri and B. Samori, *IEEE Trans. Biomed. Eng.* **52** (2005) 2074.
201. K.J. Ziegler, U. Rauwald, Z. Gu, F. Liang, W.E. Billups, R.H. Hauge and R.E. Smalley, *J. Nanosci. Nanotechnol.* **7** (2007) 2917.
202. G. Kollensperger, G. Friedbacher, M. Grasserbauer and L. Dorffner, *Fresenius J. Anal. Chem.* **358** (1997) 268.
203. S. Venkataraman, D.P. Allison, H. Qi, J.L. Morrell-Falvey, N.L. Kallewaard, J.E. Crowe and M.J. Doktycz, *Ultramicroscopy* **106** (2006) 829.
204. D.A. Chernoff, C.S. Cook and D.L. Burkhead, *Polymer Preprints* **41** (2) (2000) 1464.
205. D.A. Chernoff and D.L. Burkhead, *J. Vac. Sci. Technol. A* **17** (1999) 1457.
206. S.P. Gurden, V.F. Monteiro, E. Longo and M.M.C. Ferreira, *J. Microsc.* **215** (2004) 13.
207. S. Sheiko, M.de Silva, D. Shirvaniants, I. LaRue, S. Prokhorova, M. Moeller, K. Beers and K. Matyjaszewski, *JACS* **125** (2003) 6725.
208. R. Raiteri, H.-J. Butt, D. Beyer and S. Jonas, *Phys. Chem. Chem. Phys.* **1** (1999) 4881.
209. C. Gergely, B. Senger, J.C. Voegel, J.K.H. Horber, P. Schaaf and J. Hemmerle, *Ultramicroscopy* **87** (2001) 67.
210. D.A. Ivanov, Z. Amalou and S.N. Magonov, *Macromolecules* **34** (2001) 8944.
211. H. Jinnai, Y. Nishikawa, T. Ikehara and T. Nishi, *Adv. Polym. Sci.* **170** (2004) 115.
212. C. Thomas, P. DeVries, J. Hardin and J. White, *Science* **273** (1996) 603.
213. D. Gerlich, J. Beaudouin, M. Gebhard, J. Ellenberg and R. Eils, *Nat. Cell Biol.* **3** (2001) 852.
214. S. Schmidt, S.F. Nielsen, C. Gundlach, L. Margulies, X. Huang and D.J. Jensen, *Science* **305** (2004) 229.
215. P.C. Cheng, P.P. Hwang, J.L. Wu, G. Wang and H. Kim, Eds. *Focus on Multidimensional Microscopy* (World Scientific, Singapore, 1999).
216. D. Chernoff and S. Magonov, in *Comprehensive Desk Reference of Polymer Characterization and Analysis*, edited by R. Brady (Oxford University Press, New York, 2003).
217. M.E. Gehm and D.J. Brady. *Proc. SPIE Int. Soc. Opt. Eng.* **6090** (2006) 13.
218. X. Gong-hai, S. Rong and X. Yong-qi, *Optics and Precision Engineering* **12** (2004) 367.
219. P.G. Kotula, M.R. Keenan and J.R. Michael, *Microsc. Microanal.* **12** (2006) 36.
220. B. Minnich, H. Leeb, E.W.N. Bernroider and A. Lametschwandtner, *J. Microsc.* **195** (1999) 23.
221. *MeX; Alicona Imaging GmbH*. Available at www.alicona.com.
222. B. Javidi and F. Okano, Eds. *Three-Dimensional Television, Video, and Display Technologies* (Springer, New York, 2002).
223. T. Okosi, *Three-Dimensional Imaging Techniques* (Academic, New York, 1976).
224. N.A. Dodgson, *Appl. Optics* **35** (1996) 1705.
225. H. Tanaka, T. Kojima, H. Tsuruta, J. Chen, T. Tanji and M. Ichihashi, *J. Mater. Sci.* **41** (2006) 2659.
226. A.P. Hitchcock, T. Araki, H. Ikeura-Sekiguchi, N. Iwata and K. Tani. *J. Phys. IV, Proc* (France) **104** (2003) 509.
227. J.C. Fiala, *J. Microsc.* **218** (2005) 52.
228. W. Denk and H. Horstmann, *PLoS Biol.* **2** (2004) e329.
229. J. Alkemper and P.W. Voorhees, *J. Microsc.* **201** (2001) 388.
230. R. Magerle, *Phys. Rev. Lett.* **85** (2000) 2749.
231. R. Magerle, M. Konrad, A. Knoll, N. Rehse and G. Krausch, *Polym. Mater. Sci. Eng.* **84** (2001) 461.
232. O. Rabinovych, R. Pedrak, I.W. Rangelow, H. Ruehling and M. Maniak, *Microelectronic Engineering* **73–74** (2004) 843.
233. N. Rehse, S. Marr, S. Scherdel and R. Magerle, *Adv. Mater.* **17** (2005) 2203.
234. M.W. Phaneuf, *Micron* **30** (1999) 277–288.
235. M. Marko, C.-E. Hsieh, W.J. MoberlyChan, C.A. Mannella and J. Frank, *Microsc. Microanal.* **11** (2005) 802.
236. L.A. Giannuzzi and F.A. Stevie, *Micron* **30** (1999) 197.
237. H. White, Y. Pu, M. Rafailovich, J. Sokolov, A.H. King, L.A. Giannuzzi, C. Urbanik-Shannon, B.W. Kempshall, A. Eisenberg, S.A. Schwarz and Y.M. Strzhemechny, *Polymer* **42** (2001) 1613.
238. V.G.M. Sivel, J. van Den Brand, W.R. Wang, H. Mohdadi, F.D. Tichelaar, P.F.A. Alkemade and H.W. Zandbergen, *J. Microsc.* **214** (2004) 237.
239. T. Sakata, T. Ogiwara, H. Takahashi and T. Sekine. *8th Asian Test Symposium (ATS'99)*, (IEEE Los Alamitus, CA, Shanghai, China, 1999), p. 389.
240. E. Beach, M. Keefe, W. Heeschen and D. Rothe, *Polymer* **46** (2005) 11195.

241. M. Kato, T. Ito, Y. Aoyama, K. Sawa, T. Kaneko, N. Kawase and H. Jinnai. *55th SPSJ Annual Meeting, May 24–26, 2006*, (Soc. Polymer Science Japan, Tokyo, Nagoya, Japan, 2006), p. 903.
242. M.R. Arnison, C.J. Cogswell, N.I. Smith, P.W. Fekete and K.G. Larkin, *J. Microsc.* **199** (2000) 79.
243. J.G. McNally, T. Karpova, J. Cooper and J.A. Conchello, *Methods* **19** (1999) 373.
244. J.K. Stevens, L.R. Mills and J.E. Trogadis, Eds. *Three-Dimensional Confocal Microscopy: Volume Investigation of Biological Specimens* (Academic Press, New York, 1994).
245. R. Hamza, X.-d. Zhang, C.W. Macosko, R. Stevens and M. Listemann. *Proc. ACS Aug 25–29*, Orlando, FL, 1996, p. 803.
246. A.E. Ribbe, *Trends Polym. Sci.* **5** (1997) 333.
247. A.W. Watkins and K.S. Anseth, *Macromolecules* **38** (2005) 1326.
248. F. Fergg, F.J. Keil and H. Quader, *Colloid Polym. Sci.* **279** (2001) 61.
249. A.O. Brightman, B.P. Rajwa, J.E. Sturgis, M.E. McCallister, J.P. Robinson and S.L. Voytik-Harbin, *Biopolymers* **54** (2000) 222.
250. N.J. Zaluzec, *Microsc. Today* **6** (2003) 8.
251. J.J. Einspahr and P.M. Voyles, *Ultramicroscopy* **106** (2006) 1041.
252. J. Huisken, J. Swoger, F.D. Bene, J. Wittbrodt and E.H.K. Stelzer, *Science* **305** (2004) 1007.
253. M.A.A. Neil, R. Juskaitis and T. Wilson, *Optics Lett.* **22** (1997) 1905.
254. C.-H. Lee, H.-Y. Mong and W.-C. Lin, *Optics Lett.* **27** (2002) 1773.
255. S. Thomas, R. Zinter, Z. Cseresnyés, B. Hutter-Paier and A. Hofmeister, *Am. Biotechnol. Lab.* **21** (2003) 14.
256. J. Mitic, T. Anhut, M. Meier, M. Ducros, A. Serov and T. Lasser, *Optics Lett.* **28** (2003) 698.
257. D. Huang, E.A. Swanson, C.P. Lin, J.S. Schuman, W.G. Stinson, W. Chang and M.R. Hee, *Science* **254** (1991) 1178.
258. J.P. Dunkers, R.S. Parnas, C.G. Zimba, R.C. Peterson, K.M. Flynn, J.G. Fujimoto and B.E. Bouma, *Compos. Part A Appl. Sci. Manuf.* **30** (1999) 139.
259. J.P. Dunkers, D.P. Sanders, D.L. Hunston, M.J. Everett and W.H. Green, *J. Adhes.* **78** (2002) 129.
260. Y. Yang, R.K. Wang, E. Guyot, A.E. Haj and A. Dubois. *Proc. SPIE Int. Soc. Opt. Eng.* **5690** (2005) 18.
261. J. Kastner, E. Schlotthauer, D. Stifter and P. Burgholzer, *Comparison of Optical Coherence Tomography and X-ray Computed Tomography for Characterization of Glass-Fibre Polmer Matrix Composites*. Available at http://www.ndt.net/abstract/wcndt2004/71.htm. Accessed November 2006.
262. J. Sharpe, U. Ahlgren, P. Perry, B. Hill, A. Ross, J. Hecksher-Sorensen, R. Baldock and D. Davidson, *Science* **296** (2002) 541.
263. *OPT scanners now available*. Available at http://www.bioptonics.com/. Accessed November 2006.
264. R.E. Botto, G.D. Cody, S.L. Dieckman, D.C. French, N. Gopalsami and P. Rizo, *Solid State Nucl. Magn. Reson.* **6** (1996) 389.
265. L. Ciobanu and C.H. Pennington, *Solid State Nucl. Magn. Reson.* **25** (2004) 138.
266. C.G. Schroer, M. Kuhlmann, S.V. Roth, R. Gehrke, N. Stribeck, A. Almendarez-Camarillo and B. Lengeler, *Appl. Phys. Lett.* **88** (2006) 164102.
267. N. Stribeck, A.A. Camarilla, U. Nochel, C. Schroer, M. Kuhlmann, S.V. Roth, R. Gehrke and R.K. Bayer, *Macromol. Chem. Phys.* **207** (2006) 1139.
268. H. Sugimori, T. Nishi and H. Jinnai, *Macromolecules* **38** (2005) 10226.
269. J. Frank, Ed. *Electron Tomography: Methods for Three-dimensional Visualization of Structures in the Cell* (Springer, New York, 2006).
270. R.J. Spontak, M.C. Williams and D.A. Agard, *Polymer* **29** (1988) 387.
271. R.J. Spontak, J.C. Fung, M.B. Braunfeld, J.W. Sedat, D.A. Agard, L. Kane, S.D. Smith, M.M. Satkowski, A. Ashraf, D.A. Hajduk and S.M. Gruner, *Macromolecules* **29** (1996) 4494.
272. A. Takano, S. Wada, S. Sato, T. Araki, K. Hirahara, T. Kazama, S. Kawahara, Y. Isono, A. Ohno, N. Tanaka and Y. Matsushita, *Macromolecules* **37** (2004) 9941.
273. C. Kübel, D.P. Lawrence and D.C. Martin, *Macromolecules* **34** (2001) 9053.
274. C. Kübel, unpublished micrographs.
275. P. Sengupta and J.W.M. Noordermeer, *Macromol. Rapid Commun.* **26** (2005) 542.
276. N. Kawase, M. Kato, H. Nishioka, T. Kaneko, Y. Nishikawa and H. Jinnai. *55th SPSJ Annual Meeting, May 24–26, 2006*, (Soc. Polymer Science Japan, Tokyo, Nagoya, Japan, 2006), p. 1102.
277. S. Kohjiya, A. Katoh, J. Shimanuki, T. Hasegawa and Y. Ikeda, *Polymer* **46** (2005) 4440.

278. M. Weyland and P.A. Midgley, *Mater. Today* **7** (2004) 32.
279. P.A. Midgley, in *Handbook of Microscopy for Nanotechnology*, edited by N. Yao and Z.L. Wang (Kluwer Academic, New York, 2005), p. 601.
280. P.A. Midgley, M. Weyland, T.J.V. Yates, I. Arslan, R.E. Dunin-Borkowski and J.M. Thomas, *J. Microsc.* **223** (2006) 185.
281. P.A. Midgley and M. Weyland, *Ultramicroscopy* **96** (2003) 413.
282. K. Yamauchi, K. Takahashi, H. Hasegawa, H. Iatrou, N. Hadjichristidis, T. Kaneko, Y. Nishikawa, H. Jinnai, T. Matsui, H. Nishioka, M. Shimizu and H. Furukawa, *Macromolecules* **36** (2003) 6962.
283. C. Bathias and A. Cagnasso, *ASTM Special Technical Publication* **1128** (1992) 35.
284. L. Salvo, P. Cloetens, E. Maire, S. Zabler, J.J. Blandin, J.Y. Buffiere, W. Ludwig, E. Boller, D. Bellet and C. Josserond, *Nucl. Instrum. Methods Phys. Res. B* **200** (2003) 273.
285. P. Weiss, D.L. Nihouannen, C. Rau, P. Pilet, E. Aguado, O. Gauthier, A. Jean and G. Daculsi, *Eur. Cells Mater.* **9** (2005) 48.
286. W.C. Conner, S.W. Webb, P. Spanne and K.W. Jones, *Macromolecules* **23** (1990) 4742.
287. J.S.U. Schell, M. Renggli, G.H. van Lenthe, R. Muller and P. Ermanni, *Compos. Sci. Technol.* **66** (2006) 2016.
288. E. Bayraktar, F. Montembault and C. Bathias, *J. Mater. Sci. Technol.* **20** (2004) 27.
289. P.J. Schilling, B.R. Karedla, A.K. Tatiparthi, M.A. Verges and P.D. Herrington, *Compos. Sci. Technol.* **65** (2005) 2071.
290. O. Gauthier, R. Muller, D. Von Stechow, B. Lamy, P. Weiss, J.-M. Bouler, E. Aguado and G. Daculsi, *Biomaterials* **26** (2005) 5444.
291. A. Elmoutaouakkil, G. Fuchs, P. Bergounhon, R. Peres and F. Peyrin, *J. Phys. D* **36** (2003) 37.
292. T. Donath, F. Beckmann, R.G.J.C. Heijkants, O. Brunke and A. Schreyer, *Proc. SPIE Int. Soc. Opt. Eng.* **5535** (2004) 775.
293. T.E. Gureyev, C. Raven, A. Snigirev, I. Snigireva and S.W. Wilkins, *J. Phys. D* **32** (1999) 563.
294. C.G. Schroer, P. Cloetens, M. Rivers, A. Snigirev, A. Takeuchi and W. Yun, *MRS Bull.* **29** (2004) 157.
295. P. Cloetens, W. Ludwig, D. Van Dyck, J.P. Guigay, M. Schlenker and J. Baruchel, *Proc. SPIE Int. Soc. Opt. Eng.* **3772** (1999) 279.
296. P. Cloetens, W. Ludwig, J. Baruchel, D. van Dyck, J. van Landuyt, J.P. Guigay and M. Schlenker, *Appl. Phys. Lett.* **75** (1999) 2912.
297. A. Momose, A. Fujii, H. Kadowaki and H. Jinnai, *Macromolecules* **38** (2005) 7197.
298. J. Baruchel, J.-Y. Buffiere, P. Cloetens, M. Di Michiel, E. Ferrie, W. Ludwig, E. Maire and L. Salvo, *Scripta Mater.* **55** (2006) 41.
299. W.S. Haddad, I. McNulty, J.E. Trebes, E.H. Anderson, R.A. Levesque and L. Yang, *Science* **266** (1994) 1213.
300. Y. Wang, C. Jacobsen, J. Maseri and A. Osanna, *J. Microsc.* **197** (2000) 80.
301. W. Chao, B.D. Harteneck, J.A. Liddle, E.H. Anderson and D.T. Attwood, *Nat. Cell Biol.* **435** (2005) 1210.
302. D.I. Bower and W.F. Maddams, *The Vibrational Spectroscopy of Polymers*, Reprint Ed. (Cambridge University Press, Cambridge, 1992).
303. J.L. Koenig, *Spectroscopy of Polymers*, 2nd ed. (Elsevier Science, Amsterdam, 1999).
304. P.M. Cooke, *Anal. Chem.* **72** (2000) 169R.
305. M. Navratil, G.A. Mabbott and E.A. Arriaga, *Anal. Chem.* **78** (2006) 4005.
306. C. Marcott, R.C. Reeder, E.P. Paschalist, A.L. Boskeyt and R. Mendelsohnt. *19th International Conference—IEEE/EMBS*, (IEEE NY, NY, 1997), p. 693.
307. C.P. Schultz, *Spectroscopy* **16** (2001) 24.
308. W.J. McCarthy, *Microsc. Today* **13** (2005) 22.
309. P. Dumas and M.J. Tobin, *Spectrosc. Europe* **15** (2003) 17.
310. P. Dumas, N. Jamin, J.L. Teillaud, L.M. Miller and B. Beccard, *Faraday Discuss.* **126** (2004) 289.
311. G. Ellis, C. Marco and M. Gomez, *Infrared Phys. Technol.* **45** (2004) 349.
312. G. Ellis, C. Marco, M.A. Gomez, E.P. Collar and J.M. Garcia-Martinez, *J. Macromol. Sci. Phys.* **43** B (2004) 253.
313. R. Appel, W. Xu, T.W. Zerda and Z. Hu, *Macromolecules* **31** (1998) 5071.
314. T.W. Zerda, R. Appel and Z. Hu, *J. Appl. Polym. Sci.* **82** (2001) 1040.
315. D. Garcia, M.P. Nelson and P.J. Treado, *Technical Papers—American Chemical Society, Rubber Division, Spring Technical Program, 161st, Savannah, GA, Apr. 29-May 1, 2002* (2002) 287.
316. W. Schrof, J. Klingler, W. Heckmann and D. Horn, *Colloid Polym. Sci.* **276** (1998) 577.

317. C. Sammon, S. Hajatdoost, P. Eaton, C. Mura and J. Yarwood, *Macromol. Symp.* **141** (1999) 247.
318. N.J. Everall, *Appl. Spectrosc.* **52** (1998) 1498.
319. M. Ofuji, Y. Takano, Y. Houkawa, Y. Takanishi, K. Ishikawa, H. Takezoe, T. Mori, M. Goh, S. Guo and K. Akagi, *Jpn. J. Appl. Phys. 1* **45** (2006) 1710.
320. S. Natarajan and S. Michielsen, *Textile Res. J.* **69** (1999) 903.
321. U. Schmidt, S. Hild, W. Ibach and O. Hollricher, *Macromol. Symp.* **230** (2005) 133.
322. S.J. Stranick, D.B. Chase and C.A. Michaels, *Annual Technical Conference—Society of Plastics Engineers* 59th (SPE, Brookfield, CT, 2001) 1961.
323. W.X. Sun and Z.X. Shen, *J. Raman Spectrosc.* **34** (2003) 668.
324. D. Richards, R.G. Milner, F. Huang and F. Festy, *J. Raman Spectrosc.* **34** (2003) 663.
325. J.A. Dieringer, A.D. McFarland, N.C. Shah, D.A. Stuart, A.V. Whitney, C.R. Yonzon, M.A. Young, X. Zhang and R.P.V. Duyne, *Faraday Discuss.* **132** (2006) 9.
326. J.-X. Cheng and X.S. Xie, *J. Phys. Chem. B* **108** (2004) 827.
327. A.G. Caster, S.-H. Lim, O. Nicolet and S.R. Leone, *Spectroscopy* **21** (2006) 27.
328. A. Zumbusch, G. Holtom and X. Xie, *Phys. Rev. Lett.* **82** (1999) 4142.
329. J. Cheng, Y. Jia, G. Zheng and X. Xie, *Biophys. J.* **83** (2002) 502.
330. D.B. Williams and C.B. Carter, *Transmission Electron Microscopy: A Textbook for Materials Science* (Plenum Press, New York, 1996).
331. L. Reimer, *Transmission Electron Microscopy, Physics of Image Formation and Microanalysis*, 4th ed. (Springer, Berlin, 1997).
332. B. Fultz and J.M. Howe, *Transmission Electron Microscopy and Diffractometry of Materials*, 2nd ed. (Springer, Berlin, 2005).
333. R.F. Egerton, *Electron Energy-Loss Spectroscopy in the Electron Microscope*, 2nd ed. (Plenum Press, New York, 1996).
334. C.C. Ahn, Ed. *Transmission Electron Energy Loss Spectrometry in Materials Science and the EELS Atlas* (Wiley-VCH, Weinheim, 2005).
335. M.R. Libera and M.M. Disko, in *Transmission Electron Energy Loss Spectrometry in Materials Science and the EELS Atlas*, edited by C.C. Ahn (Wiley-VCH, Weinheim, 2005).
336. L. Reimer, Ed. *Energy-Filtering Transmission Electron Microscopy* (Springer, New York, 1995).
337. A. Du Chesne, *Macromol. Chem. Phys.* **200** (1999) 1813.
338. A.E. Ribbe, *Recent Res. Dev. Macromol.* **7** (2003) 171.
339. K. Varlot, J.M. Martin, D. Gonbeau and C. Quet, *Polymer* **40** (1999) 5691.
340. K. Varlot, J.M. Martin and C. Quet, *Polymer* **41** (2000) 4599.
341. C.A. Correa, B.C. Bonse, C.R. Chinaglia, J. Hage, E and L.A. Pessan, *Polym. Testing* **23** (2004) 775.
342. G. Schmahl, D. Rudolph, B. Niemann and O. Christ, *Q. Rev. Biophys.* **13** (1980) 297.
343. G. Schneider, T. Wilhein, B. Niemann, P. Guttman, T. Schliebe, J. Lehr, H. Aschoff, J. Thieme, D. Rudolph and G. Schmahl, *Proc. Int. Soc. Opt. Eng. SPIE* **2516** (1995) 90.
344. C. Jacobsen, S. Williams, E. Anderson, M.T. Browne, C.J. Buckley, D. Kern, J. Kirz, M. Rivers and X. Zhang, *Optics Commun.* **86** (1991) 351.
345. H. Ade and S.G. Urquhart, in *Chemical Applications of Synchrotron Radiation*, edited by T.K. Sham (World Scientific, Singapore, 2002).
346. H. Ade, *Trends Polym. Sci. (Cambridge)* **5** (1997) 58.
347. H. Ade, A.P. Smith, S. Cameron, R. Cieslinski, G. Mitchell, B. Hsiao and E. Rightor, *Polymer* **36** (1995) 1843.
348. R. Takekoh, M. Okubo, T. Araki, H.D.H. Stoever and A.P. Hitchcock, *Macromolecules* **38** (2005) 542.
349. J.I. Goldstein, D.E. Newbury, D.C. Joy, C.E. Lyman, P. Echlin, E. Lifshin, L.C. Sawyer and J.R. Michael, *Scanning Electron Microscopy and X-ray Microanalysis*, 3rd ed. (Plenum, New York, 2003).
350. A. Benninghoven, F.G. Rudenauer and H.W. Werner, *Secondary Ion Mass Spectrometry: Basic Concepts, Instrumental Aspects, Applications, and Trends* (Wiley, New York, 1987).
351. R.G. Wilson, F.A. Stevie and C.W. Magee, *Secondary Ion Mass Spectrometry: A Practical Handbook for Depth Profiling and Bulk Impurity Analysis* (Wiley-Interscience, New York, 1989).
352. *Analysis Techniques: SIMS*. Available at http://www.cameca.fr/html/sims_technique.html. Accessed December 2006.
353. D. Briggs, *Surface Analysis of Polymers by XPS and Static SIMS* (Cambridge University Press, Cambridge, 1998).

References

354. L.-T. Weng and C.-M. Chan, *Appl. Surf. Sci.* **252** (2006) 6570.
355. W.A. Lamberti, in *Handbook of Microscopy for Nanotechnology*, edited by N. Yao and Z.L. Wang (Kluwer Academic, New York, 2005), p. 207.
356. C.W.T. Bulle-Lieuwma, W.J.H. Van Gennip, J.K.J. Van Duren, P. Jonkheijm, R.A.J. Janssen and J.W. Niemantsverdriet, *Appl. Surf. Sci.* **203–204** (2002) 547.
357. M.S. Wagner, *Anal. Chem.* **77** (2005) 911.
358. K.H. Gray, S. Gould, R.M. Leasure, I.H. Musselman, J.J. Lee, T.J. Meyer and R.W. Linton, *J. Vac. Sci. Technol. A* **10** (1992) 2679.
359. J. Feng, C.-M. Chan and L.-T. Weng, *Annual Technical Conference—Society of Plastics Engineers* 58th (SPE, Brookfield, CT, 2000) 2434.
360. S. Chandra, *Appl. Surf. Sci.* **203–204** (2003) 679.
361. A.J. Eccles and T.A. Steele, *Int. J. Adhes. Adhes.* **21** (2001) 281.
362. A.J. Eccles, T.A. Steele and A.W. Robinson, *Appl. Surf. Sci.* **144–145** (1999) 106.
363. MILLBROOK, *MiniSIMS Applications*. Available at http://www.minisims.com/applications.htm. Accessed December 2006.
364. *CAMECA, NanoSIMS Materials/Geology Application Booklet*. Available at http://www.cameca.fr/doc_en_pdf/ns50_materials_application_nov2006_web200.pdf. Accessed December 2006.
365. N. Sanada, A. Yamamoto, R. Oiwa and Y. Ohashi, *Surf. Interface Anal.* **36** (2004) 280.
366. *PHI Quantera Scanning X-ray Microprobe*. Available at http://www.phi.com/products/quantera/overview.html. Accessed December 2006.
367. *PHI Quantera Scanning X-ray Microprobe*. Available at http://www.phi.com/products/quantera/quantera-brochure.pdf. Accessed December 2006.

Chapter 7
Problem Solving Summary

7.1 WHERE TO START 479
   7.1.1 Problem Solving Protocol 479
   7.1.2 Polymer Structures 480
7.2 INSTRUMENTAL
    TECHNIQUES 480
   7.2.1 Comparison of Techniques ... 480
   7.2.2 Optical Techniques 484
   7.2.3 SEM Techniques 485
   7.2.4 TEM Techniques 486
      7.2.4.1 STEM Techniques 486
   7.2.5 SPM Techniques 487
   7.2.6 Technique Selection 487
7.3 INTERPRETATION 488
   7.3.1 Artifacts 489
      7.3.1.1 Artifacts in Optical
          Microscopy 489
      7.3.1.2 Artifacts in SEM 489
      7.3.1.3 Artifacts in SPM 490
      7.3.1.4 Artifacts in TEM 491
      7.3.1.5 Artifacts in X-ray
          Microanalysis 491
   7.3.2 Summary 492
7.4 SUPPORTING
    CHARACTERIZATIONS 492
   7.4.1 X-ray Diffraction 493
   7.4.2 Thermal Analysis 495
   7.4.3 Spectroscopy 496
      7.4.3.1 X-ray Fluorescence 496
      7.4.3.2 Infrared and Raman ... 496
      7.4.3.3 Nuclear Magnetic
          Resonance 497
      7.4.3.4 X-ray Photon
          Spectroscopy 498
      7.4.3.5 Auger Spectroscopy ... 498
   7.4.4 Small Angle Scattering 499
      7.4.4.1 Small Angle Light
          Scattering 499
      7.4.4.2 Small Angle X-ray
          Scattering 499
      7.4.4.3 Small Angle Neutron
          Scattering 500
   7.4.5 Summary 500
References 501

The preceding chapters have provided a description of microscopy techniques, imaging theory, and the specimen preparation methods required to investigate polymer structures. The theme of this chapter is to put all of this together within a useful framework. This framework might be a review to experienced microscopists (who likely have developed their own protocols), but it will provide useful information regarding problem solving ideas. A problem solving protocol will be developed that permits microscopy characterizations to follow an easy and short path to a solution. These characterizations will all be classified as "problems" that require a solution. Problems can range from simple to complex and include, for example, determination of the phase structure in a polymer blend, the cause of failure of a composite or the complete and fundamental characterization of a new membrane, fiber, film, and so forth. Clearly, such problem solving will require a range of time and effort, but the protocols used to begin the characterization and to know when the problem is solved are similar overall. Generally, more than one technique is required to solve problems relating to polymer morphology, and thus complementary multidisciplinary techniques are important in conducting problem solving analyses. Interpretation of the images produced is of critical importance in evaluating polymer structures, and thus the topic of artifacts will be included in this discussion. Finally, although structural characterizations cannot generally be accomplished without microscopy methods and techniques, there are other complementary analytical techniques that are often quite important in understanding polymer structures. The last section will be devoted to a short description of these techniques, including x-ray diffraction, thermal analysis, electron spectroscopy, and others.

7.1 WHERE TO START

One of the most difficult decisions that must be made in the microscopy laboratory is how to *start* solving a problem. The difficulty is that it is not always obvious where to start and to know ahead of time the full range of techniques that will be required. It is also difficult to deal with this question when considering that there is a wide range of techniques that can be used for problem solving. However, there are some simple concepts to consider before beginning the microscopy characterization. The protocol that will be discussed here is not necessary in all cases. If a measure of the orientation of a fiber is needed, for instance, a problem solving protocol might not be needed if the investigator is aware that the birefringence can be measured using a polarizing optical microscope and a compensator. But, if the dispersed phase distribution and particle size of a polymer blend are to be correlated with impact strength, then it might not be as obvious where to start solving the problem.

7.1.1 Problem Solving Protocol

The steps involved in the problem solving protocol are outlined in Table 7.1. They are rather simple and do not take much time to consider, and such a protocol can save time in the long run. The protocol involves steps typical of scientific inquiry: collect all the currently known facts; determine the nature of the problem; state the objective of the study; obtain the correct specimen; be sure to have experimental controls; look at the sample with the naked eye and then with a stereo microscope. These provide an aid to selection of the specific microscopy techniques and preparation methods needed to begin to address the objectives. The result should be that clearly defined analyses are conducted.

It is essential to know all the facts relating to the problem to be solved. In the worst case, someone simply asks you to take a picture and not ask any questions. This sounds and is very simple; however, then you have a picture, you might even have the right picture, but you do not have a solution to the problem! The time used to take the picture is likely wasted because the problem still must be discovered and solved. It is important to gather all the relevant facts regarding the problem and the specimens that are to be characterized. It is useful to write down the objective of the experiment in order to focus on the problem itself rather than conducting a complete characterization that reveals the structure but provides more information than is needed. The next steps involve consideration of the specimens required to solve the problem. In the case of the multiphase blend described above, the correct specimens might be those that have been tested and exhibited a range of impact strength properties. Only the

TABLE 7.1. Problem solving protocol

1. Collect all the facts/data about the problem, including the chemistry of the polymer and the process used for its formation
2. Clearly define the problem and objectives
3. Select the specimen to solve the problem
4. Define scientific controls and materials to aid interpretation
5. Examine specimens with the naked eye and with a stereo binocular microscope
6. Determine the size of the morphology features to be imaged
7. Define the microscopy technique(s) and best specimen preparation methods for imaging and microanalysis
8. Take representative images and label those interesting but not representative
9. Assess potential artifacts as part of the image interpretation process

tested samples will provide the necessary data and permit comparison with specific property values, whereas using the average value from a physical test with a sample taken at random is not nearly as useful. Controls are important in any scientific study; this might involve the assessment of a specimen before and after treatment or under various conditions.

Specimens should be observed with the naked eye as the size, shape, color, and gross morphology are all important to consider when choosing microscopy techniques and preparation methods. A stereo microscope is useful to observe the specimen at low magnification as it often provides an overview of the specimen that can be critical in determining the area to be studied in more detail. One true example will show the utility of this low power, inexpensive microscope. A plastic key cap for a computer keyboard was brought into the laboratory one day with black specks on the part that the client thought required assessment via scanning electron microscopy (SEM)/energy dispersive x-ray spectroscopy (EDS). Looking at the key cap by eye, it looked like black specks were present. Observation in a stereo microscope showed the specks were on the surface and easily brushed off, saving the time it would have taken to conduct x-ray experiments.

7.1.2 Polymer Structures

The first step in the selection of a microscopy technique is to know the size of the polymer structures to be characterized. In fundamental studies, the answer might be that all of the structures present must be understood, whereas in more routine studies, a specific structure must be evaluated as part of the problem solving process. An example of a specific structure that is often evaluated is the spherulite. Spherulite sizes in semicrystalline polymers often determine the properties of the material. Monitoring this structure can be important in structure-property determinations. A listing of the most common polymer structure types, described in Chapter 1, and their characteristic sizes are shown in Table 7.2. Clearly, different microscopy techniques must be used to characterize these different structures.

TABLE 7.2. Polymer structures

Structure types	Characteristic sizes
Crystal (unit cell)	0.2–2 nm
Chain (sequence length)	2–100 nm
Lamellar crystal thickness	5–50 nm
Fibrils	5–50 nm
Spherulites	1–100 μm
Copolymer and blend phase domains	2 nm–100 μm

7.2 INSTRUMENTAL TECHNIQUES

Once the objective of the experiment is known and the specimens selected for study, the next major step is the selection of the microscopy techniques and the specimen preparation methods required to image the polymer structures of interest (Table 7.2). If lamellar crystals must be evaluated, for instance, there is no point in considering most optical techniques as they will only provide an overview of these structures. Comparisons are made in this section regarding the various techniques, in both the text and tables, as an aid in this selection process. Observations of polymer structures are limited by their size, the specimen preparations required, and the imaging techniques. If similar structures are observed by several different methods and techniques, they are more likely to be representative of the material. In this section, the advantages and disadvantages or limitations of the techniques will be compared.

Table 7.3 shows the relation of the polymer structures and their sizes, superimposed on the range of structural sizes observable by the various microscopy and scattering techniques. An important point is the overlap among the various techniques, which makes complementary analyses possible. For example, study of spherulitic structures is shown to be possible by optical, SEM, or transmission electron microscopy (TEM) methods.

7.2.1 Comparison of Techniques

Within the ranges of the general techniques shown in Table 7.3, there are many specific techniques that provide useful information

TABLE 7.3. Structural characterization

Structure	Macroscopic		Spherulitic		Domains		Lamellar		Crystal lattice
Size	10 mm	1.0 mm	0.1 mm	10 μm	1 μm	0.1 μm	10 nm	1 nm	0.1 nm
Technique	Eye ————————————								
		Stereo binocular ————————————							
		———————— Optical microscope ————————							
				———————— SEM ————————					
					———————— TEM ————————				
				———————— AFM / STM / FFM ————————					
					SAXS ————————				
								WAXS ————————	
								SAED ————————	

relating to polymer characterization. A listing of the more commonly employed microscopy techniques is shown in Table 7.4 with the type of features that are commonly imaged, the typical size range of the structures, and magnifications. This table is meant to summarize the application of these techniques, as more detailed information is found in Chapters 2, 3 and 6. Polarized light microscopy is very important in characterizing the spherulitic textures of crystalline polymers. The nature of the orientation in extruded and molded articles and the size and distribution of the spherulites have all been shown to relate to mechanical properties. An underutilized technique is phase contrast optical microscopy, which enhances the observation of small differences in refractive index between polymers, permitting the imaging of

TABLE 7.4. Microscopy techniques

Type	Features	Size range	Magnification
Optical			
Bright field	Macrostructures, microstructures, color, homogeneity	1 cm–0.3 μm	1–1,000×
Polarized light	Spherulitic textures	1 cm–0.5 μm	1–1,000×
Phase contrast	Phase variations, refractive index differences	100 μm–0.2 μm	50–1,200×
Electron			
Scanning (SEI)	Surface topography	1 mm–0.5 nm	10–60,000×
Scanning (BEI)	Atomic number contrast	1 mm–20 nm	10–10,000×
Transmission	Internal morphology, lamellar and crystalline structures	10 μm–0.2 nm	2,000–5 × 10^6×
STEM	Internal morphology, lamellar and crystalline structures	100 μm–1 nm	300–0.3 × 10^6×
Scanning probe			
STM	Surface topography	10 μm–0.3 nm	2,000–5 × 10^6×
AFM	Surface topography of insulators	10 μm–0.3 nm	2,000–1 × 10^6×
FFM	Friction and surface chemistry	10 μm–1 nm	2,000–1 × 10^6×

the dispersed phase domains. Scanning electron microscope techniques permit imaging of surface topography, by secondary electron imaging (SEI), and imaging with atomic number contrast, by backscattered electron imaging (BEI). A range of SEM techniques include variable pressure SEM, which permits study of dynamic specimens without the need for very high vacuum in the chamber. Comparison of these imaging modes is described in Chapters 2 and 3 and applications are shown in Chapter 5.

Various scattering, or diffraction, techniques also important in polymer characterization are listed in Table 7.5. The major differences among the various scattering techniques are the volume sampled and the spatial resolution of the techniques. The x-ray diffraction (XRD) techniques sample regions much larger than 100 µm, generally several millimeters across, although micro-XRD techniques have been used to sample 50 to 100 µm diameter areas. Selected area electron diffraction (SAED) techniques sample areas about 2–10 µm in diameter and thus provide crystal lattice information on very small crystallites that are difficult to consider by x-ray scattering techniques. Areas on the order of 5–100 nm are sampled by electron microdiffraction, which permits study of phase separated materials and very small local differences in the crystallinity of materials. The molecular structure of carbohydrate polymers using data from electron diffraction patterns and electron images [1] provides an excellent review of this topic.

The various general microscopy techniques are listed and compared in Tables 7.6 and 7.7. Table 7.6 compares optical, electron, and scanning probe microscope techniques, with the magnification, resolution, field of view, and imaging system listed. Useful magnifications and typical resolutions given are not the values on the knobs of the instrument, or in the instrument brochure, but these are approximate values typical of routine performance that should be considered when choosing a technique to solve a problem. The difficulty with the "best" resolution values is that often these cannot be obtained when imaging polymers due to specimen preparation and beam damage limitations. Consideration should be given to the size of the field of view of the technique (Table 7.6). One advantage of the light microscope is that large fields may be imaged and much of the specimen is observed rather than just very small areas, as in the TEM. This means that fewer samples are required to ensure that the analysis of the area studied gives information about the whole specimen. Complementary analyses provide important data regarding the uniformity of the structures in the material by analyzing both larger areas, for an overview of the structure, and smaller areas, in greater detail.

A summary and comparison of the performance of electron microscopes is found in Table 7.7. This table lists typical specimen

TABLE 7.5. Diffraction techniques

Technique	Acronym	Information	Sampled region	Sample thickness
X-ray diffraction	XRD			
Wide angle	WAXS (WAXD)	Crystal lattice Crystal size Crystal or molecular orientation	1 mm 0.1–10 mm	1 mm 1 mm
Small angle	SAXS	Lamellar thickness Domain size Fibril size and orientation	300 µm 100–600 µm	1 mm 1 mm
Electron diffraction				
Selected area	SAED	As for WAXS above, but local information	5 µm 2–10 µm	10–200 nm
Microdiffraction	µdiff	As for WAXS above, but local information	20 nm 5–100 nm	10–200 nm

Instrumental Techniques

TABLE 7.6. Comparison of various microscopies

	Stereo binocular	Compound	SEM	TEM	AFM
Useful magnifications	5–100×	30–1,500×	20–100,000×	2,000–250,000×	2,000–250,000×
Typical spatial resolution	10 μm	1 μm	5–10 nm	1 nm	2 nm
Best spatial resolution	2 μm	0.2 μm	1–5 nm	0.2 nm	0.3 nm
Field of view	Very large 5 mm, 50×	Large 2 mm, 50×	Large 20 μm, 5,000×	Small 2 μm, 50,000×	Small 2 μm, 50,000×
Imaging system	Light optical	Light optical	Scanning electron beam	Electron optical	Scanning solid probe
Lenses	Glass lenses	Glass lenses	No imaging lenses	Magnetic lenses	No lenses
Radiation damage	None	None	Serious at high voltage; minimal with conductive coatings and/or low voltage	Severe; can limit by use of replicas, cryo-TEM stage	None

thickness, viewing mode, accelerating voltage, image resolution, and x-ray analysis spatial resolution by either energy or wavelength dispersive spectroscopy. Table 7.8 focuses on these x-ray microanalysis methods and provides a comparison of the two major techniques, energy and wavelength dispersive spectroscopy. The tables provide a broad view of the techniques available in the microscopy laboratory, and they should be an aid in selection of the specific techniques required to solve materials problems. Much more detail can be found in Chapters 2 and 3.

The relatively low cost and wide availability of SEMs and more recently atomic force microscopes (AFMs) has led to concentration on these techniques, with less use of the TEM and optical microscopes. Scanning electron microscopes have great advantages; they produce easy to interpret images with excellent depth of field at high resolution. Similarly, AFMs have opened up new areas of surface studies in polymers. However, although the resolution and depth of field in an optical microscope (OM) is limited, a wealth of information becomes available when optical microscopy is used. The size

TABLE 7.7. Electron microscopy techniques

Instrument	"Regular" SEM	LVSEM	STEM	TEM
Specimen type	Thick	Thick	Thin	Ultrathin
Beam voltage (kV)	10–40	0.5–5	20–100	20–400
Useful magnifications	20–50,000×	20–100,000×	200–200,000×	3,000–250,000×
Image resolution	4–10 nm	1–5 nm	<1 nm	0.2 nm
X-ray spatial resolution	1 μm	(0.1 μm; few x-rays produced)	0.1 μm	0.1 μm

TABLE 7.8. Electron probe x-ray microanalysis

	Energy dispersive x-ray spectrometer	Wavelength dispersive x-ray spectrometer
Microscope interface	Interfaced with SEM, TEM, STEM	Interfaced with SEM, EPMA
Detection	Simultaneous detection of elements	Quantitative detection of one element at a time
Analysis time	Qualitative analysis: 10–100 s	Quantitative analysis: 100–500 s
Detectors	Single detector	Need several crystals to cover range of elements
Sensitivity	Background counts from backscattered electrons reduce sensitivity	Peak/background ratio 10 to 50 times better than EDS, good sensitivity
Energy resolution	Serious peak overlap problems; results may be ambiguous	Good energy resolution, little peak overlap
Spatial resolution	Good in TEM, AEM, STEM, poor in SEM	Poor (1–5 μm)
Resolution	130 eV at Mn Kα	8 eV resolution at Mn Kα
Atomic number limit	$Z > 11$ (regular window) $Z > 5$ (ultrathin window)	$Z > 3$

of the specimen that can be examined is tens of times the size of most SEM and AFM specimens and hundreds of times the size of a TEM specimen. Instrumental cost is much less for optical microscopes, although a research grade polarizing microscope, fully equipped for phase contrast and reflected light, can cost well above the price of a small SEM or AFM. The sections that follow consider each of the major microscopy techniques.

7.2.2 Optical Techniques

Optical microscopy is extremely useful in providing a rapid view of a relatively large area of the specimen. It should be used as a starting point in most microscopy problem solving to show the general appearance of the sample and its structure. Specialists solving problems that require the use of electron microscopy or atomic force microscopy regularly use the optical microscope to define the location of the very small area that can be imaged by these high resolution techniques and to show the relation of the high magnification images to larger scale features. Examples include using polarizing microscopy to show levels of heterogeneity in oriented films and fibers before sectioning for the TEM, reflected light microscopy of molded surfaces before profiling with the AFM, and phase contrast microscopy of multiphase polymers before study of fracture in the SEM. In the examination of spherulites, there is less chance of artifacts in the preparation of a thin section for optical study than in the preparation of an ultrathin section for TEM or an etched film for SEM or AFM. In a similar manner, dispersed phase particles can be imaged by phase contrast optical microscopy of thin sections, TEM of stained ultrathin sections, or fractured bulk samples examined in the SEM. In these examples, optical techniques are useful, rapidly providing an overview of the structures in relation to the whole specimen.

A major advantage of optical microscopy techniques is the ease of sample preparation. Thin fibers, films, or membranes can be placed directly in an appropriate immersion oil on a glass slide and information regarding the crystalline or dispersed phases, orientation, birefringence, and so forth, can be readily determined (see Section 4.1.1). Sectioning (see Section 4.3.2) of thicker materials is routinely accomplished in very short times, on the order of 30 min or so, with steel or glass knives. Observations and measurements of spherulite sizes, local orientation in molded parts, and fiber orientation are also conducted with the optical microscope. Phase contrast and Nomarski techniques provide contrast in multiphase polymers. Small differences in refractive index are enough to make the dispersed phases distinct, so the dispersed phase size and distribution can be measured. Even hard composites, such as glass fiber reinforced thermoplastics, can be thin sectioned by grinding and polishing methods (see Section 4.2.2) for reflected light microscopy. Samples for optical study are not placed in a vacuum, as with electron microscopy, and thus volatiles are not removed. Furthermore, beam damage, which is common in electron microscopy of polymers, does not occur in an optical microscope. Finally, there are many optical techniques that permit quantitative measurement of thickness, refractive indices, birefringence, roughness, and orientation that may not be applicable to study in the SEM, TEM, or AFM.

Optical microscopy has limited resolution and a decreasing depth of field with increasing magnification. The resolution limit is on the order of $0.2\,\mu m$, although $1\,\mu m$ is typical for routine analyses. Magnifications used are about 150–1,000× with a limit at about 2,000×. The poor depth of field is not a major problem if the specimen is flat, but for round specimens such as fibers this is a major problem as little of the specimen is in focus at one time. Confocal laser scanning optical microscopy (see Section 6.2.1) addresses this problem, but the instrument is not simple or inexpensive [2]. Overall, optical techniques are generally applied to the characterization of polymers as important information can be obtained in short times with minimum capital expenditure and relatively easy sample preparation methods. For many structural studies, optical techniques provide a solution to the problem. In more complex problems, these techniques provide key information that can lead to other appropriate techniques.

7.2.3 SEM Techniques

The advantages of the SEM are well known: images with a three dimensional appearance, great depth of field, ease of operation, and ease of specimen preparation. These advantages translate into micrographs that are easier to understand than the micrographs obtained by most optical and TEM techniques. The surface of even rather large samples can be imaged directly in the SEM rather than indirectly, as for TEM of replicas. Images can be formed by combinations of signals (secondary electrons, backscattered electrons, and x-rays) and the electronic signal processed to form a variety of micrographs with exceptional detail. Micrographs showing surface topography and chemical contrast are readily obtained in a short time by SEM, and they are often easy to interpret. Additionally, benchtop microscopes are relatively low in cost, and even the larger SEMs are available at half to one third the cost of a TEM. Research SEMs are also now available that are analytical instruments with combined high resolution imaging and elemental analysis by energy and wavelength dispersive x-ray techniques.

Limitations in SEM imaging include changes in the specimen caused by the vacuum and the electron beam, lack of internal detail, limited resolution and difficulties in interpreting image details. Volatiles present in a sample are removed in the vacuum, causing changes in the specimen surface as a function of time, often leaving a residue that is unrelated to the original structure. The removal of volatile components in surface finishes on textile fibers led to the need for a replication method (see Section 4.6) so the structures could be imaged without the specimen being placed in a vacuum. Alternatively, a variable pressure SEM (see Section 3.2.5) could be used. Radiation effects cause the most concern in relation to specimen damage and, in an earlier discussion (see Sections 2.6.1 and 3.4), radiation damage was shown to change the structure of some polymers and to affect imaging and resolution. Irreversible radiation effects are often responsible for the formation of structures that are easily misinterpreted. Scanning electron microscope images provide only surface detail, although samples for internal or bulk study at good resolution can be specially prepared (e.g., by fracturing or sectioning). A modern SEM with a field emission gun can be operated at low beam voltage, which limits radiation damage to a thin surface layer and reduces the need for metal coating of polymer specimens.

Interpretation of SEM images requires assessment of the data in the micrograph in light of both potential artifacts in imaging and specimen preparation and what is known regarding specimen properties. At the risk of being controversial, SEM imaging can be too easy (and the same could be said for SFM). With reasonable capital expenditure, rapid sample preparation, and minimum training, almost anyone can be taught to take a picture with the SEM. However, the esthetically pleasing SEM micrographs might have little to do with the structural problem under study. Many experienced and talented microscopists have shown that the SEM is useful for the study of polymers and as an aid to problem solving. It is important to examine specimens for structures that are representative and relate to the problem and then correlate them with the relevant properties of the material. These structure-property applications are a major contribution made possible by SEM studies during the past 35–40 years.

7.2.4 TEM Techniques

Transmission electron microscopy offers excellent resolution, down to the atomic level (e.g., [3–6]). It can provide information about molecular orientation and molecular ordering in crystals or liquid crystals, even when the ordered regions are extremely small. Combination of bright field and dark field electron microscopy with electron diffraction [7] permits the identification of the structure of ordered regions and measurement of their orientation, perfection, and size. Crystals only a few nanometers across can be detected and identified. In multiphase polymers, the dispersed phase structures can often be imaged and domains observed and quantified over a size range from less than 10 nm up to 1 μm. Attachment of EDS detectors permits the identification of local elemental composition variations at a spatial resolution of about 0.1 μm. Electron energy loss spectroscopy (EELS) (see Section 6.5.3) theoretically can provide analysis of light elements at a spatial resolution of about 20 nm. Important details of the fibrillar nature of highly oriented fibers and the transformation upon deformation of spherulites into lamellar structures have been shown by TEM characterization.

The disadvantages associated with obtaining TEM images are high capital expenditure, tedious and time consuming specimen preparation methods, the need for highly trained personnel, and "two dimensional" images that are difficult to interpret. Another serious problem for polymers is that radiation damage is often severe in the TEM, so that for many materials the high resolution structural information is very difficult to obtain directly without going through a time consuming process of preparing a replica. There are two major reasons for the time consuming nature of specimen preparation for TEM: the specimens must be extremely thin, on the order of 50 nm thick, and extra steps are often required to increase the contrast in polymers. Most of the methods for producing ultrathin sections are slow and require major capital acquisitions themselves, such as ultramicrotomes with diamond knives for sectioning and cryochambers for soft materials. Specimen preparation methods involve replication, staining, or etching due to the lack of inherent contrast. Transmission electron microscope imaging requires an understanding of image formation, the effect of the electron beam on the specimen, and knowledge of the instrument itself. Image interpretation is difficult due to changes in the specimen caused by radiation damage or exposure to vacuum. Artifacts are often caused by the specimen preparation methods. The observed sample volume is very small, and the image is not intuitively understandable in the way that an SEM image is. These difficulties notwithstanding, fundamental polymer characterization generally involves the application of TEM techniques, although AFM is playing a major role in such studies.

7.2.4.1 STEM Techniques

Scanning transmission electron microscopy (STEM) gives essentially the same type of results and has the same type of difficulties as the conventional TEM. There are two types of instruments, the "dedicated" STEMs, which generally have an ultrahigh vacuum (UHV) column, and the TEM based instruments mostly known as AEMs (analytical electron microscopes). A detailed comparison of STEM and TEM was given in Section 2.4.2. There are some advantages in using the STEM on polymer samples; in particular, it seems that thicker samples can be used. However, the added complexity and cost, combined with lower resolution in the AEM STEM mode, make it unlikely that either kind of instrument would be purchased solely for polymer studies.

7.2.5 SPM Techniques

Scanning probe microscopy (SPM) includes scanning tunneling microscopy (STM), atomic force microscopy, and frictional force microscopy, among many others (see Sections 2.5 and 3.3). A decade ago it was said to not yet be clear how large an area of application these new techniques will find in polymers, but it was clear even then that they would be very important. In fact, AFM has replaced some of the imaging conducted by all the other techniques used for polymers. Scanning tunneling microscopy was initially found useful, as samples could be metal coated as for SEM, and the images were relatively easy to interpret. Today, STM has substantially been replaced by AFM. Atomic force microscopy is being used to form high resolution images of the surface of nonconductors, in air or in water, superior to those formed in the SEM, at a lower capital cost for the instrument. The AFM surface images are like the SEM images, three dimensional, easy to understand, and they are directly quantitative, giving the absolute height profiles and roughness of the imaged surface.

It is interesting that a major advantage touted for AFM is the ease of sample preparation, and yet as shown in many sections in Chapters 4 and 5, some of the same tedious methods are now used for AFM as for TEM. Although there is no radiation damage to contend with, and some specimen preparation is trivial, many studies require cryoultramicrotomy to produce a flat block face for AFM and etching is also used, both of which are time consuming. Furthermore, the artifacts that can result from these preparation methods and from the microscope itself still suggest the use of complementary imaging to ensure interpretation is correct. Fortunately, the new microtome stages permit preparation of thin sections and the block face simultaneously and aid conducting of AFM and TEM on the same sample (see Section 5.3.4.5, Figs. 5.71–5).

Image resolution with SPM can extend to the atomic level. A great deal has been learned during the past decade about the theoretical understanding of image formation, but the technique still requires further study for a more complete understanding. As a solid scanning probe is used, the specimen can be affected by the imaging process, but the damage is now mechanical instead of radiation damage. It is comparatively easy to make pits and holes in polymer samples with the SPM, especially when using contact or tapping mode AFM, so care still has to be taken in imaging of soft polymers.

7.2.6 Technique Selection

The specific microscopy techniques and appropriate preparation methods required to solve structural problems must be selected now that the techniques and the problem solving protocol (see Table 7.1) have been considered. The advantages and limitations of these techniques have already been considered, but questions still remain. When should optical techniques be used in solving polymer structural problems? How can experiments be conducted so that it is clear if further study is required? Are there any simple "formulas" for successfully conducting microscopy studies of diverse materials in a manner that is time effective, cost effective, and really provides structural answers? The flow chart in Table 7.9 is included as an aid in selection of a technique to solve structural problems. Conducting the actual studies is an iterative process where an experimental plan is developed, the studies conducted, and further experiments planned as required. Experience has shown that certain techniques are the most likely to provide meaningful answers to certain structural problems. The structure-property applications in Chapter 5 are examples of the application of both single techniques and complementary techniques to problem solving.

TABLE 7.9. Problem solving flow chart. Questions to consider when selecting a microscopy technique

Size of polymer structure to image		Look at Tables 7.2 and 7.3
1. Known technique applied to solve this problem? ↓ No	Yes ➡	Check out examples in Chapter 5. Try it.
2. Can optical microscopy provide an overview of the structure (size)? ↓ No	Yes ➡	First try a stereo binocular microscope to select a sample for further study. Try a compound light microscope, if possible. More info? Go to 3.
3. Require surface morphology? ↓ No	Yes ➡	Try SEM or AFM study. Need higher resolution? Try high resolution SEM at low voltage with FEG or try AFM of the surface. Need more info? Go to 4.
4. Require internal bulk morphology? ↓ No	Yes ➡	Try SEM of fracture surface or try AFM of microtomed surface. For higher resolution, can use TEM of microtomed sample or AFM of microtomed block face from same sample for comparison. Go to 5.
5. Explore methods of preparation for TEM/STEM or high resolution AFM.	Yes ➡	Use preparation methods found in Chapter 4.

7.3 INTERPRETATION

There are many facets to the interpretation of images as part of structure determination and problem solving. First, it is necessary to know the effect of specimen interactions with the microscope and to understand the image formation process. Next, the effects of specimen preparation must be understood. Many of the methods of preparing specimens suitable for microscopy can deform all or part of the sample and can produce a wide range of artifacts. For the current discussion, *artifacts* are defined as any features present in a micrograph that do not correspond with structural detail in the original material. The artifacts may be introduced during specimen preparation, by radiation, thermal, or mechanical damage in the microscope, by contamination, or by some imperfection in the imaging process. In problem solving by microscopy, the only important artifacts are those that are not recognized as such and are erroneously attributed to the structure of the material. Finally, the nature of the material and its physical and mechanical properties must be considered as part of the image interpretation process. The formation of artifacts will be reviewed as the final consideration in the image interpretation process. For examples of structure determination and image interpretation the reader is directed to the examples of structure-property applications at the end of each subsection of Chapter 5.

Complementary microscopy has been stressed as being essential in understanding the nature of polymer structure, because of the common formation of artifacts that can result in misinterpretation of the image information. Carter and Harb [8] used the term *correlative microscopy*, stating that subjecting a specimen to a variety of types of microscopies can be useful to image structures not seen in one mode but seen in another. These authors refer to biological materials, but this is also relevant for polymer materials. Evaluation of the bulk structure of a polymer by TEM using thin sections is complementary to AFM of the block face resulting from the sectioning process, and each provides different types of information, the latter free of radiation effects. Polarized light microscopy provides interesting detail of the crystalline structures not seen in the SEM or AFM. Many of the applications described in Chapter 5 and from the literature give examples of complementary microscopy used in problem solving.

The use of digital imaging for many microscopy techniques brings yet another source of potential artifacts to consider when interpreting images. Koeck's [9] review of digital microscopy is worth reading. Leaving out the obvious problem of forming new structures not in the material using computerized imaging, there are many issues to consider including the recording media or whether the image is scanned or recorded directly. Naturally, such imaging can provide enhancements that aid interpretation by providing improved contrast and brightness, use of color, and image sharpening, but the microscopist should beware of making and changing image details inadvertently.

7.3.1 Artifacts

A major consideration in the selection of preparation methods for microscopy study is the nature of the potential artifacts formed, although time, cost, and the capital equipment required are also important factors. In a busy laboratory, time considerations are very important, especially if time consuming methods also have potential artifacts. The accessory equipment available must also be considered, although for this discussion it will be assumed that the laboratory has the equipment required for most general preparations. A complete discussion of specimen preparation methods can be found in Chapter 4. Typical preparations for microscopy will be outlined here with emphasis on the nature of potential artifacts.

7.3.1.1 Artifacts in Optical Microscopy

Typical preparations for optical microscopy include simple preparations, microtomy, and polishing. Placement of a thin specimen (fiber, film, etc.), as is or in an immersion oil, is a simple preparation (see Section 4.1.1) that is rapid and inexpensive. The major potential artifacts relate to the thickness of the sample or damage by the oil. Samples thicker than 10–50 μm generally do not reveal much structural detail by transmitted light techniques due to overlapping textures that may be misinterpreted or appear as a lack of structure in the specimen. Microtomy (see Section 4.3.2) is the most popular method for the preparation of specimens from fibers, films, membranes, and engineering resins. Sectioning takes a reasonable time to do, on the order of minutes to several hours per specimen, and microtomes are rather inexpensive. Potential artifacts include the deformation of the specimen, which can cause changes in the shape or the phases within the specimen or produce stress-induced structures that can be interpreted as being present in the polymer. Knife marks are found on the surface of the section, although this may be minimized by immersion of the sections in an oil with a refractive index similar to that of the embedding medium.

In grinding and polishing (see Section 4.2) of tough resins or filled composites, potential artifacts result from the deformation of the specimen, including pullout of the filler fibers or particles, undercutting of the resin, and the addition of a directionality to the specimen structures. Selection of the best polishing cloths and polishing media for a given type of specimen will minimize these effects. Polishing is an art, as is true of many microscopy preparations, and both care and experience are required for artifact-free preparations.

7.3.1.2 Artifacts in SEM

Scanning electron microscopy preparations are generally direct and rapid, yet there are quite a number of potential artifacts. Paints, glues, and tapes used to attach the specimen to the SEM stub can wick up or contaminate the specimen surfaces, adding false structures to the specimen, even far from the attachment site. The preparation of specimens from the bulk often involves deformation of the specimen (e.g., peeling, fracturing, and sectioning). Peeling (see Section 4.3.1) results in specimen fibrillation, which is a function of the force used, which is not controlled, and the sample history. Fracturing (see Section 4.8) at room temperature often causes deformation of soft or rubbery phases, whereas fracture in liquid nitrogen causes less deformation but tends to be less reproducible. Hand fracturing is not reproducible, whereas standardized testing more commonly provides fracture surfaces whose structures can be related to mechanical properties. Sections (see Section 4.3) and sectioned block faces exhibit knife damage, obvious by surface topography imaging. Etching (see Section 4.5) is one of the methods with the most potential artifacts. Chemical etching is known to remove polymer material, often redepositing it on the surface and creating new structures that might not have much to do with the structure of the specimen. Ion etching (see Section 4.5.4) is known to cause heating and melting and to create structures with directionality if great care is not taken such as with two water cooled guns at low glancing angles, whereas plasma etching has much less chance of artifact formation.

Replication (see Section 4.6) is replete with artifacts as much fine detail can be lost and new textures added, resulting in images that are often quite difficult to interpret. The application of conductive coatings (see Section 4.7) to polymers can also provide artifacts, such as the formation of grain structures. Finally, irreversible radiation damage and reversible charging effects also can cause the formation of artifacts. The picture is not as gloomy as it may seem, and if care is taken, important structural information can be obtained rather rapidly. Controls must be used to ensure the validity of the observations, and more than one specimen preparation method should be used if etching or replication are applied.

7.3.1.3 Artifacts in SPM

Sample preparation methods used for SPM imaging are often the cause of artifacts in much the same way as observed for SEM and TEM. Sectioning and formation of a block face for SPM can result in knife marks and other surface details unrelated to the material, and staining and etching can often result in changes in fine structure of contamination on the surfaces imaged. Details of potential artifacts from preparations used for SPM are discussed in Chapter 4 within the sections describing those methods. These potential artifacts due to preparation methods are similar to those for SEM by these same methods, as described in the section above. In addition, AFM, especially using contact and intermittent or tapping mode [10], the latter more commonly used for polymers [11, 12], requires significant care in image interpretation as the sample is contacted by the tip, which can cause physical damage. The environment can be an issue as humidity can result in a layer of water vapor on the specimen surface. As with any artifacts, the key issue is for the researcher to recognize the cause and effect of the image detail and interpret it appropriately.

Artifacts in SPM can also arise from many instrumental sources, such as scanner motion, tip geometry, noise (mechanical, acoustic, or electronic), drift (thermal or mechanical), signal detection, problems unique to signal detection methods, improper use of image processing (real time or postprocessed), environment (e.g., humidity), and tip-surface interaction (e.g., excessive electrostatic, adhesive, shear, and compressive forces). The most common artifacts are those that are related to scanner motion and tip geometry. In contact mode AFM or intermittent contact AFM (ICAFM), the tip is in mechanical contact with a physical surface, which will have obvious impact on image resolution. Further, current instrumentation uses piezo ceramic elements to position the physical probes relative to the test surfaces and so consideration of artifacts due to piezo element positioning are also very relevant. The ability to recognize artifacts should assist in reliable interpretation of image data and instrumental operation. These topics have been previously discussed in detail (see Section 3.3.7).

Early work reviewed by Leggett et al. [13] has shown that a number of features of highly oriented pyrolytic graphite (HOPG) surfaces, such as cracks and steps, can produce image features through thin polymer and biopolymer layers. Even worse, some graphite features and artifacts resemble molecules and are easily misinterpreted [13]. The recurrent theme, the need to conduct complementary imaging, especially when using relatively new imaging techniques, thus continues to be of utmost importance for the use of SPM techniques. West and Starostina [14] reviewed the topic of recognition and avoidance of AFM imaging artifacts noting that AFM images are a result of the probe tip shape and the shape of the feature being imaged. If the probe tip is too wide or too short, the image feature might be too small. The matching of the probe size and shape to the sample geometry can limit these artifacts. A damaged probe might result in image distortion and inaccurate shapes. Image processing also may cause artifacts with SPM images and with those collected by other techniques. The added difficulty with SPM is the depth information and the fact that the magnification in the Z direction generally differs from the X and Y directions. Until a new material is well known, use of complementary microscopy is highly recommended.

7.3.1.4 Artifacts in TEM

Ultrathin films and sections, required for TEM and STEM techniques, are produced by methods such as the formation of single crystals (see Section 4.1.3.3), dispersion (see Section 4.1.3.2), film casting (see Section 4.1.3.6), replication (see Section 4.6), and ultramicrotomy (see Section 4.3.4). The last two methods are most commonly applied to the study of industrial materials, although dispersion and sonication provide thin fragments of the specimen quite rapidly. The disadvantage of sonication is that the location of the fragment in the original specimen is unknown, whereas in ultrathin sectioning such information is known although the method is tedious. Often, ultrathin sections have no contrast, and enhancement techniques such as staining (see Section 4.4) and shadowing (see Section 4.7.2) are required to image the structures of interest. Controls must be examined as artifacts can result from stain deposits and false enhancement of shadowed structures. Replication is an alternative to sectioning a bulk material, although there are disadvantages in examining a replica when the specimen itself might be examined by a scanning technique at similar resolution (SEI). Additionally, an SEI image is much easier to interpret than a TEM image of a replica. Pretreatment of the specimen for replication can involve etching (see Section 4.5), which can also produce artifacts. Special methods (see Section 4.9) may be required for the preparation of soft or deformable microemulsion, latex, or adhesive materials that require equipment for freeze drying (see Section 4.9.2), critical point drying (see Section 4.9.3), or freeze fracture (see Section 4.9.4). These processes must be controlled and the resulting images carefully interpreted. Cryo-TEM (see Section 4.9.5) is more commonly used today and though tedious, this method provides information on fluids not previously available.

Radiation and thermal effects in the TEM and STEM can produce large scale changes in polymer specimens. These changes may be as severe as the disappearance of the specimen over time due to depolymerization and evaporation. Unless previous studies or experience show that TEM or STEM imaging is required and no other techniques provide the required information, the best approach is to conduct an optical or SEM experiment, prior to the higher resolution technique, in order to evaluate the structure or to try AFM instead.

7.3.1.5 Artifacts in X-ray Microanalysis

Many of the preparations noted for SEM and TEM also can cause artifacts for microanalysis in these same instruments. Note the various factors in Table 7.8 that describe the details and limits of energy and wavelength dispersive microanalysis. Contamination can cause problems, especially in the case of light element analysis. The source of foreign material is usually organic material on the specimen surface or in the vacuum system. The polymer microscopist has the added difficulty of working with polymers that are often the source of the contamination as the electron beam damages and reacts with the specimens, releasing smaller molecules and building up a carbon layer on the sample. Use of appropriate filters in the vacuum system, liquid nitrogen traps, and turbo pumps can all keep the system clean and attract gaseous hydrocarbons that otherwise will form on the specimen. Carbon contamination can be minimized when metal coating a specimen by use of a good vacuum during that procedure. Potential artifacts are discussed in detail in a recent text on SEM and x-ray microanalysis [15].

7.3.2 Summary

The two most important topics in the solution of structural problems are image interpretation and development of structure-property-process relations. Imaging techniques and preparative methods must be chosen that provide images of the needed structures by the most efficient experiments with a minimum of artifacts. Several major principles have been emphasized for the imaging of structures. First, the problem solving protocol (Table 7.1) should be considered prior to developing an experimental plan. As part of this protocol, the important properties of the material to be studied should be determined and the overall objective of the study defined. The size of the polymer structures required should be determined (Tables 7.2 and 7.3) as an aid in the selection of the appropriate microscopy techniques. Specimen preparation methods should be selected after considering the nature of the specimen itself, the types of structures to be imaged, and the potential artifacts. If a specimen can be examined directly, that is preferred over a less direct specimen preparation method, especially etching and replication. A key concept to keep in mind is that specimens are prepared for microscopy using deformation methods and that deformation and the other features of the preparation method, such as metal grain sizes, staining deposits, and etchants, must be carefully controlled to enable accurate image interpretation and problem solving. Finally, when more indirect and artifact prone methods are used, multidisciplinary methods and techniques should be employed to confirm the nature of the polymer structures.

7.4 SUPPORTING CHARACTERIZATIONS

Microscopy techniques provide important information about the structure of polymers. This information is often necessary in order to develop structure-property relations, but it is rarely sufficient. For example, the interpretation of anisotropic mechanical properties requires assessment of orientation and morphology as they influence these properties [16]. Morphology is described using microscopy techniques, whereas the nature of the orientation and the molecular structure must be determined by other analytical techniques. These techniques include x-ray scattering or diffraction (XRD), thermal analysis, electron spectroscopy for chemical analysis (ESCA), nuclear magnetic resonance (NMR), x-ray fluorescence (XRF), small angle light scattering (SALS), small angle neutron scattering (SANS), infrared and Raman spectroscopies, secondary ion mass spectrometry (SIMS), and the wide range of chromatographic and wet chemical analyses. The morphology and properties of block copolymers have been described by Gibson et al. [17]. Reffner [18] provides a good review of how infrared microspectroscopy (IMS) is united with microscopy for polymer science studies, with a special focus on the use of synchrotron radiation for IMS. Chalmers and Everall [19] used Fourier transform infrared (FTIR) and Raman spectroscopy to discuss quality assurance techniques for polymers such as poly(ethylene terephthalate) (PET) and ethylene-propylene copolymers. Koenig described spatially resolved spectroscopic techniques for the characterization and improvement of engineering polymer [20, 21]. The miscibility of amorphous bisphenol A polycarbonate (PC) and partially crystalline poly(butylene terephthalate) (PBT) were also studied by FTIR, differential scanning calorimetry (DSC), NMR, and polarized light microscopy giving depth into the nature of the melt blend [22]. Sperling [23] provides an excellent text on multicomponent materials, including a discussion of characterization techniques, such as microscopy, ESCA, SIMS, Auger spectroscopy, SANS, and light scattering. A text edited by Simon [24] provides a wide array of polymer characterization techniques as they are applied to polymer blends, including thermal analysis, polarized light microscopy, light, x-ray and neutron scattering, NMR, and electron microscopy.

The investigative techniques used include microscopy and some of those described in this section: x-ray diffraction, infrared and thermal analysis. Techniques that are most frequently used to complement microscopy observations will be outlined here in short summary paragraphs that briefly describe the nature of the information available, the principle of the technique, and several relevant references. Much of the current summary has been compiled from a book, *Modern Methods of Analysis*, produced by the Analytical Research Department, Summit Technical Center of the former Hoechst Celanese Corp. (now Ticona Engineering Polymers, Florence KY) [25], and my colleagues are duly and gratefully acknowledged for this useful compilation. Analytical imaging using emerging techniques not generally thought of as traditional microscopy techniques, such as Raman microscopy, x-ray microscopy, and imaging SIMS, are further described in Section 6.5.

7.4.1 X-ray Diffraction

X-ray scattering techniques are the most commonly applied complementary discipline to microscopy for structural studies. The type of information that is obtained by x-ray scattering experiments includes phase identification and quantification, crystallinity, crystallite size, lattice constants, molecular orientation and structure, molecular packing and order, and amorphous structure [26–30]. Diffraction techniques that will be described include powder diffraction, wide angle x-ray scattering (WAXS), and fiber diffraction. Small angle x-ray scattering (SAXS) will be described in Section 7.4.4.

Crystalline materials can be identified by rapid computerized powder diffraction techniques. The principle of this technique [30] is that the crystallites within a sample, placed in a collimated x-ray beam, reflect x-rays at specific angles and intensities. The diffraction pattern can be recorded photographically, using a camera (e.g., a Debye-Scherrer camera) or using a powder diffractometer. Chemical analysis depends on the fact that each chemical composition and crystallographic structure produces a unique angular distribution of diffracted intensity. Analysis is based on comparison of the diffractometer scan with known standards. Typical applications of the powder diffraction technique to polymers would be the identification of mineral fillers in engineering resins, the nature of crystalline contaminants, and determination of crystalline phases in a material.

Wide angle x-ray scattering [26, 27, 31] is used to identify the nature of crystalline phases on an atomic scale, the degree of crystallinity, the size of crystallites, and the degree of perfection and orientation of crystallites. Crystalline materials give rise to sharp diffraction rings or peaks, whereas amorphous materials produce broad, diffuse scattering of x-rays. The crystallinity of a material is a function of the amount of sharp to diffuse diffracted intensity. Analysis of the width of reflections along any crystallographic axis provides data about crystallite sizes and the degree of disorder within the crystal in that direction. Smaller crystallites result in broader reflections than those of larger crystallites.

The degree of orientation of crystallites can be computed from the arc lengths, or the angular spread of a chosen reflection. Molded specimens are known to have structural heterogeneities, including layers that can be seen visually. Wide angle x-ray scattering techniques can provide information about the degree and direction of molecular orientation within these different layers as long as the layer, or section of the specimen, is smaller than the x-ray beam. Polymer fiber bundles, aligned in the x-ray beam, provide patterns that can be analyzed for the degree of alignment of the molecules along the fiber axis [27, 32]. The uniaxial nature of fibers makes the WAXS experiment straightforward and useful in polymer structure determination. This technique has been used to determine the crystalline content of fibers as a function of process parameters, such as spinning speed.

Recent trends in x-ray diffraction are the increased use of electronic data collection and computer analysis and *in situ* or dynamic experiments. These include experiments where the x-ray patterns are obtained while the sample is stretched, heated to its melting point, or aligned with an electric field. Synchrotron radiation is several orders of magnitude more intense than regular laboratory x-ray sources, and its use allows real-time study of deformation or fiber spinning, for example. More synchrotron sources usable for polymer science are currently being built, and although they are limited to national facilities, access has improved. The increased power of computer data analysis permits "whole pattern analysis" where the crystal structure, orientation, and crystallinity are simultaneously determined [33]. More detailed numerical analysis has led to interpretation of x-ray patterns in terms of three phases in semicrystalline polymers instead of the usual two.

The ordering of polymers, determined by x-ray diffraction techniques, has been reported by many investigators. Classic methods are available [34] to determine the crystallinity in terms of the ratio of integrated peak intensity to the integrated intensity of the entire trace, although these methods depend upon good separation of crystalline peaks from the amorphous background. Hindeleh and Johnson [35] developed sophisticated calculation and peak separation techniques for the estimation of crystallinity and crystallite size in polyamide and polyester fibers. They showed that the apparent crystallite sizes increase as a function of annealing. The x-ray calculations are indirect compared with direct high resolution electron microscopy observations of true mean crystallite sizes. X-ray diffraction provides peaks that are evaluated for peak broadening, with inherent complications due to the effects that crystallite size distribution and lattice distortion have on the calculation of the average, apparent crystallite size [36]. Electron microscopy can provide lattice fringe images or replicas that are used to determine the true crystallite size distribution if unaffected by specimen preparation [37, 38]. The structure of liquid crystalline copolyester fibers has also been determined by x-ray techniques. Variations in the positions and intensities of the peaks have been used to study the randomness of the polymers [39, 40]. Liquid crystal polymers (LCPs) have also been studied by combined rheo-optical and dynamic x-ray scattering of flow induced textures to assess the underlying mechanism leading to the formation of the microstructure [41]. Polymer structure determination using electron diffraction has been described in Section 2.4.3. Although it is less precise than XRD, valuable data are available from much smaller crystals and small local areas of a specimen than by XRD. X-ray diffraction is clearly the better technique for routine structural investigations on regions of $100\,\mu$m diameter or greater. Electron diffraction and imaging is performed on samples several orders of magnitude smaller.

An example of the use of complementary techniques is a paper by Frye et al. [42]. These authors compared the structural investigation of ultrahigh modulus linear polyethylene by electron microscopy, WAXS, and combined nitric acid etching and low frequency Raman spectroscopy. Calculations of the integral breadth of the (002) reflection and dark field TEM gave the same crystal lengths. Additionally, nitric acid etching followed by Raman spectroscopy suggested that there is a broad crystal length distribution for high draw ratio, also consistent with the WAXS and TEM data. These techniques all show an increase in the crystal length in linear polyethylene (PE) with increased draw ratio. Hall [43] has several chapters on the application of x-ray and neutron scattering techniques to the study of polymers. This includes chapters on the determination of the structures of aromatic polyesters by WAXS, computer analysis of diffraction patterns, and the ability of SAXS to distinguish the morphology of crystalline polymers. The molecular relaxation processes and structure of isotactic polystyrene (iPS) films were investigated with real-time dielectric spectroscopy and simultaneous wide and small angle x-ray scattering in order to explore the restrictions imposed on molecular mobility in the vicinity of the glass transition for crystallized iPS [44]. The final example in this section is the study of the crystallization of diblock copolymers using x-ray scattering, thermal analysis, and polarized light microscopy [45]; these techniques used in isolation would not have provided a full story of the microstructure.

7.4.2 Thermal Analysis

There are a variety of thermal analysis techniques that are applied to the understanding of polymer materials. These include differential thermal analysis (DTA), differential scanning calorimetry (DSC), thermomechanical analysis (TMA), thermogravimetric analysis (TGA), and dynamic mechanical analysis (DMA) [46–51]. Differential thermal analysis techniques permit study of the thermal behavior of materials as they undergo transformations as a function of temperature. This permits evaluation of melting points, crystallinity, purity, heat of fusion, specific heat, reaction kinetics, and so forth. The principle of the technique is that a sample and a reference material are heated while both are monitored by thermocouples. The output of the instrument is the difference between the two thermocouple voltages. When there are no thermal transformations, this output voltage is zero. If the output is positive, there is an exothermic reaction, whereas a negative voltage shows an endothermic reaction. Differential thermal analysis thermograms, plots of this output as a function of the reference temperature, provide data regarding glass transition, crystallization, and melting parameters.

Differential scanning calorimetry is another thermal technique similar to DTA in the type of information available, although the experiment may be more reproducible due to the nature of the instrument. Typically, a small sample and a reference material are heated at a constant rate, and the power consumption or heat flow is measured as a function of temperature or time. The difference between the heat required by the sample and the reference is a direct measure of the thermal properties of the sample. Essentially, the technique is used to measure the energy necessary to establish a close to zero temperature difference between the material and the reference material. The DSC thermogram is a plot of the differential heat flow versus the temperature or time. Integration of peaks gives the enthalpy change of the specimen. Again, glass transitions, crystallization and melting points are determined by this important and useful technique. In the case of liquid crystalline compounds, the multiple thermal transitions often require complementary hot stage microscopy and x-ray diffraction techniques to identify the phases. One example of the technique is in the determination of the effect of the composition of copolymers on the glass transition temperature as an indicator of the degree of phase separation [17]. Advances in instrumentation have changed the sensitivity and measurements available, and these should be addressed on the Web sites of the instrument companies.

Thermomechanical analysis permits measurement of the dimensional changes of materials as a function of heating or cooling. Thermomechanical analysis measurements include expansion and contraction, degree of crosslinking, glass transitions, crystallization temperatures, and so forth. A movable core differential transformer in the TMA, in combination with specific probes, permits the various measurements to be made as a function of temperature. Thermogravimetric analysis is a technique that provides a measure of the weight change of a material as a function of temperature. Measurements by TGA include thermal stability of polymers, determination of volatiles, additives, or solvents, decomposition temperatures, and kinetics. Thermogravimetric analysis operates on a null balancing principle with a sensitive balance maintaining a reference position for comparison with the weight of the sample. A current flow is produced to balance variations in weight between the reference and the sample, and this current is proportional to the change in sample weight. The relative thermal stability of polymers is quite important in end use properties.

Dynamic mechanical analysis techniques permit measurement of the ability of materials to store and dissipate mechanical energy during deformation. Dynamic mechanical analysis is used to determine the modulus, glass transition, mechanical damping and impact resistance, and so forth, of thermoplastics, thermosets, elastomers, and other polymer materials. Information regarding the phase separation of polymers is also available by DMA [17]. In DMA, viscoelastic materials are deformed in a sinusoidal, low strain displacement, and their responses are measured. Elastic modulus and energy dissipation are the measured properties. Stress-strain relationships are determined by DMA, and temperature scans reveal glass transitions, crystallization and melting information. Blends of polypropylene and rubber have been studied by DSC where the intensity of one of the two crystallization exotherms was used as a measure of the polypropylene domains and compared with the size determined by TEM cryomicrotomy and osmium tetroxide staining methods [52]. Isothermal annealing of PET above the crystallization temperature was shown to influence the morphology and increase thermal stability by combined SAXS and DSC analysis [53]. An excellent text edited by Turi [47] describes the instrumentation and theory of thermal analysis and its application to thermoplastics, copolymers, thermosets, elastomers, additives, and fibers.

7.4.3 Spectroscopy

7.4.3.1 X-ray Fluorescence

X-ray fluorescence (XRF) spectroscopy is a technique for the determination of elemental composition of materials, for elements greater than atomic number 11, present above 0.05% concentration [54–56]. The technique is similar to the electron probe microanalyzer (EPMA), and x-ray analysis in the electron microscope (see Section 2.7.1), except that the EPMA is used for local analysis, whereas XRF is a bulk technique. Exposure of the sample to an x-ray beam causes electrons to be ejected and outer shell electrons to fall into the vacancies, emitting x-rays of discrete energy. Characteristic energies are associated with specific elements, and the x-ray intensities are related to the concentration of the element in the sample. There are problems with this direct association of x-ray intensity and concentration, due to absorption by the matrix, but standards and software programs are available to calculate elemental composition. X-ray fluorescence experiments have the advantage of being rapid and nondestructive. These techniques are usefully applied to the assessment of fillers, additives, and contaminants in polymers, and software is available for quantitation.

7.4.3.2 Infrared and Raman

Infrared and Raman are complementary vibrational spectroscopies, and both will be considered here. The absorption versus frequency characteristics of light transmitted through a specimen irradiated with a beam of infrared radiation provide a fingerprint of molecular structure. Infrared radiation is absorbed when a dipole vibrates naturally at the same frequency in the absorber. The pattern of vibrations is unique for a given molecule, and the intensity of absorption is related to the quantity of absorber. Thus, infrared spectroscopy permits the determination of components or groups of atoms that absorb in the infrared at specific frequencies, permitting identification of the molecular structure [21, 57–65]. These techniques are not limited to chemical analysis. With instruments of high spectral resolution, the tacticity, crystallinity, and molecular strain can also be measured. Copolymer dispersions can be determined as block copolymers absorb additively, and alternating copolymers deviate from this additivity due to interaction of neighboring groups.

The conventional spectrometer with a dispersive prism or grating has been largely superseded by the Fourier transform infrared (FTIR) technique. This uses a moving mirror in an interferometer to produce an optical transform of the infrared signal. Numerical Fourier analysis gives the relation of intensity and frequency, that is, the IR spectrum. The FTIR technique can be used to analyze gases, liquids, and solids with minimal preparation in short times. The FTIR technique has been applied to the study of many systems, including adsorption on polymer surfaces, chemical modification, and irradiation of polymers and oxidation of rubbers [66]. The application of infrared spectroscopy to the study of polymers has been reviewed by Bower and Maddams [62].

Infrared dichroism is a phenomenon that is used to measure the degree of orientation of the polymer chain. In this case, aligned groups, such as in a stretched polymer film, exhibit absorbance in the infrared that varies depending upon the alignment of the transition moments with respect to the polarization direction of the incident radiation [62]. Gibson et al. [17] discussed obtaining infrared dichroism measurements by uniaxially orienting a film sample and determining the absorbance of selected bands with radiation polarized parallel and perpendicular to the stretch direction. A dichroic ratio of these two absorbances is then related to an orientation function.

Infrared microspectroscopy (IMS) is widely accepted as a routine technique for polymer science, trace evidence examinations, probing biological tissues, and many other spectral analyses including polymers [18]. Imaging of the phase morphology of semicrystalline polymers and polymer blends in the melt has been done using FTIR to study different types of spherulites of poly(vinylidene fluoride), which cannot be compared in the light microscope [67]. Forensic sciences make excellent use of various microspectroscopies, especially infrared combined with SEM/EDS and Raman [18, 68], which are used to identify drugs and other materials and for prosecution of pharmaceutical patents; they are mentioned here as being of excellent use for problem solving of paints, fragments, fibers, and other materials by microscopy and spectroscopy techniques.

Laser Raman spectroscopy [21, 57, 60–62, 66, 69–71] is a light scattering process where a sample is irradiated with a laser and the inelastically scattered light collected and analyzed. Functional groups (e.g., carbon-carbon double and triple bonds and carbon bonded to sulfur and chlorine) scatter incident radiation at characteristic frequency shifts in Raman spectroscopy [69]. The vibrational frequency of the group is the amount of shift from the exciting radiation. Raman spectra depend upon polarizability, whereas infrared absorption relates to dipole moment changes so that for most groups, either the vibrational band is active in the infrared or it is observed in the Raman. The techniques are complementary for molecular structure determinations.

An important development in Raman spectroscopy has been the coupling of the spectrometer to an optical microscope. This allows the chemical and structural analysis described above to be applied to sample volumes only $1\,\mu m$ across [70, 71]. Confocal Raman spectroscopy has been used to determine the composition of binary polyfluorene composites with micro and mesoscale phase separation and the results compared with ICAFM images [71]. No more sample preparation is required than that for optical microscopy, and the microscope itself can be used to locate and record the area that is analyzed. This has obvious practical application to the characterization of small impurities or dispersed phases in polymer samples. This instrument, which may be called the "micro-Raman spectrometer," the "Raman microprobe," or the "Molecular Optics Laser Examiner" [72], has also been applied to the study of mechanical properties in polymer fibers and composites. It can act as a noninvasive strain gauge with $1\,\mu m$ resolution, as has been reviewed by Meier and Kip [73]. Even if the sample is large and homogeneous, there may be advantages in using the micro-Raman instrument. The microscope lenses are very efficient at collecting the scattered light, and because only a small volume is in the laser beam, a higher power density can be used without overheating and destroying the sample [73]. The main disadvantage is that the micro-Raman instrument is expensive. If the laser beam can be scanned, a Raman image can be formed; this is a kind of microscopy and was described in Section 6.5.2.

7.4.3.3 Nuclear Magnetic Resonance

Nuclear magnetic resonance (NMR) uses the magnetic properties of nuclei, particularly the proton ^1H and ^{13}C, to observe structural features in polymer chains. A radio frequency field and a magnetic field are applied to the sample, and at the resonance condition energy is absorbed. The exact resonance frequency depends on the local chemical environment of the atom. Measuring the small difference in resonance frequency, the *chemical shift*, and using model compounds allows the exact chemical structure to be determined [74]. Traditional NMR was a chemical technique applied to solutions to determine the chemical structure and tacticity. With Fourier transform and other techniques and today's more powerful magnets, NMR can be applied to a wider range of nuclei in solution or in the solid state [75–78]. Chemical information can be obtained not only about the average structure, but also about defects such as copolymer structure [79] chain ends and branch point density [80]. There is also a wide range of structural information available. For example, the linewidth of the resonances gives information about the mobility of the chain or of specific side groups, and thus can provide glass transitions (T_g) or crystallinity.

7.4.3.4 X-ray Photon Spectroscopy

X-ray photon spectroscopy (XPS) is also known as electron spectroscopy for chemical analysis (ESCA). Both names are descriptive, as the essential feature of the technique is bombardment of a specimen with monochromatic x-ray photons, and the energy spectrum of electrons that are emitted is measured. The binding energy of the core electrons that are emitted and the kinetic energy that they possess sum to the x-ray energy. Peaks in the plots of electron intensity versus binding energy correspond with the core energy levels that are characteristic for a given element. Small shifts in the binding energy are caused by the state of the valence electrons, that is by the local chemical environment. The spectrum thus allows elemental and chemical analysis of the top 2–10 nm of the surface [81–84]. The specimen must be in a high vacuum, but apart from that the method is nondestructive. Other advantages are that the data interpretation is relatively straightforward, and sample preparation is simple, although the equipment is expensive. The spectra can sometimes be unclear because the electron spectrometer resolution is not sufficient to prevent peaks with different chemical shifts from overlapping. A database of many spectra has been built up to aid identification [85]. X-ray photon spectroscopy has been applied to many polymer problems, particularly for the investigation of surface modification processes [66, 86] and the interaction of polymers and metal layers deposited on them [87]. Most XPS instruments have a spatial resolution of less than 1 mm. Higher resolution is possible, but a very high flux of x-rays is needed to keep the signal high, and this may destroy most polymer specimens. Briggs [84] discussed the surface analysis of polymers by the complementary techniques of XPS and SIMS. Hinder et al. [88] used a combination of ultra low angle microtomy through polymer coatings and paints to assess the morphology and chemistry using XPS, time-of-flight SIMS for compositional depth profiling, and SEM and AFM to investigate the morphology and topography of the surfaces resulting from the preparation method.

7.4.3.5 Auger Spectroscopy

Auger spectroscopy [82, 83, 89] is a surface elemental analysis technique similar to XPS although with higher spatial resolution and lower detection limits. Auger analysis is difficult to apply to polymers due to the severe sample degradation that occurs during the analysis. Scanning Auger microscopy (SAM) is a method for the elemental mapping of a surface with a rastered electron beam. It gives 5 μm spatial resolution with depth profiling possible by ion etching. An energetic beam of electrons strikes the atoms of the material in a vacuum environment, and electrons with binding energies less than the incident beam energy may be ejected from the inner atomic levels, creating a singly ionized excited atom. This inner level vacancy is filled by de-excitation of electrons from other electron energy states. The energy released can be emitted as an x-ray (fluorescence) or transferred to an electron in any atom. If this latter electron has lower binding energy than the energy from the de-excitation, then it will be ejected, with its energy related to the energy level separations in the atoms. Auger electrons are the result of de-excitation processes of these vacancies and electrons from other shells and a re-emission of an electron to carry away excess energy. The electrons emitted have a short mean free path, and thus all Auger electrons are from the first few atomic surface layers. The kinetic energy of the freed electron is detected, and these energies reflect the variations in binding energies of the levels involved in the process. The spatial resolution of Auger spectroscopy is about 20–50 nm, similar to energy dispersive spectroscopy of thin sections. Auger analysis is used in the study of adhesion on metal surfaces, in adsorption, corrosion, and oxidation studies.

7.4.4 Small Angle Scattering

Small angle scattering is a technique for the determination of morphological structures on a scale larger than that of the wavelength of the radiation used. Light, x-rays, and neutrons are used for small angle scattering, and the experimental details of these techniques are very different. They share the property of forming good numerical averages of feature size over a comparatively large volume of sample, without giving local details of the morphology. They are therefore complementary to microscopy, which can give the local details, but usually with a restricted field of view.

7.4.4.1 Small Angle Light Scattering

Small angle light scattering (SALS) uses light of wavelength 0.5 μm, so it can be used to investigate structures in the range 5–100 μm. Spherulites are structures of semicrystalline polymers that are in this size range and are studied by SALS techniques. In SALS, a monochromatic, collimated, and plane polarized laser beam passes through a thin polymer film. The scattered radiation is analyzed with a second polarizer, aligned with the first polarizer, and the scattering pattern is recorded on photographic film or by electronic detectors. The scattering of visible light is related to variations in the anisotropy and refractive index or polarizability of the specimen, and this polarizability is affected by the molecular structure [90]. Thus, light scattering techniques provide information about molecular structure and orientation. Spherulites in crystalline polymers have been found to be anisotropic scatterers, and theoretical scattering patterns have been calculated [91]. Typically, spherulitic structures are characterized by complementary SALS and polarized light microscopy techniques [90–92] where the scattering angle in the SALS pattern is used to determine the size of the spherulite. Stein et al. [90] reviewed the theory and applications of light scattering techniques, including a comparison with x-ray and neutron scattering.

7.4.4.2 Small Angle X-ray Scattering

Small angle x-ray scattering experiments are used to analyze the macrostructure of materials on a scale of about 1–200 nm [93–96]. The SAXS technique provides information regarding the electron density distribution of the material, and analysis of the angular distribution of the peak intensities (if there are peaks) reveals the periodicity and magnitude of that electron distribution. Thus, average morphological information is provided that is useful for the determination of the nature of voids or crystalline regions. Periodic structures, such as crystalline lamellae in polymers, produce small angle diffraction peaks whose measurements reflect the lamellar periodicity (i.e., thickness plus spacing). The principle of the technique is that x-rays are scattered by regions of varied electron density, such as voids or local crystalline regions, and the intensity is related to the number of such regions and their contrast. Small angle x-ray scattering is often used to define the size and shape of voids or fibrils in fibers and to measure the lamellar spacing in crystalline polymers.

Complementary studies involving SAXS and microscopy are common in polymers. They include studies of lamellar structures in polyethylene [97] and of fibrillar structures in rigid rod polymer fibers [98, 99]. Detailed analysis of the SAXS streak from high modulus fibers has been used to determine the length and orientation distribution of the fibrils [100]. To cover a wide range of feature size during phase segregation of polyethylene blends, SAXS was combined with SALS [101]. Combination of cryo-TEM (see Section 4.9.5) and SAXS was conducted on mixed solutions of poly(ethylene oxide)/sodium dodecyl sulfate systems; the former provided direct images of the microstructural building blocks, and the latter gives quantitative information not available by microscopy [102]. Two dimensional SAXS data was combined with AFM during tensile elongation and after subsequent relaxation, and DSC and birefringence of film under strain, to assess the morphology and orientation during deformation of segmented thermoplastic elastomers [103]. The deformation in lamellar and crystalline structures were studied by *in situ* simultaneous SAXS and WAXS conducted on PET fibers as a function of applied stress to understand the influence of the interactions between the crystalline and amorphous domains on the fiber properties [104]. Fibers in which the crystalline strain was large because of their strong linkages to the amorphous chains, and better load transfer, had the highest modulus and lowest ultimate elongation.

7.4.4.3 Small Angle Neutron Scattering

Neutrons are uncharged particles that may interact with a specimen by nuclear interactions resulting in a transfer of energy and momentum between the neutron and target materials [105]. Neutron scattering from hydrogen is very different to scattering from deuterium, so chemically similar deuterated molecules can be distinguished in a matrix of normal hydrogenated polymer. This allows the shape of individual molecules to be derived [96, 105, 106]. An early triumph of small angle neutron scattering (SANS) was to show that the molecules in an amorphous polymer have a random coil shape as predicted by Flory. Since then, there have been many studies of copolymers [107–109], blends [110–112], semicrystalline polymers [113, 114], and drawn fibers [115]. The emphasis in these studies has been the shape of individual molecules in these different circumstances, their diffusion and segregation to surfaces. Comparison of the radius of polybutadiene in diblock copolymers showed that the TEM values were significantly smaller than those obtained by SANS [108]. Domain boundary thicknesses, domain sizes, and domain packing order have also been determined by SANS measurements of solvent cast diblock copolymers and blends [109].

7.4.5 Summary

As was described in the first chapter in this text, and throughout the examples in various chapters, a very wide range of analytical techniques is used to determine the structure of polymeric materials. The chemical regularity, stereochemical configuration, and molecular weight distribution make up the basic molecular structure. These are unaffected by physical processing, but they define the starting material for such processing. The techniques used to determine this level of structure are essentially chemical ones.

The physical structure can be affected by processing and involves many variables, for example the molecular orientation and the distribution of dispersed phases in multiphase systems. If crystals are present, other important variables are the crystal structure, the degree of crystallinity, the crystal sizes and their arrangement into spherulites or other structures. This type of structure is determined by microscopy and by a range of scattering and spectroscopic techniques, as has been discussed in this chapter. Second phase distribution is commonly determined by electron microscopy and by x-ray or neutron scattering. The degree of crystallinity may be obtained from the density of the sample, from wide angle x-ray diffraction, from thermal analysis, and from NMR. In some cases, the two phase model of amorphous and crystalline material is insufficient, and the quantity of interfacial material must be considered. The interfacial material is associated with the fold surface of the lamellae, and the quantity of such material in PE is determined by Raman, WAXD, and NMR. Lamellar thickness is determined by small angle x-ray scattering and in some materials by Raman (low frequency longitudinal acoustic mode) and by TEM. If all are available, the combined results give a much clearer picture than is available from a single technique as each method has its advantages, but to do all of these well is time consuming.

To some polymer scientists, the word *morphology* means the assembly and relative arrangement of crystals or of second phase particles, whereas others use *supermolecular struc-*

ture to describe the same thing. Whatever the name, optical and electron microscopy and the complementary techniques of light and x-ray scattering are used to determine such structure. Common arrangements in crystalline polymers solidified from the melt are spherulites, row structures, stacks or bundles of lamellae, rods, and fibrils. Poorly ordered materials may have randomly placed single lamellae. In general, microscopy is used to define the type of arrangement, and a scattering technique, which samples a much larger volume, is used to measure average dimensions of the various structures. Without microscopy, a model must be assumed to interpret the scattering data; without the scattering, a great deal of quantitative microscopy must be done to ensure statistical sampling of the specimen structure. Thus, although our subject here is microscopy, it is apparent that at every point, complementary techniques are vital for problem solving.

References

1. S. Perez and J.F. Revol, *MSA Bull.* **23** (1993) 28.
2. G. Cox, Ed. Special Issue on Confocal Microscopy (Proceedings of a Conference held in Heidelberg, in 1999). *Micron* **32** (7) (2001).
3. D. Shindo and K. Hiraga, *High-Resolution Electron Microscopy for Materials Science* (Springer, Tokyo, 1998).
4. D.C. Martin, J. Chen, J. Yang, L.F. Drummy and C. Kübel, *J. Polym. Sci. B Polym. Phys.* **43** (2005) 1749.
5. J.C.H. Spence, *High-Resolution Electron Microscopy*, 3rd ed. (Oxford University Press, Oxford, 2003).
6. P.W. Hawkes and J.C.H. Spence, Eds. *Science of Microscopy*, 2 volume set (Springer, New York, 2006).
7. P.B. Hirsch, *Topics in Electron Diffraction and Microscopy of Materials* (CRC Press, Boca Raton, 1999).
8. H.W. Carter and J.M. Harb, *MSA Bull.* **23** (1993) 219.
9. P.J.B. Koeck, Ed. Topical Papers: Digital Electron Microscopy. *Microsc. Res. Tech.* **49** (3) (2000).
10. Q. Zhong, D. Inniss, K. Kjoller and V.B. Elings, *Surf. Sci.* **290** (1993) 688.
11. S.N. Magonov, in *Encyclopedia of Analytical Chemistry*, edited by R.A. Meyers (John Wiley & Sons Ltd, Chichester, 2000), p. 7432.
12. G.K. Bar and G.F. Meyers, *MRS Bull.* **29** (2004) 464.
13. G.J. Leggett, M.C. Davies, D.E. Jackson, C.J. Roberts and S.J.B. Tendler, *Trends Polym. Sci.* **1** (1993) 115.
14. P. West and N. Starostina, *Microsc. Today* May/June (2003) 20.
15. J.I. Goldstein, D.E. Newbury, D.C. Joy, C.E. Lyman, P. Echlin, E. Lifshin, L.C. Sawyer and J. Michael, *Scanning Electron Microscopy and X-ray Microanalysis*, 3rd ed. (Springer, New York, 2003).
16. I.M. Ward, *Mechanical Properties of Solid Polymers* (Wiley, New York, 1983).
17. P.E. Gibson, M.A. Vallance, and S.L. Cooper, *Dev. Block Copolym.* **1** (1982) 217.
18. J.A. Reffner, *Am. Lab.* **32** (2000) 36.
19. J.M. Chalmers and N.J. Everall, *Int. J. Polym. Anal. Characteriz.* **5** (1999) 223.
20. J.L. Koenig, *Microspectroscopic Imaging of Polymers* (Oxford University Press, Oxford, 1998).
21. J.L. Koenig, *Infrared and Raman Spectroscopy of Polymers* (Rapra Technology, Shropshire UK, 2001).
22. I. Hopfe, G. Pompe, K.-J. Eichhorn and L. Haussler, *J. Mol. Struct., Molecular Spectroscopy and Molecular Structure 1994. 22nd European Congress, 11–16 Sept. 1994* **349** (1995) 443–6.
23. L.H. Sperling, *Polymeric Multicomponent Materials* (John Wiley, New York, 1997).
24. G.P. Simon, Ed. *Polymer Characterization Techniques and Their Application to Blends* (Oxford University Press, Oxford, 2003).
25. D.G. Vickroy, Ed. *Modern Methods of Analysis* (Analytical Research Dept., Hoechst Celanese, Summit Technical Center, Summit, NJ, 1983).
26. H.P. Klug and L.E. Alexander, *X-ray Diffraction Procedures* (Wiley, New York, 1974).
27. L.E. Alexander, *X-ray Diffraction Methods in Polymer Science* (Krieger, Huntington, NY, 1979).
28. J.V.N. Gilfrich, R. Jenkins, T.C. Huang, R.L. Snyder, D.K. Smith, M.A. Zaitz and P.K. Predecki, Eds. *Advances in X-ray Analysis* (Springer, Colorado Springs, CO, 1995).

29. N. Kasai and M. Kakudo, *X-ray Diffraction by Macromolecules* (Springer, Berlin Heidelberg New York, 2005).
30. B.D. Cullity and S.R. Stock, *Elements of X-ray Diffraction*, 3rd ed. (Prentice-Hall, Englewood Cliffs, NJ, 2001).
31. D.L. Bish and J.E. Post, Eds. *Reviews in Mineralogy, Modern Powder Diffraction* (Mineralogical Society of America, Washington, DC, 1989).
32. H. Tadokoro, *Structure of Crystalline Polymers* (Wiley-Interscience, New York, 1979).
33. W.R. Busing, *Macromolecules* **23** (1990) 4608.
34. P.H. Hermans and A. Weidinger, *Text. Res. J.* **31** (1961) 551.
35. A.M. Hindeleh and D.J. Johnson, *Polymer* **19** (1978) 27.
36. I.H. Hall and R. Somashekar, *J. Appl. Cryst.* **24** (1991) 1051.
37. A.M. Hindeleh and D.J. Johnson, *Polymer* **21** (1980) 929.
38. D.T. Grubb and D.Y. Yoon, *Polym. Commun.* **27** (1986) 84.
39. J. Blackwell and G. Gutierrez, *Polymer* **23** (1982) 671.
40. J.B. Stamatoff, *Mol. Cryst. Liq. Cryst.* **110** (1984) 75.
41. A. Romo-Uribe and A.H. Windle, *Proc. R. Soc. London A* **455** (1999) 1175.
42. C.J. Frye, I.M. Ward, M.G. Dobb and D.J. Johnson, *J. Polym. Sci. Polym. Phys. Edn.* **20** (1982) 1677.
43. I.H. Hall, Ed. *Structure of Crystalline Polymers* (Elsevier-Applied Science, London, 1984).
44. B. Natesan, H. Xu, B. Seyhan Ince and P. Cebe, *J. Polym. Sci. B Polym. Phys.* **42** (2004) 777.
45. I.W. Hamley, V. Castelletto, R.V. Castillo, A.J. Mueller, C.M. Martin, E. Pollet and P. Dubois, *Macromolecules* **38** (2005) 463.
46. W.W. Wendlandt, *Thermal Methods of Analysis* (Wiley-Interscience, New York, 1974).
47. E. Turi, Ed. *Thermal Characterization of Polymeric Materials* (Academic Press, New York, 1981).
48. M.E. Brown, *Introduction to Thermal Analysis* (Chapman and Hall, New York, 1988).
49. B. Wunderlich, *Thermal Analysis* (Academic Press, New York, 1990).
50. P.K. Gallagher, in *Materials Science and Technology*, edited by E. Lifshin (VCH Publishers, Weinheim, 1993).
51. B. Wunderlich, *Thermal Analysis of Polymeric Materials* (Springer, Berlin Heidelberg New York, 2005).
52. A. Ghijels, N. Groesbeck and C.W. Yip, *Polymer* **23** (1982) 1913.
53. G. Groeninckx and H. Reynaers, *J. Polym. Sci. Polym. Phys. Edn.* **18** (1980) 1325.
54. E. Bertin, *Principles and Practice of X-ray Spectrometric Analysis* (Plenum, New York, 1970).
55. R. Jenkins, *X-ray Fluorescence Spectrometry* (Wiley-Interscience, New York, 1988).
56. R. Jenkins, in *Materials Science and Technology*, edited by E. Lifshin (VCH Publishers, Weinheim, 1993).
57. D.J. Cutler, P.J. Hendra and G. Fraser, in *Developments in Polymer Characterization*, edited by J.V. Dawkins (Applied Science, London, 1980), p. 71.
58. R. Zbinden, *Infrared Spectroscopy of High Polymers* (Academic Press, New York, 1964).
59. P.R. Griffiths, *Chemical Infrared Fourier Transform Spectroscopy* (Wiley, New York, 1975).
60. H.W. Siesler and K. Holland-Moritz, *Infrared and Raman Spectroscopy of Polymers* (Marcel Dekker, New York, 1980).
61. P.C. Painter, M.M. Coleman and J.L. Koenig, *The Theory of Vibrational Spectroscopy and its Application to Polymers* (John Wiley, New York, 1982).
62. D.I. Bower and W.F. Maddams, *The Vibrational Spectroscopy of Polymers* (Cambridge University Press, Cambridge, 1989).
63. E.C. Faulques, D.L. Perry and A.V. Yeremenko, Eds. *Spectroscopy of Emerging Materials* (Springer, Sudak, Crimea, Ukraine, 2004).
64. A.B. Myers, Ed. Special Issue on Raman Microscopy and Imaging. *J. Raman Spectrosc.* **27** (8) (1996).
65. G. Turrell and J. Corset, Eds. *Raman Microscopy: Developments and Applications* (Academic Press, New York, 1996).
66. L.H. Lee, Ed. *Characterization of Metal and Polymer Surfaces: Polymer Surfaces* (Academic Press, New York, 1977).
67. J. Kressler, R. Schaefer and R. Thomann, *Appl. Spectrosc.* **52** (1998) 1269.

References

68. J.A. Reffner and P.E. Leary, *Microsc. Today* March (2006) 6.
69. J.G. Grasselli, M.K. Snavely and B.J. Bulkin, *Chemical Applications of Raman Spectroscopy* (Wiley, New York, 1981).
70. J.G. Grasselli and B.J. Bulkin, *Analytical Raman Spectroscopy* (John Wiley, New York, 1991).
71. R. Stevenson, A.C. Arias, C. Ramsdale, J.D. MacKenzie and D. Richards, *Appl. Phys. Lett.* **79** (2001) 2178.
72. F. Adar and H. Noether, *Polymer* **26** (1985) 1935.
73. R.J. Meier and B.J. Kip, *Microbeam Anal.* **3** (1994) 61.
74. E.D. Becker, *High Resolution NMR - Theory and Chemical Applications* (Academic Press, New York, 1980).
75. E.A. Williams, in *Materials Science and Technology*, edited by E. Lifshin (VCH Publishers, Weinheim, 1993).
76. C.A. Fyfe, *Solid State NMR for Chemists* (CFC Press, Guelph, 1983).
77. R.A. Komoroski, Ed. *High Resolution NMR Spectroscopy of Synthetic Polymers in Bulk* (VCH Publishers, Deerfield Beach, 1986).
78. M.J. Duer, *Introduction to Solid-State NMR Spectroscopy* (Blackwell Publishing, Oxon, UK, 2005).
79. J.R. Ebdon, in *Developments in Polymer Characterization*, edited by J.V. Dawkins (Applied Science, London, 1980), p. 1.
80. A.V. Cunliffe, in *Developments in Polymer Characterization*, edited by J.V. Dawkins (Applied Science, London, 1978), p. 1.
81. V.D. Nefedov, *X-ray Photoelectron Spectroscopy of Solid Surfaces* (VSP, Utrecht, 1988).
82. D. Briggs and M.P. Seah, Eds. *Practical Surface Analysis by Auger and X-ray Photoelectron Spectroscopy* (Wiley, New York, 1990).
83. D.L. Allara and P. Zheng, in *Materials Science and Technology*, edited by E. Lifshin (VCH Publishers, Weinheim, 1993).
84. D. Briggs, *Surface Analysis of Polymers by XPS and Static SIMS* (Cambridge University Press, Cambridge, 1998).
85. NIST, *X-ray Photoelectron Database, NIST Standard Reference Database 20* (NIST, Gaithersburg, MD, 1989).
86. N.H. Turner and J.A. Schreifels, *Anal. Chem.* **64** (1992) 302R.
87. S.G. Anderson, J. Leu, B.D. Silverman and P.S. Ho, *J. Vac. Sci. Technol. A* **11** (1993) 368.
88. S.J. Hinder, C. Lowe, J.T. Maxted and J.F. Watts, *J. Mater. Sci.* **40** (2005) 285.
89. D.E. Ramaker, *Crit. Rev. Solid State Mater. Sci.* **17** (1991) 211.
90. R.S. Stein and G.P. Hadzhoannou, in *Polymer Characterization*, edited by C.C. Craver (American Chemical Society, Washington, DC, 1983), p. 721.
91. R.J. Samuels, *J. Polym. Sci.* **9** (1971) 2165.
92. G.E. Wissler and B. Crist, *J. Polym. Sci. Polym. Phys. Edn.* **23** (1985) 2395.
93. A. Guinier and G. Fournet, *Small-angle Scattering of X-rays* (Wiley, New York, 1955).
94. O. Glatter and O. Kratky, Eds. *Small Angle X-ray Scattering* (Academic Press, London, 1982).
95. F.J. Balta-Calleja and C.G. Vonk, *X-ray Scattering by Synthetic Polymers* (Elsevier, Amsterdam, 1989).
96. C. Williams and R.P. May, in *Materials Science and Technology*, edited by E. Lifshin (VCH Publishers, Weinheim, 1993).
97. G. Bloch and A.J. Owen, *Coll. Polym. Sci.* **262** (1984) 793.
98. Y. Cohen and E.L. Thomas, *Macromolecules* **21** (1988) 433.
99. Y. Cohen and E.L. Thomas, *Macromolecules* **21** (1988) 436.
100. D.T. Grubb and K. Prasad, *Macromolecules* **25** (1992) 25.
101. K. Tashiro, M.M. Satkowski, R.S. Stein, Y. Li, B. Chu and S.L. Hsu, *Macromolecules* **25** (1992) 1809.
102. D. Suss, Y. Cohen and Y. Talmon, *Polymer* **36** (1995) 1809.
103. B.B. Sauer, R.S. McLean, D.J. Brill and D.J. Londono, *J. Polym. Sci. B Polym. Phys.* **40** (2002) 1727.
104. N.S. Murthy and D.T. Grubb, *J. Polym. Sci. B Polym. Phys.* **41** (2003) 1538.
105. R.W. Richards, in *Developments in Polymer Characterization*, edited by J.V. Dawkins (Applied Science, London, 1978), p. 117.
106. J.S. Higgins and H. Benoit, *Neutron Scattering of Polymers* (Oxford University Press, Oxford, 1993).
107. G.D. Wignall, H.R. Child, F.S. Bates, R.S. Cohen, C. Berney and R.J. Samuels. *Proc.*

IUPAC 28th Macromol. Symp., (Oxford, UK, 1982), p. 654.
108. C.V. Berney, R.E. Cohen and F.S. Bates, *Polymer* **23** (1982) 1222.
109. F.S. Bates, C.V. Berney and R.E. Cohen, *Macromolecules* **16** (1982) 1101.
110. W. Wu and G.D. Wignall, *Polymer* **26** (1985) 661.
111. B.J. Bauer, R.M. Briber and C.C. Han, *Macromolecules* **22** (1989) 940.
112. E.J. Kramer and H. Sillescu, *Macromolecules* **22** (1989) 414.
113. B. Crist, J.D. Tanzer and T.M. Finerman, *J. Polym. Sci. B Polym. Phys.* **27** (1989) 875.
114. D.M. Sadler, in *Structure of Crystalline Polymers*, edited by I.H. Hall (Elsevier-Applied Science, London, 1984), p. 125.
115. D.M. Sadler and P.J. Barham, *Polymer* **31** (1990) 36.

Appendices

APPENDIX I. ACRONYMS OF POLYMER NAMES

Acrylonitrile-butadiene-styrene	ABS
Acrylonitrile-butadiene rubber	ABR
Acrylic-styrene-acrylonitrile	ASA
Cellulose acetate	CA
Cellulose nitrate	CN
Chlorinated polyethylene	CPE
Diethyl triamine	DETA
Ethylene-propylene-diene monomer rubber	EPDM
Ethylene-vinyl acetate	EVA
High density polyethylene	HDPE
High impact polystyrene	HIPS
High modulus polyethylene	HMPE
Hydroxy propyl cellulose	HPC
High temperature nylon	HTN
Liquid crystal polymer	LCP
Low density polyethylene	LDPE
Polyacetal (see Polyoxymethylene)	
Polyacrylonitrile	PAN
Polyamide (nylon)	PA
Polybutadiene	PB
Poly(butylene terephthalate)	PBT
Polycarbonate	PC
Poly(ether ether ketone)	PEEK
Polyetherimide	PEI
Polyethersulfone	PES
Polyethylene	PE
Poly(ethylene oxide)	PEO
Poly(ethylene terephthalate)	PET
Polyimide	PI
Poly(methyl methacrylate)	PMMA
Polyoxymethylene	POM
Polypropylene	PP
Poly(phenylene oxide)	PPO
Poly(phenylene sulfide)	PPS
Poly(p-phenylene benzobisoxazole)	PBZO or PBO
Poly(p-phenylene benzobisthiazole)	PBZT
Poly(p-phenylene terephthalamide)	PPTA
Poly(p-xylylene)	PPX
Polystyrene	PS
Polysulfone	PSO
Poly(tetrafluoroethylene)	PTFE
Polyurethane	PUR
Poly(vinyl acetate)	PVAC
Poly(vinyl alcohol)	PVOH
Poly(vinyl chloride)	PVC
Poly(vinylidene chloride)	PVDC
Poly(vinylidene fluoride)	PVDF
Resorcinol-formaldehyde latex	RFL
Styrene-acrylonitrile copolymer	SAN
Styrene-butadiene rubber	SBR
Styrene-butadiene-styrene	SBS

APPENDIX II. ACRONYMS OF TECHNIQUES

Note: The same acronym is often used to denote the microscope (e.g., SEM, "scanning electron microscope") or the microscopy (e.g., SEM, "scanning electron microscopy").

Technique	Acronym
Analytical electron microscope	AEM
Atomic force microscope	AFM
Backscattered electron imaging	BEI
Confocal scanning laser microscope	CSLM (also LCSM)
Confocal scanning optical microscope	CSOM
Differential interference contrast	DIC
Energy dispersive x-ray spectroscopy	EDS
Field emission scanning electron microscope	FESEM
Frictional force microscope	FFM
High pressure scanning electron microscope	HPSEM
High resolution scanning electron microscope	HRSEM
High resolution transmission electron microscope	HRTEM
Infrared spectroscopy	IR
Laser confocal scanning microscope	LCSM
Lateral force microscope	LFM
Magnetic force microscope	MFM
Microdiffraction	μdiff
Near-field optical microscope	NFOM
Nuclear magnetic resonance	NMR
Optical microscope	OM
Phase contrast microscope	PC
Polarized light microscope	PLM
Scanning electron microscope	SEM
Scanning probe microscope	SPM
Scanning thermal microscope	SThM
Scanning transmission electron microscope	STEM
Scanning tunneling microscope	STM
Scanning tunneling spectroscopy	STS
Secondary electron imaging	SEI
Selected area electron diffraction	SAED
Small angle neutron scattering	SANS
Small angle x-ray scattering	SAXS
Transmission electron microscope	TEM
Conventional TEM	CTEM
Wavelength dispersive x-ray spectroscopy	WDS
Wide angle x-ray scattering	WAXS

APPENDIX III. MANMADE POLYMER FIBERS

Fiber type	Generic name
Cellulosic	Acetate
	Rayon
	Triacetate
Noncellulosic	Acrylic
	Aramid
	Copolyester
	Fluorocarbon
	Nylon
	Polybenzimidazole (PBI)
	Polyester (PET, PEN)
	Polyethylene
	Polypropylene
	Spandex
	Ultrahigh molecular weight PE (UHMWPE)

APPENDIX IV. COMMON COMMERCIAL POLYMERS AND TRADENAMES FOR PLASTICS, FILMS, AND ENGINEERING RESINS*

Generic name	Tradename	Manufacturer	Typical end uses
Acrylonitrile-butadiene-styrene (ABS)	Novodur Magnum Cycolac	Bayer Dow SABIC	Automotive, appliance housings, furniture, construction, consumer electronics, pipes
Epoxy, rubber toughened epoxies		Dow BASF	Paints, coatings, adhesives, pipes, circuit boards
High impact polystyrene (HIPS)	Styron	Dow	Automotive, appliance housings, furniture, toys, packaging, housewares, audio and video cassettes, dinnerware, etc.
High density PE (HDPE)		Dow	Containers, pipes, fabricated parts
Low density PE (LDPE)	Ultramid	BASF Dow Eastman	Packaging, films for bags, stretch wrap
Nylon: polymer and resin	Vydyne Zytel Celanese Nylon 6,6	Monsanto DuPont Ticona	Carpet yarns, tire cords, automotive, electrical, cigarette lighters, sporting goods, brushes
Polybutadiene in copolymers and blends			Tires, rubber articles, encapsulation
Poly(butylene terephthalate) Thermoplastic polyester (PBT)	Celanex Valox	Ticona SABIC BASF	Automotive and other fabricated parts, bearings, housings
Polycarbonate	Lexan Calibre Makrolon	SABIC Dow Bayer	Bottles, safety glass, auto lenses, helmets, aircraft interiors, optical media, sheets and profiles, electrical and lighting
Toughened polycarbonates	Cycoloy Pulse Xenoy (w/EPDM)	SABIC Dow SABIC	Automotive, vacuum cleaners, computer and business machines, transportation
Poly(ether ether ketone) (PEEK)	Victrex	Victrex Manuf.	Cable insulation, coatings, composites, automotive, industrial, chemical, aerospace, electrical/electronic (E/E)
Polyetherimide) (PEI)	Ultem	SABIC	Aerospace seats, lights, wiring, films, tapes
Polyether sulfone (PES)		BASF	Electrical applications, industrial, automotive, medical
Poly(ethylene terephthalate) (PET)	Mylar THERMX	DuPont Eastman	Films for packaging, coatings, containers, bottles
PET engineering resins	Petlon Rynite Vandar Impet	Bayer DuPont Ticona	Extrudates and moldings, E/E connectors, sockets, sensors, bottles, recording tapes, electrical insulation
Polyimide (PI)	Kapton	DuPont	Printed circuit boards, insulation, films for motors, adhesives, electronics

Generic name	Tradename	Manufacturer	Typical end uses
Poly(methyl methacrylate) PMMA	Plexiglas	Arkema	Camera lenses, airplane windows, signs, molded parts, sheeting
Polyoxymethylene (POM)	Celcon Delrin	Ticona Dupont	Automotive fuel systems, E/E, plumbing, pump parts, appliances, electrical gears, zippers
Poly(phenylene oxide) (PPO) and PPO-HIPS blends	Noryl	SABIC	Appliances, housings, pumps, shields, electronic components
Polypropylene (PP)		Dow	Carpet backing, ribbons, appliance housings, automotive, consumer durables, packaging, health
Polyphenylene sulfide (PPS)	Fortron Ryton	Ticona Phillips	High temperature applications in electronics, housings, industrial, automotive, lamps, lighting fixtures
Polystyrene (PS)	Styron	Dow	Packaging, lighting, dinnerware, medical ware, toys, gloss laminations and bottles, disposables, egg cartons
Polysulfone (PSO)	Udel	Solvay	Camera bodies, electrical connectors, light sockets, food appliance coatings, cookware, membranes
Poly(tetrafluoroethylene) (PTFE)	Teflon	DuPont	Solvent resistant coatings, films, industrial, E/E parts, medical
Ethylene tetrafluoroethylene (ETFE)			Automotive, architectural, chemical and food process industries
Poly(vinylidene fluoride) (PVDF)	Kynar	Arkema	Pipe fittings, seals, lab ware, aircraft parts
Poly(vinyl acetate) (PVAC)			Paints, adhesives, coatings
Poly(vinyl alcohol) (PVOH)	Elvanol	DuPont	Coatings, adhesives, cosmetics
Poly(vinyl chloride) (PVC)		Bayer	Food wrap, furniture covers, flooring, footwear, pipes
Poly(vinylidene chloride) (PVDC)	Saran	Dow	Films, protective packaging
Saturated styrene-butadiene-styrene block copolymers	Kraton G	Kraton Polymers	Fabricated parts
Styrene-acrylonitrile (SAN)			Dentures, lenses, auto and other fabricated parts
Styrene-butadiene latex			Adhesives, coatings, binders, textile finishes
Thermoplastic polyurethanes		Dow Bayer	Automotive
Thermotropic aromatic copolyesters (liquid crystal polymers) (LCP)	Vectra Xydar Zenite	Ticona Solvay Dupont	High temperature fabricated parts, E/E interconnects, connectors, medical equipment, aerospace, automotive and industrial parts

*Table is not intended to include all common commercial polymers, their tradenames or manufacturers; for up to date manufacturers, see company Web sites.

APPENDIX V. GENERAL SUPPLIERS OF MICROSCOPY ACCESSORIES

Suppliers of microscopes are found in Appendix VI and Appendix VII. These lists of suppliers are not intended to be all inclusive of U.S. and worldwide suppliers, nor are they intended as a recommendation by the authors or the publisher.

4pi Analysis, Inc.
919-489-1757
www.4pi.com

Advanced MicroBeam, Inc.
330-394-1255
www.advancedmicrobeam.com

Aetos Technologies, Inc.
334-737-3127
www.cytoviva.com

AJA International, Inc.
781-545-7365
www.ajaint.com

Allied High Tech Products
800-675-1118
www.alliedhightech.com

Ascend Instruments
503-614-8886
www.ascendinstruments.com

BAL-TEC/RMC
800-552-2262
www.baltec-RMC.com

Buehler
847-295-6500
www.buehler.com

Cameca Instruments Inc.
203-459-0623
www.cameca.com

Delaware Diamond Knives, Inc.
800-222-5143
www.ddk.com

Denton Vacuum, USA
856-439-9100
www.dentonvacuum.com

Diatome U.S.
215-412-8390
www.emsdiasum.com

E.A. Fischione Instruments, Inc.
724-325-5444
www.fischione.com

Energy Beam Sciences, Inc.
800-992-9037
www.ebsciences.com

Electron Microscopy Sciences
215-412-8400
www.emsdiasum.com

Ernest F Fullam Inc.
518-785-5533
www.fullam.com

ETS-Lindgren
630-307-7200
www.ets-lindgren.com

Gamma Vacuum
952-445-4841
www.gammavacuum.com

Gatan, Inc.
925-463-0200
www.gatan.com

Geller MicroAnalytical Laboratory
978-887-7000
www.gellermicro.com

Hamamatsu Photonic Systems
908-231-1116
www.whatifcameras.com

IXRF Systems, Inc.
281-286-6485
www.ixrfsystems.com

Kurt J. Lesker Co.
412-387-9200
www.lesker.com

Ladd Research
802-658-4961
www.laddresearch.com

Lumenera Corporation
613-736-4077
www.lumenera.com

Mad City Labs, Inc.
608-298-0855
www.madcitylabs.com

McCrone Microscopes & Accessories
800-622-8122
www.mccrone.com/mac/

M.E. Taylor Engineering Inc.
301-774-6246
www.semsupplies.com

Micro Star Technologies
936-291-6891
www.microstartech.com

QuantomiX, Inc.
480-205-4009
www.quantomix.com

Scientific Instruments & Applications, Inc.
770-232-7785
www.sia-cam.com

SEMTech Solutions Inc.
978-663-9822
www.semtechsolutions.com

South Bay Technology, Inc.
800-728-2233
www.southbaytech.com

SPI Supplies
610-436-5400
www.2spi.com

Ted Pella, Inc.
800-237-3526
www.tedpella.com

Thermo Electron Corporation
608-276-6100
www.thermo.com/microanalysis

APPENDIX VI. SUPPLIERS OF OPTICAL AND ELECTRON MICROSCOPES, MICROANALYSIS EQUIPMENT, IMAGE ANALYSIS AND PROCESSING

These lists of suppliers are not intended to be all inclusive of U.S. and worldwide suppliers, nor are they intended as a recommendation by the authors or the publisher.

Carl Zeiss MicroImaging, Inc.
800-233-2343
www.zeiss.com/micro

EDAX Inc.
201-529-6277
www.edax.com

Evex Analytical
609-252-9192
www.evex.com

FEI Company
503-726-7500
www.feicompany.com

Hitachi High Technologies America, Inc.
925-218-2800
www.hitachi-hta.com

JEOL USA, Inc.
978-536-5900
www.jeolusa.com

Leica Microsystems Inc.
800-248-0123
www.leica-microsystems.com

Nikon Instruments Inc.
631-547-8535 x8500
www.nikonusa.com

Olympus Industrial America
845-398-9480
www.olympusamerica.com

Olympus Soft Imaging Solutions
888-FIND-SIS
www.soft-imaging.net

Oxford Instruments America, Inc.
978-369-9933
www.oxinst.com

Princeton Gamma-Tech Instruments, Inc.
609-924-7310
www.pgt.com

SII NanoTechnology USA Inc
818-280-0745
www.siintusa.com

Tescan USA Inc.
724-772-7433
www.tescan-usa.com

Thermo Electron Corporation
608-276-6100
www.thermo.com/microanalysis

APPENDIX VII. SUPPLIERS OF SCANNING PROBE MICROSCOPES AND RELATED SUPPLIES

These lists of suppliers are not intended to be all inclusive of U.S. and worldwide suppliers, nor are they intended as a recommendation by the authors or the publisher.

Agilent Technologies, Molecular Imaging
480-756-5900
www.molec.com

Ambios Technology, Inc.
831-429-4200
www.ambiostech.com

Anasys Instruments, Inc.
805-455-5482
www.anasysinstruments.com

Asylum Research
888-472-2795
www.AsylumResearch.com

BioForce Nanosciences, Inc.
515-233-8333
www.bioforcenano.com

BudgetSensors
877-521-1108
http://www.budgetsensors.co

Hysitron
952-835-6366
www.hysitron.com

Infinitesima Ltd
44-1865-811-171
www.infinitesima.com

Image Metrology
877-521-1108
www.imagemet.com

JEOL USA, Inc.
978-536-5900
www.jeolusa.com

MikroMasch
503-598-9828
www.spmtips.com

Nanonics Imaging Ltd.
866-220-6828
www.nanonics.co.il

Nanoscience Instruments, Inc.
888-777-5573
www.nanoscience.com

Nanosensors
877-521-1108
www.nanosensors.com

Nanotech America/NT-MDT
972-954-8014
www.nt-america.com

NanoWorld AG
877-521-1108
www.nanoworld.com

nPoint Inc
608-310-8770
www.npoint.com

Pacific Nanotechnology, Inc.
800-246-3704
www.pacificnano.com

PSIA
408-986-1110
www.psiainc.com

Team Nanotec
805-696-9002
www.team-nanotec.de

Veeco Instruments
805-967-1400
www.veeco.com

Veeco Probes
805-696-9002
www.veecoprobes.com

WITec Instruments Corp.
877-948-3201; 217-351-9705
www.witec-instruments.com

Index

A

Abbe offset errors   115
Abbe theory of imaging   73
Aberration(s)   73–75
  chromatic   40, 73–75, 439
  in SEM   86–87
  spherical   73–75, 439
  in TEM   74–77
Aberration corrected EM   41, 438–440
Abrio   438
ABS. *See* Acrylonitrile-butadiene-styrene
Acid etching   168, 183–184
  literature review   183
Acrylics   200, 303
  embedding media   151
  staining of   178
Acrylonitrile-butadiene-styrene (ABS)   2, 8, 151, 309, 345
  etching of   182, 184, 196
  staining methods   167–168
Acrylonitrile-chlorinated polyethylene-styrene (ACS)   309
Acrylonitrile-styrene-acrylate (ASA)   167
  with polyurethanes   345
  staining methods for   167–168
Additives   11, 217. *See also* Composites
Adhesion   8. *See also* Adhesives
  AFM of   47, 102–105
  in composites   216–218, 354, 362
  failure   367, 396
  in multiphase polymers   323, 338

Adhesive(s)   2, 380–398
  interfaces by EFTEM   157
  "Post-it"   387
  RFL   268–269, 387
Adhesive forces in AFM   109–114
AEM. *See* Analytical electron microscopy
Aerospace applications   7, 8, 387, 408
AFM. *See* Atomic force microscopy
AM-AFM. *See* Amplitude modulation AFM
Amorphous polymers   4, 8, 316–318
  diffraction from   69, 122
  oriented   14, 35, 70
  structure in TEM   281–282
Amplitude modulation AFM (AM-AFM)   110
Analysis. *See* Analytical microscopy, Analytical Imaging, Electron Probe, Failure analysis, Image analysis, Microanalysis; Thermal analysis
Analytical electron microscopy (AEM)   44, 55–56
Analytical imaging   459–468
  EELS   461–462
  FTIR microscopy   459–460
  imaging surface analysis   464–468
  Raman microscopy   460–461
  x-ray microscopy   462–464
Analytical microscopy   53–56, 459–468

Anisotropic materials   6, 35, 81–83, 270, 315, 403–404
Annealing   280, 288, 343
Antistatic
  additives   356
  devices   103, 154
  sprays   222
Anti-tank missiles   459
Apertures
  in illumination   78–80
  objective lens   30, 71–2
  optical sectioning and   454
  SAED   44, 71–2
  SEM final lens   36–42, 86–87
Aplanatic lenses   78
Apochromats (Apo)   32
aPP. *See* Atactic polypropylene
Applications of microscopy   47, 248–434
  adhesives   381–387
  composites   354–380
  emulsions   380–385
  engineering resins and plastics   308–354
  fibers   250–276
  films   276–294
  liquid crystalline polymers   398–418
  membranes   294–308
Aramid(s)   8, 270–273, 399
  etching   189–193
  fibers   255, 270, 272, 399–403, 409–411
  fractures in   255
  hollow fiber membranes from   305
  liquid crystalline polymers   270, 272, 399–402

Argon
  etching 189–193
  in sputter coating 202–206, 232
Aromatic copolyesters 8, 399, 412
  high modulus fibers 409–412
  LCPs 399–412
Aromatic polyamides, See Aramids
Aromatic polymers 1, 119
  beam damage in 78
Artifacts 488–492
  in amorphous films 282
  from beam damage 121–123, 209–210
  charging effects and 207–209
  etching and 181–194, 490
  in FESEM 210–211
  in microtomy 160
  in OM 489
  polishing and 142–143
  replication and 197, 490
  in SEM 207–211, 489–490
  in SPM 114–118, 490–491
  in TEM 121–123, 282, 491
  from AFM tips 117–118, 490–491
  in x-ray microanalysis 491–492
ASA. See Acrylonitrile-styrene-acrylate
Ashing 357–358
Atactic polymers, definition 3
Atactic polypropylene (aPP) 3, 159, 167
Atactic polystyrene (aPS) 4, 141, 186, 281
Atomic force microscopy (AFM) 46–51, 97–118, 140–142, 481. See also Contact mode AFM; Noncontact AFM; Intermittent contact mode AFM
  Adhesive forces in 109–114
  applied to
    fibers 271–272
    films 286–290
    latex 142, 389–393
    membranes 297
    multiphase polymers 110–111, 159, 340
    nanocomposites 377, 380
    resins and plastics 311, 331–337
    single crystal 136
  calibration 58–59
  cantilevers 46–50, 97–99, 101–102
  conductive 48
  contact mode 47–48, 102–105, 442
  elasticity and 104–105, 330, 451
  feedback in 47, 113
  force-separation curves 101–102
  imaging 97–118
  indentation and 47, 223
  in-situ deformation 223–225
  non-contact mode 47, 49–50, 101, 112
  probe-specimen interactions in 100–102
  properties summary 29
  Raman microscopy with 461
  image resolution 114
  and SAXS 500
  scanners 99–100, 114–115
  Introduction 46–51
  staining for 166
  video rate 445
Atomic number contrast (Z-contrast) 36–38, 217, 219, 350, 452, 456. Also Compositional contrast
Atomic resolution 46, 103
Attenuated total reflection (ATR) 184, 459
Auger spectroscopy 498–499
Automated SPM 449–451
Axial fiber splitting 255

B

Back focal plane 30, 33, 71, 78–79
Backscattered electron imaging (BEI) 37–38, 54, 217, 481, 482
  characteristics of 37–38, 55, 88
  contrast and 254
  compared to SEI 38–39
  fiber studies 254, 268
  mineral filled composites 351
  multiphase polymers 350
Backscattered electrons (BSE) 37–38, 88–92
  detectors for 37, 95
  directionality of 91
Backscattering coefficient 37, 89
Bakelite 354
Banded structures (liquid crystalline polymers) 402, 405, 412–413
Barrel temperature effects 312
BCB. See Benzocyclobutene
Beam damage 118–124
  artifacts in SEM 209–210
  and thermal stability 78, 120
Becke line method 34, 252
BEI. See Backscattered electron imaging
Bend contours 43
Bending, in situ 59
Benzocyclobutene (BCB)
  as coating 15
  photodefinable 291
Berek (rotary) compensator 84
Bertrand lens 35
BF. See Bright field
Biaxial materials 14, 81–82, 118, 277
  LVSEM of blown film 287
Binocular stereomicroscopes 31, 132, 481, 483
Biocompatibility 346
Biodegradable polymers 347–349
Biopol 347
Biostability 346
Birefringence 11, 35, 81–84, 438
  biaxial 82
  definition of 35
  in fibers 251–253
  in films 283, 288
  measurement 83–85, 253
  negative, positive 82
  refractive index and 35, 81
  uniaxial 82
Birefringence imaging 438–439
Bisphenol A (BPA) 329
  epoxy resins 180
  Polybutylene terephthalate blends 168
  Polycarbonate/polyethylene blends 178
Blends 3, 11, 16, 141, 331–338, 496
  AFM 111–113
  freeze fracture 231
  by layering 16

phase domains 329, 369, 412, 482
specimen preparation of 153–196
toughening 309, 323–326, 329, 339
staining of 310, 329–331, 339–350
with LCPs 329, 409
Block copolymers 2–3, 141, 337–345
amorphous, crystalline 337
examples of 159, 170, 177, 190, 223, 337–339
Blow molding 9, 14, 311–312
Boil-in bags 373
Bond breaking 414, 442
mass loss 121
in radiation damage 118–123
BOPP. *See* Biaxially oriented polypropylene
Bottle films 283
Bottles 373
Bowing in SPM scanners 115
BPA. *See* Bisphenol A
Bräce-Köhler compensator 84
Bragg's law 69, 77
Bright field (BF) imaging 30, 32–33, 481
in TEM 42–43, 53, 341
in OM 32, 150
in STEM 154, 286
Bright field defocus phase contrast 280
Brittle fracture 5, 214–215, 231, 255, 327, 349, 361
Brittle matrix polymers 325
Bromine 179
Bromobenzene 135
BSE. *See* Backscattered electrons
Bulletproof vests 270, 272

C

CA. *See* Cellulose acetate
Cables 270, 352–353
Calcium carbonate filler 217, 370
Calibration 57–59
in AFM 58–59
by thermal tuning 59
Canadian Balsam 145
Cantilever(s) in AFM 46–51, 58, 97–101
deflection 102
force modulation imaging 104
harmonic imaging 443
oscillation in ICAFM 49, 106–110
Quality factor (Q) 98
Capillary forces
in contact mode AFM 48, 103
in IC-AFM 108, 112
Carbon
amorphous, electron diffraction 70
coatings 19, 202–204, 207
support films 134–135, 138, 198, 203
Carbon black filled polymers 8, 32, 150, 284, 354
AFM of 336–337
OM of 368
TEM of 369
Carbon black filled rubber 368–369
Carbon fiber composites 8, 356–357, 365–366
OM of 144, 365
SEM of 366–377
specimen preparation method for 142–144
Carbon fibers, etching of 189–190
Carbon nanotubes 50–51, 219, 375–376, 380
Carbon replicas 198–200, 314, 386
Carboxyl terminated butadiene-acrylonitrile (CTBN) 327
modified epoxy 389
CARS. *See* Coherent anti-Stokes Raman scattering
Cast films 137–139
CB. *See* Chlorobutyl rubber
CCD. *See* Charge-coupled device
Celgard 165, 206, 303–305
stained with OsO_4 304
Cellulose acetate (CA) 200, 301
etching of 181, 185, 271
hollow fiber membranes 301
in replica formation 200, 314
RO membrane 300
Cellulose nitrate 300
Cellulose fibers
as filler 273, 354
staining 161, 178–179
Ceramic, as filler 354
Cerium hexaboride 40
CFE. *See* Cold field emission
CFM. *See* Chemical force microscopy
Chain folded structure 5–6
Channel plate electron multiplier detector 95
Chaotic advection 16, 293–294, 326, 376
Characteristic x-rays 53, 88, 91
Characterization techniques summary 17–21
microscopy techniques 18, 480–488
non-microscopy techniques 492–500
spatial resolution 18, 29, 72–74, 114, 123
Characterizers in AFM 117
Charge-coupled device (CCD) 28, 30, 124
for TEM 59
Charging effects 90, 94, 201, 207
SEM artifacts and 207–209
Charpy tester. *See* Impact
Chemical force microscopy (CFM) 441–442
polymer applications 442
Chemical microscopy 459
Chemical/solvent etching 181–183
Chemical vapor deposition (CVD) 15
Chlorinated polyolefin (CPO) 394
Chlorobutyl rubber (CB) 177
Chlorosulfonic acid staining 162, 173–175, 179–180
applications 162, 173–174, 185, 281, 284
literature review 173
Chromatic aberration 73–75
with EFTEM 75
in HREM 77
Chromic acid 183–184, 196
Chromium oxide 143
Circuit boards 291, 373
Circular polarization 80
Clay filler 217, 358
in nanocomposites 374–379
Cleavage plane splitting 146

Clouding temperature 135
CLSM. *See* Confocal laser scanning microscopy
Coatings. *See* Conductive coatings for EM specimens; Polymer coatings
COC. *See* Cycloolefin copolymers
CoContinuity 309
Coherent anti-Stokes Raman scattering (CARS) 461
Coherent light, definition 68
Cohesive failure 367
Cold drawing process 10
Cold FEG 40, 87–88, 304. *Also* Cold field emission (CFE) source
Cold stage, examples of use 176, 228, 232–233, 385
Collodion support films 134, 197, 200–201
Colloids 380, 394–398
Compact disks (CDs) 317
Compensators, polarized light 35, 84–85, 252–253
  Babinet, Berek, Elliptic, Senarmont 84
  first order red plate 84
  quartz wedge 84
Competitive analysis 349–353
Complementary techniques 1
  in microscopy 138–141, 184, 266, 268, 301, 394, 481
  non-microscopy 17–18, 250, 492–500
Composite membranes 295
Composites 8, 354–380. *See also* Nanocomposites
  adhesion in 354, 360
  applications 354–355
  carbon black filled rubber 368–369
  carbon fiber 356–357, 365–366
  characterization of 357–363
  conductive fillers 356, 373
  cryofracture 227
  fillers in 354
  fracture 217, 354
  graphite fiber 356–357, 365–366
  hybrids 357
  OM of 355, 357–359
  particle filled 366–370
  processing of 354–355
  SEM of 355, 359–362
  TEM of 357, 362–364
Compositional contrast 36–38, 217, 219, 350, 452, 456. *Also* Atomic number contrast
Compositional mapping 464
Compounding 11–12
Compound microscopes 31, 483
Compression stages 59
Compression molding 14, 152, 311
Compressive strength for LCPs 178
  kink bands and 416
Condenser lenses 30, 68, 78
  for DIC 79
  for phase contrast 79
Conductive AFM 48
Conductive coatings for EM specimens 201–211
  artifacts 204–211
  carbon 202–204, 207
  coating devices 202–203
  high resolution 202–203, 205–206
  and LVSEM, VPSEM 202
  metals for 203, 317
  produced with IBS 204–206, 317
  shadowing with 203
  for SEM (and STM) 203–207
  sputtered 202, 204–206, 284
  for TEM 203
  vacuum evaporators and 202
Confocal laser scanning microscopy (CLSM) 21, 143, 232, 358, 436–441
  for nanocomposites 374
  optical sectioning and 454
Conoscopic view 35
Constant height mode AFM 444
Constant signal mode SPM 45
Contact microradiography 356
Contact mode AFM 47–48, 102–105. *See also* Noncontact AFM, IC-AFM
  capillary forces 103
  electrostatic forces 103
  force modulation imaging 48, 104–105
  force volume imaging 104, 442
  tips 115–118, 490
Continuous chaotic advection blender (CCAB) 16, 17, 294
Continuum x-rays 54
Contrast 18–19, 28–29, 72 *See also* Differential interference contrast; Phase contrast
  atomic number (or compositional) 36–38, 217, 219, 350, 452, 456
  crystallographic 43
  diffraction 43, 112
  Hoffman modulation 33–34
  in SEM 38–39, 254
  in TEM 43
  topographic 36, 38, 254
Contrast transfer function (CTF) 72
Controlled environment vitrification system (CEVS) 233
Conventional transmission electron microscopy (CTEM). *See* Transmission electron microscopy
Copolyesters. *See* Aromatic copolyesters
Copolymers 2–3, 8, 292, 309. *See also* Blends; Block copolymers; Graft copolymers
  HAADF of 345
  LVSEM of 310
  random 8, 308, 340
  TEM of 337
Corona discharge 387
Correlative microscopy 488–489
Cosmetics, emulsions for 380
Coverage, of fabrics 258
CPD. *See* Critical point drying
CPO. *See* Chlorinated polyolefin
Crazes and crazing 4, 212, 217–222
  definition 4
  fractography 212
  in HIPS 218, 222

microtomy and 220
real time study 219
sample deformation methods 212, 221
SAXS and 221
preparation for TEM 219–221
Creep in AFM scanners 115
Critical fiber length 355, 360
Critical point 230
Critical point drying (CPD) method 20, 230–231
Critical pressure 230
Crossed-beam FIB microscopes 453
Crossed polarizers (polars) 34–35, 83–85, 253
Cross-linkable epoxy thermoplastics (CET) 327
Crosslinking reaction 3, 119
of polybutadiene 381
in radiation damage 119–120
Crossover energies in LVSEM 90, 94, 208
Cryodeformation 219
Cryo-FESEM 232. *See also* Cryomicroscopy
Cryogenic specimen preparation 226–232
Cryomicroscopy 53, 232–234
Cryomicrotomy 146, 154–157, 169
cryoultramicrotomy 146, 154–157, 292, 352
knives 154
of latex 382
of nanocomposites 376–378
for SPM 158, 450, 487
Cryo-polishing 328
Cryo-SEM 232–233
Cryo-TEM 233–234, 394–398
artifacts and 491
of nanoparticles 395–398
with SAXS 499
Cryoultramicrotomy 146, 151, 154–157, 292, 352
Crystalline melting temperature 4–5, 53, 113, 495
Crystalline polymers *See* Semicrystalline polymers
Crystallinity 4, 283, 500
and deformation in films 283

loss of, due to radiation 121–122
and Raman microscopy 460
transcrystallinity 321
Crystallographic contrast 43
Crystals 4–5. *See also* Single Crystals
AFM of 141
Diffraction from 69–71
Extended chain 7, 272
HREM of 45, 137
Liquid Crystals 399
Optical properties 81–82
in PE 5, 135–138, 174, 280
in TLCPs 186
Crystallization 5–7, 496
in-situ AFM 47, 159, 224
in-situ PLM 133, 278
in LVSEM 271
CTBN. *See* Carboxyl terminated butadiene-acrylonitrile
CTEM. *See* Transmission electron microscopy
CTF. *See* Contrast transfer function
Cyano-acrylate glue 141, 145
Cycloolefin copolymers (COC) 170

D

Dark field (DF) 30, 32–33. *See also* High angle annular dark field
optical microscopy 32
TEM 43, 53, 414
Deflection
of AFM cantilever 46–48, 99, 101–106
SEM display mode 254
Defocus
imaging 43, 139, 280
and phase contrast 43, 76–77, 280
optimum TEM 75–76
Deformation 4–5, 47, 212–213. *See also* In situ deformation, shear bands
in AFM 47, 102, 223–224
in copolymers 340
crazing and 221
cryodeformation 219
crystallinity and 283
and fracture 221–226
in PE 285

orientation and 35, 139, 250
of spherulites 283
stages for 59–60
Degradation
aminolysis 182
biodegradation 348
hydrolysis 13, 15, 123
radiation induced 118–122, 310
Delamination 356, 394, 396
Delustrant 165, 262
Depth of field 30, 42, 74–75, 87, 440, 485
Depth of focus 30, 74–75
DETA. *See* Diethylene triamine
Detachment replicas 201
Detector resolution 72–74
Detectors
AFM photo diode 99
BSE 38
CCD 28, 30, 59, 124, 458
channel plate 95
E-T 39, 92
IR 459
for LVSEM 95–96
Robinson 38
for VPSEM 96–97
x-ray 54–55
DF. *See* Dark field
Diamond knives for (cryo)ultramicrotomy 152–156, 223, 331, 382
oscillating 153, 160
Diamond-like carbon (DLC) 50–51
DIC. *See* Differential interference contrast
Dichloroacetic acid 182
Dichroic 34, 438
Dichroism 461, 497
Diethylene triamine (DETA) 168
Differential interference contrast (DIC) 33–34
condenser lenses 79
of etched surfaces 182, 185
for optical sectioning 454
of polished sections 144
video enhanced 384
Differential scanning calorimetry (DSC) 60, 492, 495–496
Differential thermal analysis (DTA) 495

Diffraction 44, 68, 482 *See also* Electron diffraction; Small angle x-ray scattering; X-ray diffraction
  in amorphous polymers 69
  contrast, in TEM 43
  electron diffraction 44, 69–71
  limit to resolution 72
  microdiffraction 44, 412–414
  SAED 44, 122, 275, 412, 481
  SAXS 5, 186, 297, 481, 499–500
  techniques listed 482
  X-ray diffraction 374, 482, 493–495
Digital
  imaging 31, 233, 489
  image analysis 56, 234, 275
Dimensional changes, radiation-induced 42, 122–123
Disintegration, as specimen preparation 137
Dispersed phases 321–348
  morphology 329, 369
  size of 311, 480
Dispersion, as specimen preparation 135
Distortion
  of AFM image 58, 114–115, 491
  of SEM image 87, 94–95
  of specimens 122, 147, 151
DLC. *See* Diamond-like carbon
DMA. *See* Dynamic mechanical analysis
Drying methods 151, 226–234. *See also* Critical point drying; Freeze drying
DSC. *See* Differential scanning calorimetry
DTA. *See* Differential thermal analysis
Dual-beam FIB microscopes 453
Dual-pass technique, NCAFM 112–113
Ductile fracture 215, 230, 255, 328
Ductile matrix polymers 145, 216, 325
DVDs 317–319
Dwell time 15
Dyes 161
Dynamic charging 208

Dynamic mechanical analysis (DMA) 495–496
Dynamic microscopy 59–60
  cold stage in 60, 233–234
  hot stage in 60
  tensile stage in 59–60, 87

E

Eastman 910 glue 148, 197
EBA. *See* Ethylene butylacrylate
Ebonite method for specimen preparation 177–178
  and tire cords 269–270
EDS. *See* Energy dispersive x-ray spectroscopy
EELS. *See* Electron energy loss spectroscopy
EFI. *See* Energy filtered image
EFM. *See* Electric force microscopy
EFTEM. *See* Energy filtering electron microscopy
E-GMA. *See* Ethylene-glycidyl methacrylate
Elastic properties
  AFM and 104–105, 224
  of silicon 98
  viscoelasticity 49, 104, 223
Elastic scattering 37, 42, 88, 285
Elastomers 3, 309
  with multiphase polymers 323, 326
  spherulites and 155, 329
Electric force microscopy (EFM) 50, 113
Electrons. *See* Backscattered electrons; Secondary electrons
Electron beam (E Beam) sputtering technique 51, 202
Electron diffraction 18, 44–45, 69–71, 121–122
  advantages of 486
  of films 280, 410
  example patterns 69–71, 122, 280, 412
  interpretation 69–70
  microdiffraction 44, 412–414
  selected area 44, 275, 481
Electron energy loss spectroscopy (EELS) 53, 461–462
  and EFTEM 157, 462

  parallel 461–462
  with STEM 138, 461–462
Electron microscopy (EM) 35–45, 69–78, 85–97, 438–440. *See also* Analytical EM; High resolution EM; Scanning EM
Electron probe microanalyzer (EPMA) 54–55, 484, 496
  compared to XRF 496
Electron sources 39–41
Electron spectroscopy for chemical analysis (ESCA). *See* X-ray photon spectroscopy
Electron yield 90, 208
Electrostatic forces in AFM 103, 108, 114
Elemental mapping by x-ray analysis 55–56, 464
  examples with SEM imaging 350–351, 394, 396
Elliptical polarized light 80
Elliptic compensator 84
EM. *See* Electron microscopy
EMAA. *See* Polyethylene-ran-methacrylic acid
Embedding
  in acrylic 151
  in epoxy 151, 154
  in GMA 154
  media for OM 149
  media for TEM and AFM 151–153
  in polyesters 151
Emulsion(s)
  2, 380–398. *See also* Latexes
  and latexes 381–385
  microemulsions 380–382
  polymerization 381–382
End point dose 120
Energy dispersive x-ray spectroscopy (EDS) 54–55, 484
  AEM 55
  elemental mapping 55–56, 351, 361–362, 396
  of metal loaded fibers 265
  with SEM, TEM 53
  WDS comparison 55, 484
Energy filtered image (EFI) 462
Energy filtering electron microscopy (EFTEM) 70, 461–462

Index

with EELS 157, 462
and interface adhesion 157
of PMMA/SAN blend 462
for thick samples 75
of unstained samples 161
Energy spread
of electron sources 40
and resolution 86
Engineering plastics and resins 2, 308–353
characterization of 309–311
extrudates 311–312
failure analysis of 349–350
molded parts 311–315
multiphase polymers 321–348
Environmental SEM (ESEM, HPSEM or VPSEM) 21, 41, 96–97, 452
applications 348–349
of blends 232, 329
conductive coatings and 202
detectors for 96–97
of hydrated materials 123, 348
of latex 382–383
EO. See Ethylene octane copolymer
EPDM. See Ethylene-propylene-diene monomer
EPMA. See Electron probe microanalyzer
Epoxy 3, 15, 309
as adhesive 386
as brittle matrix polymer 325
Cross-linkable epoxy thermoplastics (CET) 166, 327
crazing in 220
as embedding medium 143, 151, 154
Epotek 197
fracture 366
Liquid crystalline epoxy (LCE) 186
rubber toughened 174, 326–328, 389
Equatorial reflection 69, 415
ESCA. See X-ray photon spectroscopy
ESEM. See Environmental SEM
Etching 20, 52, 181–196
with acids 183–184
artifacts and 490
with FIB 194–195

freeze fracture 20, 231–232
ion/plasma 188–194
with permanganate acid 184–188
plasma 188–194
summary table 195–196
solvents for 181–183
with xylene 168
E-T detector. See Everhart-Thornley detector
Ethene-co-1-butene (PEB) 311, 330
Ethylene butylacrylate (EBA) 352
Ethylene octane (EO) copolymer 335, 393
Ethylene-propylene-diene monomer (EPDM) 154, 309
in blends 168, 336
Ethylene-propylene rubber (EPR) 331–334, 377
Ethylene vinyl acetate (EVA) 335, 393
Ethylene-vinyl alcohol (EVOH) 374
EVA. See Ethylene vinyl acetate
Evaporative coatings 202–204
and vacuum evaporators 198, 202
Everhart-Thornley (E-T) detector 39, 92
LVSEM and 95
Exfoliation 373
Extinction, in PLM
incomplete 402
positions 35, 83–84
Extraction replicas 201
Extrudates 11–17
of films 283, 288
of LCPs 403–409
of PBZT 411–412
Extrusion processes 12, 311–312

F
Fabric
coverage 258
hand 258
nonwoven 222, 258–259
OM of 251
protective 270
SEM of 258–259
woven 258
Failure analysis 349–353

False coloring 57, Color plate XIII
Fast Fourier transforms (FFTs) 234, 275, 297
Fast scanning SPM 444–445
Fatigue fracture
in composite polymers 354
of fibers 255, 257
Feedback control
in AFM 47, 110, 113
in SPM 45
FEG. See Field emission gun
FESEM. See Field emission scanning electron microscopy
FFM. See Frictional force microscopy
FFTs. See Fast Fourier transforms
FIB. See Focused ion beam
Fiber composites 355–357, 365–366
contact microradiography of 356
critical fiber length 355, 360
interfacial bond failure 217–218
OM of 357–359
SEM of 359–362
single polymer, PE in PE 271
specimen preparation 143–145, 217
Fiber finishes 197–198, 254, 360
Fibers 2, 250–276. See also High modulus fibers; Hollow fiber membranes; Microfibers; Textile fibers
aramid 193, 409–410, 417
formation of 9–11
fractography 213–214, 254–258
fracture summary 255
high performance 270–276
industrial 267–270
metal loaded 265–267
nanofibers 273–275
optical retardation of 253
peelback of 146–147
replication and 200
spider silk 275–276
TEM of 259–260
textile 251–265
with titanium dioxide 262
wood pulp 258, 270, 272–273

Fibrils. *See also* Microfibrils
  in crazes 4, 220–222
  in LCPs 417–418
  in membranes 304–306
  and SAXS 499
Field curvature 73
Field emission gun (FEG) 40–41, 44, 80, 304, 331
  in HRSEM 41
  properties summary 40
Field emission scanning electron microscopy (FESEM) 40–41, 87, 196, 288
  artifacts at low voltage 210–211
  of membranes 297, 304, 307
  of microfibrils 414
  of PVC 258
Field of view 32–33
Filled LCP moldings 407–408
Fillers 8, 11, 32, 53, 118, 217, 354. *See also* Particle filled composites
  mica 354, 366, 369–371
  minerals as 366
Films 2, 276–308. *See also* Langmuir-Blodgett films
  amorphous "structure" 281–282
  OM of anisotropic 283
  birefringence of 277, 283, 288
  blown 11, 277, 283–288
  bottle 283
  casting for TEM 137–139
  dichroic 34
  drawing for TEM 139–140
  extrusion of 283, 288
  formation of 9–11
  Formvar 134
  HAADF of 285
  industrial 282–294
  model studies of 278–281
  multilayered 292–294
  orientation classes of 277
  peelback of 146–147
  polyester 283
  polyimide 282
  refractive index of 283
  SEM of 284
  semicrystalline 280–281
  spherulites in 138
  SPM of 286–287
  surfaces of 140
  wettability of 287

Filter membranes 295
First-order red plate 35, 83–85, Color plate II
Flat film membranes 294–305
Fluor lens 32
Fluorescence microscopy 28, 454–456
Fluorescence yield 91
FMM. *See* Force modulation microscopy
Focal plane array 459
Focus 30, 85. *See also* Defocus
  Gaussian or geometric 75–76, 86
  Scherzer 75–76
  underfocus 76, 80
Focused ion beam (FIB) 440, 453–454
  etching with 194–195
Force modulation imaging *or* Force modulation microscopy (FMM) 48, 104, 279, 330
Force-separation curves in AFM 101–102
Force spectroscopy 47–48, 441, 444
Force volume imaging 105, 442
Form birefringence 35
Formvar films 134
Fountain (flow field) model 356
Fourier's theorem 69
Fourier transform infrared (FTIR) microscopy 459–460, 492, 497
  resolution 459
Fourier transform infrared (FTIR) spectroscopy 18, 273, 303, 383, 496–497
  dichroism with 497
Fractography 212–213
  of fibers 213–214, 254–258
Fracture 212–217. *See also* Fatigue fracture
  brittle 255, 349
  of carbon fibers 366
  of composite polymers 217, 354–357, 366
  crazing and 217–221
  of fibers, summary 255
  hackles 214, 350, 360
  of plastics 214–216
  SEM of 212, 214, 257

  of semicrystalline polymers 215
  standard physical testing 213–217
  types of 213–214
  at weld line 316
Free amplitude 49, 111, 450
Freeze drying sample preparation 20, 227
  examples of use 227–230
  TEM and 227
Freeze fracture-etching 20, 231
  examples 231–232
Freezing methods 226, 227–233
Freon 230–232
Frequency modulated detection in NC-AFM 112–113
Frequency sweep data 107–108
Frictional force microscopy (FFM) 47, 50, 103–104, 481
FTIR. *See* Fourier transform infrared
Fuel cells 297

G
G values 118–120
  definition of 118
  table of values 119
Gas path length in VPSEM 96–97
Gastric balloons 346
Gaussian focal plane 75–76, 86
Gelatin 197–201, 300
Geometric focal plane. *See* Gaussian focal plane
Glass fiber composites 354
  SEM of 218, 355, 359–364
  OM of 357, 360, 408
Glass fibers
  microscopy of 358, 392
  plasma etching of 190–192
  coating on 392
Glass knives 146, 152, 154, 292
  crazing and 220
  cryomicrotomy and 169, 187
Glass transition temperature 4, 113, 447, 495
Glycerol 138, 143, 279
Glycol methacrylate (GMA) 154
GMA. *See* Glycol methacrylate

Index

Gold
  backscattering coefficient 37
  -coated AFM tips 441
  colloid 117
  conductive coating for EM 203–205, 211
  decoration 19, 174, 211, 278
  for shadowing 203
Graft copolymers 3, 8, 337–345
  HIPS as 337
Grafted rubber concentrate (GRC) 327
Graphite. *See also* Highly oriented pyrolytic graphite
  fiber composites 356–357, 365–366
  substrates 137, 140–141
Gray (Gy), definition 120
GRC. *See* Grafted rubber concentrate

H

HAADF *See* High angle annular dark field
Hackle (fracture) morphology 213, 217, 256
  in composite matrix 217, 360–361, 367
  in engineering resins and plastics 349–350
  in fibers 256
Hand, of fabrics 258
Hard elastic polypropylene (HEPP) 223
Hardy microtome 147
Harmonic imaging in IC-AFM 443–444
  of PMMA/PS 444
  Low Quality factor (Q) and 443
  resonant cantilevers for 444
HDPE. *See* High density polyethylene
Heat aging 157, 254
Heat conduction 120, 356
Heated tip thermal microscopy 47, 447
HEPP. *See* Hard elastic polypropylene
Heteropolymers, definition 3
Hexacyanoferrate (HCF) 142
Hexafluoroisopropanol (HFIP) 137, 198

Heptane etching 346
HFIP. *See* Hexafluoroisopropanol
High angle annular dark field (HAADF) 44, 233, 456
  of nanoparticles 345
  of ionomer films 285
High density polyethylene (HDPE), *see also* Polyethylene (PE) 151, 186, 279
  deformation in 285
  end uses listed 508
  etching of 181, 184–186, 190
  fibers 271
  films 235, 271, 284
  microporous membranes 300
  replication methods 195
  single crystals 135
  spherulites 84
  staining of 167–168
  thin film specimens 83, 139
High impact polystyrene (HIPS) 3, 8, 94, 151, 155, 171
  3D imaging of 453
  AFM and TEM compared 155–156
  crazing in 218, 222
  end uses listed 508
  etching of 196
  graft copolymer 337
  multiphase polymer 309
  staining methods for 168, 171–172
Highly oriented pyrolytic graphite (HOPG) 133, 140, 286, 491
High modulus fibers 7, 137, 270–272, 409–412
  aromatic copolyesters 412
  aromatic polyamides 409–411
  LCPs 399
  PE 270–272
  rigid rod polymers 411–412
High modulus low shrink yarns (HMLS) 272
High molecular weight polymers 135, 288. *See also* Ultrahigh molecular weight PE
High performance polymers 398–418. *See also* Liquid

crystalline polymers (LCPs)
  extrusion of 403–409
  high modulus fibers 409–412
  high performance fibers 270–276
  LCPs 399–400
  moldings 403–409
High pressure SEM (HPSEM or VPSEM or ESEM) 21, 41, 59–60, 96–97, 452
  applications 348–349
  of blends 232, 329
  conductive coatings and 202
  detectors for 96–97
  of hydrated materials 123, 348
  of latex 382–383
High resolution coating devices 202–204
High resolution scanning electron microscopy (HRSEM) 21, 41, 86. *See also* Field Emission SEM,
  aberrations in 86
  compared to AFM 289, 305–307, 341
  FEG in 41
High resolution (transmission) electron microscopy (HREM, also HRTEM) 21, 45, 77
  of dispersed crystals 137
  lens aberrations and 77
  low dose (LD) 78, 275, 414–415
  nanofiber example 273–275
  specimen preparation for 137–138
  of PE 285
  of PVC 281
High speed spin-draw fiber process 10, 262–264
High temperature ashing 357
HIPS. *See* High impact polystyrene
HMLS. *See* High modulus low shrink yarns
High voltage electron microscope 53, 136, 292
Hoffman modulation contrast 33
  compared to DIC 34

Hollow fiber membranes 305–308
  of modified PEEK 307
  of PE 308
  of polyimide 307
  of polysulfone 305
  of PTFE 307
Homeotropic orientation 400
Homopolymers 2, 52, 313
  examples 309
HOPG. *See* Highly oriented pyrolytic graphite
Hot compaction process 15, 270, 271
Hot stage microscopy 60, 221, 329
  in AFM 224–225, 344, 393
  in polarized light microscopy 132, 284, 399
  in SEM 197, 222
HPSEM. *See* High pressure SEM
HREM. *See* High resolution electron microscopy
HRSEM. *See* High resolution scanning electron microscopy
HRTEM. *See* High resolution transmission electron microscopy
Humidity
  and forces in AFM 103–104, 490
  in CFM 442
Hybrid composite polymers 357
Hyperspectral imaging 451
Hysteresis in AFM
  in cantilever deflection 101
  in IC-AFM 109
  in scanner motion 115

I

IBS. *See* Ion beam sputter coating
IC-AFM. *See* Intermittent contact mode AFM
Illumination 29–30, 33–34, 73
  in CSLM 436
  in optical sectioning 454–455
Illumination systems 78–80
  for OM 78–79
  for TEM 80

Image analysis 19, 56–59, 309
  examples 145, 161, 227, 284, 331, 375
  in automated SPM 450–452
Image formation 28–31
  in AFM 97–118
  with lenses(OM, TEM) 29–30, 68–85
  radiation and 121–123
  in SEM 92–94
Image processing 19, 53, 56–57, 93, 455, 490
  examples 271, 329
Imaging. *See also* Analytical imaging; Backscattered electron imaging; Birefringence imaging, Force modulation imaging; Harmonic imaging; Image formation; Three dimensional imaging
  in AFM 97–118
  BEI 37–39
  force-volume imaging 105, 451
  with lenses (OM, TEM) 68–85
  lenticular, in 3D display 452
  phase, in AFM 49, 442
  SEI 39
  in SEM 35–39, 85–97
  in SPM 45–51
  structured light imaging 455
  in TEM 42–44
  tomographic spectral imaging 451
Impact (Charpy or Izod) test 190, 212, 216, 323
  on composite 360–361
Impact modified thermoplastics 328–337
Impact strength 2–3, 11, 16, 316, 479
Incident beam voltage 483
  in SEM 37, 41, 52, 88–90, 92, 94
  in TEM 53
Incident dose 120
Incident light techniques 28
Incoherent radiation, definition 68
Indentation and AFM 47, 223, 393. *See also* Nanoindentation
Index ellipsoid 81–82

Indicatrix 81–82
Indium-tin-oxide (ITO) coating 140, 141
Industrial fibers 267–270
Industrial films 282–294
Inelastic scattering 70, 88, 161
Infrared (IR) spectroscopy 18, 496–497. *See also* Fourier transform infrared spectroscopy
  dichroism 497
Infrared microspectroscopy (IMS). *See* Fourier transform infrared microscopy
Injection molding 12–14, 311–316
  reaction IM (RIM) 15
  of semicrystalline polymers 316
In-lens SEM design 86
In situ deformation 221–223, 311
  in AFM 223–225
  in SEM 222–223
  in TEM 223
Instron tensile tester 212–213, 255
Interaction volume
  in AFM 104
  in SEM 37–38, 88–89, 92, 94, 123
  in STEM 91
  in X-ray microanalysis 54–56, 91
Interference 68
  colors 35, 221
  contrast 33, 34. *See also* Differential interference contrast microscopy 33–34
  in X-ray microscopy 458
Interferometric optical profiler 436
Intermediate aperture 44, 71
Intermediate lens 30
Intermittent contact mode AFM (IC-AFM) *or* Tapping mode AFM 47–49, 105–106, 140
  applied to
    blends 334–337
    block copolymer films 290, 343

Index

cellulose fibers 273
etched spherulites 315–316
lithography process 291
microporous membranes 305–306
toughened thermoset 328
cantilevers for 98
cantilever oscillation 106–110
harmonic imaging 443
imaging parameters 110–111
interaction distance regime 101
modeling of 112
probe tips 51, 490
Quality factor (Q) 99, 109
Interpretation of images 488–492
Inversion walls (LCPs) 406
Iodine staining 52, 179, 348
Ion beam sputter coating (IBS) 206, 317
Ion etching 188–194
Ion microscopy 440–441
Ionomers 8, 285, 328
Ion Tech microsputter gun 191
iPP. See Isotactic polypropylene
IR. See Infrared
Isogyres 399, 401
Isoprene inclusion (and staining) method 165
  examples 178, 264–265
Isopropanol 141, 182
Isotactic polymers, definition of 3
Isotactic polypropylene (iPP) 3, 184–189, 224, 330
  chemical force microscopy of 442
ITO. See Indium-tin-oxide
Izod. See Impact

J
Jamin-Lebedeff interference microscope 33

K
Kapton 277
Kevlar 187, 272, 415–417
Kink bands 174, 255, 412
  compression and 272, 416
  HREM of 415
  in LCPs 413–416
  STM of 416
$KMnO_4$. See Permanganate acid

Knee replacement, UHMWPE for 224
Knives. See Diamond knives; Glass knives
Köhler illumination system 78–79
Kraton 187, 271, 339

L
LaB_6. See Lanthanum hexaboride
Lamellae or Lamellar crystals 4–6, 52, 480
  in block copolymers 399–343
  in PE 136, 280, 284–285
  and SAXS 499–500
  thickening of 279
Langmuir-Blodgett films (LB)
  AFM 140, 286, 450
  NSOM 449
Lanthanum hexaboride (LaB_6)
  source 40, 41
  SEM 41, 85, 87, 331
  TEM 44, 302
Laser confocal scanning microscopy (LCSM). See Confocal laser scanning microscopy
Laser-induced fluorescence spectroscopy of nanocomposites 374
Lateral force microscopy (LFM) 47, 103–104, 441
Lateral forces 104
Latex(es) 381–386. See also Resorcinol-formaldehyde-latex
  for calibration 57, 117
  characterization
    AFM 111, 389–393
    OM, SEM 383–384
    TEM 172, 176, 384–385
    VPSEM 383
  dispersion for TEM 135
  cryomicrotomy and 157, 382
  film coalescence 386
  film formation 9, 393
  freeze drying of 227
  particle size measurement 385–386
  replication methods for 386
  staining methods for 167, 170–173, 176

Lattice imaging 43, 45, 72, 76–78, 494
  in high modulus fibers 275, 409–410
Layer multiplying coextrusion 16, 292
LB. See Langmuir-Blodgett films
LCE. See Liquid crystalline epoxy
LCPs. See Liquid crystalline polymers
LCSM. See Confocal laser scanning microscopy
LDPE. See Low density polyethylene
Lead zirconate titanate (PZT) 99
Lennard-Jones potential 100–101
Lens(es). 28 See also Condenser lenses; Objective lenses
  aberrations in 40, 73
  acceptance angle 72–74
  aplanatic 78
  Bertrand 35
  chromatic aberration 40, 75, 86
  condenser lenses 29–30, 44, 78–80
    in SEM 36, 86–87
  contrast with 72–76
  diffraction in 68–69
  glass for OM 29, 32
  illumination systems 78–80
  imaging with 29–30, 68–85
  intermediate 30
  objective lenses 32, 75
  projector lenses 30, 74
  phase contrast with 76–78
  resolution 72–76
Lens-imaging microscopes 29–30
Lenticular imaging, in 3D display 452
LFM. See Lateral force microscopy
LFRTs. See Long fiber reinforced thermoplastics
Light emitting diodes 446, 449
Light microscopy. See Optical microscopy
Light scattering techniques 499
Linear low density polyethylene (LLDPE) 186

Linear polarization of light 80
Liquid crystalline epoxy (LCE) 186
Liquid crystalline polymers (LCPs) 7–8, 141, 185, 399–400, 494
 AFM of 406
 aromatic copolyester 400
 aromatic polyamide 400
 banded structure 402
 blends 408–409
 chemistry of 399–400
 domain texture 401–402, 405, 408–409
 films from 282
 high modulus fibers 399, 409–417
  formation of 9
  microfibrils in 415–417
  structural model 417–418
 lyotropic liquid crystals 398–399
 microstructure 400–403, 413–417
 nematic crystals 9, 399–401, 406, 408
 optical textures 400–402
 PLM 400
 rigid rod polymers 399
 smectic crystals 45, 399
 TLCPs 185–186, 270
Liquid crystals (LC) 399
Liquid nitrogen 53, 149, 226, 233, 382
Liquid sulfur 220
Lithography
 of AFM tips 50
 AFM for 287, 444
LLDPE. *See* Linear low density polyethylene
Local thermal analysis (LTA) 352–353
Long fiber reinforced thermoplastics (LFRTs) 11
Low density polyethylene (LDPE)
 AFM 315
 blends 16, 293–295, 348
 chemical force microscopy 442
Low dose, high resolution electron microscopy (LD-HREM) 45, 78, 137, 273

Low dose TEM 52–53
Low temperature RF plasma asher (LTA) 191, 357
Low voltage SEM (LVSEM) 52, 94–96, 133
 applied to
  blown PE film 287
  copolymers 310
  crystallization 271
 detectors for 95–96
 specimen charging and 94–95, 202, 207–208
 image comparison 93
 techniques compared 483
LTA. *See* Local thermal analysis; Low temperature RF plasma asher
LVSEM. *See* Low voltage SEM
Lyotropic 8

M

Macroemulsions 380
Magnetic force microscopy (MFM) 50, 112–113
Magnetic resonance imaging (MRI) 456
Magnetron sputtering 206, 232
Maleated polypropylene (MAH-g-PP) 375
Mapping
 elemental, by x-ray analysis 55, 266, 351, 372, 396, 464
 IR 460
Mass loss, by radiation damage 121
Mass spectroscopy (MS) 18
Mass thickness contrast 42–43, 221
Material safety data sheet (MSDS) 161, 162, 170
Matrix cleavage of composites 217, 357
Matrix cracking 217
Matrix polymers 33, 325
MCT. *See* Mercury cadmium telluride
Mechanical deformation. *See* Deformation
MEK. *See* Methyl ethyl ketone
Melting point 4–5, 9–11, 45, 53, 78, 120, 495

Membranes 2, 276–308. *See also* Hollow fiber membranes; Reverse osmosis membranes; Microporous membranes
 applications of 276, 289, 297–298, 301–302, 304, 307
 cast 276–277
 composite 295, 299–300
 flat film 294–305
 freeze fractured 300
 hollow fiber 230–231, 305–308
 microporous 295–296, 303–305
 reverse osmosis 296, 302, 305
 types of 277
Mercury cadmium telluride (MCT) 459
Mercuric trifluoroacetate, staining with 178–179
Meridional reflections 69
Metal decoration 19. *See also* Gold decoration
Metal loaded fibers 265–267
Metal shadowing 19, 199, 203
Methyl ethyl ketone (MEK) as etchant 182
Metripol 438
Mettler hot stage 279
MF. *See* Microfilters
MFM. *See* Magnetic force microscopy
Mica
 test object in AFM 109
 flakes 217, 279, 290
 as filler 354, 366, 369–370
 as substrate 133, 136–138, 278
Micelles 382, 395
 definition of 380
Michel-Levy (polarization color) chart 84
Microanalysis. *See* X-ray microanalysis
Microdiffraction 44, 412–414, 482
Microemulsion, definition of 380–381
Microfibrils, *or* microfibers 6, 10, 250, 418
 in fiber structure 250, 271–271
 examples of 262, 265, 414

Index

in LCPs 412, 414–418
in composites 157, 228–230
STM, SEM compared 416
TEM, Electron diffraction 414
Microfilters (MF) 295
Micro mar resistance (MMR) 224
Microporous membranes 295–296, 303–306
 examples 205–206, 211, 289, 303–306
Microscopes 28–31. *See also* Binocular stereomicroscopes; Microscopy
 compared 483
 compound 31
 crossed-beam *or* dual-beam FIB 453
 image formation in 68–124
 lens-imaging 29–30
 radiation damage in 118–124
 scanning-imaging 30–31
 simple 31
Microscopy 18–19, 21, 28–31
 analytical 53–56, 459–468
 calibration in 57–59
 cryomicroscopy 53, 60, 233–234
 dynamic 59–60
 quantitative 56–59, 386
 techniques compared 480–488
Microtomes 147–150
 rotary 148
 sledge 148
Microtomy 19, 146–160. *See also* Cryomicrotomy; Ultramicrotomy
 for 3D imaging 57, 452–454
 artifacts in 160
 block trimming for 152
 embedding for 149, 151
 mounting for 152
 microtomes 147–150
 for OM 147–150
 for SEM 150
 specimen mounting for 148–149
 for SPM 150–154, 158–159, 315
 for TEM 150–154

Microwave oven technique 151, 170–171
Minerals, as fillers 366
MiniSIMS 466
Mirror, as fracture surface region 29, 33, 213, 256
Mist, as fracture surface region 256
MMR. *See* Micro mar resistance
Modulation transfer function (MTF) 72
Modulus 2, 9–11, 223–224. *See also* High modulus fibers
Molding processes 11–17, 311, 403–409. *See also* Injection molding; Moldings, microstructure of
 blow molding 14
 compression molding 14, 311
 mold temperature 11, 356
 reaction injection molding (RIM) 15
 structure-property relations for 4–5, 8, 17
 thermoforming 14–15
Moldings, microstructure of
 spherulitic textures 5–7
 in fiber composites 355–356, 365–366
 in filled LCP resins 407–409
 cryofracture 227
 in LCPs 403–409
 skin-core morphology 12–13, 315–316
Molecular orientation 5, 13
 by birefringence 35, 251
 in fibers 250–253
 by NSOM 449
 in radiation damage 121
Molecular weight distribution 18
Monomer, definition 2
Morphology, definition of 1, 3–4
 introduction to 1–21
Mounting, of specimens 148–149, 152, 213
MPDI. *See* M-phenylene isophthalamide
M-phenylene isophthalamide (MPDI) 45, 137, 273
MRI. *See* Magnetic resonance imaging

MS. *See* Mass spectroscopy
MSDS. *See* Material safety data sheet
MTF. *See* Modulation transfer function
Muffle furnace 357
Multilayered films 292–294
 nano thermal analysis of 447
 PC/PET 159, 292
Multiphase polymers 3, 8, 47, 309, 321–349
 AFM of 159, 330, 332, 338
 biodegradable 347–349
 block copolymers and 337–345
 carbon black filled 368
 copolymers 337–344
 elastomers with 323, 345
 examples listed 309
 EPDM 336–337
 etching techniques for 185, 193–196
 FTIR and IC-AFM of 389
 graft copolymers and 337–345
 HIPS 155, 171
 impact modified thermoplastics 217, 328–337
 OM, PLM of 330, 332, 338, 484
 particle size in 8
 polyurethanes in 345–347
 processing of 326
 random copolymers and 337–345
 resins of 321–349
 SAN 462–463
 SBS 339, 343
 SEM of 216, 331, 318, 346, 350
 staining techniques for 167, 175, 181, 329, 341
 TEM of 159, 332–333, 339–341, 486
 toughened resins 323–326
 toughened thermoset resins 326–327
Multiwalled carbon nanotubes (MWCNTs) 219, 372, 376
MWCNTs. *See* Multiwalled carbon nanotubes

Mylar 277
  as extensilble substrate 138, 201, 278

N

NA. *See* Numerical aperture
Nanocomposites 8, 47, 354, 370–380
  AFM of 373, 377–378, 380
  carbon nanotubes 375–376, 380
  clay in 373, 376
  characterization of 374
  HRTEM of 376
  laser-induced fluorescence spectroscopy 374
  processing of 17
  SEM of 373, 376
  TEM of 373, 377, 379
Nanofibers 273–275. *See also* microfibrils
Nanofilters (NF) 295
Nanofoams 154
Nanoindentation 47, 223–224
  probes for 50, 116
Nanolayers 16
Nanotechnology 21, 219
Nanotubes. *See* Carbon nanotubes
National Institute for Standards and Technology (NIST) 58
Natural rubber 4, 166, 201, 382
NC-AFM. *See* Noncontact AFM
Near edge x-ray absorption fine structure (NEXAFS) 463–464
Near field scanning optical microscopy (NSOM) 449
  polarized, for molecular orientation 449
  with Raman microscopy 461
Negative birefringence 82
Negative staining 161, 168, 172
Negative phase contrast 76
Negative replicas 197
Nematic liquid crystal 399
  texture in LCPs 401–402
Neutron scattering techniques 4, 492, 500
NEXAFS. *See* Near edge x-ray absorption fine structure
NF. *See* Nanofilters

NIST. *See* National Institute for Standards and Technology
Nitric acid etching 183, 494
N-methylpyrrolidone 233
NMR. *See* Nuclear magnetic resonance
Noise *See also* Signal-to-noise ratio
  in AFM 58, 114
  in HREM 78
  in SEM imaging 86, 93–94, 210
  limit to resolution in EM 123–124
Nomarski interference contrast. *See* Differential interference contrast
Nomex 275
Noncontact AFM (NCAFM) 47, 49–50, 101, 112–113
  dual-pass technique with 112–113
  frequency modulated detection in 112, 113
Nonlinear geometric mixing 116
Nonlinear motion of AFM scanner 115
Nonperiodic layer (NPL) crystal 415
Nonwoven fabrics 222, 258–259
Noryl GTX 351
NSOM. *See* Near field scanning optical microscopy
Nuclear magnetic resonance (NMR) 18, 492, 497–498
Nucleation density 5, 313
Numerical aperture (NA) 32, 73
Nylon (*or* Polyamide). *See also* Aromatic polyamides
  composite with glass fiber 11, 217, 358, 362–364
  etching of 181–182
  fibers 175, 250
    fatigue failure of 256–257
  molded specimens 313
  nanocomposites 374–377
  polishing of 144
  PTA and 176
  particle filled 150, 367
  rubber toughening 157, 296
  PLM of spherulites in 6–7, 150, 313

SEM of fracture 215
  staining of 161–164, 167–169, 175, 180
Nyquist criterion 73

O

Objective aperture 30, 71, 80
Objective lenses 30–32, 71
  image formation by 71–73
  in TEM 71
OCT. *See* Optical coherence tomography
Off-axis aberrations 73–4
Off-axis reflections in fiber diffraction 69
OM. *See* Optical microscopy
Optical coherence tomography (OCT) 438–439
  for optical sectioning 455
  polarization sensitive 438
Optical microscopy (OM) 18–19, 31–35, 484–485. *See also* Polarized light microscopy
  applied to
    composite polymers 355, 360, 365, 368
    extracted filler particles 357–359
    fibers 251
    latex 384
    multiphase polymers 332, 338
  artifacts in 489
  basic optics of 29, 68–69
  birefringence imaging in 438
  calibration 57
  compound microscope 31
  confocal scanning microscope (CLSM) 21, 358, 436–437
  diffraction limit for 72–73
  dynamic hot stage 60, 132, 399
  fluorescence microscopy 28, 454–456
  illumination systems for 29–33, 78–79
  imaging modes 32–33, 481
  Köhler illumination 78–79
  microtomy for 147–150
  near field scanning 449
  phase contrast 33, 76
  polarized light 34, 83–85

resolution 18, 29, 72–74, 481, 483
specimen preparation methods for 132, 143–145, 147–149
stereo microscopes 28, 31, 132–133, 479–481
Optical path difference 68, 455
Optical sectioning 357, 454–455
apertures and 454
DIC for 454
Optical coherence tomography (OCT) for 455
STEM and 454–455
wide field 455
Optical texture, of LCPs 400–402
Optic axis
of birefringent object 81–84
of instrument 437, 454
Ortho-phosphoric acid. See phosphoric acid
Oscillating
cantilever, in AFM 49, 98–99, 106–110
knife in microtomy 153, 450
Osmium tetroxide (OsO₄)
staining 162–165, 180–181
examples of use 161–163, 166, 304
inclusion methods 164–165
of multiphase polymers 155, 329, 339
practical details 164
safety precautions when using 162
for SEM and SPM 165
for TEM 163–165
two step reactions 163
OsO₄. See Osmium tetroxide
Oversampling 74
Overvoltage 91

P
PA6 (Polyamide-6). See Nylon
PAA. See Polyacrylic acid
Packaging 11, 14, 276, 329
electronic 287, 296
food 373
Paints 140, 151
emulsions for 380
PAN. See Polyacrylonitrile
Paper, emulsions for 380

Parallel EELS (PEELS) 461–462
Parison 14
Particle Atlas 358
α-particle emitters, anti-static 103
Particle filled composites 366–370
carbon-black filled 368–369
Particle size 56, 159
distribution 114, 141, 177, 309
effect on film formation 393
measurement of 385–386
PBI. See Polybenzimidazole
PBO. See Polybenzobisoxazole
PBT. See Polybutylene terephthalate
PBZO. See Poly-p-phenylene-benzobisoxazole
PBZT. See Poly-p-phenylene benzobisthiazole
PC. See Polycarbonate
PCL. See Poly-ε-caprolactone
PDAC. See Polydiallyldimethylammoniumchloride
PDMS. See Polydimethylsiloxane
PE. See Polyethylene
PEB. See Ethene-co-1-butene
PEDOT. See Poly-3,4-ethylene-dioxythiophene
PEEK. See Poly ether ether ketone
Peelback method (for fibers/films) 146–147, 200
cleavage plane splitting technique 146
examples 147, 414
PEELS. See Parallel EELS
PEI. See Polyetherimide
Penning sputtering 203, 205–206
Pentacene 288
PEO. See Polyethylene oxide
Pepper and salt texture 282, 401–402
Perfluorodecalin 442
Permanganate-acid etching 159, 168, 184–188, 315
Permanganic acid 184
PET. See Polyethylene terephthalate
Peterlin model (of drawing process) 10, 250, 278

PETG. See Polyethylene terephthalate-co-1,4-cyclohexanedimethylene terephthalate
Phase contrast 33, 76–77, 481
AFM 106–109
in OM 19, 33, 76
condenser lens for 79
in TEM 43, 76–77
Phase (contrast) imaging 49
AFM examples 158, 332–337
OM examples 161, 331–332, 338, 360
TEM examples 169, 280
PHB. See Poly-3-hydroxybutyrate
PHBA. See p-hydroxybenzoic acid
PHIC. See Polyhexyl isocyanate
Phosphoric acid
as etchant 184
with potassium permanganate 184–187
as substrate 138
Phosphotungstic acid (PTA) stain 161, 171, 175–177
examples of use 173, 176, 339–340, 385
literature review 175
practical details 176
PhotoBCB. See Photodefinable benzocyclobutene
Photodefinable benzocyclobutene (PhotoBCB) 291–292
Photodiode detector 99
Photon tunneling microscope 449
Photosensitizers, by NSOM 449
PHV. See Poly 3-hydroxyvalerate
p-hydroxybenzoic acid (PHBA) 400, 413
PID. See Proportional-integral-differential
Piezoelectric elements in SPM 46, 99–100, 106
Pigments 11, 32, 172, 262
Pinhole lens, for SEM 87
Piperazine 303
Pipes, from PVC 349
Pixel (picture element) 30, 73
size, effect on resolution 73–74, 114

Plan apo 32
Plasma etching 188–194
Plastics 3, 308–353. *See also* Thermoplastic(s)
  amorphous polymers 316–318
  characterization of 309–311
  competitive analysis 349–353
  extrusion of 311–312
  failure analysis of 349–353
  fracture of 214–216
  multiphase polymers 321–349
  semicrystalline polymers 318–321
  skin-core structures 315–316
  spherulites 313–315
  toughening 8, 323–326
Platinum/carbon 138, 198, 203, 382
Platinum coating 204, 317, 400
PLM. *See* Polarized light microscopy
PMMA. *See* Polymethyl methacrylate
PMMA/PS, *See* Polystyrene/polymethyl methacrylate
Point-to-point resolution 72, 77
Polarization 34
  colors 35, 83–84
  plane of 80, 82–83
  state 34, 80–81, 83, 438
Polarization sensitive optical coherence tomography (PS-OCT) 438
Polarized light 80–85
  circularly polarized light 35, 80, 405
  elliptically polarized light 80
  extinction positions 35, 83–84, Color plate II
  lenses for 32
  linear or plane polarized light 80
Polarized light microscopy (PLM) 34–35, 83–85, 132, 154, 278
  applied to
    fibers 253
    LCPs 400–403
    molded bar 313, 320
    spherulites 6, 7, 283
  hot stage 399
Polarizer (Polar) 34

Polarizing microscopy. *See* Polarized light microscopy
Polishing methods 142–145
  artifacts caused by 142–143
  examples 142–146, 365–366, 403–404, Color plate XI
  followed by etching 144–145
  for thin sections 145–146
PolScope 438
Poly-3,4-ethylenedioxythiophene (PEDOT) 298, 393
Polyacetal. *See* Polyoxymethylene (POM)
Poly(acrylic acid) (PAA) 198, 200–201
Polyacrylonitrile (PAN) 297
  hollow fiber membranes from 305
Polyamide. *See* Nylon. *See also* Aromatic polyamides
Polybenzimidazole (PBI) 230, 296, 300–301
  drying method 230
  hollow fiber membranes from 305
  membranes of 296, 305
Polybenzobisoxazole (PBO) 45, 153, 272, 325
  fibers 153, 272
Polybenzyl L-glutamate 291
Polybutadiene (PB)
  crosslinking of 381
  staining methods for 161
Poly(butylene terephthalate) (PBT) 137, 157, 492
  blends 168, 182
  ductile matrix polymer 325
  etching of 191
  specimen preparation 137, 316, 336
  spherulites in 314
  zones of 321
Polycaprolactone 85
Polycarbonate (PC) 151, 153, 281, 492
  compact disks 317–319
  crazing in 218
  in multilayered film 292
  probes for 447
  staining method 329
  SAXS of 282

Poly(diallyldimethylammonium chloride) (PDAC) 233, 395
  cryo-TEM of 395
Polydimethylsiloxane (PDMS) 330
  AFM of 311
Poly-ε-caprolactone (PCL) AFM of 225
Polyesters. *See also* Aromatic copolyesters
  as embedding media 151–152
  etching method 196
  fatigue fracture of 257
  fibers 165, 251–257, 267, 269
  films 191, 283
  heat aging of 254
  high speed spun 264
  in polarized light 253
  SEM of 257
  staining for 164, 183
Poly(etherether ketone) (PEEK) 138, 185, 187
Polyetherimide (PEI) 206, 356
Polyethersulfone (PES) 297
Polyethylene (PE) 5, 122, 280. *See also* High density polyethylene (HDPE); Low density polyethylene (LDPE); Ultrahigh molecular weight PE (UHMWPE)
  blends 335
  radiation damage in 119, 122–123
  electron diffraction of 70, 122, 280
  etching of 184, 188, 200
  fibers 200, 252–253
  high modulus fibers 270–272
  films 271, 279–281, 376
  hollow fiber membranes 308
  industrial films 282–284, 287–288, 293
  HREM of 285
  lamellae in 52, 184, 280, 284
  melt cast film 83, 281
  melt extruded 287
  microporous membranes 289, 298
  PLM of 253, Color Plate II
  microtomy of 155

Index

replication methods for 201
single crystals 123, 135–136
staining methods for 170, 173–175
thin film preparation 135, 139
Poly(ethylene-*block*-polyferrocenyldimethylsilane) (PS-PFS) 279
Poly(ethylene oxide) (PEO) 183
  AFM 224
Poly(ethylene-polypropylene) 8
Poly(ethylene-ran-methacrylic acid) (EMAA) 285
Poly(ethylene terephthalate) (PET) 11, 84, 105, 138, 182, 277, 282
  Chemical force microscopy (CFM) for 442
  blends of 335, 409
  etching of 182, 185, 189–191, 195
  fibers 9, 147, 165, 214, 253, 163–265
  films 192, 277–287
  fracture of 255–256
  melt crystallization of 224
  in multilayered films 292
  SAXS for 282, 500
  SEM for 265, 278–279
  staining method for 163–165, 177
  TEM for 163, 165, 265, 352, 409
  thin film preparation 138
  WAXS for 500
Poly(ethylene terephthalate-co-1,4-cyclohexanedimethylene terephthalate) (PETG) 292
Poly(ethylene-co-vinyl alcohol) (EVOH) 94, 155, 348
  blends 155, 348
  nanocomposites 374
Poly(hexyl isocyanate) (PHIC) 291
Poly-3-hydroxybutrate (PHB) 183, 199
  biodegradable 347
  etching of 187
Poly-3-hydroxyvalerate (PHV) 347
Polyimide (PI) films 282
Polyisoprene 4
Poly(lactic acid) 230, 348

Polymer blends. *See* Blends
Polymer coatings 2, 380–398
  list of polymers used 15
  adhesion of 15
  processes with 15
  surfaces of 140
  wettability and 388–398
Polymeric light emitting diodes 449
Polymers. *See also* Amorphous polymers; Aromatic polymers; Blends; Composite polymers; High performance polymers; Multiphase polymers; Rigid rod polymers; Semicrystalline polymers; Single phase polymers
  acronyms for 521
  applications of 2, 250–418
  biodegradable 347–349
  characterization of 17–21
  classes of 3
  colloids and 394–398
  crosslinking and 119
  crystallinity loss of 121–122
  definitions of 2–3
  deformation in 340
  degradation of 13
  etching of (summary) 195
  high melting point 78
  high molecular weight 288
  LCPs 8
  mass loss in 120
  materials of 1–3
  measurement values of 208
  morphology of 3–8
  processes with 8–17
  radiation and 118–120
  staining for (summary) 179–180
  starburst 117
  structure of (summary) 480
  surfactants and 394
  viscosity of 13
Poly(methyl methacrylate) (PMMA) 37, 110, 111, 119, 151, 154, 191, 282
  amorphous polymer 316
  beam damage of 209
  blends 330, 335, 458, 466, 518
  block copolymers 119, 345

brittle matrix polymer 325
EFTEM for 462
electron interaction volume in 38
etching methods 195–196
films 110, 191, 448
harmonic imaging of 444
nanocomposites 376
replication method 197
with SAN 462
SAXS of 282
staining method for 164, 167, 172–173, 465
Polymethyl methacrylate/polystyrene. *See* Polystyrene/ polymethyl methacrylate
Polyolefins 356. *See also* Thermoplastic polyolefin
  chlorosulfonic acid for 281
  staining of 281
  surface pretreatment of 387
Polyoxymethylene (POM), (acetal) 119, 152, 187
  polarized light micrography of 6
  single crystals of 135
  molded bar 320
  spherulites in 370
Poly(p-1,2-dihydrocyclobuta phenylene terephthalamide) (PPXTA) 272
Poly(phenylene oxide) (PPO) 171, 374
  crazes 220
  etching of 374
  nanocomposites 374
  staining methods for 171–172, 179
Poly(phenylene sulfide) (PPS) 355, 409
Poly(phenyl ether) (PPE) 389
Poly(p-phenylene-benzobisoxazole) (PBZO) 297
Poly(p-phenylene benzobisthiazole) (PBZT) 45, 272, 297, 400
  extrusion of 411–412
Poly(p-phenylene terephthalamide) (PPTA) 45, 399, 402, 409

Polypropylene (PP) 16, 151. *See also* Biaxially oriented PP
  atactic (aPP) 3
  blends 158, 326, 334–335, 338
  etching of 182–185, 195
  film 200, 278, 287, 442
  HEPP 223
  isotactic (iPP) 3, 184–189, 224, 330
  hollow fiber membranes from 305
  MAH-g-PP 375
  microporous membranes 206
  nanocomposites 374, 377
  reinforcement of 367
  replication methods for 200
  sectioning of 154–155, 158
  staining methods for 165, 167–170, 175, 180
  syndiotactic PP (sPP) 3
Polypropylene/ethylene propylene rubber (PP/EPR) 332
Polypyrrole (PPy) 298
Polysaccharide(s) 141, 348
Polystyrene (PS) 2, 45, 282. *See also* High impact polystyrene
  amorphous polymer 70, 316
  atactic 4, 141
  brittle matrix polymer 325–326
  crazing studies 219–221
  etching 191, 195
  films 219
  freeze drying of latex 229
  isotactic 122, 281–282
  latex 173, 385–386, 390, 392
  nanocomposites 374–375
  optical properties of 68
  radiation effects on 121
  rubber toughening of 292, 309
  rigid 151
  SAXS of 282
  staining methods for 165, 168–170, 181
  syndiotactic 4, 168
Poly(styrene)-*block*-poly(ethene-*co*-but-1-ene)-*block*-poly(styrene) (SEBS) 141, 159, 331–333
Polystyrene-*block*-*t*-butylacrylate (PS-*b*-PtBA) 442

Polystyrene-*block*-ferocenyldimethylsilane (PS-PFS) 279
Polystyrene-*block*-isoprene-*block*-styrene (SIS) 157
Poly(styrene-butadiene). *See* Styrene-butadiene rubber
Polystyrene-polybutadiene-polystyrene (PS/PB/PS). *See* styrene block coplymers
Polystyrene-polyethylene oxide 337
Polystyrene-poly(methyl methacrylate) (PS-PMMA) 458
  FFM of 104
  staining method 167
  TEM and STXM of 464–465
  X-ray tomography of 458
Polystyrene-poly(phenylene oxide) (PS-PPO), single phase thermoplastic 309
Polysulfides, as coating 15
Polysulfone 296, 299, 300
  hollow fiber membranes from 305
Polytetrafluoroethylene (PTFE) 119, 151, 155
  AFM of 225, 286
  extraction replicas of 201
  Chemical force microscopy (CFM) of 442
  FESEM of membrane 305
  membranes 305–307
  radiation damage in 119
  wear deposited films 286
Polyurethanes (PUR)
  as coating 15
  etching method 183, 196
  fracture 345
  as multiphase polymers 345–347
  OCT of 455
  with polyacetal 346
  RIM processing 15, 345
  staining methods 163
  thermoplastic (TPU) blends 182, 183, 345
Poly(vinyl acetate) (PVAC) 142, 385, 309
Poly(vinyl alcohol) (PVOH) 142, 201

Poly(vinyl chloride) (PVC) 151, 258, 329
  chlorinated polyethylene multiphase polymer 328
  as ductile matrix polymer 325
  FESEM for 258
  fracture of 232
  pipes from 349
  SAXS of 282
  staining method for 167, 181
  TEM and 281, 328
Poly(vinylidene chloride) (PVDC), latex particles 384
Poly(vinylidene fluoride) (PVDF) 185, 297
POM. *See* Polyoxymethylene
Porosity 141, 260, 297
  of hollow fibers 305
Positive birefringence 82
Post-it® 387–388
Potassium permanganate (KMnO$_4$). *See* Permanganate-acid
PP. *See* Polypropylene
PPE. *See* Poly(phenyl ether)
PP/EPR. *See* Polypropylene/ethylene propylene rubber
PP-LDPE blends 16, 294
PPO. *See* Poly(phenylene oxide)
PPTA. *See* Poly(p-phenylene terephthalamide)
PPXTA. *See* Poly(p-1,2-dihydrocyclobuta phenylene terephthalamide)
PPy. *See* Polypyrrole
Prepreg 146, 365
Pressure-sensitive adhesives 387
Probes in SPM 50–51. *See also* Cantilevers; Scanning probe microscopy
  artifacts due to probe goemetry 115–117, 491
  artifacts due to tip wear 117–118
  microfabrication of 50
  probe specimen interaction in AFM 100–102
  for scanning thermal microscopy 445–448
  silicon 50
  silicon nitride 50

Index

Probes in SEM 85–87
  BSE 88
  probe specimen interaction in SEM 88–91
  probe size 86–88
  Secondary electrons 89–91
  Probes in STEM 85–86, 91
Problem solving 478–501
  adhesives 387
  coatings 389
  composites 362–364
  engineering resins and plastics 349–354
  fiber studies 260–267
  films 287–292
  flow chart for 488
  instrumentation techniques 481–484
  interpretation considerations 488–491
  nanocomposites 376–379
  protocol for 478–480
  starting point for 478–480
  supporting characterizations 492–500
Processes 8–15, 311–315, 326, 403
  coating 15
  compounding 11, 294, 360, 376
  extrusion 12
  fiber extrusion 8–9, 270
  film extrusion 11, 277, 284
  molding 12–14, 311–315, 319, 326
Projector lens 30, 73
Proportional-integral-differential (PID) 99, 113
Prostheses 224, 346
Protective fabrics 270, 272
PS. See Polystyrene
PS-OCT. See Polarization sensitive optical coherence tomography
PS/PB/PS. See Polystyrene-polybutadiene-polystyrene
PS-PFS. See Polyethylene-block-polyferrocenyl-dimethylsilane
PS/PMMA. See Polystyrene/polymethyl methacrylate
PS-PPO. See Polystyrene-poly(phenylene oxide)
PTA. See Phosphotungstic acid

PTFE. See Polytetrafluoroethylene
Pull-off force 48, 442
PVA. See Poly(ethylene-vinyl alcohol); Poly(vinyl acetate); Poly(vinyl alcohol)
PVC. See Poly(vinyl chloride)
PVDF. See Poly(vinylidene fluoride)
PVF_2 See Poly(vinylidene fluoride)
PZT. See Lead zirconate titanate

Q

Quality factor (Q) 98
  for AFM cantilever oscillation 106–109
  effect of environment 98
  in fast scanning SPM 444
  harmonic imaging and 443
  in IC-AFM 99
  in NCAFM 112
Quantitative microscopy 56–59
  calibration techniques 54, 57–59, 62, 121
  fundamentals of 56–59, 386
  image analysis 56–57, 114, 369, 450
  latex particle size analysis 309, 385–386
  stereology 57
Quartz crystal monitor 203
Quartz wedge 84

R

Radial distribution function (RDF) 69
Radial growth rate, of spherulites 5
Radiation
  coherent 68
  incoherent 68
  synchrotron 457–459, 494
  ultraviolet 387
Radiation dose 52, 120, 123, 281
Radiation effects in EM 118–123
  artifacts 122
  chemical changes 118–120
  crosslinking 119
  crystallinity loss 121
  dimensional changes 122

mass loss 121
scission 118
specimen heating 120–121
Radiation effects in SEM 52, 123
Radiation induced contrast 196, 310
Radiation sensitive materials 51–53
  low dose TEM operation 53
  SEM operation for 52
Raman microscopy 18, 53, 376, 449, 460–461, 492, 497
  with AFM 461
  analytical imaging with 460–461
  Coherent anti-Stokes Raman scattering (CARS) 461
  dichroism in 461
  NSOM with 461
  Surface enhanced Raman scattering (SERS) 461
  Tip-enhanced Raman scattering (TERS) 461
Raman spectroscopy 18, 187, 281, 460, 492, 497
Random copolymers 337–345
Rayleigh criterion 72–74
RDF. See Radial distribution function
Reaction injection molding (RIM) 15, 311, 345
Reflection electron microscopy (REM) 28
Reflected light microscopy 28, 33–34, 132, 484
  examples 144, 145, 353, 365, 404
  specimen preparation for 142–145, 183, 485
Refractive index 68, 76
  birefringence and 35, 81–82, 283
  measurement of 34, 283, 438
  and phase contrast 33
Reinforcement 8, 11, 354–357; See also Composites
  fiber lengths of 360
  with fibers 144, 354–357, 360, 367
  with microfibers 157
  with mineral fibers 367
  with particles 369–370

Relative humidity (RH). *See* Humidity
REM. *See* Reflection electron microscopy
Replication 19, 195–201
 artifacts in 490
 carbon replicas 184, 198–200, 314, 386
 detachment replicas 201
 direct replicas 198
 extraction replicas 201
 of fibers 200
 for OM 197
 for SEM 197
 using silicone rubber (Silastic or Xantopren Blue) 197
 for TEM 198–201
 two stage replicas 199–201
Resin-rich regions 357, 366
Resins 308–353
 amorphous polymers 316–318
 characterization of 309–311
 competitive analysis of 349–353
 engineering 2, 308–353
 extrusion of 311–312
 failure analysis of 349–353
 hackle in 349–350
 of multiphase polymers 321–349
 photodefinable 291
 semicrystalline polymers 318–321
 single phase polymers 316–321
 skin-core structures in 315–316
 spherulites in 312–315
 thermoset 291, 308
 toughened 323–326
 toughened thermoset 326–327
Resolution 18, 28, 68. *See also* High resolution SEM; High resolution TEM
 in AFM 47, 51, 103, 114, 342–343
 in analytical microscopy 53–56
 definition of 28
 of detector 73
 in lens-based systems 72–76
  chromatic aberration 75

diffraction limit 72–73
focus and 75
spherical aberration 75–76
of microscopy techiques, compared 29, 484–487.
noise limited 123–124
in OM 29, 31, 73, 147
point-to-point 72
Rayleigh criterion, defined 72
in SEM 37, 41–42, 52, 85–88, 90
in STEM 44, 91
in SThM 48
in TEM 45, 74–76, 438–439
x-ray 53–56, 91
Resorcinol-formaldehyde-latex (RFL) 177
 as adhesive 268–269, 387
 in tire cords 268–269
Retardation 35, 438
 of biaxial films 283
 definition of 35, 81
 of fibers 253
 measurement of 84, 438
Reverse osmosis (RO) membranes 295, 300–303
 CA as 300
RFL. *See* Resorcinol-formaldehyde-latex
RH. *See* Humidity
Rigid rod polymers
 beam damage in 78, 120
 as high modulus fibers 411–412
 as LCPs 399
 Diffraction from 297, 412–414
RIM. *See* Reaction injection molding
RO. *See* Reverse osmosis membranes
Robinson detectors 38, 97
Rotary microtomes 148, 159
Row nucleated structure 5–7
Rubber 3
 carbon black-filled 368–369
 chlorobutyl (CB) 177
 ebonite method for 177
 EPR 334, 377
 GRC 327
 natural 4
 PP/EPR 332

staining methods for 161, 164, 329
vulcanization of 3
Rubber toughened polymers 8, 150, 223, 327–328
RuO_4. *See* Ruthenium tetroxide
Ruthenium dioxide 170
Ruthenium tetroxide (RuO_4) staining method 138, 158, 166–173
 artifacts and 169
 EELS and 169
 examples of use 171, 379, 396, 465
 Literature review 167–169
 PC and 329
 practical details 169–170
Ruthenium trichloride 170

S
Saddle field ion gun 203, 204
SAED. *See* Selected area electron diffraction
SALS. *See* Small angle light scattering
SAMs. *See* Self-assembled monolayers
SAN. *See* Styrene-acrylonitrile
SANS. *See* Small angle neutron scattering
Santovac-5 389
SAXS. *See* Small angle x-ray scattering
SBR. *See* Styrene-butadiene rubber
SBS. *See* Styrene-butadiene-styrene
Scan generator, SEM 36
Scan speed in SPM 113–114, 444–445
Scanned sample SPM 115
Scanned tip SPM 115
Scanners for SPM 99–100
 artifacts due to 114–115, 490
Scanning Auger spectroscopy (SAM) 498–499
Scanning capacitance microscopy (SCM) 48
Scanning electron microscopy (SEM) 18, 35–42, 85–97. *See also* Field emission SEM; High resolution

Index

SEM; Low voltage SEM;
    Variable pressure SEM
aberration corrected   440
applied to
    blends   327, 338, 346
    composites   218, 355, 359–362, 370, 408
    fabrics   258–259
    fibers   147, 214, 253–258, 261–263, 416
    films   279, 285, 288, 294–295
    fractures   212, 214
    membranes   231, 298–300, 302, 306–307
    multiphase polymers   216, 327, 338, 324
    impact test samples   360–361
    latex particles   383–385, 454
    artifacts in   489–490
calibration   57–58
charging effects   94–95, 207–209
conductive coatings for   203–207, 317
EDS with   55, 363, 497
electron sources   39–41, 85
elemental mapping in   55–56, 351, 372
examples
    BEI and elemental mapping   350–351
    complementary study with AFM   287, 306, 406
    complementary study with OM & TEM   301
with FIB   194, 440, 453–454
fundamentals of   35–42
specimen heating in   121
IBS coating for   199, 204–205, 317–318
image formation in   92–94
image interpretation   36, 39, 485, 488
imaging signals   85–97
final lens design for   86–87
in situ deformation in   222–223
low voltage operation   52, 94–96, 133. *See also* Low voltage SEM
microtomy for   150

noise   39, 42, 86, 93–94, 123, 208
optimization   42
peelback for   146–147
probe formation   85–87
probe-specimen interactions in   88–91
properties of   29
radiation damage in   123
replication methods for   197
resolution   37, 41–42, 52, 85–88, 90
schematic of   36
specimen preparation methods for   133, 144, 146, 150, 165, 170, 197
staining methods for   165–167, 170
types of   41, 483
Scanning-imaging microscopes   30–31
Scanning ion microscope (SIM)   440
Scanning probe microscopy (SPM)   45–51, 114, 279, 441–451, 487
    artifacts in   114–118, 490–491
    automated   449–451
    CFM   441–442
    cryomicrotomy for   154–157
    fast scanning   444–445
    feedback in   45
    of films   286–288
    microtomy for   150–154, 158–159
    motion control in   46
    of nanocomposites   373
    NSOM   449
    probes for   50–51
    probe tips   50–51
    schematic of   45
    specimen preparation for   140–142
    staining for   165–166
    SThM   445–449
    for surface analysis   464
    ultramicrotomy for   154
Scanning reflection electron microscopy (SREM)   28
Scanning thermal microscopy (SThM)   48, 291, 445–449

Scanning transmission electron microscopy (STEM)   28, 43–44, 85–86, 486–487
    aberration corrected   439
    dedicated   44, 486
    with EELS   138, 461–462
    electron sources for   40
    FEG with   44
    HAADF in   44, 145, 233, 285, 345, 456
    of ionomer   285
    optical sectioning and   454–455
    probe-specimen interactions in   91
    radiation sensitivity and   123–124
    staining for   170
    techniques   486–487
    x-ray analysis in   55
Scanning transmission x-ray microscopy (STXM)   463–464
Scanning tunneling microscopy (STM)   46, 116, 317, 481, 487
    applied to
        CD surface features   318
        LCP microstructure   414–417
        microporous membranes   289
    conductive coatings for   203–207
    sample preparation for   140–141
Scherzer focus position   75, 76
Schlieren texture in PLM   401
Schottky field emission gun   40
Scission   118–119, 121, 209
SCM. *See* Scanning capacitance microscopy
SDS. *See* Sodium dodecyl sulfate
SE. *See* Secondary electrons
Secondary electron imaging (SEI)   39, 93, 191–193, 482
    examples   191–192, 205, 263, 396
    in LVSEM   39, 362–363
    optimization of   41–42
    in VPSEM   41

Secondary electrons (SE) 39, 89–91
  types of 39, 90
  produced by BSE 39
Secondary ion mass spectrometry (SIMS) 18, 464–466, 498
Sectioning *See* Microtomy
  cryosectioning. *See* Cryomicrotomy
  optical 357, 454–455
  physical, for 3D imaging 452–454
  serial 452
SEI. *See* Secondary electron imaging
Selected area aperture 44, 71–72
Selected area electron diffraction (SAED) 44, 122, 275, 412, 481
Selective plane illumination microscopy (SPIM) 455
Self-assembled monolayers (SAMs) 287, 441
Self-reinforcing LCPs 403
SEM. *See* Scanning electron microscopy
Semicrystalline polymers 4–7, 318–321
  chlorosulfonic acid staining 173–175
  crystallization 5–7
  melt-drawn films 280–281
  SEM of 215, 284, 322, 324
  injection molding of 316
  lamellae 4–7, 10, 136, 174, 175, 206, 280, 284
  spherulites 7, 83–85, 163, 184, 189, 313–315, 320, 322
  OsO_4 staining 163
  PLM of 6, 7, 83, 85, 253
  TEM of 163, 174, 175, 189, 280
  toughening of 339–340
  microstructure of molded part 14, 313, 315, 320, 330
Senarmont compensator 84
Sensitive tint plate 84
Serial sectioning 452
SERS. *See* Surface enhanced Raman scattering
Set point 49, 102, 110, 445
  ratio 110, 327, 450

Shadowing 19, 183, 186, 199, 203, 285, 384
Shear bands 4–5, 212, 412–413
Shish kebab structure *See* Row nucleated structure
Shrinkage 13, 253, 366
  in amorphous polymers 14
Signal-to-noise ratio (SNR)
  resolution and 123–124, 439
  visibility limit 93
Silastic
Silicon SPM probes 50–51, 98
Silicone replica method 197
Silicon nitride 50
Silver nitrate staining method 178
Silver sulfide insertion method 178
SIM. *See* Scanning ion microscopy
Simple microscopes 31
SIMS. *See* Secondary ion mass spectrometry
Single crystals 5, 136
  AFM 136–137
  diffraction from 122
  formation 5, 135
  of PE 136
  TEM 135–136
Single phase polymers 308–309
Single phase plastics 316–321
  examples 309
Single walled carbon nanotubes (SWCNTs) 372, 376–378, 380
SIS. *See* Polystyrene-*block*-isoprene-*block*-styrene
Skin-core morphology (texture) 12–13, 168
  examples 320, 403–407, 411
  in plastics and resins 315–316
Sledge microtome 148
Small angle light scattering (SALS) 499
Small angle neutron scattering (SANS) 500
Small angle x-ray scattering (SAXS) 5, 186, 297, 481, 499–500
  AFM with 188, 346, 451, 500
  of amorphous films 282
  crazing and 221

TEM with 186, 233, 499
STEM with 285
Smectic liquid crystals 45, 183, 399
Snorkel lens, for SEM 86
SNR. *See* Signal-to-noise ratio
Sodium dodecyl sulfate (SDS) 233, 395
Sodium hypochlorite 169
Sodium (meta)periodate 169, 170
Sonication, sample preparation 137, 187
  examples 403, 409–412
Solvents for etching 181–183
SPB. *See* Spherical polyelectrolyte brush
Specimen preparation methods 132–234
  conductive coatings 201–211
  CPD 230–231
  cryogenic methods 226–234
  and cryomicroscopy 232–234
  drying 151, 226–234
  embedding 149
  etching 181–196
  fracture studies 212–217
  freeze drying 227–230
  freeze fracture-etching 231
  microtomy 147–160
    mounting for 148–149, 152
  for OM 132–133, 144
  peelback 146
  polishing 142–146
  replication 196–201
  for SEM 133, 144
  single crystal formation 135–137
  staining 160–181
  for SPM 140–142, 154–159, 165, 223–226
  for TEM 134–140, 150–157, 219, 223, 233
Specimen support films 134
Spectroscopy techniques 496–499. *See also* Raman spectroscopy; X-ray photoelectron spectroscopy
  Auger 498–499
  FTIR 496–497
  MS 18
  NMR 497–498
  XRF 496

Index

Spherical aberration 73–75, 439
Spherical polyelectrolyte brush (SPB) 395
Spherulites 4–7, 83–85, 480
  AFM of 189, 314, 315
  deformation of 10, 283
  elastomer addition and 329
  permanganate etching 184–186
  in cast thin films 138
  SEM of 322
  PLM of 6, 7, 83–85, 163, 253, 313, 320
  in POM 370, 372
  structure of 6
  TEM of 163, 184
Spider silk 270, 275–276
SPIM. *See* Selective plane illumination microscopy
Spinneret 9
  contamination of 267
SPM. *See* Scanning probe microscopy
Spray spun nonwoven fabrics 258–259
Sputter coaters and coating 202, 204–206
  examples of sputter coating 205–206, 307, 416
SREM. *See* Scanning reflection electron microscopy
Staining methods 160–181. *See also* Vapor staining
  for ABS 167
  with bromine 179
  with chlorosulfonic acid 162, 173–175, 281
  with ebonite 177–178
  compared to EFTEM 161
  with iodine 179
  with mercuric trifluoroacetate 178–179
  negative staining 161
  for nylon 161–162
  with osmium tetroxide (OsO_4) 162–166, 180–181, 183, 339
    for AFM 166
    for SEM 165–167
  positive staining 161
  for polybutadiene 162, 167, 171–173
  for polyester 164, 183
  summary of 180–181
  for PS/PMMA 167
  with phosphotungstic acid (PTA) 161, 175–177
  for fiber indentification 258
  for rubber 161, 164
  with ruthenium tetroxide (RuO_4) 154, 166–173, 182
    for SEM 170
    for STEM 170
    of SBS 155, 164, 339
  with silver sulfide 178
  with Sudan Black B 161
  for TEM 160–161, 170–173
  with uranium salts 161
  with uranyl acetate 173–175
Starburst polymers 117
Starches 347–349
  degradation of 348
STEM. *See* Scanning transmission electron microscopy
Stereo binocular microscopes. *See* Binocular stereomicroscopes
Stereology 57
SThM. *See* Scanning thermal microscopy
STM. *See* Scanning tunneling microscopy
Structured light imaging 455
STXM. *See* Scanning transmission x-ray microscopy
Styrene-acrylonitrile (SAN) copolymers 381
  as brittle matrix polymer 325
  EFTEM of blend 462
  a multiphase polymer 309
  staining 163, 168
Styrene block copolymers
  styrene-butadiene diblock copolymer 190, 339, 382
    in situ deformation 223
  styrene-butadiene-styrene (SBS) triblock copolymer 2, 155, 290, 342
    staining methods 162, 164, 178, 181
  non-butadiene blocks 141, 157, 159
Styrene-butadiene rubber (SBR) 161, 177, 368, 382
Sudan Black B 161
Sulfur
  as embedding media 152
  liquid 220
Sulfuric acid, etchant 159, 183–188
Suppiers listed in Appendices
  accessories 510–511
  microscopes and microanalysis equipment 512
  scanning probe microscopes and accessories 513
Surface enhanced Raman scattering (SERS) 461
Surfactants 380
  in cryo-EM 233
  in etching 165, 166, 304
  polymer interaction 394–398
Supporting characterization techniques 492–500
  spectroscopies 496–498
  thermal analysis 495
  x-ray diffraction 493–495
SWCNTs. *See* Single walled carbon nanotubes
Synchrotron radiation 447, 457–459, 494
Syndiotactic polymer, definition of 3

T

Talc particles 217, 354, 358–359
Tantalum-tungsten alloy 185, 199–200
Tapping mode AFM (TMAFM). *See* Intermittent contact mode AFM
TEM. *See* Transmission electron microscopy
Tensile properties 2, 10, 270, 271, 316, 323, 376, 417
Tensile stress 10
Tensile tester 212–213, 255
  in situ 47, 59–60, 222–223, 346
Ternary blends 311, 330–331
TERS. *See* Tip-enhanced Raman scattering
Tetrafluoromethane (CF_4) plasma 190
Tetrahydrofuran (THF)
  and $Os O_4$ 163, 327
  etchant 345

Textile fibers 251–260
  birefringence studies 35, 251–253
  characterization of 262–265
  OM of 251–253
  SEM of 253–254, 263
  TEM of 259–263
Textile World Manmade Fiber Chart 251
TGA. *See* Thermogravimetric analysis
Thermal analysis 4, 17, 47, 495–496
  of nanocomposites 373
  of UHMWPE 271
Thermal field emission gun 41
Thermal infrared detection 447
Thermally cured resins 291
Thermal tuning in AFM 59
Thermionic emission source 40
Thermoforming 14–15, 311
Thermogravimetric analysis (TGA) 495
Thermomechanical analysis (TMA) 495
Thermoplastic(s) 3, 309
  cross-linkable epoxy thermoplastics (CET) 327
  cold drawing 10
  impact modified 328–337
Thermoplastic polyolefin (TPO) 154, 347, 377–379, 394
Thermoplastic polyurethane (TPU) 183
Thermoplastic vulcanizate (TPV) 336
Thermosets 3, 14, 309
  as matrix 354
  toughened 326–328
Thermotropic, definition of 8
Thermotropic liquid crystalline polymers (TLCPs) 185–186, 270
  PLM of 400–402
THF. *See* Tetrahydrofuran
Three dimensional imaging 57, 436, 451–459, 466
  by optical sectioning 454–455
  by physical sectioning 452–454, 466
  by tomography 57, 455–459

Time-of-flight secondary ion mass spectrometry (ToF-SIMS) 159, 464
Tip-enhanced Raman scattering (TERS) 461
Tips of SPM probes 50–51
  artifacts due to 115–118, 490
  chemically modified 35, 441–442
  coated 50
  wear 117, 450
Tire cords 267–270
  characterization of 267
  ebonite with 269–270
  Kevlar in 399
  RFL in 268–269
Titanium dioxide in fibers 262–265
TLCPs. *See* Thermotropic liquid crystalline polymers
TMAFM, Tapping mode AFM. *See* Intermittent contact mode AFM
ToF-SIMS. *See* Time-of-flight secondary ion mass spectrometry
Tomographic spectral imaging 451
Tomography 57, 275, 451, 455–459
  Optical coherence tomography (OCT) 438, 455
  x-ray microtomography 457–459
Topographic contrast 36, 38, 254, 331
Toughening
  of plastics and resins 8, 323–326, 339
  with rubber 8, 327–328
  of thermoset resins 326–327
Tows 366
TPO. *See* Thermoplastic polyolefin
TPU. *See* Thermoplastic polyurethane
TPV. *See* Thermoplastic vulcanizate
Transcrystallinity 321

Transmission electron microscopy (TEM) 42–45, 71–78, 80, 486. *See also* Cryo-TEM; Energy filtering EM; Scanning transmission EM
  abberations in 74–76, 438–439
  applied to
    blends 330–333
    filled rubber 368
    copolymers 337, 339, 341
    composites 357, 362–364
    crazes 219–222
    fibers 259–260, 274–275, 409–416
    films 280, 284–285, 294–296
    membranes 301–304
    multiphase polymers 159, 339–341
    nanocomposites 373, 377, 379
    latexes 172, 176, 383–386
    semicrystalline films 280–281
    PET fibers 165, 265
  artifacts in 491
  basic optics of 29, 68–69, 71
  calibration of 57–58
  CCD for 59
  cold stage 233–234
  comparison of techniques 29, 44, 480–487
  contrast modes 42–43
    BF and DF 42–43, 341, 402
    crystallographic *or* diffraction 43, 122, 456
    mass thickness 42–43, 221
    phase 43, 76–78
  diffraction techniques 18, 44–45, 69–71. *See also* Electron diffraction
  with EELS 53, 461–462, 486
  EFTEM 70, 75, 157, 461–462
  electron sources for 40
  electron beam heating in 120
  illumination systems for 80
  intermediate aperture 44, 71. *See also* Selected area ED
  introduction to 28–30, 42, 486
  lattice imaging 76–78, 274–275, 410, 415. *See also* High resolution TEM
  lenses in 29–30

Index

low dose technique 53
objective aperture 71
radiation damage 51–52, 118–124
resolution 72–78, 438–440
Selected area electron diffraction (SAED) 44, 122, 275, 412, 481
   in situ deformation in 59–60, 223
specimen preparation for 19, 134–140
   carbon coating 203
   cryomicrotomy 154–157
   freeze drying 227–229
   microtomy 150–157
   replication 198–201
   shadowing 203
   staining, examples 170–173
   staining, methods 160–181
   ultramicrotomy 152–154
x-ray analysis in 44, 55–56, 486
Transmission x-ray microscopy (TXM) 463
Transmitted light microscopy 147, 358
Tube length 32
Tungsten filament electron source 29, 40
   in SEM 41, 88
TXM. See Transmission x-ray microscopy

U
UFs. See Ultrafilters
UHMWPE. See Ultrahigh molecular weight PE
UHV. See Ultrahigh vacuum
ULAM. See Ultra-low-angle microtomy
Ultrafiltration membranes (UFs) 295–297
   PAN 296
   PES 297
   PVDF 297
Ultrahigh molecular weight PE (UHMWPE) 175, 188
   fibers 270–272
   nanoindentation of 224
   self composite (with HDPE) 271
   composite with carbon nanotubes 376

Ultrahigh vacuum (UHV) 40, 49, 90
   NCAFM 49, 112
   STEM 44, 486–487
   XPS 467
Ultra-low-angle microtomy (ULAM) 159
Ultramicrotomy 146, 152, 154–157
   cryoultramicrotomy 146, 151, 154–157, 292, 352
   knives for 154
Ultrasonic bath
   for cleaning 143, 197
   for disintegration 137
Ultrathin window (UTW) detectors 54
Ultrathin sections. See Ultramicrotomy
Ultraviolet (UV) irradiation 59, 387
Underfocusing technique 34, 76, 342. See also Defocus
Unfilled LCP moldings 406–407
Uniaxial birefringence 82
Uniaxially oriented material 10, 81–82, 250, 493
Uranium salts, staining with 161
   uranyl acetate 173–175
UTW. See Ultrathin window detectors

V
Vacuum. See also Ultrahigh vacuum
   artifacts 492
   for electron sources 40
   evaporators 198–199, 202
   in freeze drying 227
   low vacuum SEM 41, 96, 363, 383. See also Variable pressure SEM
Vacuum in a bottle 442
Vapor staining 138, 154–157, 164–168, 170
Variable pressure SEM (VPSEM or HPSEM or ESEM) 21, 41, 59–60, 96–97, 452
   applications 348–349
   of blends 232, 329
   conductive coatings and 202
   detectors for 96–97

of hydrated materials 123, 348
of latex 382–383
Vectra® LCP resins 400, 408
Vectran® LCP fibers
   peelback of 414
   SEM, STM of 416
Virtual aperture 86
Viscoelasticity, lateral forces and 104
Viscosity
   of embedding resins (low) 151, 154, 178, 197
   of polymer melts 11–13, 323–326
Visibility limits 93
Void growth 217
Voxel 451, 466
VPSEM. See Variable pressure SEM

W
Warpage 13
Waves, interference of 68
Wavelength dispersive x-ray spectrometer (WDS) 54–55, 206, 265
   compared to EDS 55, 4, 84
   mapping in metal loaded fibers 265
WAXD. See X-ray diffraction
WAXS. See X-ray diffraction
WDS. See Wavelength dispersive x-ray spectroscopy
Weld line 14
   fractures at 316
Wettability 141, 190, 254
   and coatings 388–398
   of etched films 287
Wide-angle x-ray diffraction (WAXD) See X-ray diffraction
Wide-angle x-ray scattering (WAXS) See X-ray diffraction
Wide field optical sectioning 455
Wollaston wire 446–447
Wollastonite 217, 446
   as filler 354
Wood pulp fiber 173, 258, 273
Work function 40
Working distance 41–42, 86–87, 96, 440

WORM (write once optical data storage disks) 317
Woven fabrics 258, 300

X

X7G polymers 400–401, 412–413
Xantopren Blue 197
Xenoy 335, 336
XPS. *See* X-ray photon spectroscopy
X-ray(s). *See also* Energy dispersive x-ray spectroscopy; Small angle x-ray scattering; Wavelength dispersive x-ray spectrometer
  characteristic 53, 88, 91
  continuum 54
X-ray diffraction (XRD) 4–5, 17, 374, 482, 493–495
  of fibers 250, 263, 275, 409, 417, 500
  of membranes 297
  of nanocomposites 374, 375
X-ray fluorescence (XRF) 492, 496
X-ray microanalysis 19, 38, 53–56, 350, 438, 484
  artifacts in 491–492
  elemental mapping by 53–56, 349–351, 394–396, 464
  energy dispersive spectrometer (EDS) 53–54
  in failure analysis 349–351
  during in-situ deformation in SEM 222
  at low beam voltage 95
  of metal loaded fibers 266
  of multiphase polymers 332, 350, 372
  SEM compared to AEM 55
  of paint layers 394–396
  wavelength dispersive spectrometer (WDS) 53–54
X-ray microanalyzer (XRM). *See* Electron probe microanalyzer
X-ray microscopy 21, 462–464
X-ray microtomography 457–459
X-ray photoelectron spectroscopy (XPS) 18, 159, 184, 190, 273, 498
  with CFM 441
  of fibers 250, 273
  high resolution spatial imaging 466–468
XRD. *See* X-ray diffraction
XRF. *See* X-ray fluorescence
XRM. *See* X-ray microanalyzer
Xydar 400
Xylene 135, 233, 279
  etching with 168, 181
  in replication 198

Y

Yarns 258
  HMLS 272
Yielding and fracture 212–218, 325
Young's modulus 10, 316

Z

Zero net force, contact AFM 101, 103, 105
Zirconium oxide electron emitter 40
Z-piezo position 102, 105, 111, 115
Z-contrast. *See* Atomic number contrast
Z-sweep 111